Reagents for Organic Synthesis

Fieser and Fieser's

Reagents for Organic Synthesis

VOLUME NINE

Mary Fieser
Harvard University

Rick L. Danheiser
Massachusetts Institute of Technology

William Roush
Massachusetts Institute of Technology

A WILEY-INTERSCIENCE PUBLICATION
JOHN WILEY & SONS
NEW YORK · CHICHESTER · BRISBANE · TORONTO

Library of Congress Cataloging in Publication Data:

Library of Congress Catalog Card Number: 66-27894

ISBN 0-471-05631-6

Printed in the United States of America

10 9 8 7 6 5 4 3 2 1

PREFACE

This volume of reagents includes references to papers published during the period from July 1978 through 1979. We are very grateful for the advice and help from many colleagues. Paul B. Hopkins and Alan E. Barton read the entire manuscript. We are especially grateful for invaluable proofreading by Dr. Mark A. Wuonola, Dr. William Moberg, Dr. James V. Heck, Professor Bruce Lipshutz, Dr. Ving Lee, Dr. May Lee, Professor Janice Smith, Professor Dale L. Boger, Dr. John Secrist, Dr. Stephen Benner, Dr. Anthony Marfat, Dr. Donald R. Deardorff, Daniel Brenner, Thomas M. Echrich, Anthony Feliu, Katherine Brighty, Jeffrey C. Hayes, Jay W. Ponder, John E. Munroe, Donald Wolanin, Charles Manly, Stuart Schreiber, Craig Shaefer, William P. Roberts, Steven Freilich, Robert E. Wolf, Jr., J. Jeffry Howbert, William McWhorter, Marco Pagnotta, Marifaith Hackett, Steven E. Hall, Thomas E. D'Ambra, David J. Carini, John M. Maher, John M. Morin, Howard Sard, and Dr. Francis Brion.

MARY FIESER

RICK L. DANHEISER

WILLIAM R. ROUSH

Cambridge, Massachusetts
February 1981

CONTENTS

Reagents for Organic Synthesis

A

Acetaldehyde N-*t*-butylimine, $CH_3CH{=}NC(CH_3)_3$ (**1**). Mol. wt. 99.18, n_D 1.4005. Preparation.[1]

α,β-Unsaturated aldehydes. Meyers and co-workers[2] have simplified the phosphonate imine method for synthesis of enals (**2**, 131–132; **3**, 96) by generation of the reagent *in situ* from **1** and diethyl chlorophosphate to form **2**, which reacts with carbonyl compounds to form α,β-unsaturated aldehydes (equation I).

$$\text{(I)}\quad \mathbf{1} \xrightarrow[\text{2) }(C_2H_5O)_2POCl]{\text{1) LDA}} \left[(C_2H_5O)_2\overset{O}{\overset{\|}{P}}CH{=}CHN\begin{matrix}C(CH_3)_3\\ \\ Li\end{matrix}\right] \xrightarrow[55\text{–}95\%]{\text{1) }R^1COR^2\text{; 2) }H_3O^+} \begin{matrix}R^1\\ \\ R^2\end{matrix}\!\!>C{=}CHCHO$$

2

[1] H. R. Snyder and D. S. Matterson, *Am. Soc.*, **79**, 2217 (1957).
[2] A. I. Meyers, K. Tomioka, and M. P. Fleming, *J. Org.*, **43**, 3788 (1978).

Acetic anhydride, 1, 3; **2**, 7–10; **5**, 3–4; **6**, 1–2; **7**, 1; **8**, 1–2.

Phenoxycoumarins.[1] 3-Phenoxycoumarins have been obtained by a Perkin reaction of salicylaldehyde with the sodium salt of a substituted phenoxyacetic acid in

$$\text{(I)}\quad \text{salicylaldehyde (o-}HOC_6H_4CHO) + R\text{-}C_6H_4OCH_2COONa \xrightarrow[\sim 40\%]{Ac_2O} \text{3-(R-phenoxy)coumarin}$$

$(R = CH_3, OCH_3, NO_2)$

acetic anhydride at 170–180° (equation I). The reaction fails in the case of the parent acid: Only the triacetate of salicylaldehyde is formed.

[1] J. R. Merchant and A. S. Gupta, *Chem. Ind.*, 628 (1978).

Acetic anhydride–Pyridine.

Acetylenes; allenes. Nitrimines, prepared from ketoximes and nitrous acid, on acylation with acetic anhydride and pyridine with catalysis by 4-dimethylamino-pyridine fragment to alkynes and/or allenes.[1]

Examples:

$$(CH_3)_3C\overset{NNO_2}{\overset{\|}{C}}(CH_2)_3CH_3 \xrightarrow[65\%]{Ac_2O,\ Py} (CH_3)_3CC{\equiv}C(CH_2)_2CH_3$$

[1] G. Büchi and H. Wüest, *J. Org.*, **44**, 4116 (1979).

$$CH_3\overset{NNO_2}{\overset{\|}{C}}(CH_2)_2C_7H_{15} \xrightarrow[28\%]{} HC{\equiv}C(CH_2)_2C_7H_{15} + H_2C{=}C{=}CHCH_2C_7H_{15} + CH_3C{\equiv}C(CH_2)_2C_6H_{13} \quad (31:20:49 \text{ ratio})$$

$$C_6H_5CH_2C(CH_3)_2{-}\overset{NNO_2}{\overset{\|}{C}}CH_3 \xrightarrow[71\%]{} C_6H_5CH_2C(CH_3)_2{-}C{\equiv}CH$$

$$\text{Cyclododecanone nitroimine } (={NNO_2}) \xrightarrow[76\%]{} \text{cyclododecyne} + \text{1,2-cyclododecadiene} \quad 81:19$$

Acetyl methanesulfonate, 5, 5.

Cleavage of cyclopropyl ketones. Cyclopropyl ketones are cleaved under neutral conditions at 20° by the combined reaction of AcOMs and the nucleophile tetramethylammonium bromide or iodide (as a source of Br^- or I^-). CH_3CN is somewhat superior to CH_2Cl_2 as solvent.[1]

Examples:

$$\text{Tricyclic cyclopropyl ketone} \xrightarrow[87\%]{AcOMs,\ Br^-,\ CH_3CN,\ 20°} \text{bicyclic enol acetate (OAc, Br)} + \text{bridged enol acetate (OAc, Br)} \quad 4:1$$

$$\text{Bicyclo[4.1.0]heptan-2-one} \xrightarrow[72\%]{} \text{3-(bromomethyl)cyclohexenyl acetate (OAc, } CH_2Br)$$

$$CH_3\overset{O}{\overset{\|}{C}}{-}\text{cyclopropyl} \xrightarrow[81\%]{} \underset{CH_3\qquad H}{\overset{AcO\qquad CH_2CH_2Br}{C{=}C}} + \underset{AcO\qquad H}{\overset{CH_3\qquad CH_2CH_2Br}{C{=}C}} \quad 3:1$$

[1] M. Demuth and P. R. Raghavan, *Helv.*, **62**, 2338 (1979).

3-Acetyl-4-oxazoline-2-one (1). Mol. wt. 111.10, b.p. 108–112°/24 mm.

Preparation[1]:

$$\text{3-acetyl-2-oxazolidinone } (N{-}COCH_3) \xrightarrow{Cl_2,\ h\nu,\ CCl_4} \text{Cl-substituted 3-acetyl-2-oxazolidinone} \xrightarrow[60\text{–}90\%]{120\text{–}150°} \underset{\mathbf{1}}{\text{3-acetyl-4-oxazolin-2-one}}$$

Diels–Alder reactions. Two laboratories[2,3] have reported use of **1** in Diels-Alder reactions and transformation of the resulting adducts into β-amino alcohols. The reagent is only a moderately active dienophile, somewhat more reactive than vinylene carbonate. The parent heterocycle, 4-oxazoline-2-one, is less useful because of thermal instability at the temperatures usually required for cycloaddition. The adducts from **1** can be deacetylated by treatment with KOH or ammonium hydroxide in CH_3OH at 20°. On hydrolysis with KOH or NaOH in refluxing methanol or ethanol the adducts give *cis*-β-amino alcohols in good yield.

Example:

1 + CH_3, CH_2, CH_3, CH_2 —(160°, ~50%)→ adduct (CH_3, CH_3, H, O, =O, N, H, $COCH_3$) —(KOH, CH_3OH, Δ, 95%)→ (CH_3, OH, CH_3, NH_2)

adduct —(95%, KOH, CH_3OH, 20°)→ (CH_3, H, O, =O, CH_3, N, H, H)

The 4-oxazoline-2-one **1** also undergoes photochemical [2 + 2] cycloaddition with olefins. Unfortunately this reaction shows no regioselectivity. The adducts can also be hydrolyzed to β-aminocyclobutanols.[4]

[1] K.-H. Scholz, H.-G. Heine, and W. Hartmann, *Ann.*, 1319 (1976).
[2] *Idem, ibid.*, 2027 (1977).
[3] J. A. Deyrup and H. L. Gingrich, *Tetrahedron Letters*, 3115 (1977).
[4] K.-H. Scholz, H.-G. Heine, and W. Hartmann, *ibid.*, 1467 (1978).

(S)-1-Alkoxymethyl-2-aminopropane, polymeric (1).

Enantioselective α-alkylation of cyclohexanone. A polymeric form of this chiral amine (**1**) has been prepared as shown in equation (I). The reaction of **1** with cyclohexanone leads to the polymer-bound chiral alkoxyimine (**2**). Alkylation of the anion of **2** followed by mild acid cleavage results in an (S)-2-alkylcyclohexanone (**4**). When methyl iodide is the alkylating reagent, the optical yield is 95%; it is somewhat less when isopropyl iodide is used.[1] These results compare favorably with those obtained by Enders and Eichenauer by alkylation of a chiral hydrazone of cyclohexanone (**7**, 10–11). For a related reaction, *see* Benzyl(methoxymethyl)methylamine, this volume.

[1] P. M. Worster, C. R. McArthur, and C. C. Leznoff, *Angew. Chem., Int. Ed.*, **18**, 221 (1979).

(I) Ⓟ—$C_6H_4CH_2Cl$ + $KOCH_2C^*H(CH_3)$—N(phthalimido) → Ⓟ—$C_6H_4CH_2OCH_2C^*H(CH_3)$—N(phthalimido)

$\xrightarrow{H_2NNH_2}$ Ⓟ—$C_6H_4CH_2OCH_2C^*H(CH_3)$—$NH_2$ (**1**) $\xrightarrow{\text{cyclohexanone}}$ Ⓟ—$C_6H_4CH_2OCH_2C^*H(CH_3)$—N=cyclohexylidene (**2**)

$\xrightarrow{\text{1) LDA, 2) RI}}$ Ⓟ—$C_6H_4CH_2OCH_2C^*H(CH_3)$—N=(2-R-cyclohexylidene)* (**3**) $\xrightarrow[\sim 80\%]{H_3O^+}$ **1** + 2-R-cyclohexanone*

4 [$R = CH_3$, ee 95%; $R = CH(CH_3)_2$, ee 60%]

2-Alkyl-1,3-benzodithioles, (2-H, 2-R-1,3-benzodithiole). These benzodithioles are prepared by reaction of Grignard reagents with 2-alkoxy-1,3-benzodithioles, readily available from the reaction of benzyne with carbon disulfide in the presence of an alcohol (**6**, 96).[1]

Synthesis of ketones. These benzodithioles are converted into rather stable anions on treatment with n-butyllithium at −30°. They react with various electrophiles (alkyl halides, carbonyl compounds, epoxides) to give adducts, which are hydrolyzed to ketones by red HgO and BF_3 etherate in aqueous THF.[2]

Example:

2-Li-2-$CH_2CH_2CH_3$-1,3-benzodithiole (**1**) $\xrightarrow[93\%]{n\text{-}C_4H_9I}$ 2-(C_4H_9-n)-2-$CH_2CH_2CH_3$-1,3-benzodithiole $\xrightarrow[89\%]{HgO,\ BF_3\cdot(C_2H_5)_2O,\ THF,\ H_2O}$ $O{=}C(C_4H_9\text{-}n)(C_3H_7\text{-}n)$

The anions **1** are particularly useful for synthesis of hindered ketones. Thus they react with even hindered trialkylboranes to form organoboranes, which are converted into ketones by alkaline hydrogen peroxide.[3] A typical procedure is formulated.

$$\mathbf{1} + (\text{C}_5\text{H}_9)_3\text{B} \xrightarrow[-30\to 20°]{\text{THF}} \left[o\text{-LiS-C}_6\text{H}_4\text{-S-C}(\text{CH}_2\text{CH}_2\text{CH}_3)(\text{C}_5\text{H}_9)\text{-B}(\text{C}_5\text{H}_9)_2 \right] \xrightarrow[73\%]{\text{H}_2\text{O}_2,\ \text{OH}^-} \text{CH}_3(\text{CH}_2)_2\text{C(=O)-C}_5\text{H}_9$$

If the intermediates are treated with mercury(II) chloride to induce a second migration from boron to carbon and then oxidized, tertiary alcohols can be obtained. The very hindered cyclohexyldicyclopentylcarbinol was prepared in this way in 80% yield from **1** (R = cyclohexyl) and dicyclopentylhexylborane.[4]

$$\mathbf{1} + \text{R}^1{}_2\text{R}^2\text{B} \rightarrow [\text{R}^1\text{R}^2\text{B—CR}^1\text{R}^3\text{SC}_6\text{H}_4\text{SLi}] \xrightarrow{\text{HgCl}_2} \text{R}^2\text{B(X)—CR}^1{}_2\text{R}^3 \xrightarrow[60\text{–}85\%]{\text{H}_2\text{O}_2,\ \text{OH}^-} \text{R}^1{}_2\text{R}^3\text{COH}$$

[1] S. Ncube, A. Pelter, and K. Smith, *Tetrahedron Letters*, 255 (1977).
[2] S. Ncube, A. Pelter, K. Smith, P. Blatcher, and S. Warren, *ibid.*, 2345 (1978).
[3] S. Ncube, A. Pelter, and K. Smith, *ibid.*, 1893 (1979).
[4] *Idem*, *ibid.*, 1895 (1979).

Alkyllithium reagents.

ortho-Lithiation of substituted benzenes.[1] It has been recognized for some time that some substituents on the benzene ring can direct metalation primarily to the *ortho*-position. The most effective groups include CH_2NR_2, $CONR_2$, SO_2NR_2, and OR.[2] This directed lithiation of N-substituted benzamides has been used by two laboratories[3,4] for synthesis of unsymmetrical anthraquinones, as shown in equation (I).

(I) $m\text{-CH}_3\text{OC}_6\text{H}_4\text{CONHC}_6\text{H}_5 \xrightarrow[\text{TMEDA}]{n\text{-BuLi}} [\text{2-Li-3-CH}_3\text{O-C}_6\text{H}_3\text{C(OLi)=NC}_6\text{H}_5] \xrightarrow[97\%]{\text{2-CH}_3\text{O-5-C}_6\text{H}_5\text{CH}_2\text{O-C}_6\text{H}_3\text{CHO}}$ lactone (OCH$_3$, OCH$_2$C$_6$H$_5$, OCH$_3$) $\xrightarrow[\text{quant.}]{\text{Zn, KOH, H}_2\text{O, Py}}$ COOH / OCH$_2$C$_6$H$_5$ / OCH$_3$ / OCH$_3$ diarylmethane $\xrightarrow{\text{Several steps}}$ anthraquinone (OAc, OCH$_3$, OCH$_3$)

Directed lithiation has also proved useful in synthesis of phthalideisoquinoline alkaloids, as shown in equation (II).[4]

(II)

$(C_2H_5)_2NC(=O)$–, OCH_3, OCH_3 $\xrightarrow[\text{2) } CH_3I]{\text{1) } sec\text{-BuLi}}$ CH_3, $(C_2H_5)_2NC(=O)$–, OCH_3, OCH_3 $\xrightarrow[\text{47\% overall}]{\text{1) } sec\text{-BuLi; 2) } CO_2}$

$HOOCCH_2$, $(C_2H_5)_2NC(=O)$–, OCH_3, OCH_3 $\xrightarrow{\text{Several steps}}$ O, O, O, OCH_3, OCH_3 $\xrightarrow{Py \cdot HBr \cdot Br_2}$

Br, O, O, O, OCH_3, OCH_3 **1**

1 + CH_3O, CH_3O, $NHCH_3$ $\xrightarrow[56\%]{}$ CH_3O, CH_3O, O, N, CH_3, O, O, OCH_3, OCH_3 $\xrightarrow{POCl_3}$

CH_3O, CH_3O, N, CH_3, O, O, OCH_3, OCH_3

Reactions with thioesters ($R\overset{S}{\overset{\|}{C}}OC_2H_5$); monothioacetals and vinyl sulfides.[5] Alkyllithiums react with thioesters (**1**) to form adducts (**a**) with the >C=S group exclusively. The adducts can be trapped with methyl iodide to form monothioacetals (**2**), which are hydrolyzed to ketones **3** by chloramine-T (**4**, 75).

Examples:

$$C_6H_5\overset{S}{\overset{\|}{C}}OC_2H_5 + CH_3Li \xrightarrow[-78^\circ]{THF} \left[C_6H_5\underset{OC_2H_5}{\overset{SLi}{C}}-CH_3 \right] \xrightarrow[97\%]{CH_3I}$$

1 **a**

$$C_6H_5\underset{OC_2H_5}{\overset{SCH_3}{C}}-CH_3 \xrightarrow[80\%]{\text{Chloramine-T}, CH_3OH, H_2O, 20^\circ} C_6H_5\overset{O}{\overset{\|}{C}}CH_3$$

2 **3**

$$C_6H_{11}\overset{S}{\overset{\|}{C}}OC_2H_5 \xrightarrow[91\%]{} C_6H_{11}-\underset{OC_2H_5}{\overset{SCH_3}{C}}-CH_3 \xrightarrow[88\%]{} C_6H_{11}\overset{O}{\overset{\|}{C}}CH_3$$

The adducts (**a**) are converted by LDA, with elimination of C_2H_5OH, preferentially into the less substituted thioenolates (**b**), which afford vinyl sulfides (**4**) on alkylation. When the substitution pattern is similar (*i.e.*, $R^2 = H$), a mixture of vinyl sulfides is formed. The (Z)-isomer of **4** is usually formed.

$$(R^1)(R^2)CH\overset{S}{\overset{\|}{C}}OC_2H_5 \xrightarrow{R^3CH_2Li} \left[(R^1)(R^2)CH\underset{OC_2H_5}{\overset{SLi}{C}}-CH_2R^3 \right] \xrightarrow{LDA}$$

a

$$\left[(R^1)(R^2)CH\overset{SLi}{C}=CHR^3 \right] \xrightarrow[60-85\%]{CH_3I} (R^1)(R^2)CH\overset{SCH_3}{C}=CHR^3$$

b **4**

Alkynes.[6] Arylsulfonylalkynes react with organolithium reagents to give products of substitution rather than of addition. The reaction is complete in less than 1 minute, even at -78°. Grignard reagents can be used in the same way, but this reaction is much slower and yields are lower.

$$ArSO_2C\equiv CR + R^1Li \xrightarrow[60-95\%]{THF, -78^\circ} R^1C\equiv CR + ArSO_2Li$$

1,4-*Addition reactions.* α,β-Unsaturated carbonyl compounds **1** and **3** undergo exclusive 1,4-addition reactions with organolithium (and organomagnesium) compounds owing to steric hindrance of the carbonyl group. The adducts **2** and **4** are converted into carboxylic acids by cleavage with potassium *t*-butoxide and H_2O in the case of **2** (*see* potassium *t*-butoxide, this volume) and by treatment with ethanolic

$$(C_6H_5)_3C\text{-CO-CH=CH-R} \xrightarrow[57\text{–}90\%]{R'\text{-Li}} (C_6H_5)_3C\text{-CO-CH}_2\text{-CHR'-R}$$

1 **2**

$$\mathbf{3} \xrightarrow{67\text{–}98\%} \mathbf{4}$$

HCl in the case of **4**. Thus the triphenylmethyl and the substituted piperidyl groups of **1** and **3** function as protecting groups of carboxylic acids.[7]

[1] Review: H. W. Gschwend and H. R. Rodriguez, *Org. React.*, **26**, 1 (1979).
[2] D. W. Slocum and C. A. Jennings, *J. Org.*, **41**, 3653 (1976).
[3] J. E. Baldwin and K. W. Bair, *Tetrahedron Letters*, 2559 (1978).
[4] S. O. de Silva, J. N. Reed, and V. Snieckus, *ibid.*, 5099 (1978); S. O. de Silva and V. Snieckus, *ibid.*, 5103 (1978); S. O. de Silva, I. Ahmad, and V. Snieckus, *ibid.*, 5107 (1978).
[5] L. Narasimhan, R. Sanitra, and J. S. Swenton, *J.C.S. Chem. Comm.*, 719 (1978).
[6] R. L. Smorada and W. E. Truce, *J. Org.*, **44**, 3444 (1979).
[7] D. Seebach and R. Locher, *Angew. Chem., Int. Ed.*, **18**, 957 (1979).

Allyltrimethyltin, $(CH_3)_3SnCH_2CH{=}CH_2$ (**1**). Mol. wt. 204.86.

Homoallylic alcohols.[1] The reagent reacts with carbonyl compounds in the presence of BF_3 etherate to form homoallylic alcohols. In general, aldehydes are more reactive than ketones, and methyl ketones are more reactive than internal ketones.

Examples:

$$C_6H_5CHO + (CH_3)_3SnCH_2CH{=}CH_2 \xrightarrow[92\%]{BF_3\cdot(C_2H_5)_2O,\ CH_2Cl_2, -78°} C_6H_5CH(OH)CH_2CH{=}CH_2$$

$$[CH_3(CH_2)_2]_2C{=}O + (C_4H_9)_3SnCH_2CH{=}CH_2 \xrightarrow[95\%]{} [CH_3(CH_2)_2]_2C(OH)CH_2CH{=}CH_2$$

For a related reaction, *see* Bis(diethylaluminum) sulfate, this volume.

[1] Y. Naruta, S. Ushida, and K. Maruyama, *Chem. Letters*, 919 (1979).

Alumina, 1, 19–20; **2**, 17; **3**, 6; **4**, 8; **6**, 16–17; **7**, 5–7; **8**, 9–13.

Cyclization of vinyl ether epoxides.[1] Epoxy dihydropyranes undergo cyclization in the presence of Lewis acids. Thus treatment of **1** with basic alumina affords **2** in almost quantitative yield. The product can be converted into the keto aldehyde **5** by treatment with a mineral acid followed by oxidation. Another useful reaction of **2** is conversion to the keto lactone **3** by singlet oxygen.

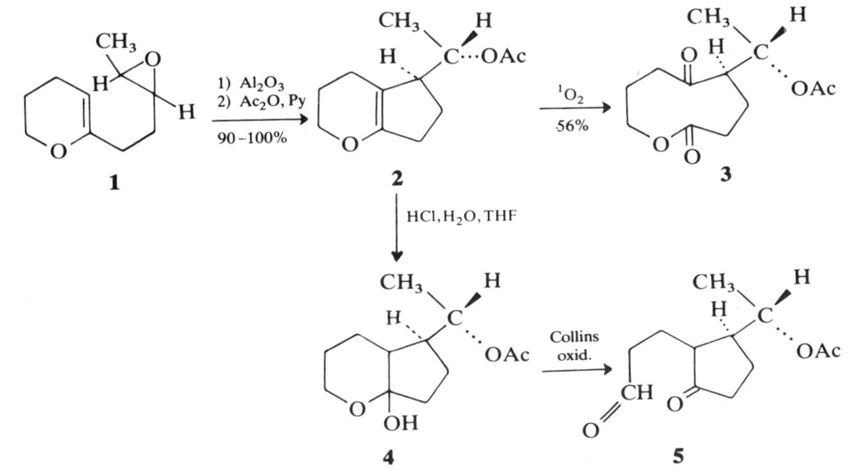

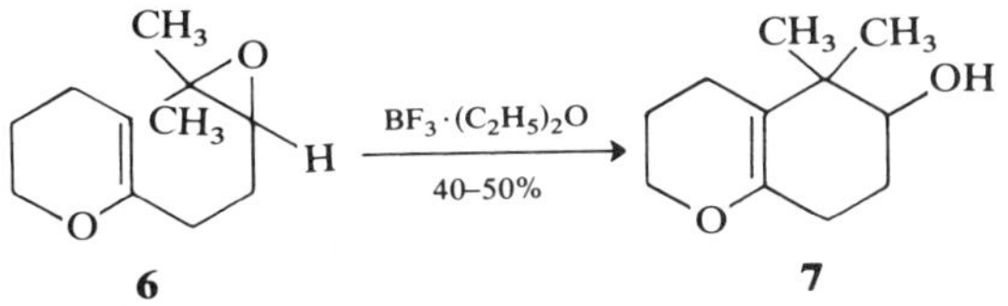

1) Al_2O_3
2) Ac_2O, Py
90–100%

$^1O_2$
56%

HCl, H_2O, THF

Collins oxid.

$BF_3 \cdot (C_2H_5)_2O$
40–50%

The course of the cyclization depends on the substitution pattern of the epoxide; thus **6** cyclizes to **7** only.

Cyclization of a germacrene. Epoxygermacrene-D (**1**) in the presence of basic alumina (20°, 2.5 hours) rearranges to **2**, **3**, and **4** in the yields indicated. Two of these

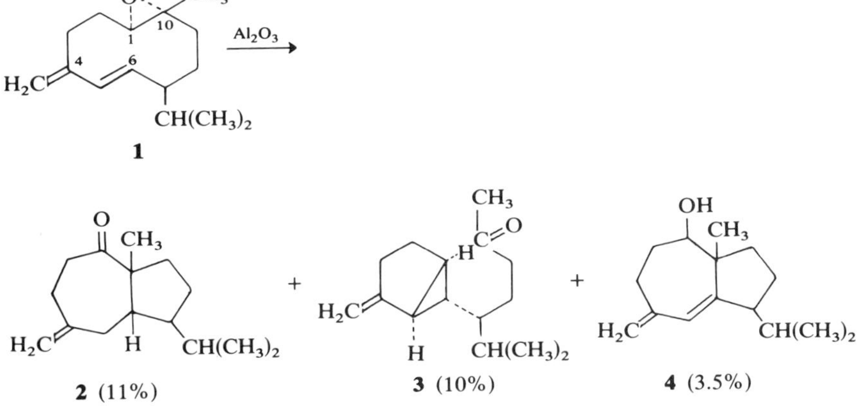

Al_2O_3

2 (11%) + **3** (10%) + **4** (3.5%)

products, **2** and **4**, contain the same carbon skeleton as periplanone-A (**5**), a sexual stimulant of the American cockroach.[2]

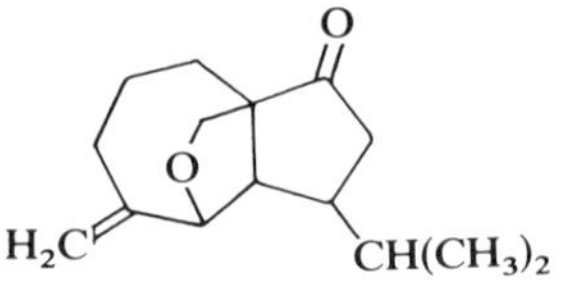

O
H$_2$C
CH(CH$_3$)$_2$
5

These transannular reactions contrast with those observed on acid-catalyzed cyclization of **1**, which results in C_5–C_{10} bond formation (equation I).[3]

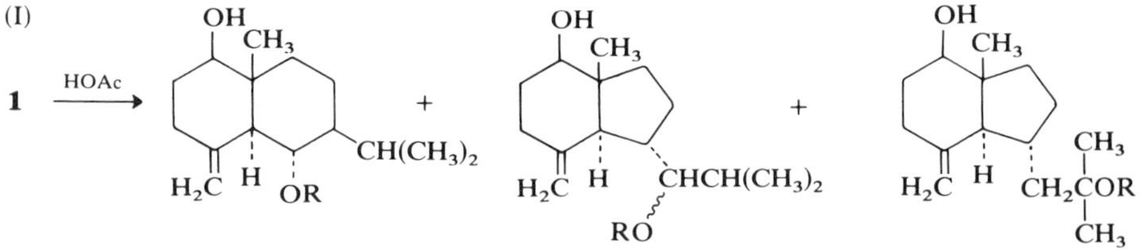

(I) **1** $\xrightarrow{\text{HOAc}}$ **6**/**7** + **8**/**9** + **10**/**11**

6 (R = Ac, 11.6%)
7 (R = H, 9.3%)

8 (R = Ac, 9.3%)
9 (R = H, 9.6%)

10 (R = Ac, 5.9%)
11 (R = H, 5.1%)

Alumina-catalyzed cleavage of epoxides (**6**, 16–17; **8**, 10–12). The nucleophilic opening of epoxides by alcohols can also be promoted by alumina used in catalytic amounts.[4] The reaction then requires higher temperatures and longer reaction periods, but is economical for large-scale preparation. The reaction of cyclohexene oxide with allyl alcohol (equation I) can be carried out on 0.3 moles of the oxide with only 3 g. of W-200-N alumina as purchased. The yield is the same as that obtained with large quantities of alumina doped with 4 weight % of the alcohol.

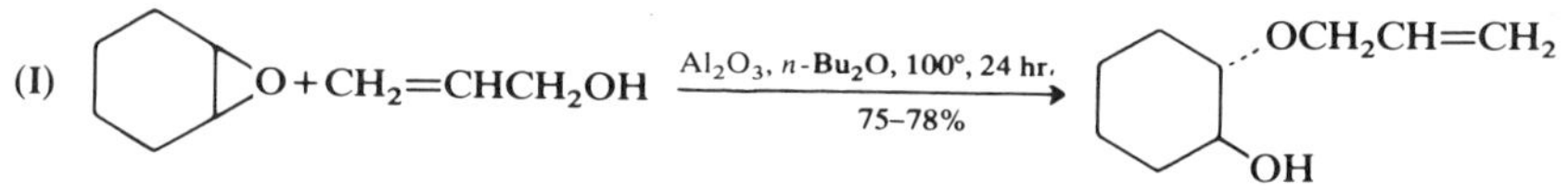

(I) cyclohexene oxide + $CH_2{=}CHCH_2OH$ $\xrightarrow[75\text{–}78\%]{Al_2O_3,\ n\text{-}Bu_2O,\ 100°,\ 24\ hr.}$ 2-($OCH_2CH{=}CH_2$)cyclohexanol (OH)

An intramolecular nucleophilic epoxide cleavage of this type was used to convert **1** into **2**, an intermediate in a synthesis of **3**, which was originally believed to be the structure of the alkaloid cannivonine of the New Brunswick cranberry.[5]

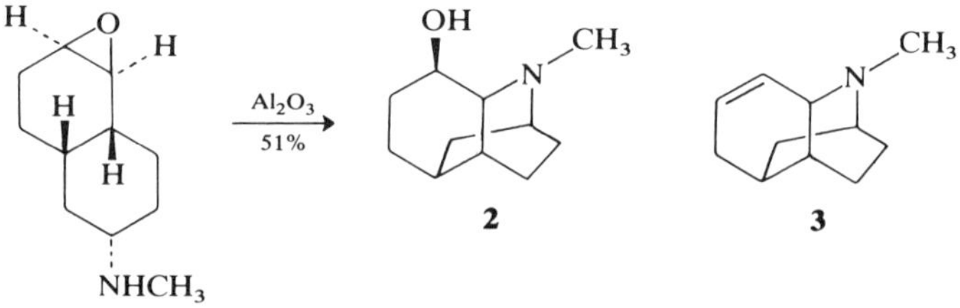

1 (NHCH$_3$) $\xrightarrow[51\%]{Al_2O_3}$ **2** (OH, N–CH$_3$)   **3** (N–CH$_3$)

Dehydrofluorination (**8**, 11–12). The ability of neutral alumina to convert a *gem*-difluorocycloalkane to a cyclic vinyl fluoride was first reported by Strobach and Boswell.[6] The conversion of the difluorosteroid **1** to **2** is particularly facile.

$$\mathbf{1} \xrightarrow[95\text{–}98\%]{Al_2O_3,\ C_6H_{14},\ \Delta,\ 1\ hr} \mathbf{2}$$

Oxidation of thiols. Thiols are oxidized to disulfides in 90–95% yield by oxygenation in the presence of Merck basic aluminum oxide. Benzene is the most satisfactory solvent.[7]

Triphase catalyst.[8] Commercial neutral alumina can function as a triphase catalyst in solid–liquid–solid systems. It can catalyze displacement reactions as well as the permanganate oxidation of alcohols.

[1] R. K. Boeckman, Jr., K. J. Bruza, and G. R. Heinrich, *Am. Soc.*, **100**, 7101 (1978).
[2] M. Niwa, M. Iguchi, and S. Yamamura, *Tetrahedron Letters*, 4291 (1979).
[3] *Idem, ibid.*, 4043 (1978).
[4] G. H. Posner and D. Z. Rogers, *Am. Soc.*, **99**, 8208, 8214 (1977); G. H. Posner and A. Romero, *Org. Syn.*, submitted (1979).
[5] D. A. Evans, A. M. Golob, N. S. Mandel, and G. S. Mandel, *Am. Soc.*, **100**, 8170 (1978).
[6] D. R. Strobach and G. A. Boswell, Jr., *J. Org.*, **36**, 818 (1971).
[7] K.-T. Liu and Y.-C. Tong, *Synthesis*, 669 (1978).
[8] S. Quici and S. L. Regen, *J. Org.*, **44**, 3436 (1979).

Aluminum bromide, 1, 22–23; **2**, 19–21; **3**, 7; **4**, 10; **5**, 10; **6**, 17.

***Bromination of adamantanone* (1).**[1] Bromination of **1** with Br_2 as solvent and a large excess of $AlBr_3$ under the best conditions (55°, 10 hours) gives 1-bromo-4-adamantanone (**2**) in 25% yield. However, if *t*-butyl bromide is added, **2** is obtained in 72–89% yield from a reaction conducted at 20° for 3 days (compare the "sludge" catalyst of Schleyer, **2**, 20; **3**, 7).

$$\mathbf{1} \xrightarrow[72\text{–}89\%]{Br_2,\ AlBr_3,\ BrC(CH_3)_3} \mathbf{2}$$

[1] H. Klein and R. Wiartalla, *Syn. Comm.*, **9**, 825 (1979).

Aluminum chloride, 1, 24–34; **2**, 21–23; **3**, 7–9; **4**, 10–15; **5**, 10–13; **6**, 17–19; **7**, 7–9; **8**, 13–15.

Ene reaction (**5**, 11; **7**, 7–8). Complete details of the ene reaction of methyl propiolate with alkenes catalyzed by aluminum chloride have been published. Since the original report, Snider *et al.*[1] have found that dichloroethylaluminum is superior to $AlCl_3$, probably because it can serve as a proton scavenger as well as a Lewis acid. Optimal yields are obtained with *ca.* 1 equiv. of the catalyst and reaction periods of *ca.* 1 day in CH_2Cl_2 at room temperature. Ene products are formed as the major product in reactions with 1,1-di-, tri-, and tetrasubstituted alkenes. Monosubstituted alkenes give mixtures of ene adducts and [2+2] cycloadducts. 1,2-Disubstituted alkenes form [2+2] cycloadducts stereospecifically; cyclohexene is unusual in that it also gives an ene adduct. Ester, nitrile, ether, and nitro groups do not interfere with the reaction.

Examples:

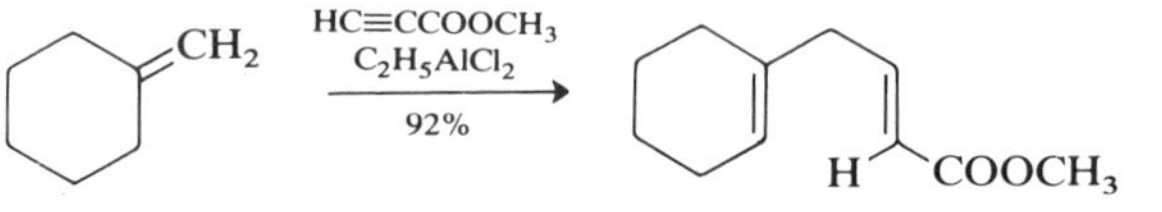

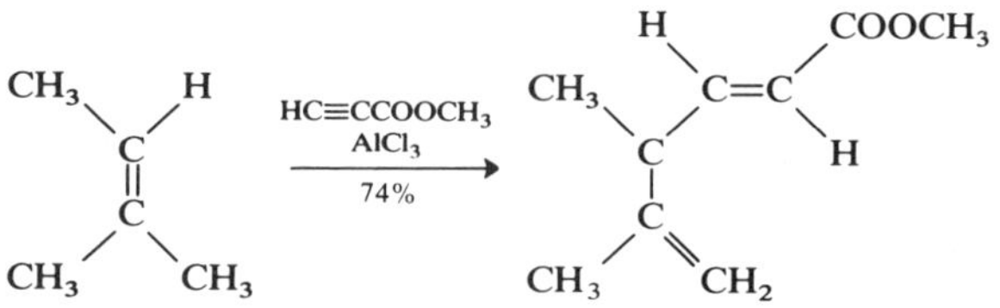

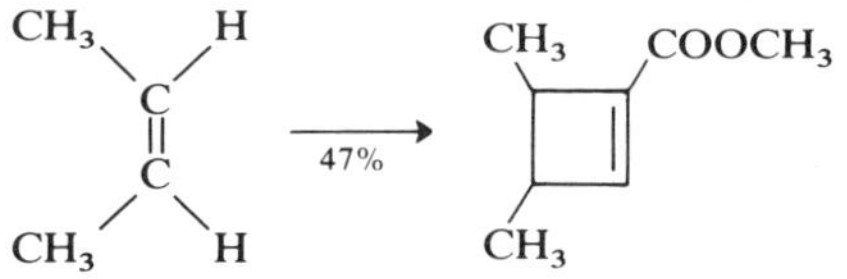

$HC{\equiv}CCOOCH_3$, $C_2H_5AlCl_2$; 92%

$HC{\equiv}CCOOCH_3$, $AlCl_3$; 74%

47%

The catalyzed ene reaction has been used for isoprenylation of monoterpenes with an isopropenyl side chain to form sesquiterpenes of the bisabolane group. An example is the three-step conversion of (+)-limonene (**1**) into β-bisabolene (**2**).[2]

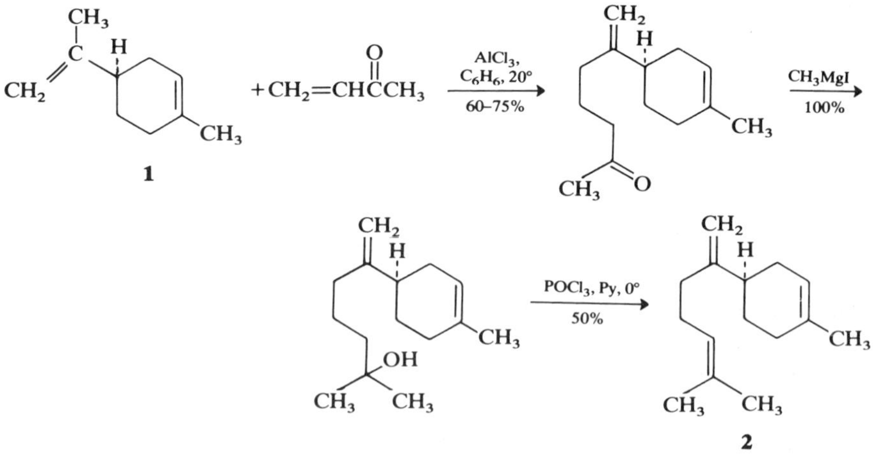

1 $+ CH_2{=}CHCCH_3$ (C=O) — $AlCl_3$, C_6H_6, 20°; 60–75% → — CH_3MgI; 100% →

— $POCl_3$, Py, 0°; 50% → **2**

Diels–Alder catalyst. Aluminum chloride catalyzes the reaction of 2-cycloalkenones (**1**) with 1,3-butadiene to give bicyclic ketones in satisfactory yields. The rate of the addition and the yields are strongly dependent on the quantity of catalyst. When less than 0.3 equiv. of $AlCl_3$ is used, the rate and yield are low; when more than 1 equiv. is used, the desired reaction fails. When R = H, the initial *cis*-adducts isomerize to the *trans*-isomers.[3]

$$\mathbf{1}\ (R = H, CH_3;\ n = 1, 2, 3) + H_2C{=}CH{-}CH{=}CH_2 \xrightarrow{AlCl_3} \mathbf{2}$$

[1] B. B. Snider, D. J. Rodini, R. S. E. Conn, and S. Sealfon, *Am. Soc.*, **101**, 5283 (1979).
[2] G. Mehta and A. V. Reddy, *Tetrahedron Letters*, 2625 (1979).
[3] F. Fringuelli, F. Pizzo, A. Taticchi, and E. Wenkert, *Syn. Comm.*, **9**, 391 (1979).

Aluminum chloride–Ethanethiol.

Dealkylation of esters; cleavage of lactones. The combination of $AlCl_3$ or $AlBr_3$ and a thiol as the nucleophile can cleave esters in high yield under mild conditions (equation I). The reaction is not subject to steric hindrance. Debenzylation of benzyl esters is also possible.

$$\text{(I)}\qquad RC(=O)OR^1 + AlCl_3 + R^2SH \rightarrow RC(OAlCl_2){=}O + R^1SR^2 + HCl;\quad \downarrow\ RCOOH$$

Under these conditions some lactones are cleaved to ω-ethylthiocarboxylic acids.[1]

$$\gamma\text{-butyrolactone} \xrightarrow[88\%]{C_2H_5SH,\ AlBr_3} HOOCCH_2CH_2CH_2SC_2H_5$$

The new method has proved valuable in the case of the very sensitive β-lactam antibiotics, particularly for cleavage of benzyl, benzhydryl, and *t*-butyl esters.[2]

[1] M. Node, K. Nishide, M. Sai, and E. Fujita, *Tetrahedron Letters*, 5211 (1978).
[2] T. Tsuji, T. Kataoka, M. Yoshioka, Y. Sendo, Y. Nishitani, S. Hirai, T. Maeda, and W. Nagata, *ibid.*, 2795 (1976).

Aluminum chloride–Potassium chloride–Sodium chloride. A eutectic mixture of $AlCl_3$ (5.85 g.), KCl (0.848 g.), and NaCl (0.803 g.) is obtained when these salts are heated until a clear solution forms.

Ene catalyst.[1] Aluminum chloride does not catalyze an ene reaction of methyl acrylate with 1-alkenes. However, this modified Lewis acid does catalyze the reaction of methyl acrylate with 1-octene to give the ene product 5-undecenoic acid methyl ester as a mixture of (E)- and (Z)-isomers, **1a** and **1b**, in the ratio 86 : 14. The mixture was converted into **1b** mainly by addition and subsequent elimination of

$$CH_2{=}CHCOOCH_3 + CH_2{=}CH(CH_2)_5CH_3 \xrightarrow[40\%]{AlCl_3,\ KCl,\ NaCl}$$

$$CH_3(CH_2)_4CH{=}CH(CH_2)_3COOCH_3\ (\mathbf{1a},\ E) + CH_3(CH_2)_4CH{=}CH(CH_2)_3COOCH_3\ (\mathbf{1b},\ Z) \quad 86:14$$

$$\mathbf{1a} \xrightarrow[2)\ NaI,\ DMF]{1)\ NBS,\ HCl} \mathbf{1b}$$

BrCl. Hydrolysis of **1b** to the acid and reaction with decyllithium results in **2**, the sex pheromone of the Douglas fir tussock moth.

$$CH_3(CH_2)_9\overset{O}{\overset{\|}{C}}(CH_2)_3\overset{H}{C}{=}\overset{H}{C}(CH_2)_4CH_3$$

2

[1] B. Åkermark and A. Ljunggvist, *J. Org.*, **43**, 4387 (1978).

Aluminum chloride–Potassium phenoxide.

Isoprenylation of phenols. The reaction of phenols and isoprene with this combination of $AlCl_3$ and C_6H_5OK (prepared *in situ* from phenol and K pellets) provides a highly selective synthesis of 2,2-dimethylchromanes (equation I). No other combinations of a phenoxide and a Lewis acid were found to be comparable with respect to selectivity and yield.[1]

(I) $$R^1,R^2,R^3,R^4\text{-phenol} + H_2C{=}C(CH_3)CH{=}CH_2 \xrightarrow[50\text{–}85\%]{AlCl_3\text{–}C_6H_5OK\ (1:1),\ C_6H_6,\ 20^\circ} R^1,R^2,R^3,R^4\text{-2,2-dimethylchromane}$$

[1] L. Bolzoni, G. Casiraghi, G. Casnati, and G. Sartori, *Angew. Chem., Int. Ed.*, **17**, 684 (1978).

Aluminum isopropoxide, 1, 35–37; **3**, 10; **4**, 15–16; **5**, 14; **6**, 19.

Allylic alcohols. A few years ago Eschinasi[1] reported that 3,4-epoxy-*cis*-*p*-menthane (**1**) is rearranged to **2** by aluminum isopropoxide in isopropanol at 120–130°. The *trans*-isomer of **1** rearranges to a mixture of products including **2**.

$$\mathbf{1} \xrightarrow[\sim quant.]{Al(OC_3H_7\text{-}i)_3,\ 120\text{–}130^\circ} \mathbf{2}$$

This reaction has also been used to prepare allylic alcohols (**5**) in high yield from isoprenoid compounds such as **3** by selective epoxidation followed by rearrangement.[2]

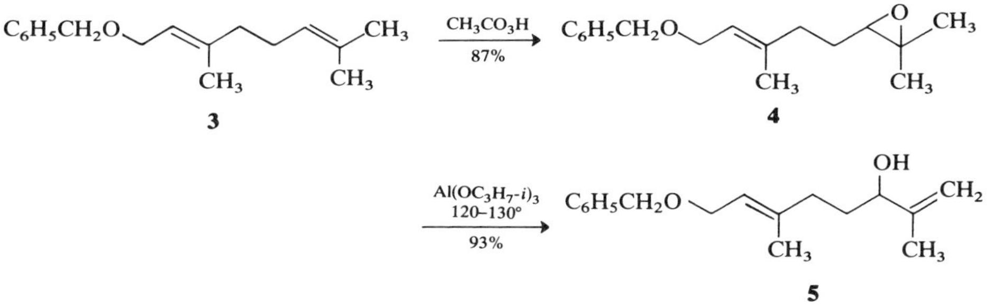

$C_6H_5CH_2O$–(chain)–CH_3 (**3**, with CH_3, CH_3 substituents) $\xrightarrow[87\%]{CH_3CO_3H}$ $C_6H_5CH_2O$–(chain, epoxide O)–CH_3 (**4**, with CH_3, CH_3 substituents)

$\xrightarrow[93\%]{Al(OC_3H_7\text{-}i)_3,\ 120\text{–}130°}$ $C_6H_5CH_2O$–(chain, OH)–CH_2 (**5**, with CH_3, CH_3 substituents)

[1] E. H. Eschinasi, *J. Org.*, **35**, 1598 (1970).
[2] S. Terao, M. Shiraishi, and K. Kato, *Synthesis*, 467 (1979); *idem*, *J. Org.*, **44**, 868 (1979).

Aluminum thiophenoxide, $Al(SC_6H_5)_3$. Mol. wt. 354.49. The reagent is obtained as a white solid by reaction of $Al(CH_3)_3$ with C_6H_5SH.

Phenyl thiolesters. This reagent (2.5 equiv.) converts simple esters into phenyl thiolesters at 5° (4 hours) or at 25° (1 hour).[1]

Examples:

$$(CH_3)_2CHCOOCH_3 \xrightarrow[100\%]{Al(SC_6H_5)_3,\ C_6H_6,\ 25°} (CH_3)_2CHCOSC_6H_5.$$

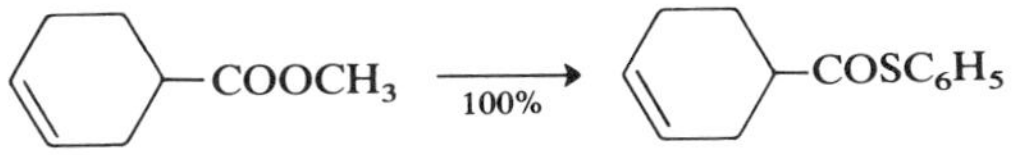

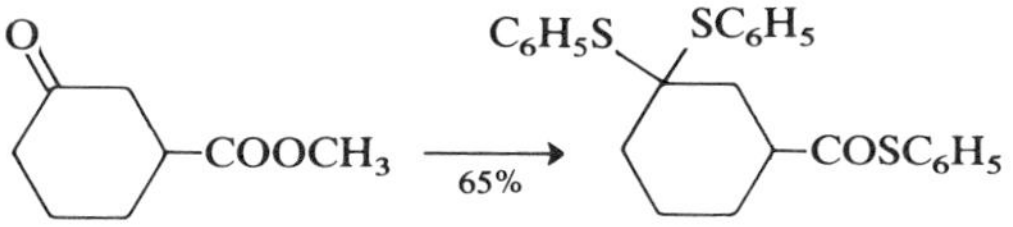

(cyclohexenyl)–$COOCH_3$ $\xrightarrow{100\%}$ (cyclohexenyl)–$COSC_6H_5$

(3-oxocyclohexyl)–$COOCH_3$ $\xrightarrow{65\%}$ (3,3-bis(SC_6H_5)cyclohexyl)–$COSC_6H_5$

Boron thiophenoxide (**6**, 409–410) can also be used for this transformation, but requires a temperature of 140° (refluxing xylene). However, this reagent is useful for conversion of α,β-unsaturated esters to thiolesters.

Examples:

$$(CH_3)_2C{=}CHCOOCH_3 \xrightarrow[90\%]{B(SC_6H_5)_3,\ 140°} (CH_3)_2C{=}CHCOSC_6H_5$$

$$C_6H_5C(H){=}C(H)COOCH_3 \xrightarrow{75\%} C_6H_5C(H){=}C(H)COSC_6H_5$$

Dimethylaluminum phenylthiolate[2] has been used previously for this reaction; dimethylaluminum *t*-butylthiolate has been used to prepare *t*-butyl thiolesters.[3]

Ketene bis(phenylthio)acetals.[4] Reaction of this reagent with carboxylic acids or esters results in simple ketene bis(phenylthio)acetals. Yields are almost quantitative if the acid or ester is branched at the α- or β-position.

Examples:

$$\text{cyclohexyl(H)COOCH}_3\ (\mathbf{1}) \xrightarrow[100\%]{Al(SC_6H_5)_3,\ \text{xylene},\ \Delta} \text{cyclohexylidene}{=}C(SC_6H_5)_2\ (\mathbf{2})$$

$$CH_3CH_2COOCH_3 \xrightarrow[56\%]{} CH_3CH{=}C(SC_6H_5)_2$$

Some of the useful applications of ketene thioacetals of this type are shown in Scheme (I) for some reactions of **2**. For other examples, see lithium naphthalenide, this volume.

1) *sec*-BuLi TMEDA 2) $(C_6H_5S)_2$ ← **2** → 1) *sec*-BuLi, TMEDA 2) CH_3OH

68%

$C(SC_6H_5)_2$ / SC_6H_5 — $CH(SC_6H_5)_2$ — BF_3, 89% from **2** → $CHSC_6H_5$ / SC_6H_5

Cu^+, base ↓ — 73% Cu^+ ↓ — 45% overall, several steps ↓

$C(SC_6H_5)_2$ — $CHSC_6H_5$ — $CHCH_3$ / O

Scheme (I)

Reaction of $Al(SC_6H_5)_3$ with an α,β-unsaturated ester results in a 1,1,3-tris(phenylthio)-1-alkene such as **3**, which was converted into the isoprene derivative **5**, as shown in equation (I).

$$\text{(I)}\quad (CH_3)_2C{=}CHCOOCH_3 \xrightarrow[85\%]{Al(SC_6H_5)_3,\ C_6H_6,\ \Delta} (CH_3)_2C(SC_6H_5)CH{=}C(SC_6H_5)_2\ (\mathbf{3}) \xrightarrow{[C_{10}H_8]^{\dot{-}}Li^+,\ -70^\circ}$$

$$(CH_3)_2C{\cdots}\overset{Li^+}{\bar{C}H}{\cdots}C(SC_6H_5)_2\ (\mathbf{a}) \xrightarrow[71\%]{CH_3OH} (CH_3)_2C{=}CHCH(SC_6H_5)_2\ (\mathbf{4}) \xrightarrow[76\%]{Cu^+,\ \text{base}}$$

$$CH_2{=}C(CH_3){-}CH{=}CHSC_6H_5\ (\mathbf{5})$$

[1] T. Cohen and R. E. Gapinski, *Tetrahedron Letters*, 4319 (1978).
[2] E. J. Corey and D. J. Beames, *Am. Soc.*, **95**, 5829 (1973).
[3] R. H. Hatch and S. M. Weinreb, *J. Org.*, **42**, 3960 (1977).
[4] T. Cohen, R. E. Gapinski, and R. R. Hutchins, *ibid.*, **44**, 3599 (1979).

(S)-1-Amino-2-(methoxymethyl)pyrrolidine, (1), **8**, 16–17.

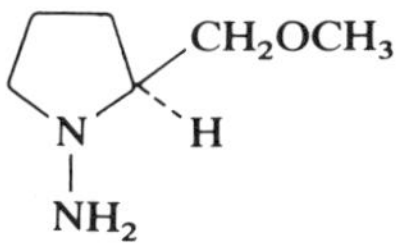

CH_2OCH_3 H N NH_2

Asymmetric α-alkylation of ketones (**7**, 10–11; **8**, 16–17). Enders and Eichenauer[1] have now obtained almost complete asymmetric α-alkylation of acyclic ketones by way of hydrazones formed with **1**. An example is the synthesis of (+)-(S)-4-methyl-3-heptanone (**4**), an ant alarm pheromone, from diethyl ketone. The hydrazone **2** was metalated and alkylated with n-propyl iodide in ether to give **3**.

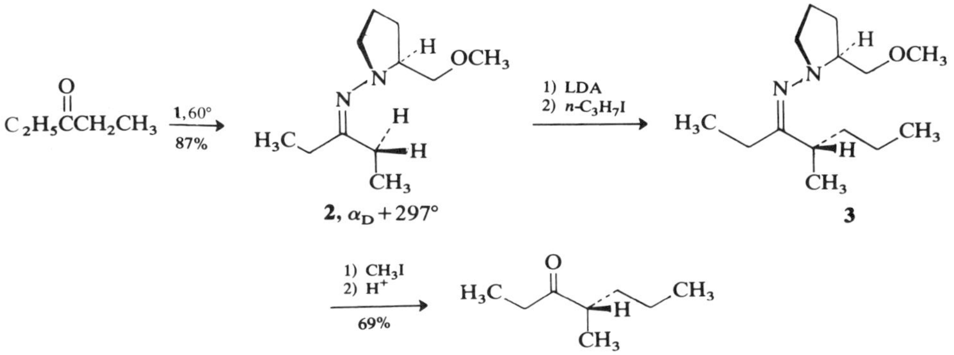

$C_2H_5\overset{O}{\overset{\|}{C}}CH_2CH_3$ $\xrightarrow[87\%]{1, 60°}$ **2**, α_D +297° $\xrightarrow{1)\ LDA\ \ 2)\ n\text{-}C_3H_7I}$ **3**

$\xrightarrow[69\%]{1)\ CH_3I\ \ 2)\ H^+}$ **4**, α_D +22°, 99.5% ee

The crude product was cleaved by acid via the N-methyl iodide. The optical purity of the final product was consistently >99%. The extent of asymmetric induction varies with the solvent; it is highest in ether and DME, and only 20% in THF–HMPT.

The same method was used successfully for synthesis of a number of chiral acyclic ketones.

Enantioselective aldol reaction. The (R)-enantiomer **2** (m.p. 154–155°, α_D + 11.3°) of **1** is prepared in 35% overall yield from (R)-glutamic acid (**3**). Reagent **2** has

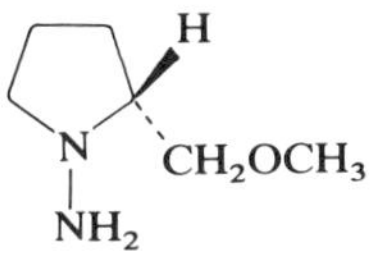

2 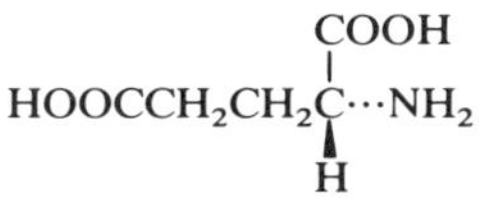 **3** ($HOOCCH_2CH_2C(COOH)(H)\cdots NH_2$)

been used as the chiral auxiliary reagent for synthesis of (+)-(S)-[6]-gingerol (**4**), the major pungent principle of ginger (Scheme I).[2]

[1] D. Enders and H. Eichenauer, *Angew. Chem., Int. Ed.*, **18**, 397 (1979).
[2] D. Enders, H. Eichenauer, and R. Pieter, *Ber.*, **112**, 3703 (1979).

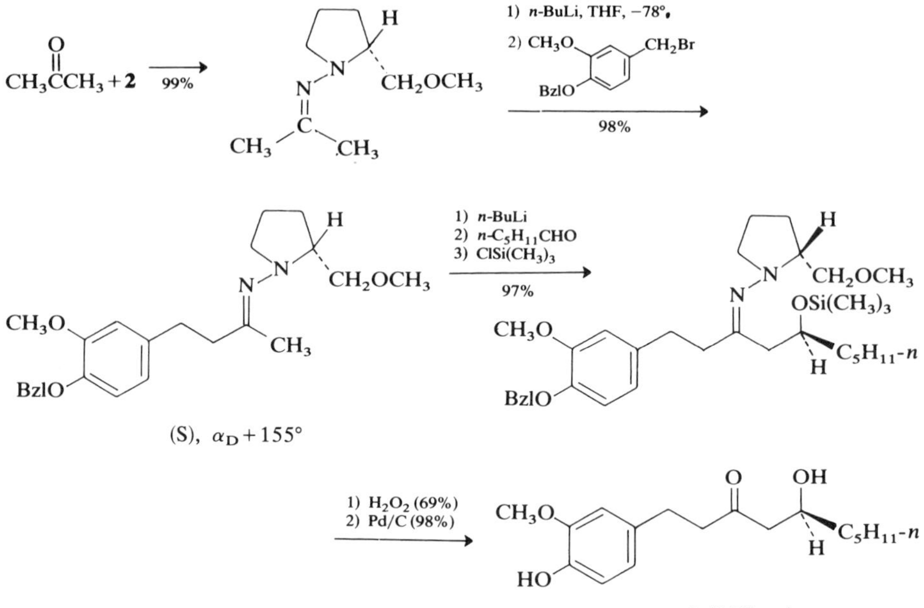

$CH_3COCH_3 + 2 \xrightarrow{99\%}$ [hydrazone of acetone with (S)-2-(methoxymethyl)pyrrolidin-1-amine] $\xrightarrow[98\%]{\text{1) } n\text{-BuLi, THF, } -78°\text{; 2) } CH_3O(BzlO)C_6H_3CH_2Br}$

(S), $\alpha_D + 155°$ $\xrightarrow[97\%]{\text{1) } n\text{-BuLi; 2) } n\text{-}C_5H_{11}CHO\text{; 3) } ClSi(CH_3)_3}$ $\xrightarrow{\text{1) } H_2O_2 (69\%)\text{; 2) Pd/C (98\%)}}$

4 (39% ee)

Scheme I

2-Aminopyridine, (**1**). Mol. wt. 94.12, m.p. 60°, b.p. 208–210°.

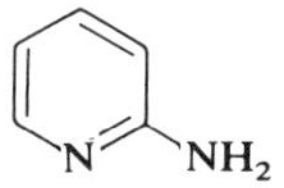

Decarboxylative dimerization of maleic anhydride (**2**). When maleic anhydride (2 equiv.) and **1** (1 equiv.) are heated in HOAc at 110°, CO_2 is evolved, and on acidification dimethylmaleic anhydride (**3**) is obtained in 75% yield.[1]

(I) 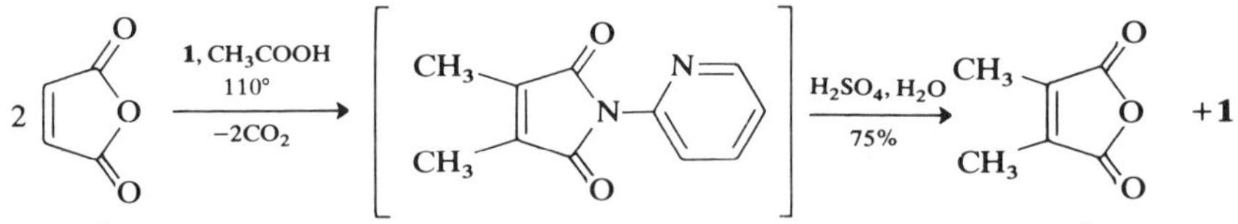 2 **2** $\xrightarrow[-2CO_2]{\mathbf{1}, CH_3COOH, 110°}$ [**a**] $\xrightarrow[75\%]{H_2SO_4, H_2O}$ **3** + **1**

[1] M. E. Baumann and H. Bosshard, *Helv.*, **61**, 2751 (1978).

Ammonium peroxydisulfate–Silver nitrate, 5, 16–17; **8**, 18.

Alkoxycarbonylation of quinones.[1] This couple converts monoesters of oxalic acid in part into the radical ·COOR, which alkoxycarbonylates quinones in satisfactory yield. Naphthoquinone itself is extremely reactive in this reaction and is converted into the 2,3-disubstituted derivative (**1**). However, 2-alkyl-1,4-naphthoquinones are converted into monoalkoxycarbonylated products; this reaction has been used as a route to the naphthacenequinone **4** (equation II).

(I) 1,4-naphthoquinone $+ 2C_2H_5OOCCOOH \xrightarrow[51\%]{(NH_4)_2S_2O_8,\ AgNO_3}$ 2,3-bis($COOC_2H_5$)-1,4-naphthoquinone (**1**)

(II) 2-benzyl-1,4-naphthoquinone (**2**) $\xrightarrow{\sim 50\%}$ 3-($COOC_2H_5$)-2-benzyl-1,4-naphthoquinone (**3**) $\xrightarrow[30\%]{PPA,\ 100°}$ hydroxynaphthacenequinone (**4**)

[1] S. C. Sharma and K. Torssell, *Acta Chem. Scand.*, **B32**, 347 (1978).

(S)-2-(Anilinomethyl)pyrrolidine (1). Mol. wt. 176.25, b.p. 111–112°/0.55 mm., $\alpha_D + 18.5°$.

Preparation from (S)-proline[1]:

(S)-proline (COOH) $\xrightarrow{97\%}$ N-$COOCH_2C_6H_5$, COOH $\xrightarrow{79\%}$ N-$COOCH_2C_6H_5$, $CONHC_6H_5$ $\xrightarrow[94\%]{H_2,\ Pd}$ NH, $CONHC_6H_5$ $\xrightarrow[66\%]{LiAlH_4}$ NH, $CH_2NHC_6H_5$ (**1**)

Optically active α-hydroxy aldehydes.[2] A synthesis of chiral α-hydroxy aldehydes involves reaction of **1** with methyl hydroxymethoxyacetate to form the methoxycarbonyl aminal **2**. The product is treated with a Grignard reagent in the presence of $MgCl_2$ to form the keto aminal **3**. A second Grignard reaction with **3** forms an intermediate (**a**), which is hydrolyzed to **1** and an α-hydroxy aldehyde (**4**);

optical yields are high, 80–100%. The configuration of **4** can be reversed by a change in the order in which the two R groups are introduced.

$$\mathbf{1} + (CH_3O)(HO)CHCOOCH_3 \xrightarrow{C_6H_6} \mathbf{2}\ (\text{COOCH}_3) \xrightarrow[\text{30–80\% from } \mathbf{1}]{R^1MgX,\ MgCl_2} \mathbf{3}\ (\text{COR}^1)$$

$$\xrightarrow{R^2MgX} [\mathbf{a}] \xrightarrow{\text{40–80\%}} \underset{\mathbf{4}}{OHC{-}C(R^1)(R^2)OH} + \mathbf{1}$$

The high optical yields depend at least in part on complexation of the magnesium atom with the ring nitrogen and the carbonyl group to form a rigid structure.

One advantage of this synthesis is that the chiral auxiliary is recovered.

The method has been used to prepare (S)-frontalin (**5**), a pheromone of certain beetles, from (S)-2-hydroxy-2,6-dimethyl-6-heptenal.[3]

5

[1] T. Mukaiyama, M. Asami, J. Hanna, and S. Kobayashi, *Chem. Letters*, 783 (1977).
[2] T. Mukaiyama, Y. Sakito, and M. Asami, *ibid.*, 1253 (1978).
[3] *Idem, ibid.*, 705 (1979).

Antimony(V) fluoride, 5, 9; **6**, 23–24.

Cyclopropyl carbonyl compounds. SbF_5 promotes condensation of α,β-enones with α-diazo carbonyl compounds to form 1,2-disubstituted cyclopropyl carbonyl compounds. BF_3 etherate or TiF_4 is much less efficient. The reaction is subject to steric hindrance by substituents on the vinyl group. SbF_5 also promotes isomerization of the initial *trans–cis* mixture of products to the more stable isomer.[1]

Examples:

$$RCOCHN_2 + H_2C{=}CHCOCH_3 \xrightarrow[\text{70–80\%}]{SbF_5,\ CH_2Cl_2} RCO{-}C_3H_4{-}COCH_3$$

$[R = C_6H_5, OC_2H_5, C(CH_3)_3]$ (*trans/cis* = ~10)

$$RCOCHN_2 + H_2C{=}CHCHO \xrightarrow{20\text{–}70\%} RC(=O)\text{-cyclopropyl-}CHO$$

[1] M. P. Doyle, W. E. Buhro, and J. F. Dellaria, Jr., *Tetrahedron Letters*, 4429 (1979).

Arenechromium tricarbonyl reagents, 6, 27–28, 103–104, 125–126; **7,** 71; **8,** 21–22.

1-Methoxy-1,3-cyclohexadienes (*cf.*, **7**, 21). π-Anisolechromium tricarbonyl reacts with nucleophiles selectively at the *meta*-position (*see also* chromium carbonyl, this volume); the products on protonation give 1-methoxy-1,3-cyclohexadiene derivatives. A typical example is formulated in equation (I) for reaction with 2-lithio-2-methylpropionitrile.

(I) CH_3O-C_6H_5·$Cr(CO)_3$ + $LiC(CH_3)_2CN$ → [Li^+ anionic CH_3O-cyclohexadienyl-$C(CH_3)_2CN$ · $Cr(CO)_3$] $\xrightarrow{CF_3CO_2H}$ CH_3O-cyclohexadiene-$C(CH_3)_2CN$ $\xrightarrow[70\%\ \text{overall}]{H_3O^+}$ cyclohexenone-$C(CH_3)_2CN$

An intramolecular version of this reaction as a route to spirocyclohexenones has also been reported (equation II).[1]

(II) CH_3O-C_6H_4-$(CH_2)_4CN$ $\xrightarrow[70\%]{Cr(CO)_6}$ CH_3O-C_6H_4-$(CH_2)_4CN$ · $Cr(CO)_3$ $\xrightarrow[70\%]{1)\ LDA,\ -78^\circ;\ 2)\ CF_3CO_2H}$ CH_3O-spirodiene-CN $\xrightarrow[96\%]{H_3O^+}$ spirocyclohexenone-CN

[1] M. F. Semmelhack, J. J. Harrison, and Y. Thebtaranonth, *J. Org.*, **44**, 3275 (1979).

Azidotrimethylsilane, 1, 1236; **3,** 316; **4,** 542; **5,** 719–720; **6,** 632. The reagent can be generated conveniently *in situ* from chlorotrimethylsilane and sodium azide in DMF.[1]

Examples:

Cyclohexanone $\xrightarrow[58\%]{ClSi(CH_3)_3,\ NaN_3,\ DMF}$ 1-azido-1-$OSi(CH_3)_3$-cyclohexane (N_3)

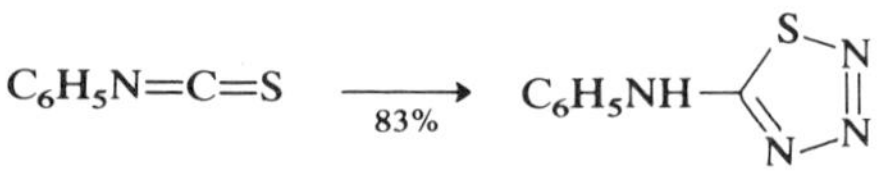

$$C_6H_5N{=}C{=}S \xrightarrow[83\%]{} C_6H_5NH\text{-}(\text{1,2,3,4-thiatriazol-5-yl})$$

[1] H. Vorbrüggen and K. Krolikiewicz, *Synthesis*, 35 (1979).

1,1′-(Azodicarbonyl)dipiperidine (1), 8, 19–20.

Oxidation. Meyers *et al.*[1] used Saigo's reagent (**1**) for oxidation of the primary alcohol **2** to **3** (70.5% yield). The reaction was one step in the total synthesis of maysine (**4**).

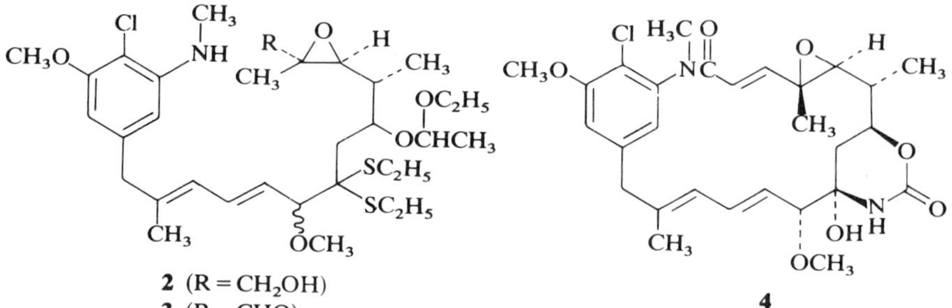

2 ($R = CH_2OH$)
3 ($R = CHO$)

4

[1] A. I. Meyers, D. L. Comins, D. M. Roland, R. Henning, and K. Shimizu, *Am. Soc.*, **101**, 7104 (1979).

B

Barium manganate, 8, 21.

Oxidation of allylic alcohols. Barium manganate is also effective for this oxidation. Thus **1** is oxidized to the ketone **2** in 80–85% yield.[1] The yield with MnO_2 as oxidant is 38%. Two-phase Jones oxidation gives a complex mixture.

$Ba(MnO_4)$, CH_2Cl_2 80–85%

[1] P. J. DeClercq and L. A. Van Royen, *Syn. Comm.*, **9**, 771 (1979).

Barium oxide, BaO.

Alkyl aryl ethers. *o*-Bromophenols are converted into *o*-alkoxyphenols by reaction with an alcohol, BaO, and $CuCl_2$ in DMF at 110° for 3 hours followed by removal of the solvents under reduced pressure. An alternative method is to treat the bromophenol with an alcohol, NaOH, CaO, and $CuCl_2$ in DMF. Yields are in the range 85–95%. The same reaction with a *p*-bromophenol gives *p*-alkoxyphenols in much lower yield. The reaction with 2-bromoanisole fails.[1]

[1] S. Torii, H. Tanaka, T. Siroi, and M. Akada, *J. Org.*, **44**, 3305 (1979).

Benzeneboronic acid, 1, 833–834; **2**, 317; **3**, 221–222; **5**, 513–514.

ortho-*Substitution of phenol by aldehydes.* Some time ago Peer[1] reported that boric acid is a useful catalyst for condensation of phenol with formaldehyde to give saligenol (**2**). Benzeneboronic acid is a much more efficient catalyst particularly in combination with propionic acid. Thus the reaction in the presence of a slight excess of benzeneboronic acid gives the dioxaborin **1** in high yield. This product is converted into saligenin (**2**) by an exchange reaction or by oxidation with hydrogen peroxide.

+ $HCHO + C_6H_5B(OH)_2$ $\xrightarrow[97\%]{C_2H_5COOH,\ C_6H_6,\ \Delta}$ **1** $\xrightarrow[90\%]{}$ **2**

The reaction is generally applicable to other aldehydes and to phenols except those with *ortho* electron-donating groups (OCH_3, $OCOCH_3$)[2].

[1] H. G. Peer, *Rec. trav.*, **79**, 825 (1960).

[2] W. Nagata, K. Okada, and T. Aoki, *Synthesis*, 365 (1979).

Benzeneselenenic acid (1), 8, 24–25.

Cyclic ethers from dienes.[1] The reagent **1** reacts with 1,5-cyclooctadiene in CH_2Cl_2 at 25° to give **2** in 90% yield. Presumably the first step is addition of **1** to one double bond to form a β-hydroxy selenide, which undergoes further reaction with **1** to form the cyclic ether **2**. The product can be converted to **3** or to **4** by oxidation or reduction.

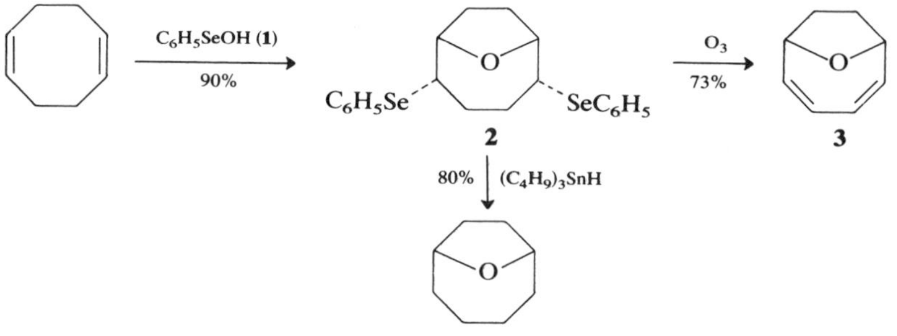

This new cyclization was used for synthesis of eucalyptole (equation I) and linalool (equation II).

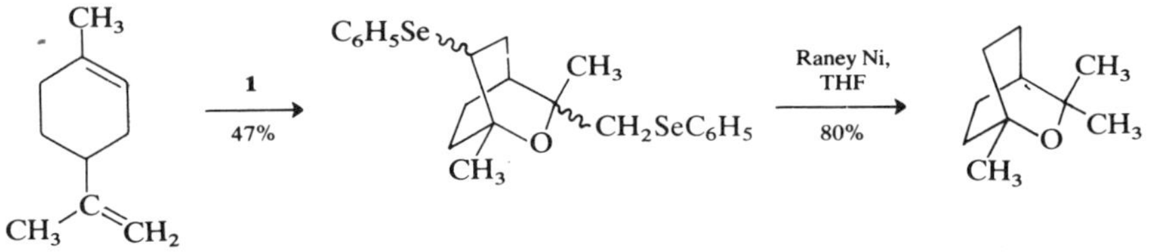

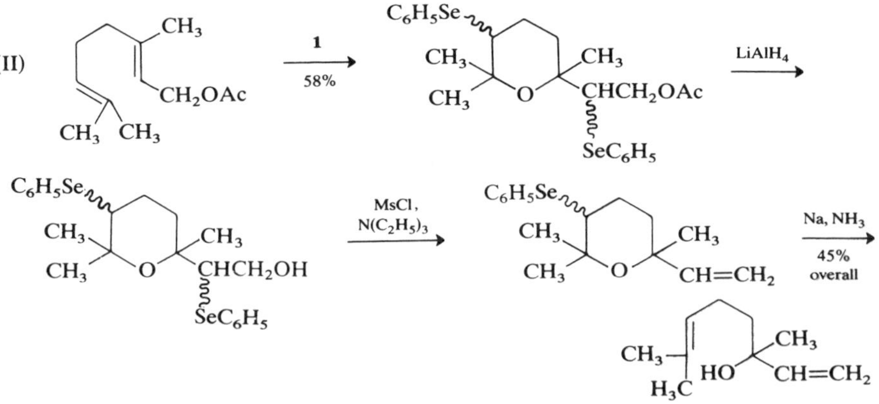

[1] R. M. Scarborough, Jr., A. B. Smith III, W. E. Barnette, and K. C. Nicolaou, *J. Org.*, **44**, 1743 (1979).

Benzeneselenenic anhydride, $(C_6H_5Se)_2O$ (**1**). Mol. wt. 328.13. The anhydride can be generated *in situ* from diphenyl diselenide and *t*-butyl hydroperoxide or from a mixture of diphenyl diselenide and benzeneseleninic anhydride:

$$2(C_6H_5Se)_2 + [C_6H_5\overset{\overset{O}{\|}}{Se}]_2O \rightarrow \underset{\mathbf{1}}{3(C_6H_5Se)_2O}$$

α-Phenylseleno ketones.[1] 1-Alkenes react with **1** to form α-phenylseleno ketones and α-phenylseleno aldehydes. The ratio of these two products is markedly dependent on the solvent. The former products predominate in reactions conducted in DMSO.

Benzyl ethers of allylic alcohols also show a marked regioselectivity in reactions with this reagent, as shown in the second example.

Examples:

$$C_8H_{17}CH{=}CH_2 \xrightarrow[70\%]{(C_6H_5Se)_2O,\ DMSO} C_8H_{17}\overset{\overset{O}{\|}}{C}CH_2SeC_6H_5 + C_8H_{17}\underset{SeC_6H_5}{\underset{|}{C}}HCHO \quad (91:9)$$

$$C_6H_5(CH_2)_2\overset{\overset{C_6H_5CH_2O}{|}}{C}H{-}CH{=}CH{-}CH_3 \xrightarrow{79\%} C_6H_5(CH_2)_2\overset{\overset{C_6H_5CH_2O}{|}}{C}H\underset{SeC_6H_5}{\underset{|}{C}}H\overset{\overset{O}{\|}}{C}CH_3 + C_6H_5(CH_2)_2\overset{\overset{C_6H_5CH_2O}{|}}{C}H\underset{O}{\underset{\|}{C}}\overset{\overset{SeC_6H_5}{|}}{C}HCH_3 \quad (95:5)$$

α-Phenylseleno aldehydes.[2] Oxidation of allylic *t*-butyldimethylsilyl ethers results in α-phenylseleno aldehydes, which on desilylation give α-phenylseleno-α,β-unsaturated enals. A typical example is formulated in equation (I).

$$\text{(I)}\quad C_7H_{15}\overset{\overset{OSi(CH_3)_2C(CH_3)_3}{|}}{C}HCH{=}CH_2 \xrightarrow[87\%]{(C_6H_5Se)_2O} C_7H_{15}\overset{\overset{OR}{|}}{C}H\underset{SeC_6H_5}{\underset{|}{C}}HCHO \xrightarrow[71\%]{KF,\ 18\text{-crown-6}} \underset{C_7H_{15}}{\overset{H}{\;}}C{=}C\underset{SeC_6H_5}{\overset{CHO}{\;}}$$

Allyl methyl ethers of the type shown in equation (II) are also oxidized to α-phenylseleno aldehydes.

$$\text{(II)}\quad \text{cyclohexane-1-(OCH}_3\text{)-1-(CH=CH}_2\text{)} \xrightarrow[89\%]{(C_6H_5Se)_2O} \text{cyclohexane-1-(OCH}_3\text{)-1-}\underset{SeC_6H_5}{\underset{|}{C}}HCHO$$

[1] M. Shimizu, R. Takeda, and I. Kuwajima, *Tetrahedron Letters*, 419 (1979).
[2] *Idem, ibid.*, 3461 (1979).

Benzeneselenenyl bromide and chloride, 5, 518–522; **6**, 459–460; **7**, 286–287.

Phenylselenolactonization. Full experimental details for phenylsulfeno- and phenylselenolactonization are now available.[1]

Reaction with β,γ-unsaturated acids. Phenylselenolactonization (**8**, 26–28) is the usual reaction of benzeneselenenyl chloride with γ,δ- or δ,ε-unsaturated acids. The reaction with acyclic β,γ-unsaturated acids results in adducts that are converted by base into allyl phenyl selenides or phenylselenolactones. The first reaction is favored by a β-substituent, and the second by a γ-substituent.[2]

Examples:

Phenylselenoetherification (**8**, 26–28). A related reaction leading to a potential precursor to deoxy sugars has been reported by Current and Sharpless.[3] Thus the reaction of *p*-chlorobenzeneselenenyl bromide (**6**, 421) with (E)-4-hexenal (**1**) in CCl_4 containing benzyl alcohol at reflux leads to **2** in about 65% yield. This product is converted into **3** on selenoxide elimination. When benzeneselenenyl bromide itself is used, this elimination reaction is very slow. The unsaturated compound can be functionalized in various ways. Of course, alcohols other than benzyl alcohol can be used.

4 (25%) **5** (13%)

Reversal of phenylseleno cyclizations. Sodium in liquid ammonia reacts with phenylselenolactones and ethers to give olefinic products in good yield.[4]

Addition to alkenes. Liotta and Zima[5] have examined the regioselectivity of the addition of C_6H_5SeCl to olefins as well as the stability of the adducts. Even though some of the adducts are thermally and solvolytically unstable, most can be obtained by a reaction conducted below −50°. In general, high regioselectivity obtains. Terminal olefins react initially to form anti-Markownikoff adducts, which isomerize on standing to the more stable Markownikoff adduct (example I). Disubstituted olefins usually react to form the *trans* adducts (example II). The reagent does not react with a conjugated double bond (example III). Adducts with tri- and tetrasubstituted double bonds are unstable at 20°, but the Markownikoff adducts are produced at low temperatures (examples IV–VI). Terpineol (example V) forms an adduct at low temperatures which cyclizes in the presence of triethylamine.

Examples:

(I) $$n\text{-}C_4H_9CH{=}CH_2 \xrightarrow[94\%]{C_6H_5SeCl,\ -10^\circ} n\text{-}C_4H_9CH(SeC_6H_5){-}CH_2Cl \xrightarrow[100\%]{25^\circ,\ 100\text{ hr.}} n\text{-}C_4H_9CH(Cl){-}CH_2SeC_6H_5$$

(II) cyclopentene $\xrightarrow[100\%]{25^\circ}$ *trans*-1-chloro-2-(SeC_6H_5)cyclopentane

(III) $\xrightarrow{}$ [adduct with SeC_6H_5, Cl] $\xrightarrow[100\%]{30\text{ min.}}$ OH, OH, SeC_6H_5, Cl

(IV) 1-methylcyclohexene (CH_3) $\xrightarrow[96\%]{25^\circ}$ CH_3, Cl, SeC_6H_5

(V) CH_3, $HOC(CH_3)_2$ $\xrightarrow[100\%]{-50^\circ}$ H_3C, Cl, SeC_6H_5, $HOC(CH_3)_2$ $\xrightarrow[90\%]{N(C_2H_5)_3}$ CH_3, O, CH_3, CH_3, SeC_6H_5

(VI) $$(CH_3)_2C{=}C(CH_3)_2 \xrightarrow[100\%]{-50^\circ} (CH_3)_2C(Cl){-}C(CH_3)_2(SeC_6H_5)$$

β-Hydroxy selenides. Benzeneselenenyl bromide also adds very readily to terminal alkenes; in fact this reaction can be carried out selectively in the case of polyolefins. The resulting β-bromo selenides can be converted into β-hydroxy selenides by reaction with water in trifluoroethanol, which is superior to DMF or HMPT. β-Azido selenides can be prepared in the same way by use of sodium azide.[6]

$$RCH{=}CH_2 \xrightarrow{C_6H_5SeBr} \underset{\displaystyle Br}{RCH}{-}CH_2SeC_6H_5 \xrightarrow[\text{75–80\% overall}]{H_2O,\ CF_3CH_2OH} \underset{\displaystyle OH}{RCH}{-}CH_2SeC_6H_5$$

$$\underset{\displaystyle Br}{RCH}{-}CH_2SeC_6H_5 \xrightarrow{NaN_3,\ CF_3CH_2OH} \underset{\displaystyle N_3}{RCH}CH_2SeC_6H_5$$

Olefin cyclization. (Z,Z)-1,5-Cyclononadiene (**1**) reacts with benzeneselenenyl chloride in acetic acid buffered with NaOAc with transannular participation by the remote double bond to afford **2** in ~68% yield.[7] Under these conditions, geranyl acetate (**3**) does not afford **6**, but rather the usual olefin addition products **4** and **5**. Cyclization of **4** to **6** does occur on treatment with CF_3CO_2H in CH_2Cl_2; the seleniranium ion **a** is probably an intermediate, since the compound analogous to **4** but lacking the C_6H_5Se— group fails to cyclize under these conditions.[8]

C_6H_5SeCl, HOAc, NaOAc; ~68%

1 → **2** (AcO, H, H, SeC_6H_5)

3 (CH_3, OAc, CH_3, CH_3) → **4** (R = H), **5** (R = Ac) (C_6H_5Se, CH_3, OAc, RO, CH_3, CH_3)

CF_3COOH, CH_2Cl_2

[**a**: C_6H_5—Se⊕, CH_3, OAc, CH_3, CH_3] → 65% → **6** (OH, CH_3, OAc, C_6H_5Se, H_3C, CH_3)

Reaction with α,β-unsaturated ketones. α-Phenylseleno-α,β-unsaturated ketones can be prepared by addition of the enone to a methylene chloride solution of benzeneselenenyl chloride and pyridine (about 1.10 equiv. of each). The reaction at room temperature proceeds slowly (1–6 hours for unhindered enones). As the number of substituents in the γ- and δ-positions is increased, the reaction time increases substantially until no reaction occurs even on prolonged heating. The reaction is believed to involve conjugate addition of pyridine (triethylamine is ineffective) followed by α-selenylation of the enolate and subsequent loss of pyridine and a proton.[9]

Examples:

$$\text{cyclohexenone} \xrightarrow[71\%]{C_6H_5SeCl,\ Py} \text{2-}(SeC_6H_5)\text{cyclohexenone}$$

$$CH_3CH{=}CHCHO \xrightarrow[75\%]{} CH_3CH{=}C(SeC_6H_5)CHO$$

α-Phenylseleno aldehydes and ketals. These substances can be prepared by reaction of C_6H_5SeCl with enol ethers (equation I) and with ethylene ketals (equation II).[10]

$$\text{(I)}\quad R^1R^2C{=}CHOCH_3 + C_6H_5SeCl \xrightarrow{CH_2Cl_2,\ -78^\circ} \left[R^1R^2C(SeC_6H_5){-}CHClOCH_3\right] \xrightarrow[85\text{–}95\%]{NaHCO_3,\ H_2O} R^1R^2C(SeC_6H_5)CHO$$

$$\text{(II)}\quad R^1CH_2(R^2)C(OCH_2CH_2O) + C_6H_5SeCl \xrightarrow[20\text{–}70\%]{CH_2Cl_2,\ 25^\circ} R^1CH(SeC_6H_5)(R^2)C(OCH_2CH_2O)$$

Ester Claisen rearrangement. The first step in a new version of this rearrangement is addition of C_6H_5SeBr to an enol ether such as ethyl vinyl ether (**1**) to give an adduct (**a**) that reacts with an allylic alcohol such as β-methylallyl alcohol (**2**) in the presence of diisopropylamine to form **3** in high yield. The corresponding selenoxide (**4**) is stable at 20°, but decomposes at 140° to form, after hydrolysis, the δ-unsaturated acid **5** in high yield. The conversion of **4** to **5** involves a Claisen rearrangement via **b**. Five examples of this procedure were recorded.[11]

$$\underset{\mathbf{1}}{C_2H_5OCH{=}CH_2} \xrightarrow{C_6H_5SeBr} \underset{\mathbf{a}}{\left[C_2H_5OCH(Br)CH_2SeC_6H_5\right]} + \underset{\mathbf{2}}{HOCH_2C(CH_3){=}CH_2} \xrightarrow[99\%]{\text{Base}}$$

$$\underset{\mathbf{3}}{C_2H_5OCH(CH_2SeC_6H_5)OCH_2C(CH_3){=}CH_2} \xrightarrow[100\%]{NaIO_4}$$

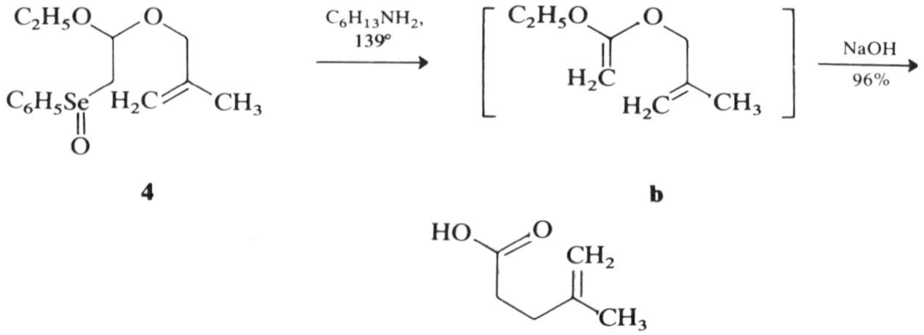

C_2H_5O, O, C_6H_5Se(=O), H_2C, CH_3 (**4**) $\xrightarrow{C_6H_{13}NH_2,\ 139°}$ [C_2H_5O, O, H_2C, H_2C, CH_3] (**b**) $\xrightarrow{NaOH,\ 96\%}$ HO, O, CH_2, CH_3 (**5**)

This sequence is equally applicable to cyclic and bicyclic allylic alcohols.[12]

Example:

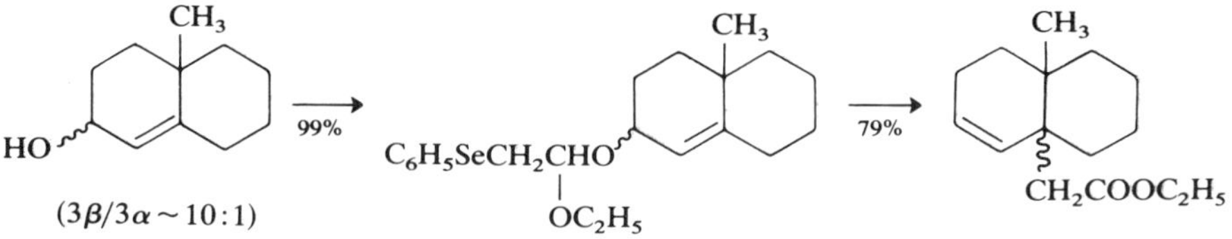

CH_3, HO ($3\beta/3\alpha \sim 10:1$) $\xrightarrow{99\%}$ $C_6H_5SeCH_2CHO$(OC_2H_5), CH_3 $\xrightarrow{79\%}$ CH_3, $CH_2COOC_2H_5$ ($cis/trans \sim 10:1$)

An intramolecular version of this Claisen rearrangement was used for the synthesis of a natural 10-membered lactone, phoracantholide J (**8**).[13] The synthesis involves preparation of **6** by known reactions. Treatment of **6** with C_6H_5SeBr in the presence of Hünig's base gives the cyclic acetal **7** as a mixture of diastereoisomers in 71% yield. Decomposition of the selenoxide results in the rearranged γ,δ-unsaturated lactone **8** as the major product. The ready formation of a 10-membered ring is noteworthy because yields of rings of this size are low on lactonization of ω-hydroxy acids.

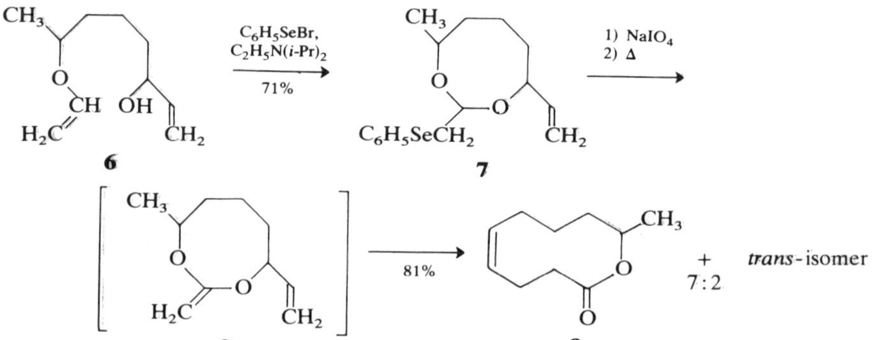

CH_3, O, CH, OH, H_2C, CH_2 (**6**) $\xrightarrow{C_6H_5SeBr,\ C_2H_5N(i\text{-}Pr)_2,\ 71\%}$ CH_3, O, O, $C_6H_5SeCH_2$, CH_2 (**7**) $\xrightarrow{1)\ NaIO_4\ \ 2)\ \Delta}$

[CH_3, O, O, H_2C, CH_2] (**a**) $\xrightarrow{81\%}$ CH_3, O, O (**8**) + *trans*-isomer (7:2)

α,β-Enones. A route to acyclic *trans*-α,β-unsaturated ketones (**4**) from acetylenes involves as the second step reaction of a trialkylalkynylborate salt (**1**), prepared as shown, with benzeneselenenyl chloride. Selective oxidation of **2** at boron is possible with trimethylamine oxide (**3**, 309; **6**, 624), and leads to an isolable seleno ketone (**3**). The last step is hydrogen peroxide oxidation. Overall yields are in the range 60–75% (five examples). One advantage of this approach is that specific enolate formation is not involved; consequently the method is particularly useful for synthesis of unsymmetric acyclic enones.[14]

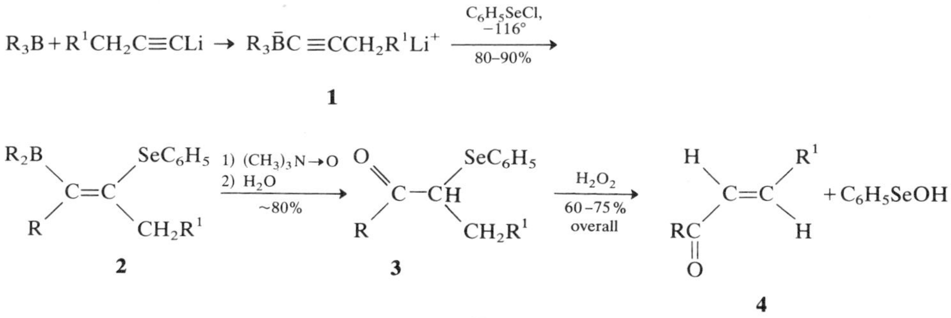

$$R_3B + R^1CH_2C{\equiv}CLi \rightarrow R_3\bar{B}C{\equiv}CCH_2R^1Li^+ \xrightarrow[80\text{–}90\%]{C_6H_5SeCl,\ -116°}$$

1

$$\underset{\mathbf{2}}{(R_2B)(R)C{=}C(SeC_6H_5)(CH_2R^1)} \xrightarrow[\sim 80\%]{1)\ (CH_3)_3N{\rightarrow}O\quad 2)\ H_2O} \underset{\mathbf{3}}{RC(=O){-}CH(SeC_6H_5)(CH_2R^1)} \xrightarrow[60\text{–}75\%\ \text{overall}]{H_2O_2} \underset{\mathbf{4}}{RC(=O)(H)C{=}C(H)(R^1)} + C_6H_5SeOH$$

Cleavage of a cyclopropyl ketone.[15] Reaction of **1** with benzeneselenenyl chloride at room temperature gives a mixture of **2** and **3**. The latter compound was converted into (±)-erysotramidine (**4**) by reaction with CH_3OH and $AgNO_3$ followed by desulfurization with Raney nickel (50% overall yield).

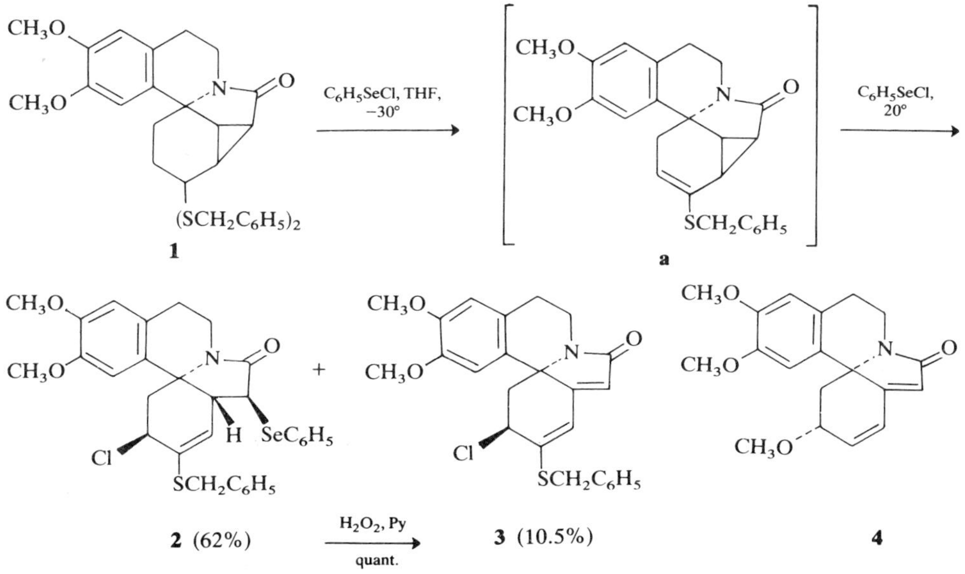

CH_3O, CH_3O, N, O, $(SCH_2C_6H_5)_2$ — **1** $\xrightarrow{C_6H_5SeCl,\ THF,\ -30°}$ [CH_3O, CH_3O, N, O, $SCH_2C_6H_5$ — **a**] $\xrightarrow{C_6H_5SeCl,\ 20°}$

CH_3O, CH_3O, N, O, Cl, H, SeC_6H_5, $SCH_2C_6H_5$ — **2** (62%) + CH_3O, CH_3O, N, O, Cl, $SCH_2C_6H_5$ — **3** (10.5%); CH_3O, CH_3O, N, O, CH_3O — **4**

2 (62%) $\xrightarrow[\text{quant.}]{H_2O_2,\ Py}$ **3** (10.5%)

[1] K. C. Nicolaou, S. P. Seitz, W. J. Sipio, and J. F. Blount, *Am. Soc.*, **101**, 3884 (1979).
[2] D. Goldsmith, D. Liotta, C. Lee, and G. Zima, *Tetrahedron Letters*, 4801 (1979).

[3] S. Current and K. B. Sharpless, *ibid.*, 5075 (1978).
[4] K. C. Nicolaou, W. J. Sipio, R. L. Magolda, and D. A. Claremon, *J.C.S. Chem. Comm.*, 83 (1979).
[5] D. Liotta and G. Zima, *Tetrahedron Letters*, 4977 (1978).
[6] J. N. Denis, J. Vicens, and A. Krief, *ibid.*, 2697 (1979).
[7] D. L. J. Clive, G. Chittattu, and C. K. Wong, *J.C.S. Chem. Comm.*, 441 (1978).
[8] T. Kametani, K. Suzuki, H. Kurobe, and H. Nemoto, *ibid.*, 1128 (1979).
[9] G. Zima and D. Liotta, *Syn. Comm.*, **9**, 697 (1979).
[10] K. C. Nicolaou, R. L. Magolda, and W. J. Sipio, *Synthesis*, 982 (1979).
[11] M. Petrzilka, *Helv.*, **61**, 2286 (1978).
[12] R. Pitteloud and M. Petrzilka, *ibid.*, **62**, 1319 (1979).
[13] *Idem, ibid.*, **61**, 3075 (1978).
[14] J. Hooz and R. D. Mortimer, *Can. J. Chem.*, **56**, 2786 (1978).
[15] K. Ito, F. Suzuki, and M. Haruna, *J.C.S. Chem. Comm.*, 733 (1978).

Benzeneseleninic acid, 8, 28–29.

Selenol esters, $R\overset{O}{\overset{\|}{C}}SeC_6H_5$. Oxidation of N-acylhydrazines with benzeneseleninic acid in CH_2Cl_2 leads to selenol esters, presumably via a diazene. Yields are usually increased by addition of triphenylphosphine and are then in the range 75–95%.[1]

$$R\overset{O}{\overset{\|}{C}}NHNH_2 + C_6H_5SeO_2H \rightarrow \left[R\overset{O}{\overset{\|}{C}}N{=}NH + C_6H_5SeOH \rightarrow R\overset{O}{\overset{\|}{C}}N{=}NSeC_6H_5\right] \xrightarrow{-N_2} R\overset{O}{\overset{\|}{C}}SeC_6H_5$$

[1] T. G. Back and S. Collins, *Tetrahedron Letters*, 2661 (1979).

Benzeneseleninic anhydride, 6, 240–241; **7**, 139; **8**, 29–32.

Oxidation of alcohols. The anhydride oxidizes benzylic alcohols to aldehydes rapidly and in high yield. Allylic and saturated alcohols are oxidized more slowly. 3-Hydroxy steroids can be oxidized to $\Delta^{1,4}$-3-ketones by use of 4 equiv. of the reagent (60% yield). Δ^1-3-Ketones can be obtained from 3-hydroxy-4,4-dimethyl steroids in similar yields.[1]

Benzylic oxidation. Barton *et al.*[2] have reported several examples of oxidation of benzylic hydrocarbons to aldehydes and ketones by benzeneseleninic anhydride. Phenylselenylated by-products are also obtained.

Examples:

$$C_6H_5CH_2C_6H_5 \xrightarrow[90\%]{C_6H_5\overset{O}{\overset{\|}{Se}}O\overset{O}{\overset{\|}{Se}}C_6H_5} C_6H_5\overset{O}{\overset{\|}{C}}C_6H_5$$

$$CH_3\text{-}C_6H_4\text{-}CH_3 \xrightarrow[40\text{–}65\%]{} CH_3\text{-}C_6H_4\text{-}CHO$$

Selective oxidation.[3] The reagent was used to oxidize the unstable phenol **1** to 1-hydroxy-4-isopropyl-7-methoxy-1,6-dimethyl-2(1*H*)-naphthalenone (**2**), a cytokinetic constituent of cotton implicated in a syndrome associated with inhalation of cotton dust.

trans-*3,4-Dihydroxy-3,4-dihydro-7,12-dimethylbenz[a]anthracene* (*3*). Oxidation of phenol **1** with benzeneseleninic anhydride affords the *o*-quinone **2** in 80% yield. Reduction of **2** with $LiAlH_4$ affords the *trans*-diol **3** in 43% yield. This substance is a highly carcinogenic metabolite of 7,12-dimethylbenz[*a*]anthracene.[4]

Angular hydroxylation. The reagent can be used for angular hydroxylation adjacent to enolizable tricyclic ketones such as **1**, **4**, and **6**.[5]

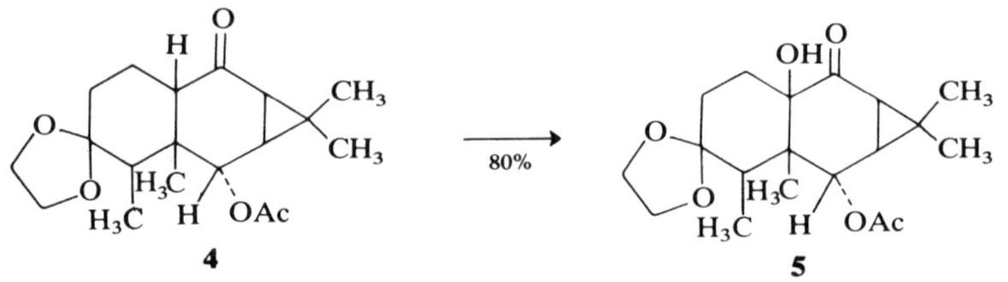

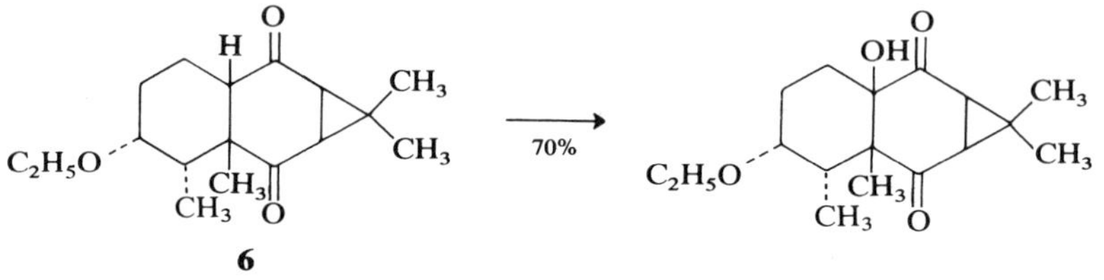

4 $\xrightarrow{80\%}$ **5**

6 $\xrightarrow{70\%}$

Cleavage of selenoacetals and selenoketals. Krief *et al.*[6] have compared several methods for effecting this reaction. Oxidative cleavage with benzeneseleninic anhydride is apparently the preferred method for selenoacetals and for selenoketals of open-chain ketones (80% yields). Cleavage of selenoketals of cycloalkanones is effected in highest yields with $CuCl_2 \cdot 2H_2O$ and CuO in acetone–water. In general, these substrates are more readily cleaved than thioacetals and thioketals, but they are more prone to elimination resulting in vinyl selenides.

[1] D. H. R. Barton, A. G. Brewster, R. A. H. F. Hui, D. J. Lester, S. V. Ley, and T. G. Back, *J.C.S. Chem. Comm.*, 952 (1978).
[2] D. H. R. Barton, R. A. H. F. Hui, D. J. Lester, and S. V. Ley, *Tetrahedron Letters*, 3331 (1979).
[3] P. W. Jeffs and D. G. Lynn, *ibid.*, 1617 (1978).
[4] K. B. Sukumaran and R. G. Harvey, *Am. Soc.*, **101**, 1353 (1979).
[5] K. Yamakawa, T. Satoh, N. Ohba, and R. Sakaguchi, *Chem. Letters*, 763 (1979).
[6] A. Burton, L. Hevesi, W. Dumont, A. Cravador, and A. Krief, *Synthesis*, 877 (1979).

Benzeneselenocyanate–Copper(II) chloride.

Oxyselenation of dienes. Reactions of several dienes with this combination of reagents affords cyclic ethers substituted with two C_6H_5Se groups.[1]

Examples:

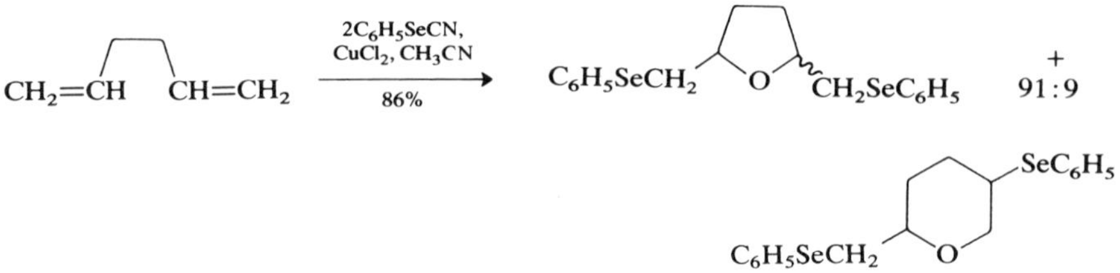

$$CH_2{=}CH\ \ CH{=}CH_2 \xrightarrow[86\%]{2C_6H_5SeCN,\ CuCl_2,\ CH_3CN} C_6H_5SeCH_2\text{–(tetrahydrofuran)–}CH_2SeC_6H_5 + \text{(tetrahydropyran, } C_6H_5SeCH_2,\ SeC_6H_5) \quad 91:9$$

88%

C_6H_5Se … SeC_6H_5 + C_6H_5Se … SeC_6H_5 44:56

[1] S. Uemura, A. Toshimitsu, T. Aoai, and M. Okano, *Chem. Letters*, 1359 (1979).

Benzeneselenol, 6, 28–29. This reagent has been prepared by reaction of selenium with phenylmagnesium bromide (57–71% yield)[1] and by reduction of diphenyl diselenide with hypophosphorous acid[2] or with $NaBH_4$ (86% yield).[3]

Dealkylation of amines.[3] Most amines (primary, secondary, and tertiary) form crystalline salts with benzeneselenol. When heated (150°) some of these salts decompose to form the dealkylated amine and an alkyl phenyl selenide. Salts of primary alkylamines for unknown reasons undergo dealkylation very slowly; hindered, tertiary amines are dealkylated most rapidly.

Deoxygenation of an S-oxide.[4] The Δ^3-cephem S-oxide **1** is reduced to the corresponding sulfide (**2**) by benzeneselenol. If the less hazardous *p*-phenylbenzeneselenol[5] is used, the reaction time can be reduced from *ca.* 3 days to 3–4 hours with little change in yield.

$C_6H_5CH_2CONH$, H, O, S, N, O, CH_3, $COOCH_2CCl_3$ **1** $\xrightarrow[89\%]{C_6H_5SeH,\ CH_2Cl_2,\ CH_3OH}$ $C_6H_5CH_2CONH$, H, S, N, O, CH_3, $COOCH_2CCl_3$ **2**

[1] D. G. Foster, *Org. Syn. Coll. Vol.*, **3**, 771 (1955).
[2] W. H. H. Günther, *J. Org.*, **31**, 1202 (1966).
[3] H. J. Reich and M. L. Cohen, *ibid.*, **44**, 1979 (1979).
[4] M. J. Perkins, B. V. Smith, B. Terem, and E. S. Turner, *J. Chem. Res.* (*S*), 341 (1979).
[5] D. L. J. Clive, W. A. Kiel, S. N. Menchen, and C. K. Wong, *J.C.S. Chem. Comm.*, 657 (1977).

Benzenesulfenyl chloride, 5, 523–524; **6**, 30–32; **8**, 32–34.

Unsaturated sulfoxide–sulfenate rearrangements. This [2,3] sigmatropic rearrangement (**6**, 30–32) has been used twice in an efficient method for conversion of a steriod 17-ketone into a 17α,21-dihydroxy-20-ketone, typical of cortical hormones.[1] The two-carbon side chain is introduced by reaction with dipotassium acetylide to give **2**. This reaction is possible in the presence of a Δ^4-3-keto group, which is converted into the unreactive enolate. The unnatural configuration is inverted to the desired configuration by conversion to the 17-sulfenate ester (**a**), which undergoes [2,3] sigmatropic rearrangement at −40° to the allene sulfoxide **3**. Reaction of **3** with sodium methoxide in methanol results in 1,4-addition to give the enol ether **4**. When heated in methanol, **4** undergoes a [2,3] sigmatropic rearrangement to the sulfenate (**b**), which is converted into **5** and then into **6**.

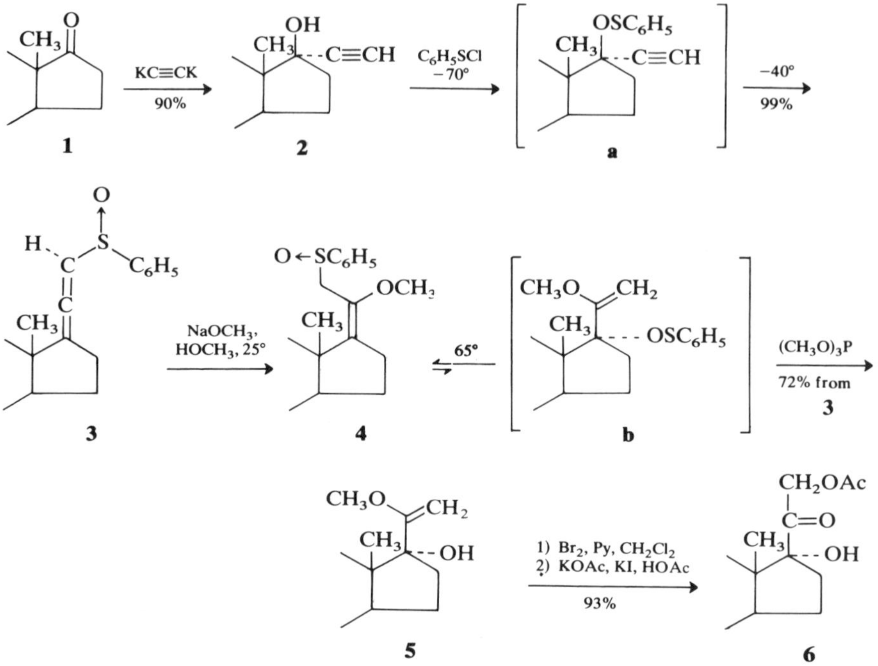

The paper includes a discussion of the stereochemical aspects of these rearrangements.

This sequence was developed for conversion of 9α-hydroxy-4-androstene-3,17-dione, a biodegradation product of the common plant sterol β-sitosterol, to corticosteroids. The transposition of the 9α-hydroxy group to an 11β-hydroxy group can be accomplished readily by dehydration to the 9(11)-ene, addition of HOBr, and reduction of the 9α-bromine with chromium(II) chloride.

γ-Sulfenylation of O-silylated dienolates.[2] The trimethylsilyl enol ethers of α,β- or β,γ-unsaturated esters, ketones, and aldehydes undergo γ-sulfenylation on reaction with this reagent in CH_2Cl_2 at $-78°$. In nine cases, the product of α-attack was observed only once and then only as a minor product.

Examples:

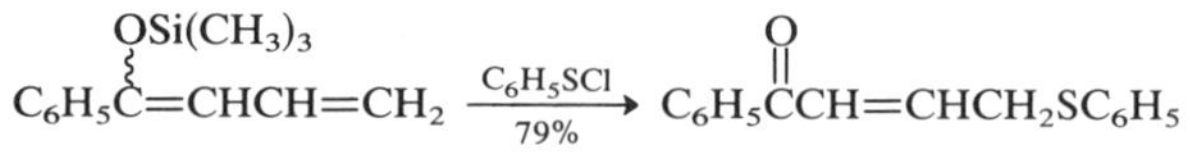

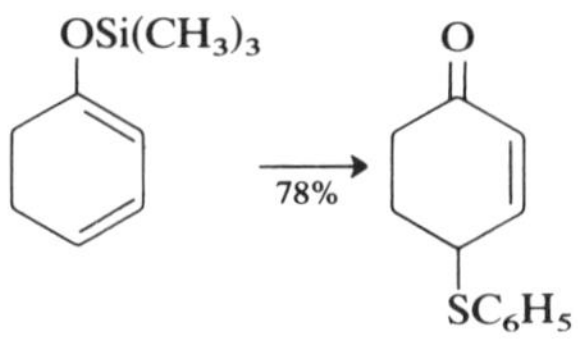

$$C_6H_5C(OSi(CH_3)_3){=}CHCH{=}CH_2 \xrightarrow[79\%]{C_6H_5SCl} C_6H_5C(O)CH{=}CHCH_2SC_6H_5$$

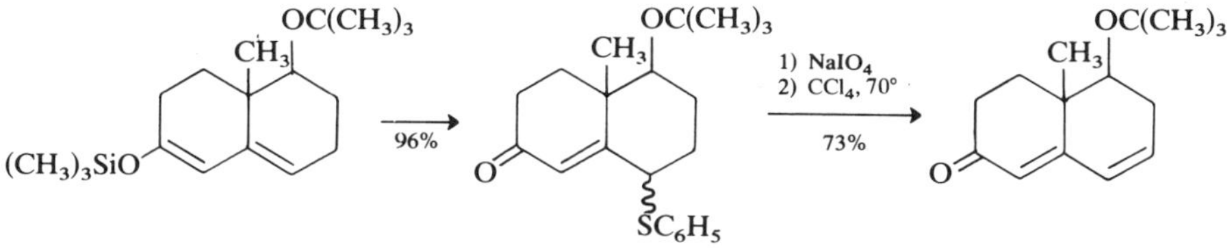

Homoconjugative addition. Benzenesulfenyl chloride adds to 5-methylene-2-norbornene (**1**) to give 1-chloromethyl-3-nortricyclyl phenyl sulfide (**2**) in good yield.[3]

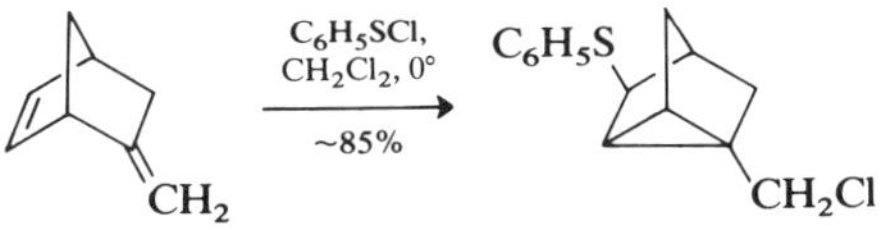

This reaction has been used for a synthesis of tricyclo-*eka*-santalol (**3**), a minor constituent of sandalwood oil (equation I).

Stereoselective syntheses of di- and trisubstituted olefins, (**5**, 400–402; **6**, 30–31). The stereochemistry of [2,3] sigmatropic rearrangements has been reviewed (103 references). The review covers the rearrangements of allyl sulfoxides and allyl selenoxides, as well as Stevens and Wittig rearrangements.[4]

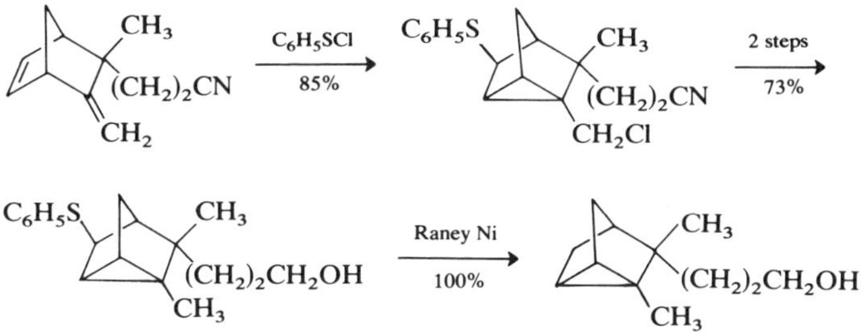

Functionalization of the isopropylidene group. Benzenesulfenyl chloride adds selectively to an isopropylidene terminus of isoprenoids to give a pair of regioisomeric adducts (**1a** and **1b**) in high yield. Separation of the isomers is not necessary because subsequent reactions can involve an intermediate episulfonium cation (**a**). For example, dehydrochlorination of the mixture with $N(C_2H_5)_3$ in DMF gives the allylic sulfide **2** in good yield. This can be converted into the *trans* allylic alcohol **3** by the procedure of Evans (**5**, 400–401). The mixture of **1a** and **1b** is converted into a terminal β-hydroxy sulfide, **4**, by treatment with silica gel.[5] This product can also be converted into **2**.

The addition of benzenesulfenyl chloride occurs selectively to the terminal double bond of farnesol and related polyisoprenoids; it has been used in a stereoselective

$RCH_2(H)C{=}C(CH_3)_2$ —(C_6H_5SCl, quant.)→ $RCH_2CH(SC_6H_5)C(CH_3)_2Cl$ (**1a**) + $RCH_2CHClC(CH_3)_2SC_6H_5$ (**1b**)

1a, **1b** —($DMF, N(C_2H_5)_3$)→ [episulfonium ion **a**: $RCH_2CH{-}C(CH_3)_2$ bridged by $\overset{+}{S}C_6H_5$] —(75–90%)→ $RCH_2CH(SC_6H_5)C(CH_3){=}CH_2$ (**2**)

1b —(65–85%, SiO_2)→ $RCH_2CH(SC_6H_5)C(CH_3)_2OH$ (**4**) —(75%, TsOH, C_6H_6, 50°)→ **2**

2 —(1) $NaIO_4$; 2) $P(OCH_3)_3$; 75–85%)→ $RCH_2(H)C{=}C(CH_3)CH_2OH$ (**3**)

synthesis of solanesol (**5**), an all-*trans* alcohol containing nine isoprene units, from three farnesol units.[6]

$HO{-}[CH_2(CH_3)C{=}CH{-}CH_2]_9{-}OH$

5

The adducts can be converted into allylic alcohols. Thus the regioisomeric adducts **6** can be converted via an episulfonium ion into either **7** or **9** and then into **8** or **11**.[7]

[1] V. VanRheenan and K. P. Shephard, *J. Org.*, **44**, 1582 (1979).
[2] I. Fleming, J. Goldhill, and I. Patterson, *Tetrahedron Letters*, 3205 (1979).
[3] D. Heissler and J.-J. Riehl, *ibid.*, 3957 (1979).
[4] R. W. Hoffmann, *Angew. Chem., Int. Ed.*, **18**, 563 (1979).
[5] Y. Masaki, K. Hashimoto, and K. Kaji, *Tetrahedron Letters*, 4539 (1978).
[6] *Idem, ibid.*, 5123 (1978).
[7] Y. Masaki, K. Hashimoto, K. Sakuma, and K. Kaji, *J.C.S. Chem. Comm.*, 855 (1979).

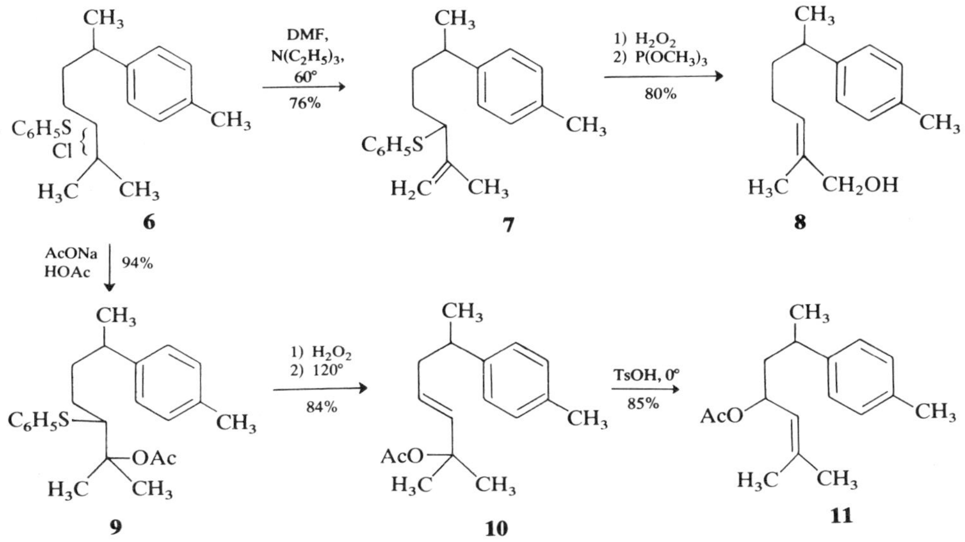

Benzenesulfonylnitrile oxide, $C_6H_5SO_2C{\equiv}N{\rightarrow}O$ (**1**). Mol. wt. 183.10.

Preparation:

$$C_6H_5SO_2CH_2NO_2 \xrightarrow{Br_2,\ NaOAc} C_6H_5SO_2CH(Br)NO_2 \xrightarrow[30\text{–}40\%]{1)\ CH_2N_2,\ 0^\circ;\ 2)\ 40^\circ,\ -CH_2O} C_6H_5SO_2C(Br){=}NOH\ (\mathbf{2}) \xrightarrow{Na_2CO_3} \mathbf{1}$$

Since the reagent dimerizes readily, it is prepared *in situ*.

1,3-Dipolar cycloaddition to olefins; cyanohydrins. Slow addition of the bromo oxime **2** to a mixture of an alkene and aqueous sodium carbonate affords [1.3]cycloadducts in satisfactory yield. With the exception of reactive olefins, a large excess of olefin is required. Tetrasubstituted olefins react sluggishly. The products of these cycloaddition reactions are transformed into *syn*-β-cyano alcohols on reduction with 2% sodium amalgam. The reagent thus effects *syn*-addition of a hydroxy and a cyano group to a double bond.[1]

Examples:

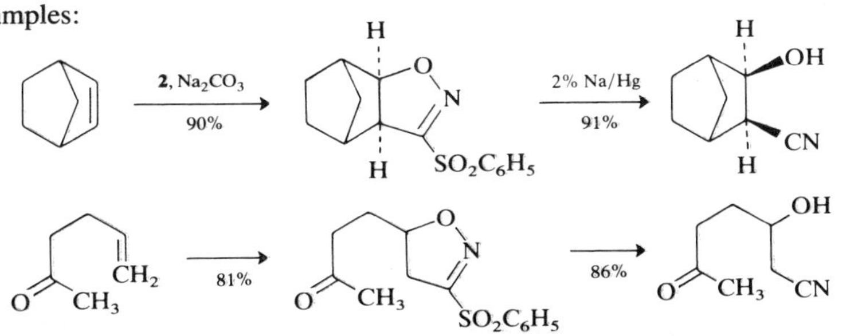

[1] P. A. Wade and H. R. Hinney, *Am. Soc.*, **101**, 1319 (1979).

Benzocyclobutenedione monoethylene ketal, 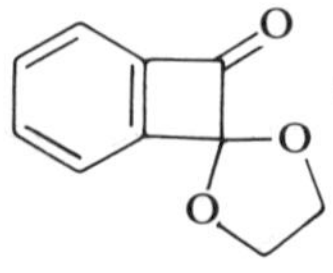 (**1**). Mol. wt. 144.2, m.p. 143–147°. The monoketal is prepared from the corresponding dione by reaction with ethylene glycol with azeotropic removal of water (C_6H_6, reflux, TsOH, 77% yield).[1]

Anthracyclinone synthesis.[2] Reaction of **1** with 1-lithio-3,3,6,6-tetramethoxy-1,4-cyclohexadiene (**7**, 191) affords the quinone monoketal **3** (70% yield). Similarly, **1** reacts with **4** and **5** to afford **6** (47% yield) and **7** (63% yield), respectively. The mechanism of this transformation is unknown. In any event, **1** functions as an equivalent of the 1,4-dipole **8**.

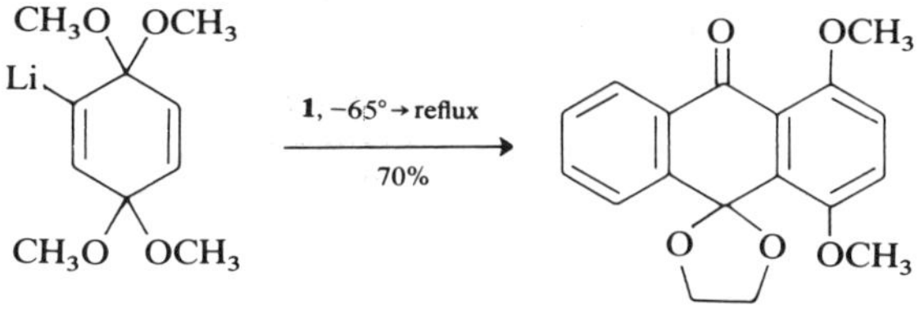

1, −65° → reflux, 70%

2 → **3**

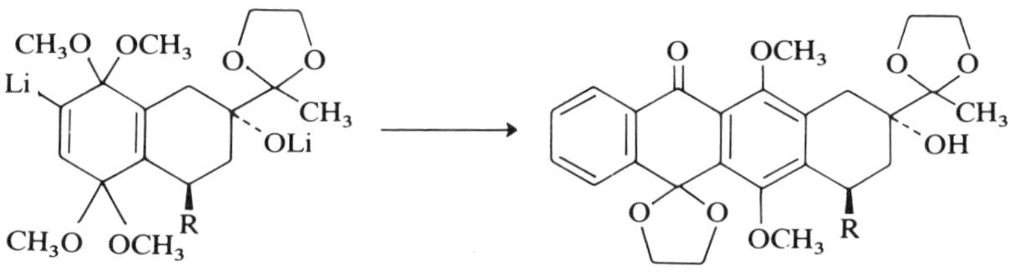

4 (R = H)
5 (R = OCH_3)

6 (47%)
7 (63%)

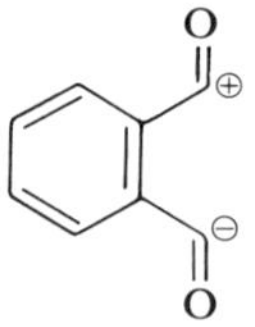

8

[1] M. P. Cava and R. P. Stein, *J. Org.*, **31**, 1866 (1966).
[2] D. K. Jackson, L. Narasimhan, and J. S. Swenton, *Am. Soc.*, **101**, 3989 (1979).

2-Benzoylthio-1-methylpyridinium chloride (1). Mol. wt. 265.76, hygroscopic. Treatment of 1-methyl-2(1*H*)-pyridothione (**2**) with benzoyl chloride in refluxing acetonitrile for 2 hours affords **1** as a hygroscopic precipitate. The reagent can be generated *in situ* for benzoylation reactions by the combination of **2** and C_6H_5COCl in chloroform–water.[1]

$$\text{2 (1-methyl-2(1H)-pyridinethione)} \xrightarrow[CH_3CN]{C_6H_5COCl} \text{1 (1-methyl-2-(benzoylthio)pyridinium chloride)}$$

Benzoylation. Reaction of **1** with carboxylic acids, amines, and phenols produces mixed anhydrides, benzamides, and benzoate esters, respectively, in good yield. These reactions can also be carried out using C_6H_5COCl and 0.3 equiv. of **2** in place of preformed **1**.[1]

$$C_6H_5CH_2CH_2COOH \xrightarrow[NaHCO_3,\ H_2O,\ CHCl_3,\ 82\%]{\mathbf{2},\ C_6H_5COCl} C_6H_5CH_2CH_2COOCOC_6H_5$$

$$O_2N\text{-}C_6H_4\text{-}OH \xrightarrow{81\%} O_2N\text{-}C_6H_4\text{-}OCOC_6H_5$$

[1] M. Yamada, Y. Watabe, T. Sakakibara, and R. Sudoh, *J.C.S. Chem. Comm.*, 179 (1979).

Benzylchlorobis(triphenylphosphine)palladium(II). $C_6H_5CH_2PdCl[P(C_6H_5)_3]_2$ (**1**). Mol. wt. 757.56, m.p. 147–151°, stable to air. The complex is formed from benzyl chloride and $Pd[P(C_6H_5)_3]_4$.[1]

Ketone synthesis.[2] Ketones can be obtained in notably high yield by the reaction of acid chlorides with organotin compounds in the presence of this palladium(II) complex (equation I). The reaction is faster in HMPT, but THF can be used. The reaction is compatible with many functional groups: aldehyde, nitrile, nitro,

$$\text{(I)} \quad R^1COCl + R^2{}_4Sn \xrightarrow{\mathbf{1},\ HMPT,\ 65°} R^1COR^2 + R^2{}_3SnCl$$

methoxy, ester, and aryl halo. Oxygen has an activating effect, triphenylphosphine a deactivating effect. The active catalyst is probably bis(triphenylphosphine)palladium(0), generated *in situ*. The paper discusses a probable catalytic cycle.

Examples:

$$(E)\text{-}C_6H_5CH{=}CHCOCl + (CH_3)_4Sn \xrightarrow[91\%]{\mathbf{1},\ HMPT} (E)\text{-}C_6H_5CH{=}CHCOCH_3$$

$$CH_3COCl + (C_6H_5)_4Sn \xrightarrow{76\%} C_6H_5COCH_3$$

$$CH_3OOC(CH_2)_3COCl + (n\text{-}C_4H_9)_3SnCH{=}CH_2 \xrightarrow[92\%]{} CH_3OOC(CH_2)_3COCH{=}CH_2$$

[1] P. Fitton, J. E. McKeon, and B. C. Ream, *Chem. Commun.*, 370 (1969).
[2] D. Milstein and J. K. Stille, *J. Org.*, **44**, 1613 (1979).

N-Benzylhydroxylamine, $C_6H_5CH_2NHOH$ (**1**). Mol. wt. 123.16, m.p. 58°. The amine is formed by hydrolysis of N-benzylbenzaldoxime or N-benzylacetoxime with hydrochloric acid.[1]

Primary amines.[2] A new method for preparation of primary amines involves N-alkylation of this derivative of hydroxylamine, dehydration of the product, **2**, to **4**, and finally acid hydrolysis of the N-benzylidenealkylamine (**4**) to the amine **5**. HMPT is the solvent of choice for the alkylation step. 2-Fluoro-1-methylpyridinium *p*-toluenesulfonate (**3**) is superior to tosyl chloride for the second step.

$$\text{(I)}\quad \underset{\mathbf{1}}{C_6H_5CH_2NHOH} + RX \xrightarrow[\text{HMPT}]{N(C_2H_5)_3,\ (C_2H_5)_4NI} \underset{\mathbf{2}}{C_6H_5CH_2N(OH)R} \xrightarrow{\mathbf{(3)},\ N(C_2H_5)_3}$$

$$\text{1-methyl-2-pyridone} + \underset{\mathbf{4}}{C_6H_5CH{=}NR} \xrightarrow[\text{65–85\% overall}]{H_3O^+} \underset{\mathbf{5}}{RNH_2} + C_6H_5CHO$$

(**3**: 2-fluoro-1-methylpyridinium TsO⁻)

An alternate method involving a Michael addition of **1** is shown in equation (II).

$$\text{(II)}\quad \mathbf{1} + CH_2{=}CHCOOC_2H_5 \xrightarrow{CH_3OH} C_6H_5CH_2N(OH)CH_2CH_2COOC_2H_5$$

$$\xrightarrow[82\%]{\mathbf{3},\ N(C_2H_5)_3} C_6H_5CH{=}NCH_2CH_2COOC_2H_5 \xrightarrow[\text{quant.}]{H_3O^+} H_2NCH_2CH_2COOC_2H_5 \cdot HCl$$

[1] L. W. Jones and C. N. Sneed, *Am. Soc.*, **39**, 674 (1917).
[2] T. Mukaiyama, T. Tsuji, and Y. Watanabe, *Chem. Letters*, 1057 (1978).

(Benzylmethoxymethyl)methylamine (1), 7, 17.

Asymmetric synthesis of ketones (**7**, 17). Meyers and Williams[1] have extended the asymmetric alkylation of cyclohexanones via the imines formed from **1** to acyclic ketones. Initially optical yields were only 3–44%, but they can be increased to 20–98% by heating the lithioenamines to reflux (THF) prior to alkylation at −78°. Evidently the lithioenamines formed at −20° are mixtures of (E)- and (Z)-isomers. The optical yields are lowered as the size of substituents on the ketone increases.

Example:

$$\underset{(R)\text{-}\mathbf{1}}{C_6H_5CH_2CH(NH_2)CH_2OCH_3} + O{=}C(C_2H_5)CH_2CH_3 \longrightarrow C_6H_5CH_2CH(CH_2OCH_3)N{=}C(C_2H_5)CH_2CH_3 \xrightarrow{\text{1) LDA; 2) } C_2H_5I}$$

$$C_6H_5CH(CH_2OCH_3)N{=}C(C_2H_5)C(CH_3)(H)(C_2H_5) \xrightarrow[48\%]{H_3O^+} O{=}C(C_2H_5)C(CH_3)(H)(C_2H_5)$$

(S, 76% ee)

[1] A. I. Meyers and D. R. Williams, *J. Org.*, **43**, 3245 (1978).

Benzyl(triethyl)ammonium permanganate, $C_6H_5CH_2(C_2H_5)_3\overset{+}{N}MnO_4^-$. Mol. wt. 311.26. The reagent precipitates on reaction of the corresponding chloride with potassium permanganate. It is soluble in organic solvents such as CH_2Cl_2.

Caution[1,2]: This reagent is much less stable than originally suggested. It decomposes violently on drying the crystalline form at 80°/1 mm. The reagent should not be stored, and room temperature should not be exceeded when it is dried *in vacuo*.

Oxidation.[3] Benzylic methylene and methine groups are oxidized in reasonable yields with this reagent in CH_2Cl_2. Some aliphatic methine groups require higher temperatures and use of acetic acid as solvent. In general, higher yields are obtainable by dry ozonolysis.

$$C_6H_5CH_2CH_2CH_2CH_3 \xrightarrow[61\%]{MnO_4^-} C_6H_5\overset{O}{\overset{\|}{C}}CH_2CH_2CH_3$$

$$C_6H_5CH(CH_3)_2 \xrightarrow{98\%} C_6H_5C(OH)(CH_3)_2$$

Decalin (H, H) → 9-hydroxydecalin (OH, H) (37%) + 1-decalone (H, O; H) (43%)

Ethers are oxidized to esters, generally in high yield.

$$CH_3(CH_2)_3O(CH_2)_3CH_3 \xrightarrow[80\%]{} CH_3(CH_2)_2\overset{O}{\overset{\|}{C}}O(CH_2)_3CH_3$$

$$C_6H_5CH_2OC_6H_5 \xrightarrow[94\%]{} C_6H_5\overset{O}{\overset{\|}{C}}OC_6H_5$$

[1] H. Jager, J. Lütolf, and M. W. Meyer, *Angew. Chem., Int. Ed.*, **18**, 786 (1979).
[2] H.-J. Schmidt and H. J. Schafer, *ibid.*, **18**, 787 (1979).
[3] *Idem, ibid.*, **18**, 68 (1979).

Bis(acetonitrile)dichloropalladium(II), 7, 21–22; **8**, 39.

Rearrangement of allylic acetates. This soluble Pd(II) salt is very effective for equilibration of allylic acetates at room temperature. The (E)-isomer is obtained as the major isomer.[1]

Examples:

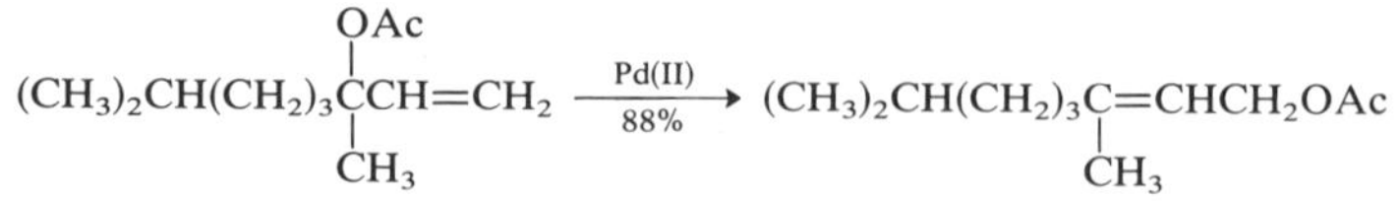

$$(CH_3)_2CH(CH_2)_3\underset{CH_3}{\overset{OAc}{C}}CH{=}CH_2 \xrightarrow[88\%]{Pd(II)} (CH_3)_2CH(CH_2)_3\underset{CH_3}{C}{=}CHCH_2OAc$$

(E/Z = 78:22)

$$\text{1-acetoxy-1-vinylcyclohexane} \xrightarrow[93\%]{} \text{cyclohexylidene}{=}\overset{H}{C}CH_2OAc$$

$$C_3H_7\underset{OAc}{C}HCH{=}CH_2 \rightleftharpoons C_3H_7CH{=}CHCH_2OAc$$

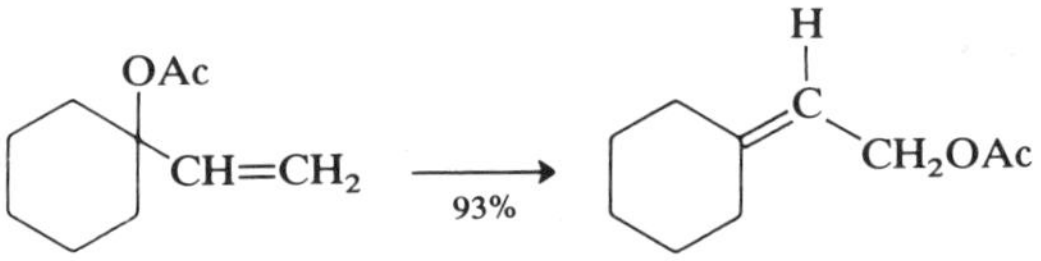

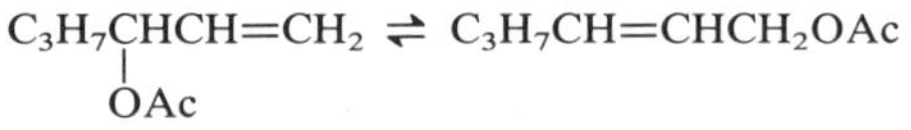

(41%) (59%)

[1] L. E. Overman and F. M. Knoll, *Tetrahedron Letters*, 321 (1979).

Bis(benzonitrile)dichloropalladium(II), 6, 45–47.

Chlorination of steroids.[1] Chlorination of cholesterol with this reagent gives cholesteryl chloride, with the usual retention of configuration shown by Δ^5-3β-ol steroids in displacement reactions. However in other cases, the chlorination is effected with an unusual stereospecificity. Inversion of configuration obtains when the OH group is equatorial, but substitution with retention obtains when the OH group is axial. Thus 3α- and 3β-hydroxysteroids are converted to the same product.

Examples:

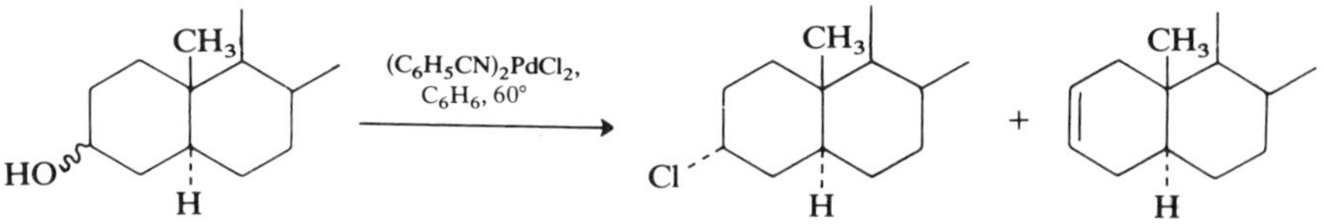

CH3, HO, H (both isomers) $\xrightarrow{(C_6H_5CN)_2PdCl_2,\ C_6H_6,\ 60°}$ CH3, Cl, H (80%) + CH3, H (20%)

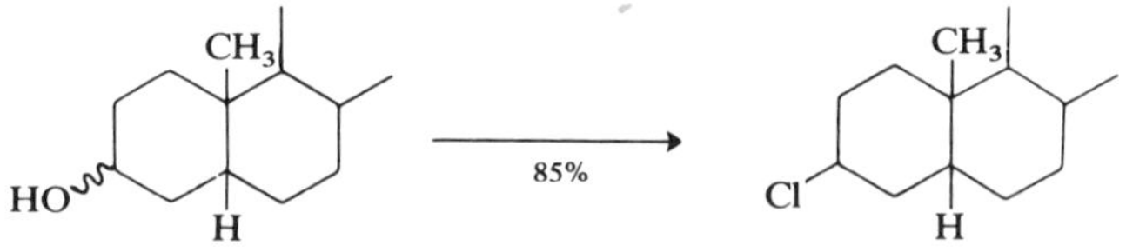

CH3, HO, H (both isomers) $\xrightarrow[85\%]{}$ CH3, Cl, H

Chlorides from hindered sterols are formed in low yield, probably because a bulky coordination complex is an intermediate.

Chlorohydrins from epoxides. Steroid epoxides are converted into chlorohydrins by this reagent (benzene, 20°). The ring opening is of the usual *trans*, diaxial type; conditions are so mild that side reactions observed with hydrochloric acid are avoided.[2]

Examples:

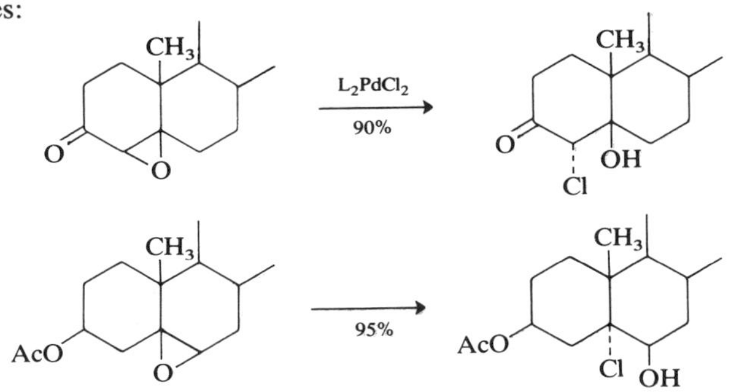

CH_3 ... $\xrightarrow[90\%]{L_2PdCl_2}$ ... CH_3, O, Cl, OH

CH_3, AcO, O $\xrightarrow{95\%}$ CH_3, AcO, Cl, OH

[1] E. Mincione, G. Ortaggi, and A. Sirna, *Tetrahedron Letters*, 4575 (1978).
[2] *Idem, J. Org.*, **44**, 1569 (1979).

Bis(1,5-cyclooctadiene)nickel(0), 4, 33–35; **5**, 34–35; **7**, 428–429.

α-Methylene-γ-lactones. Semmelhack *et al.*[1] have used the recent one step cyclization–lactonization method (**7**, 428–429) for a synthesis of the pseudoguaianolide confertin (**4**). The key step involved cyclization–lactonization of the sulfonium salt **1**, obtained as an E/Z (1 : 4) mixture. Treatment of **1** with zinc–copper gave the *cis*-fused lactone **3**, but with the unnatural configuration. Use of $(COD)_2Ni$

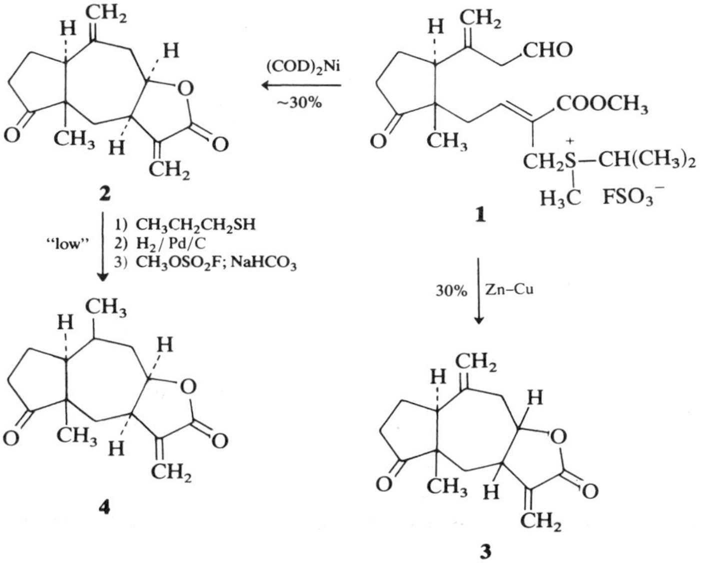

1 (CH_2, CHO, $COOCH_3$, CH_3, $CH_2\overset{+}{S}-CH(CH_3)_2$, H_3C FSO_3^-) $\xrightarrow[\sim 30\%]{(COD)_2Ni}$ **2**

2 $\xrightarrow{\text{"low"}}$ 1) $CH_3CH_2CH_2SH$; 2) $H_2/Pd/C$; 3) CH_3OSO_2F; $NaHCO_3$ → **4**

1 $\xrightarrow[30\%]{Zn\text{–}Cu}$ **3**

gave a mixture of two lactones (2 : 1 ratio). The major product was assumed to be the desired **2**. This was converted into confertin (**4**) in low yield by protection of the α-methylene-γ-lactone group, hydrogenation, and deprotection.

[1] M. F. Semmelhack, A. Yamashita, J. C. Tomesch, and K. Hirotsu, *Am. Soc.*, **100**, 5565 (1978).

Bis(dialkylamino)alanes–Dichlorobis(cyclopentadienyl)titanium, $HAl(NR_2)_2$–Cp_2TiCl_2. The alanes are prepared in quantitative yield by the reaction shown in equation (I).

(I) $$Al + H_2 + 2R_2NH \rightarrow HAl(NR_2)_2 + \tfrac{3}{2}H_2$$

Hydroalumination.[1] Cp_2TiCl_2 is a particularly efficient catalyst for hydroalumination of alkenes with these alanes (equation II). The intermediate adducts can also be trapped with carbonyl compounds, oxygen, or carbon dioxide, but yields are not impressive.

(II) $$RCH{=}CH_2 + HAl(N\text{-}i\text{-}Pr_2)_2 \xrightarrow[90\%]{1)\ Cp_2TiCl_2\ \ 2)\ H_2O} RCH_2CH_3$$

[1] E. C. Ashby and S. A. Noding, *J. Org.*, **44**, 4364 (1979).

Bis(dibenzylideneacetone)palladium(0), $(C_6H_5CH{=}CHCOCH{=}CHC_6H_5)_2Pd$ (**1**). Mol. wt. 575.0, m.p. 150° dec. Preparation.[1]

α-Alkylation of cyclohexenone.[2] Cyclohexenone reacts with methylenecyclopropane in the presence of **1** and methyldiphenylphosphine (1 : 1) as catalyst to afford **2** and **3** in 60 and 19% yields, respectively. The ratio **2** : **3** is improved (*ca.* 9 : 1) by use of excess cyclohexenone. With 3-methylcyclohexenone (**4**), the 1 : 1 adduct **5** is obtained in 81% yield, together with only a trace of a dialkylated product.

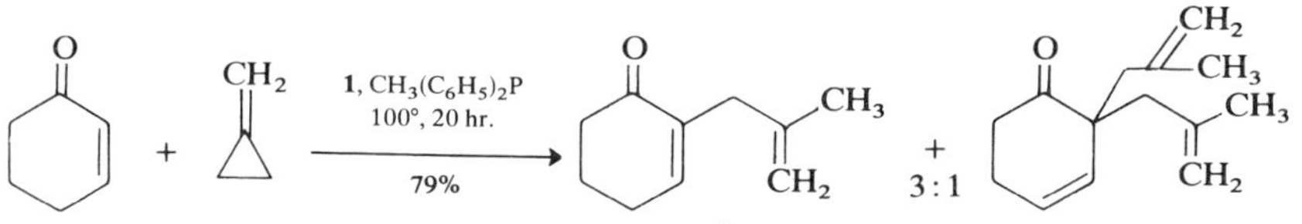

O CH₂ 1, $CH_3(C_6H_5)_2P$, 100°, 20 hr. 79% → **2** + **3** (3 : 1)

4 + CH₂=cyclopropane → 81% → **5**

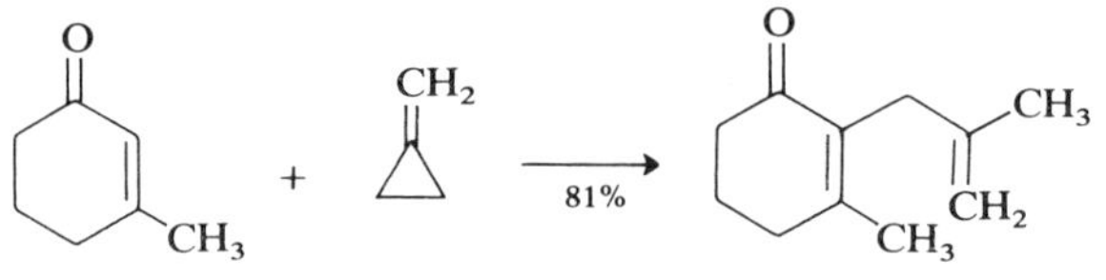

[1] Y. Takahashi, Ts. Ito, S. Sakai, and Y. Ishii, *J.C.S. Chem. Comm.*, 1065 (1970).
[2] G. Balavoine, C. Eskenazi, and M. Guillemont, *ibid.*, 1109 (1979).

Bis(diethylaluminum) sulfate, $[(C_2H_5)_2Al]_2SO_4$ (**1**). Mol. wt. 266.27.

Allylation. This compound is a useful Lewis acid for activation of electrophiles containing oxygen to allylation with allylstannanes. It is usually more effective than BF_3 etherate.[1]

Examples:

$$CH_3(CH_2)_2CHO + (CH_3)_3SnCH_2CH{=}CH_2 \xrightarrow[89\%]{\mathbf{1},\ 80°,\ 16\ hr.} CH_3(CH_2)_2CH(OH)CH_2CH{=}CH_2$$

$$C_6H_5(CH_2)_2CHO + (CH_3)_3SnCH_2CH{=}C(CH_3)_2 \xrightarrow[77\%]{BF_3\cdot(C_2H_5)_2O} C_6H_5(CH_2)_2CH(OH){-}C(CH_3)_2CH{=}CH_2$$

$$\text{2-cyclohexenone} + (CH_3)_3SnCH_2CH{=}CH_2 \xrightarrow[46\%]{\mathbf{1},\ 80°,\ 16\ hr.} \text{3-}(CH_2CH{=}CH_2)\text{cyclohexanone}$$

[1] A. Hosomi, H. Iguchi, M. Endo, and H. Sakurai, *Chem. Letters*, 977 (1979).

(S,S)-(+)-1,4-Bis(dimethylamino)-2,3-dimethoxybutane (DDB),

$$(CH_3)_2NCH_2CH(OCH_3){-}CH(OCH_3)CH_2N(CH_3)_2.$$

Mol. wt. 204.31, b.p. 38°/0.01 mm., α_D +14.7°. Supplier: Aldrich. Preparation.[1]

Enantioselective reactions in chiral solvents. Previous studies of addition of organometallic reagents to carbonyl compounds in chiral solvents indicated generally low stereoselectivities; the highest optical yields were observed with the alkaloid sparteine (one case 18% ee) and with (+)-2,3-dimethoxybutane (one case 17% ee).[2] Seebach *et al.*[1] have prepared chiral compounds from tartaric acid, lactic acid, malic acid, and proline and tested their ability to induce enantioselective addition of *n*-butyllithium to benzaldehyde. Of these, the most effective are DDB and (S,S)-(−)-1,2,3,4-tetramethoxybutane, both derived from natural tartaric acid. The optical yields depend on the temperature, being higher the lower the temperature. The highest optical yield reported in the paper was 40%, observed in the reaction shown in equation (I).

$$\text{(I)}\qquad C_6H_5CHO + n\text{-}C_4H_9Li \xrightarrow[60\text{–}70\%,\ 30\text{–}40\%\ ee]{DDB,\ isopentane,\ -140°} C_6H_5C(OH)(H)C_4H_9\text{-}n\ \ (R)$$

Grignard and Reformatsky reagents also react in this way to give one enantiomer in 10–20% optical yield.[3] Use of the chiral solvent also permits enantioselective 1,4-additions of organometallic reagents to α,β-unsaturated carbonyl compounds, nitro compounds, and ketene thioacetals.[4] In one case, an enantiomeric excess of 43% was observed (equation I).

(I) $CH_3CH{=}CHNO_2$ + 2-lithio-1,3-dithiane (H, Li) $\xrightarrow[72\%]{\text{DDB, pentane, }-78°}$ 1,3-dithiane–CH(H)$CH_3\overset{|}{C}HCH_2NO_2$ (43% ee)

[1] D. Seebach, H.-O. Kalinowski, B. Bastani, G. Crass, H. Daum, H. Dörr, N. P. DuPreez, V. Ehrig, W. Langer, C. Nüssler, H.-O. Oei, and M. Schmidt, *Helv.*, **60**, 301 (1977).
[2] Review: J. D. Morrison and H. S. Mosher, *Asymmetric Organic Reactions*, A.C.S., Washington, D.C., 1976, pp. 415–417.
[3] D. Seebach and W. Langer, *Helv.*, **62**, 1701 (1979).
[4] W. Langer and D. Seebach, *ibid.*, **62**, 1710 (1979).

Bis(N,N-dimethylformamido)oxodiperoxomolybdenum(VI), $MoO(O_2)_2(DMF)_2$ (**1**). Mol. wt. 322.13, m.p. 100–102°, water soluble, stable on storage. The complex is prepared (65% yield) by reaction of H_2MoO_4 with 30% H_2O_2 and 2 equiv. of DMF. Several molybdenum pentoxide complexes are known. Two useful ones are $MoO_5 \cdot HMPT$ (**4**, 203–204) and $MoO_5 \cdot Py \cdot HMPT$ (**5**, 269–270). These HMPT complexes are more easily prepared because they are insoluble in water, but the released HMPT can interfere with work-up.

Hydroxamic acids (**4**, 202). This oxidant is preferred for oxidation of trimethylsilylated amides to hydroxamic acids.[1] Although the yields on oxidation of silylated secondary aliphatic amides with **1** are usually only moderate (15–40%), hydroxylation of acetanilides proceeds consistently in yields of 40–50%. The reaction is efficient for oxidation of benzoxazinones, for example, **2** → **3**. The products are useful herbicides because of inhibition of auxin.

(I) **2** (2-$OSi(CH_3)_3$, N-$Si(CH_3)_3$ benzoxazinone) $\xrightarrow{\text{1) 1; 2) EDTA}}$ **3** (2-OH, N-OH benzoxazinone)

[1] S. A. Matlin, P. G. Sammes, and R. M. Upton, *J.C.S. Perkin I*, 2481 (1979).

***trans*-2,3-Bis(diphenylphosphino)-[2.2.1]-5-bicycloheptene (Norphos) (1).** Mol. wt. 428.41.

Preparation[1]:

$$\text{cyclopentadiene} + (C_6H_5)_2P(O)CH=CHP(O)(C_6H_5)_2 \longrightarrow \text{bicyclo[2.2.1]heptene-}2,3\text{-}(PO(C_6H_5)_2)_2$$

$$\xrightarrow[\text{2) } HSiCl_3]{\text{1) resolve with L-(}-\text{)-dibenzoyltartaric acid}} \text{bicyclo[2.2.1]heptene-}2,3\text{-}(P(C_6H_5)_2)_2 \quad (-)\text{-}\mathbf{1}$$

Both enantiomers of **1** are available from the resolution of the intermediate phosphine oxide.

Asymmetric hydrogenation.[1] The chiral bisphosphine **1** is an easily prepared optically active chelating reagent. The hydrogenation of **2** using the rhodium complex prepared from (−)-**1** and $Rh_2Cl_2(COD)_2$ (COD = 1,5-cyclooctadiene) results in one of the highest optical yields reported for asymmetric hydrogenation. In contrast to several other chelating phosphines, both enantiomers of **1** are available and therefore both configurations of the hydrogenation products can be obtained.

$$\underset{\mathbf{2}}{H_5C_6CH=C(NHAc)COOH} \xrightarrow[Rh_2Cl_2(COD)_2]{(-)\text{-}\mathbf{1},\ H_2,\ C_2H_5OH} \underset{(R)}{C_6H_5CH_2CH(NHAc)COOH} + \underset{(S)}{C_6H_5CH_2CH(NHAc)COOH} \quad 2:98$$

[1] H. Brunner and W. Pieronczyk, *Angew. Chem., Int. Ed.*, **18**, 620 (1979).

2,4-Bis(4-methoxyphenyl)-1,3,2,4-dithiadiphosphetane-2,4-disulfide (1). Mol. wt. 202.23, m.p. 228°, hygroscopic. The reagent is prepared in 80% yield by reaction of anisole with P_4S_{10} at reflux for 1 hour (*caution*: H_2S is formed).[1]

$$CH_3O-C_6H_4-P(=S)(\mu\text{-}S)_2P(=S)-C_6H_4-OCH_3$$

Thiation. This dimer (**1**) of *p*-methoxyphenylthionophosphine sulfide is an efficient reagent for thiation of different classes of carbonyl compounds: ketones, amides, esters,[2] lactones,[1,3] lactams, and imides.[4] Reactions are typically conducted in refluxing xylene or toluene for 2–10 hours and usually proceed in yields >80%. A typical reaction is that with butyrolactone to form thionobutyrolactone.

$$\text{butyrolactone} \xrightarrow[74\text{–}83\%]{\mathbf{1},\ C_6H_5CH_3,\ 110°,\ 3\ \text{hr.}} \text{thionobutyrolactone}$$

[1] B. S. Pedersen, S. Scheibye, N. H. Nilsson, and S.-O. Lawesson, *Bull. Soc. Chim. Belg.*, **87**, 223 (1978); I. Thomsen, K. Clausen, S. Scheibye, and S.-O. Lawesson, *Org. Syn.*, submitted (1979).
[2] S.-O. Lawesson and co-workers, *Bull. Soc. Chim. Belg.*, **87**, 223, 229, 293, 299 (1978).
[3] S. Scheibye, J. Kristensen, and S.-O. Lawesson, *Tetrahedron*, **35**, 1339 (1979).
[4] R. Shabana, S. Scheibye, K. Clausen, S. O. Olesen, and S.-O. Lawesson, *Nouv. J. Chim.*, **4**, 47 (1980).

Bis(*p*-methoxyphenyl)telluroxide, $CH_3O{-}C_6H_4{-}Te(=O){-}C_6H_4{-}OCH_3$ (**1**). Mol. wt., 357.86, m.p. 190–191°.

Preparation[1]:

$$CH_3O{-}C_6H_5 \xrightarrow[90\%]{TeCl_4,\ 160°} (CH_3O{-}C_6H_4)_2{-}TeCl_2 \xrightarrow{NaOH} \mathbf{1}$$

Oxidation.[2] Bis(*p*-methoxyphenyl)telluroxide (**1**) is a mild and selective oxidant for conversion of xanthates, thiocarbamates, thioamides, and nonenolizable thiones into the corresponding oxo derivatives, and also of thiols into disulfides. Typically these reactions afford products in 70–100% yield. 1,2- and 1,4-Hydroquinones are oxidized by **1** to *o*- and *p*-quinones, respectively. Phenylhydroxylamine is oxidized to nitrosobenzene (90% yield). Phenols, amines, enamines, alcohols, oximes, dithiolanes, isonitriles, and 2,4-dinitrophenylhydrazones are unreactive.

[1] K. Lederer, *Ber.*, **49**, 1076 (1916); G. T. Morgan and R. E. Kellett, *J. Chem. Soc.*, 1080 (1926); J. Bergman, *Tetrahedron*, **28**, 3323 (1972).
[2] D. H. R. Barton, S. V. Ley, and C. A. Merholz, *J.C.S. Chem. Comm.*, 755 (1979).

Bis(*o*- or *p*-nitrophenyl)phenylphosphonate, $C_6H_5P(=O)(OC_6H_4NO_2)_2$ (**1**).

Peptide synthesis.[1] The condensation of the tetrabutylammonium salt of N-benzyl-L-leucine (**2**) with glycine ethyl ester (**3**), the components in the Young test,[2] in DMF in the presence of **1** as coupling reagent proceeds in 75–80% yield without racemization.

$$\underset{\mathbf{2}}{(CH_3)_2CHCH_2CH(NHBzl)COO^-\overset{+}{N}(C_4H_9)_4} + \underset{\mathbf{3}}{NH_2CH_2COOC_2H_5} \xrightarrow[75\text{–}80\%]{\mathbf{1},\ DMF}$$

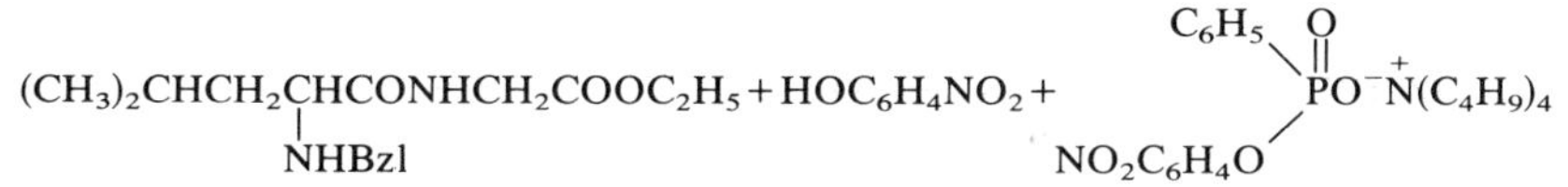

$$(CH_3)_2CHCH_2CH(NHBzl)CONHCH_2COOC_2H_5 + HOC_6H_4NO_2 + (C_6H_5)(NO_2C_6H_4O)P(O)O^- \overset{+}{N}(C_4H_9)_4$$

(75–80%)

4

[1] T. Mukaiyama, N. Morito, and Y. Watanabe, *Chem. Letters*, 1305 (1979).
[2] M. W. Williams and G. T. Young, *J. Chem. Soc.*, 881 (1963).

Bis(2,4-pentanedionato)copper, $Cu(C_5H_7O_2)_2$, **2**, 81; **3**, 62–63. Preparation.[1]

Δ^2-Cycloalkene-1,4-diones.[2] α,ω-Bisdiazoketones couple intramolecularly to cyclic ene-1,4-diones in the presence of $Cu(acac)_2$ in benzene at 60°. Other copper catalysts are unsatisfactory. The procedure fails when n is 5 or 6, but is satisfactory when n is 4 and >6. Both *cis*- and *trans*-isomers are formed, with the latter predominating except when n is 4. The products are reduced by sodium dithionite to cycloalkane-1,4-diones, which can be cyclized by NaOH to fused-ring cyclopentenones when n is 7, 9, 10, and 16.

Under these conditions intermolecular coupling of diazoketones is also possible.

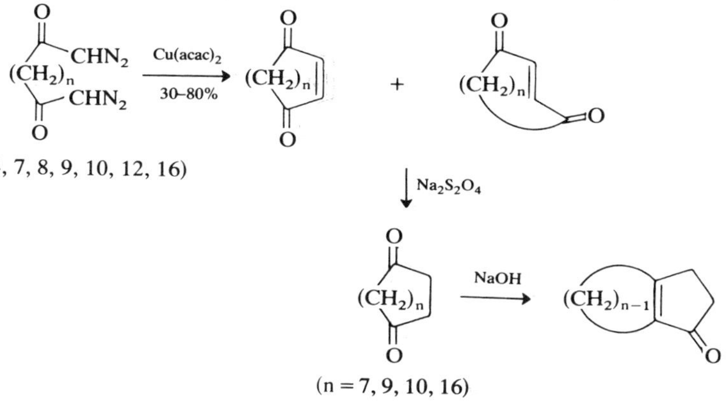

$$N_2HC\text{-}CO(CH_2)_nCO\text{-}CHN_2 \xrightarrow[30\text{–}80\%]{Cu(acac)_2} cis\text{-}\Delta^2\text{-cycloalkene-1,4-dione} + trans\text{-isomer}$$

(n = 4, 7, 8, 9, 10, 12, 16)

$$\downarrow Na_2S_2O_4$$

$$\text{cycloalkane-1,4-dione } (CH_2)_n \xrightarrow{NaOH} \text{fused cyclopentenone } (CH_2)_{n-1}$$

(n = 7, 9, 10, 16)

[1] M. Takebayashi, T. Ibata, H. Kohara, and B. H. Kim, *Bull. Chem. Soc. Japan*, **40**, 2392 (1967).
[2] S. Kulkowit and M. A. McKervey, *J.C.S. Chem. Comm.*, 1069 (1978).

Bis(2,4-pentanedionato)nickel, 5, 471; **6**, 417; **7**, 250.

Chromans.[1] $Ni(acac)_2$ catalyzes the cycloaddition of phenols and 1,3-dichloro-3-methylbutane (**1**) to form 2,2-dimethylchromans (**2**). The synthesis involves

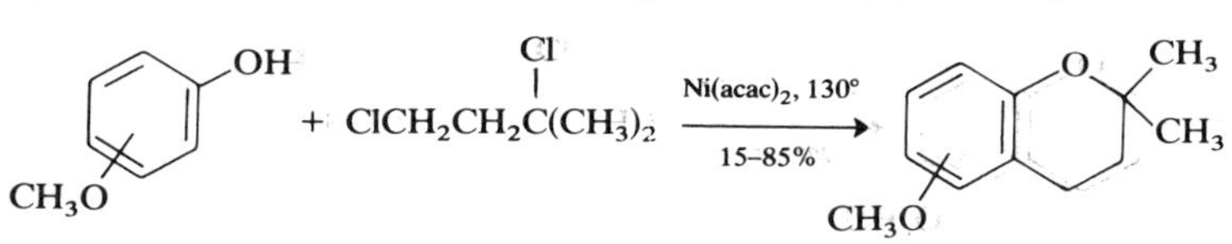

$$CH_3OC_6H_4OH + ClCH_2CH_2CCl(CH_3)_2 \xrightarrow[15\text{–}85\%]{Ni(acac)_2,\ 130^\circ} \text{CH}_3\text{O-2,2-dimethylchroman}$$

1 **2**

o-alkylation and etherification. Prenyl chloride, $ClCH_2CH{=}C(CH_3)_2$, can be used in place of **1**. *See also* Aluminum chloride–Potassium phenoxide, this volume.

[1] F. Camps, J. Coll, A. Messeguer, M. A. Pericás, and R. Ricart, *Synthesis*, 126 (1979).

Bis(2,4-pentanedionato)nickel–Trimethylaluminum, $Ni(C_5H_7O_2)_2$–$(CH_3)_3Al$.

Carbometalation of silylalkynes. In the presence of this nickel–aluminum catalyst, methylmagnesium bromide adds to 1-trimethylsilyl-1-octyne (**1**) to give the *cis*-addition product **2**, which isomerizes slowly to **3**. The use of ethylmagnesium bromide results in reduction mainly to *cis*-1-trimethylsilyl-1-octene. The products **2** and **3**, as expected, react with a variety of electrophiles to give di- and trisubstituted

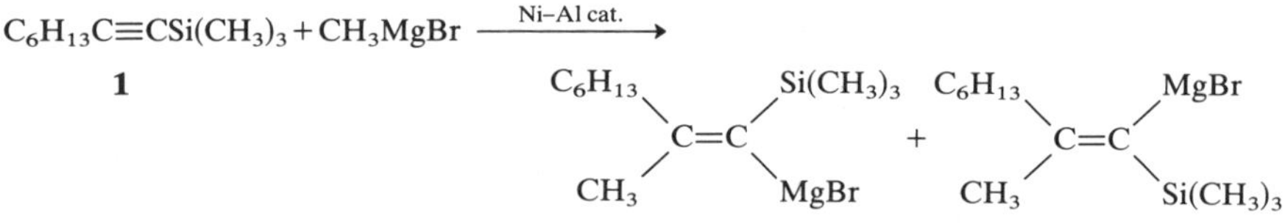

$$\underset{\mathbf{1}}{C_6H_{13}C{\equiv}CSi(CH_3)_3} + CH_3MgBr \xrightarrow{\text{Ni–Al cat.}} \underset{\mathbf{2}}{(C_6H_{13})(CH_3)C{=}C(Si(CH_3)_3)(MgBr)} + \underset{\mathbf{3}}{(C_6H_{13})(CH_3)C{=}C(MgBr)(Si(CH_3)_3)}$$

vinylsilanes, as shown for the reaction with formaldehyde (equation I). Similar reactions are observed with CO_2 and iodine. In addition, **2** and **3** undergo coupling reactions, probably nickel catalyzed, with vinyl and allyl bromide to give dienes. Since desilylation proceeds with retention of stereochemistry, these reactions provide stereoselective syntheses of substituted alkenes.[1]

Similar reactions are observed with 1-silyl-1-alkynes containing the following terminal groups: $N(C_2H_5)_2$, OTHP, and $OCH(CH_3)OC_2H_5$.[2] As before, the *cis*-adduct is formed initially and isomerizes to the *trans*-isomer (equation II).

The carbometalation reaction is also compatible with the presence of a double bond, as shown by a synthesis of geraniol (equation III).

$$\text{(I)}\quad \mathbf{2+3} \xrightarrow[66\%]{HCHO} (C_6H_{13})(CH_3)C{=}C(Si(CH_3)_3)(CH_2OH) + (C_6H_{13})(CH_3)C{=}C(CH_2OH)(Si(CH_3)_3) \quad 9:1$$

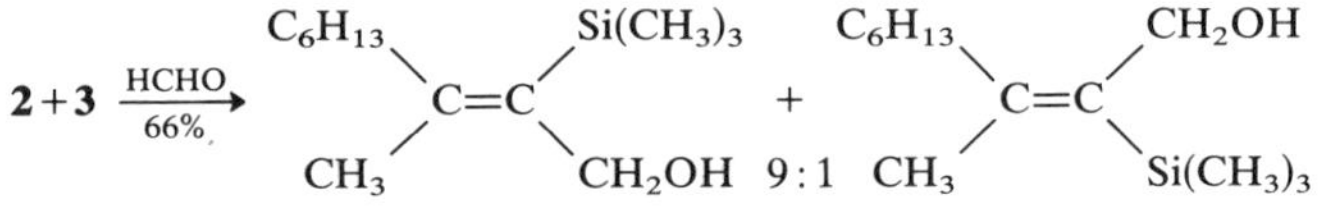

$$\text{(II)}\quad THPOCH_2CH_2C{\equiv}CSi(CH_3)_3 \xrightarrow{CH_3MgBr,\ \text{Ni cat.}} (THPOCH_2CH_2)(CH_3)C{=}C(Si(CH_3)_3)(MgBr) + (THPOCH_2CH_2)(CH_3)C{=}C(MgBr)(Si(CH_3)_3) \quad 4:1$$

$$\text{(III)}\quad (CH_3)_2C{=}CH(CH_2)_2C{\equiv}CSi(CH_3)_3 \xrightarrow[38\%]{1)\ CH_3MgBr,\ \text{cat.};\ 2)\ HCHO} ((CH_3)_2C{=}CH(CH_2)_2)(CH_3)C{=}C(CH_2OH)(Si(CH_3)_3) \xrightarrow[80\%]{F^-} ((CH_3)_2C{=}CH(CH_2)_2)(CH_3)C{=}C(CH_2OH)(H)$$

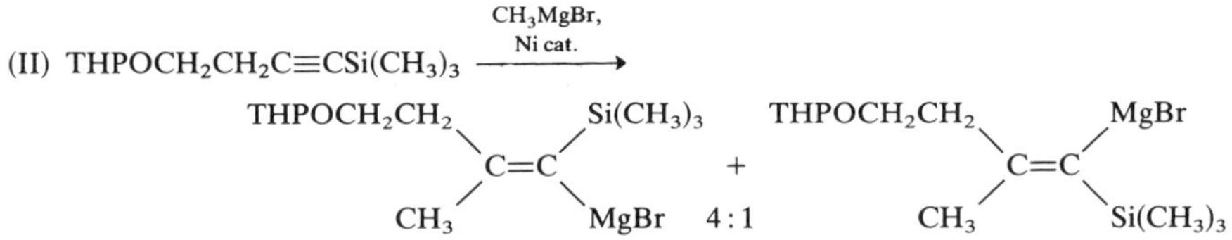

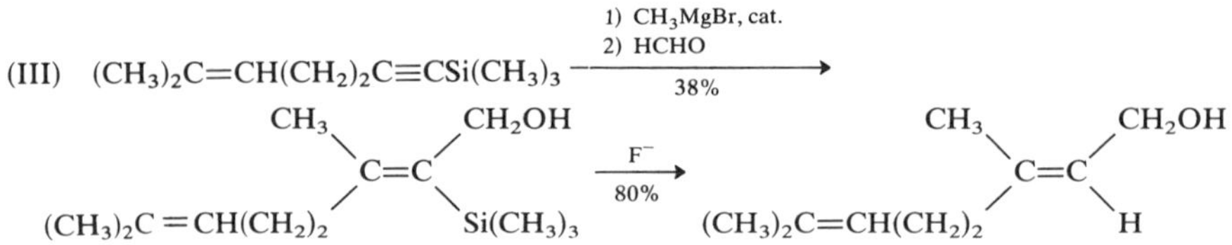

[1] B. B. Snider, M. Karras, and R. S. E. Conn, *Am. Soc.*, **100**, 4626 (1978).
[2] *Idem, Tetrahedron Letters*, 1679 (1979).

Bis(tetra-*n*-butylammonium) dichromate, $[(C_4H_9)_4N]_2Cr_2O_7$. Mol. wt. 700.91, orange crystals, m.p. 129–133°. This solubilized form of potassium dichromate is prepared by treatment of potassium chromate with tetrabutylammonium hydrogen sulfate.

Oxidation of alkyl halides. This reagent oxidizes benzyl and allyl halides to the corresponding carbonyl compounds in high yield. Secondary alkyl bromides afford ketones in good yield, but primary alkyl bromides are oxidized in low yield to aldehydes.[1]

[1] D. Landini and F. Rolla, *Chem. Ind.*, 213 (1979).

Bis(tri-*n*-butyltin) oxide, 6, 56–57; **7**, 26–27; **8**, 43–44.

Selective alkylation and acylation of a vic-diol. Selective acylation or alkylation of cyclohexanoid, axial-equatorial *vic*-diols is possible by way of the dibutylstannylene derivative, obtained by reaction with bis(tri-*n*-butyltin) oxide.[1] A typical reaction is formulated for a *myo*-inositol derivative in equation (I). The method

(I) [structures: BzlO, OBzl, OH; $[Bu_3Sn]_2O$ →; O—$SnBu_2$; $C_6H_5CH_2Br$, 60% overall →; OBzl]

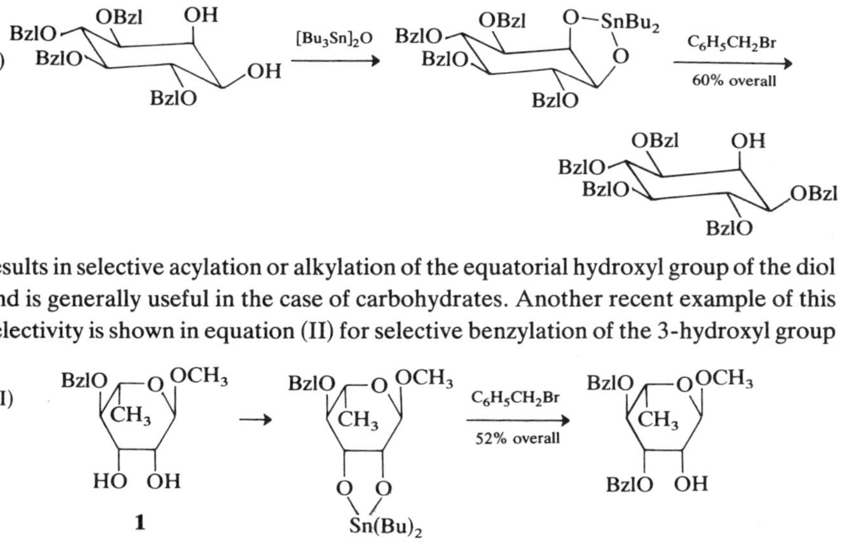

results in selective acylation or alkylation of the equatorial hydroxyl group of the diol and is generally useful in the case of carbohydrates. Another recent example of this selectivity is shown in equation (II) for selective benzylation of the 3-hydroxyl group

(II) [structures: BzlO, OCH_3, CH_3, HO OH (**1**) →; O O, $Sn(Bu)_2$; $C_6H_5CH_2Br$, 52% overall → BzlO OH]

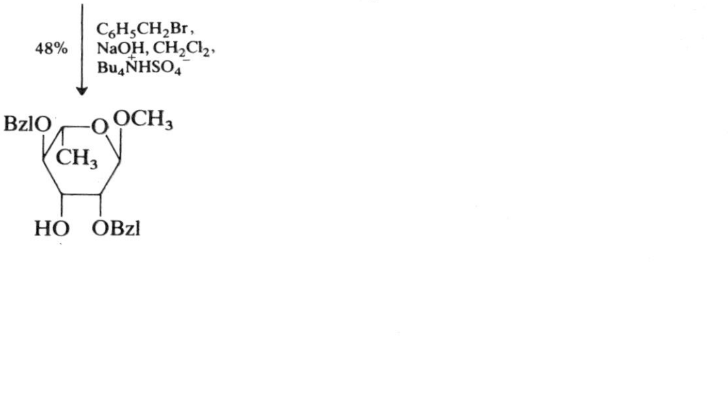

1 ↓ 48%, $C_6H_5CH_2Br$, NaOH, CH_2Cl_2, $Bu_4\overset{+}{N}HSO_4^-$ → BzlO, OCH_3, CH_3, HO OBzl of a rhamnopyranoside (**1**).[2] The 2-hydroxyl group of **1** can be benzylated selectively by use of the phase-transfer catalyzed technique (48% yield) or by selective tritylation, benzylation, detritylation (yield not given).

[1] M. A. Nashed and L. Anderson, *Tetrahedron Letters*, 3503 (1976); M. A. Nashed, *Carbohydrate Res.*, **60**, 200 (1978).
[2] V. K. Handa, C. F. Piskorz, J. J. Barlow, and K. L. Matta, *ibid.*, **74**, C5 (1979).

Bis(trifluoroacetoxy)iodobenzene [Iodobenzene bis(trifluoroacetate)], $C_6H_5I(OCOCF_3)_2$. Mol. wt. 430.05, m.p. 124–126°. The reagent is prepared by reaction of iodobenzene diacetate (**1**, 508; **4**, 266) with trifluoroacetic acid (53% yield).[1]

Hofmann rearrangement. Primary amides are converted into amines in high yield by reaction with this oxidant at room temperature in aqueous acetonitrile (equation I). The reaction is related to a recent method using lead tetraacetate (**6**, 316); however, the intermediate isocyanate in this case is not trapped as the carbamate, but is hydrolyzed directly to the amine.

$$\text{(I)}\quad R\overset{O}{\overset{\|}{C}}NH_2 + C_6H_5I(OCOCF_3)_2 + H_2O \xrightarrow[85\text{–}95\%]{CH_3CN} RNH_2 + CO_2 + C_6H_5I + 2CF_3COOH$$

One limitation is that this reaction cannot be used for rearrangement of amides of aryl carboxylic acids because anilines are subject to further oxidation.[2]

[1] S. Spyroudis and A. Varvoglis, *Synthesis*, 445 (1975).
[2] A. S. Radhakrishna, M. E. Parham, R. M. Riggs, and G. M. Loudon, *J. Org.*, **44**, 1746 (1979).

Bis(trifluoromethyl)methylenetriphenylphosphorane, $(C_6H_5)_3P{=}C(CF_3)_2$ (**1**). Mol. wt. 412.32.

Preparation:

$$4\,P(C_6H_5)_3 + \text{[2,2,4,4-tetrakis(trifluoromethyl)-1,3-dithietane: } (CF_3)_2C\langle S_2\rangle C(CF_3)_2] \xrightarrow{\text{Ether, }-78°} 2\ \mathbf{1} + 2(C_6H_5)_3P{=}S$$

Bis(trifluoromethyl)alkenes.[1] This ylide (**1**), prepared *in situ*, reacts with aldehydes to form bis(trifluoromethyl)alkenes in 30–65% yield (isolated). Note that the related phosphoranes $(RO)_3P{=}C(CF_3)_2$[2] react with ketones but not with aldehydes.

[1] D. J. Burton and Y. Inouye, *Tetrahedron Letters*, 3397 (1979).
[2] W. J. Middleton and W. H. Sharkey, *J. Org.*, **30**, 1384 (1965).

1,3-Bis(trimethylsilyloxy)-1-methoxybuta-1,3-diene (1). Mol. wt. 260.49.

Preparation[1]:

$$((CH_3)_3SiO)(CH_3)C{=}CHCOOCH_3 \xrightarrow[2)\ (CH_3)_3SiCl]{1)\ LDA,\ THF,\ TMEDA} H_2C{=}C(OSi(CH_3)_3){-}CH{=}C(OCH_3)(OSi(CH_3)_3)$$

1

Reaction with electrophiles. This enol ether functions as a dianion equivalent of methyl acetoacetate in reactions with electrophiles.[1]

$$C_6H_5CH(OH)CH_2COCH_2COOCH_3 \xleftarrow[72\%]{C_6H_5CHO,\ TiCl_4} \mathbf{1} \xrightarrow[89\%]{Br_2} BrCH_2COCH_2COOCH_3$$

$$\mathbf{1} \xrightarrow[66\%]{CH_3COCH{=}C(OSi(CH_3)_3)CH_3} \text{methyl 2-hydroxy-4,6-dimethylbenzoate (OH, } COOCH_3\text{, } H_3C\text{, } CH_3\text{)}$$

[1] T.-H. Chan and P. Brownbridge, *J.C.S. Chem. Comm.*, 578 (1979).

Bistrimethylsilyl ether, $[(CH_3)_3Si]_2O$ (**1**). Mol. wt. 162.38, b.p. 98–100°. The ether is prepared in high yield by reaction of $(CH_3)_3SiCl$ with $NaHCO_3$.

Trimethylsilyl ethers. These ethers can be prepared by reaction of an alcohol with **1** in refluxing benzene with pyridinium *p*-toluenesulfonate (**8**, 427–428) as the acid catalyst. *p*-Toluenesulfonic acid is a satisfactory catalyst for acid-stable alcohols. The water formed is removed by 4 Å molecular sieves (Soxhlet extractor). The reaction requires several days for completion. Yields of trimethylsilyl ethers are 80–90%.[1]

[1] H. W. Pinnick, B. S. Bal, and N. H. Lajis, *Tetrahedron Letters*, 4261 (1978).

Bis(trimethylsilyl) lithiomalonate, $LiCH[COOSi(CH_3)_3]_2$ (**1**). Mol. wt. 254.35. The anion is prepared by reaction of bis(trimethylsilyl) malonate[1] with *n*-butyllithium in ether at −60°.

β-*Keto acids.*[2] The reagent converts acid chlorides into products (**a**) that undergo hydrolysis and decarboxylation in the presence of water to form β-keto acids (**2**). The yields usually are improved by use of a 100% excess of **1**.

$$RCOCl + \mathbf{1} \xrightarrow{\text{Ether, } 0^\circ} \underset{\mathbf{a}}{RCOCH[COOSi(CH_3)_3]_2} \xrightarrow[60\text{–}90\%]{H_2O,\ 20^\circ} \underset{\mathbf{2}}{RCOCH_2COOH}$$

[1] U. Schmidt and M. Schwochau, *Monatsh.*, **98**, 1492 (1967).
[2] J. W. F. K. Barnick, J. L. van der Baan, and F. Bickelhaupt, *Synthesis*, 787 (1979).

Bis(trimethylsilyl)mercury (1). Mol. wt. 346.98, m.p. 102–104°.

Preparation[1]:

$$2\,(CH_3)_3SiBr + Hg/Na \rightarrow \underset{\mathbf{1}}{[(CH_3)_3Si]_2Hg}$$

Debromination of 1,2-dibromoalkanes. Bis(trimethylsilyl)mercury debrominates 1,2-dibromoalkanes to olefins.[2] This reaction has been applied to 1,2-dibromo- and 1,2-diiodoadamantane (**2**) as a route to adamantene (**a**).[3] When performed in the presence of 1,3-diphenylisobenzofuran, the hydrocarbon **3** is obtained in 30–35% yield. In principle, this route could be extended to isolation of adducts of other anti-Bredt alkenes.

X X 1 C_6H_5 O C_6H_5 30–35% C_6H_5 C_6H_5

2 (X = Br, I) **a** **3**

[1] E. Wiberg, O. Stecher, H. H. Andrascheck, L. Kreuzbichler, and E. Staude, *Angew. Chem., Int. Ed.*, **2**, 507 (1963).
[2] S. W. Bennett, C. Eaborn, R. A. Jackson, and R. W. Walsingham, *J. Organomet. Chem.*, **27**, 195 (1971).
[3] J. I. G. Cadogan and R. Leardini, *J.C.S. Chem. Comm.*, 783 (1979).

Bis(trimethylsilyl) monoperoxysulfate (1). Mol. wt. 258.44.
Preparation[1]:

$$(CH_3)_3SiOOSi(CH_3)_3 \xrightarrow[CH_2Cl_2,\ -30^\circ]{SO_3,} (CH_3)_3SiOS(=O)_2OOSi(CH_3)_3$$

1

Baeyer–Villiger oxidation.[2] This reagent is generally superior to Caro's acid (monoperoxysulfuric acid) for Baeyer–Villiger oxidation. Limitations are that double bonds are attacked, that α,β-enones react sluggishly, and that aryl alkyl ketones are hydrolyzed to phenols.

[1] B. Bressel and A. Blaschette, *Z. Anorg. Allg. Chem.*, **377**, 182 (1970).
[2] W. Adam and A. Rodriguez, *J. Org.*, **44**, 4969 (1979).

1,3-Bis(trimethylsilyloxy)-1,3-butadiene, 7, 27–28.

Anthracyclinones.[1] A key step in a synthesis of 4-desmethoxydaunomycinone (**5**) is the Diels–Alder reaction of naphthazarin with this diene (**1**); the initial adduct is unstable, so it is treated with dilute acid to give the monosilyl ether **2**, obtained in 90% yield. Further treatment with acid gives **3**. The advantage of this route is that the desired two oxygen functions in ring A are introduced. Reaction of **3** with ethynylmagnesium bromide introduces the two-carbon side chain. The D ring is formed by a Diels–Alder reaction of **4** with 1-acetoxybutadiene, which proceeds in almost quantitative yield. The final step involves hydration of the triple bond. Overall yield of **5** from naphthazarin is 29%.

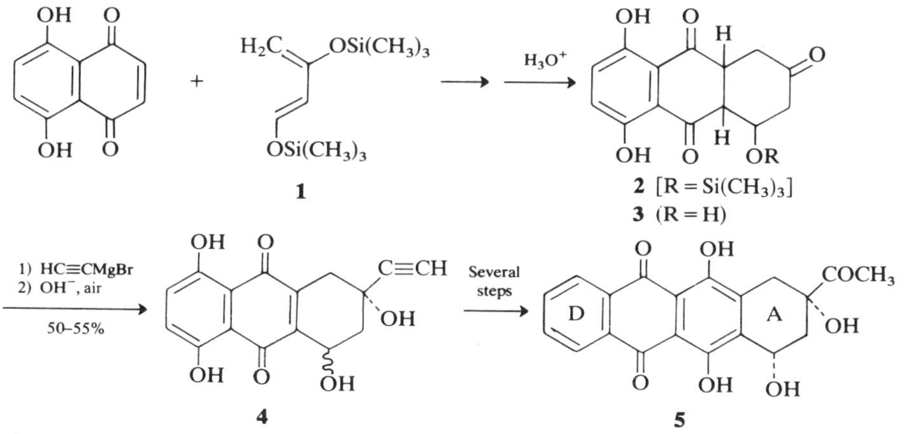

OH O + H_2C $OSi(CH_3)_3$ $OSi(CH_3)_3$ → H_3O^+ →

1

2 [$R = Si(CH_3)_3$]
3 (R = H)

1) HC≡CMgBr
2) OH^-, air
50–55%

4

Several steps

5

[1] K. Krohn and K. Tolkiehn, *Tetrahedron Letters*, 4023 (1978).

Bis(triphenylphosphine)copper tetrahydroborate, 8, 47. An improved preparation from copper(I) chloride, triphenylphosphine, and $NaBH_4$ in the ratio 1 : 2 : 1 has been described. The paper includes examples of reduction of a wide range of acid chlorides to aldehydes. The rate of reduction is increased by added triphenylphosphine. The reduction is compatible with various functional groups. The reagent is converted to $(C_6H_5)_3PCuCl$, which can be recycled by treatment with $NaBH_4$.[1]

[1] G. W. J. Fleet and P. J. C. Harding, *Tetrahedron Letters*, 975 (1979).

9-Borabicyclo[3.3.1]nonane (9-BBN), 2, 31; **3**, 24–29; **4**, 41; **5**, 46–47; **6**, 62–64; **7**, 29–31; **8**, 47–49.

Hydroboration of acetylenes.[1] Internal acetylenes are generally hydroborated by 9-BBN (**1**) at 25° in THF with 1.0 equiv. of **1**, whereas terminal acetylenes are hydroborated with 0.5 equiv. of **1** at 0°. The use of larger amounts of **1** or higher reaction temperatures in reactions of terminal acetylenes leads to substantial quantities of *gem*-dibora derivatives. The regioselectivity of and the sensitivity to electronic factors by 9-BBN in acetylene hydroboration are greater than those with thexylborane, disiamylborane, or dicyclohexylborane. B-Vinylic-9-BBN derivatives undergo protonolysis in methanol, oxidation to aldehydes or ketones by inverse addition to buffered hydrogen peroxide, 1,2 addition to aldehydes,[2] and 1,4 addition to methyl vinyl ketone and related derivatives.[3]

Examples:

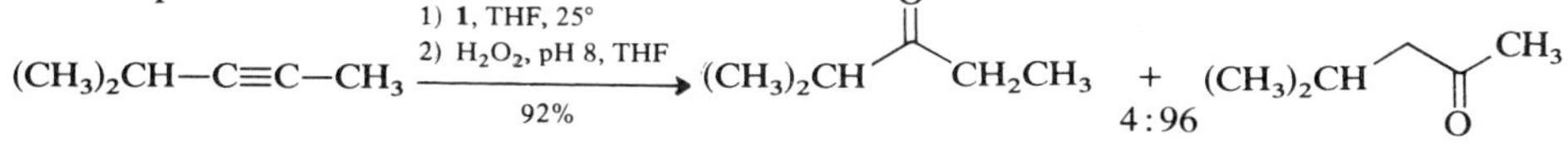

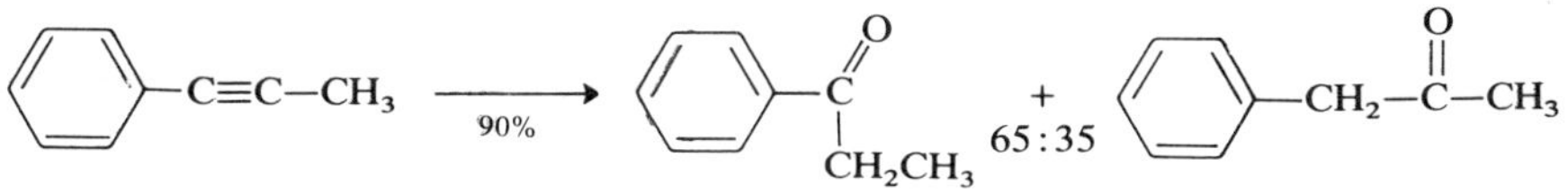

$(CH_3)_2CH{-}C{\equiv}C{-}CH_3$ —(1) **1**, THF, 25°; 2) H_2O_2, pH 8, THF; 92%)→ $(CH_3)_2CHCOCH_2CH_3$ + $(CH_3)_2CHCH_2COCH_3$ (4 : 96)

$C_6H_5{-}C{\equiv}C{-}CH_3$ —(90%)→ $C_6H_5{-}CO{-}CH_2CH_3$ + $C_6H_5{-}CH_2{-}CO{-}CH_3$ (65 : 35)

Hydroboration of allenes.[4] With few exceptions, the 9-BBN hydroboration of allenes affords B-allylic-9-BBN derivatives. In contrast, the reactions of allenes with disiamylborane or dicyclohexylborane afford predominantly vinylic boranes. With unsymmetrical allenes, 9-BBN binds to the less substituted carbon atom. Allene itself affords a 1,3-dibora derivative. The B-allylic-9-BBN derivatives are useful reagents for the allylic boration of carbonyl compounds.[5]

Example:

$$H_2C{=}C{=}CHCH_3 \xrightarrow[\text{THF}]{\text{9-BBN}} \left[\text{9-BBN}{-}CH_2CH{=}CHCH_3 \right] \xrightarrow[\text{oxid.}]{(CH_3)_2C{=}O,} H_2C{=}CHCH(CH_3)C(CH_3)_2OH + \left[HOCH_2CH_2CH(OH)CH_3\right]$$

(*cis–trans* mixture)

74% 12%

Selective hydroboration of alkenes.[6] Alkenes react more readily with 9-BBN than structurally similar alkynes. This behavior contrasts with that of other dialkylboranes. The difference permits selective hydroboration of allylic acetylenes (equations I and II).

$$\text{(I)}\quad RC{\equiv}CCH_2CR^1{=}CH_2 \xrightarrow[\text{THF}]{\text{9-BBN,}} RC{\equiv}CCH_2CHR^1CH_2B\text{(9-BBN)} \xrightarrow[85\text{–}95\%]{H_2O_2,\ OH^-} RC{\equiv}CCH_2CHR^1CH_2OH$$

$(R^1 = H, CH_3)$

$$\text{(II)}\quad H_2C{=}CHC_6H_{13}\text{-}n + HC{\equiv}CC_5H_{11}\text{-}n \rightarrow H_2C(B)CH_2C_6H_{13}\text{-}n + (H)(B)C{=}C(H)(C_5H_{11}\text{-}n)$$

5.7 : 1

Hydroboration of trimethylsilyl enol ethers.[7] This reaction can be used to effect reductive 1,2-transposition of acyclic ketones (equation I). Highest yields of the transposed alcohol are obtained when $R^2 = H$.

$$\text{(I)}\quad R^1C(O)CH_2R^2 \rightarrow (CH_3)_3SiO(R^1)C{=}CHR^2 \xrightarrow{\text{9-BBN}} [R^1CH{=}CHR^2] \xrightarrow[65\text{–}80\%\ \text{overall}]{1)\ \text{9-BBN}\quad 2)\ H_2O_2,\ OH^-} R^1CH_2CH(OH)R^2$$

[1] H. C. Brown, C. G. Scouten, and R. Liotta, *Am. Soc.*, **101**, 96 (1979).
[2] P. Jacob III and H. C. Brown, *J. Org.*, **42**, 579 (1977).
[3] *Idem, Am. Soc.*, **98**, 7832 (1976).
[4] H. C. Brown, R. Liotta, and G. W. Kramer, *ibid.*, **101**, 2966 (1979).
[5] G. W. Kramer and H. C. Brown, *J. Org.*, **42**, 2292 (1977).
[6] C. A. Brown and R. A. Coleman, *ibid.*, **44**, 2328 (1979).
[7] G. L. Larson and L. M. Fuentes, *Syn. Comm.*, **9**, 841 (1979).

Borane–Pyridine, 1, 963–964; **8**, 50–51.

Reductions in trifluoroacetic acid. Borane–pyridine in refluxing benzene is a rather weak reducing agent. When used in trifluoroacetic acid it has much stronger reducing properties. It has been used to effect a number of useful reductions, such as those shown.[1]

Examples:

$$2RCHO \xrightarrow[60\text{–}85\%]{Py\cdot BH_3,\ CF_3COOH,\ 5\text{ min., }23^\circ} RCH_2OCH_2R$$

$$R^1CHO + R^2CH_2OH \xrightarrow[65\text{–}80\%]{} R^1CH_2OCH_2R^2$$

$$C_6H_5CH_2COCH_2C_6H_5 \xrightarrow[99\%]{} C_6H_5CH_2CH(OH)CH_2C_6H_5$$

$$C_6H_5CH(OH)C_6H_5 \xrightarrow[97\%]{12\text{ hr., }23^\circ} C_6H_5CH_2C_6H_5$$

[1] Y. Kikugawa and Y. Ogawa, *Chem. Pharm. Bull. Japan*, **27**, 2405 (1979).

Boric acid, H_3BO_3. Mol. wt., 61.83.

Decarboalkoxylation. Malonic esters undergo this reaction in yields of 90–95% when heated with boric acid at 170–190° for 4 hours.[1]

[1] T.-L. Ho, *Syn. Comm.*, **9**, 609 (1979).

Boronic acid resins, Ⓟ—$B(OH)_2$, **7**, 33.

Acylated carbohydrate derivatives.[1] Treatment of glycosides with polymer-supported boronic acid (**1**) with azeotropic removal of water affords the more stable five- or six-membered boronate. The unprotected hydroxyl groups of the protected glycosides can be acylated, often in yields higher than those realized by classical methods.

Examples:

OCH3; CH3; O; HO; HO; OH → 1) **1**; 2) C_6H_5COCl → OCH3; H_3C; O; C_6H_5COO; O; B; O; Ⓟ → H^+, 85% → OCH3; CH3; O; C_6H_5COO; OH; OH

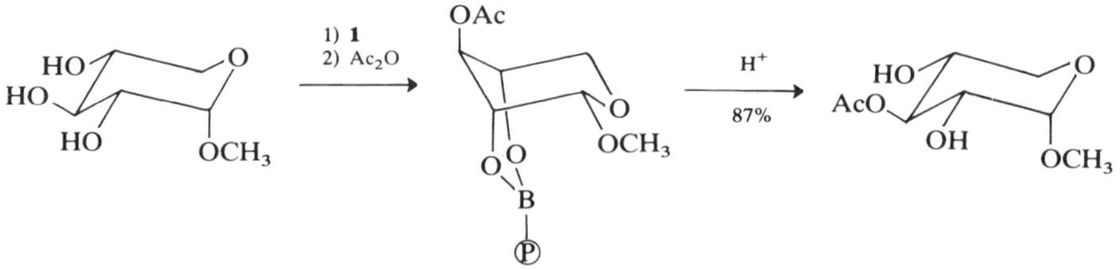

[1] J. M. J. Fréchet, L. J. Nuyans, and E. Seymour, *Am. Soc.*, **101**, 432 (1979).

Boron triacetate, $B(OCOCH_3)_3$. Mol. wt. 187.95. Supplier: Alfa. Boron triacetate can be prepared by heating boric acid with excess acetic anhydride. It is generally less destructive than BF_3 etherate.

Diels–Alder catalyst. Kelly and Montury[1] have examined the effect of three Lewis acids on the Diels–Alder reaction of *peri*-hydroxylated naphthoquinones with 1-methoxy-3-methyl-1,3-butadiene: magnesium iodide, boron trifluoride etherate, and boron triacetate. All three increase regioselectivity. BF_3 etherate and $B(OAc)_3$

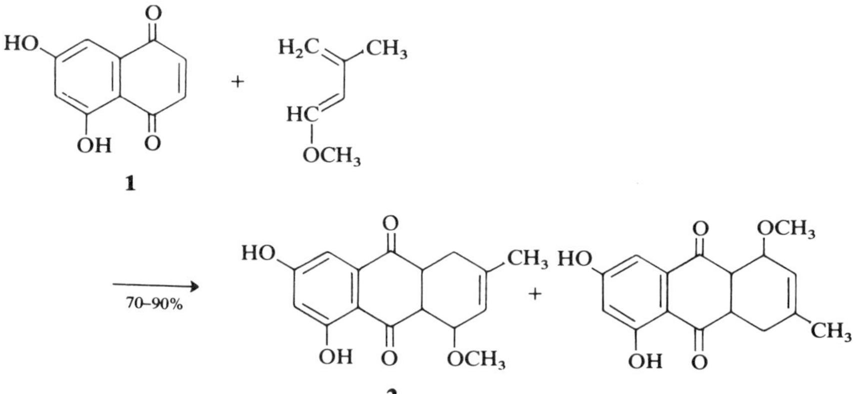

	2	3
25°	45%	55%
$BF_3 \cdot (C_2H_5)_2O$	≥95%	≤5%
$B(OAc)_3$	≥95%	≤5%
MgI_2	≤5%	≥95%

exert a similar effect; magnesium iodide has an equal, but opposite effect. In the case of the boron compounds, the effect may be the result of complexes formed with the *peri*-hydroxyl group. The situation with MgI_2 is less clear.

The catalytic effect of boron triacetate was used in a stereocontrolled synthesis of altersolanol B (**6**).[2]

Unexpectedly boron triacetate induces a marked regioselectivity on the cycloaddition of isoprene to the anthracene analog (**7**) of juglone. The uncatalyzed

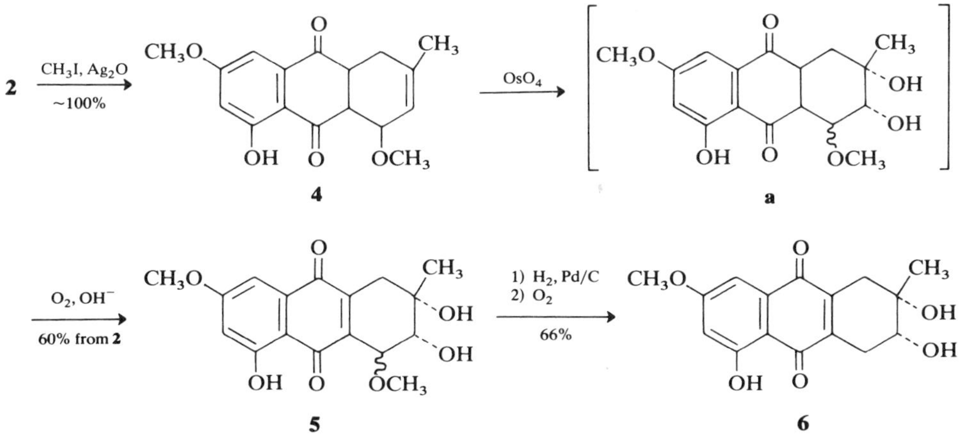

2 $\xrightarrow[\sim 100\%]{CH_3I,\ Ag_2O}$ **4** (O, CH_3O, CH_3, OH, O, OCH_3) $\xrightarrow{OsO_4}$ [**a** (O, CH_3O, CH_3, OH, OH, OH, O, OCH_3)]

$\xrightarrow[60\%\ from\ 2]{O_2,\ OH^-}$ **5** (O, CH_3O, CH_3, OH, OH, OH, O, OCH_3) $\xrightarrow[66\%]{1)\ H_2,\ Pd/C\quad 2)\ O_2}$ **6** (O, CH_3O, CH_3, OH, OH, OH, O)

reaction, conducted at 150° for 2 hours, leads to **8** and **9** in the ratio 11 : 10. When $B(OAc)_3$ is present (large excess), the addition proceeds at 20° and **8** is formed exclusively in 60% yield.[3]

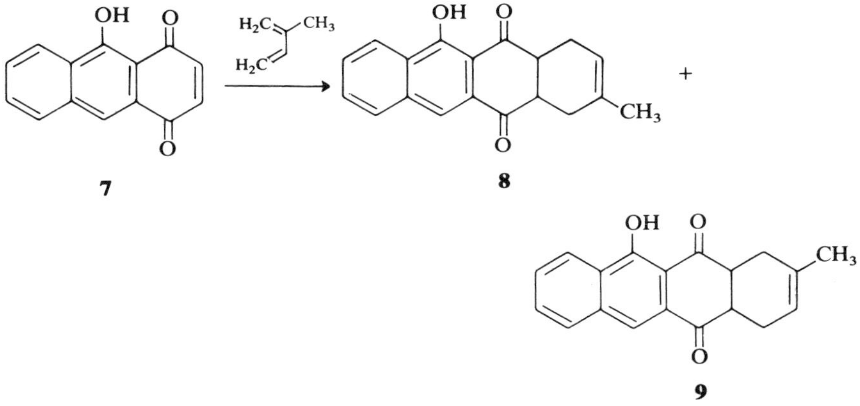

7 (OH, O, O) $\xrightarrow{H_2C=C(CH_3)-CH=CH_2}$ **8** (OH, O, O, CH_3) + **9** (OH, O, O, CH_3)

[1] T. R. Kelly and M. Montury, *Tetrahedron Letters*, 4311 (1978).
[2] *Idem, ibid.*, 4309 (1978).
[3] R. A. Russell, G. J. Collin, M. Sterns, and R. N. Warrener, *ibid.*, 4229 (1979).

Boron tribromide, 1, 66–67; **2**, 33–34; **3**, 30–31; **4**, 42; **5**, 49; **6**, 64–65.

Deethylation. Boron tribromide cleaves the ethyl ethers of phenols at 20° in yields consistently greater than 90%. The yields of phenols on demethylation of the corresponding methyl ethers with BBr_3 are erratic (15–90%).[1]

[1] J. B. Press, *Syn. Comm.*, **9**, 407 (1979).

Boron trichloride, 1, 67–68; **2**, 34–35; **3**, 31–32; **4**, 42–43; **5**, 50–51; **6**, 65.

ortho-*Substitution of anilines.*[1] N-Methylaniline is converted by BCl_3 into N-methylanilinodichloroborane (**1**), which reacts with a benzaldehyde regiospecifically to form **2**. The yield in this reaction is improved considerably by addition of a tertiary amine to trap the hydrogen chloride formed. Actually the reaction can be conducted in a one-pot procedure. The aniline is refluxed with BCl_3 in benzene to form **1**; then the aldehyde and tertiary amine are added to the solution of **1**. This method is not applicable to aniline itself.

X–C₆H₄–NHCH₃ $\xrightarrow{BCl_3}$ [X–C₆H₄–N(CH₃)–BCl₂ (**1**) $\xrightarrow[-HCl]{ArCHO,\ NR_3}$ cyclic N(CH₃)–BCl–O–CH(Ar) adduct (**a**)] $\xrightarrow[65\text{–}90\%]{H_2O}$ X–C₆H₄(NHCH₃)–CHOH–Ar (**2**)

The borane **1** can also be used for *ortho*-acylation of anilines by reaction with nitriles. The yields in this reaction are considerably improved by addition of aluminum chloride, but other Lewis acids are ineffective. This reaction is also successful for primary anilines. No other metal halides are effective in these two substitution reactions.

1 + RC≡N $\xrightarrow{AlCl_3}$ [X–C₆H₃(NHCH₃)–C(=NH)R (**b**)] $\xrightarrow[40\text{–}80\%]{H_3O^+}$ X–C₆H₃(NHCH₃)–C(=O)R (**3**)

Aldol condensation. Boron trichloride converts imines into vinylaminodichloroboranes, which react at room temperature with carbonyl compounds to form aldols in reasonable yield. An example is shown in equation (I). This aldol condensation can be carried out with comparable yield in a one-pot procedure without isolation of the vinylaminodichloroborane.[2]

The method has also been used for asymmetric aldol condensation of acetophenone and benzaldehydes.[3] The ketone is converted into a chiral imine by condensation with isobornylamine. The imine is then treated with BCl_3 and then

(I) [Reaction scheme: cyclohexanone N-cyclohexylimine (NC_6H_{11}) → (BCl_3, $N(C_2H_5)_3$, CH_2Cl_2; 75%) → N-cyclohexyl-N-(cyclohexenyl)aminodichloroborane ($NBCl_2$) → (R^1COR^2) → [imine adduct with $OBCl_2$, $C(R^1)(R^2)$] → (H_3O^+; 30–70%) → 2-($R^1R^2C(OH)$)cyclohexanone]

benzaldehyde. The aldol product is found to have the (R)-configuration and is obtained in optical purity of 47.7%. Lower enantioselectivity has been observed in several cases.

A related condensation between benzaldehydes and acetonitrile was effected with diethylaminodichloroborane[4] and triethylamine.[5] In this reaction, the reactive intermediate is considered to be the diethylaminochloroketeniminoborane (**a**) in equation (II).

$$\text{(II)}\quad CH_3CN \xrightarrow[N(C_2H_5)_3]{(C_2H_5)_2NBCl_2} \left[CH_2{=}C{=}NB(N(C_2H_5)_2)(Cl) \right] \xrightarrow{C_6H_5CHO}$$

a

$$\left[C_6H_5CH(CH_2CN){-}OB(N(C_2H_5)_2)(Cl) \right] \xrightarrow[45\text{–}50\%]{H_3O^+} C_6H_5CH(N(C_2H_5)_2)CH_2CN$$

b

[1] T. Sugasawa, T. Toyoda, M. Adachi, and K. Sasakura, *Am. Soc.*, **100**, 4842 (1978).
[2] *Idem, Syn. Comm.*, **9**, 515 (1979).
[3] T. Sugasawa and T. Toyoda, *Tetrahedron Letters*, 1423 (1979).
[4] K. Niedenzu and J. W. Dawson, *Am. Soc.*, **81**, 3561 (1959).
[5] T. Sugasawa and T. Toyoda, *Syn. Comm.*, **9**, 553 (1979).

Boron trifluoride–Ethanethiol.

Cleavage of ethers. This combination cleaves methyl ethers of primary and secondary alcohols readily. With the latter substrates, retention of configuration obtains. Aryl methyl ethers are cleaved very slowly by this method.[1] Benzyl ethers of both alcohols and phenols are also cleaved and in high yield; the former ethers are cleaved more readily. Presumably BF_3 coordinates with the ethereal oxygen to form an oxonium intermediate that is then attacked by the thiol (a soft nucleophile).[2]

[1] M. Node, H. Hori, and E. Fujita, *J.C.S. Perkin I*, 2237 (1976).
[2] K. Fuji, K. Ichikawa, M. Node, and E. Fujita, *J. Org.*, **44**, 1660 (1979).

Boron trifluoride–Methanol. Supplier: Fluka.

Detritylation. Trityl ethers are rapidly cleaved at room temperature in 85–95% yield by the complex of BF_3 with methanol or by BF_3 etherate and methanol in aprotic, anhydrous solvents (CH_2Cl_2, C_6H_6). O- and N-Acyl groups, O-sulfonyl, N-alkoxycarbonyl, O-methyl, O-benzyl, and acetal groups are stable under these conditions.[1]

[1] K. Dax, W. Wolflehner, and H. Weidmann, *Carbohydrate Res.*, **65**, 132 (1978).

Boron trifluoride etherate, 1, 70–72; **2**, 35–36; **3**, 33; **4**, 44–45; **5**, 54–55; **6**, 65–70; **7**, 31–32; **8**, 51–52.

Cyclization of humulene-4,5-epoxide. Treatment of this epoxide (**1**) with BF_3 etherate results in cyclization to equal amounts of the two isomeric alcohols **2** and **3**, related to the sesquiterpene africanol **4**, of marine origin. Such a reaction could be involved in biogenesis of **4**.[1]

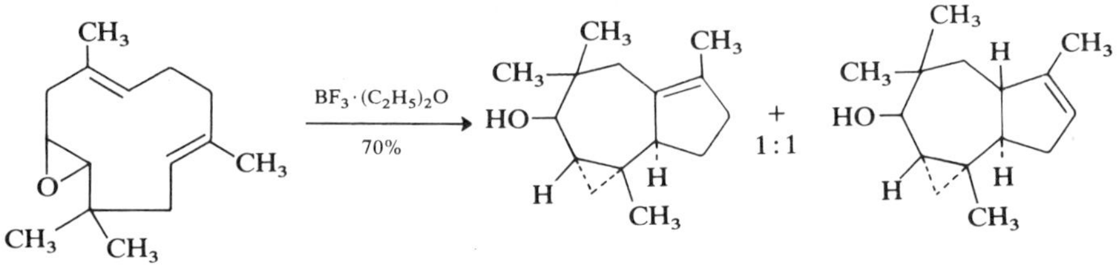

CH$_3$ CH$_3$ O CH$_3$ CH$_3$ **1** $\xrightarrow[70\%]{BF_3\cdot(C_2H_5)_2O}$ **2** + **3** (1:1)

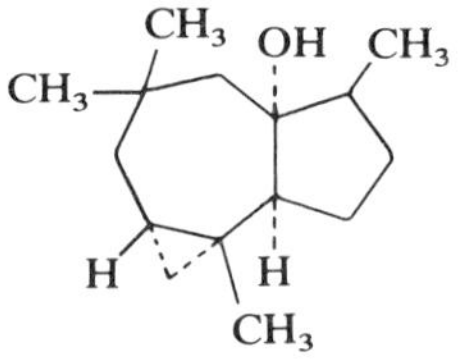

4

Cyclization of epoxy β-keto esters. Treatment of **1** with BF_3 etherate effects cyclization to the 6,8-dioxabicyclo[3.2.1]octane **2** in high yield. This basic skeleton had been encountered in several sex pheromones of bark beetles such as frontalin (**3**), which can be synthesized readily by the β-keto ester cyclization, since the carboxylic acid corresponding to **2** is readily decarboxylated when heated at 220°.[2]

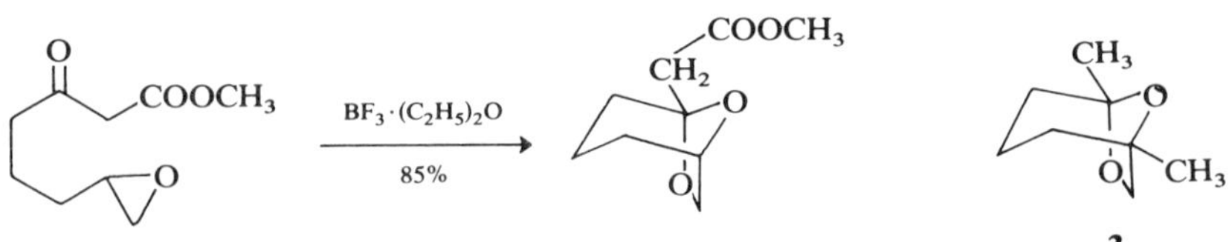

1 $\xrightarrow[85\%]{BF_3\cdot(C_2H_5)_2O}$ **2**   **3**

Cleavage of* t-*butyldimethylsilyl ethers (**4**, 176–177). These ethers, usually cleaved by acid or F^-, can also be cleaved in 85–90% yield by BF_3 etherate in $CHCl_3$ or CH_2Cl_2 at 0–25°.[3]

[1] J. A. Mlotkiewicz, J. Murray-Rust, P. Murray-Rust, W. Parker, F. G. Riddell, J. S. Roberts, and A. Sattar, *Tetrahderon Letters*, 3887 (1979).
[2] P.-E. Sum and L. Weiler, *Can. J. Chem.*, **57**, 1475 (1979).
[3] D. R. Kelly, S. M. Roberts, and R. F. Newton, *Syn. Comm.*, **9**, 295 (1979).

Boron triiodide, BI_3. Mol. wt. 391.52, m.p. 43–44°. Supplier: Alfa.

Demethylation of methoxybenzaldehydes. BI_3 is more reactive than BCl_3 (**2**, 34–35; **4**, 43) and cleaves methoxyl groups of aromatic aldehydes regardless of the position. BBr_3 is also effective but this reagent can convert benzaldehydes into benzal bromides.[1]

[1] J. M. Lansinger and R. C. Ronald, *Syn. Comm.*, **9**, 341 (1979).

Bromine, 3, 34; **4,** 46–47; **5,** 55–57; **6,** 70–73; **7,** 33–35; **8,** 52–53.

Bromolactonization. Corey and Hase[1] have observed high stereoselectivity in halolactonization of an acylic β,γ-unsaturated acid. Thus reaction of bromine with the thallium salt of *trans*-2-methyl-3-pentenoic acid (**1**) with Br_2 in CH_2Cl_2 at −78° results in one halolactone (**2**) in about 90% yield, even though two additional asymmetric centers are introduced. The reaction was also conducted with the resolved acid.

CH₃ CH₃ H COOTl **1** —(Br_2, CH_2Cl_2, −78°; ~90%)→ Br H CH₃ CH₃ H H O O **2** —(DIBAH; 94%)→ Br CH₃ CH₃ O OH **3** (two epimers)

3 + HS(CH₂)₃SH —($BF_3\cdot(C_2H_5)_2O$)→ Br CH₃ CH₃ HO S S —($KOC(CH_3)_3$)→ H O CH₃ CH₃ H S S **4**

3 —($LiBH_4$)→ Br CH₃ CH₃ HO CH₂OH —($KOC(CH_3)_3$)→ H₃C O H HOCH₂ CH₃ H **5**

The lactol **3** has been converted into the protected epoxy aldehyde **4** and into the hydroxy epoxide **5**, both with established configuration. These particular six-carbon units were required for synthesis of a natural product, rifamycin. A number of other transformations are also possible.

Cleavage of ketoximes. Ketoximes can be cleaved by aqueous bromine buffered with $NaHCO_3$. Sodium hypobromite may be the actual reagent. The method can be used for cleavage of tosylhydrazones; in this case yields are improved by added HMPT.[2]

Sulfoxides. The complexes of bromine with tertiary amines have been used to convert sulfides to sulfoxides.[3] The reaction can be conducted more conveniently with bromine and potassium hydrogen carbonate as the hydrogen bromide acceptor under two phase conditions (H_2O–CH_2Cl_2). Yields are in the range 80–100%.[4]

$$R^1SR^2 \xrightarrow{Br_2} \left[R^1\overset{\displaystyle Br}{\underset{\displaystyle Br}{S}}R^2 \right] \xrightarrow[80\text{–}100\%]{H_2O} R^1\overset{\displaystyle O}{\overset{\|}{S}}R^2 + 2HBr$$

[1] E. J. Corey and T. Hase, *Tetrahedron Letters*, 335 (1979).
[2] G. A. Olah, Y. D. Vankar, and G. K. S. Prakash, *Synthesis*, 113 (1979).
[3] S. Oae, Y. Onishi, S. Kozuka, and W. Tagaki, *Bull. Chem. Soc. Japan*, **39**, 364 (1966).
[4] J. Drabowicz, W. Midura, and M. Mikołajczyk, *Synthesis*, 39 (1979).

Bromine–Dibenzo-18-crown-6 (DBC·Br_2).

Stereoselective bromination of olefins.[1] Exclusive *trans*-addition of bromine to alkenes can be effected with this bromine–crown ether complex. Thus *cis*- and *trans*-β-methylstyrene react with DBC·Br_2 to yield *threo*- and *erythro*-1-phenyl-1,2-dibromopropane, respectively. Other brominating agents (Br_2, Py·Br_2) are significantly less selective for this reaction.[1]

[1] K. H. Pannell and A. Mayr, *J.C.S. Chem. Comm.*, 132 (1979).

Bromoacetylmethylenetriphenylphosphorane, $BrCH_2COCHP(C_6H_5)_3$ (**1**). Mol. wt. 397.26, m.p. 222–224°. Preparation.[1]

Cyclopentenone annelation.[2] Enolates condense with **1** in an alkylation–intramolecular Wittig reaction sequence to afford cyclopentenones (yields, 14–39%).

Example:

[2-methylcyclohexane-1,3-dione → (1, NaH, DMF; 31%) → bicyclic enone with O, CH_3, =O]

[1] M. LeCorre, *Compt. Rend., C*, **273**, 81 (1971); K. Issleib and R. Lindner, *Ann.*, **173**, 12 (1968).
[2] H.-J. Altenbach, *Angew. Chem., Int. Ed.*, **18**, 940 (1979).

Bromodimethylsulfonium bromide, 4, 174–175.

Dethioacetalization.[1] Bromodimethylsulfonium bromide is generally superior to bromine[2] for cleavage of thioacetals (yields 55–90%).

Oxidation of sulfides.[3] The reagent, in combination with triethylamine, oxidizes sulfides to disulfides in 80–95% yield.

α-Bromo ketones.[4] The reagent in combination with triethylamine converts oxides of alkenes and oxides of five- to seven-membered cycloalkenes into α-bromo ketones in 70–80% yield. Morpholinoenamines under these conditions are also converted into α-bromo ketones.

Chlorodimethylsulfonium chloride (generated from chlorine and dimethyl sulfide) can be used in the same way for preparation of α-chloro ketones.

[1] G. A. Olah, Y. D. Vankar, M. Arvanaghi, and G. K. S. Prakash, *Synthesis*, 720 (1979).
[2] W. E. Truce and F. E. Roberts, *J. Org.*, **28**, 961 (1963).
[3] G. A. Olah, M. Arvanaghi, and Y. D. Vankar, *Synthesis*, 721 (1979).
[4] G. A. Olah, Y. D. Vankar, and M. Arvanaghi, *Tetrahedron Letters*, 3653 (1979).

1-Bromo-2-ethoxycyclopropyllithium, C_2H_5O–cyclopropane–(Li)(Br) (**1**). Mol. wt. 170.96. The reagent is prepared[1] by treatment of 1,1-dibromo-2-ethoxycyclopropane[2] with n-butyllithium in THF at −95°.

α,β-Unsaturated aldehydes.[1,3] The reagent can serve as the equivalent of 2-lithiopropenal. Thus the reaction of **1** with heptanal gives, after ring opening, the acetal **3** in 75% overall yield. The acetal **3** can be converted to α,β-unsaturated aldehydes of the type found in terpenoids. For example, reaction of the acetate of **3** with $(CH_3)_2CuLi$ leads to the homoterpenoid **4**. Or **3** after acetylation and hydrolysis to the aldehyde **5** can be reduced to **6**. The most efficient reducing agent is $Fe(CO)_5$; $NaBH_3CN$ can be used, but the yield is considerably lower.

$$\mathbf{1} + CH_3(CH_2)_5CHO \longrightarrow \underset{\mathbf{2}}{C_2H_5O\text{-}c\text{-}C_3H_3(Br)CH(OH)C_6H_{13}\text{-}n} \xrightarrow[\sim 75\%]{C_2H_5OH,\ K_2CO_3,\ \Delta}$$

$$\underset{\mathbf{3}}{(C_2H_5O)_2CHC(=CH_2)CH(OH)C_6H_{13}\text{-}n} \xrightarrow[\sim 50\%]{1)\ Ac_2O,\ Py;\ 2)\ (CH_3)_2CuLi;\ 3)\ H_3O^+} \underset{\mathbf{4}}{(CH_3CH_2)(OHC)C{=}C(H)C_6H_{13}\text{-}n}$$

$$\mathbf{3} \xrightarrow[80\%]{1)\ Ac_2O;\ 2)\ H_2SO_4} \underset{\mathbf{5}}{OHCC(=CH_2)CH(OAc)C_6H_{13}\text{-}n} \xrightarrow[90\%]{Fe(CO)_5,\ Dabco} \underset{\mathbf{6}}{(CH_3)(OHC)C{=}C(H)C_6H_{13}\text{-}n}$$

These last sequences were used in syntheses of the juvenile hormone (**7**) and of β-sinensal (**8**).

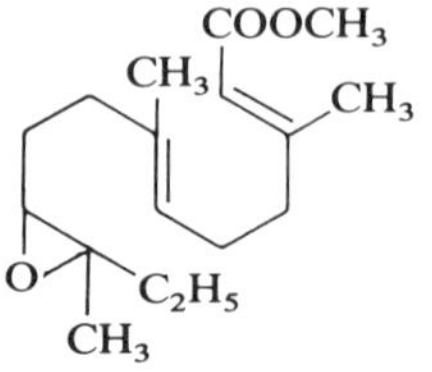

COOCH3, CH3, CH3, O, C2H5, CH3 — **7**

CH3, H2C, CH=CH2, OHC, CH3 — **8**

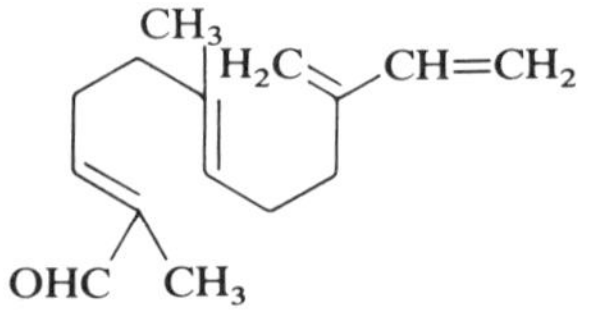

[1] T. Hiyama, A. Kanakura, H. Yamamoto, and H. Nozaki, *Tetrahedron Letters*, 3047 (1978).
[2] E. J. Corey and P. Ulrich, *ibid.*, 3685 (1975).
[3] T. Hiyama, A. Kanakura, H. Yamamoto, and H. Nozaki, *ibid.*, 3051 (1978).

Bromomethyl phenyl sulfoxide, $C_6H_5\overset{O}{\overset{\|}{S}}CH_2Br$. Mol. wt. 219.12, b.p. 113°/0.1 mm. The compound is prepared by reaction of methyl phenyl sulfoxide with NBS–Br_2 in the presence of pyridine.[1]

Bromomethyl ketones. Aldehydes can be converted into bromomethyl ketones by reaction with the anion of bromomethyl phenyl sulfoxide to form a β-hydroxy-α-bromo sulfoxide followed by pyrolysis in xylene or diglyme (equation I).[2]

$$\text{(I)}\qquad RCHO + C_6H_5\overset{O}{\overset{\|}{S}}\underset{Li}{\underset{|}{C}}HBr \xrightarrow[70-85\%]{} R\underset{H}{\overset{HO}{C}}-\underset{S(=O)C_6H_5}{\overset{Br}{C}}-H \xrightarrow[85-95\%]{\Delta} R\overset{O}{\overset{\|}{C}}CH_2Br$$

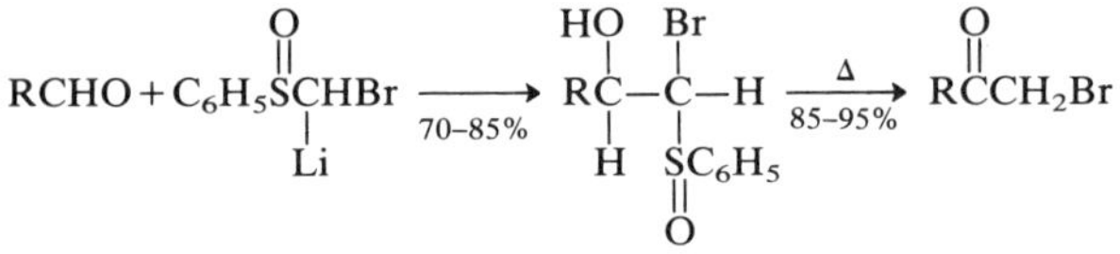

[1] S. Iriuchijima and G. Tsuchihashi, *Synthesis*, 588 (1970).
[2] V. Reutrakul, A. Tiensripojamarn, K. Kusamran, and S. Nimgirawath, *Chem. Letters*, 209 (1979).

3-Bromo-3-methyl-2-trimethylsilyloxy-1-butene (1). Mol. wt. 237.23, b.p. 63°/9 mm.

Preparation:

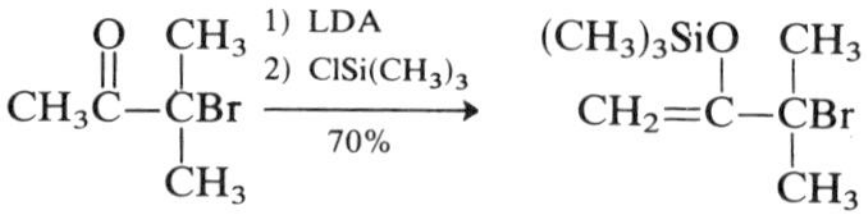

$$CH_3\overset{O}{\overset{\|}{C}}-\underset{CH_3}{\overset{CH_3}{C}}Br \xrightarrow[70\%]{1)\ LDA\quad 2)\ ClSi(CH_3)_3} CH_2=\overset{(CH_3)_3SiO}{C}-\underset{CH_3}{\overset{CH_3}{C}}Br$$

1

Cycloadditions.[1] The reaction of **1** in CH_2Cl_2 with 1,3-dienes with zinc chloride as catalyst results in [3 + 4]cycloaddition to give cycloheptenones as major products. In some cases [3 + 2]cycloadducts are formed as minor products. Styrene derivatives react with **1** to form mainly cyclopentanones by [3 + 2]cycloaddition.

Examples:

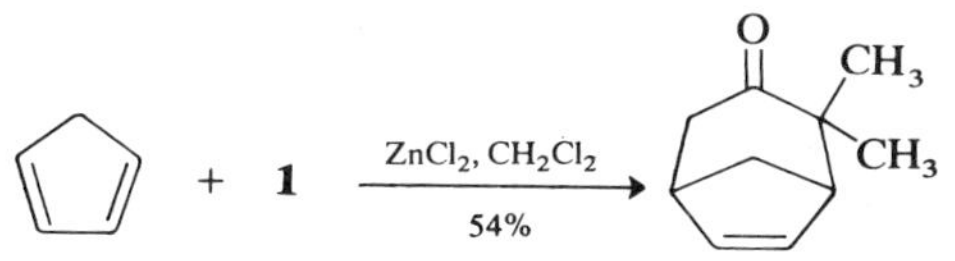

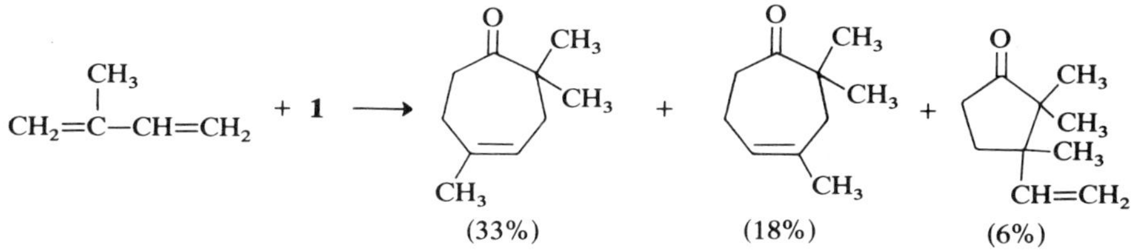

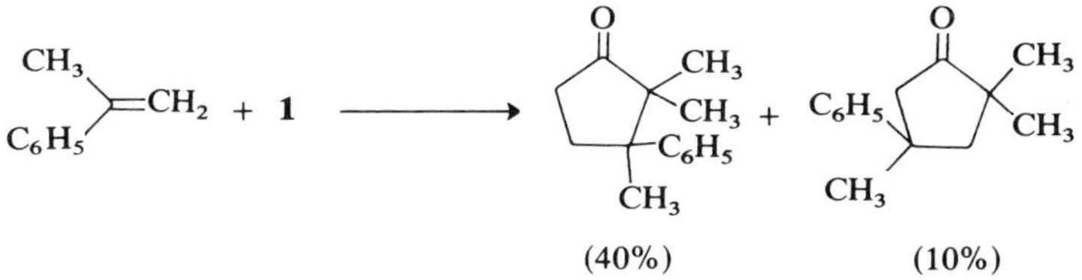

Reactions of this type have been reported for reactions of α,α'-dibromoketones with iron carbonyl compounds or zinc–copper couple as catalysts (**4**, 157–158; **5**, 221–224).

[1] H. Sakurai, A. Shirahata, and A. Hosomi, *Angew. Chem., Int. Ed.*, **18**, 163 (1979).

2-Bromopropane (Isopropyl bromide), $BrCH(CH_3)_2$. Mol. wt. 123, b.p. 59°.

Protection of enols of β-keto aldehydes.[1] The reaction of 2-hydroxymethylenecyclohexanone (**1**) with 2-bromopropane and K_2CO_3 in DMF gives **2** in 72% yield.

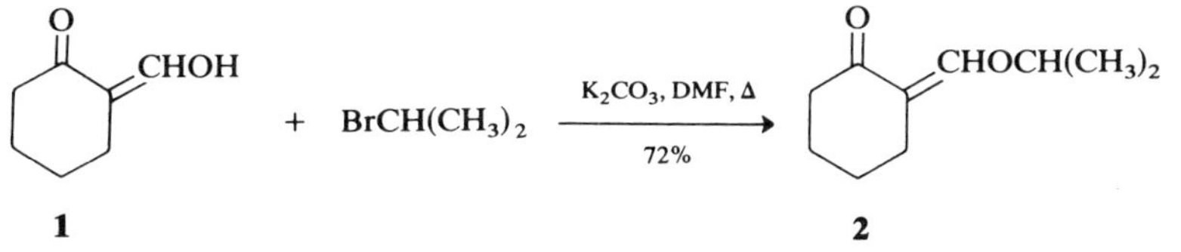

Protection of phenols.[2] Aryl isopropyl ethers are also obtained by reaction of phenols with 2-bromopropane in DMF. The protective group is cleaved by BCl_3 in CH_2Cl_2 at 0°, but it is stable to $SnCl_4$ or $TiCl_4$ under these conditions. It is recommended for protection of phenolic groups during formylation with dichloromethyl methyl ether in the presence of $SnCl_4$.

[1] T. M. Cresp and M. V. Sargent, *J.C.S. Perkin I*, 2145 (1974).
[2] T. Sala and M. V. Sargent, *ibid.*, 2593 (1979).

β-Bromopropionaldehyde ethylene acetal, $BrCH_2CH_2CH\langle O_2C_2H_4 \rangle$ (**1**). Mol. wt. 181.04, b.p. 68–70°/8 mm. Preparation.[1]

Cyclopentene annelation.[2] The Grignard derivative (**2**) of **1**, prepared efficiently with Reike's magnesium powder (**5**, 419), undergoes conjugate addition to enones in the presence of a copper(I) salt, particularly with $CuBr \cdot S(CH_3)_2$ (**6**, 235). The adducts are converted to annelation products on treatment with aqueous THF containing HCl.

Examples:

$CuBr \cdot S(CH_3)_2$, THF, −75→0°; 85%; H_3O^+; 89%

77%; H_3O^+; 54%; OH

[1] G. Büchi and H. Wüest, *J. Org.*, **34**, 1122 (1969).
[2] A. Marfat and P. Helquist, *Tetrahedron Letters*, 4217 (1978).

N-Bromosuccinimide, 1, 78–80; **2,** 40–42; **3,** 34–36; **4,** 49–53; **5,** 65–66; **6,** 74–76; **7,** 37–40; **8,** 54–56.

Acid bromides. Aldehydes can be oxidized to acid bromides by NBS under irradiation. For example, the reaction with benzaldehyde results in benzoyl bromide (48% yield). In general, the acid bromide is converted into an ester or an amide *in situ*.[1]

Examples:

$$C_6H_{11}CHO \xrightarrow{NBS,\ CCl_4,\ h\nu} [C_6H_{11}COBr] \xrightarrow[84\%]{CH_3OH} C_6H_{11}COOCH_3$$

$$n\text{-}C_3H_7CHO \rightarrow [n\text{-}C_3H_7COBr] \xrightarrow[66\%]{C_6H_5CH_2NH_2} n\text{-}C_3H_7CONHCH_2C_6H_5$$

Bromohydrins.[2] NBS in aqueous DMSO is useful for preparation of bromohydrins (**2**, 41; **3**, 34–35). Cyclopenta[*cd*]pyrene, a highly mutagenic environmental pollutant, is oxidized specifically to bromohydrin **2** by this method. Sodium methoxide treatment of **2** affords the acid sensitive and highly mutagenic epoxide **3**, which possibly is the ultimate mutagen derived from **1**.

OH
Br
NBS, H_2O, DMSO
66%
$NaOCH_3$, THF
33%
O

1 **2** **3**

Epoxidation. Precocene (**1**) causes larvae of certain insects to moult precociously to premature adults. The highly reactive epoxide **3**, probably the active metabolite of **1**, has now been synthesized. NBS oxidation of **1** in aqueous THF affords bromohydrin **2**, which when treated with NaH affords **3** in 88% yield. Attempts to prepare **3** by MCPBA, $VO(acac)_2$–*t*-butylhydroperoxide, or $Mo(CO)_6$–*t*-butylhydroperoxide oxidations were unsuccessful.[3]

CH_3O O CH_3 CH_3
NBS, THF, H_2O
OH Br CH_3O O CH_3 CH_3
NaH
88% overall

1 **2**

O
CH_3O O CH_3 CH_3

3

Selective oxidation. Oxidation of the diol **1** with PCC and other Cr(VI) reagents results mainly in oxidation of the primary alcohol group to give lactone **2**. In contrast, aqueous NBS (1.5 equiv.) oxidizes the secondary alcohol group selectively to a carbonyl group in essentially quantitative yield. As a result of this differing behavior, the diol **1** can serve as a precursor to both 2- and 9-isocyanopupukeanane (**6** and **7**), tricyclic sesquiterpene isocyanides produced from a species of sponge (Hymeniacidon).[4]

[1] Y.-F. Cheung, *Tetrahedron Letters*, 3809 (1979).
[2] A. Gold, J. Brewster, and E. Eisenstadt, *J.C.S. Chem. Comm.*, 903 (1979).
[3] R. C. Jennings and A. P. Ottridge, *ibid.*, 920 (1979).
[4] E. J. Corey and M. Ishiguro, *Tetrahedron Letters*, 2745 (1979).

CH_2OH, CH_3, H, OH, CH_3, H, $(CH_3)_2CH$ **1** — NBS, H_2O, DME, >98% → CH_2OH, CH_3, H, O, CH_3, H, $(CH_3)_2CH$ **3** — PCC, CH_2Cl_2, 90% → CHO, CH_3, H, O, CH_3, H, $(CH_3)_2CH$ **4**

1 ⇅ LiAlH$_4$ (100%) / PCC, CH_2Cl_2 (~100%) **2** (O, C, O, CH_3, H, CH_3, H, $(CH_3)_2CH$)

2 → Several steps → **7** (CH_3, NC, H, CH_3, $CH(CH_3)_2$)

4 → 98% K_2CO_3, CH_3OH → **5** (CH_3, O, OH, H, CH_3, $CH(CH_3)_2$) → Several steps → **6** (CH_3, NC, H, CH_3, H, $CH(CH_3)_2$)

N-Bromosuccinimide–Dimethylformamide.

Bromination of arenes. NBS in DMF is useful for monobromination of reactive arenes and also of phenols and anilines. The reaction is not useful for monobromination of *o*- and *p*-hydroxyphenols or for dianilines. Some compounds (and the yields) prepared in this way are listed.[1]

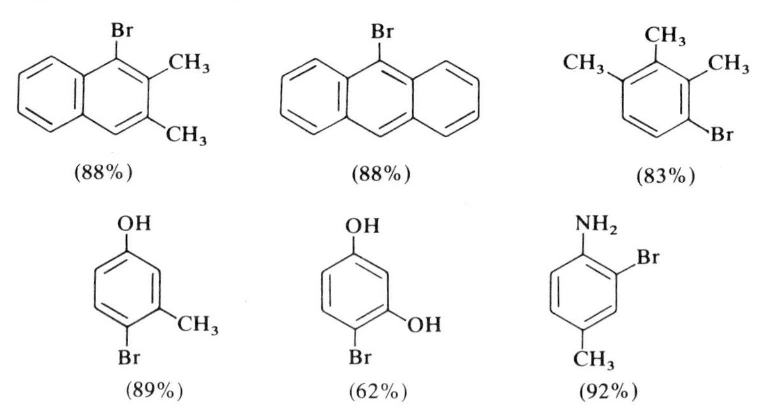

[1] R. H. Mitchell, Y.-H. Lai, and R. V. Williams, *J. Org.*, **44**, 4733 (1979).

N-Bromosuccinimide–Dimethyl sulfoxide.

α-Bromo ketones.[1] One useful route to bromohydrins is reaction of olefins with NBS in aqueous DMSO (Dalton reaction, **2**, 41–42; **3**, 34–35). If the reaction is carried out in dry DMSO, α-bromo ketones are obtained. This variation has been reported for conversion of cyclohexene into α-bromocyclohexanone (63% yield) and of dihydropyrane into α-bromo-δ-valerolactone (equation I).

(I) [dihydropyran] —NBS, DMSO, 62%→ [α-bromo-δ-valerolactone, Br]

This new reaction was first observed as one step in the conversion of the hexaacetate (**1**) of aucubin, an iridoid glucoside, into the lactone **3**, a possible chiral intermediate to prostaglandins.[1]

1 (OAc, $AcOCH_2$, $OGlu(Ac)_4$) —NBS, DMSO, 90%→ **2** (two isomers) (OAc, Br, O, $AcOCH_2$, $OGlu(Ac)_4$) —Zn, HOAc, 90%→ **3** (OAc, O, $AcOCH_2$, $OGlu(Ac)_4$)

[1] W. F. Berkowitz, I. Sasson, P. S. Sampathkumar, J. Hrabie, S. Choudhry, and D. Pierce, *Tetrahedron Letters*, 1641 (1979).

Bromotrimethylsilane, $BrSi(CH_3)_3$. Mol. wt. 153.10, b.p. 79°. Suppliers: Aldrich, Alfa. Preparation.[1]

Alkyl bromides.[2] Alcohols are converted into bromides by reaction with bromotrimethylsilane (1.5–4 equiv.) in $CHCl_3$ at 25–50° (equation I). The reaction occurs with inversion. Tertiary and benzylic alcohols react more rapidly than primary or secondary alcohols. Bromides are not formed under the same conditions from trimethylsilyl ethers of alcohols. However, trimethyl orthoformate is converted into methyl formate, $HC(OCH_3)_3 \rightarrow HCOOCH_3$. Unlike iodotrimethylsilane, the bromosilane does not dealkylate esters, ethers, or carbamates.

$$\text{(I)}\quad ROH + BrSi(CH_3)_3 \rightarrow \underset{(85\text{–}100\%)}{RBr} + HOSi(CH_3)_3 \xrightarrow{BrSi(CH_3)_3} [(CH_3)_3Si]_2O + HBr$$

Cleavage of lactones and carbonates.[3] Lactones and carbonates react with bromotrimethylsilane to afford ω-bromocarboxylic acid derivatives (equation I) and bromohydrin trimethylsilyl ethers (equation II), respectively; acyclic, aliphatic esters do not react with bromotrimethylsilane. Iodotrimethylsilane reacts in an analogous fashion with lactones, but in reaction with ethylene carbonate the main product is 1,2-diiodoethane (equation III). The ω-bromocarboxylate derivatives are converted into acid chlorides by reaction with $SOCl_2$ (equation I).

(I) $$\text{lactone } (CH_2)_n\text{–O–C(=O)} \ (n = 2–5) \xrightarrow[79–96\%]{BrSi(CH_3)_3,\ 100°C} Br(CH_2)_nCOOSi(CH_3)_3 \xrightarrow[87–95\%]{SOCl_2} Br(CH_2)_nCOCl$$

(II) $$\text{propylene carbonate} \xrightarrow[79\%]{BrSi(CH_3)_3} CH_3CH(OSi(CH_3)_3)CH_2Br + CO_2$$

(III) $$\text{ethylene carbonate} \xrightarrow{ISi(CH_3)_3} ICH_2CH_2I$$

Selective phosphonate ester dealkylation.[4] Alkyl phosphonate esters are selectively and nearly quantitatively cleaved by bromotrimethylsilane in the presence of alkyl carboxylate esters, carbamates, acetylenes, ketones, and halides. Alkyl iodides do not exchange under the reaction conditions. The resulting bis(trimethylsilyl) phosphonates are hydrolyzed in acetone by a small excess of water.

Examples:

$$C_2H_5OOCCH_2PO(OC_2H_5)_2 \xrightarrow[\sim 100\%]{1)\ BrSi(CH_3)_3;\ 2)\ H_2O,\ (CH_3)_2C{=}O} C_2H_5OOCCH_2PO(OH)_2$$

$$CH_3NHCOCH_2PO(OC_2H_5)_2 \xrightarrow[\sim 100\%]{} CH_3NHCOCH_2PO(OH)_2$$

[1] B. O. Pray, L. H. Somer, G. M. Goldberg, G. T. Kerr, P. A. DiGiorgio, and F. C. Whitmore, *Am. Soc.*, **70**, 433 (1948).
[2] M. E. Jung and G. L. Hatfield, *Tetrahedron Letters*, 4483 (1978).
[3] H. R. Kricheldorf, *Angew. Chem., Int. Ed.*, **18**, 689 (1979).
[4] C. E. McKenna and J. Schmidhauser, *J.C.S. Chem. Comm.*, 739 (1979).

1-Butadienyldimethylsulfonium tetrafluoroborate,

$CH_2{=}CHCH{=}CH\overset{+}{S}(CH_3)_2BF_4^-$ (**1**). Mol. wt. 202.02.

Preparation[1]:

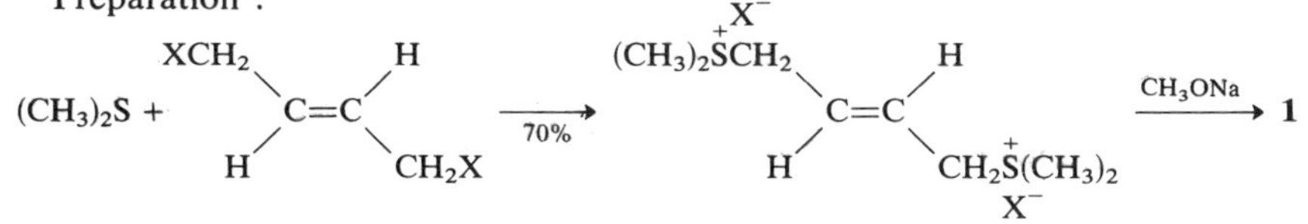

$$(CH_3)_2S + XCH_2CH{=}CHCH_2X \xrightarrow{70\%} (CH_3)_2\overset{+}{S}CH_2CH{=}CHCH_2\overset{+}{S}(CH_3)_2\ 2X^- \xrightarrow{CH_3ONa} \mathbf{1}$$

Epoxyannelation.[2] The enolate of cyclohexanone reacts with this reagent (**1**) to form the α,β-unsaturated epoxide **2**, with displacement of dimethyl sulfide, and a mixture of two sulfur-containing products (**3**). The reaction to give an epoxide is general for enolates of unhindered, cyclic ketones, but is not observed with enolates of acyclic ketones. However, it is possible with enolates of aldehydes. An example is the conversion of phenylacetaldehyde into **4** and **5**, in somewhat low yield.

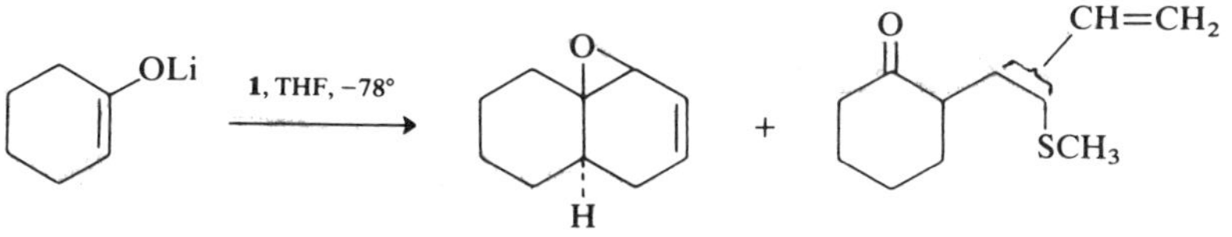

OLi; **1**, THF, −78°; O; H; **2** (58%); +; O; CH=CH$_2$; SCH$_3$; **3** (18%)

$C_6H_5CH{=}CHOLi \xrightarrow[27\%]{\mathbf{1}}$ O; C_6H_5; **4**; + 4:1; C_6H_5; CHO; CH=CH$_2$; **5**

Epoxyannelation has been effected previously by intramolecular Darzens condensation of enolates of ketones with 1,4-dichlorobutane-2-one[3] and chloroacrylonitrile,[4] as shown (equations III and IV).

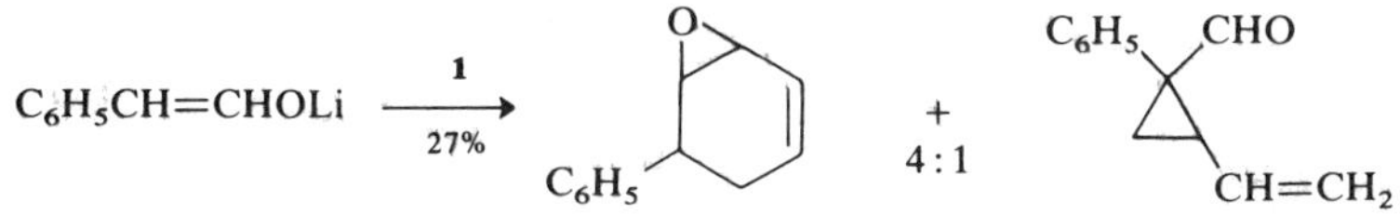

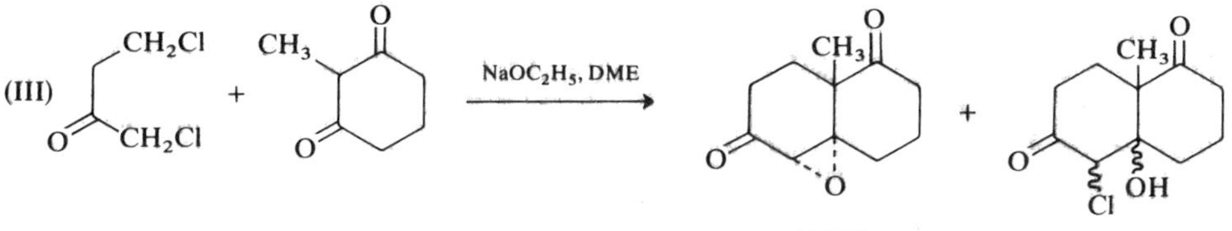

(III) CH$_2$Cl; O; CH$_2$Cl; + CH$_3$; O; O; $\xrightarrow{NaOC_2H_5,\ DME}$ CH$_3$; O; O; O; (43%) + CH$_3$; O; O; Cl; OH; (11%)

(IV) CH$_3$; C=O; CH$_2$; CH(COOC$_2$H$_5$)$_2$ + Cl; CN; C; CH$_2$ $\xrightarrow[61\%]{KOC(CH_3)_3,\ C_6H_6}$ CH$_3$; O; CN; C$_2$H$_5$OOC; COOC$_2$H$_5$

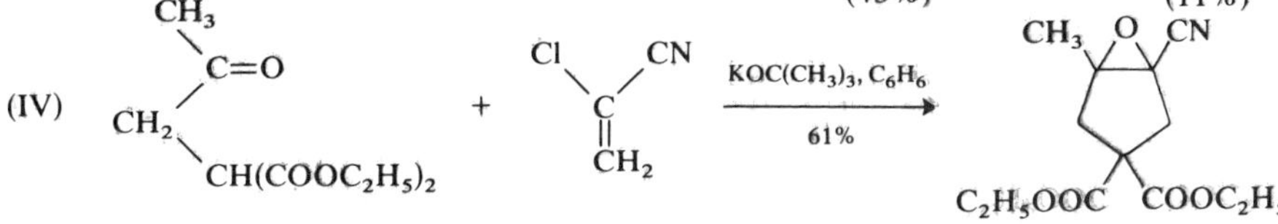

[1] H. Braun, N. Meyer, and G. Kresze, *Ann.*, **762**, 111 (1972).
[2] M. E. Garst, *J. Org.*, **44**, 1578 (1979).
[3] S. Danishefsky and G. A. Koppel, *J.C.S. Chem. Comm.*, 367 (1971).
[4] D. R. White, *ibid.*, 95 (1975).

***t*-Butoxybis(dimethylamino)methane, 5**, 71–73; **7**, 41.

Carboxylation. Bredereck's reagent has been used in a simple synthesis of L-γ-carboxyglutamic acid (**5**), an unusual natural amino acid present in prothrombin and believed to be involved in clotting. The starting material is the lactam **1** derived from L-glutamic acid and available commercially. It is converted by reaction with this reagent into the enamide **2**. Reaction with 2,2,2-trichloroethoxycarbonyl chloride transforms **2** into the trichloroethyl ester **3** in moderate yield. The synthesis of **5** is completed by reaction with benzyl alcohol and triethylamine followed by hydrogenolysis.[1]

$COOCH_2C_6H_5$ $COOCH_2C_6H_5$ **1** $\xrightarrow[95\%]{[(CH_3)_2N]_2CHOC(CH_3)_3}$ $COOCH_2C_6H_5$ $(CH_3)_2N$ $COOCH_2C_6H_5$ **2**

$\xrightarrow[41\%]{ClCOCH_2CCl_3}$ $COOCH_2C_6H_5$ Cl_3CCH_2OC $COOCH_2C_6H_5$ **3** $\xrightarrow[62\%]{C_6H_5CH_2OH,\ N(C_2H_5)_3,\ 104°}$

$C_6H_5CH_2OOC$ $C_6H_5CH_2OOC$ $NHCO_2CH_2C_6H_5$ $COOCH_2C_6H_5$ **4** $\xrightarrow[93\%]{H_2,\ Pd/C}$ HOOC HOOC NH_2 COOH **5**

[1] S. Danishefsky, E. Berman, L. A. Clizbe, and M. Hirama, *Am. Soc.*, **101**, 4385 (1979).

B-*n*-Butyl-9-borabicyclo[3.3.1]nonane (1). Mol. wt. 178.13. Supplier: Aldrich.

1,5-Dienes. Regioselective head-to-tail coupling of alkylthioallyllithiums and allylic halides can be effected by prior conversion of the allyllithium to the lithium allyl borate via reaction with **1**.[1] Reaction of the alkylthioallyllithium compounds

Bu-*n* B **1** $\equiv R_3B$

i-PrS CH_2 $\xrightarrow{1)\ sec\text{-}BuLi;\ 2)\ \mathbf{1}}$ [*i*-PrS H $\bar{B}R_3Li^+$] $\xrightarrow[77\%]{H,\ CH_2Br,\ CH_3,\ H}$ *i*-PrS CH_2 CH_2 CH_3

directly, affords head-to-head coupling products,[2] whereas the alkylthioallylcopper species yield tail-to-tail coupling products.[3]

[1] Y. Yamamoto, H. Yatagai, and K. Maruyama, *J.C.S. Chem. Comm.*, 157 (1979).
[2] J. F. Biellmann and J. B. Ducep, *Tetrahedron Letters*, 5629 (1968).
[3] K. Oshima, H. Yamamoto, and H. Nozaki, *Am. Soc.*, **95**, 7926 (1973).

***t*-Butylchlorodimethylsilane, 4**, 57–58, 176–177; **5**, 74–75; **6**, 78–79; **7**, 59.

Selective protection of primary alcohols. Two recent syntheses of warburganal (**1**) required selective monoprotection of a primary diol. In both cases success was realized by use of this bulky silane. Thus the diol **2** is converted into the monosilyl ether **4** in quantitative yield by reaction with *t*-butyldimethylchlorosilane and imidazole in DMF.[1] Similarly, the triol **3** can be selectively protected to give **5**.[2]

$ClSi(CH_3)_2C(CH_3)_3$

2 (R = H)
3 (R = OH)

4 (R = H)
5 (R = OH)

Protection of* vic*-diols. Several laboratories have recently reported migration of *t*-butyldimethylsilyl protective groups of ribonucleosides.[3-5] Thus the 2′- and 3′-O-*t*-butyldimethylsilyl derivatives of ribonucleosides can rearrange in the presence of base (pyridine) by an intramolecular reaction. Obviously care must be exercised in the use of silicon-containing reagents for partial protection of polyols.

[1] S. P. Tanis and K. Nakanishi, *Am. Soc.*, **101**, 4398 (1979).
[2] T. Nakata, H. Akita, T. Naito, and T. Oishi, *ibid.*, **101**, 4400 (1979).
[3] K. K. Ogilvie, S. L. Beaucage, A. L. Schifman, N. Y. Theriault, and K. L. Sadana, *Can. J. Chem.*, **56**, 2768 (1978).
[4] K. L. Sadana and P. C. Loewen, *Tetrahedron Letters*, 5095 (1978).
[5] S. S. Jones and C. B. Reese, *J.C.S. Perkin I*, 2762 (1979).

***t*-Butyl chloromethyl ether,** $ClCH_2OC(CH_3)_3$. The ether can be prepared as a solution in CCl_4 by reaction of NCS with *t*-butyl methyl ether (*hν*, 35°). The ether decomposes on attempted isolation.

Protection of hydroxyl groups. *t*-Butoxymethyl ethers of alcohols can be obtained by reaction with *t*-butyl chloromethyl ether and triethylamine in THF at room temperature. The acetals are obtained in 55–80% yield. Yields are poor in the case of phenols. Deprotection is accomplished with TFA at room temperature. The acetals are stable to hot acetic acid; therefore selective deprotection of other acid-sensitive groups is possible.[1]

[1] H. W. Pinnick and N. H. Lajis, *J. Org.*, **43**, 3964 (1978).

t-Butyldimethylsilyl perchlorate (1). Mol. wt. 214.73, b.p. 35°/0.06 mm.

Protection of hydroxyl groups. The *t*-butyldimethylsilyl group has been exceedingly useful for protection of hydroxyl groups (**4**, 57–58; **5**, 176–177; **6**, 75) because of the greater stability to solvolysis than is shown by the trimethylsilyl group. They have been prepared using *t*-butyldimethylsilyl chloride, but the etherification can be slow owing to steric hindrance. Barton and Tully[1] reasoned that the silyl perchlorate (**1**) should be more reactive and developed a method for the preparation shown in equation (I). The perchlorate has no ionic character and is a covalent ester of perchloric acid.

$$\text{(I)}\quad (CH_3)_2SiHCl + (CH_3)_3CLi \xrightarrow[90\%]{\text{Pentane}} (CH_3)_2SiHC(CH_3)_3 \xrightarrow[91\%]{(C_6H_5)_3C^+ClO_4^-,\ CH_2Cl_2} (CH_3)_3CSi(CH_3)_2OClO_3 \quad \mathbf{1}$$

The method was extended to a synthesis of di-*t*-butylmethylsilyl perchlorate (**2**), b.p. 65°/0.1 mm., and to tri-*t*-butylsilyl perchlorate (which proved too unreactive to be of value).

Both (**1**) and (**2**) convert alcohols, even tertiary ones, into the corresponding silyl ethers in the presence of pyridine without formation of alkenes; the reaction is more rapid than that with *t*-butyldimethylsilyl chloride. The ethers derived from **2** are unusually stable; they are not cleaved by CsF in DMSO even at elevated temperatures. But they can be cleaved by BF_3 etherate in CH_2Cl_2 at 0° to give borate ethers, which can be hydrolyzed by $NaHCO_3$ at 0°.

Caution: All perchlorates are potentially explosive.

[1] T. J. Barton and C. R. Tully, *J. Org.*, **43**, 3649 (1978).

t-Butyl hydroperoxide (TBHP), 1, 88–89; **2**, 49–50; **3**, 37–38; **5**, 75–77; **6**, 81–82; **7**, 43–45; **8**, 62–64.

Review.[1] In comparison with related sources of oxygen atoms such as H_2O_2 and CH_3COOOH, TBHP is actually safer to handle; it is one of the most stable known organic peroxides. However, it should be handled with respect. For some reactions, the 70% TBHP available commercially (Lucidol, Aldrich) has adequate reactivity. When anhydrous TBHP is required, it can be obtained in an aprotic organic solvent (dichloroethane) by azeotropic drying. The procedure is described in detail. This anhydrous reagent is required for epoxidation of isolated double bonds in a reasonable time (5–24 hours at reflux). It is equivalent in reactivity to 90% TBHP.

Epoxidation. The last steps in a total synthesis of pentalenolactone (**3**) required epoxidation of **1**. The reaction of alkaline hydrogen peroxide with **1** afforded mainly the undesired β-epoxide, epimeric with **2**. The desired reaction was effected by reduction of **1** with diisobutylaluminum hydride to the allylic hemiacetal and application of the Sharpless epoxidation reaction. After reoxidation (Jones reagent), the desired epoxide **2** was obtained in 45% yield. Alkaline hydrolysis then gave the natural product (**3**).[2]

1) DIBAL
2) $(CH_3)_3COOH$, $VO(acac)_2$
3) CrO_3

1

2 ($R = CH_3$, 45%)
3 ($R = H$)

[1] K. B. Sharpless and T. R. Verhoeven, *Aldrichim. Acta.*, **12**, 63 (1979).
[2] S. Danishefsky, M. Hirama, K. Gombatz, T. Harayama, E. Berman, and P. Schuda, *Am. Soc.*, **100**, 6536 (1978).

t-Butyl hydroperoxide–Diaryl diselenides.

Oxidation of alcohols.[1] Alcohols can be oxidized to carbonyl compounds by *t*-butyl hydroperoxide (slight excess) and a diaryl diselenide (0.1–0.5 equiv.). Diphenyl diselenide is satisfactory, but bis(2,4,6-trimethylphenyl) diselenide is generally more satisfactory. The method is particularly efficient for benzylic and primary allylic alcohols (1–2 hours, 87–100% yield); oxidation of saturated alcohols, primary and secondary, requires 4–17 hours, but still proceeds in excellent yield.

The mechanism is not clear. Possibly ArSe(=O)OH or ArSeOSeAr is the true catalyst. In any case the diaryl diselenide is recoverable in about 70% yield.

[1] M. Shimizu and I. Kuwajima, *Tetrahedron Letters*, 2801 (1979).

t-Butyl hydroperoxide–Selenium dioxide, 8, 64–65.

Propargylic oxidation. α,α'-Dioxygenation is the normal pattern observed in oxidation of alkynes with *t*-butyl hydroperoxide catalyzed by selenium dioxide. The reactivity sequence is $CH_2 \simeq CH > CH_3$. Alkynes with one CH_2 and one CH group afford enynones as the major product.[1]

Examples:

$$CH_3(CH_2)_3C{\equiv}C(CH_2)_3CH_3 \xrightarrow[70\%]{(CH_3)_3COOH,\ SeO_2,\ 23^\circ}$$

$$\underset{OH}{CH_3(CH_2)_3C{\equiv}CCH(CH_2)_2CH_3} + \underset{OH\qquad OH}{CH_3(CH_2)_2CHC{\equiv}CCH(CH_2)_2CH_3} + \text{ketone}(7\%) + \text{ketol}(3\%)$$

30:60

$$CH_3(CH_2)_6C{\equiv}CCH_3 \xrightarrow[58\%]{} \underset{OH}{CH_3(CH_2)_5CHC{\equiv}CCH_3} + \underset{OH}{CH_3(CH_2)_5CHC{\equiv}CCH_2OH}$$

70:20

$$(CH_3)(CH_2{=})CC{\equiv}CCH_2CH_3 \xrightarrow[55\%]{} (HOCH_2)(CH_2{=})CC{\equiv}CCH_2CH_3$$

$$C_6H_{11}{-}C{\equiv}CCH_2CH_3 \xrightarrow[78\%]{} C_6H_{10}(OH){-}C{\equiv}C\underset{OH}{CHCH_3} + C_6H_9{-}C{\equiv}CCOCH_3$$

74:26

[1] B. Chabaud and K. B. Sharpless, *J. Org.*, **44**, 4202 (1979).

t-Butyl hydroperoxide–Triton B.

Epoxidation of α,β-unsaturated ketones. Some years ago, *t*-butyl hydroperoxide in the presence of a basic catalyst (Triton B) was reported to epoxidize unhindered *α,β*-unsaturated ketones in yields comparable to those obtained with alkaline hydrogen peroxide (**1**, 88). This reaction results in Michael adducts in the case of acrylonitrile or methyl acrylate.

More recently, Grieco *et al.*[1] have reported a successful epoxidation with basic *t*-butyl hydroperoxide in contrast to no reaction with basic hydrogen peroxide. Thus **1** can be epoxidized to **2** in 83% yield by a reaction conducted in THF at 25° for 15

CH_2OCH_3, $CH_2CH{=}C(CH_3)_2$, H, O

1

$$\xrightarrow[83\%]{(CH_3)_3COOH,\ \text{Triton B},\ THF,\ 25^\circ}$$

CH_2OCH_3, $CH_2CH{=}C(CH_3)_2$, H, O

2

hours. The method was also used by Still[2] in his synthesis of (±)-periplanone-B (**6**), the female sex excitant of the American cockroach. Actually two selective epoxidations were used. The first epoxidation, **3** → **4**, was carried out with *t*-butyl hydroperoxide catalyzed by KH in THF at −20°. This step was followed by reaction with dimethylsulfonium methylide (**1**, 314) to give a single bisepoxide **5**. The pheromone **6** was obtained from **5** by deprotection and oxidation.

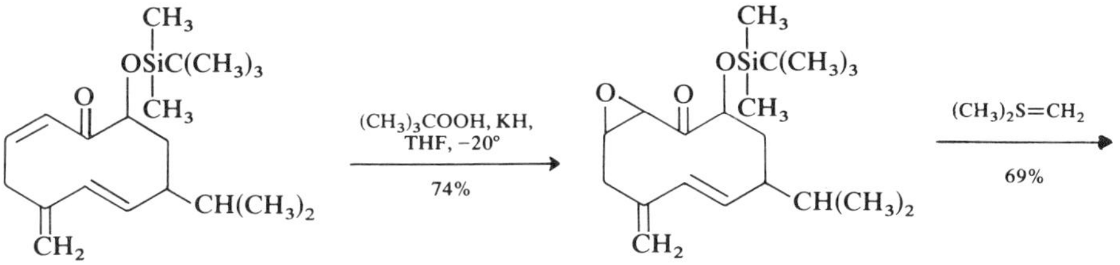

$(CH_3)_3COOH$, KH, THF, −20°, 74%; $(CH_3)_2S{=}CH_2$, 69%; 1) $(C_4H_9)_4NF$ 2) $CrO_3 \cdot 2Py$, 81%

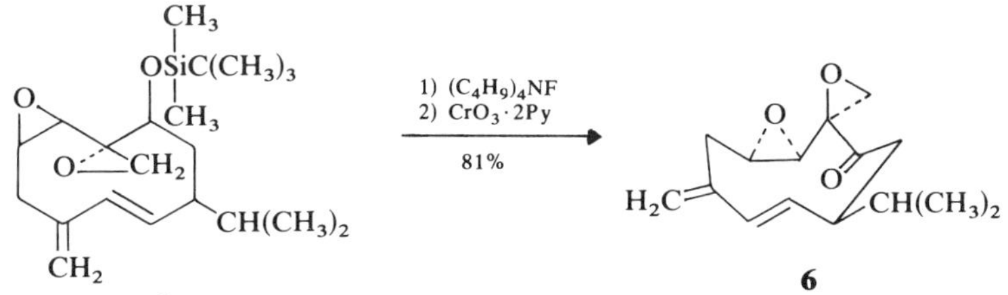

3 **4** **5** **6**

[1] P. A. Grieco, M. Nishizawa, T. Oguri, S. D. Burke, and N. Marinovic, *Am. Soc.*, **99**, 5773 (1977).
[2] W. C. Still, *ibid.*, **101**, 2493 (1979).

***t*-Butyl hydroperoxide–Vanadyl acetylacetonate, 2,** 287; **4,** 346; **5,** 75–76; **7,** 43–44.

Epoxidation of 2-cyclohexene-1-ols (**5**, 75–76). If the epoxidation of 2-cyclohexenol with *t*-butyl hydroperoxide catalyzed by $VO(acac)_2$ is carried to completion (toluene, 75°, 5 hours), *cis*-2,3-epoxycyclohexanol can be obtained in 89% yield; 2-cyclohexenone is also obtained in a significant amount (11% yield). This oxidation to an enone becomes the major pathway (91% yield) in the reaction of *cis*-5-*t*-butylcyclo-2-hexene-1-ol.[1] Fortunately this anomalous reaction is not observed in epoxidation of this substrate with *m*-chloroperbenzoic acid. On the basis of these results, peracid epoxidation should be used for preparation of *cis*-epoxy alcohols from pseudoequatorial alcohols and hydroperoxide–$VO(acac)_2$ epoxidation for pseudoaxial alcohols.

Epoxidation of allylic alcohols (**5**, 75–76). Corrected values for the *erythro/threo* ratios obtained on epoxidation of allylic alcohols with TBHP catalyzed with $VO(acac)_2$ or $Mo(CO)_6$ have been reported.[2,3] In general greater selectivity is observed with the vanadium catalyst. In simple systems formation of *erythro*-epoxy alcohols is markedly favored over the *threo*-isomers, but substitution of a *cis*-alkyl group on the double bond favors formation of *threo*-epoxy alcohols.

Examples:

$$CH_2{=}CHCH(OH)R \xrightarrow{\text{TBHP, VO(acac)}_2} \text{(erythro)} + \text{(threo)} \quad 80:20$$

(*erythro*) (*threo*)

$$\text{cis-}CH_3CH{=}CHCH(OH)CH_3 \longrightarrow \text{(threo)} + \text{(erythro)} \quad 71:29$$

(*threo*) (*erythro*)

Reasonable selectivities are also observed with some homoallylic and bishomoallylic alcohols. The most common competing reaction is observed with cyclohexenols with a quasiequatorial group and a hindered double bond.

[1] R. B. Dehnel and G. H. Whitham, *J.C.S. Perkin I*, 953 (1979).
[2] E. D. Mihelich, *Tetrahedron Letters*, 4729 (1979).
[3] B. E. Rossiter, T. R. Verhoeven, and K. B. Sharpless, *ibid.*, 4733 (1979).

***t*-Butyl isocyanide, 2,** 50–51.

Peptides. In the presence of an aliphatic isocyanide, amino acids and amines condense to form amides. Yields are increased by addition of $ZnCl_2$; CH_2Cl_2 is preferable to alcoholic solvents. Yields of dipeptides range from 25 to 90%.[1]

$$R^1COOH + (CH_3)_3CNC \rightarrow \left[(H)(R^1COO)C{=}NC(CH_3)_3 \right] \xrightarrow{R^2NH_2} R^1CONHR^2 + HCONHC(CH_3)_3$$

[1] H. Aigner and D. Marquarding, *Tetrahedron Letters*, 3325 (1978); D. Rehn and I. Ugi, *J. Chem. Res.* (*S*), 119 (1977).

***t*-Butyl lithioacetate, 5,** 371.

β-Ketoadipic acid esters.[1] These 1,4-dicarbonyl compounds are available by reaction of succinic anhydrides with enolates of acetic acid esters, which are prepared from the acetates with lithium hexamethyldisilazide. The *gem*-dimethyl anhydride **2** reacts preferentially at the more substituted carbonyl function. This reaction is the first example of regioselective addition of a carbon nucleophile in a system where two modes of attack are possible. The same regioselectivity is observed in reductions of *gem*-disubstituted succinic anhydrides and succinimides.[2]

Examples:

$$\mathbf{1} + CH_2{=}C(OLi)OR \xrightarrow{\text{THF, }-78^\circ} HOOC(CH_2)_2COCH_2COOR$$

$R = C(CH_3)_3$ 78%
$R = CH_2C_6H_5$ 62%

H_3C, CH_3, O, O, 2 + $CH_2{=}C(OLi)OC(CH_3)_3$ → 1) THF, −78° 2) CH_2N_2, 96% → CH_3OOC–CH_2–$C(CH_3)_2$–CO–CH_2–$COOC(CH_3)_3$ + CH_3OOC–$C(CH_3)_2$–CH_2–CO–CH_2–$COOC(CH_3)_3$ (71:29)

[1] F.-P. Montforts and S. Ofner, *Angew. Chem., Int. Ed.*, **18**, 632 (1979).
[2] D. M. Bailey and R. E. Johnson, *J. Org.*, **35**, 3574 (1970); J. B. P. A. Wijneberg, H. E. Schoemaker, and W. N. Speckamp, *Tetrahedron*, **34**, 179 (1978).

***n*-Butyllithium, 1**, 95–96; **2**, 51–53; **4**, 60–63; **5**, 78; **6**, 85–91; **7**, 45–47; **8**, 65–66.

3-Cyclopentenols. Danheiser and co-workers[1] have developed an efficient method for conversion of 1,3-dienes to 3-cyclopentenols that involves an alkoxy-accelerated vinylcyclopropane rearrangement. Treatment of conjugated dienes [**1**, R = CH_3, C_6H_5, $C(CH_3)_3$] with $ClCH_2OCH_2CH_2Cl$ and LiTMP (**7**, 214) generates mixtures of cyclopropyl ethers. Ether cleavage and rearrangement of the resulting lithium salts can be effected in a single step by reaction at 25° with *n*-BuLi in THF, DME, or diethyl ether–HMPT.

1 (R–$C(=CH_2)$–$CH{=}CH_2$) → $ClCH_2OCH_2CH_2Cl$, LiTMP, $(C_2H_5)_2O$, 43–54% → **2** (OCH_2CH_2Cl) + **3** (OCH_2CH_2Cl) → *n*-BuLi, 0° → 25°, 69–95% → **4** (R, OH)

Application of this procedure to 1,3-cyclohexadiene (**5**) yields only the *anti*-7-norborneol **6**.

5 → 24% → **6** (HO)

Stereo- and regiospecific synthesis of trimethylsilyl enol ethers.[2] Addition of vinylmagnesium bromide to acyltrimethylsilanes, $R^1COSi(CH_3)_3$, affords 1-trimethylsilylallylic alcohols (**1**).[3] Lithium alkoxides of **1**, prepared by treatment with *n*-butyllithium in hexane at −78°, rearrange to the homoenolates **2**. These anions are alkylated by primary alkyl iodides to afford (Z)-trimethylsilyl enol ethers in 63–72% yield and with 99% regio- and stereospecificity.

$$R^1COSi(CH_3)_3 \xrightarrow{CH_2{=}CHMgBr} R^1C(OH)(Si(CH_3)_3){-}CH{=}CH_2 \;(\mathbf{1}) \xrightarrow[-78\%]{n\text{-}C_4H_9Li,\ hexane}$$

$$\left[R^1{-}C(OLi)(Si(CH_3)_3){-}CH{=}CH_2\right] \longrightarrow (CH_3)_3SiO(R^1)C{=}CH{-}CH_2Li \;(\mathbf{2}) \xrightarrow[63\text{–}72\%\ \text{from}\ \mathbf{1}]{R^2I} (CH_3)_3SiO(R^1)C{=}CH{-}CH_2R^2 \;(\mathbf{3})$$

$R^1 = n\text{-}C_3H_7,\ C_6H_5CH_2CH_2,\ C_9H_{19}$
$R^2 = CH_3,\ n\text{-}C_4H_9$

Another regiospecific preparation of trimethylsilyl enol ethers involves treatment of acyltrimethylsilanes with the lithium anions of alkyl sulfones or nitriles.[4] In this case, the sulfone or nitrile group is eliminated during the silyl alkoxide rearrangement (*e.g.*, **5** → **6**). Mixtures of olefin stereoisomers are obtained. Note that **4** and **8** give complementary regiochemical results.

$$H_5C_6CH_2CH_2COSi(CH_3)_3 \;(\mathbf{4}) \xrightarrow{CH_3CH(Li)CN,\ -78°} \left[C_6H_5CH_2CH_2C(OLi)(Si(CH_3)_3)CH(CH_3)CN\right] (\mathbf{5}) \xrightarrow{80\%}$$

$$C_6H_5CH_2CH_2C(OSi(CH_3)_3){=}C(H)CH_3 \;\; \mathbf{6}\ (Z/E = 32{:}68) \;+\; H_5C_6CH_2CH_2COCH_2CH_3 \;\; 5\% \;+\; H_5C_6CH_2CH{=}C(OSi(CH_3)_3)CH_2CH_3 \;\; \mathbf{7}\ <4\%$$

$$H_5C_6CH_2COSi(CH_3)_3 \xrightarrow[70\%]{C_6H_5SO_2CH(Li)CH_3} H_5C_6CH_2C(OSi(CH_3)_3){=}C(H)CH_3$$

$$H_5C_6CH_2CH(SO_2C_6H_5)COSi(CH_3)_3 \;(\mathbf{8}) \xrightarrow[70\%]{C_2H_5Li,\ (C_2H_5)_2O,\ -78°} C_6H_5CH_2CH{=}C(OSi(CH_3)_3)CH_2CH_3 \;(\mathbf{7})$$

Alkylidenecyclopropanes.[5] *n*-Butyllithium in THF or *t*-butyllithium in ether effect CH_3Se–Li exchange reactions (**6**, 85–86) with 1,1-dimethylselenocyclopropane. The resulting anion **2** condenses with aldehydes or ketones to afford **3** in 64–75% yield. These β-hydroxycyclopropyl selenides afford alkylidenecyclopropanes **4** in 58–75% yield on reaction with N,N'-carbonyldiimidazole in toluene at 110° (for adducts prepared from aldehydes) or on treatment with PI_3 (for adducts derived from ketones). When treated with *p*-TsOH in refluxing benzene, **3** rearranges to cyclobutanones (**5**) in good yield.

SeCH₃/SeCH₃ cyclopropane **1** —RLi→ **2** (SeCH₃, Li) —R^1COR^2, 58–65%→ **3** (SeCH₃, C(OH)R^1R^2) —58–75%→ **4** (cyclopropylidene=CR^1R^2)

3 —TsOH, 73%→ **5** (2-R^1,2-R^2-cyclobutanone)

Addition to ketenes. *n*-Butyllithium adds to diethylketene to form an enolate (**a**), which can be alkylated (equation I). The product is not available by conventional base-catalyzed alkylation.[6]

$$\text{(I)}\quad (C_2H_5)_2C{=}C{=}O \xrightarrow[\text{THF, }-70^\circ]{n\text{-BuLi,}} \left[(C_2H_5)_2C{=}C(OLi)(Bu\text{-}n)\right]_{\mathbf{a}} \xrightarrow[55\%]{CH_3I} C_2H_5C(C_2H_5)(CH_3){-}C({=}O){-}Bu\text{-}n$$

α-Ketovinyl anion equivalents. Smith and co-workers[7] have found that α-bromo ketals react with *n*-butyllithium in THF at −76° to form vinyl anions that react with a variety of electrophiles in good yield. Hydrolysis of the resulting ketals can be conveniently effected by treatment with oxalic acid in a two-phase (CH_2Cl_2–H_2O)

1 (2-bromocyclohexenone ethylene ketal, Br) —*n*-BuLi, THF→ [**a** (Li)] —$ClCOOC_2H_5$→ **2** ($COOC_2H_5$ ketal) —$(COOH)_2$, H_2O, CH_2Cl_2, 81%→ **3** (2-$COOC_2H_5$-cyclohex-2-enone)

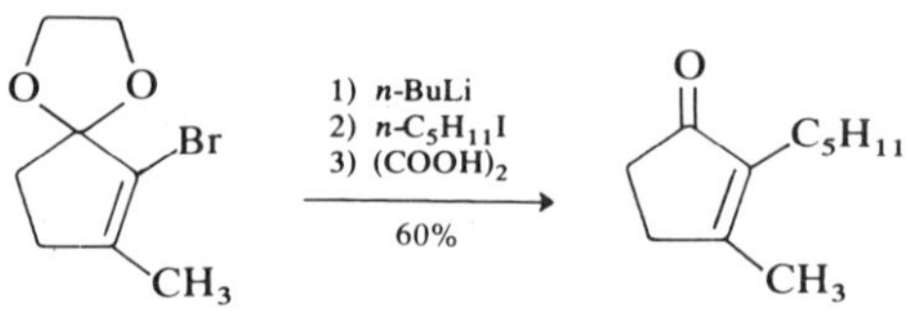

4 → 5: 1) *n*-BuLi; 2) *n*-$C_5H_{11}I$; 3) $(COOH)_2$; 60%

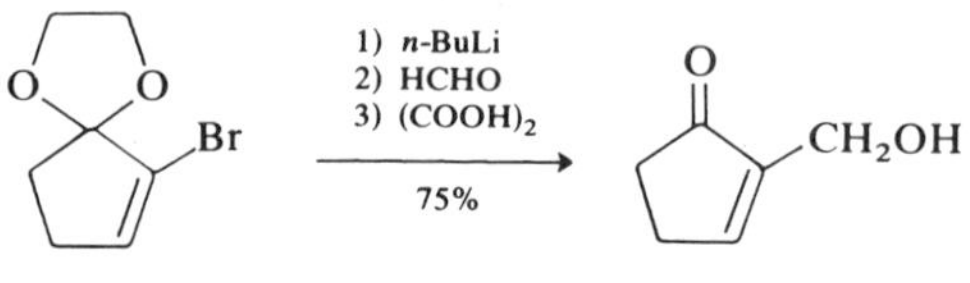

6 → 7: 1) *n*-BuLi; 2) HCHO; 3) $(COOH)_2$; 75%

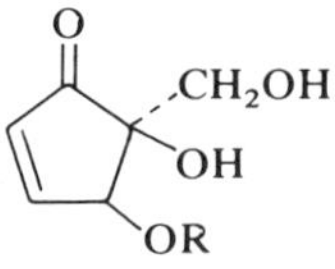

8 (R = H)
9 (R = Ac)

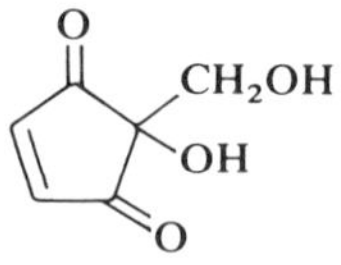

10

system. The product **7**[8] was used as a precursor to the antibiotics pentenomycin I (**8**) and II (**9**) and dehydropentenomycin I (**10**).[9]

Lithium o-lithiobenzylate.[10] This dianion (**1**) is prepared by treatment of benzyl alcohol with *n*-butyllithium (2 equiv.) and TMEDA at 20° in about 70% yield. It reacts with a variety of electrophiles. Typical reactions are shown in equations (I)–(III).

(I) **1** + CH_3I → (30%)

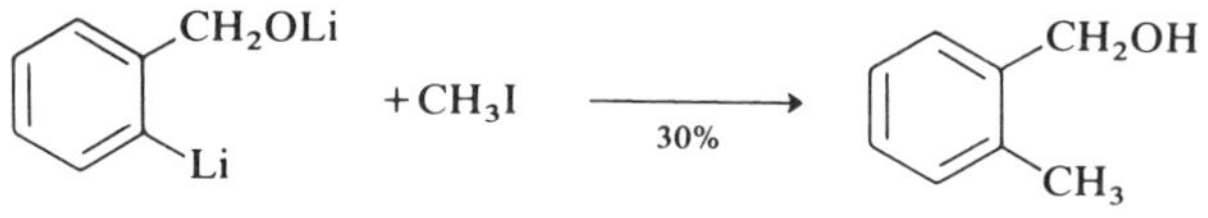

(II) **1** + cyclohexanone → (71%); HCl, Ether (95%)

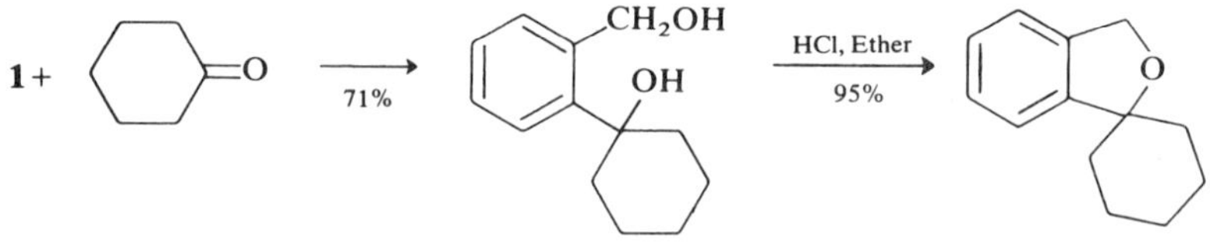

(III) **1** + CO_2 → (50%)

[1] R. L. Danheiser, C. Martinez-Davila, and J. M. Morin, *J. Org.*, **45**, 1340 (1980).
[2] I. Kuwajima and M. Kato, *J.C.S. Chem. Comm.*, 708 (1979).
[3] I. Kuwajima, T. Abe, and N. Minami, *Chem. Letters*, 993 (1976); I. Kuwajima, M. Arai, and T. Sato, *Am. Soc.*, **99**, 4181 (1977); *idem.*, *J.C.S. Chem. Comm.*, 478 (1978).
[4] H. J. Reich, J. J. Rusek, and R. E. Olson, *Am. Soc.*, **101**, 2225 (1979).
[5] S. Halazy and A. Krief, *J.C.S. Chem. Comm.*, 1136 (1979).
[6] T. T. Tidwell, *Tetrahedron Letters*, 4615 (1979).
[7] M. A. Guaciaro, P. M. Wovkulich, and A. B. Smith, *ibid.*, 4661 (1978).
[8] A. B. Smith, S. J. Branca, M. A. Guaciaro, P. M. Wovkulich, and A. Korn, *Org. Syn.*, submitted (1979).
[9] S. J. Branca and A. B. Smith, *Am. Soc.*, **100**, 7767 (1978).
[10] N. Meyer and D. Seebach, *Angew, Chem., Int. Ed.*, **17**, 521 (1978).

***n*-Butyllithium–Hexamethylphosphoric triamide.**

2-Alkenyl-3-methyl-2-cyclohexene-1-ones. The kinetic enolate of 3-methyl-3-cyclohexene-1-one is alkylated only by reactive halides such as allyl bromide. An alternate route to compounds of type **2** starts with the 2-diethylaminoethyl ether of 2,5-dihydro-*m*-cresol (**1**), obtained by Birch reduction. (The corresponding methyl ether is not readily metalated.) *n*-Butyllithium–HMPT in THF can metalate **1**. The resulting carbanion is alkylated by a variety of alkyl bromides to give, after acid hydrolysis, cyclohexenones **2** in yields of 80–90%.[1]

$O(CH_2)_2N(C_2H_5)_2$; CH_3 (**1**) — 1) *n*-BuLi, HMPT, THF, −78°; 2) $CH_2{=}CH(CH_2)_2Br$; 3) HCl, $(CH_3)_2C{=}O$; 86% → O; $(CH_2)_2CH{=}CH_2$; CH_3 (**2**)

[1] J. Amupitan and J. K. Sutherland, *J.C.S. Chem. Comm.*, 852 (1978).

***n*-Butyllithium–Potassium *t*-butoxide.**

Deprotonation. Preparation of the dianion (**2**) of isobutylene (**1**) with *n*-BuLi and TMEDA requires several days, but can be conducted in pentane with *n*-BuLi and $KOC(CH_3)_3$ in a few minutes.[1]

$CH_2{=}C(CH_3)_2$ (**1**) — *n*-BuLi, $KOC(CH_3)_3$ → $C(CH_2)_3^{2-}\ 2Li^+$ (**2**)

[1] R. B. Bates, W. A. Beavers, B. Gordon III, and N. S. Mills, *J. Org.*, **44**, 3800 (1979).

***sec*-Butyllithium, 5**, 78–79.

Debenzylation. Evans *et al.*[1] were unable to cleave the acid-sensitive benzyl ether **1** by catalytic hydrogenolysis. However, efficient debenzylation to **2** was accomplished by benzylic metalation with *sec*-BuLi followed by oxidation.

$$(CH_3)_2\text{-}t\text{-BuSiO}\ldots\ OCH_2C_6H_5\ (\mathbf{1}) \xrightarrow[\text{2) } B(OCH_3)_3,\ H_2O_2]{\text{1) } sec\text{-BuLi, THF}} (CH_3)_2\text{-}t\text{-BuSiO}\ldots\ HO\ (\mathbf{2})$$

[1] D. A. Evans, C. E. Sacks, W. A. Kleschick, and T. R. Taber, *Am. Soc.*, **101**, 6789 (1979).

***sec*-Butyllithium–Tetramethylethylenediamine** (TMEDA), **5**, 78–79; **8**, 69.

α-Lithio derivatives of primary alcohols. These useful intermediates can be obtained indirectly by metalation of an ester of a 2,4,6-trisubstituted benzoate (**1**) with *sec*-butyllithium–TMEDA adjacent to oxygen to give **a**, which can be trapped by electrophiles. Thus reaction of acetone with **a** leads to **2**, which can be reduced to **3**

$$2,4,6\text{-}R_3C_6H_2COOCH_2CH_3\ (\mathbf{1}) \xrightarrow[\text{TMEDA}]{sec\text{-BuLi,}} ArC(=O)OCH(Li)CH_3\ (\mathbf{a}) \xrightarrow[41\%]{CH_3COCH_3,\ HMPT}$$

$$ArC(=O)OCH(CH_3)C(OH)(CH_3)_2\ (\mathbf{2}) \xrightarrow[38\%\ \text{overall}]{LiAlH_4} (CH_3)_2C(OH)\text{—}CH(OH)CH_3\ (\mathbf{3})$$

by lithium aluminum hydride. Another example of this process is conversion of 1-iodopentane to 2-heptanol (equation I).[1]

(I) $$CH_3(CH_2)_4I \xrightarrow{38\%} CH_3(CH_2)_4CH(OH)CH_3$$

ortho-***Lithiation of aryl amides.***[2] Secondary and tertiary amide groups are superior to sulfonamide,[3] methoxy, (dimethylamino)methyl, chloro, carboxyl, or methyl groups in directing lithiation to an adjacent position. Since an amide is convertible into an aldehyde, amine, alcohol, or alkyl group, this selectivity is useful for many *ortho*-substitutions. *n*-BuLi cannot be used because it displaces dialkylamino groups.[4]

[1] P. Beak, M. Baillargeon, and L. G. Carter, *J. Org.*, **43**, 4255 (1978).
[2] P. Beak and R. A. Brown, *ibid.*, **44**, 4463 (1979).
[3] D. W. Slocum and C. A. Jennings, *ibid.*, **41**, 3653 (1976).
[4] L. Barsky, H. W. Gschwend, J. McKenna, and H. R. Rodriguez, *ibid.*, **41**, 3651 (1976).

***t*-Butyllithium,** **1**, 96–97; **5**, 79–80; **7**, 47; **8**, 70–72.

Metalation of 2-alkynyl and 1,2-alkadienyl tetrahydropyranyl ethers; furane synthesis. *t*-Butyllithium metalates the lithium alkoxide **1** to afford the allenyllithium compound **a** quantitatively. This anion reacts with alkyl halides or CH_3OH to afford **2**. Another metalation–alkylation protonation sequence proceeds via **b** to afford **3**. Hydrolysis of the latter intermediate affords furanes directly. The overall sequence can be performed in one pot from a propargyl tetrahydropyranyl ether, *t*-butyllithium, and an aldehyde.[1]

$$\underset{\mathbf{1}}{LiOCH_2C{\equiv}CCH_2{-}OTHP} \xrightarrow{t\text{-}C_4H_9Li} \underset{\mathbf{a}}{\left[LiO{-}CH_2C{\equiv}\overset{Li^+}{\bar{C}}{-}CHOTHP\right]} \xrightarrow{RX}$$

$$\underset{\mathbf{2}}{LiOCH_2{-}\overset{R}{C}{=}C{=}CH{-}OTHP} \xrightarrow{t\text{-}C_4H_9Li} \underset{\mathbf{b}}{\left[LiOCH_2{-}\overset{R}{C}{=}C{=}C(Li)(OTHP)\right]}$$

$$\xrightarrow{R^1X} \underset{\mathbf{3}}{LiOCH_2{-}\overset{R}{C}{=}C{=}\overset{R^1}{C}{-}OTHP} \xrightarrow{H_3O^+} \underset{\mathbf{4}}{\text{3-R-5-}R^1\text{-furan}}$$

Example:

$$HC{\equiv}C\overset{CH_3}{C}HOTHP \xrightarrow[64\%]{1)\ t\text{-}C_4H_9Li;\ 2)\ n\text{-}C_3H_7CHO;\ 3)\ t\text{-}C_4H_9Li;\ 4)\ C_2H_5I;\ 5)\ H_3O^+} \text{2-}(n\text{-}C_3H_7)\text{-3-}C_2H_5\text{-5-}CH_3\text{-furan}$$

A related furane synthesis is outlined in equation (I).[2]

$$\text{(I)}\quad C_5H_{11}{-}C{\equiv}C{-}CH_2{-}OTHP \xrightarrow{1)\ n\text{-}C_4H_9Li,\ -70°;\ 2)\ C_6H_5CHO;\ 3)\ H_3O^+} \text{3-}C_5H_{11}\text{-2-}C_6H_5\text{-furan}$$

[1] M. Stähle and M. Schlosser, *Angew. Chem. Int. Ed.*, **18**, 875 (1979).
[2] F. Mercier, R. Epstein, and S. Holand, *Bull. soc.*, 690 (1972).

t-Butyl perbenzoate, 1, 98–101; **2,** 54–55; **4,** 66.

Oxidation of pyrrolizines. Reaction of t-butyl perbenzoate with bicyclic pyrrole derivatives such as **1** results in introduction of the benzoyloxy group at C_1 in 40–45% yield. Lead tetraacetate, selenium dioxide, DDQ, and NBS are not useful for oxidation of **1**.[1]

$C_6H_5COOC(CH_3)_3$, C_6H_5Cl, 100°; 40–45%

1 ($R = H_2, O$) → **2** ($OCOC_6H_5$ at C_1)

[1] F. Bohlmann, W. Klose, and K. Nichisch, *Tetrahedron Letters*, 3699 (1979).

***trans*-4-(t-Butylthio)-3-butene-2-one,** $CH_3COCH{=}CHSC(CH_3)_3$ (**1**). Mol. wt. 158.26, m.p. 37°. The reagent is obtained by reaction of *trans*-4-chloro-3-butene-2-one with t-butylmercaptan in aqueous NaOH (85% yield).

Polyunsaturated aldehydes. The reagent is useful for synthesis of various unsaturated aldehydes as shown in Scheme (I).[1]

$$R^1CH{=}CH{-}\overset{O}{\overset{\|}{C}}CH{=}CHSC(CH_3)_3 \xleftarrow[60\text{–}85\%]{NaOH,\ R^1CHO} \mathbf{1} \xrightarrow{RLi} \left[R\overset{OH}{\underset{CH_3}{C}}CH{=}CHSC(CH_3)_3\right] \xrightarrow[50\text{–}75\%]{H_3O^+,\ -(CH_3)_3CSH} R\underset{CH_3}{C}{=}CHCHO$$

LiAlH4 branch:

$$\left[R^1CH{=}CH\underset{OH}{C}HCH{=}CHSC(CH_3)_3\right] \xrightarrow[70\text{–}80\%]{H_3O^+,\ -(CH_3)_3CSH} R^1CH{=}CHCH{=}CHCHO$$

R^2Li branch:

$$\left[R^1CH{=}CH\overset{OH}{\underset{R^2}{C}}CH{=}CHSC(CH_3)_3\right] \xrightarrow[60\text{–}90\%]{H_3O^+,\ -(CH_3)_3CSH} R^1CH{=}CH\underset{R^2}{C}{=}CHCHO$$

(E and Z isomers)

Scheme (I)

The reagent has been used for synthesis of the natural carotenoid isorenieratene (**2**) (equation I).[2]

(I) 2,3,6-trimethylbenzaldehyde (CH_3, CH_3, CHO, CH_3) + **1** $\xrightarrow[65\%]{NaOC_2H_5}$ $Ar{-}CH{=}CH{-}CO{-}CH{=}CH{-}SC(CH_3)_3$ $\xrightarrow{1)\ CH_3Li;\ 2)\ H_3O^+;\ 3)\ I_2}$

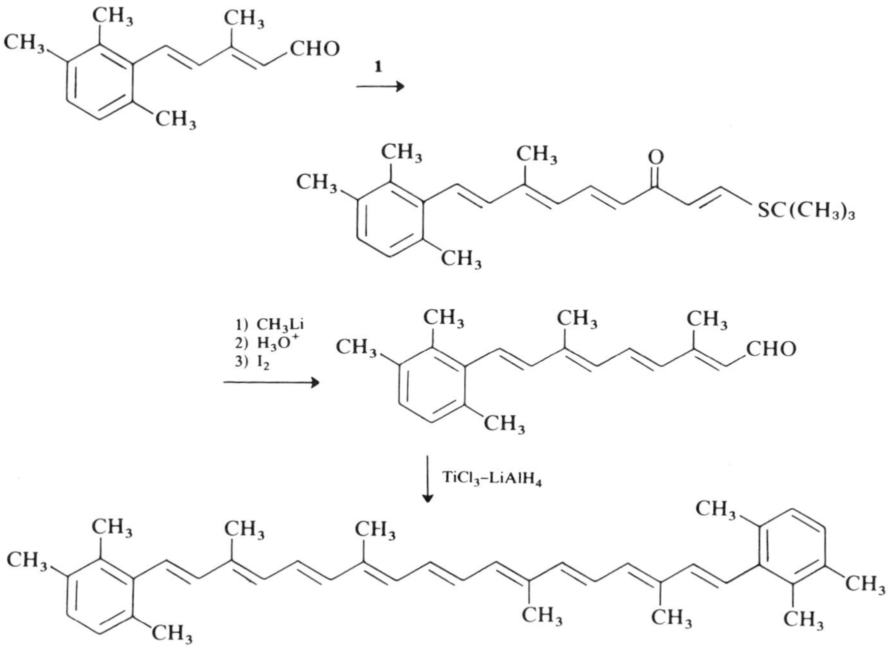

$$\xrightarrow{\mathbf{1}}$$

$$\xrightarrow{1)\ CH_3Li;\ 2)\ H_3O^+;\ 3)\ I_2}$$

$$\downarrow TiCl_3\text{-}LiAlH_4$$

2

[1] S. Akiyama, S. Nakatsuji, T. Hamamura, M. Kataoka, and M. Nakagawa, *Tetrahedron Letters*, 2809 (1979).
[2] S. Akiyama, S. Nakatsuji, S. Eda, M. Kataoka, and M. Nakagawa, *ibid.*, 2813 (1979).

***t*-Butyl thionitrate,** $(CH_3)_3CSNO_2$ **(1).** Mol. wt. 135.19, b.p. 55°/13 mm., stable. The reagent is obtained in 98% isolated yield by reaction of $(CH_3)_3CSH$ with N_2O_4 (25°, 2–5 minutes).[1]

Deaminative substitution.[2] Arylamines undergo various substitution reactions in the presence of **1** (slight excess). The reactions proceed through a diazonium salt to form an aryl radical.

Examples:

$$C_6H_5NH_2 + C_6H_6 \xrightarrow[62\%]{\mathbf{1},\ 80°} C_6H_5C_6H_5$$

$$p\text{-}CH_3C_6H_4NH_2 + (CH_3S)_2 \xrightarrow[62\%]{\mathbf{1},\ 110°} p\text{-}CH_3C_6H_4SCH_3$$

$$p\text{-}NO_2C_6H_4NH_2 + CHBr_3 \xrightarrow[80\%]{\mathbf{1},\ 120°} p\text{-}NO_2C_6H_4Br$$

$$p\text{-}CH_3OC_6H_4NH_2 + I_2 \xrightarrow[51\%]{\mathbf{1},\ C_6H_6,\ 25°} p\text{-}CH_3OC_6H_4I$$

N*-(t-*Butylthio*)-p-*benzoquinonimines. These derivatives of quinonimines are obtained by reaction of *p*-aminophenols with **1** in CH_3CN. Addition of $CuCl_2$ generally improves the yield (6–45%).[3]

$$HO-C_6H_4-NH_2 + \mathbf{1} \xrightarrow[40\%]{CH_3CN} O{=}C_6H_4{=}NSC(CH_3)_3$$

[1] Y. H. Kim, K. Shinhama, D. Fukushima, and S. Oae, *Tetrahedron Letters*, 1211 (1978).
[2] S. Oae, K. Shinhama, and Y. H. Kim, *Chem. Letters*, 939 (1979).
[3] *Idem, ibid.*, 1077 (1979).

1-*t*-Butylthio-1-trimethylsilyloxyethylene, $CH_2{=}C(OSi(CH_3)_3)SC(CH_3)_3$ **(1).** Mol. wt., 204.4, b.p. 65°/12 mm. The reagent is prepared by conversion of the S-*t*-butyl ester of thioacetic acid into the enolate (LDA) followed by reaction with $ClSi(CH_3)_3$; yield 81%.[1,2]

Michael additions. Gerlach and Künzler[1] report that the lithium enolate of S-*t*-butyl thioacetate undergoes 1,4-addition to cyclopentenone. They have extended this Michael reaction to a synthesis of methyl jasmonate (**5**), based on the similar conjugate addition of the trimethylsilyl enolate **1** promoted by tetra-*n*-butylammonium fluoride. The adduct **2** was alkylated by 1-bromo-2-pentyne in the presence of tetra-*n*-butylammonium fluoride to give **3** in rather low yield. Remaining steps to **5** were methanolysis and partial hydrogenation of the triple bond.

Cyclopentenone + **1** $\xrightarrow[72\%]{Bu_4NF,\ THF}$ 1-$OSi(CH_3)_3$-cyclopentene, 3-$CH_2COSC(CH_3)_3$ (**2**) $\xrightarrow[-FSi(CH_3)_3,\ 20\%]{CH_3CH_2C{\equiv}CCH_2Br,\ Bu_4NF,\ THF}$

cyclopentanone, 2-$CH_2C{\equiv}CC_2H_5$, 3-$CH_2COSC(CH_3)_3$ (**3**) $\xrightarrow{CH_3OH,\ Hg(NO_3)_2}$ cyclopentanone, 2-$CH_2C{\equiv}CC_2H_5$, 3-CH_2COOCH_3 (**4**) $\xrightarrow{H_2,\ Lindlar\ cat.}$

cyclopentanone, 2-$CH_2CH{=}CHC_2H_5$, 3-CH_2COOCH_3 (**5**)

The adduct **2** was also used as a precursor to jasmone ketolactone **8**, another component of the essential oil of *Jasminum grandiflorum*. The final step involved lactonization, carried out by the 2,2′-dipyridyl disulfide–triphenylphosphine method (**5**, 286; **6**, 246).

$$\mathbf{2} + BrCH_2C{\equiv}CCH_2CH_2OTHP \xrightarrow[25\%]{Bu_4NF}$$ 2-oxocyclopentane with $CH_2C{\equiv}CCH_2CH_2OTHP$ and $CH_2COSC(CH_3)_3$ (**6**)

$$\xrightarrow{\text{1) } H_3O^+ \text{ 2) } H_2/Pd \text{ 3) } Hg(NO_3)_2}$$ 2-oxocyclopentane with $CH_2\overset{H}{C}{=}\overset{H}{C}CH_2CH_2OH$ and CH_2COOH (**7**) $$\xrightarrow[71\%]{\text{1) } (C_5H_4NS)_2, P(C_6H_5)_3 \text{ 2) } AgClO_4}$$ **8**

[1] H. Gerlach and P. Künzler, *Helv.*, **61**, 2503 (1978).
[2] G. Simchen and W. West, *Synthesis*, 247 (1977).

C

Calcium–Methylamine/ethylenediamine.

Reduction of arenes.[1] Calcium in a mixed solvent composed of methylamine and ethylenediamine is comparable to lithium and an alkylamine for reduction of aromatic hydrocarbons to monoenes.[2] Thus naphthalene is reduced by this newer system to Δ^9-octalin (82%) and $\Delta^{1(9)}$-octalin (18%). The method is more efficient than one using calcium and liquid ammonia, although the latter method is considerably improved in rate and selectivity by small amounts of HMPT.

[1] R. A. Benkeser and J. Kang, *J. Org.*, **44**, 3737 (1979).
[2] E. M. Kaiser, *Synthesis*, 391 (1972).

Carbon dioxide, 3, 40–41; 5, 93–94; 6, 94–95; 7, 52.

Cyclic carbonates. Carbon dioxide reacts with epoxides (autoclave, 60–120°) in the presence of Sb(V) compounds [$(C_6H_5)_5Sb$, $(C_6H_5)_3SbBr_2$] to form cyclic carbonates in yields of 70–95% (equation I).[1]

(I) CO_2 + R-epoxide —Sb(V), 70–95%→ R-cyclic carbonate (O, O, C=O)

[1] H. Matsuda, A. Ninagawa, and R. Nomura, *Chem. Letters*, 1261 (1979).

Carbon disulfide–Methyl iodide.

S-Methylation. The known reaction of Grignard reagents with carbon disulfide and then with methyl iodide[1] has been extended successfully to the more reactive allylic Grignard reagents **1a** and **1b** to give dithioesters **2a** and **2b**. These can be converted to the isoprenic ketene dimethyl dithioacetals **3a** and **3b**. These

$R^1(H)C{=}C(R^2)CH_2MgCl$ —1) CS_2, 2) CH_3I→ $CH_2{=}C(R^2){-}CH(R^1){-}C({=}S)SCH_3$ —1) LDA, 2) CH_3I→ $CH_2{=}C(R^2){-}C(R^1){=}C(SCH_3)_2$

1a ($R^1 = H$, $R^2 = CH_3$) — **2a** (47%) — **3a** (83%)
1b ($R^1 = CH_3$, $R^2 = H$) — **2b** (86%) — **3b** (80%)

compounds are stable at −30° for a few months; they can be utilized as isoprene synthons. Thus they undergo clean conjugate addition with organolithium reagents. An example is shown in equation (I).[2]

(I) 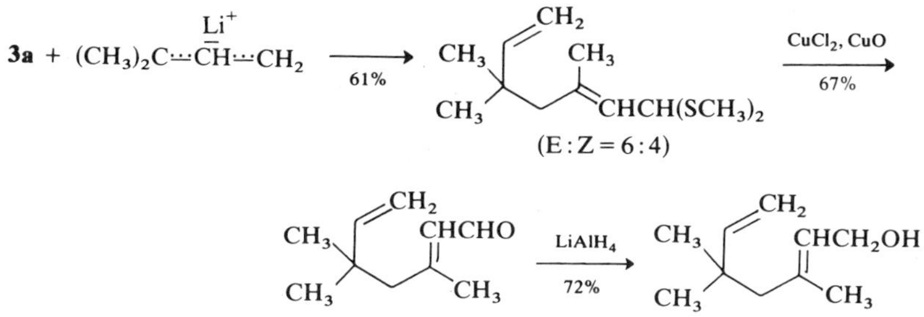

[1] J. M. Beiner and A. Thuillier, *Compt. Rend.*, **274C**, 642 (1972); J. Meijer, P. Vermeer, and L. Brandsma, *Rec. trav.*, **92**, 601 (1973).
[2] B. Cazes and S. Julia, *Tetrahedron Letters*, 4065 (1978).

Carbon monoxide, 2, 60, 204; **3,** 41–43; **5,** 96; **7,** 53; **8,** 76–77.

α,β-*Unsaturated esters.* Organopentafluorosilicates (**1**), prepared as shown from alkynes, are cabonylated in the presence of palladium salts, sodium acetate, and an alcohol at room temperature to form (E)-α,β-unsaturated esters usually in high yield (equation I). The reaction fails when R^1 and R^2 are C_6H_5, but ester and ether groups are compatible with this carbonylation.[1]

(I) 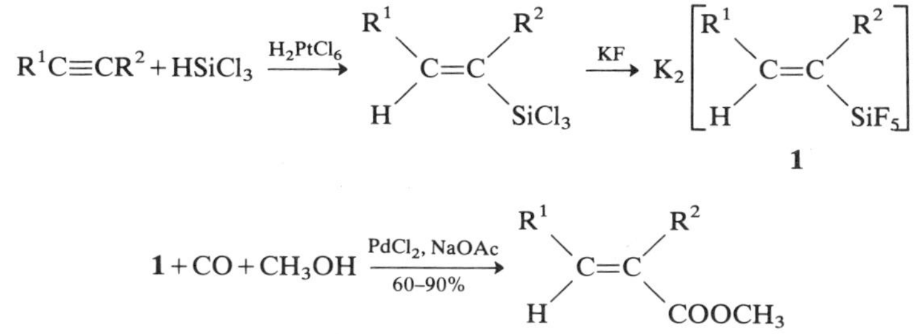

Berbines.[2] The reaction of **1** with CO in the presence of $Pd(OAc)_2$ and $P(C_6H_5)_3$ leads to **2** with insertion of CO.

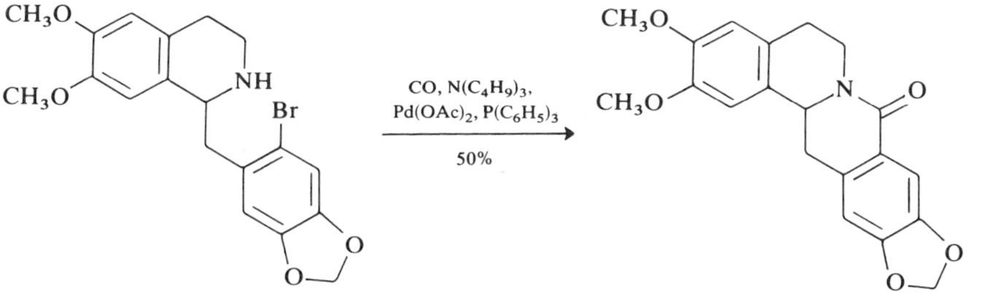

1 **2**

[1] K. Tamao, T. Kakui, and M. Kumada, *Tetrahedron Letters*, 619 (1979).
[2] G. D. Pandey and K. P. Tiwari, *Syn. Comm.*, **9**, 895 (1979).

N,N'-Carbonyldiimidazole, 1, 114–116; **2**, 61; **5**, 97–98; **6**, 97; **8**, 77.

C-Acylation. C-Acylation of active methylene compounds is usually conducted under basic conditions. Masamune *et al.*[1] have developed a new method for conducting this reaction under neutral conditions that is patterned on the enzymic synthesis of fatty acids. The acylating reagent is the imidazolide of a carboxylic acid (**1**) prepared *in situ*. The substrate is the neutral magnesium salt of a mono ester or thioester of a malonic acid (**2**), prepared with magnesium ethoxide. The reaction of **2** with **1** is conducted in THF at 25–35° for 18–24 hours; the yield of products (**3**) is generally >85%.[1]

$$R^1C(=O)\text{–N(imidazole)} \; (\mathbf{1}) + Mg[OC(=O)CH(R^2)C(=O)SR^3]_2 \; (\mathbf{2}, R^2 = H, CH_3) \xrightarrow[85\text{–}100\%]{THF} R^1C(=O)CH(R^2)C(=O)SR^3 \; (\mathbf{3})$$

A somewhat similar biomimetic acetyl transfer reaction has been reported.[2] Thus reaction of mono-*n*-butyl thiolmalonate (**4**) with phenyl thiolacetate (**5**) in the presence of imidazole and magnesium acetate in THF at room temperature gives *n*-butyl acetothiolacetate (**6**) in 60% yield.

$$HOC(=O)CH_2C(=O)SC_4H_9\text{-}n \; (\mathbf{4}) + CH_3C(=O)SC_6H_5 \; (\mathbf{5}) \xrightarrow[60\%]{\text{imidazole}, \; Mg(OAc)_2} CH_3C(=O)CH_2C(=O)SC_4H_9\text{-}n \; (\mathbf{6}) + CO_2 + C_6H_5SH$$

Tetronic acids. N,N'-Carbonyldiimidazole (**1**) can serve as an equivalent of $^{2+}C{=}O$ in a synthesis of tetronic acids from dianions of α-hydroxy ketones or α-diketones.[3]

Examples:

$$(CH_3)_2C(OH)C(=O)CH_3 \xrightarrow[48\%]{1)\ 2LDA;\ 2)\ \mathbf{1}} \text{4-hydroxy-5,5-dimethyl-2(5H)-furanone (HO, } CH_3, CH_3\text{)}$$

$$C_6H_5CH_2C(=O)C(=O)CH_2C_6H_5 \xrightarrow{92\%} \text{5-benzylidene(H, } C_6H_5\text{)-4-hydroxy(OH)-3-}C_6H_5\text{-2(5H)-furanone}$$

[1] D. W. Brooks, L. D.-L. Lu, and S. Masamune, *Angew. Chem., Int. Ed.*, **18**, 72 (1979).
[2] Y. Kobuke and Y. Yoshida, *Tetrahedron Letters*, 367 (1978).
[3] P. J. Jerris, P. M. Wovkulich, and A. B. Smith III, *Tetrahedron Letters*, 4517 (1979).

6-Carboxy-5-carbomethoxy-7-oxabicyclo[2.2.1]heptene-2 (1). Mol. wt. 198.2, m.p. 112–113°. Preparation:

$$\text{furan} + \text{maleic anhydride} \longrightarrow \text{adduct} \xrightarrow[92\%]{CH_3OH,\ 45°} \mathbf{1}\ (COOH,\ COOCH_3)$$

5,5-Dialkylbutenolides. The reagent is converted by reaction with Grignard reagents into the lactone **2**, which loses furane when heated to 150° to form 5,5-dialkylbutenolides (**3**).

$$\mathbf{1} \xrightarrow[60\text{–}100\%]{1)\ RMgX,\ THF;\ 2)\ H_3O^+} \mathbf{2} \xrightarrow[95\text{–}100\%]{150°} \mathbf{3} + \text{furan}$$

[1] J.-C. Grandguillot and F. Rouessac, *Synthesis*, 607 (1979).

(2-Carboxy-2-propenyl)triphenylphosphonium bromide, $CH_2{=}C(COOH)CH_2\overset{+}{P}(C_6H_5)_3Br^-$ **(1).** Mol. wt. 427.3, m.p. 214–217°. The Wittig reagent **1** is prepared in 90% yield by reaction of triphenylphosphine with 2-(bromomethyl)acrylic acid.

1,3-Butadiene-2-carboxylic acids.[1] The ylide of **1**, prepared with NaH (toluene), reacts with aldehydes in DMSO–toluene (1:1) to give 4-alkyl-1,3-butadiene-2-carboxylic acids (15–40% yield). The products have the (Z)-configuration.

$$RCHO \xrightarrow[15\text{–}40\%]{1)\ \mathbf{1};\ 2)\ CH_2N_2} CH_2{=}C(COOCH_3)CH{=}CHR$$

[1] H. Duttmann and P. Weyerstahl, *Ber.*, **112**, 3480 (1979).

Catecholborane, 4, 25, 69–70; **5**, 100–101; **6**, 33–34, 98; **7**, 54–55; **8**, 79–80.

Amines; lactams. Simple acyloxyboranes are known to react with amines to form amides, but in low yield. This reaction has been improved by use of catecholborane (equation I). It can be extended to synthesis of lactams. Thus addition of catecholborane to a suspension of an ω-amino acid in pyridine (80°) results in a lactam. Yields are high (85–95%) for 3- and 5-membered lactams. Yields are low (~6%) in the case of medium-size lactams, but improve to 10–15% in the synthesis of 14- and 16-membered lactams. (Dimers are formed preferentially.) Yields are

(I) 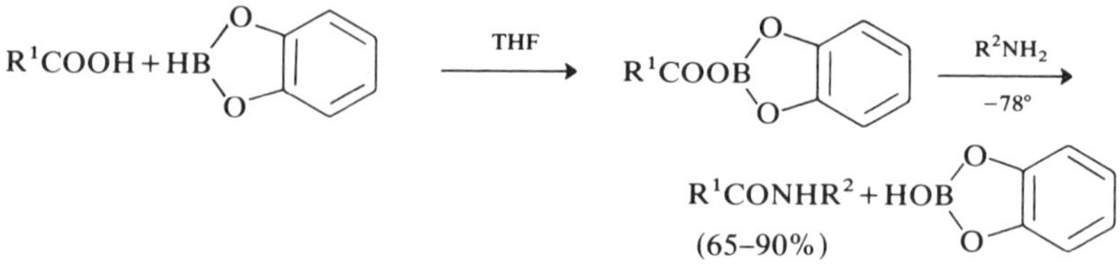

somewhat improved when the soluble tetra-*n*-butylammonium salts of ω-amino acids are used with B-chlorocatecholborane as the condensing agent, and dimeric lactams are not formed.

The boranes derived from 3-methoxy- and 4-nitrocatechol are somewhat superior to catecholborane.[1]

Hydroboration of allenes.[2] Monohydroboration of cyclic allenes with catecholborane (**1**) followed by oxidation results in a single ketone arising from regiospecific attack of boron at the central carbon atom of the allene (equation I). The reaction with diborane or disiamylborane is not selective.

(I) 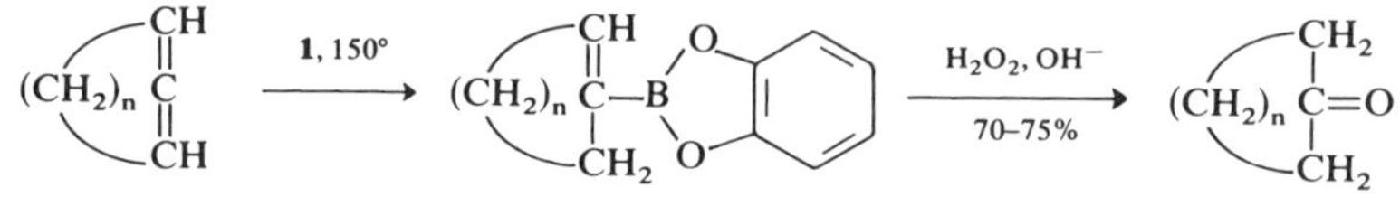

(n = 6, 7, 10)

Deoxygenation. Greene[3] has applied Kabalka's method for reduction of an enone (**6**, 98; **7**, 54) to the tosylhydrazone of a cross-conjugated ketone, 6-*epi*-α-santonin (**2**). Unexpectedly only one olefin (**3**) was obtained in 50% yield. This product was converted into the diene **4** by allylic oxidation, reduction to an allylic alcohol, and dehydration. The product was converted into (−)-dictyolene (**5**) by a method developed previously.[4]

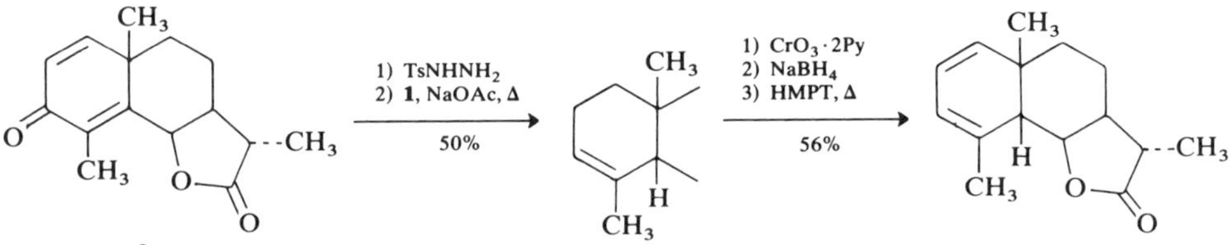

2 **3** **4**

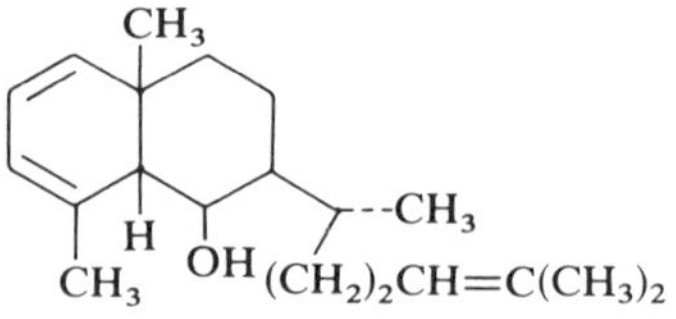

5

Reduction of tosylhydrazones of conjugated acetylenic ketones with catecholborane (**1**) proceeds with migration of the triple bond to give allenes.[5]

Examples:

$$CH_3C{\equiv}C{-}\overset{\overset{\displaystyle NNHTs}{\|}}{C}{-}C_6H_5 \xrightarrow[75\%]{\text{1) } \mathbf{1}, CHCl_3, 20^\circ \text{ 2) } NaOAc} CH_3CH{=}C{=}CHC_6H_5$$

$$HC{\equiv}C\overset{\overset{\displaystyle NNHTs}{\|}}{C}CH_3 \xrightarrow[49\%]{} CH_2{=}C{=}CHCH_3$$

[1] D. B. Collum, S.-C. Chen, and B. Ganem, *J. Org.*, **43**, 4393 (1978).
[2] V. V. Ramana Rav, S. K. Agarwal, D. Devaprobhakara, and S. Chandrasekaran, *Syn. Comm.*, **9**, 437 (1979).
[3] A. E. Greene, *Tetrahedron Letters*, 63 (1979).
[4] *Idem, ibid.*, 851 (1978).
[5] G. W. Kabalka, R. J. Newton, Jr., and J. H. Chandler, *J.C.S. Chem. Comm.*, 726 (1978).

Cerium(II) ammonium nitrate (CAN), 1, 120–121; **2**, 63–65; **3**, 44–45; **4**, 71–74; **5**, 101–102; **6**, 99; **7**, 55–56.

Oxidative demethylation. Both silver(II) oxide (**4**, 431–432)[1] and CAN (**7**, 55) have been used for oxidative demethylation of dimethyl ethers of 1,4-hydroquinones to give *p*-quinones. A direct comparison of the two reagents indicates that yields are higher with CAN. With either reagent, yields are improved when the N-oxide of pyridine-2,6-dicarboxylic acid is added as catalyst. The oxide is prepared by oxidation of pyridine-2,6-dicarboxylic acid with H_2O_2 and Na_2SO_4.[2]

[1] C. D. Snyder and H. Rapoport, *Am. Soc.*, **96**, 8046 (1974).
[2] L. Syper, K. Kloc, and J. Mlochowski, *Synthesis*, 521 (1979).

Cerium(III) chloride; erbium(III) chloride.

Selective ketalization.[1] Ketalization with lanthanoid catalysts can distinguish between aliphatic and aromatic ketones; the latter remain unaffected. Also selective ketalization of aldehydes in the presence of ketones is usually possible. $CeCl_3$ and $ErCl_3$ are more efficient than the heavier rare earth ions.

[1] A. L. Gemal and J.-L. Luche, *J. Org.*, **44**, 4187 (1979).

Cerium(III) sulfate, $Ce_2(SO_4)_3$. Mol. wt., 568.42. Supplier: Alfa.

Dehalogenation.[1] Cerium(III) sulfate in combination with sodium iodide reduces α-halo ketones at room temperature in 30 minutes to the parent ketone in 80–90% yield. Presumably CeI_3 is the effective reagent.

[1] T.-L. Ho, *Syn. Comm.*, **9**, 241 (1979).

Cesium carbonate, 7, 57; **8**, 81.

Crown ethers. The dicesium salts of catechol, resorcinol, and salicylic acid react with dibromopolyethylene glycols in DMF to afford crown ethers.[1] Similar reactions employing other alkali metal salts proceed less cleanly in significantly lower yield.

Example:

$$\text{catechol} \xrightarrow[\text{50–78\%}]{\text{1) } Cs_2CO_3\text{, DMF; 2) } Br(CH_2CH_2O)_nCH_2CH_2Br} \text{benzo crown ether} \quad (n = 3\text{–}5)$$

Lactonization. Heating the cesium salt of ω-bromo- or ω-iodocarboxylic acids at 40° in DMF produces mixtures of lactones and diolides.[2]

$$X(CH_2)_nCOOH \xrightarrow{Cs_2CO_3\text{, DMF}} (CH_2)_n\text{-lactone} + \text{diolide}$$

	Lactone	Diolide
n = 4	70%	0%
n = 5, 8	0%	88, 95%
n = 10	33%	54%
n = 14	83%	17%

[1] B. J. van Keulen, R. M. Kellogg, and O. Piepers, *J.C.S. Chem. Comm.*, 285 (1979).
[2] W. H. Kruizinga and R. M. Kellogg, *ibid.*, 286 (1979).

Cesium fluoride, 7, 57–58; **8**, 81–82.

Desilylation. Anhydrous cesium fluoride desilylates trimethylsilylmethylsulfonium, trimethylsilylmethylammonium, and trimethylsilylmethylphosphonium salts at room temperature to produce ylides, which undergo various useful transformations.[1] Use of potassium fluoride–18-crown-6 or a tetraalkylammonium fluoride gives products in low yield in these reactions. The trimethylsilylmethylonium salts are prepared by alkylation of sulfides, amines, imines, and phosphines with (trifluoromethanesulfonylmethyl)trimethylsilane (**1**).

$$C_5H_{11}CH(SCH_3)COOCH_3 \xrightarrow{(CH_3)_3SiCH_2OSO_2CF_3 (\mathbf{1}),\ CH_3CN} \left[C_5H_{11}CH(COOCH_3)\overset{+}{S}(CH_3)CH_2Si(CH_3)_3\ OTf^-\right] \xrightarrow[81\%]{CsF} C_4H_9CH{=}CHCOOCH_3$$

$$C_6H_5CH{=}CHCH_2N(\text{piperidine}) \xrightarrow{\mathbf{1},\ CH_3CN} \left[C_6H_5CH{=}CHCH_2{-}\overset{+}{N}(\text{piperidine})CH_2Si(CH_3)_3\ OTf^-\right] \xrightarrow[54\%]{CsF} C_6H_5CH(CH{=}CH_2)CH_2N(\text{piperidine})$$

[1] E. Vedejs and G. R. Martinez, *Am. Soc.*, **101**, 6452 (1979).

Chloramine-T, 4, 75, 445–446; **5**, 104; **7**, 58; **8**, 83.

Tosylamination of alkenes.[1] The reaction of diphenyl disulfide with chloramine-T in acetone gives, in 91% yield, an adduct of the composition $(C_6H_5SSC_6H_5)(NTs)_2$ shown to have the structure (**1**) by virtue of subsequent reactions with alkenes. Thus it reacts with cyclohexene to give (**2**), which was reduced by sodium borohydride to (**3**), whose structure was established by an independent synthesis.

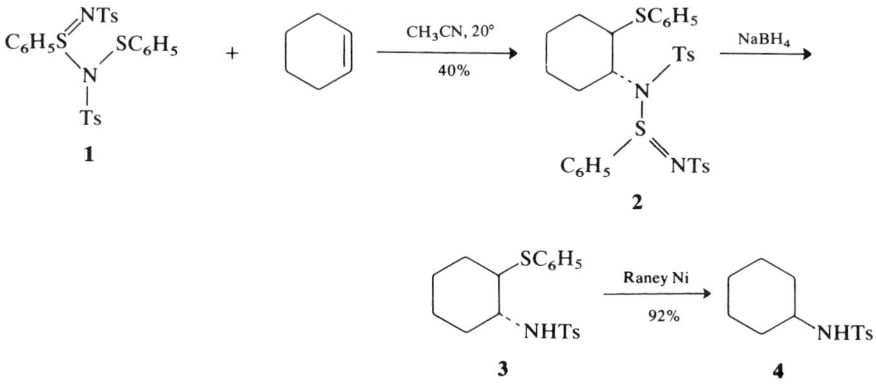

A complex similar to **1** could not be isolated from the reaction of chloramine-T with diphenyl diselenide, but its formation was demonstrated by the reaction of the diselenide and chloramine-T in the presence of an olefin. The initial product of addition to cyclohexene is also unstable and not isolable, but reduction with $NaBH_4$ gave a *trans*-1,2-amidoselenide (**5**).

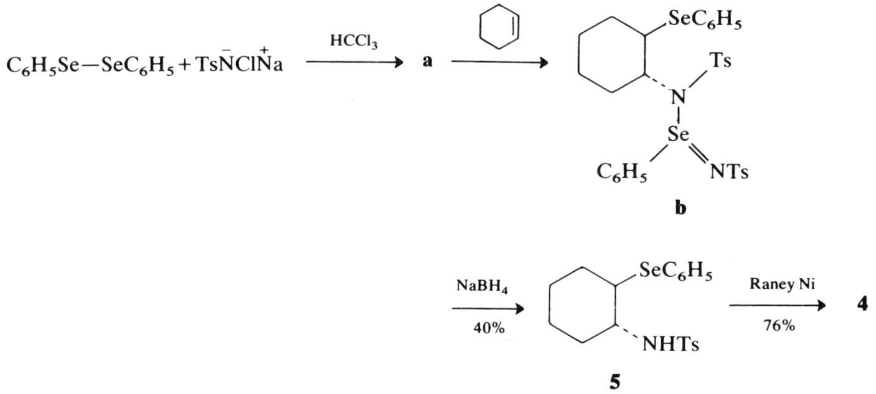

α-Amino aldehydes.[2] Enamines react with chloramine-T trihydrate in CH_2Cl_2 to afford α-amino aldehydes in 50–84% yield. The probable pathway is formulated in equation (I). Analogous oxidation of enol ethers gives N-tosyl derivatives of α-amino aldehydes (equation II). The latter reaction can also be performed in the presence of OsO_4 (**7**, 256).[3]

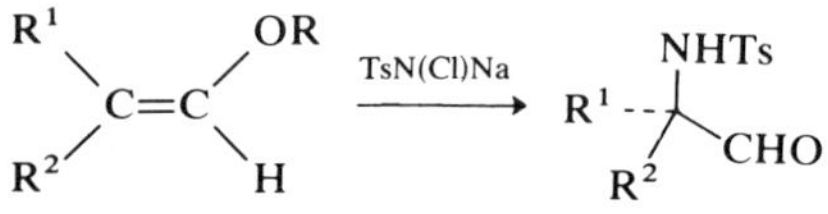

[1] D. H. R. Barton, M. R. Britten-Kelly, and D. Ferreira, *J.C.S. Perkin I*, 1090 (1978).
[2] I. Dyong and Q. Lam-Chi, *Angew. Chem., Int. Ed.*, **18**, 933 (1979).
[3] I. Dyong, G. Schulte, Q. Lam-Chi, and H. Friege, *Carbohydrate Res.*, **68**, 257 (1979).

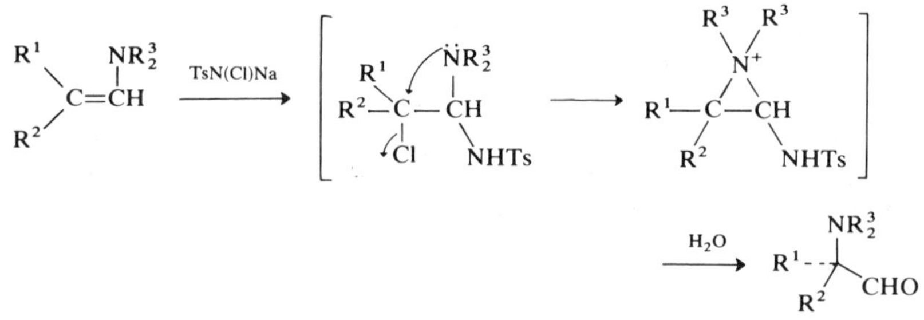

***o*-Chloranil, 1**, 128–129; **2**, 345; **4**, 76; **7**, 355–356.

Polycyclic phenols.[1] Polycyclic aromatic ketones such as **1** can be converted into phenols by dehydrogenation of the corresponding enol acetates with *o*-chloranil (equation I).

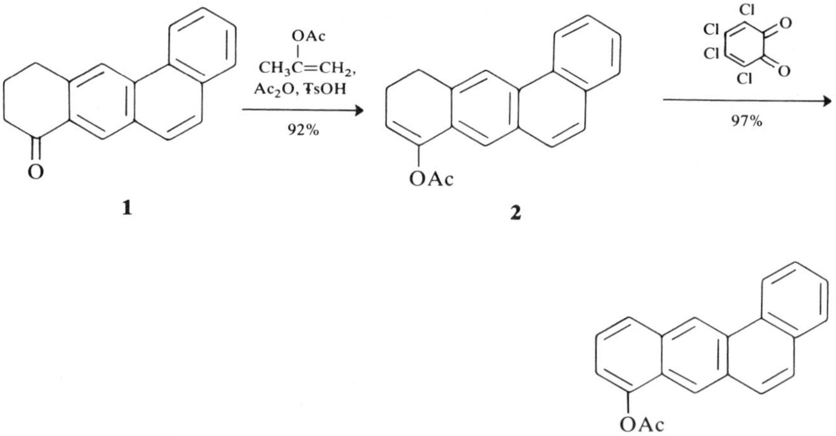

[1] P. P. Fu, C. Cortez, K. B. Sukumaran, and R. G. Harvey, *J. Org.*, **44**, 4265 (1979).

4-Chlorobenzenediazonium tetrafluoroborate, $p\text{-ClC}_6\text{H}_4\text{N}_2{}^+\text{BF}_4{}^-$ **(1).** Slightly hygroscopic, stable at 20°.

Sulfoxides. Sulfones can be reduced to sulfoxides as outlined in equation (I).[1]

$$\text{(I)}\quad R^1SO_2R^2 + \mathbf{1} \xrightarrow{\Delta} \left[\overset{BF_4^-}{p\text{-}ClC_6H_4O\overset{+}{S}(=O)(R^1)R^2} \right] \xrightarrow{NaBH_4\text{-}Al_2O_3} \left[\overset{BF_4^-}{H\overset{+}{S}(=O)(R^1)R^2} \right] \xrightarrow[45\text{-}90\%]{-H^+} R^1SOR^2$$

[1] I. W. J. Still and S. Szilagyi, *Syn. Comm.*, **9**, 923 (1979).

Chlorobis(cyclopentadienyl)tetrahydroboratozirconium(IV), $Cp_2Zr(Cl)BH_4$. Mol. wt. 271.70. An early preparation has been reported without details.[1] A convenient preparation involves reaction of $Cp_2Zr(H)Cl$ and $BH_3 \cdot S(CH_3)_2$; yield 70–80%.[2]

Reduction of carbonyl groups.[2] Aldehydes and ketones are reduced by this complex in high yield; esters, carboxylic acids, nitriles, and nitro compounds are reduced very slowly. The reagent thus resembles $NaBH_4$, but can be used for reductions in benzene.

α,β-Unsaturated ketones are reduced in low yield to a mixture of saturated and unsaturated alcohols. 4-*t*-Butylcyclohexanone is reduced to a 2:1 mixture of the *trans*- and *cis*-alcohols, respectively.

[1] R. K. Nanda and M. G. H. Wallbridge, *Inorg. Chem.*, **3**, 1978 (1964).
[2] T. N. Sorrell, *Tetrahedron Letters*, 4985 (1978).

Chlorocyanoketene, 8, 88.

Cycloaddition to benzaldehydes.[1] Chlorocyanoketene (**1**) undergoes [2+2] cycloaddition to benzaldehydes to form β-lactones (**a**), detected by IR and ^1H NMR spectroscopy, which then lose CO_2 to form (E)-1-chloro-1-cyano-2-phenylethenes (**2**). The rate and yield are dependent on the nature of the R group. They are

$$\mathbf{1} + p\text{-}RC_6H_4CHO \xrightarrow{C_6H_6,\ \Delta} \left[\underset{\mathbf{a}}{\text{Cl(CN)C–C(=O)–O–CH}(C_6H_4R\text{-}p)} \right] \xrightarrow[30\text{-}92\%]{} \underset{\mathbf{2}}{(Cl)(CN)C{=}C(H)(C_6H_4R\text{-}p)}$$

increased by electron-releasing groups (OCH_3, $N(CH_3)_2$, OAc) and decreased by electron-attracting groups (Cl, NO_2). This trend is the opposite of that observed in cycloadditions of dichloroketene.[2] The paper presents evidence for a zwitterionic mechanism by way of $(Cl)(NC)C{=}C(\bar{O})\text{–}\overset{+}{O}{=}CHC_6H_4R$, with the ketene as an electrophile rather than a nucleophile, as with other ketenes.

[1] H. M. Moore, F. Mercer, D. Kunert, and P. Albaugh, *Am. Soc.*, **101**, 5435 (1979).
[2] H. O. Krabbenhoft, *J. Org.*, **43**, 1305 (1978).

Chlorodi(η^5-cyclopentadienyl)hydrozirconium(IV), **6**, 177–179; **7**, 101–102; **8**, 84–87. The definitive paper for use of the reagent for synthesis of saturated and α,β-unsaturated ketones has been published.[1]

[1] D. B. Carr and J. Schwartz, *Am. Soc.*, **101**, 3521 (1979).

1-Chloro-1-dimethylaminoisoprene (1). Mol. wt. 161.68, b.p. 55°/12 mm.

Preparation:

$$(CH_3)_2NC(=O)C(CH_3)=CHCH_3 \; (E+Z) \xrightarrow[80\%]{1)\ COCl_2;\ 2)\ N(C_2H_5)_3} (CH_3)_2N(Cl)C=C(CH_3)CH=CH_2 \; \mathbf{1}\ (E+Z)$$

Diels–Alder reactions.[1] The amino group at C_1 renders this position electrophilic, whereas C_4 is nucleophilic (push–pull diene). The diene (**1**) can react as a vinyl keteniminium chloride (**1a**) in reactions with simple nitriles such as acetonitrile or acrylonitrile. After treatment with base a pyridine derivative (**2**) is obtained. The

$$\mathbf{1} \rightleftarrows \underset{\mathbf{1a}}{CH_2=CH-C(CH_3)=C=\overset{+}{N}(CH_3)_2\ Cl^-} \xrightarrow{CH_3C\equiv N} [\mathbf{a}]\ Cl^- \xrightarrow[62\%]{N(C_2H_5)_3} \mathbf{2}$$

diene **3**, in which chlorine is replaced by fluorine, undergoes normal cycloaddition to acrylonitrile and methyl acrylate to yield 6-substituted-2-methylcyclohexenones **4**.

$$\mathbf{1} \xrightarrow[75\%]{KF} \underset{\mathbf{3}}{CH_2=CH-C(CH_3)=C=\overset{+}{N}(CH_3)_2\ F^-} \xrightarrow{CH_2=CHCN} [\mathbf{b}] \xrightarrow[90\%]{H_3O^+} \mathbf{4}$$

Yet another mode of cycloaddition obtains with diene **5** and dimethyl acetylenedicarboxylate to yield an aromatic product **6**, possibly by way of a cyclobutene (**c**).

$$\mathbf{1} \xrightarrow[80\text{–}85\%]{NaOCH_3,\ KI} \underset{\mathbf{5}}{CH_2=CH-C(CH_3)=C(OCH_3)N(CH_3)_2} \xrightarrow{CH_3OOCC\equiv CCOOCH_3} [\mathbf{c}]$$

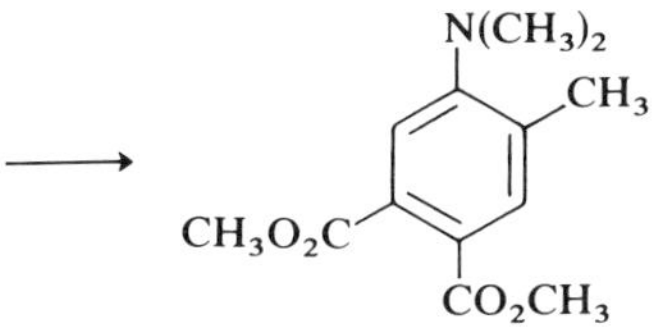

$\longrightarrow$ $N(CH_3)_2$, CH_3, CH_3O_2C, CO_2CH_3

6

[1] M. Gillard, C. T'Kint, E. Sonveaux, and L. Ghosez, *Am. Soc.*, **101**, 5837 (1979).

Chlorodimethylsulfonium chloride, 4, 191–192.

Methylthiomethyl esters (**4**, 84–85).[1] These esters can also be prepared by reaction of the triethylammonium salts of carboxylic acids with the Corey-Kim reagent and excess triethylamine initially at −70° and then at 20° for 30 minutes. Yield 80–85%.

[1] T.-L. Ho, *Syn. Comm.*, **9**, 267 (1979).

2-Chloro-3-ethylbenzoxazolium tetrafluoroborate, (C_2H_5, N^+, O, Cl, BF_4^-) **(1), 8,** 90–92.

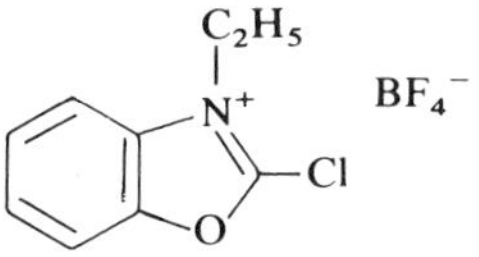

Dehydration. The reagent, in combination with 2 equiv. of triethylamine, converts formamides into isocyanides at 20° in 70–85% yield (equation I).[1] This

(I) $$RNHCHO + \mathbf{1} + 2N(C_2H_5)_3 \xrightarrow[70\text{–}85\%]{CH_2Cl_2} RNC + (C_2H_5, N, O, {=}O) + (C_2H_5)_3N \cdot HCl + (C_2H_5)_3N \cdot HBF_4$$

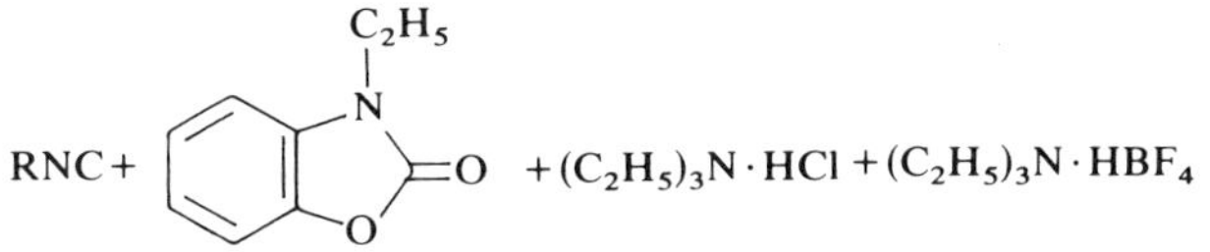

dehydration can be used to convert ketones of the type $ArCOCH_2R$ into alkynes in fair yields (equation II).[2]

(II) $$ArCOCH_2R + \mathbf{1} + 2\,N(C_2H_5)_3 \xrightarrow[40\text{–}80\%]{} ArC{\equiv}CR + (C_2H_5, N, O, {=}O) + (C_2H_5)_3N \cdot HCl + (C_2H_5)_3N \cdot HBF_4$$

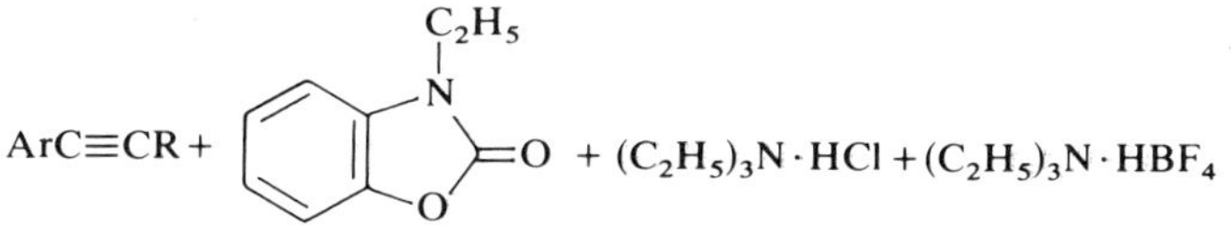

A similar reaction has been used to convert α-hydroxy carboxylic acids into ketones by decarbonylation and dehydration (equation III).[3]

(III) 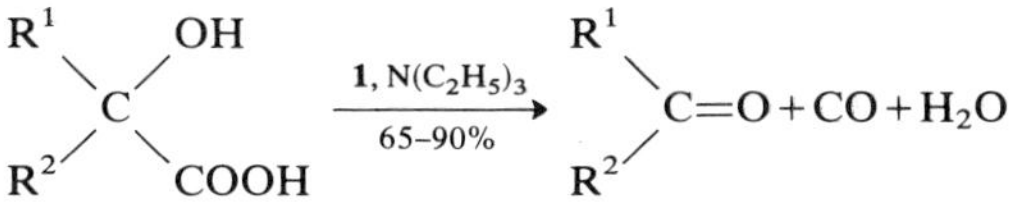 $$R^1R^2C(OH)COOH \xrightarrow[65\text{–}90\%]{\mathbf{1},\ N(C_2H_5)_3} R^1R^2C{=}O + CO + H_2O$$

α-Chloro nitriles. α-Chloro nitriles are formed by reaction of cyanohydrins with **1** and tetraethylammonium chloride (equation I).[4]

(I) RCH(OH)CN + **1** $\xrightarrow{(C_2H_5)_3N,\ CH_2Cl_2}$ [RCH(CN)O–(3-ethylbenzoxazolium) BF_4^-] $\xrightarrow[70\text{–}95\%]{(C_2H_5)_4\overset{+}{N}Cl^-}$ RCH(Cl)CN + 3-ethylbenzoxazol-2-one

[1] Y. Echigo, Y. Watanabe, and T. Mukaiyama, *Chem. Letters*, 697 (1977).
[2] T. Tsuji, Y. Watanabe, and T. Mukaiyama, *ibid.*, 481 (1979).
[3] T. Mukaiyama and Y. Echigo, *Tetrahedron Letters*, 49 (1978).
[4] T. Mukaiyama, K. Kawata, A. Sasaki, and M. Asami, *Chem. Letters*, 1117 (1979).

2-Chloroethyl thiocyanate, $ClCH_2CH_2SCN$ **(1).** Mol. wt. 121.58, b.p. 87–88°/18 mm. This substance is obtained by reaction of potassium thiocyanate with 1-bromo-2-chloroethane in refluxing methanol (77–88% yield).

Vinylic sulfides; 1-vinylthio-1-alkynes.[1] These useful compounds can be prepared as shown in equations (I) and (II).

$$\text{(I)}\quad C_6H_5MgBr \xrightarrow{\mathbf{1}} C_6H_5SCH_2CH_2Cl \xrightarrow{KOC(CH_3)_3,\ THF} C_6H_5SCH{=}CH_2$$

$$\text{(II)}\quad (CH_3)_3SiC{\equiv}CLi \xrightarrow{\mathbf{1}} (CH_3)_3SiC{\equiv}CSCH_2CH_2Cl \xrightarrow{1)\ LiNH_2,\ NH_3;\ 2)\ H_2O} HC{\equiv}CSCH{=}CH_2$$

[1] W. Verboom, J. Meijer, and L. Brandsma, *Synthesis*, 577 (1978).

1-Chloro-2-iodoethane (Ethylene iodochloride), $ClCH_2CH_2I$. Mol. wt. 190.41, b.p. 140°. This reagent is prepared by addition of iodine monochloride to ethylene.[1]

Iodination of lithium aryls. The reagent reacts with **2** to provide the aryl iodide **3**, a key intermediate for synthesis of the antibiotic frustulosin.[2] The iodination is

1 (1,4-dimethoxy-2-($CH_2OCH{=}CH_2$)benzene) $\xrightarrow{n\text{-BuLi, pentane}}$ **2** (3-Li derivative) $\xrightarrow[51\%]{ClCH_2CH_2I,\ (C_2H_5)_2O}$ **3** (3-I derivative; OCH_3, $CH_2OCH{=}CH_2$, I, OCH_3)

rapid and specific at room temperature; excess reagent is easily removed by distillation; and ethylene and LiCl are the only by-products. Other sources of electrophilic iodine attack the vinylic ether of **2**.

[1] M. Simpson, *Ann.*, **127**, 372 (1863).
[2] R. C. Ronald and J. M. Lansinger, *J.C.S. Chem. Comm.*, 124 (1979).

Chloromethyl methyl ether, 1, 132–135; **5,** 120; **7,** 61–62. The conventional synthesis produces the potent carcinogen bis(chloromethyl) ether as a by-product. A new preparation (equations I and II) is free from this drawback.[1] Methanol is required only in a catalytic amount, since it is regenerated as the ether is formed. The by-product is methyl acetate.

$$\text{(I)} \qquad CH_3COCl \xrightarrow{CH_3OH} CH_3\overset{O}{\overset{\|}{C}}OCH_3 + HCl$$

$$\text{(II)} \qquad CH_2(OCH_3)_2 \underset{}{\overset{HCl}{\rightleftharpoons}} CH_3OCH_2Cl + CH_3OH$$

[1] J. S. Amato, S. Karady, M. Sletzinger, and L. M. Weinstock, *Synthesis*, 970 (1979).

Chloromethyl methyl sulfide, 6, 109–110; **8,** 94.

Methylthiomethyl ethers. MTM ethers of a wide range of alcohols can be prepared in 60–80% yield by reaction of primary or secondary alcohols with chloromethyl methyl sulfide and silver nitrate in the presence of triethylamine (benzene, 20–80°).[1]

Protection of carboxyl groups (**8**, 94). Two new methods for deprotection of methylthiomethyl esters have been reported; one involves oxidation to the sulfone followed by alkaline hydrolysis, and the other involves methylation to a sulfonium iodide and hydrolysis.[2]

$$R\overset{O}{\overset{\|}{C}}{-}OCH_2SCH_3 \xrightarrow{H_2O_2,\ (NH_4)_6Mo_7O_{24}} R\overset{O}{\overset{\|}{C}}OCH_2\underset{O}{\underset{\|}{\overset{O}{\overset{\|}{S}}}}CH_3 \xrightarrow[90\text{–}100\%]{NaOH} R\overset{O}{\overset{\|}{C}}OH$$

$$R\overset{O}{\overset{\|}{C}}{-}OCH_2SCH_3 \xrightarrow{CH_3I} R\overset{O}{\overset{\|}{C}}OCH_2\overset{+}{S}(CH_3)_2I^- \xrightarrow[85\text{–}95\%]{NaOH} R\overset{O}{\overset{\|}{C}}OH$$

[1] K. Suzuki, J. Inanaga, and M. Yamaguchi, *Chem. Letters*, 1277 (1979).
[2] J. M. Gerdes and L. G. Wade, Jr., *Tetrahedron Letters*, 689 (1979).

Chloromethyl phenyl sulfide, $C_6H_5SCH_2Cl$ **(1).** Mol. wt. 158.66, b.p. 66°/0.2 mm. The sulfide is prepared by chlorination of thioanisole with sulfuryl chloride in CH_2Cl_2.[1]

α-Methylenation of carbonyl compounds. Silyl enol ethers of lactones and esters are converted into α-phenylthiomethyl derivatives in good yield by reaction with **1** catalyzed by anhydrous zinc bromide. These products are converted into α-methylene lactones and esters by oxidation and sulfoxide elimination.[2]

Examples:

$$\xrightarrow[98\%]{\text{LDA, ClSi(CH}_3)_3} \text{—OSi(CH}_3)_3 \xrightarrow[96\%]{\text{1, ZnBr}_2}$$

$$\text{CH}_2\text{SC}_6\text{H}_5 \xrightarrow[96\%]{\text{1) NaIO}_4;\ \text{2) } \Delta} \text{CH}_2$$

$$\text{CH}_3\text{CH}_2\text{CH=C(OSi(CH}_3)_3)(\text{OC}_2\text{H}_5) \xrightarrow[96\%]{} \text{CH}_3\text{CH}_2\text{CH(CH}_2\text{SC}_6\text{H}_5)\text{COOC}_2\text{H}_5 \xrightarrow[95\%]{} \text{CH}_3\text{CH}_2\text{C(=CH}_2)\text{COOC}_2\text{H}_5$$

The same sequence is applicable to ketones, but in this case $TiCl_4$ is superior to $ZnBr_2$ as the catalyst. The method can be used with both cyclic and acyclic ketones and is applicable to both kinetic and thermodynamic silyl enol ethers for control of regiospecificity.[3]

[1] B. M. Trost and R. A. Kunz, *J. Org.*, **39**, 2648 (1974).
[2] I. Paterson and I. Fleming, *Tetrahedron Letters*, 993 (1979).
[3] *Idem, ibid.*, 995 (1979).

Chloro(norbornadiene)rhodium dimer, $[RhCl(C_7H_8)]_2$. Mol. wt. 461.00, air-stable, golden yellow solid. Supplier: Strem.

Quinolines. Aminoarenes react with aldehydes in the presence of this rhodium complex at 180° to form quinolines in fair to good yield.[1]

$$\text{X–C}_6\text{H}_4\text{–NH}_2 + \text{RCH}_2\text{CHO} \xrightarrow{\text{Rh(I), C}_6\text{H}_5\text{NO}_2\text{, 180°}} \text{X–quinoline(3-R, 2-CH}_2\text{R)}\ (30\text{–}80\%) + \text{X–C}_6\text{H}_4\text{–NH(CH}_2)_2\text{R}$$

[1] Y. Watanabe, M. Yamamoto, S. C. Shim, T. Mitsudo, and Y. Takegami, *Chem. Letters*, 1025 (1979).

***m*-Chloroperbenzoic acid, 1**, 135–139; **2**, 68–69; **3**, 49–50; **6**, 110–114; **7**, 62–64; **8**, 97–102.

Stereocontrolled epoxidation of acyclic systems. Stereoselectivity in the epoxidation of homoallylic alcohols is usually low. Nonetheless Kishi *et al.*[1] have observed that the MCPBA epoxidation of **1** in two-phase CH_2Cl_2–aqueous $NaHCO_3$ at room temperature affords **2** in nearly quantitative yield.

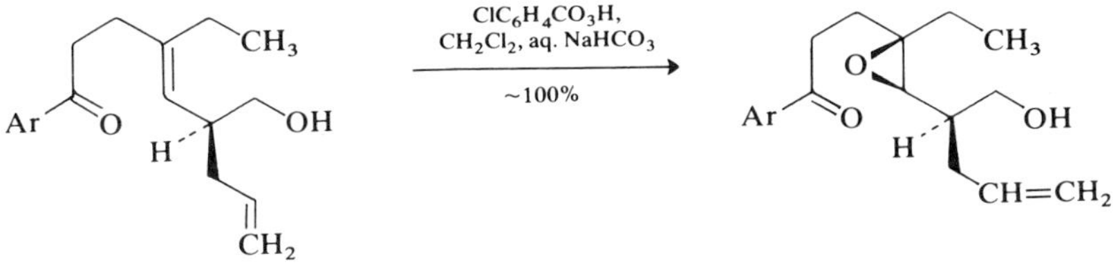

$ClC_6H_4CO_3H$, CH_2Cl_2, aq. $NaHCO_3$; ~100%

1 → **2**

The stereoselectivity of this reaction is probably related to the preference of **1** for adopting conformation **A**, in which steric interactions between the C_3-ethyl and the C_1-substituents are minimized. Hydroxyl-assisted epoxidation (**5**, 76) then leads to **2**.

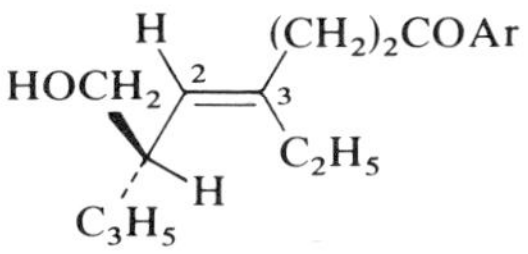

A

Stereoselective epoxidation of allylic alcohols (**5**, 76; **7**, 62–63). The effect of an allylic hydroxyl group on the stereoselectivity of epoxidation of a double bond is reinforced by a neighboring ether group.[2] Typical examples are shown in equations (I) and (II).[3]

(I) $ClC_6H_4CO_3H$, CH_2Cl_2, 0°

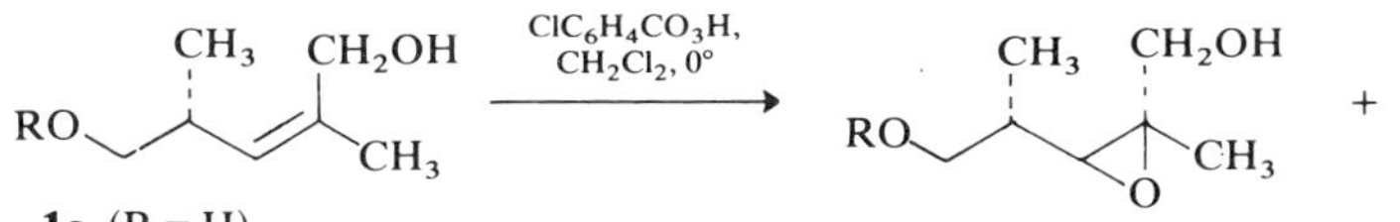

1a (R = H)
1b (R = $CH_2C_6H_5$)

2a (R = H) 4:1
2b (R = $CH_2C_6H_5$) 15:1

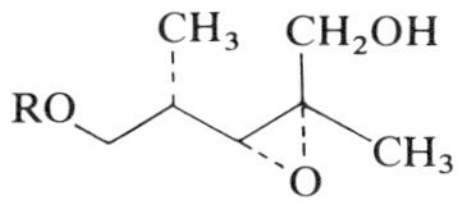

3a (R = H)
3b (R = $CH_2C_6H_5$)

(II) 97%; $(CH_3)_2CuLi$, 95%

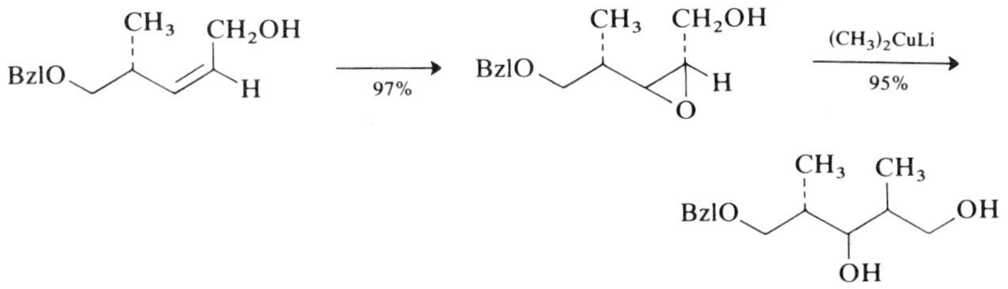

Epoxidation. Isobe *et al.*[4] report that epoxidation of **1** with this reagent can only be achieved in methylene chloride–water. Reaction under other conditions produces the alcohol **3**.

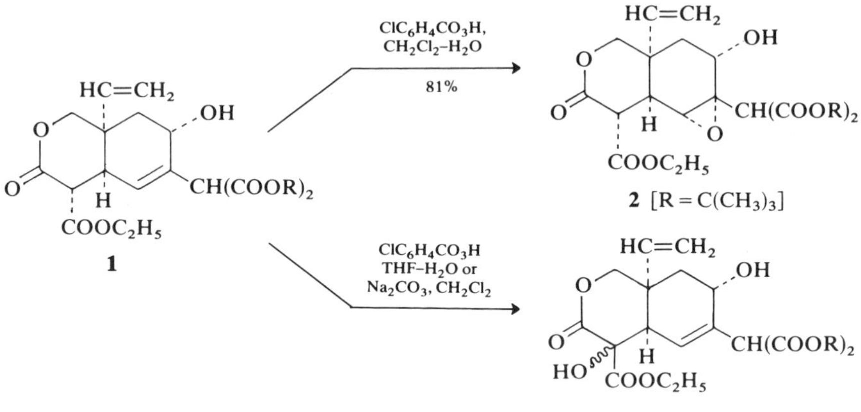

HC=CH₂ ... OH ... CH(COOR)₂ ... H ... COOC₂H₅

1

$ClC_6H_4CO_3H$, CH_2Cl_2–H_2O, 81% → **2** [R = C(CH₃)₃]

$ClC_6H_4CO_3H$, THF–H_2O or Na_2CO_3, CH_2Cl_2 → **3** (R = C_2H_5)

[1] T. Fukuyama, C.-L. J. Wang, and Y. Kishi, *Am. Soc.*, **101**, 260 (1979).
[2] M. R. Johnson and Y. Kishi, *Tetrahedron Letters*, 4347 (1979).
[3] M. R. Johnson, T. Nakata, and Y. Kishi, *ibid.*, 4343 (1979).
[4] H. Iio, M. Isobe, T. Kawai, and T. Goto, *Am. Soc.*, **101**, 6076 (1979).

9-Chloro-9-phenylxanthene, (C_6H_5, Cl) **(1).** Mol. wt. 292.75, m.p. 105–106°. Preparation.[1]

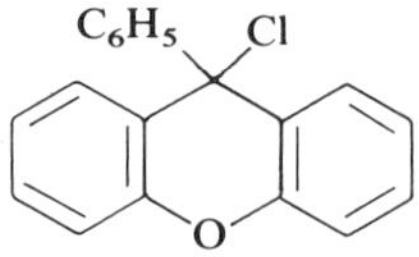

Protection of 5′-hydroxyl groups of 2′-deoxyribonucleosides.[2] The 9-phenyl-9-xanthyl (pixyl) derivatives of 2′-deoxyribonucleosides have better crystallization properties than the corresponding, usually used, dimethoxytrityl ethers, but also have similar acid lability, being removable in 8–15 minutes by 80% HOAc at 20°.

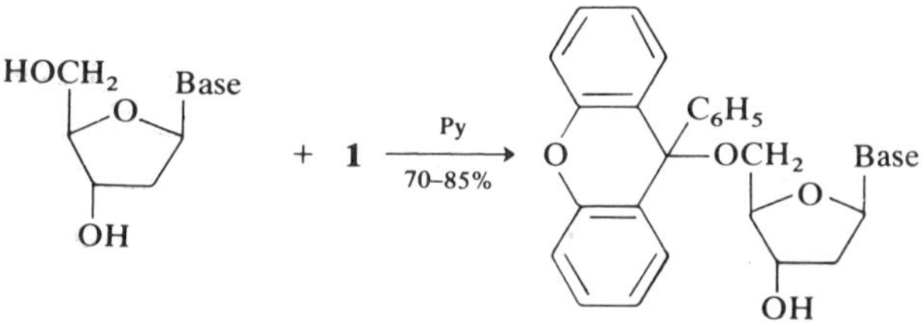

$HOCH_2$ Base, OH + **1** —Py, 70–85%→ C_6H_5, OCH_2 Base, OH

[1] M. Gomberg and L. H. Cone, *Ann.*, **370**, 142 (1909).
[2] J. B. Chattopadyaya and C. B. Reese, *J.C.S. Chem. Comm.*, 639 (1978).

N-Chlorosuccinimide, 1, 139; **2**, 69–70; **5**, 127–129; **6**, 115–118; **7**, 65; **8**, 103–105.

1-Chloro-1-alkynes. Arylchloroacetylenes can be prepared in about 65% yield by conversion of an arylacetylene into the lithium acetylide followed by reaction with NCS in THF and HMPT at −50° initially and eventually at 25°.[1]

$$\mathrm{ArC{\equiv}CH} \xrightarrow{n\text{-BuLi}} [\mathrm{ArC{\equiv}CLi}] \xrightarrow[60\text{–}70\%]{\text{NCS, THF–HMPT}} \mathrm{ArC{\equiv}CCl}$$

[1] W. Verboom, H. Westmijze, L. J. De Noten, P. Vermeer, and H. J. T. Bos, *Synthesis*, 296 (1979).

N-Chlorosuccinimide–Benzeneselenenyl chloride.

Chlorination of alkenes. Allylic chlorination is usually conducted by a free-radical reaction with NCS. Chlorination can be effected with catalysis by C_6H_5SeCl, in which case the main product is usually the rearranged allylic chloride. However, the formation of rearranged allylic chloride is sensitive to the structure of the alkenes and also to the particular selenium compound used as catalyst.[1] *See also* N-phenylselenosuccinimide, this volume.

Examples:

NCS, $C_6H_5SeSeC_6H_5$; 87%; products 96:4

30%; products 74:12:15

[1] T. Hori and K. B. Sharpless, *J. Org.*, **44**, 4204 (1979).

N-Chlorosuccinimide–Dimethyl sulfide, 4, 87–90; **5**, 129–132.

Enaminones. The usual method for conversion of a β-diketone such as **1** results in low yields (10–40%). The reaction is improved considerably by use of the Corey–Kim reagent in the presence of trifluoroacetic acid.[1] Undoubtedly an intermediate sulfonium salt is involved, as in Corey–Kim oxidation (**5**, 129–130).

1) NCS, $(CH_3)_2S$, CH_2Cl_2; 2) CF_3COOH; 3) RNH_2; 25–60%

1 ($C_6H_5CH_2$–N) → ($C_6H_5CH_2$–N, NHR) ($R = CH_3, CH_2C_6H_5, C_6H_5$)

[1] Y. Tamura, L. C. Chen, M. Fujita, H. Kiyokawa, and Y. Kita, *Chem. Ind.*, 668 (1979).

2-Chlorotetrahydrofurane, (**1**). Mol. wt. 106.5, b.p. 35–36°/15 mm. A convenient preparation involves reaction of THF (excess) with sulfuryl chloride (85% yield).

Protection of hydroxyl groups.[1] Alcohols and phenols react with **1** and triethylamine to form tetrahydro-2-furanyl (THF) ethers (85–98% yield). The reaction of acids with **1** results in THF esters. These derivatives are stable to base and nucleophilic reagents; they are readily removed by acid-catalyzed hydrolysis or methanolysis. One synthetic application is conversion of the ethers into alkyl bromides by reaction with triphenyphosphine dibromide (**1**, 1247–1248), a reaction that is faster than that with the free alcohols.

$$\text{RO-THF} \xrightarrow[\text{60–85\% isolated}]{(C_6H_5)_3PBr_2} \text{RBr}$$

[1] C. G. Kruse, F. L. Jonkers, V. Dert, and A. van der Gen, *Rec. trav.*, **98**, 371 (1979).

Chlorotrichloroethylketene, $CCl_3CH_2(Cl)C{=}C{=}O$ (**1**). Mol. wt. 207.89.

Preparation *in situ*:

$$CH_2{=}CHCOOH \xrightarrow[76\%]{1)\ CCl_4,\ Cu_2Cl_2,\ 140^\circ;\ 2)\ SOCl_2} \underset{\mathbf{2}}{CCl_3CH_2CHClCOCl} \xrightarrow{N(C_2H_5)_3,\ 65^\circ} \mathbf{1}$$

Cycloaddition. This ketene undergoes [2+2]cycloaddition with olefins.[1] Examples:

$$(CH_3)_2C{=}CH_2 \xrightarrow[67\%]{\mathbf{2},\ N(C_2H_5)_3} \text{3,3-dimethyl-2-chloro-2-(CH}_2\text{CCl}_3\text{)cyclobutanone}$$

$$\text{cyclopentadiene} \xrightarrow[89\%]{} \text{bicyclic cyclobutanone (Cl, CH}_2\text{CCl}_3\text{)}$$

[1] P. Martin, H. Greuter, and D. Bellus, *Am. Soc.*, **101**, 5853 (1979).

Chlorotrimethylsilane, 1, 1232; **2,** 435–438; **3,** 310–312; **4,** 537–539; **5,** 709–713; **6,** 626–628; **7,** 66–67; **8,** 107–109.

Silylation. Cooper[1] has reviewed this reaction, particularly as applied to large scale production of pharmaceuticals such as prostaglandins and semisynthetic penicillins and cephalosporins. For this purpose, bis(trimethylsilyl)urea $[(CH_3)_3SiNH)_2CO]$ and monotrimethylsilyl acetamide $[(CH_3)_3SiNHCOCH_3]$ are the preferred reagents because they are more reactive than chlorotrimethylsilane and give neutral by-products.

[1] B. E. Cooper, *Chem. Ind.*, 794 (1978).

Chlorotris(triphenylphosphine)rhodium(I), 1, 1252; **2**, 248–253; **3**, 325–329; **4**, 559–562; **5**, 736–740; **6**, 652–653; **7**, 68; **8**, 109.

Stereoselective hydrogenation.[1] A stereoselective synthesis has been reported of the naturally occurring form of α-multistriatin (**5**), the aggregation pheromone of the European elm beetle, the vector of Dutch elm disease in North America. The synthesis is another example of the value of carbohydrates as chiral precursors to natural products.[2] In this case, the known epoxide **1**, derived from D-glucose, was converted in several steps into **2**. The crucial next step required hydrogenation to **3** with the 1,3-diaxial configuration of the two methyl groups. The desired selectivity was attained by use of Wilkinson's catalyst.

The final step in the synthesis required cyclization of **4** to **5**. This reaction was accomplished without epimerization at C_4 by mercuric chloride and mercuric oxide[3] in anhydrous acetonitrile. The pheromone **5** was obtained in greater than 99.7% optical purity.

C_6H_5 O O O O OCH_3 **1** —Several steps→ H_2C $OC(C_6H_5)_3$ CH_3 O OCH_3 **2** —H_2, C_6H_6, $[(C_6H_5)_3P]_3RhCl$, 81%→

CH_3 $OC(C_6H_5)_3$ CH_3 O OCH_3 **3** —Several steps→ CH_3 OH CH_3 OH 4 CH_2CH_3 S S **4** —$HgCl_2$, HgO, CH_3CN, 80%→

O O CH_2CH_3 CH_3 CH_3 **5**

Catalytic hydroacylation.[4] Aldimines of 3-methyl-2-aminopyridine and aromatic aldehydes react with chlorotris(triphenylphosphine)rhodium(I) (**1**) in THF at 55° to afford products of imine C—H insertion (**3**). Aminals of 2-aminopyridine and aldehydes with α-hydrogens (**4**) similarly react with **1** to give **5**; presumably, the aminals are in equilibrium with the corresponding imines under these conditions. These complexes undergo hydroacylation reactions with ethylene as illustrated for **2**. The overall reaction can be performed with catalytic quantities of **1**, as indicated for the reaction of **4**.

2 $\xrightarrow[>90\%]{1}$ **3** $\xrightarrow[80\%]{1)\ CH_2{=}CH_2,\ 150\ psi,\ 170°;\ 2)\ H^+}$ $CH_3CH_2\overset{O}{\overset{\|}{C}}Ar$

4 $\xrightarrow{0.05\ equiv.\ 1}$ [**5**] $\xrightarrow[10\text{–}45\%\ overall]{1)\ CH_2{=}CH_2,\ 150\ psi,\ 160°;\ 2)\ H^+}$ $CH_3CH_2\overset{O}{\overset{\|}{C}}R$

[1] P.-E. Sum and L. Weiler, *Can. J. Chem.*, **56**, 2700 (1978).
[2] B. Fraser-Reid, *Accts. Chem. Res.*, **8**, 192 (1975).
[3] H. Tanaka and S. Torii, *J. Org.*, **40**, 462 (1975).
[4] J. W. Suggs, *Am. Soc.*, **101**, 489 (1979).

Chromic acid, 1, 142–144; **2**, 70–72; **3**, 54; **4**, 95–96; **5**, 138–140; **6**, 123–124; **7**, 68–69.

Medium effects on oxidations with H_2CrO_4. The oxidation of an alcohol such as *t*-butylphenylmethanol (**1**) with chromic acid in aqueous acetic acid results in significant amounts of the cleavage products benzaldehyde and *t*-butyl alcohol. This cleavage is reduced to 6% when the reaction is conducted in aqueous acetone; it is completely suppressed on addition of oxalic acid (6 equiv.), which is known to accelerate oxidations with chromic acid (**5**, 538).

$$\underset{\mathbf{1}}{C_6H_5\overset{OH}{\overset{|}{C}}H{-}C(CH_3)_3} \xrightarrow[100\%]{H_2CrO_4,\ CH_3COCH_3,\ H_2O,\ (HOOC)_2} \underset{\mathbf{2}}{C_6H_5\overset{O}{\overset{\|}{C}}{-}C(CH_3)_3}$$

The same method also results in quantitative yields from the oxidation of 1,2-diphenylethanol and in much improved yields from oxidation of 7-norbornanol.[1]

[1] P. Müller and J. Blanc, *Helv.*, **62**, 1980 (1979).

Chromic anhydride, 1, 144–147; **2**, 72–75; **3**, 54–57; **4**, 96–97; **5**, 140–141; **7**, 70.

Remote oxidation. The major products of oxidation of (−)-bornyl acetate (**1**) with CrO_3 are the C_5- and C_6-ketones, **2** and **3**. Preferential hydroxylation of **1** at C_5 is also observed with microorganisms. The correspondence in regioselectivity probably is a result of greater accessibility of the C_5-position.[1]

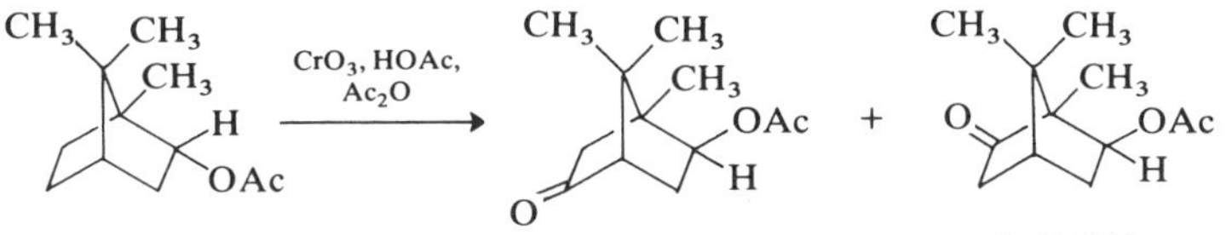

CrO_3, HOAc, Ac_2O: **1** → **2** (40%) + **3** (16%)

The selectivity was useful in a synthesis of nojigku alcohol (**8**) from (−)-isobornyl acetate (**4**). Oxidation of **4** with CrO_3 gave a mixture of **5** and **6**, which was converted into **7** in high yield by Riley oxidation and reduction. The conversion of **7** into **8** involves elimination of the 5-keto group and Wagner-Meerwein rearrangement.[2]

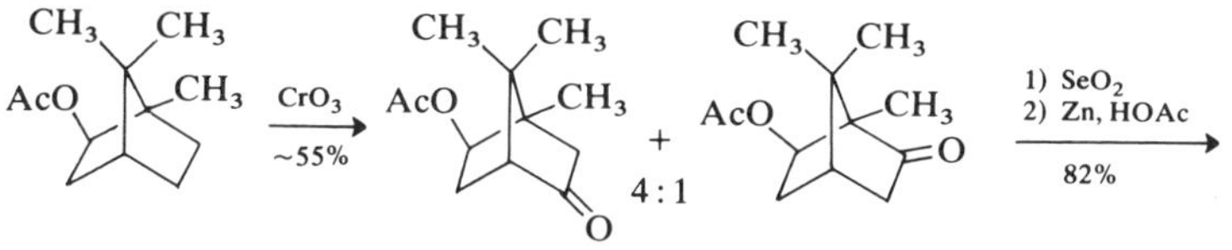

4 → (CrO_3, ~55%) → **5** + **6** (4:1) → (1) SeO_2; 2) Zn, HOAc; 82%) → **7** → (Several steps) → **8**

[1] M. S. Allen, N. Darby, P. Salisbury, E. R. Sigurdson, and T. Money, *Can. J. Chem.*, **57**, 733 (1979).

[2] N. Darby, N. Lamb, and T. Money, *ibid.*, **57**, 742 (1979).

Chromic anhydride–Diethyl ether.

Oxidation of alcohols. CrO_3 (1.0–2.0 equiv.) in a mixture of diethyl ether and dichloromethane (1:3) is useful for oxidation of alcohols to carbonyl compounds. Yields are higher and work-up simpler if Celite is added. Aldehydes undergo further oxidation to carboxylic acids, but ketones and α-diketones are stable. The reagent is not suitable for allylic oxidation.[1]

Examples:

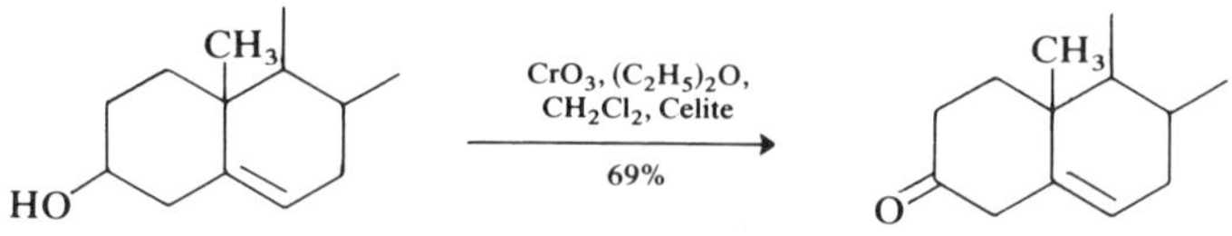

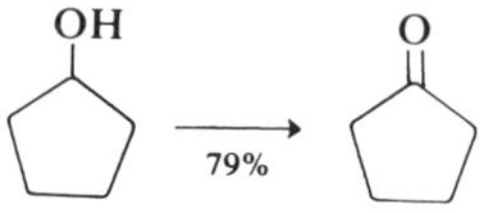

$$\xrightarrow[69\%]{CrO_3, (C_2H_5)_2O, CH_2Cl_2, Celite}$$

$$\xrightarrow[79\%]{}$$

$$CH_3CH(OH)-C(=O)CH_3 \xrightarrow[78\%]{} CH_3C(=O)-C(=O)CH_3$$

$$C_6H_5CH_2OH \xrightarrow[75\%]{} C_6H_5CHO$$

[1] S. J. Fleet, G. W. J. Fleet, and B. J. Taylor, *Synthesis*, 815 (1979).

Chromic anhydride–3,5-Dimethylpyrazole complex (1), 5, 142; **8,** 110.

Allylic oxidation (**8**, 110). The complex **1** (15 equiv.) was used to oxidize **2** to **3** in a total synthesis of (+)-carpesiolin (**4**), an antibacterial helenanolide.

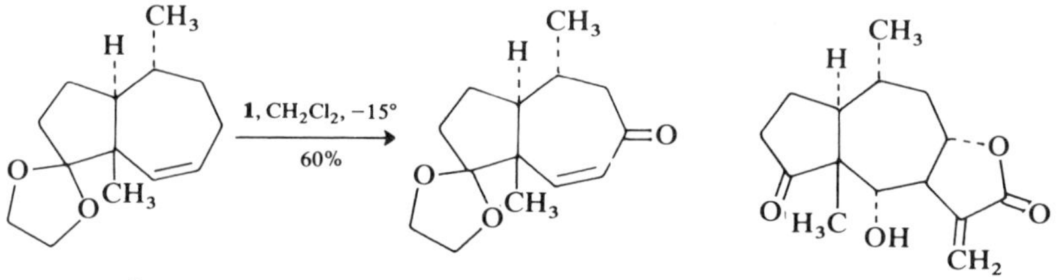

$$\mathbf{2} \xrightarrow[60\%]{\mathbf{1}, CH_2Cl_2, -15^\circ} \mathbf{3} \qquad \mathbf{4}$$

[1] P. Kok, P. J. De Clereq, and M. E. Vandewalle, *J. Org.*, **44**, 4553 (1979).

Chromic anhydride–Hexamethylphosphoric triamide, 7, 71.

Selective oxidation of a primary alcohol. The final step in a recent synthesis of strophanthidin required selective oxidation of the primary hydroxyl group of strophanthidol (**1**) to an aldehyde group.[1] Oxidations with chromic acid, N-haloamides,

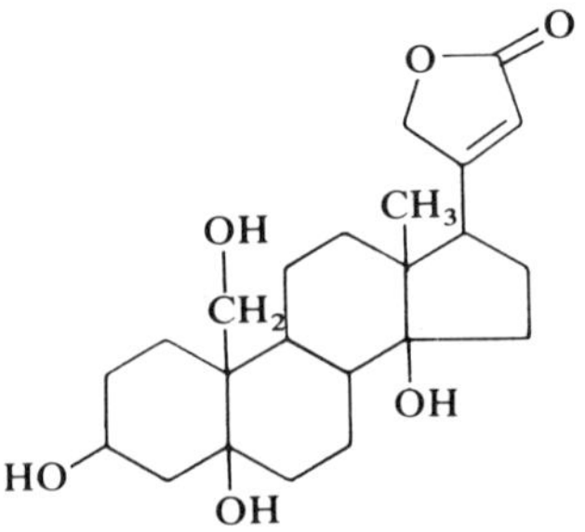

1

and PtO_2 are known to favor oxidation of the 3-hydroxyl group. This preference is also shown by pyridinium chlorochromate and DMSO–SO_2–pyridine. However, chromic trioxide in HMPT is an excellent reagent for this oxidation. The rate is slow, but strophanthidin is obtained in about 35% yield with no ketonic products being formed.

[1] E. Yoshii, T. Oribe, K. Tumura, and T. Koizumi, *J. Org.*, **43**, 3946 (1978).

Chromium(II)–Amine complexes, 3, 57–59.

Reduction of acetylenes.[1] Alkyl phenylacetylenes are reduced by chromium(II) perchlorate–triethylamine to *cis*-olefins in high yield. Terminal acetylenes are also reduced readily, but dialkylacetylenes are not reduced.

[1] J. K. Crandall and W. R. Heitmann, *J. Org.*, **44**, 3471 (1979).

Chromium carbonyl, $Cr(CO)_6$, **5**, 142–143; **6**, 125–126; **7**, 71–72; **8**, 110.

Addition of carbanions to π-arenechromium tricarbonyl complexes. Experimental details for addition of carbanions to η^6-benzenechromium tricarbonyl and for subsequent transformations (oxidation, reaction with electrophiles) of the resulting addition compounds have been published.[1]

Substituent effects in addition of carbanions to π-arenechromium tricarbonyl complexes have been examined.[2] Methyl[3] and chloro substituents give mixtures of *ortho*- and *meta*-substitution, with the amount of *meta*-isomer increasing with bulk of the anions. Methoxyl and dimethylamino substituents are strongly *meta*-directing. The more hindered 3-position of 1,2-dimethoxybenzene is substituted even by a tertiary carbanion. Naphthalene shows 99% α-substitution. A trimethylsilyl group is strongly *para*-directing.

OCH_3 — 1) $Cr(CO)_6$, dioxane, Δ 2) $LiCH_2CO_2R$ 3) I_2 — 93% → OCH_3 / COOR + OCH_3 / COOR (4:96)

CH_3 / OCH_3 / $Si(CH_3)_3$ — 1) $Cr(CO)_6$ 2) $LiC(OR)(CN)CH_3$ 3) I_2 4) H_3O^+ — 82% → O / C / CH_3 / CH_3 / OCH_3

Complexed Aryllithiums. Arenechromium tricarbonyl complexes are metalated with *n*-butyllithium and TMEDA in THF at −78° to afford chromium tricarbonyl complexes of aryllithiums. Metalation occurs specifically *ortho* to methoxyl, fluoro, and chloro substituents; and ring-hydrogens are abstracted in preference to benzylic ones. The complexed aryllithiums react with a variety of reactive electrophiles (Me_3SiCl, $MeOSO_2F$, PhCHO, CH_3COCH_3) in good to excellent yield. With carbonyl compounds, an acid–base reaction sometimes competes with 1,2-addition, whereas with less reactive electrophiles such as CH_3I rapid proton transfer competes with alkylation to give mixtures of polyalkylated products.[4]

Example:

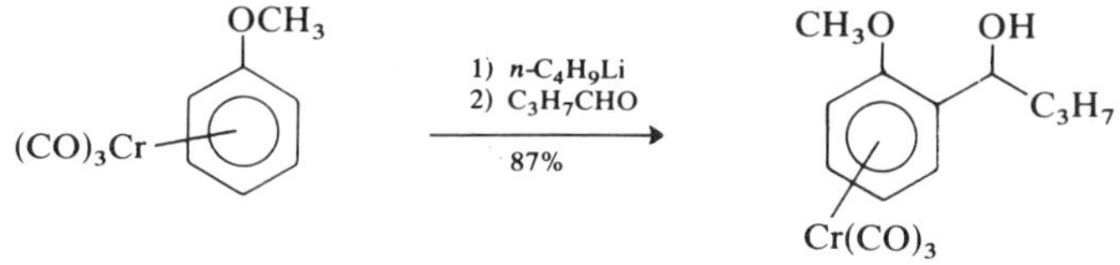

OCH3
(CO)3Cr
1) n-C4H9Li
2) C3H7CHO
87%
CH3O
OH
C3H7
Cr(CO)3

Decomplexation of ArCr(CO)₃. The chromium carbonyl complexes of arenes are useful for activation of the aryl group to nucleophilic attack (**6**, 28, 125–126; **7**, 71–72). Decomplexation has been effected with iodine or by photochemical oxidation with destruction of the expensive $Cr(CO)_3$ unit. A more recent method involves reflux with pyridine to form Py_3–$Cr(CO)_3$ in yields of 70–100%. The pyridine complex in the presence of BF_3 can be reused for preparation of $ArCr(CO)_3$.[5]

Isomerization of 1,3-dienes. Ergosteryl acetate (**1**) is isomerized by chromium carbonyl to ergosteryl B_2 acetate (**2**) in 81% yield. Under the same conditions ergosteryl B_3 acetate (**3**) is isomerized to ergosteryl B_1 acetate (**4**). Both reactions involve isomerization of a *cisoid* diene to a *transoid* diene.[6] In contrast iron carbonyl isomerizes steroidal *transoid* 3,5- and 4,6-dienes to 2,4-dienes.[7]

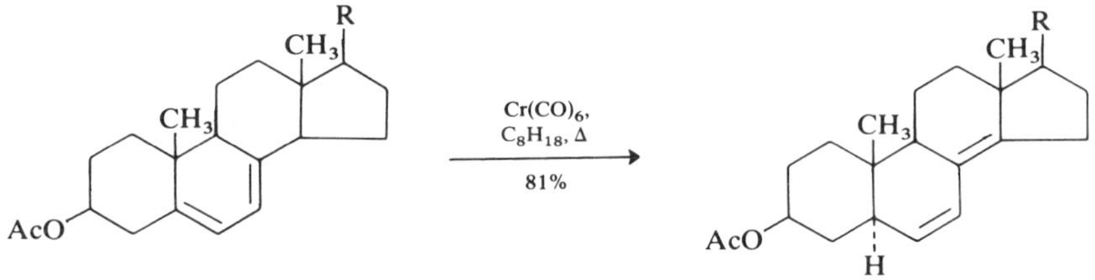

R
CH3
CH3
AcO
Cr(CO)6,
C8H18, Δ
81%
R
CH3
CH3
AcO
H

1 **2**

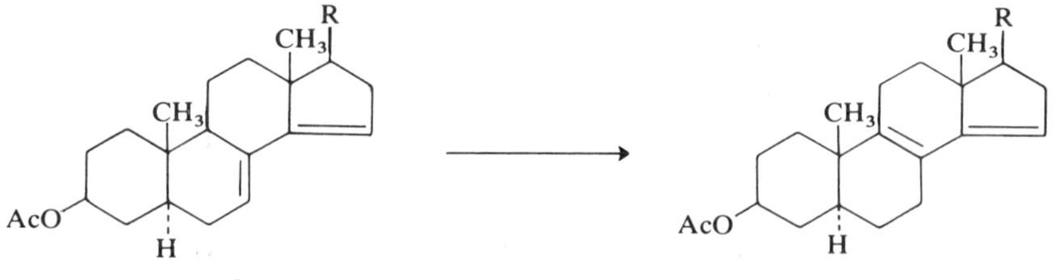

R
CH3
CH3
AcO
H
R
CH3
CH3
AcO
H

3 **4**

[1] M. F. Semmelhack, H. T. Hall, Jr., R. Farina, M. Yoshifuji, G. Clark, T. Bargar, K. Hirotsu, and J. Clardy, *Am. Soc.*, **101**, 3535 (1979).
[2] M. F. Semmelhack, G. R. Clark, R. Farina, and M. Saeman, *ibid.*, **101**, 217 (1979).
[3] M. F. Semmelhack and G. Clark, *ibid.*, **99**, 1675 (1977).
[4] M. F. Semmelhack, J. Bisaha, and M. Czarny, *ibid.*, **101**, 768 (1979).
[5] G. Carganico, P. Del Buttero, S. Maiorana, and G. Riccardi, *J.C.S. Chem. Comm.*, 989 (1978).
[6] D. H. R. Barton, S. G. Davies, and W. B. Motherwell, *Synthesis*, 265 (1979).
[7] H. Alper and J. T. Edward, *J. Organomet. Chem.*, **14**, 411 (1968).

Chromium(III) chloride–Lithium aluminum hydride, 8, 110–112.

Allenes. The propargylic bromide **1** reacts with aldehydes and ketones in the presence of this Cr(II) reagent to give only allenic alcohols (**2**) (equation I).[1]

(I) $$\underset{\mathbf{1}}{C_5H_{11}C{\equiv}C{-}CH_2Br} + R^1COR^2 \xrightarrow[75\text{–}80\%]{CrCl_3\text{–}LiAlH_4} \underset{\mathbf{2}}{H_2C{=}C{=}C(C_5H_{11}){-}C(R^1)(R^2)OH}$$

[1] P. Place, F. Delbecq, and J. Gore, *Tetrahedron Letters*, 3801 (1978).

Cobalt(III) acetate, 6, 127.

Oxidation of alkenes. Two different reactions have been observed in the oxidation of alkenes with $Co(OAc)_3$. α-Methylstyrene is oxidized exclusively to the 1,2-adduct (equation I), whereas either (E)- or (Z)-β-methylstyrene is oxidized mainly at the allylic position (equation II).[1]

(I) $$(C_6H_5)(CH_3)C{=}CH_2 \xrightarrow[HOAc]{Co(OAc)_3,} CH_3C(C_6H_5)(OH){-}CH_2OAc$$

(II) $$C_6H_5CH{=}CHCH_3 \rightarrow C_6H_5CH{=}CHCH_2OAc$$

[1] M. Hirano and T. Morimoto, *J. Chem. Res.* (*M*), 1069 (1979).

Cobalt(II) phthalocyanine (1). Mol. wt. 571.45, violet crystals. This very insoluble compound is prepared by heating phthalic anhydride, urea, cobalt(II) chloride, and a trace of ammonium molybdate in nitrobenzene at 190–200° (45% yield). It can be converted in high yield into the crystalline anions **2** or **3** by reaction with lithium, dilithium benzophenone, sodium naphthalenide, or sodium amalgam. The anion is sensitive to acid, but fairly stable to oxygen. Of greater significance, it behaves as a nucleophile. Thus it is readily alkylated even by primary alkyl halides to form stable cobalt(III) phthalocyanine derivatives.

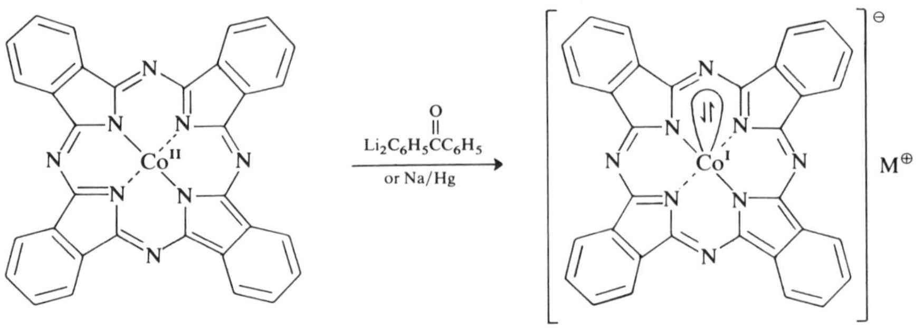

$$\mathbf{1} \xrightarrow[\text{or Na/Hg}]{Li_2C_6H_5\overset{O}{\overset{\|}{C}}C_6H_5} \left[\text{Co}^{I} \right]^{\ominus} M^{\oplus}$$

1

2 (M = Li)
3 (M = Na)

Cleavage of β-halourethanes. Eckert and Ugi[1] have found that urethanes of this type can be cleaved to the amine by treatment with the anion of **1** in acetone or methanol at room temperature. The reaction involves alkylation followed by reductive fragmentation (equation I). The rate of cleavage depends on the structure of the blocking group; 2,2,2-trichloroethoxycarbonyl derivatives ($R^1 = R^2 = Cl$) are cleaved within 1–5 minutes. Cobalt(II) phthalocyanine (**1**) is recovered in almost quantitative yield in reusable form.

$$\text{(I)} \quad ClC(R^1)(R^2)-C(R^3)(R^4)-OCONHR \xrightarrow[70\text{–}90\%]{2\,\mathbf{2},\ CH_3OH,\ 20^\circ} RNH_2 + 2\,\mathbf{1} + R^1R^2C{=}CR^3R^4 + LiCl + LiHCO_3$$

The 2,2,2-trichloroethoxycarbonyl group is stable to acid, but somewhat labile to base. A superior protective group is the 2,2,2-trichloro-*t*-butyloxycarbonyl group, which is stable to both acid and base. The group is introduced by means of the stable chloroformate, prepared from chloretone (equation II). This protecting group is cleaved by the anion **2** in about 1 minute in approximately 90% yield. It is also

$$\text{(II)} \quad Cl_3C-C(CH_3)_2-OH \xrightarrow{COCl_2} Cl_3C-C(CH_3)_2-OCOCl \xrightarrow[70\text{–}95\%]{RNH_2} Cl_3C-C(CH_3)_2-OCONHR$$

cleaved by zinc in acetic acid, but this cleavage requires several hours and the amine is obtained in lower yield.[2] Trichloro-*t*-butyl chloroformate, m.p. 32–34°, is available commercially (Fluka).

Protection of carboxyl groups.[3] The deblocking reaction discussed above with **2** is also applicable to a halogenated protected derivative of the terminal carboxyl group of peptides. The N-protected amino acid is converted by the Passerini reaction with an α-halo aldehyde into a protected derivative such as **4**. The derivative is cleaved by reaction with **2** at 20° in acetonitrile or methanol (equation III).

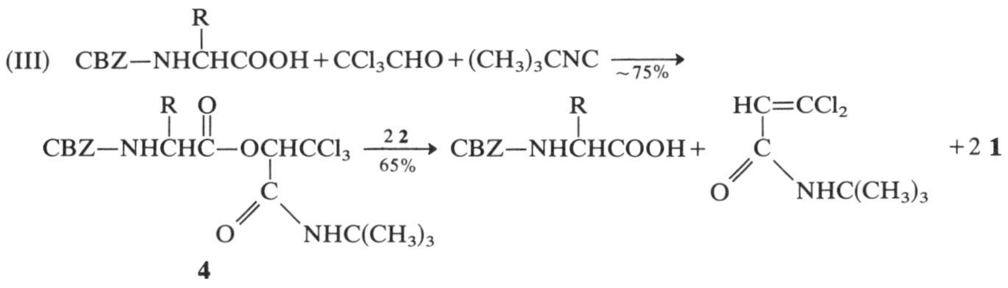

(III) $CBZ{-}NHCH(R)COOH + CCl_3CHO + (CH_3)_3CNC \xrightarrow[\sim 75\%]{}$

$CBZ{-}NHCH(R)C(=O){-}OCH(CCl_3)C(=O)NHC(CH_3)_3$ (**4**) $\xrightarrow[65\%]{\textbf{2}}$ $CBZ{-}NHCH(R)COOH + HC{=}CCl_2{-}C(=O)NHC(CH_3)_3 + 2\ \textbf{1}$

[1] H. Eckert and I. Ugi, *Ann.*, 278 (1979).
[2] H. Eckert, M. Listl, and I. Ugi, *Angew. Chem., Int. Ed.*, **17**, 361 (1978).
[3] H. Eckert, *Synthesis*, 332 (1977).

Collins reagent (Dipyridine chromium(VI) oxide), 2, 74–75; **3,** 55–56; **4,** 216–217.

α,β-Ynones. This chromium(VI) reagent and pyridinium chlorochromate are the most efficient oxidants for conversion of alkynes, $RC{\equiv}CCH_2R$, to conjugated ynones, $RC{\equiv}CCOR$. Chromium trioxide on graphite (Seloxcette reagent, Alpha) is completely unreactive to alkynes. Chromic acid attacks the triple bond almost exclusively.[1]

Oxidation of carbohydrates. Collins reagent has been used to oxidize primary hydroxyl groups of sugars to aldehydes in 50–75% yield, but has not been particularly useful for oxidation of isolated secondary hydroxyl groups.[2] However, in a more recent publication, Garegg and Samuelsson[3] found that excellent results can be obtained if acetic anhydride is added. The highest yields result when the molar ratios of substrate/$CrO_3 \cdot Py_2$/Ac_2O is 1:4:4. If the reaction is extended beyond 5–10 minutes, the yield of the carbonyl product is lowered owing to further oxidation. Some typical results are shown in the formulation of the substrates with yields of product.

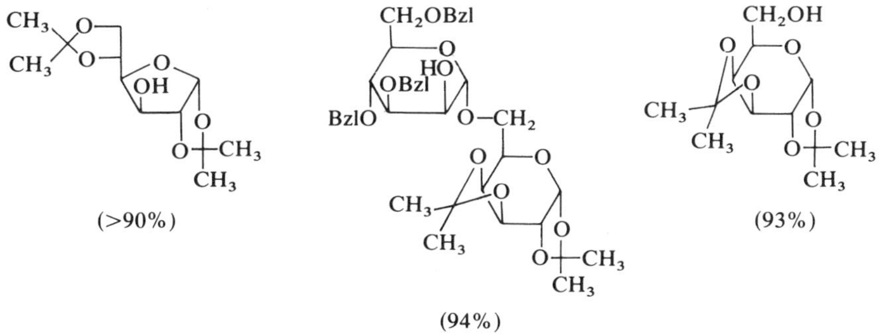

(>90%) (94%) (93%)

[1] W. B. Sheats, L. K. Olli, R. Stout, J. T. Lundeen, R. Justus, and W. G. Nigh, *J. Org.*, **44**, 4075 (1979).
[2] R. F. Butterworth and S. Hanessian, *Synthesis*, 70 (1971).
[3] P. J. Garegg and B. Samuelsson, *Carbohydrate Res.*, **67**, 267 (1978).

Copper, 1, 157–158; **2**, 82–84; **3**, 63–65; **4**, 102–103; **5**, 146–148; **7**, 73–74; **8**, 113–114.

Active copper powder. A highly reactive copper powder can be prepared by reduction of CuI with potassium naphthalenide in DME (8 hours). The slurry should be used immediately. It is particularly effective for Ullmann reactions; high yields of biaryls are obtained even at 85°. It also promotes cross-coupling with allyl halides.[1]

The same paper reports preparation of an activated uranium by reduction of uranium tetrachloride with Na/K alloy and naphthalene.

[1] R. D. Rieke and L. D. Rhyne, *J. Org.*, **44**, 3445 (1979).

Copper(0)–Isonitrile complexes, 4, 101; **5**, 150; **6**, 128–129.

Reduction of halides.[1] Benzylic and allylic halides are dimerized in good yield by the complex of copper metal and cyclohexylisonitrile (Wurtz reaction). Secondary halides are dimerized in low yield, and primary halides do not react. The system is less active than Cu(I) salts and resembles potassium graphite.

The complex reduces *gem*-dibromocyclopropanes to allenes. The *gem*-dichloro compounds are not reduced.

Br Br — Cu-RNC, 53% →

$(CH_3)_3C$ Br Br — 58% → $(CH_3)_3CCH{=}C{=}CH_2$

[1] A. Ballatore, M. P. Crozet, and J.-M. Surzur, *Tetrahedron Letters*, 3073 (1979).

Copper(I) *t*-butoxide, 4, 109; **6**, 144–145. Supplier: Alfa.

Ullmann reaction of 1,3-dinitrobenzene. 2,6-Dinitrobiphenyls can be prepared in good yield by the Cu(I) *t*-butoxide mediated reaction of aryl iodides and 1,3-dinitrobenzene.[1] The reagent is conveniently generated *in situ* from CuCl and potassium *t*-butoxide in DME.

ArI + O_2N NO_2 — $CuOC(CH_3)_3$, Py, DME, 50–95% → O_2N NO_2 Ar

Urethanes.[2] Methyl carbamates (**1**) can be prepared from primary or secondary amines, alkyl halides, and carbon dioxide in a reaction promoted by copper(I) *t*-butoxide (equation I). The ligand *t*-butyl isocyanide can be replaced with tri-*n*-butylphosphine. Copper(I) *t*-butoxide is more effective than other copper salts. In the case of diethylamine, the intermediates **a** and **b** were isolated and **b** was converted to the methyl carbamate in 86% yield.

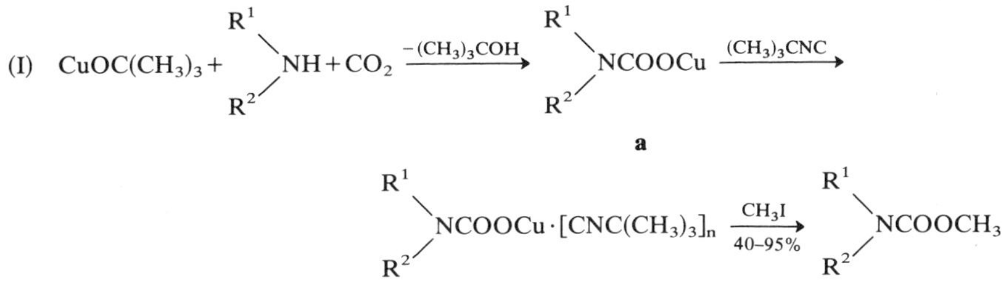

$$\text{(I)}\quad \text{CuOC(CH}_3)_3 + \text{R}^1\text{R}^2\text{NH} + \text{CO}_2 \xrightarrow{-(\text{CH}_3)_3\text{COH}} \underset{\mathbf{a}}{\text{R}^1\text{R}^2\text{NCOOCu}} \xrightarrow{(\text{CH}_3)_3\text{CNC}}$$

$$\underset{\mathbf{b}}{\text{R}^1\text{R}^2\text{NCOOCu}\cdot[\text{CNC(CH}_3)_3]_n} \xrightarrow[40\text{–}95\%]{\text{CH}_3\text{I}} \underset{\mathbf{1}}{\text{R}^1\text{R}^2\text{NCOOCH}_3}$$

[1] J. Cornforth, A. F. Sierakowski, and T. W. Wallace, *J.C.S. Chem. Comm.*, 294 (1979).
[2] T. Tsuda, H. Washita, K. Watanabe, M. Miwa, and T. Saegusa, *ibid.*, 815 (1978).

Copper(I) chloride, 1, 166–169; **2**, 91–92; **3**, 67–69; **4**, 109–110; **5**, 164–165; **6**, 145–146; **7**, 80–81; **8**, 118–119.

Spiroannelation of phenolic α-diazoketones (**8**, 118–119). The Cu(I) decomposition of the phenolic α-diazo ketone **1** has been used for a synthesis of α-chamigrene (**3**).[1]

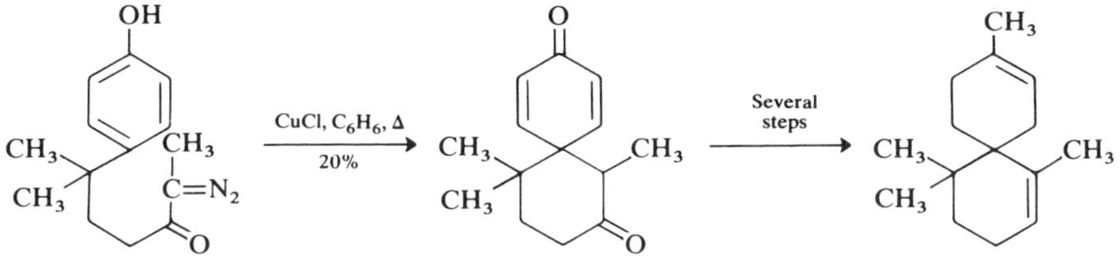

CuCl, C_6H_6, Δ, 20% ; Several steps

1 2 3

[1] C. Iwata, M. Yamada, and Y. Shinoo, *Chem. Pharm. Bull. Japan*, **27**, 274 (1979).

Copper(II) chloride, 1, 163; **2**, 84–85; **3**, 66; **4**, 105–107; **5**, 158–160; **6**, 139–141; **7**, 79; **8**, 119–120.

1,4-Cyclohexanediones. Paquette and co-workers[1] have applied Saegusa's method (**6**, 139) to the intramolecular oxidative coupling of **1**. Addition of the dienolate obtained from **1** to a solution of anhydrous $CuCl_2$ in THF and DMF at −78° affords the polycyclic diketone **2** in 58% yield.

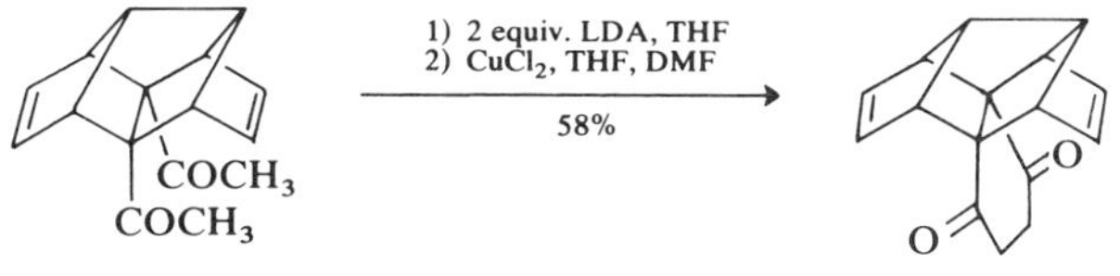

1) 2 equiv. LDA, THF
2) $CuCl_2$, THF, DMF
58%

1 2

[1] L. A. Paquette, R. A. Snow, J. L. Muthard, and T. Cynkowski, *Am. Soc.*, **101**, 6991 (1979).

Copper(I) cyanide, 5, 166–167; **6**, 146–147.

Alkylation of β-arylcyclopentanones. Addition of 10 mole% of CuCN to the lithium enolate prepared from β-arylcyclopentanones and LDA increases the amount of the less stable product of alkylation. Polyalkylation is also suppressed. Similar results are obtained when methyl- or phenylcopper is added to the enolate prepared by alkyllithium cleavage of trimethylsilyl enol ethers. The mechanism by which Cu(I) influences these alkylations is not as yet understood. The regiospecificity of enolate formation in the example illustrated in equation (I) has been attributed to a directing effect of the proximate phenyl group. This effect is also observed in the deprotonation of β-arylcyclohexanones. Quantitative, but not qualitative, differences exist between five- and six-membered rings, probably because of conformational differences.[1]

(I)

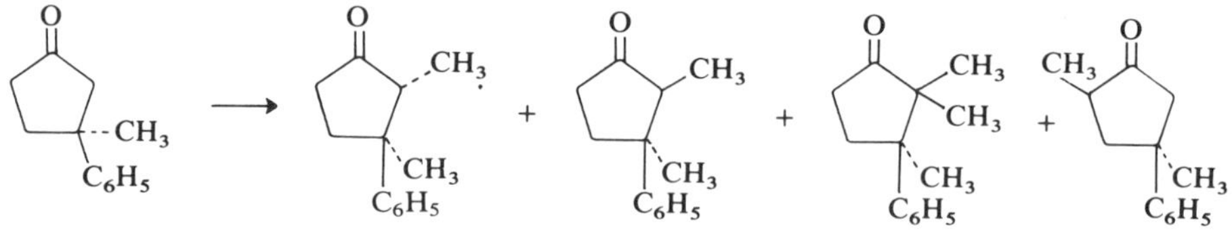

1) LDA, then CH_3I	30.6%	3.4%	47%	10%
2) LDA, CuCN, then CH_3I	28%	52%	5%	10%

[1] G. H. Posner and C. M. Lentz, *Am. Soc.*, **101**, 934 (1979).

Copper(I) iodide, 1, 169; **2**, 92; **3**, 69–71; **5**, 167–168; **6**, 147; **7**, 81–83; **8**, 121–122.

Reaction of Grignard reagents with epoxides. Epoxides are known to undergo ring opening with Grignard reagents, but the reaction can proceed under milder conditions and in higher yield in the presence of copper(I) iodide. The salt is also beneficial in the reaction with the less reactive oxetanes.[1]

Examples:

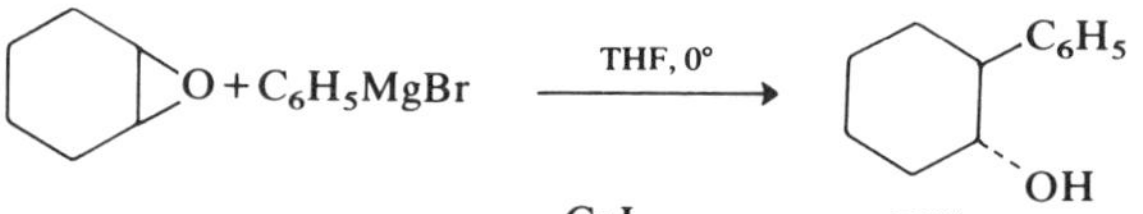

$O + C_6H_5MgBr \xrightarrow{\text{THF, 0°}}$ (product with C_6H_5, OH)

CuI	82%
no catalyst	3%

$+ n\text{-}C_4H_9MgCl \xrightarrow{\text{THF, 0°}} n\text{-}C_4H_9CH_2CH_2OH$

CuI	88%
no catalyst	55%

$+ n\text{-}C_4H_9MgCl \xrightarrow{\text{Ether, 39°}} n\text{-}C_4H_9(CH_2)_3OH$

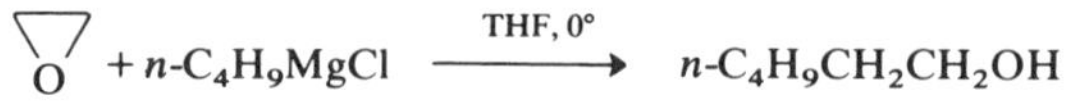

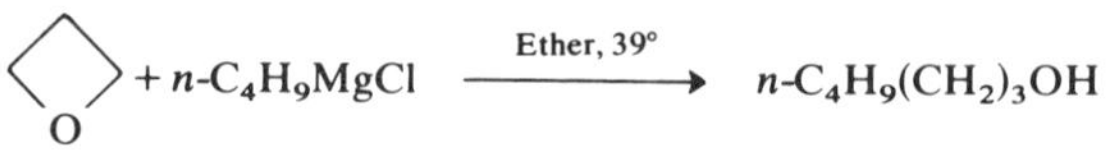

CuI	75%
no catalyst	<10%

Addition of 10% CuI to the reaction of allylic Grignard reagents with epoxides exerts a marked effect on the regioselectivity. The catalyst also increases the rate of reaction. The "α-products" are also the main products when sufficient CuI is added to form the stoichiometric dialkyl cuprate of the Grignard reagent, but yields are lower than in the copper-catalyzed reaction.[2]

Examples:

$$(CH_3)_2\overset{\gamma}{C}{=}CH\overset{\alpha}{C}H_2MgCl \rightleftharpoons (CH_3)_2C(MgCl)CH{=}CH_2$$

70%, propylene oxide (CH_3-oxirane)

(γ-product) + (α-product)

no catalyst 91:9
10% CuI 1:99

1-cyclohexenyl-CH_2MgBr + ethylene oxide $\xrightarrow{\sim 85\%}$ 2-methylenecyclohexyl-CH_2CH_2OH + 1-cyclohexenyl-$CH_2CH_2CH_2OH$

no catalyst 99:1
CuI 3:97

[1] C. Huynh, F. Derguini-Boumechal, and G. Linstrumelle, *Tetrahedron Letters*, 1503 (1979).
[2] G. Linstrumelle, R. Lorne, and H. P. Dang, *ibid.*, 4069 (1978).

Copper(II) sulfate–Pyridine.

Hydration of aryl-substituted epoxides. Treatment of aryl-substituted oxiranes with copper(II) sulfate and pyridine in an aqueous phosphate buffer (pH 7) affords *cis*-diols by reaction of water at the benzylic position. Less than 15% of *trans* cleavage products obtain. In the absence of pyridine, the oxides are rapidly destroyed. Triethylamine cannot be substituted for pyridine. *cis*-Chlorohydrins are obtained when aryl-substituted oxides are treated with copper(II) sulfate, pyridine, and LiCl in THF. When methanol is used as solvent, the corresponding *cis*-glycol monomethyl ether is obtained. In all cases substitution occurs at the benzylic position. Alkyl oxiranes do not react under these conditions.[1]

Examples:

$$\text{indene oxide} \xrightarrow[>95\%]{CuSO_4,\ Py,\ H_2O} \text{cis-indane-1,2-diol (OH, OH)}$$

$$C_6H_5\text{-epoxide}(H, CH_3) \longrightarrow \text{threo-diol} + \text{erythro-diol} \quad 90:10$$

[1] M. Imuta and H. Ziffer, *Am. Soc.*, **101**, 3990 (1979).

Crown ethers, 4, 142–145, **5**, 152–155; **6**, 133–137; **7**, 76–79; **8**, 128–130.

Dehydrohalogenation. Alkyl halides can be converted into alkenes by solid potassium *t*-butoxide and a catalytic amount of 18-crown-6. Petroleum ether with a suitable boiling point range is used as solvent. No reaction is observed in the absence of a crown ether; other bases (KOH, K_2CO_3) are ineffective. The method is particularly useful for dehydrohalogenation of primary alkyl halides, which are converted into ethers by potassium *t*-butoxide under usual conditions.[1]

Examples:

$$n\text{-}C_6H_{13}Cl \xrightarrow[86\%]{KOC(CH_3)_3,\ 18\text{-crown-6},\ \text{petroleum ether},\ 60°} CH_3(CH_2)_3CH{=}CH_2$$

$$CH_3C(CH_3)_2CH_2CH_2Br \xrightarrow[95\%]{25°} CH_3C(CH_3)_2CH{=}CH_2$$

$$(CH_3)_2C(Br)C{\equiv}CC(Br)(CH_3)_2 \xrightarrow[86\%]{} CH_2{=}C(CH_3){-}C{\equiv}C{-}C(CH_3){=}CH_2$$

Synthesis from polyethylene glycol dibromides. Treatment of a poly(ethylene glycol) dibromide with $Ba(OH)_2 \cdot 8H_2O$ as a source of OH^- and Ba^{2+} followed by pyrolysis results in a crown ether. The Ba^{2+} ion acts as a template. The present method requires only one building block (with one less oxygen than the crown ether formed). Highest yields (50%) are obtained with 18-crown-6.[2]

$$BrCH_2(CH_2OCH_2)_nCH_2Br \xrightarrow{OH^-} BrCH_2(CH_2OCH_2)_nCH_2O^- \xrightarrow[-Br^-]{Ba^{2+},\ \Delta} (CH_2OCH_2)_{n+1}$$

(n = 4, 5, 6)

Ferrocene derivatives. Tetrahydrofuran solutions of alkylcyclopentadienes, Fe(II) chloride, potassium hydroxide, and 18-crown-6 as phase transfer catalyst afford substituted ferrocenes in 45–65% yields. Less than 12% of ferrocene is obtained from the reaction of cyclopentadiene in the absence of the crown ether.

$$2C_5H_5R + 2KOH + FeCl_2 \xrightarrow[45\text{–}65\%]{18\text{-crown-6}} (RC_5H_4)_2Fe$$

R = H, CH_3, $CH_2C_6H_5$, n-C_3H_7, c-C_6H_{11}

Other crown ethers (e.g., dibenzo-18-crown-6) can be used, but the reaction work-up is not as simple as with 18-crown-6.[3]

Arenediazo cyanides. Arenediazonium tetrafluoroborates react with solid KCN and 18-crown-6 (0.05 equiv.) in CH_2Cl_2 to give arenediazo cyanides in 75–95% yield. In the absence of 18-crown-6, only low yields of the arenediazo cyanides are obtained after prolonged reaction times. Quaternary ammonium salts are not useful as phase-transfer catalysts. Arenediazo cyanides undergo various reactions, including the Diels–Alder cycloaddition.[4]

$$ArN_2^+BF_4^- \xrightarrow[75\text{–}95\%]{\text{KCN, 18-crown-6}} ArN{=}N{-}CN$$

[1] E. V. Dehmlow and M. Lissel, *Synthesis*, 372 (1979).
[2] L. Mandolini and B. Masci, *Syn. Comm.*, **9**, 851 (1979).
[3] M. Sališová and H. Alper, *Angew. Chem., Int. Ed.*, **18**, 792 (1979).
[4] M. F. Ahern and G. W. Gokel, *J.C.S. Chem. Comm.*, 1019 (1979).

Cryptates, 5, 156; **6**, 137–138; **8**, 130.

Williamson ether synthesis. Cryptate [2.2.2] is an efficient catalyst for the Williamson ether synthesis. Thus reaction of the potassium salt of cyclooctanol with benzyl bromide at 20° for 5 hours in the presence of the cryptate gives the ether in 98% yield.[1]

Reduction of enones. Cyclohexenone is reduced by $LiAlH_4$ in ether almost exclusively by 1,2-addition to give 2-cyclohexenol. However in the presence of cryptate [2.2.1], which binds Li^+, 1,4-addition is the predominant reaction. The same effect is observed with $LiBH_4$, but is less pronounced.[2]

[1] J. F. Biellmann, H. D'Orchymont, and M. P. Goeldner, *Tetrahedron Letters*, 4209 (1979).
[2] A. Loupy and J. Seyden-Penne, *ibid.*, 2571 (1978).

Cyanogen bromide, 1, 174–176; **2**, 93; **4**, 110; **5**, 169–170; **6**, 148–149; **8**, 131.

Cyanatobenzene. This reagent can be prepared in a one-flask reaction by cyanation of phenol with cyanogen bromide prepared *in situ.*[1]

$$Br_2 + NaCN \xrightarrow[-NaBr]{H_2O} BrCN \xrightarrow[75\text{–}85\%]{C_6H_5OH,\ N(C_2H_5)_3,\ CCl_4,\ -5\text{–}10°} C_6H_5OCN + (C_2H_5)_3\overset{+}{N}HBr^-$$

[1] D. Martin and M. Bauer, *Org. Syn.* submitted (1979).

Cyanotrimethylsilane, 4, 542–543; **5**, 720–722; **6**, 632–633; **7**, 397–399; **8**, 133. Two laboratories[1,2] have published independently a new simple, inexpensive preparation of this useful reagent that avoids hydrogen cyanide or silver cyanide (expensive). The reaction of either KCN or NaCN with $ClSi(CH_3)_3$ is conducted in N-methylpyrrolidone (b.p. 202°) for several hours at a temperature of 90–100°. In one case, Adogen 464 was added as a catalyst. Yields are 60–70%.

Cyanohydrins. Cyanohydrins, even of hindered ketones, can be prepared by reaction with a slight excess of cyanotrimethylsilane in the presence of ZnI_2 as catalyst in CH_2Cl_2 at 65°. The trimethylsilyl cyanohydrin as obtained is hydrolyzed with 3*N* hydrochloric acid at 25–45° (equation I). Yields are generally >90%.[3]

$$\text{(I)}\quad R^1COR^2 + (CH_3)_3SiCN \xrightarrow{ZnI_2} R^1C(OSi(CH_3)_3)(CN)R^2 \xrightarrow[90\text{–}99\%]{HCl} R^1R^2C(OH)CN$$

Alkyl aryl ketones.[4] The adducts (**2**) of aromatic aldehydes and cyanotrimethylsilane can be deprotonated with LDA (THF or ether, −78°). The resulting anions can be alkylated by alkyl iodides, bromides, and tosylates in good yield. The products are usually not isolated but are hydrolyzed by dilute hydrochloric acid or benzyltrimethylammonium fluoride to ketones in yields mainly of 80–90%.

$$\text{(I) } \underset{\mathbf{1}}{ArCHO} \xrightarrow[75\text{–}95\%]{(CH_3)_3SiCN,\ ZnCl_2,\ 100°} \underset{\mathbf{2}}{Ar{-}C(OSi(CH_3)_3)(CN){-}H} \xrightarrow{1)\ LDA,\ -78°;\ 2)\ RX} \underset{\mathbf{a}}{\left[Ar{-}C(OSi(CH_3)_3)(CN){-}R\right]} \xrightarrow[50\text{–}90\%]{H_3O^+} \underset{\mathbf{3}}{ArC(O)R}$$

Acyloins.[5] The anion of the adduct of cyanotrimethylsilane with benzaldehyde reacts with aldehydes and ketones to form acyloin silyl ethers (**2**), by way of a 1,4-O-silyl rearrangement, **a** → **b**. On hydrolysis acyloins (**3**) are obtained in high yield.

$$\underset{\mathbf{1}}{C_6H_5C(OSi(CH_3)_3)(CN){-}H} \xrightarrow{1)\ LDA;\ 2)\ R^1COR^2,\ -78° \to 20°} \left[\underset{\mathbf{a}}{C_6H_5C(OSi(CH_3)_3)(CN){-}C(OLi)(R^1){-}R^2} \to \underset{\mathbf{b}}{C_6H_5C(OLi)(CN){-}C(OSi(CH_3)_3)(R^2){-}R^1}\right] \xrightarrow[90\text{–}100\%]{-LiCN}$$

$$\underset{\mathbf{2}}{C_6H_5C(O){-}C(OSi(CH_3)_3)(R^2){-}R^1} \xrightarrow[80\text{–}90\%]{H_3O^+} \underset{\mathbf{3}}{C_6H_5C(O){-}C(OH)(R^2){-}R^1}$$

Thiocyanates.[6] Thiocyanates are obtained in yields usually over 80% by reaction of sulfenyl chlorides with this silane in dry acetonitrile (1 hour, 25°). The advantage of this method is that isothiocyanates are not formed and that even aromatic thiocyanates are easily prepared. The paper reports that the most satisfactory method for preparation of sulfenyl chlorides is reaction of sulfuryl chloride with a thiol or disulfide (80–90% yield).

$$\left.\begin{matrix} RSH \\ \text{or} \\ RSSR \end{matrix}\right\} \xrightarrow[80\text{–}90\%]{SO_2Cl_2} RSCl \xrightarrow{N{\equiv}CSi(CH_3)_3} \underset{(75\text{–}95\%)}{RSCN} + ClSi(CH_3)_3$$

[1] S. Hünig and G. Wehner, *Synthesis*, 522 (1979).
[2] J. K. Rasmussen and S. M. Heilmann, *ibid.*, 523 (1979).
[3] P. G. Gassman and J. J. Talley, *Tetrahedron Letters*, 3773 (1978); *Org. Syn.*, submitted (1978).

[4] K. Deuchert, W. Hertenstein, S. Hünig, and G. Wehner, *Ber.*, **112**, 2045 (1979).
[5] S. Hünig and G. Wehner, *ibid.*, **112**, 2062 (1979).
[6] D. N. Harpp, B. T. Freidlander, and R. A. Smith, *Synthesis*, 181 (1979).

Cyclodextrins, 6, 151–152; **8**, 133–135.

***Photolysis of chlorpromazine* (1).** The photolysis of **1** results mainly in oxidation to the corresponding sulfoxide. Irradiation in the presence of a cyclodextrin (CD), however, results mainly in dechlorination to promazine (**2**), which is also converted into a sulfoxide, (**3**), on photolysis. The rate of conversion of **1** to **2** depends on the cyclodextrin used, the rates following the order β-cyclodextrin > γ-cyclodextrin > α-cyclodextrin. Since these cyclodextrins differ in the size of the cavity, it appears that the conversion of **1** to **2** takes place within the inclusion complex.[1]

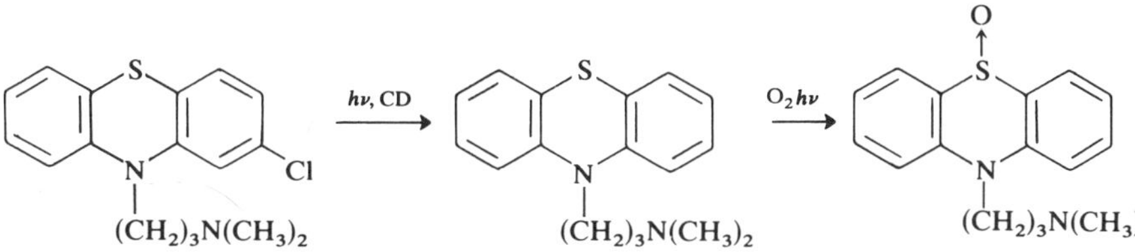

S, N, Cl, $(CH_2)_3N(CH_3)_2$ **1** —$h\nu$, CD→ S, N, $(CH_2)_3N(CH_3)_2$ **2** —$O_2 h\nu$→ O↑S, N, $(CH_2)_3N(CH_3)_2$ **3**

Acylation of β-cyclodextrin. The acylation of β-cyclodextrin is modestly accelerated by bound *m*-nitrophenyl acetate, $m\text{-}NO_2C_6H_4OCOCH_3$. Recently acylation of β-cyclodextrin at a rate comparable to acylation of chymotrypsin has been reported.[2] The acylating reagent is the *p*-nitrophenyl ester (**1**) of ferrocinnamic acid.[3] This reagent was chosen because ferrocene is strongly bound within the cavity of β-cyclodextrin. The acylation is accelerated by >50,000-fold compared to hydrolysis of **1** in DMSO–H_2O alone buffered at pH 6.8. Thus cyclodextrins can behave as artificial enzymes.

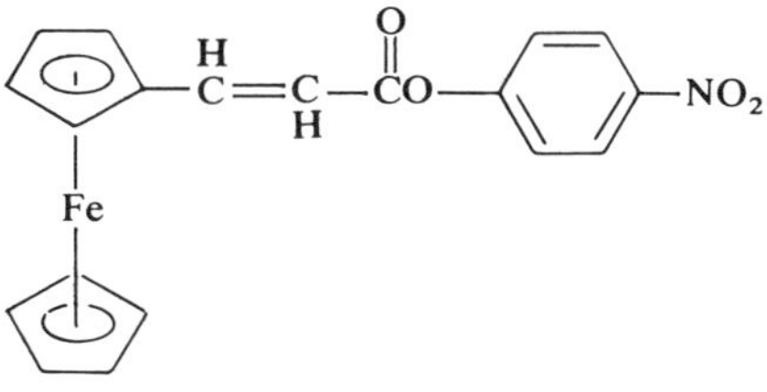

H, C=C, H, O, CO, NO_2, Fe

1

[1] K. Uekama, T. Irie, and F. Hirayama, *Chem. Letters*, 1109 (1978).
[2] M. F. Czarniecki and R. Breslow, *Am. Soc.*, **100**, 7771 (1978).
[3] C. R. Hauser and J. K. Lindsay, *J. Org.*, **22**, 905 (1957).

D

Dessicants for solvents. The efficiency of various dessicants for the commonly used dipolar aprotic solvents has been examined.[1] One finding is that no one drying agent is satisfactory for all solvents. HMPT is very resistant to dessication; but distillation from calcium hydride followed by storage over 3 Å molecular sieves gives reasonably dry solvent. Molecular sieves are also recommended for elimination of water from DMF. Distillation of DMSO with elimination of the first 20% of the distillate provides fairly dry solvent; further dessication can be accomplished with molecular sieves. In the case of acetone, boric anhydride was found to be the dessicant of choice. Oddly enough, this chemical is not satisfactory for the other solvents.

[1] D. R. Burfield and R. H. Smithers, *J. Org.*, **43**, 3966 (1978).

(E,E)-1,4-Diacetoxybutadiene, 1, 183–185.

Naphthalene oxides.[1] The Diels–Alder reaction of 1,4-diacetoxybutadiene (**1**) with benzyne (generated from *o*-benzenediazonium carboxylate, **1**, 46) gives the diacetate of 1,4-dihydronaphthalene-*cis*-1,4-diol (**2**) in 53% yield. The product is converted into **3** by reaction with hydrogen peroxide and catalytic amounts of osmium tetroxide. The monotosylate of **3** is converted into **4** by treatment with base. This epoxy diol has the stereochemistry found in the carcinogenic epoxy diols formed as metabolites of aromatic hydrocarbons. The dimesylate of **3** is converted into the *syn*-diepoxide **5**.

[1] R. R. Schmidt and R. Angerbauer, *Angew. Chem., Int. Ed.*, **18**, 304 (1979).

Dialkylboron trifluoromethanesulfonates, R_2BOTf. These triflates are prepared by reaction of trialkylboranes with triflic anhydride.

Stereoselective aldol condensations. Mukaiyama *et al.*[1] have reported that ketones are converted into vinyloxyboranes by reaction with dialkylboron triflates and a tertiary amine and that these enolates undergo regioselective aldol reactions with aldehydes (equation I). Mukaiyama used di-n-butylboryl triflate in combination

$$\text{(I)}\quad R^1\overset{O}{\overset{\|}{C}}CH_2R^2 \xrightarrow[\text{base}]{R_2BOTf,} (R_2BO)(R^1)C{=}C(H)(R^2) \xrightarrow{R^3CHO} R^1\overset{O}{\overset{\|}{C}}\underset{R^2}{\underset{|}{C}H}\overset{OH}{\overset{|}{C}}HR^3$$

with ethyldiisopropylamine and 9-BBN triflate in combination with 2,6-lutidine. He also noted that the two reagents result in formation of isomeric aldols from the same condensation. Actually the products are stereoisomers.[2] Thus (E)-vinyloxyboranes derived from the reaction of α-diazo ketones and tri-n-butylborane are converted into *threo*-aldols selectively and the (Z)-isomers are converted selectively into *erythro*-aldols. The same relationship obtains in the case of dialkylvinyloxyboranes derived from cyclohexyl ethyl ketone (**1**). By a proper selection of the borane triflate and use of ethyldiisopropylamine as base, the ketone can be converted into either the (E)- or the (Z)-vinyloxyborane (equation II).[3,4]

$$\text{(II)}\quad \underset{\mathbf{1}}{C_6H_{11}\overset{O}{\overset{\|}{C}}CH_2CH_3} \xrightarrow[C_2H_5N(i\text{-}C_3H_7)_2]{\text{9-BBN-OTf}} \left[\underset{\mathbf{a}}{(CH_3)(H)C{=}C(OB\text{-9-BBN})(C_6H_{11})}\right] \xrightarrow[79\%]{C_6H_5CHO} \underset{\mathbf{2}\ (erythro)}{C_6H_{11}\overset{O}{\overset{\|}{C}}{-}\underset{CH_3}{CH}{-}\overset{OH}{C}HC_6H_5} + \mathbf{3}\quad >97:<3$$

$$\mathbf{1} \xrightarrow[C_2H_5N(i\text{-}C_3H_7)_2]{(C_5H_9)_2BOTf} \left[\underset{\mathbf{b}}{(H)(CH_3)C{=}C(OB(C_5H_9)_2)(C_6H_{11})}\right] \xrightarrow[88\%]{C_6H_5CHO} \underset{\mathbf{3}\ (threo)}{C_6H_{11}\overset{O}{\overset{\|}{C}}{-}\underset{CH_3}{CH}{-}\overset{OH}{C}HC_6H_5} + \mathbf{2}\quad 86:14$$

Similar stereoselective syntheses have been obtained in the aldol condensation of vinyloxyboranes derived from S-t-butyl propanethioate (equations III and IV).[5]

$$\text{(III)}\quad CH_3CH_2\overset{O}{\overset{\|}{C}}SC(CH_3)_3 \xrightarrow[C_2H_5N(i\text{-}C_3H_7)_2]{(C_5H_9)_2BOTf} \left[(H)(CH_3)C{=}C(OB(C_5H_9)_2)(SC(CH_3)_3)\right] \xrightarrow[83\%]{C_6H_5CHO}$$

$$\underset{\mathbf{4}\ (threo)}{C_6H_5\overset{OH}{C}H\underset{CH_3}{C}H\overset{O}{\overset{\|}{C}}SC(CH_3)_3} + \mathbf{5}\quad 95:5$$

$$\text{(IV)}\quad \underset{H}{\overset{CH_3}{}}\!\!>C{=}C{=}O \xrightarrow{[(CH_3)_3C]_2BSC(CH_3)_3} \left[\underset{H}{\overset{CH_3}{}}\!\!>C{=}C<\!\!\underset{SC(CH_3)_3}{\overset{OB[C(CH_3)_3]_2}{}}\right] \xrightarrow[78\%]{C_6H_5CHO} \underset{\mathbf{5}\ (erythro)}{C_6H_5\overset{OH}{C}H\underset{CH_3}{C}H\overset{O}{\overset{\|}{C}}SC(CH_3)_3} + \mathbf{4}\quad 93:7$$

[1] T. Mukaiyama and T. Inone, *Chem. Letters*, 559 (1976); T. Inone, T. Uchimaru, and T. Mukaiyama, *ibid.*, 153 (1977).
[2] S. Masamune, S. Mori, D. Van Horn, and D. W. Brooks, *Tetrahedron Letters*, 1665 (1979).
[3] D. E. Van Horn and S. Masamune, *ibid.*, 2229 (1979).
[4] *See also* Di-*n*-butylborontrifluoromethanesulfonate, this volume.
[5] M. Hirama and S. Masamune, *Tetrahedron Letters*, 2225 (1979).

Di-π-allylpalladium chloride, $(C_3H_5PdCl)_2$. Mol. wt. 365.85, m.p. 120°. Supplier: Alfa.

Allylic sulfones. The reaction of 1,3-dienes with arenesulfinic acids or sodium arenesulfinates in THF in the presence of this palladium catalyst and triphenylphosphine results in allylic sulfones. The more substituted isomer is formed predominately[1]

Examples:

$$CH_2{=}CHCH{=}CH_2 + C_6H_5SO_2H \xrightarrow{\text{Pd cat., } P(C_6H_5)_3} \underset{(76\%)}{CH_3\underset{SO_2C_6H_5}{C}HCH{=}CH_2} + \underset{(19\%)}{CH_3CH{=}CH\underset{SO_2C_6H_5}{C}H_2}$$

$$CH_2{=}\overset{CH_3}{C}{-}CH{=}CH_2 + C_6H_5SO_2H \xrightarrow[96\%]{} CH_3\overset{CH_3}{\underset{SO_2C_6H_5}{C}}{-}CH{=}CH_2$$

[1] M. Julia, M. Nel, and L. Saussine, *J. Organometal. Chem.*, **181**, C17 (1979).

1,5-Diazabicyclo[5.4.0]undecene-5 (DBU), 2, 101, **4,** 16–18; **5,** 177–178; **6,** 158; **7,** 87–88; **8,** 141.

Cyclohexadienes. The last steps in a recent novel synthesis of β-damascenone (**3**) from dimedone[1] by way of **1** required oxidation of an allylic hydroxyl group and 1,4-elimination of acetic acid. Activated MnO_2 proved to be superior to pyridinium chlorochromate (40% yield) and to NCS–$S(CH_3)_2$ (**4,** 87–89), which caused elimination of the hydroxyl group. The last step was effected by DBU at reflux for 20 seconds in 79% yield.[1]

$$\text{AcO-cyclohexenyl}(CH_3, CH_3)\text{-CH(OH)-CH=CH-CH}_3 \ (\mathbf{1}) \xrightarrow[87\%]{MnO_2,\ CH_2Cl_2} \text{AcO-cyclohexenyl}(CH_3, CH_3)\text{-CO-CH=CH-CH}_3 \ (\mathbf{2}) \xrightarrow[79\%]{DBU,\ \Delta} \mathbf{3}$$

Esterification. Carboxylic acids can be esterified by reaction with alkyl halides and DBU in benzene at 25 or 80° (equation I). The reaction is widely applicable to hindered or unstable acids and to BOC- or CBZ-protected amino acids. Yields are invariably high. Triethylamine is of less value as the base. Presumably the hydrogen-bonded complex of DBU and the acid plays a significant role in the reaction. The DBU can be recovered by treatment of the hydrohalide with sodium hydroxide.[2]

(I) $$R^1COOH + R^2X + DBU \xrightarrow{C_6H_6} R^1COOR^2 + DBU \cdot HX$$

[1] S. Torii, T. Inokuchi, and H. Ogawa, *J. Org.*, **44**, 3412 (1979).
[2] N. Ono, T. Yamada, T. Saito, K. Tanaka, and A. Kaji, *Bull. Chem. Soc. Japan*, **51**, 2401 (1978).

Diazoacetyl chloride, $N_2CHCOCl$ (**1**). Mol. wt. 104.50, b.p. 15°/0.5 mm., decomposes only slowly below 0°. The acid chloride is prepared from phosgene and diazomethane (1 : 2).

Diazoacetylation. Diazoacetyl chloride (**1**) reacts with alcohols, phenols, thiols, and amines in the presence of triethylamine to afford the diazoacetic ester or amide derivatives **2** in high yield.[1]

$$\underset{\mathbf{1}}{N_2{=}CH{-}C({=}O){-}Cl} \xrightarrow[86\text{–}94\%]{R{-}XH,\ Et_3N} \underset{\mathbf{2}}{H{-}C({=}N_2){-}C({=}O){-}XR}$$

X = O, S, NR
R = alkyl or aryl

[1] H. J. Bestmann and F. M. Saliman, *Angew. Chem. Int. Ed.*, **18**, 947 (1979).

Diazomethane, 1, 191–195; **2**, 102–104; **3**, 74; **4**, 120–122; **5**, 179–182; **6**, 158; **7**, 88–89.

Cyclopentanones. Treatment of 2,2-dichlorocyclobutanones, products of [2+2]cycloaddition of dichloroketene with olefins (**1**, 221–222; **2**, 118; **3**, 87–88;

4, 134–135), with ethereal diazomethane affords 2,2-dichlorocyclopentanones with high regiospecificity; the diazomethane ring expansion reactions of the corresponding monochlorocyclobutanones and the non-chlorinated cyclobutanones are much less regioselective. Reduction of the resulting dichlorocyclopentanones with zinc in acetic acid affords cyclopentanones in 54–72% overall yield based on starting olefin.[1]

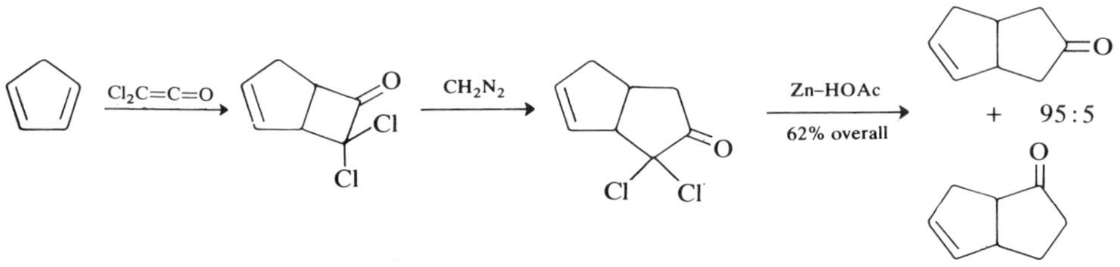

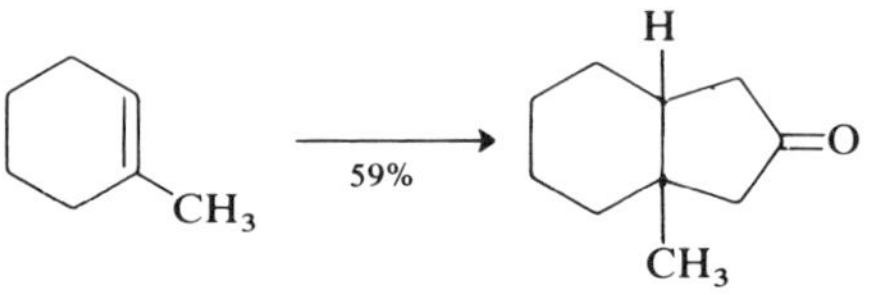

$Cl_2C{=}C{=}O$ → CH_2N_2 → Zn–HOAc, 62% overall; 95:5

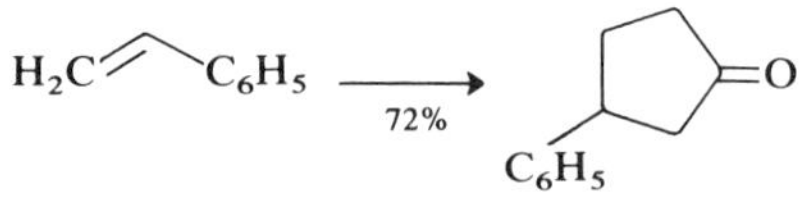

$H_2C{=}CHC_6H_5$ → 72%

→ 59%

[1,3]Cycloaddition to anti-Bredt olefins. Trisubstituted olefins usually exhibit low reactivity to the usual 1,3-dipolar reagents.[2] However, introduction of strain renders such olefins more reactive for 1,3-dipolar cycloadditions. Swiss chemists[3] have examined the reaction of anti-Bredt olefins with diazomethane, phenyl azide, and mesitonitrile oxide. Actually the Bredt olefin **1** reacts with diazomethane more rapidly than a comparable monocyclic olefin. Two regioisomers, **2** and **3**, are obtained, but no stereoisomers. Similar results were obtained with phenyl azide and mesitonitrile oxide. Note that cycloaddition to **4** is fairly regioselective. The results are not completely in accord with frontier orbital theory.

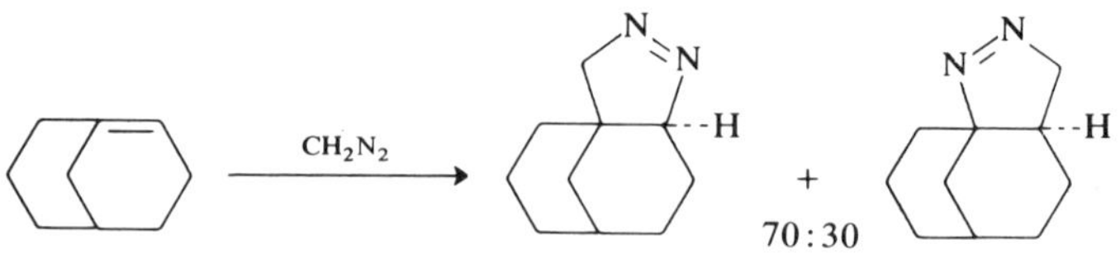

1 → CH_2N_2 → **2** + **3** (70:30)

4 → >80% → **5**

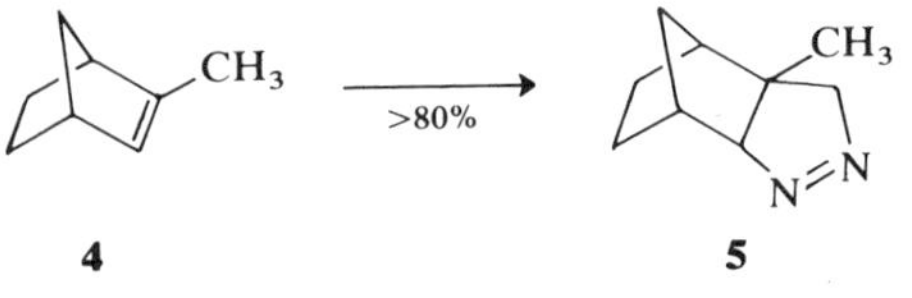

Selective methylation of α,α′α″-trihydroxyanthraquinones. Quinones of this type (*e.g.* **1**) are mono-methylated by reaction with excess diazomethane in CH_2Cl_2. This reaction can be used to convert carminomycinone into daunomycinone (**3**) in 45% yield.[4]

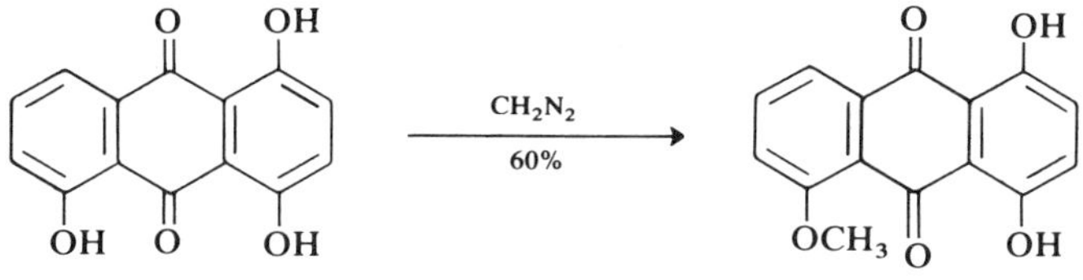

CH_2N_2, 60%

[1] A. E. Green and J.-P. Duprés, *Am. Soc.*, **101**, 4003 (1979).
[2] R. Husigen, *Angew Chem., Int. Ed.*, **2**, 565, 633 (1963).
[3] K. B. Becker and M. K. Hohermuth, *Helv.*, **62**, 2025 (1979).
[4] R. J. Blade and P. Hodge, *J.C.S. Chem. Comm.*, 85 (1979).

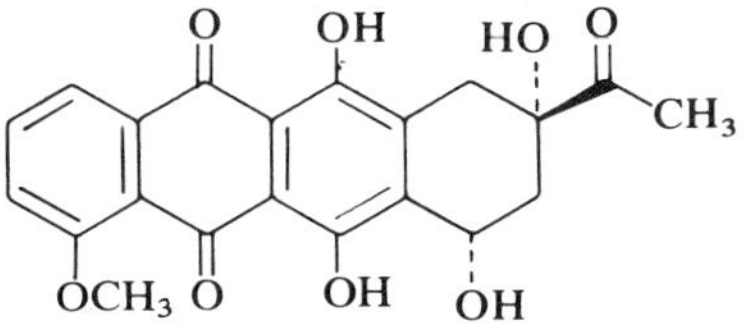

Diazomethane–Silica gel.

Methylation of alcohols.[1] Alcohols can be methylated by diazomethane in the presence of silica gel (30–100 wt. equiv.), silicic acid, neutral alumina, or zeolite. The methylation is faster as the polarity of the alcohol increases. The method is useful for sensitive substrates such as prostaglandins and carbohydrates.

Methylation of amides.[2] Diazomethane in the presence of silica gel forms methyldiazonium silicate, which can methylate amides. Thus this reagent converts caprolactam (**1**) into O-methylcaprolactim (**2**) in 95% yield in 15 minutes.

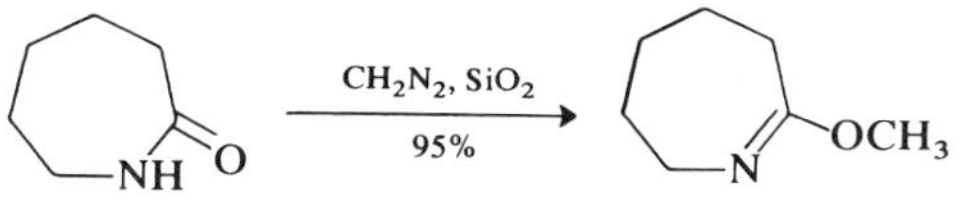

CH_2N_2, SiO_2, 95%

CH_2N_2, SiO_2, 90%; 3:2

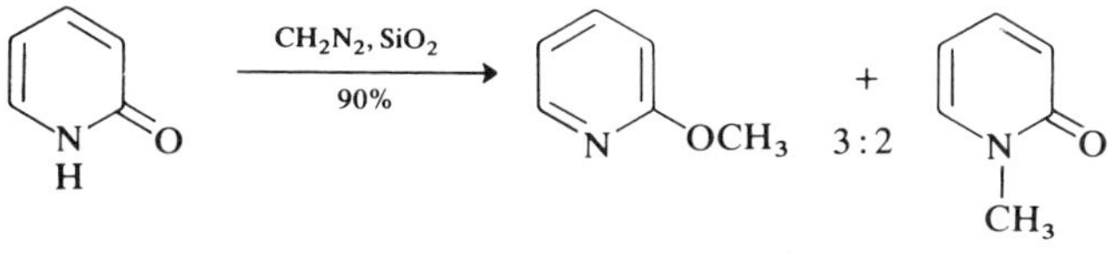

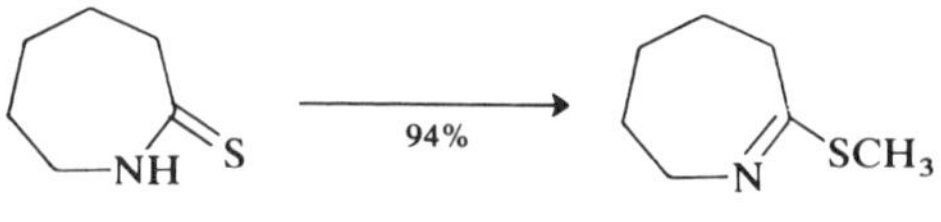

NH S —94%→ N SCH$_3$

[1] K. Ohno, H. Nishiyama, and H. Nagase, *Tetrahedron Letters*, 4405 (1979).
[2] H. Nishiyama, H. Nagase, and K. Ohno, *ibid.*, 4671 (1979).

Dibenzoyl peroxide–Nickel(II) bromide.

Oxidation of alcohols.[1] Secondary alcohols are oxidized to ketones in 85–95% yield by dibenzoyl peroxide in the presence of catalytic amounts of $NiBr_2$ or the soluble complex with dimethoxyethane, $NiBr_2 \cdot (CH_2OCH_3)_2$ (Alfa), in acetonitrile. Under the same conditions, primary alcohols are oxidized to esters. However if 1.5–2.5 equiv. of $NiBr_2$ is used, aldehydes can be obtained in reasonable yield.

[1] M. P. Doyle, W. J. Patrie, and S. B. Williams, *J. Org.*, **44**, 2955 (1979).

Dibenzylammonium trifluoroacetate. Mol. wt. 311.31 m.p. 139°. This salt is obtained in 78% yield by reaction of trifluoroacetic acid and dibenzylamine in ether at room temperature for 30 minutes.

Aldol condensation. This reagent is uniquely effective in catalyzing the aldol cyclization of the dialdehyde **1**.[1] Elimination of the tetrahydropyranyloxy group with other amine–acid combinations is a complication.

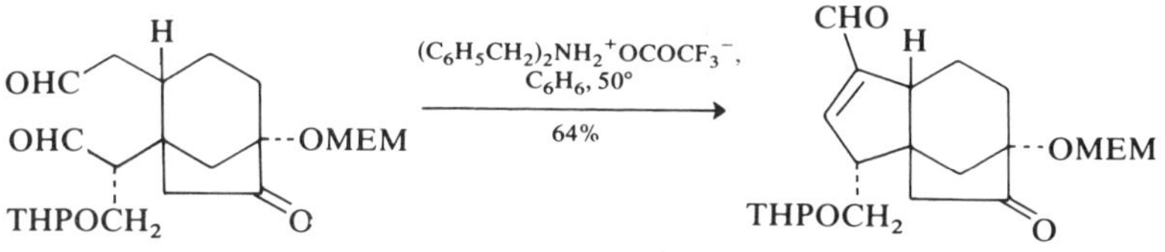

H, OHC, OHC, ···OMEM, THPOCH$_2$, O
1

$(C_6H_5CH_2)_2NH_2^+OCOCF_3^-$, C_6H_6, 50°
64%

CHO, H, ···OMEM, THPOCH$_2$, O
2

[1] E. J. Corey, R. L. Danheiser, S. Chandrasekaran, P. Siret, G. E. Keck, and J.-L. Gras, *Am. Soc.*, **100**, 8031 (1978).

Diborane, 1, 199–207; **2**, 106–108; **3**, 76–77; **4**, 124–126; **5**, 184–186; **6**, 161–162; **7**, 89; **8**, 141–143.

Control of acyclic stereochemistry. The five contiguous stereocenters of aldehyde **5**, an intermediate in Kishi's total synthesis of monensin (**6**),[1] were established by the hydroboration reactions of **1** and **3**, which afforded **2** and **4**, respectively, with high stereoselectivity (8 : 1 diastereomer ratio for **1**; 12 : 1 for **3**). Kishi argues that the

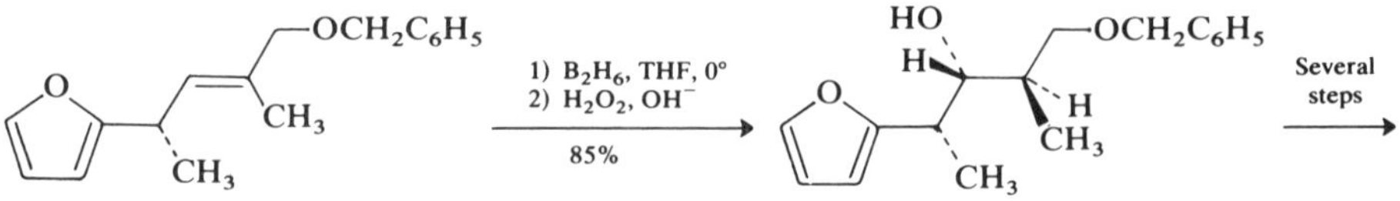

O, OCH$_2$C$_6$H$_5$, CH$_3$, CH$_3$
1

1) B_2H_6, THF, 0°
2) H_2O_2, OH^-
85%

HO, H, O, OCH$_2$C$_6$H$_5$, H, CH$_3$, CH$_3$
2

Several steps →

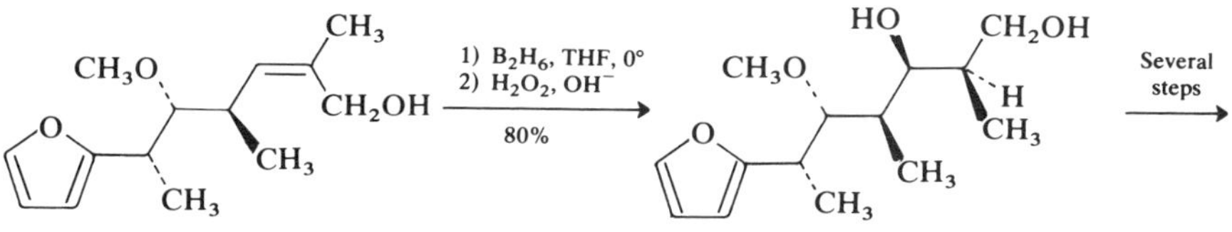

CH_3O ... CH_3 ... CH_2OH ... CH_3 ... CH_3

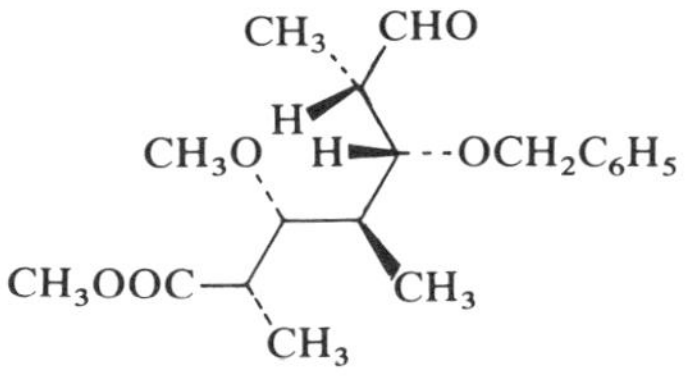

3

1) B_2H_6, THF, 0°
2) H_2O_2, OH^-
80%

HO ... CH_2OH ... CH_3O ... H ... CH_3 ... CH_3 ... CH_3

4

Several steps

CH_3 CHO ... H ... CH_3O H ... $OCH_2C_6H_5$ ... CH_3OOC ... CH_3 ... CH_3

5

HO ... H_3C ... H_3C ... CH_3 ... CH_3 ... CH_3O H ... O ... O ... O ... O ... O ... CH_2OH ... H_3C H ... H ... H H ... OH ... C_2H_5 ... HOOC ... CH_3 ... CH_3

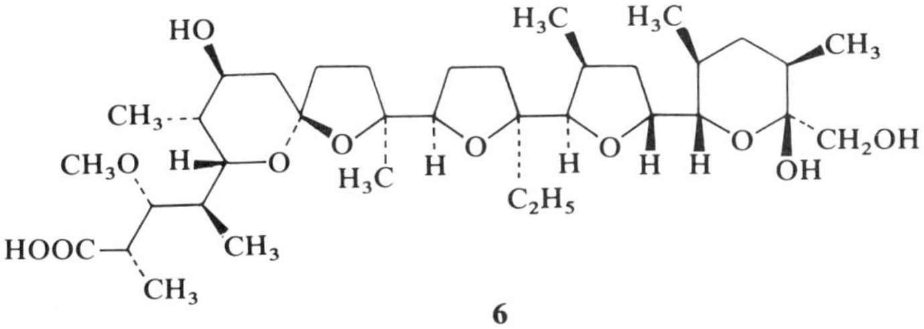

6

remarkable stereoselectivity of these transformations is related to the preference for these systems to adopt conformations such as **A** for **1** in which hydroboration occurs preferentially from the less hindered side. An analogous result has been reported previously by Matsumoto *et al.*[2]

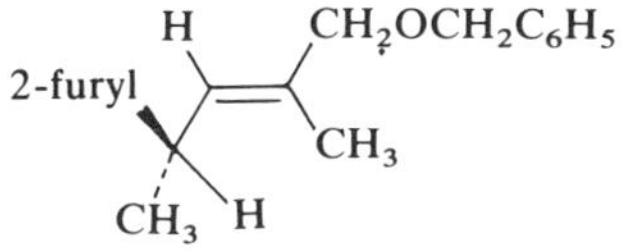

H ... $CH_2OCH_2C_6H_5$ ... 2-furyl ... CH_3 ... CH_3 H

A

Reduction of α,β-unsaturated epoxides. Allylic alcohols can be prepared by reduction of α,β-unsaturated epoxides. Although yields are only moderate (40–65%), the method is stereoselective in that the (Z)-isomer is formed. The reaction is believed to involve conjugate reduction.[3]

Examples:

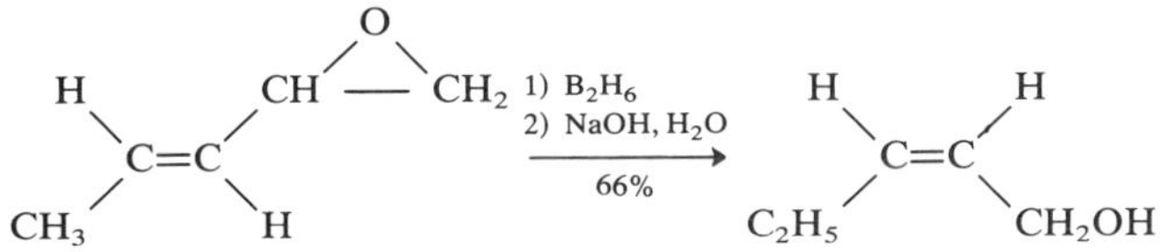

$$\underset{CH_3}{\overset{H}{}}C{=}C\overset{H}{\underset{H}{}}\!-\!CH\overset{O}{—}CH_2 \xrightarrow[66\%]{1)\ B_2H_6;\ 2)\ NaOH,\ H_2O} \underset{C_2H_5}{\overset{H}{}}C{=}C\underset{CH_2OH}{\overset{H}{}}$$

$$CH_3(CH_2{=}CH)C\overset{O}{-}CH(CH_3) \xrightarrow{43\%} (H)(CH_3)C{=}C(CH_3)CH(OH)CH_3$$

[1] G. Schmid, T. Fukuyama, K. Akasaka, and Y. Kishi, *Am. Soc.*, **101**, 259 (1979); T. Fukuyama, K. Akasaka, D. S. Karanewsky, C.-L. J. Wang, G. Schmid, and Y. Kishi, *ibid.*, **101**, 262 (1979).
[2] T. Matsumoto, Y. Hosoda, K. Mori, and K. Fukui, *Bull. Chem. Soc. Japan*, **45**, 3156 (1972).
[3] M. Zaidlewicz, A. Uzarewicz, and R. Sarnowski, *Synthesis*, 62 (1979).

1,1-Dibromoalkyllithium reagents, $RCBr_2Li$ (**1**).

Preparation:

$$RX + Br_3CLi \xrightarrow{\text{HMPT, } -110°} RCBr_3 \xrightarrow[\text{THF–ether, } -100°]{n\text{-BuLi, LiBr,}} \mathbf{1}$$

Homologation of aldehydes or ketones.[1] These reagents react with aldehydes or ketones to form homologated ketones.

Examples:

$$C_6H_5CHO + C_2H_5CBr_2Li \rightarrow \left[C_6H_5\underset{OLi}{CH}{-}CBr_2C_2H_5 \right] \xrightarrow[86\%]{n\text{-BuLi}} C_6H_5COC_3H_7\text{-}n + (C_6H_5)(C_2H_5)CHCHO$$

86 : 14

$$\text{cyclohexanone} + n\text{-}C_4H_9CBr_2Li \xrightarrow{60\%} \text{2-}(C_4H_9\text{-}n)\text{cycloheptanone}$$

[1] J. Villieras, P. Perriot, and J. F. Normant, *Synthesis*, 968 (1979).

2-Dibromomethylbenzoyl chloride, $o\text{-}(CHBr_2)C_6H_4COCl$ (**1**). Mol. wt. 312.40, m.p. 53.5–54.5°. The preparation[1] involves bromination of *o*-toluic acid[2] and then reaction with thionyl chloride.

Alcohol protection.[1] Primary and secondary hydroxyl groups are esterified by **1** in the presence of pyridine (71–92% yield). The usual selectivity for acylation of primary versus secondary hydroxyl functions obtains. The resulting 2-dibromomethylbenzoates **2** are deprotected under neutral conditions by silver perchlorate mediated hydrolysis to 2-formylbenzoate esters **3** (2,6-lutidine or 2,4,6-collidine is added to maintain a virtually neutral reaction medium). Addition of morpholine then leads to rapid deacylation with release of the alcohol in high yield

(87–90% yield; two examples). This step probably proceeds by way of the morpholine adduct **a**,[3] as formulated in equation I.

(I) **2** (o-$CHBr_2$, o-COOR benzene) $\xrightarrow[\text{THF-}H_2O\ (19:1)]{AgClO_4,\ 2,4,6\text{-collidine}}$ **3** (o-CHO, o-COOR benzene) $\xrightarrow{\text{morpholine}}$ [**a**: o-(morpholino)CH(O^-), o-COOR benzene] $\longrightarrow$ 3-morpholinophthalide + ROH

[1] J. B. Chaltopadhyaya, C. B. Reese, and A. H. Todd, *J.C.S. Chem. Comm.*, 987 (1979).
[2] E. L. Eliel and D. E. Rivard, *J. Org.*, **17**, 1252 (1952).
[3] M. L. Bender and M. S. Silver, *Am. Soc.*, **84**, 4589 (1962); M. L. Bender, J. A. Reinstein, M. S. Silver, and R. Mikulak, *ibid.*, **87**, 4545 (1965).

1,3-Di-*n*-butoxypropene (1). Mol. wt. 186.29.

Preparation:

$$CH_2Cl\text{-}CHOH\text{-}CH_2Cl \xrightarrow[n\text{-BuOH}]{n\text{-BuONa,}} CH_2OC_4H_9\text{-}CHOH\text{-}CH_2OC_4H_9 \rightarrow CH_2OC_4H_9\text{-}CHOTs\text{-}CH_2OC_4H_9 \xrightarrow[DMSO]{KOC(CH_3)_3,} C_4H_9OCH{=}CHCH_2OC_4H_9$$

1 (Z/E = 4 : 1)

Deprotonation. This compound is both a vinyl and an allyl ether and conceivably could be deprotonated to a vinyl or an allyl anion or a mixture of both. However, on deprotonation with *t*-butyllithium–HMPT only the vinyl anion is formed. Therefore **1** can serve as the equivalent of $O{=}\bar{C}CH{=}CH_2$.[1]

$$\mathbf{1} \xrightarrow[HMPT]{t\text{-BuLi,}} \left[C_4H_9O\underset{}{\overset{Li}{C}}{=}CHCH_2OC_4H_9\right] \xrightarrow[90\%]{CH_3I} C_4H_9O\underset{CH_3}{C}{=}CHCH_2OC_4H_9$$

a **2**

[1] S. J. Gould and B. D. Remillard, *Tetrahedron Letters*, 4353 (1978).

2,6-Di-*t*-butyl-*p*-benzoquinone, (2,6-$[C(CH_3)_3]_2$-1,4-benzoquinone) (**1**). Mol. wt. 220.31, m.p. 65–67°. Supplier: Aldrich. Preparation (**1**, 218).

Oxidation of amines to amides. It is usually difficult to oxidize selectively the α-methylene group of an amine to obtain the corresponding amide. However, arylmethylamines (**2**) condense with **1** to afford **3** in satisfactory yield. Base-catalyzed oxygenation of **3** results in amides (**4**) with regeneration of **1**. Two conditions for oxygenation can be used. Potassium *t*-butoxide in *t*-butanol gives **4** in 36–85% yield, but **1** is recovered in only 10% yield (**1** is unstable under these conditions). Alternatively, the combination of KOH in ethanol leads to **4** in 15–79% yield with better recovery of **1** (70%).[1]

$ArCH_2NH_2$ (**2**) $\xrightarrow[42–82\%]{1}$ $HO-C_6H_2[C(CH_3)_3]_2-N{=}CHAr$ (**3**) $\xrightarrow{O_2,\ base}$ $Ar-C(=O)-NH_2$ (**4**) + **1**

[1] A. Nishinaga, T. Shimizu, and T. Matsuura, *J.C.S. Chem. Comm.*, 970 (1979).

Di-*n*-butylboryl trifluoromethanesulfonate, $(n\text{-}C_4H_9)_2BOSO_2CF_3$ (**1**), **7**, 91–92.

Stereoselective aldol condensations. Ketones react with **1** in the presence of ethyldiisopropylamine to form (Z)-boron enolates with high stereoselectivity. These derivatives react with aldehydes to afford primarily *erythro*-β-ketols after oxidation and alkaline hydrolysis.[1]

$C_2H_5COCH_2CH_3$ $\xrightarrow[i\text{-}Pr_2EtN,\ ether]{1,}$ $C_2H_5C(OBBu_2){=}CHCH_3$ (>99% Z) $\xrightarrow[77\%]{1)\ C_6H_5CHO;\ 2)\ MoO_5\cdot Py\cdot HMPA;\ 3)\ NaOH}$ $C_2H_5COCH(CH_3)CH(OH)C_6H_5$ (*erythro*, 97%) + $C_2H_5COCH(CH_3)CH(OH)C_6H_5$ (*threo*, 3%)

In the case of cyclohexanone the cyclopentylthexyl boron enolate must be employed to achieve high diastereoselectivity. *See also* Dialkylboron trifluoromethanesulfonates, this volume.

Cyclohexanone $\xrightarrow{R^1R^2BOTf}$ 1-(OBR^1R^2)cyclohexene ($R^1 = C_5H_9$, R^2 = thexyl) $\xrightarrow{73\%}$ 2-[$CH(OH)C_6H_5$]cyclohexanone (erythro) + (threo), 96:4

[1] D. A. Evans, E. Vogel, and J. V. Nelson, *Am. Soc.*, **101**, 6120 (1979).

2,6-Di-*t*-butyl-4-methylpyridine (1). Mol. wt. 205.35, m.p. 31–32°, b.p. 148–153°/95 mm. Supplier: Fluka.

Preparation[1]:

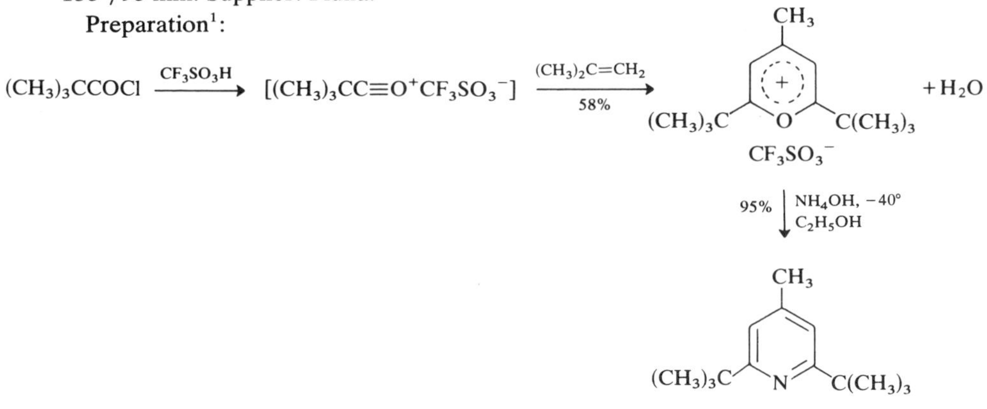

$(CH_3)_3CCOCl \xrightarrow{CF_3SO_3H} [(CH_3)_3CC{\equiv}O^+CF_3SO_3^-] \xrightarrow[58\%]{(CH_3)_2C{=}CH_2}$ 2,6-di-t-butyl-4-methylpyrylium $CF_3SO_3^-$ $+ H_2O$

$\xrightarrow[C_2H_5OH]{NH_4OH,\ -40°;\ 95\%}$ **1**

This hindered, non-nucleophilic base has been used in the alkylation of arenes with vinyl triflates. This reaction proceeds via vinyl cations.[2]

Example:

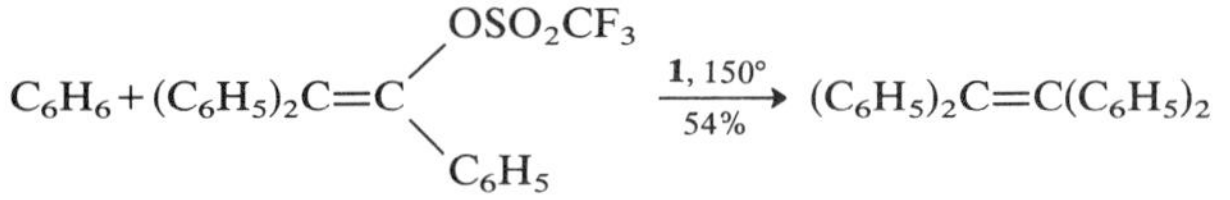

$$C_6H_6 + (C_6H_5)_2C{=}C(OSO_2CF_3)(C_6H_5) \xrightarrow[54\%]{\mathbf{1},\ 150°} (C_6H_5)_2C{=}C(C_6H_5)_2$$

The combination of alkyl trifluoromethanesulfonates and **1** is recommended for methylation of carbohydrates containing acid or base labile groups.[3]

[1] A. G. Anderson and P. J. Stang, *J. Org.*, **41**, 3034 (1976); *idem*, *Org. Syn.* submitted (1979).

[2] P. J. Stang and A. G. Anderson, *Am. Soc.*, **100**, 1520 (1978).

[3] J. Arnarp, L. Kenne, B. Lindberg, and J. Lonngren, *Carbohydrate Res.*, **44**, C5 (1975); J. Arnarp and J. Lonngren, *Acta Chem. Scand.*, **B32**, 465 (1978).

Di-*n*-butyltin oxide, 5, 188.

Selective oxidation of alcohols. The procedure of David and Thieffry[1] has been employed for selective oxidation of the 4-hydroxyl group in the tetraol **1**,[2] a key intermediate in the synthesis of the aminoglycoside antibiotic spectinomycin (**3**).

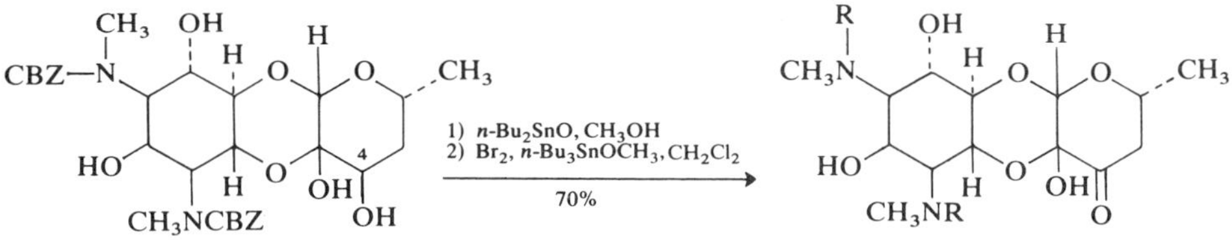

1 $\xrightarrow[70\%]{1)\ n\text{-}Bu_2SnO,\ CH_3OH;\ 2)\ Br_2,\ n\text{-}Bu_3SnOCH_3,\ CH_2Cl_2}$ **2** ($R = CO_2CH_2C_6H_5$), **3** ($R = H$)

[1] S. David and A. Thieffry, *J.C.S. Perkin I*, 1568 (1979).

[2] S. Hanessian and R. Roy, *Am. Soc.*, **101**, 5839 (1979).

Dicarbonyl cyclopentadienylcobalt, 5, 172–173; **6,** 153–154; **7,** 94–95; **8,** 146–147.

Phenanthrene derivatives. As part of a program to determine the relationship between strain and carcinogenicity of polycyclic hydrocarbons, Perkins and Vollhardt[1] have prepared **3** and **4** from 1,5-hexadiyne. Trimerization of the diyne with $CpCo(CO)_2$ (high dilution) gives **1** in 55% yield. This product is converted into the *trans*-stilbene **2**, which is converted into **3** and **4** by oxidative photocyclization. The electronic spectra of **3** and **4** are indicative of strained ring fusion, particularly for **3**. The hydrocarbons also differ in chemical behavior: **3** is hydrogenated twice as readily as **4**. The latter hydrocarbon has no mutagenic properties, but **3** is a weak carcinogen.

C≡C·H, C≡C·H; $CpCo(CO)_2$, 55% → **1**; $h\nu$, NBS, CCl_4 → H, H **2**

$h\nu$, O_2, I_2 → **3** + **4** (2 : 1)

Estrone synthesis. Funk and Vollhardt's work on cooligomerization of bis(trimethylsilyl)acetylene and a 1,5-hexadiyne (**8**, 147) has culminated in three syntheses of *dl*-estrone,[2] one of which shows the desired regioselectivity. The synthesis of **3** by condensation of **1** and **2** had been reported earlier,[3] and the problem remained of finding a suitable modification for estrone synthesis. Use of methoxytrimethylsilylacetylene rather than **1** was explored, but resulted in the undesired regioselectivity and led to 2-methoxyestratrienone as the major final product. Bromination of **3** is regioselective, but the major product results from displacement of the 2- rather than the 3-trimethylsilyl group. The increased reactivity of the 2-position was then exploited by selective reaction of **3** with trifluoroacetic acid at −30°C to give **4** and **5** in a 1:9 ratio. Oxidative cleavage of **5** with lead tetrakis(trifluoroacetate) (**4**, 282–283) proceeds in almost quantitative yield to give estrone (**6**). This sequence is the shortest synthesis to date of *dl*-estrone: five steps from 2-methylcyclopentenone in 24% overall yield and seven steps from 1,5-hexadiyne in 17% overall yield.

$$Si(CH_3)_3-C\equiv C-Si(CH_3)_3 \ (\mathbf{1}) + \mathbf{2} \xrightarrow[71\%]{CpCo(CO)_2} \mathbf{3} \xrightarrow[90\%]{TFA, -30°} \mathbf{4} + \mathbf{5} \ (1:9)$$

$$\mathbf{5} \xrightarrow[\sim 100\%]{Pb(OCOCF_3)_4} \mathbf{6}$$

[1] P. Perkins and K. P. C. Vollhardt, *Angew. Chem., Int. Ed.*, **17**, 615 (1978).
[2] R. L. Funk and K. P. C. Vollhardt, *Am. Soc.*, **101**, 215 (1979).
[3] *Idem, ibid.*, **99**, 5483 (1977).

Dicarbonylcyclopentadienylironmethyldimethylsulfonium tetrafluoroborate (1). Mol. wt. 339.93. This yellow solid is stable in air at room temperature for several months.

Preparation:

$$[CpFe(CO)_2]_2 \xrightarrow[93\%]{1)\ Na(Hg)\quad 2)\ ClCH_2SCH_3} CpFe(CO)_2CH_2SCH_3 \xrightarrow[82\%]{(CH_3)_3O^+BF_4^-,\ CH_2Cl_2} CpFe(CO)_2CH_2\overset{+}{S}(CH_3)_2BF_4^- \ (\mathbf{1})$$

Cyclopropanation. Cyclopropanes can be prepared by reaction of olefins with 2 equiv. of this reagent in dioxane at reflux. The reaction is stereospecific.[1]

Examples:

$$\text{cyclooctene} \xrightarrow[92\%]{\mathbf{1},\ dioxane} \text{cis-bicyclo[6.1.0]nonane}$$

$$\text{trans-}C_4H_9CH=CHC_4H_9 \xrightarrow[\text{67\% (49\% conversion)}]{} \text{trans-1,2-di-}C_4H_9\text{-cyclopropane}$$

[1] S. Brandt and P. Helquist, *Am. Soc.*, **101**, 6473 (1979).

Di-μ-carbonylhexacarbonyldicobalt, $Co_2(CO)_8$, **1**, 224–225; **3**, 89; **4**, 139; **5**, 204–205; **6**, 172; **7**, 99–100; **8**, 148–150.

Acylation of alkenes and alkynes. Dienes and alkynes can be acylated under phase-transfer conditions by carbon monoxide and methyl iodide in the presence of catalytic amounts of $Co_2(CO)_8$. The reaction with alkynes results in a 4-hydroxy-2-butenolide; the probable mechanism is shown in equation (I).

(I) $CH_3I + CO + Co_2(CO)_8 \xrightarrow{\text{Cat., NaOH, } H_2O,\ C_6H_6} [CH_3\overset{O}{\overset{\|}{C}}Co(CO)_4] \xrightarrow{RC{\equiv}CH} [(CO)_4Co(R)C{=}C(H)COCH_3]$

$\xrightarrow{CO,\ OH^-} [HO(O{=})C(R)C{=}C(H)COCH_3] \xrightarrow{20\text{–}70\%}$ (4-R-5-CH₃-5-OH-2(5H)-furanone: O, R, CH_3, OH)

1,3-Dienes are acylated under the same conditions at the least substituted carbon atom to give exclusively an (E)-conjugated enone.[1]

Examples:

1,3-Cyclohexadiene $\xrightarrow[54\%]{CH_3I,\ CO,\ Co_2(CO)_8}$ 1-acetyl-1,3-cyclohexadiene ($COCH_3$)

1-Vinylcyclohexene ($CH{=}CH_2$) $\xrightarrow[60\text{–}63\%]{}$ (E)-1-(cyclohexenyl)-$CH{=}CH$-$COCH_3$ (H, $COCH_3$, C=C, H)

$CH_2{=}CH{-}C(CH_3){=}CH_2 \xrightarrow[40\%]{} CH_3CO(H)C{=}C(H)C(CH_3){=}CH_2$

$C_6H_5(H)C{=}C(H)CH{=}CHCH{=}CH_2 \xrightarrow[43\%]{} C_6H_5(H)C{=}C(H)\text{–}(H)C{=}C(H)\text{–}(H)C{=}C(H)COCH_3$

Catalytic reactions with hydrosilanes and carbon monoxide. The reactions of olefins, cyclic ethers, and aldehydes with a hydrosilane and carbon monoxide catalyzed by $Co_2(CO)_8$ or Ru or Rh complexes have been reviewed (48 references). A number of the more useful transformations are summarized in equations (I)–(IV).[2]

I: cyclohexene $\xrightarrow[89\%]{HSiCH_3(C_2H_5)_2,\ CO,\ Co_2(CO)_8,\ 140°}$ cyclohexylidene–$CH{=}OSi(CH_3)(C_2H_5)_2$

II: tetrahydrofuran $\xrightarrow[53\%]{140°}$ $(C_2H_5)_2CH_3SiO(CH_2)_4CHO$

III: cyclohexene oxide $\xrightarrow[51\%]{140°}$ 2-$OSiCH_3(C_2H_5)_2$-cyclohexane-1-CHO

IV: RCHO (excess) $\xrightarrow[49\%]{P(C_6H_5)_3,\ 100°}$ $RCH(OSiCH_3(C_2H_5)_2)CHO$

[1] H. Alper, J. K. Currie, and H. des Abbayes, *J.C.S. Chem. Comm.*, 311 (1978); H. Alper and J. K. Currie, *Tetrahedron Letters*, 2665 (1979).
[2] S. Murai and N. Sonoda, *Angew. Chem., Int. Ed.*, **18**, 809 (1979).

gem-Dichloroallyllithium, 5, 188–189.

Carbenoid equivalent.[1] Generation of *gem*-dichloroallyllithium (**1**) from 3,3-dichloropropene and lithium 2,2,6,6-tetramethylpiperidide[2] in the presence of an alkene results in formation of *gem*-chlorovinylcyclopropanes (**2**) in 20–40% yield (crude). The paper also reports reduction of one of these products to a vinylcyclopropane (**3**) with sodium–ammonia.[3]

$$CH_2{=}CHCHCl_2 \xrightarrow{LiTMP,\ ether,\ -5°} \underset{\mathbf{1}}{CH_2{=}CHCCl_2Li} \xrightarrow{(CH_3)_2C{=}CH_2} \underset{\mathbf{2}}{(CH_3)_2C\!-\!CH_2\!-\!C(Cl)(CH{=}CH_2)}$$

$$\xrightarrow[53\%]{Na,\ NH_3,\ -50°} (CH_3)_2C\!-\!CH_2\!-\!C(H)(CH{=}CH_2)$$

[1] A. Moss and R. C. Munjal, *Synthesis*, 425 (1979).
[2] The base should be generated from the corresponding amine with methyllithium (Alfa) freed from LiBr by treatment with dry triglyme, which precipitates most of the salt.
[3] Procedure of R. F. Heldeweg and H. Hogeveen, *J. Org.*, **43**, 1916 (1978).

Dichlorobis(cyclopentadienyl)titanium, $(C_5H_5)_2TiCl_2$. Mol. wt. 249.00. Suppliers: Alfa, Strem.

Lactone synthesis.[1] This Ti(II) complex can serve as catalyst for hydromagnesiation of allylic or homoallylic alcohols, prepared by addition of vinyl- or allyl-Grignard reagents to ketones. Ethylmagnesium bromide is used as the source of magnesium hydride. The carbonylation of the organomagnesium intermediate results in a γ- or a δ-lactone (equation I).

$$\text{(I)}\quad \begin{matrix} R^1 \\ \diagdown \\ C{=}O \\ \diagup \\ R^2 \end{matrix} + CH_2{=}CH(CH_2)_nMgX \rightarrow R^1{-}\overset{OH}{\underset{R^2}{C}}{-}(CH_2)_nCH{=}CH_2 \xrightarrow[-CH_2=CH_2]{2CH_3CH_2MgBr,\ Cp_2TiCl_2}$$

$$(n = 0, 1)$$

$$\left[R^1, R^2 \text{-substituted cyclic } O{-}Mg{-}(CH_2)_n \right] \xrightarrow[55\text{–}60\%]{CO_2} R^1, R^2 \text{-substituted lactone } O{-}C(=O){-}CH_2{-}(CH_2)_n$$

[1] J. J. Eisch and J. E. Galls, *J. Organometal. Chem.*, **160**, C8 (1978).

Dichlorobis(cyclopentadienyl)titanium–Lithium borohydride.

Hydroboration.[1] When these two reagents are mixed in THF, hydrogen is evolved and a solution of a violet titanium–boron complex forms. This complex catalyzes the hydroboration of alkenes with $LiBH_4$ to form lithium alkylborohydrides, which are converted to alcohols by $NaOCH_3$ and H_2O_2. The relative reactivity of alkenes is terminal > cyclic > internal. The reaction involves anti-Markownikoff addition.

[1] K. Isagawa, H. Sano, M. Hattori, and Y. Otsuji, *Chem. Letters*, 1069 (1979).

Dichlorobis(cyclopentadienyl)titanium–Trialkylaluminum, $Cp_2TiCl_2{-}AlR_3$ **(1).**

1,1-Dialkylalkenes. Monoalkylalkenes of structure **2** react with 2 equiv. of reagent **1** in CH_2Cl_2 at room temperature to form alkenes of type **3**. The reaction is

$$\text{(I)}\quad \underset{\mathbf{2}}{(R^1)(H)C{=}CH_2} \xrightarrow{2\,\mathbf{1}} \underset{\mathbf{3}}{(R^1)(R)C{=}CH_2}$$

subject to steric effects: the yield of **3** decreases as the size of R is increased. It is compatible with some functional groups: bromo, hydroxyl, and ester.[1]

Examples:

$$n\text{-}C_8H_{17}CH{=}CH_2 + Cp_2TiCl_2{-}Al(CH_3)_3 \xrightarrow[92\%]{} (n\text{-}C_8H_{17})(CH_3)C{=}CH_2$$

$$\text{(cyclohexenyl)}CH{=}CH_2 + Cp_2TiCl_2{-}Al(C_2H_5)_3 \xrightarrow[10\%]{} \text{(cyclohexenyl)}C(C_2H_5){=}CH_2$$

[1] J. J. Barber, C. Willis, and G. M. Whitesides, *J. Org.*, **44**, 3603 (1979).

Dichloro[1,1′-bis(diphenylphosphine)ferrocene]palladium(II) [Pd(dppf)Cl$_2$]. This palladium(II) complex is prepared by reaction of bis(acetonitrile)dichloropalladium(II) with 1,1′-bis(diphenylphosphine)ferrocene in benzene.

Cross-coupling of Grignard reagents and organic halides. This palladium complex is an efficient catalyst for this cross-coupling reaction, particularly in the case of *sec*-alkyl Grignard reagents. Use of tetrakis(triphenylphosphine)-palladium(0) or dichlorobis(triphenylphosphine)palladium(II) is not useful because of formation of various side products resulting from isomerization, elimination, and reduction.[1]

$$CH_3CH_2CH(CH_3)MgCl + C_6H_5CH{=}CHBr \xrightarrow[97\%]{L_2PdCl_2} CH_3CH_2CH(CH_3){-}CH{=}CHC_6H_5$$

[1] T. Hayashi, M. Konishi, and M. Kumada, *Tetrahedron Letters*, 1871 (1979).

Dichlorobis(triphenylphosphine)nickel(II), $[(C_6H_5)_3P]_2NiCl_2$ (**1**). Mol. wt. 654.21. Supplier: Alfa. Preparation.[1]

Coupling of Grignard reagents with unsaturated sulfur compounds. This nickel-phosphine complex catalyzes coupling of Grignard reagents with alkenyl or aryl sulfides with retention of configuration.[2,3]

Examples:

$$(C_6H_5S)HC{=}CH(C_6H_5) \xrightarrow[97\%]{1,\ C_6H_5MgBr,\ THF} (C_6H_5)HC{=}CH(C_6H_5)$$

$$(CH_3S)HC{=}CH(C_6H_5) \xrightarrow[85\%]{1,\ C_6H_5MgBr} (C_6H_5)HC{=}CH(C_6H_5)$$

$$p\text{-}(CH_3)_3CC_6H_4SCH_3 \xrightarrow[53\%]{1,\ C_6H_5MgBr} p\text{-}(CH_3)_3CC_6H_4C_6H_5$$

$$CH_3SO_2C_6H_5 \xrightarrow[53\%]{1,\ p\text{-}CH_3C_6H_4MgBr} C_6H_5C_6H_4CH_3\text{-}p$$

The cross-coupling between allylic sulfides and Grignard reagents proceeds more readily.[4] In the case of hindered allylic sulfides, $NiCl_2[P(C_6H_5)_3]_2$ is more effective than $NiCl_2[(C_6H_5)_2PCH_2CH_2P(C_6H_5)_2]_2$, but the latter catalyst is more effective for coupling with cinnamyl methyl sulfide.

Examples:

$$(CH_3)_2CHSCH_2CH{=}CH_2 \xrightarrow[99\%]{Ni(II),\ C_6H_5MgBr} C_6H_5CH_2CH{=}CH_2$$

$$CH_3SCH(CH_3)CH{=}CH_2 \rightarrow \underset{(55\%)}{C_6H_5CH_2CH{=}CHCH_3} + \underset{(38\%)}{C_6H_5CH(CH_3)CH{=}CH_2}$$

$$CH_3SCH_2CH{=}CHC_6H_5 \xrightarrow[96\%]{} (C_6H_5CH_2)HC{=}CH(C_6H_5)$$

[1] F. A. Cotton, O. D. Faut, and D. M. L. Goodgame, *Am. Soc.*, **83**, 344 (1961).
[2] H. Okamura, M. Miura, and H. Takei, *Tetrahedron Letters*, 43 (1979).
[3] E. Wenkert, T. W. Ferreira, and E. L. Michelotti, *J.C.S. Chem. Comm.*, 637 (1979).
[4] H. Okamura and H. Takei, *Tetrahedron Letters*, 3325 (1979).

Dichlorobis(triphenylphosphine)platinum(II)–Tin(II) chloride dihydrate, $[(C_6H_5)_3P]_2PtCl_2{-}SnCl_2\cdot 2H_2O$.

Methoxylation of allylic alcohols. This complex catalyzes the methoxylation of steroidal allylic alcohols to rearranged allyl methyl ethers.[1]

Examples:

HO, CH$_3$ → (CH$_3$OH, cat., 15°; ~100%) → CH$_3$, OCH$_3$

HO, CH$_3$ → CH$_3$, OCH$_3$ (75%) + CH$_3$, OCH$_3$ (25%)

CH$_3$, OH → CH$_3$, OCH$_3$ (60%) + CH$_3$, OCH$_3$ (20%) + CH$_3$ (20%)

[1] Y. Ichinone, H. Sakamaki, and N. Kameda, *Chem. Letters*, 835 (1978).

2,3-Dichloro-5,6-dicyano-1,4-benzoquinone (DDQ), 1, 215–219; **2**, 112–117; **3**, 83–84; **4**, 130–134; **5**, 193–194; **6**, 168–170; **7**, 96–97; **8**, 153–156.

Oxidative cyclization of arylidenebutyrolactone lignans. Reaction of **1** with 2 equiv. of DDQ in refluxing benzene for 18 hours produces the 4-aryl-2,3-naphthalide lignan **2**.[1]

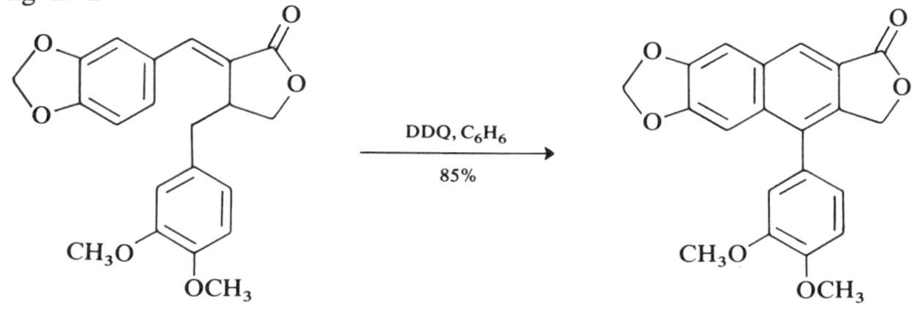

DDQ, C_6H_6
85%

CH_3O OCH_3 **1** CH_3O OCH_3 **2**

Oxidative ring closure to thiadiazoles. Modified cephalosporins of type **3** have been prepared from 3-formyl-3-cephems **1** as shown in equation (I).[2]

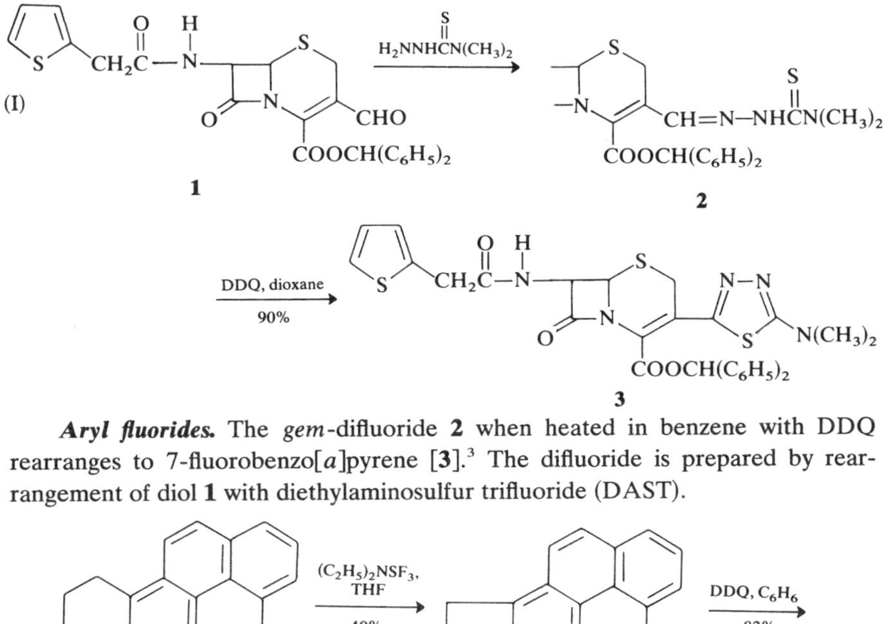

(I) CH_2C—N— CHO $COOCH(C_6H_5)_2$ **1** $\xrightarrow{H_2NNHC(S)N(CH_3)_2}$ $CH{=}N{-}NHC(S)N(CH_3)_2$ $COOCH(C_6H_5)_2$ **2**

DDQ, dioxane
90%

CH_2C—N— N—N $N(CH_3)_2$ $COOCH(C_6H_5)_2$ **3**

Aryl fluorides. The *gem*-difluoride **2** when heated in benzene with DDQ rearranges to 7-fluorobenzo[*a*]pyrene [**3**].[3] The difluoride is prepared by rearrangement of diol **1** with diethylaminosulfur trifluoride (DAST).

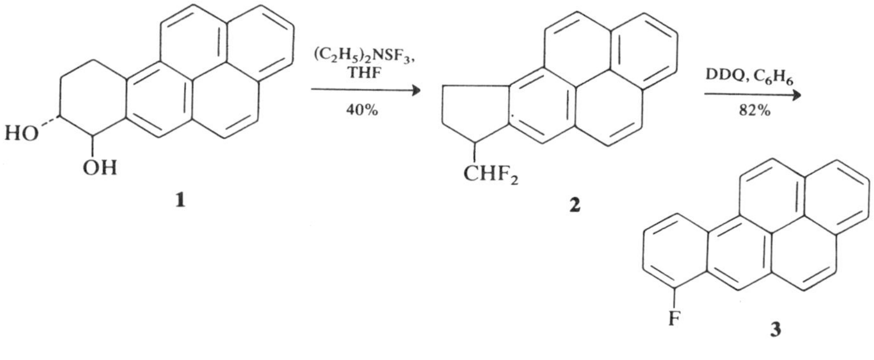

HO OH **1** $\xrightarrow[40\%]{(C_2H_5)_2NSF_3,\ THF}$ CHF_2 **2** $\xrightarrow[82\%]{DDQ,\ C_6H_6}$ F **3**

α,β-*Enones.* Two groups[4,5] have reported the oxidation of silyl enol ethers to α,β-unsaturated ketones using DDQ. This method is most suitable for cyclohexanone derivatives. Fleming and Paterson[6] report that yields are generally improved by addition of collidine (1.5 equiv.), which presumably removes the by-product $DDQH_2$. This last communication also reports improvements in the preparation of either the kinetic or thermodynamic silyl enol ethers by House's method (**3**, 310–311). By a combination of the two reactions, carvone (**6**) can be prepared from 2-methylcyclohexanone (**1**) in an overall yield of 28% (equation I).

(I) **1** → (1) LDA; 2) $ClSi(CH_3)_3$; 97%) → **2** → (DDQ, collidine, C_6H_6, 20°; 73%) → **3** → ($CH_2{=}C(CH_3)MgBr$, CuCl; 85%) → **4** → ($ClSi(CH_3)_3$, $N(C_2H_5)_3$, DMF, 130°, 68 hr.; 87%) → **5** → (DDQ, collidine, C_6H_6, 20°; 52%) → **6**

Other examples[6]:

4-tert-butyl-1-(trimethylsilyloxy)cyclohexene → 4-tert-butylcyclohex-2-enone (84%)

1-(trimethylsilyloxy)cyclopentene → cyclopentenone (20%)

1-(trimethylsilyloxy)cycloheptene → cycloheptenone (19%)

Oxidation of allylic alcohols.[7] A new method for this reaction involves use of a catalytic amount of DDQ in a slightly acidic benzene–water system with periodic acid for regeneration of DDQ from $DDQH_2$.

$$\text{(I)} \quad R^1CH{=}CH\overset{OH}{\overset{|}{C}}HR^2 \xrightarrow[60\text{–}90\%]{DDQ,\ H_5IO_6,\ C_6H_6\text{–}H_2O,\ HCl} R^1CH{=}CH\overset{O}{\overset{\|}{C}}R^2$$

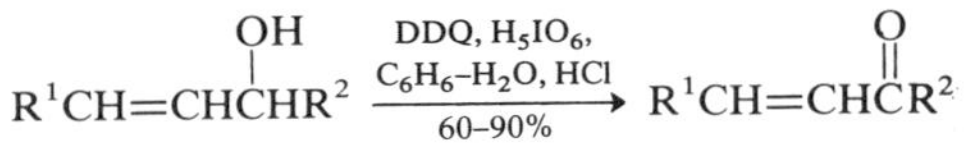

p-Benzoquinone monoketals. McKillop and Taylor have prepared these substances recently by oxidation of 4-alkoxy- or 4-aryloxyphenols with TTN (**7**, 363). Yields in this oxidation can be improved by addition of $KHCO_3$. DDQ or $FeCl_3$ in CH_3OH can also be used, and in some cases is successful where TTN fails.[8]

Examples:

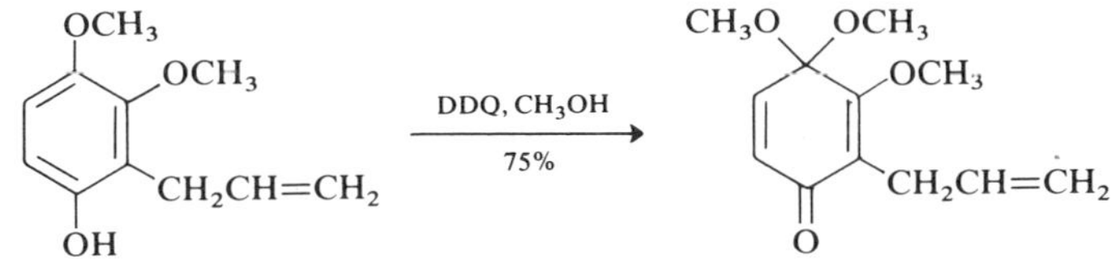

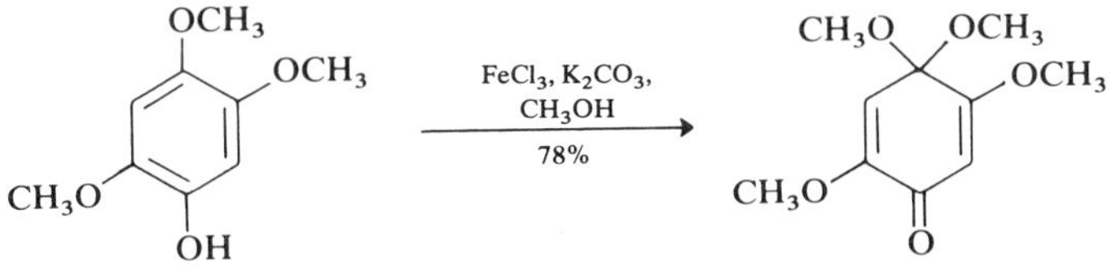

2,3-Dimethoxy-4-allylphenol (ring substituents OCH_3, OCH_3, $CH_2CH{=}CH_2$, OH) $\xrightarrow[75\%]{DDQ,\ CH_3OH}$ 4,4,3-trimethoxy-2-allyl cyclohexadienone (CH₃O OCH₃ ketal; OCH_3; $CH_2CH{=}CH_2$; O)

OCH_3, OCH_3, CH_3O, OH substituted phenol $\xrightarrow[78\%]{FeCl_3,\ K_2CO_3,\ CH_3OH}$ CH_3O OCH_3 ketal; OCH_3; CH_3O; O

Dehydrogenation of Δ^2-isoxazolines. The Δ^2-isoxazoline **1** is transformed into the isoxazole **2** by reaction with 3 equiv. of DDQ in refluxing benzene or dioxane in 83–90% yield.[9] The yields are higher than those reported using NBS, but comparable to those with MnO_2.[10] However, DDQ is more available than MnO_2.

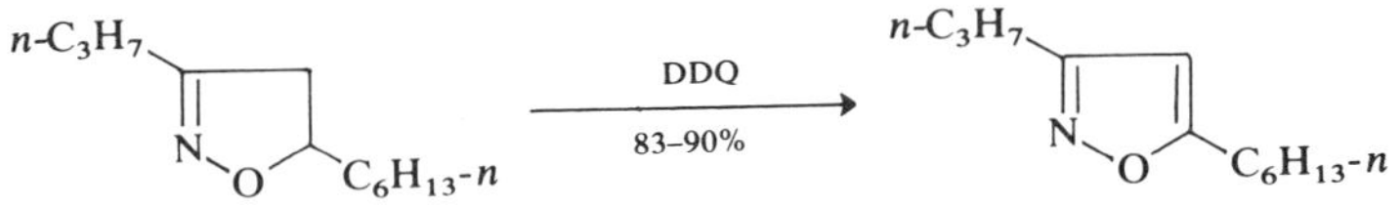

3-(n-C_3H_7)-5-(C_6H_{13}-n)-Δ^2-isoxazoline (**1**) $\xrightarrow[83\text{–}90\%]{DDQ}$ 3-(n-C_3H_7)-5-(C_6H_{13}-n)-isoxazole (**2**)

[1] S. Ghosal and S. Banerjee, *J.C.S. Chem. Comm.*, 165 (1979).
[2] T. Sugawara, H. Masuya, T. Matsuo, and T. Miki, *Chem. Pharm. Bull. Japan*, **27**, 2544 (1979).
[3] M. S. Newman, V. K. Khanna, and K. Kanakarajan, *Am. Soc.*, **101**, 6788 (1979).
[4] M. E. Jung, Y.-G. Pan, M. W. Rathke, D. F. Sullivan, and R. P. Woodbury, *J. Org.*, **42**, 3961 (1977).
[5] I. Ryu, S. Murai, Y. Hatayama, and N. Sonoda, *Tetrahedron Letters*, 3455 (1978).
[6] I. Fleming and I. Paterson, *Synthesis*, 736 (1979).
[7] S. Cacchi, F. La Torre, and G. Paolucci, *ibid.*, 848 (1978).
[8] G. Büchi, P.-S. Chu, A. Hoppmann, C.-P. Mak, and A. Pearce, *J. Org.*, **43**, 3983 (1978).
[9] G. Bianchi and M. De Amici, *J. Chem. Res.* (*S*), 311 (1979).
[10] A. Barco, S. Benetti, G. Pollini, P. G. Baraldi, M. Guarneri, and C. B. Vincenti, *J. Org.*, **44**, 105 (1979).

Dichlorodiethylaminoborane, $(C_2H_5)_2NBCl_2$. Mol. wt. 153.85.

Directed aldol condensation. This borane converts ketones into vinyloxyaminochloroboranes, some of which are stable, isolable compounds such as **1**, formed from cyclohexanone.[1] These products react with aldehydes at room temperature to form a mixture of aldols (**2**), in which the *erythro*-isomer generally predominates.

The reaction can be carried out as a one-pot procedure. The ketone, an aldehyde, the borane, and 2 equiv. of triethylamine are stirred for 20 hours in CH_2Cl_2.[2]

$(C_2H_5)_2NBCl_2$, $N(C_2H_5)_3$, 56%; **1**; RCHO, CH_2Cl_2; **a**; $N(C_2H_5)_3$; **b**; H_3O^+, 50–75%; **2**

[1] K. Niedenzu and J. W. Dawson, *Am. Soc.*, **81**, 3561 (1959).
[2] T. Sugaswa, T. Toyoda, and K. Sasakura, *Syn. Comm.*, **9**, 583 (1979).

Dichloroketene, 1, 221–222; **2**, 118; **3**, 87–88; **4**, 134–135; **8**, 156.

Cycloaddition to silyl enol ethers. As expected, dichloroketene, generated from trichloroacetyl chloride and activated zinc, reacts with silyl enol ethers to form 3-silyloxy-2,2-dichlorocyclobutanones, usually in good yield. Cycloaddition does not obtain with dichloroketene generated from dichloroacetyl chloride and triethylamine. In some cases only acyclic products are formed, but these may arise by ring opening of intermediate cyclobutanones.[1]

Examples:

$Cl_2C{=}C{=}O$, 92%; H_3O^+, 83%

81%; 85%

Cycloaddition to alkynes.[2] Dichloroketene, generated from trichloroacetyl chloride by zinc (**8**, 156), adds to alkynes to give cyclobutenones. The reaction occurs preferentially with the triple bond of enynes.

Examples:

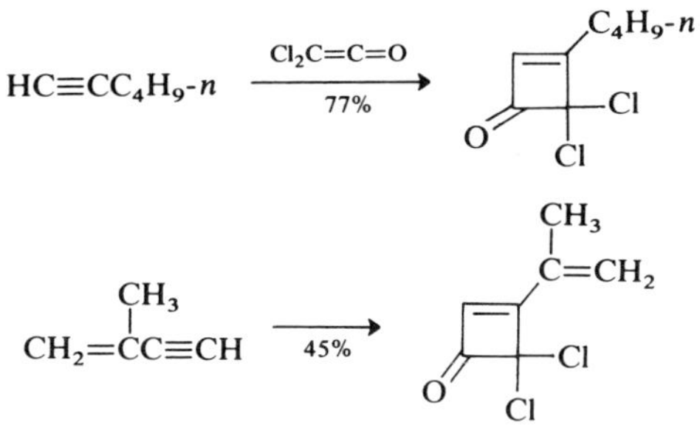

$$HC{\equiv}CC_4H_9\text{-}n \xrightarrow[77\%]{Cl_2C{=}C{=}O}$$ 3-n-butyl-4,4-dichlorocyclobut-2-enone (C_4H_9-n, Cl, Cl, O)

$$CH_2{=}C(CH_3)C{\equiv}CH \xrightarrow{45\%}$$ 4,4-dichloro-3-(CH_3)C$=CH_2$-cyclobut-2-enone

Ketene Claisen rearrangement.[3] Dichloroketene (**1**) reacts with the allylic sulfide **2** to give the expected [2+2] cycloadduct (**3**) and the S-phenyl ester of the unsaturated acid (**4**). The latter product is evidently formed by a Claisen rearrangement of a 1,3-dipolar intermediate (**a**). Products related to **4** are also obtained from allylic ethers and selenides; in fact, in the case of allylic ethers only the rearranged esters are usually isolable.

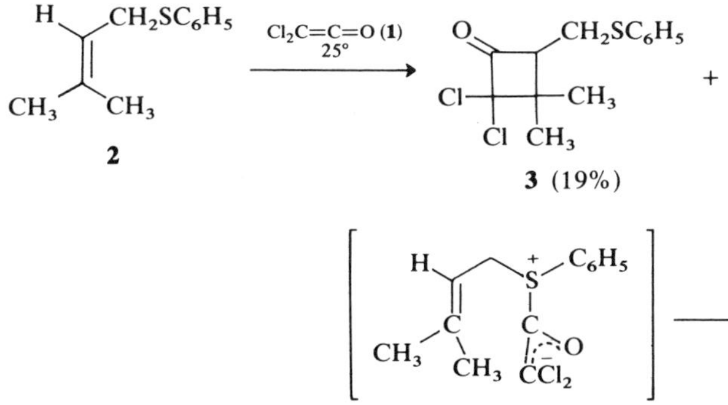

2 ($(CH_3)_2C{=}CHCH_2SC_6H_5$) $\xrightarrow[25°]{Cl_2C{=}C{=}O\ (\mathbf{1})}$ **3** (19%) +

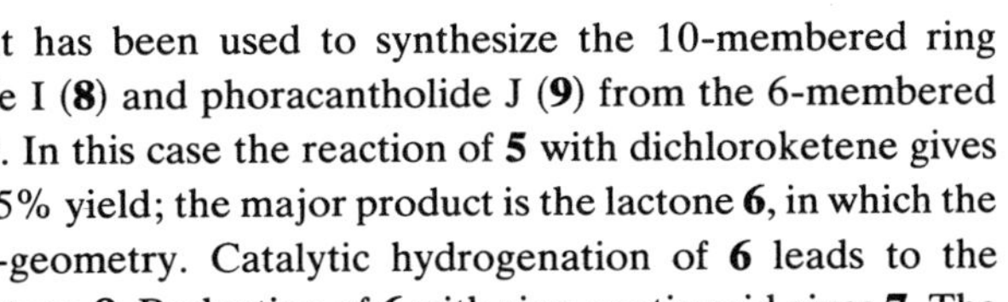

a $\longrightarrow$ **4** (26%)

This new rearrangement has been used to synthesize the 10-membered ring macrolides phoracantholide I (**8**) and phoracantholide J (**9**) from the 6-membered α-vinyl substituted ether **5**. In this case the reaction of **5** with dichloroketene gives the [2+2] cycloadduct in 15% yield; the major product is the lactone **6**, in which the double bond has the (E)-geometry. Catalytic hydrogenation of **6** leads to the saturated dechlorinated lactone **8**. Reduction of **6** with zinc–acetic acid gives **7**. The final step in the synthesis of **9** involved (E)- to (Z)-isomerization, conducted by irradiation in the presence of a catalytic amount of diphenyl disulfide.[4]

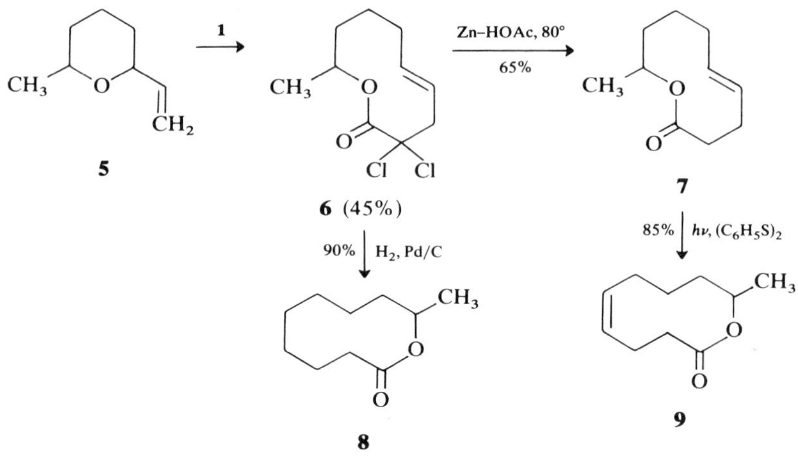

5 →(1) 6 (45%) →(Zn–HOAc, 80°; 65%) 7

6 →(90%; H_2, Pd/C) 8

7 →(85%; $h\nu$, $(C_6H_5S)_2$) 9

[1] L. R. Krepski and A. Hassner, *J. Org.*, **43**, 3173 (1978).
[2] A. Hassner and J. Dillon, *Synthesis*, 689 (1979).
[3] R. Malherbe and D. Belluš, *Helv.*, **61**, 3096 (1978).
[4] K. H. Schulte-Elte and G. Ohloff, *Helv.*, **51**, 548 (1968).

Dichloromethyllithium, 1, 223–224; **2**, 119; **3**, 89; **4**, 138–139; **5**, 199–200; **6**, 170–171.

1-Alkynes. Primary alkyl halides react with dichloromethyllithium in the presence of HMPT (1 equiv.) to form homologated 1,1-dichloroalkanes in 60–90% yield. Several methods have been used for dehydrochlorination of the products. A new method is the reaction with 3 equiv. of n-butyllithium to form a lithium acetylide.[1]

$$RCH_2X \xrightarrow[60\text{–}90\%]{LiCHCl_2} RCH_2CHCl_2 \xrightarrow[30\text{–}95\%]{1)\ n\text{-BuLi};\ 2)\ H_3O^+} RC{\equiv}CH$$

[1] J. Villieras, P. Perriot, and J. F. Normant, *Bull. soc.*, 765 (1977); *idem, Synthesis*, 502 (1979).

Dichloromethyl methyl ether, 2, 120; **5**, 200–203; **7**, 98–99.

Formylation of unsaturated silanes. α,β-Unsaturated aldehydes can be prepared by reaction of vinylsilanes with dichloromethyl methyl ether in the presence of 1 equiv. of $TiCl_4$ (equation I).[1] The enals have the (E)-configuration, regardless of

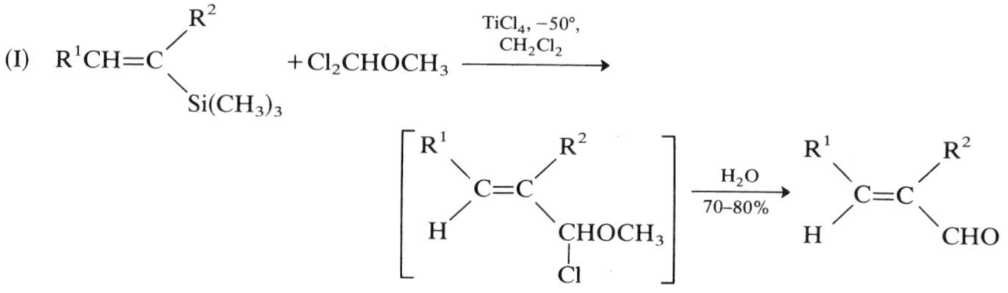

$$\text{(I)}\quad R^1CH{=}C(R^2)Si(CH_3)_3 + Cl_2CHOCH_3 \xrightarrow{TiCl_4,\ -50°,\ CH_2Cl_2} \left[(R^1)(H)C{=}C(R^2)CHClOCH_3\right] \xrightarrow[70\text{–}80\%]{H_2O} (R^1)(H)C{=}C(R^2)CHO$$

stereochemistry of the starting material. This reaction was used in a synthesis of nuciferal (**1**), a sesquiterpene aldehyde.

$$\text{(p-tolyl)CH(CH}_3\text{)CH}_2\text{CH}_2\text{CH=C(Si(CH}_3)_3\text{)CH}_3 \xrightarrow[48\%]{Cl_2CHOCH_3,\ TiCl_4} \text{(p-tolyl)CH(CH}_3\text{)CH}_2\text{CH}_2\text{CH=C(CHO)CH}_3$$

1

The reaction is not applicable to allyltrimethylsilanes, but it can be used to obtain conjugated dienals from 1-trimethylsilyl-1,3-dienes (equation II). These dienes can be prepared from reaction of the anion of allyltrimethylsilane with aldehydes followed by dehydration (equation III). They probably have the (E,E)-configuration.[2]

(II) $$RCH=CH-CH=CHSi(CH_3)_3 + Cl_2CHOCH_3 \xrightarrow{TiCl_4,\ CH_2Cl_2} \xrightarrow[50-70\%]{H_2O} RCH=CH-CH=CHCHO$$

(III) $$(CH_3)_3SiCH_2CH{=}CH_2 \xrightarrow{n\text{-BuLi, TMEDA}} \left[(CH_3)_3SiCH{\cdots}\bar{C}H{\cdots}CH_2\ Li^+\right] \xrightarrow{RCHO}$$

$$(CH_3)_3SiCH{=}CHCH_2CH(OH)R \xrightarrow[45-55\%]{1)\ CH_3SO_2Cl;\ 2)\ DBU} (CH_3)_3SiCH{=}CH-CH{=}CHR$$

α-Keto acid chlorides (**6**, 172). Detailed directions for preparation of pyruvoyl chloride (equation I) are available.[3]

(I) $$CH_3COCOOH + Cl_2CHOCH_3 \xrightarrow[-HCl]{20-50^\circ} \underset{42-54\%}{CH_3COCOCl} + HCOOCH_3$$

A two-step synthesis of this acid chloride has been reported (equation II).[4]

(II) $$CH_3COCOOH \xrightarrow[87\%]{ClSi(CH_3)_3,\ N(C_2H_5)_3} CH_3COCOOSi(CH_3)_3 \xrightarrow[50-80\%]{(COCl)_2} CH_3COCOCl$$

Formylation of arenes (**1**, 220; **2**, 120). Yields in this reaction can be improved by addition of the arene to a mixture of the ether (25% excess) and stannic chloride (2 equiv.). This inverse addition decreases formation of by-products of the type

Ar_2CH_2 and Ar_3CH. In one case the yield of the aldehyde by the original method of 50% was increased to 75%.[5]

[1] K. Yamamoto, O. Nunokawa, and J. Tsuji, *Synthesis*, 721 (1977); K. Yamamoto, J. Yoshitake, N. T. Qui, and J. Tsuji, *Chem. Letters*, 859 (1978).
[2] K. Yamamoto, M. Ohta, and J. Tsuji, *ibid.*, 713 (1979).
[3] H. C. J. Ottenheijm and M. W. Tijhuis, *Org. Syn.*, submitted (1979).
[4] J. Hausler and U. Schmidt, *Ber.*, **107**, 145 (1974).
[5] A. H. Lewin, S. R. Parker, N. B. Fleming, and F. I. Carroll, *Org. Prep. Proc. Int.*, **10**, 201 (1978).

N,N-Dichlorourethane, 2, 121–122; **4,** 139; **8,** 161.

Chlorination of tertiary carbon atoms (**8,** 161). Radical chlorination of the trifluoroacetate **1** with N,N-dichlorourethane gives as one product the 25-chloro derivative **2,** which can be converted into the 25-hydroxy derivative **3** by refluxing aqueous ethanol.[1] The corresponding 25-hydroxy-8-ketone has been converted into 25-hydroxyvitamin D_3.[2] This method for hydroxylation of C_{25} is somewhat more selective than dry ozonation or peracetic acid oxidation, both of which hydroxylate all four tertiary carbons, but the yield of **3** is about the same as that obtained directly by dry ozonation.

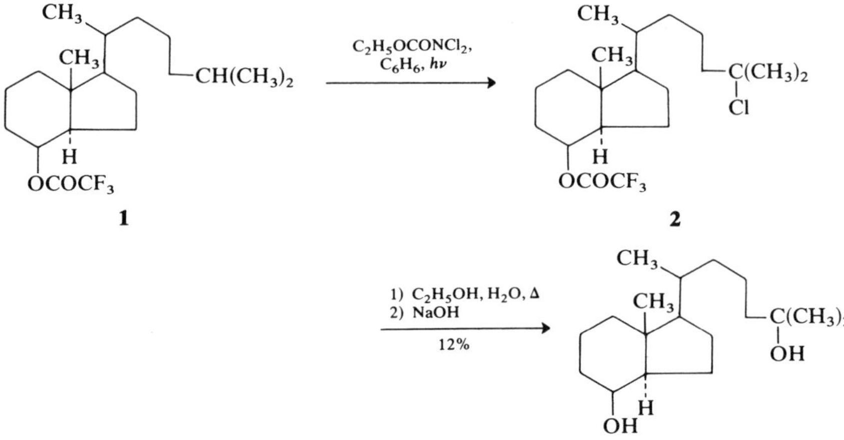

1 → ($C_2H_5OCONCl_2$, C_6H_6, $h\nu$) → **2** → (1) C_2H_5OH, H_2O, Δ; 2) NaOH; 12%) → **3**

[1] Z. Cohen, E. Berman, and Y. Mazur, *J. Org.*, **44**, 3077 (1979).
[2] B. Lythgoe, D. A. Rogerts, and I. Waterhouse, *J.C.S. Perkin I*, 1608 (1977).

Dicyclohexylcarbodiimide (DCC), 1, 231–236; **2,** 126; **3,** 91; **4,** 141; **5,** 206–207; **6,** 174; **7,** 100–101; **8,** 162–163.

Esterification. DCC has been satisfactory only for esterification of phenols and thiophenols because of variable yields. However, if 4-dimethylaminopyridine (**3,** 118–119; **5,** 26) or 4-pyrrolidinopyridine (this volume) is added as catalyst, alcohols and thiols are esterified readily at room temperature in satisfactory yields. Sterically hindered esters can be formed; however, *t*-butyl esters of very hindered acids are obtained, if at all, in low yield. Some racemization is observed on esterification of CBZ-amino acids.[1,2]

Use of pyridine as solvent also promotes esterification with DCC. Further increase of yield is observed in the presence of a catalytic amount of a strong acid such as *p*-TsOH. Esters of primary and secondary alcohols as well as of phenols are generally obtainable in high yields.[3]

[1] B. Neises and W. Steglich, *Angew. Chem., Int. Ed.*, **17**, 522 (1978).
[2] A. Hassner and V. Alexanian, *Tetrahedron Letters*, 4475 (1978).
[3] K. Holmberg and B. Hansen, *Acta Chem. Scand.*, **B33**, 410 (1979).

2,2-Diethoxyethylidenetriphenylphosphorane (**1**). Mol. wt. 376.44.

Preparation:

$$(C_6H_5)_3P{=}CHCHO + C_2H_5Br \rightarrow (C_6H_5)_3\overset{+}{P}{-}C(H){=}C(H)(OC_2H_5)\ Br^-$$

$$(C_6H_5)_3\overset{+}{P}{-}C(H){=}C(H)(OC_2H_5)\ Br^- \xrightarrow{NaNH_2,\ NH_3} (C_6H_5)_3P{=}C{=}C(H)(OC_2H_5) \xrightarrow{C_2H_5OH,\ (C_2H_5)_2O} \underset{\mathbf{1}}{(C_6H_5)_3P{=}CHCH(OC_2H_5)_2}$$

$$(C_6H_5)_3\overset{+}{P}{-}C(H){=}C(H)(OC_2H_5)\ Br^- \xrightarrow{NaOC_2H_5,\ THF} \mathbf{1}$$

The yields of Wittig olefination products are generally higher when **1** is prepared by the three-step procedure formulated.

(Z)-α,β-Unsaturated aldehydes.[1] The reagent (**1**) condenses with aldehydes in a (Z)-stereoselective Wittig reaction to afford diethyl acetals of (Z)-α,β-unsaturated aldehydes in 57–86% yield. Hydrolysis of the acetals with *p*-TsOH as catalyst in acetone–water or with moist silica gel (2 days at 23°) affords the corresponding (Z)-unsaturated aldehydes in 47–98% yield. The product usually contains 4–14% of the (E)-isomer. These results contrast with those of Wittig reactions of formylmethylenetriphenylphosphorane and 1,3-dioxolan-2-ylmethylenetriphenylphosphorane (**5**, 269), which afford (E)-unsaturated aldehydes.

Examples:

$$CH_3\overset{O}{\overset{\|}{C}}O{-}(CH_2)_8{-}CHO \xrightarrow[70\%]{\mathbf{1},\ THF} CH_3\overset{O}{\overset{\|}{C}}O{-}(CH_2)_8{-}C(H){=}C(H){-}CH(OC_2H_5)_2 \xrightarrow[87\%]{H_3O^+,\ 0^\circ}$$

$$CH_3\overset{O}{\overset{\|}{C}}O{-}(CH_2)_8{-}C(H){=}C(H){-}CHO\ (Z) + CH_3\overset{O}{\overset{\|}{C}}O{-}(CH_2)_8{-}C(H){=}C(H){-}CHO\ (E)$$

97 : 3

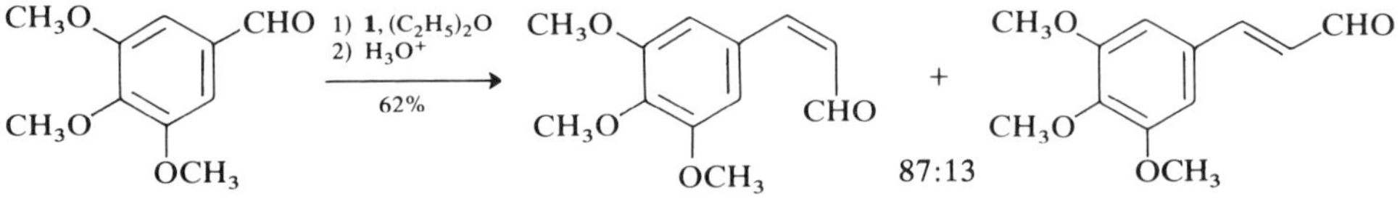

[1] H. J. Bestmann, K. Roth, and M. Ettlinger, *Angew. Chem., Int. Ed.*, **18**, 687 (1979).

3,3-Diethoxy-1-propene. Mol. wt. 130.19, b.p. 125°. Suppliers: Aldrich, Fluka.

Ethoxyallylation. Treatment of the cyano ketone **1** with this acetal in refluxing benzene furnishes **2** in 76% yield.[1] This transformation probably involves a Claisen rearrangement of the mixed acetal **a**.

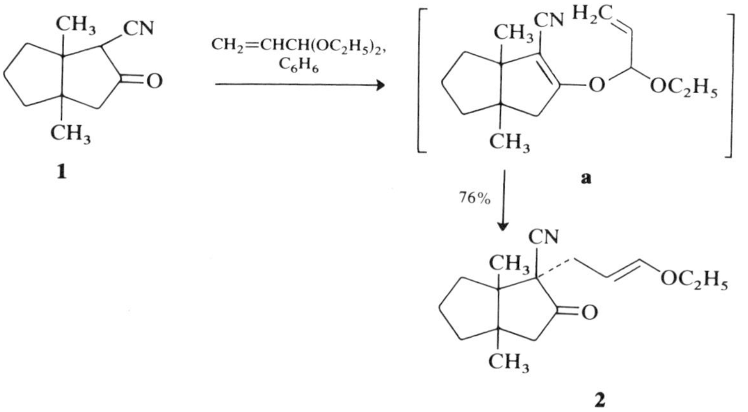

[1] R. M. Coates, S. K. Shah, and R. W. Mason, *Am. Soc.*, **101**, 6765 (1979).

Diethylaluminium cyanide, 1, 244; **2**, 127–128; **4**, 146–147; **6**, 180–181.

Hydrocyanation. Although 2-carbomethoxy-2-cyclopentene-1-one (**2**) is prone to polymerization, it reacts smoothly with diethylaluminium cyanide to give the adduct **3**. This product has the carbon skeleton of sarkomycin (**7**), an antibiotic active against ascites-type tumors. Selective manipulation of the functional groups resulted in the first regiospecific synthesis of **7**.[1]

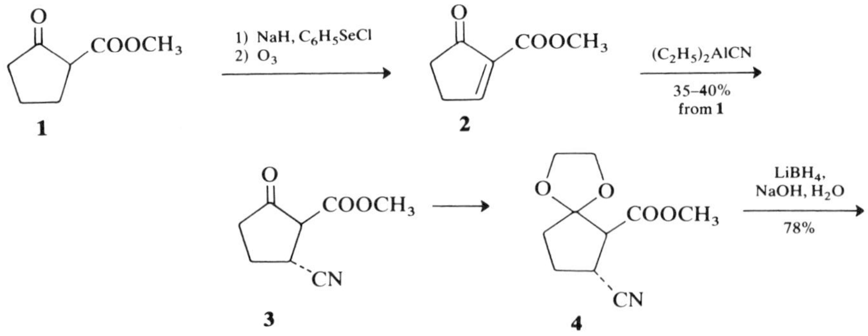

$$\mathbf{5} \xrightarrow[91\%]{\text{NaOH}} \mathbf{6} \xrightarrow{H_3O^+} \mathbf{7}$$

[1] J. N. Marx and G. Minaskanian, *Tetrahedron Letters*, 4175 (1979).

N,N-Diethylaminoacetonitrile (**1**). Mol. wt. 112.18, b.p. 61–63°/14 mm.

Preparation[1]:

$$(C_2H_5)_2NH + CH_2O + NaCN \xrightarrow[90\%]{NaHSO_3} \underset{\mathbf{1}}{(C_2H_5)_2NCH_2CN}$$

Synthesis of aldehydes and ketones.[2] This substance can function as a protected cyanohydrin of formaldehyde in an extension of Stork and Maldonado's synthesis of ketones from aldehydes, RCHO → RCOR′, by way of cyanohydrins (**4**, 300–301). Thus the anion of **1**, generated with LDA, does not undergo self-condensation, but can be alkylated; the product on hydrolysis gives the homologous aldehyde of the alkyl halide. Thus the reagent serves as the latent anion of formaldehyde, $H\bar{C}{=}O$. An example is shown in equation (I).

$$\text{(I)}\quad \mathbf{1} \xrightarrow[90\%]{\text{1) LDA, THF; 2) } CH_3(CH_2)_7I} CH_3(CH_2)_7CH(N(C_2H_5)_2)(CN) \xrightarrow[90\%]{H_2O,\ HOOCCOOH} CH_3(CH_2)_7CHO$$

Ketones can be prepared by dialkylation of **1**; in this case the resulting aminonitrile is hydrolyzed by $CuSO_4$ in 95% ethanol. They can also be prepared by monoalkylation of **1** followed by Michael addition to an enone, as illustrated in equation (II).

$$\text{(II)}\quad \mathbf{1} \xrightarrow{\text{1) LDA; 2) } ICH_2CH_2CH{=}CHC_2H_5} \left[C_2H_5CH{=}CHCH_2CH_2CH(N(C_2H_5)_2)(CN)\right] \xrightarrow{\text{1) LDA; 2) } CH_3COCH{=}CH_2}$$

$$C_2H_5CH{=}CHCH_2CH_2C(N(C_2H_5)_2)(CN)CH_2CH_2COCH_3 \xrightarrow[\text{72\% overall}]{CuSO_4,\ C_2H_5OH} C_2H_5CH{=}CHCH_2CH_2\overset{O}{\overset{\|}{C}}CH_2CH_2\overset{O}{\overset{\|}{C}}CH_3$$

This synthesis is compatible with several variations. Since acetal groups are stable to the conditions, keto aldehydes can be prepared in this way. The second alkylation can also be carried out with tosylates and with epoxides.

[1] C. F. H. Allen and J. A. VanAllen, *Org. Syn. Coll. Vol.*, **3**, 275 (1955).
[2] G. Stork, A. A. Ozorio, and A. Y. W. Leong, *Tetrahedron Letters*, 5175 (1978).

Diethyl azodicarboxylate (DAD), 1, 245–247; **2,** 128–129; **4,** 148–149; **5,** 212–213; **6,** 185.

5-Deazaflavins.[1] A new synthesis of 5-deazaflavins (**2**), in which N_5 of a flavin is replaced by CH, involves oxidative cyclization of 5-benzylidene-6-(N-substituted amino)uracils (**1**) with DAD at 150°.

$$\mathbf{1} \xrightarrow[45\text{–}70\%]{\text{DAD, }150^\circ} \mathbf{2}$$

[1] K. Mori, K. Shinozuka, Y. Sakuma, and F. Yoneda, *J.C.S. Chem. Comm.*, 764 (1978).

Diethyl 2,2-dichloro-1-ethoxyvinyl phosphate, $Cl_2C{=}C(OP(O)(OC_2H_5)_2)(OC_2H_5)$ (**1**). Mol. wt. 293.09. The phosphate is prepared by the Perkow reaction of ethyl trichloroacetate and triethyl phosphite (83% yield).

α-Chloro-α,β-unsaturated esters. These products can be prepared from **1** by lithiation followed by reaction with aldehydes and ketones (equation I).[1]

$$\text{(I)}\quad Cl_2C{=}C(OP(O)(OC_2H_5)_2)(OC_2H_5)\ \mathbf{1} \xrightarrow[50\text{–}85\%]{1)\ n\text{-BuLi},\ -78^\circ;\ 2)\ R^1COR^2} R^1R^2C{=}C(Cl)COOC_2H_5\ \mathbf{2}$$

[1] F. Karrenbrock and H. J. Schäfer, *Tetrahedron Letters*, 2913 (1979).

S,S′-Diethyl dithiomalonate, $CH_2(COSC_2H_5)_2$ (**1**). Mol. wt. 192.30, b.p. 135°/10 mm. The reagent is prepared in 95% yield by reaction of malonyl dichloride and ethanethiol in ether.

Hydroxyethylation.[1] This reaction can be carried out by substitution of the anion of **1** with electrophiles to give β-keto thiolesters, which undergo dealkylthiocarbonylation on treatment with Raney nickel (equation I).

$$\text{(I)}\quad \mathbf{1} + RX \xrightarrow{NaH} RCH(COSC_2H_5)_2 \xrightarrow{\text{Raney Ni}} RCH_2CH_2OH$$

Examples:

$$CH_3(CH_2)_4CH_2I \xrightarrow[78\%]{\mathbf{1},\ NaH} CH_3(CH_2)_5CH(COSC_2H_5)_2 \xrightarrow[84\%]{\text{Raney Ni}} CH_3(CH_2)_7OH$$

$$Br(CH_2)_5Br \xrightarrow{91\%} \text{(cyclohexane)}(COSC_2H_5)_2 \xrightarrow{89\%} \text{(cyclohexyl)}-CH_2OH$$

$$C_6H_5CH_2CH(COSC_2H_5)_2 \xrightarrow[100\%]{CH_3I} (C_6H_5CH_2)(CH_3)C(COSC_2H_5)_2 \xrightarrow{66\%} (C_6H_5CH_2)(CH_3)CHCH_2OH$$

[1] H.-J. Liu and H. K. Lai, *Can. J. Chem.*, **57**, 2522 (1979).

Diethyl ethylidenemalonate, $CH_3CH{=}C(COOC_2H_5)_2$ (**1**). Mol. wt. 186.21, b.p. 115–118°/17 mm. Preparation.[1] Supplier: Aldrich, Fluka.

5-Arylpenta-2,4-dienoates.[2] The reagent can be used for stereospecific synthesis of (2Z, 4E)- and (2E, 4E)-isomers of this system (scheme I).

$$ArCHO + \mathbf{1} \xrightarrow{C_6H_5CH_2\overset{+}{N}(CH_3)_3OH^-} \left[\text{6-Ar-3-COOCH}_3\text{-dihydropyran-2-one}\right] \xrightarrow{66\%} Ar{-}CH{=}CH{-}CH{=}C(COOH)(COOCH_3)$$

$[Ar = C_6H_3(OCH_3)_2\text{-}3,5]$

$$\xrightarrow[66\%]{KOH} Ar{-}CH{=}CH{-}CH{=}C(COOH)_2 \xrightarrow[79\%]{\text{Quinoline, }\Delta} Ar{-}CH{=}CH{-}CH{=}CH{-}COOH \xrightarrow{h\nu,\ I_2} Ar{-}CH{=}CH{-}CH{=}CH{-}COOH$$

$$\xrightarrow[52\%]{Py,\ \Delta} Ar{-}CH{=}CH{-}CH{=}CH{-}COOCH_3$$

Scheme (I)

[1] W. S. Fones, *Org. Syn. Coll. Vol.* **4**, 293 (1963).
[2] L. Crombie, W. M. L. Crombie, G. W. Kilbee, and P. Tuchinda, *Tetrahedron Letters*, 4773 (1979).

Diethyl lithio-N-benzylideneaminomethylphosphonate, $(C_2H_5O)_2\overset{O}{\overset{\|}{P}}\overset{-}{C}HN{=}CHC_6H_5\ Li^+$ (**1**). The phosphonate is obtained by lithiation of the corresponding Wittig–Horner reagent.[1]

***gem*-Substitution of C=O.**[2] The reagent reacts with carbonyl compounds to form 2-azadienes (**2**), which can be isolated but usually are converted directly into metalloenamines (**3**), which react readily with electrophiles, as shown in the examples.

$$R^1R^2C{=}O \xrightarrow{\mathbf{1}} R^1R^2C{=}C(H)N{=}CHC_6H_5 \ (\mathbf{2}) \xrightarrow{n\text{-BuLi}} R^1R^2C{=}C(H)N(Li){-}CH(n\text{-Bu})C_6H_5 \ (\mathbf{3})$$

Examples:

$$\text{cyclohexanecarboxaldehyde (CHO)} \longrightarrow \mathbf{3} \xrightarrow[\text{43\% overall}]{CH_3I} \text{cyclohexyl-}CH(CH_3)CHO$$

$$\text{cyclohexanone} \longrightarrow \mathbf{3} \xrightarrow[\text{51\% overall}]{CH_2{=}CHCH_2Br} \text{1-(}CH_2CH{=}CH_2\text{)cyclohexane-1-CHO}$$

A variation of this sequence involves use of the homologous reagent **4**, as shown in equation (I).

$$\text{(I)} \quad C_6H_{11}(H)C{=}O + (C_2H_5O)_2\overset{O}{\overset{\|}{P}}\bar{C}(CH_3)N{=}CHC_6H_5 \ Li^+ \ (\mathbf{4}) \longrightarrow$$

$$C_6H_{11}(H)C{=}C(CH_3)N{=}CHC_6H_5 \ (\mathbf{5}) \xrightarrow[\text{41\% overall}]{\text{1) } n\text{-BuLi; 2) } CH_3I} \text{cyclohexyl-}CH(CH_3)COCH_3 \ (\mathbf{6})$$

[1] R. W. Ratcliffe and B. G. Christensen, *Tetrahedron Letters*, 4645 (1973).
[2] S. F. Martin and G. W. Phillips, *J. Org.*, **43**, 3792 (1978).

Diethyl methoxyethoxymethylphosphonate, $(C_2H_5O)_2P(O)CH_2OCH_2CH_2OCH_3$ (**1**). Mol. wt. 226.21, b.p. 90–91°/0.2 mm. The phosphonate is obtained in 77% yield by the reaction of methoxyethoxymethyl chloride (**7**, 227–229) with triethyl phosphite (3 hours at 155°).

Homologated aldehydes.[1] Unlike diethyl methoxymethylphosphonate, $(C_2H_5O)_2P(O)CH_2OCH_3$, which does not undergo Wittig–Horner reactions,[2] the anion of **1** reacts readily with both aldehydes and ketones to give 1,2-adducts (**2**) in high yield. These products can be converted into enol ethers of aldehydes (**3**) either by heat or by treatment with potassium *t*-butoxide.

$$\mathbf{1} \xrightarrow[65-90\%]{1)\ LDA, THF;\ 2)\ R^1COR^2} R^1-\underset{R^2}{\overset{HO}{C}}-\overset{O=P(OC_2H_5)_2}{CH}OCH_2CH_2OCH_3 \xrightarrow[60-85\%]{(CH_3)_3COK}$$

2

$$\underset{\mathbf{3}}{(R^1)(R^2)C=CH(OCH_2CH_2OCH_3)} \xrightarrow[20-80\%]{H_2O,\ HClO_4} \underset{\mathbf{4}}{(R^1)(R^2)CHCHO}$$

$$(C_4H_9O)_2P(O)CH_2OTHP$$

6

Kluge has also prepared the related phosphonate **6** by tetrahydropyranylation of di-*n*-butyl hydroxymethylphosphonate. This reagent has the advantage over **1** in that the enol ethers corresponding to **3** are very easily hydrolyzed by acid.

[1] A. F. Kluge, *Tetrahedron Letters*, 3629 (1978).
[2] M. Green, *J. Chem. Soc.*, 1324 (1963).

Diethyl phosphorocyanidate, 5, 217; **6**, 192–193; **7**, 107.

C-Acylation. Active methylene compounds can be C-acylated directly by carboxylic acids with diethyl phosphorocyanidate (**1**) in combination with triethylamine. DMF is the preferred solvent; TMEDA or DBU can replace triethylamine.[1]

Examples:

$$C_6H_5COOH + CH_2(CN)(COOC_2H_5) \xrightarrow[93\%]{(C_2H_5O)_2P(O)CN\ (\mathbf{1}),\ N(C_2H_5)_3} C_6H_5COCH(CN)(COOC_2H_5)$$

$$CH_3C(OCH_2CH_2O)CH_2CH_2COOH + H_2C(COOC_2H_5)_2 \xrightarrow[58\%]{} CH_3C(OCH_2CH_2O)CH_2CH_2COCH(COOC_2H_5)_2$$

$$\xrightarrow{1)\ CH_2{=}CHCH_2I,\ Bu_4NOH;\ 2)\ NaCl,\ DMSO;\ 3)\ HCl,\ CH_3OH} CH_3\overset{O}{\overset{\|}{C}}CH_2CH_2\overset{O}{\overset{\|}{C}}CH_2CH_2CH{=}CH_2$$

$$C_6H_5CH_2\underset{}{CH}(NHCOOCH_2C_6H_5)COOH + H_2C(CN)(COOC_2H_5) \xrightarrow[87.8\%]{} C_6H_5CH_2CH(NHCOOCH_2C_6H_5)COCH(CN)(COOC_2H_5)$$

Strecker α-amino nitrile synthesis.[2] This synthesis can be conducted by reaction of ketones and an amine and this reagent for introduction of the CN group (equation I).[3]

$$\text{(I)}\quad R^1R^2C{=}O + R^3NHR^4 + (C_2H_5O)_2P(O)CN \xrightarrow[75\text{–}100\%]{\text{THF, 20–60°}} R^1R^2C(CN)NR^3R^4$$

[1] T. Shioiri and Y. Hamada, *J. Org.*, **43**, 3631 (1978).
[2] A. Strecker, *Ann.*, **75**, 27 (1850); **91**, 349 (1854).
[3] S. Harusawa, Y. Hamada, and T. Shioiri, *Tetrahedron Letters*, 4663 (1979).

N,N-Diethyl-1-propynylamine (N,N-diethylaminopropyne), 2, 133–134; **3,** 98; **5,** 217–219; **7,** 107–108.

Cycloaddition to an α,β-unsaturated lactam. Ficini *et al.*[1] have extended cycloaddition reactions of diethylaminoalkynes (**5**, 219; **7**, 107–108) to include the α,β-unsaturated lactam **1**. N,N-Diethylaminobutyne condenses with lactam **1** in the presence of $MgBr_2$ to afford enamine **2**. This product is hydrolyzed and esterified to afford **3** in ~50% overall yield from **1**. The intermediate **3** is converted into the indole alkaloid (±)-dihydroantirhine (**4**) by $LiAlH_4$ reduction and debenzylation.

$$\mathbf{1} \xrightarrow{H_5C_2{-}C{\equiv}C{-}N(C_2H_5)_2,\ MgBr_2,\ THF} \mathbf{2} \xrightarrow[\sim 50\%\ \text{overall from } \mathbf{1}]{1)\ 10\%\ HCl\quad 2)\ CH_2N_2} \mathbf{3} \xrightarrow[80\%]{1)\ LiAlH_4\quad 2)\ Na,\ NH_3} \mathbf{4}$$

Cycloadditions of ynamines to nitroalkenes.[2] 1-Nitrocyclopentene condenses with ynamine **1** in the expected manner (**5**, 219; **7**, 107–108) to afford cyclobutene **2** in 40% yield. This reaction is not general, however, since nitroalkenes **3** and **4**

$$\text{1-nitrocyclopentene} + \underset{\mathbf{1}}{\text{N-pyrrolidinyl}-C\equiv C-C_6H_5} \xrightarrow{40\%} \mathbf{2}\ (\text{bicyclic cyclobutene: } NO_2,\ \text{N-pyrrolidinyl},\ C_6H_5)$$

$$\underset{\mathbf{3}\ (R=CH_3),\ \mathbf{4}\ (R=H)}{H_5C_6CH=C(R)NO_2} + \underset{\mathbf{5}}{(C_2H_5)_2N-C\equiv C-CH_3} \xrightarrow{25\text{–}30\%} \mathbf{6}\ (R=CH_3),\ \mathbf{7}\ (R=H)\ (\text{nitrone: } R,\ =N\rightarrow O,\ CH_3,\ C_6H_5,\ CON(C_2H_5)_2)$$

condense with **5** to afford nitrones **6** and **7** in 25–30% yield. Nitrones are also obtained from the condensation of **3** and **4** with **1**. The report includes a possible mechanism.

[1] J. Ficini, A. Guingant, and J. d'Angelo, *Am. Soc.*, **101**, 1318 (1979).
[2] A. D. DeWit, M. L. M. Pennings, W. P. Trompenaars, D. N. Reinhoudt, S. Harchema, and C. Nevestveit, *J.C.S. Chem. Comm.*, 993 (1979).

Diethyl 1-trimethylsilyloxyethylphosphonate, $CH_3CH(OSi(CH_3)_3)-P(O)(OC_2H_5)_2$ (**1**). The reagent is prepared by reaction of acetaldehyde and diethyl trimethylsilyl phosphite.

Ketone synthesis. Phosphonates of this type serve as acyl anion equivalents.[1] Thus lithiation followed by alkylation gives **2**, which can be hydrolyzed to ketones **3** by base. However, the alkylation must be conducted at $-78\rightarrow 20°$ to prevent a Brook–Wittig rearrangement[2] to **4**.

$$\mathbf{1} \xrightarrow{\text{LDA, }-78°} \underset{\mathbf{a}}{CH_3C(OSi(CH_3)_3)(Li)P(O)(OC_2H_5)_2} \xrightarrow[-78\rightarrow 20°]{RX} \underset{\mathbf{2}}{CH_3C(OSi(CH_3)_3)(R)P(O)(OC_2H_5)_2}$$

$$\mathbf{a} \xrightarrow{20°} \underset{\mathbf{4}}{CH_3C(OH)(Si(CH_3)_3)HP(O)(OC_2H_5)_2}$$

$$\mathbf{2} \xrightarrow[80\text{–}85\%]{NaOH,\ C_2H_5OH} \underset{\mathbf{3}}{CH_3\overset{O}{\overset{\|}{C}}R}$$

The related reagent diethyl trimethylsilyl phosphite, $(C_2H_5O)_2POSi(CH_3)_3$ (**5**), has been used to prepare ketones from aldehydes (equation I).

$$\text{(I)}\quad C_6H_5CHO + \mathbf{5} \xrightarrow[91\%]{} C_6H_5CH(OSi(CH_3)_3)P(O)(OC_2H_5)_2 \xrightarrow[90\%]{1)\ LDA\quad 2)\ CH_3I} C_6H_5C(CH_3)(OSi(CH_3)_3)-P(O)(OC_2H_5)_2 \xrightarrow[86\%]{NaOH,\ C_2H_5OH} C_6H_5C(O)CH_3$$

[1] M. Sekine, M. Nakajima, A. Kume, and T. Hata, *Tetrahedron Letters*, 4475 (1979); *see also* T. Hata, A. Hashizume, M. Nakajima, and M. Sekine, *ibid.*, 363 (1978).
[2] A. G. Brook, *Accts. Chem. Res.*, **7**, 77 (1974).

Difluoromethylenetriphenylphosphorane, 2, 443.

gem-*Difluoroalkenes.* The phosphorane can be generated from CBr_2F_2 and $P(C_6H_5)_3$.[1] It converts aldehydes and activated ketones into *gem*-difluoroalkenes. The reaction, however, is extremely sensitive to moisture. The addition of zinc dust improves the yield considerably over that from reactions with aldehydes and permits use of N,N-dimethylacetamide as solvent. Under these conditions isolated yields of 1,1-difluoroalkenes are ~60%. The products can be reduced selectively by sodium bis(2-methoxyethoxy)aluminium hydride (SMEAH) to 1-monofluoroalkenes in 80–95% yield (equation I).[2]

$$\text{(I)}\quad RCHO \xrightarrow[\sim 60\%]{(C_6H_5)_3P{=}CF_2,\ Zn,\ DMA} RCH{=}CF_2 \xrightarrow[80\text{–}95\%]{SMEAH} RCH{=}CHF \quad (E/Z = 3\text{–}13:1)$$

[1] D. G. Naae and D. J. Burton, *J. Fluorine Chem.*, **1**, 123 (1971/72).
[2] S. Hayashi, T. Nakai, N. Ishikawa, D. J. Burton, D. G. Naae, and H. S. Kesling, *Chem. Letters*, 983 (1979).

2,2-Difluoro-1-tosyloxyvinyllithium, $CF_2{=}CLiOSO_2C_6H_4CH_3$-*p* (**1**). Mol. wt. 242.17. The reagent is prepared *in situ* by reaction of 2,2,2-trifluoroethyl *p*-toluenesulfonate (Aldrich) with LDA (2 equiv.) at −78° (THF).

α-Keto acids. The reagent has been used for bishomologation of carbonyl compounds as a route to α-keto acids (equation I).[1]

$$\text{(I)}\quad \mathbf{1} + R_2C{=}O \xrightarrow{THF,\ 0^\circ} R_2C(OH)-C(OTs){=}CF_2 \xrightarrow[60\text{–}80\%]{H_3O^+} R_2C{=}C(OTs)COOH \xrightarrow[90\text{–}98\%]{OH^-} R_2CHCOCOOH$$

[1] K. Tanaka, T. Nakai, and N. Ishikawa, *Tetrahedron Letters*, 4809 (1978).

Dihydridotetrakis(triphenylphosphine)ruthenium, $H_2Ru[P(C_6H_5)_3]_4$. Mol. wt. 1152.25. Preparation.[1]

Silyl enol ethers.[2] Silyl ethers of primary allyl alcohols are isomerized by this hydride complex of ruthenium at 100–150° (1–20 hours) to a mixture of (Z)- and (E)-isomers of silyl enol ethers in which the former isomer usually predominates.

$$R^1CH{=}C(R^2)CH_2OSi(CH_3)_3 \xrightarrow[60-100\%]{H_2RuL_4} R^1CH_2C(R^2){=}CHOSi(CH_3)_3$$

[1] J. J. Levison and D. S. Robinson, *J. Chem. Soc. A*, 2947 (1970).
[2] H. Suzuki, Y. Koyama, Y. Moro-Oka, and T. Ikawa, *Tetrahedron Letters*, 1415 (1979).

1,2-Dihydro-4,6-dimethyl-2-thioxo-3-pyridinecarbonitrile (1). Mol. wt. 164.2, m.p. 250–264° dec.

Preparation:[1,2]

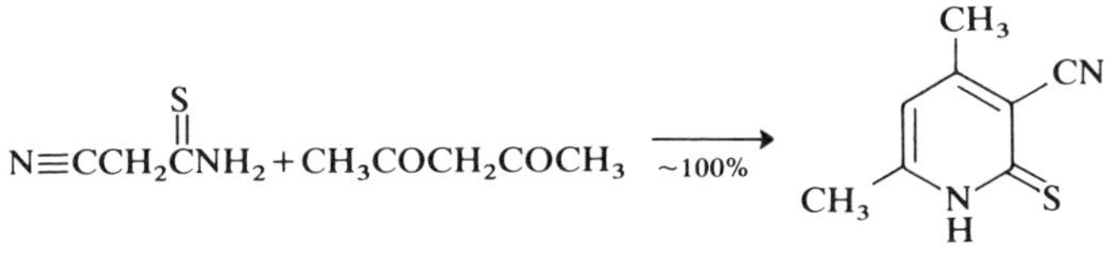

$$N{\equiv}CCH_2C(=S)NH_2 + CH_3COCH_2COCH_3 \xrightarrow{\sim 100\%} \mathbf{1}$$

(**1**: pyridine ring with CH_3 at 4, CN at 3, =S at 2, NH at 1, CH_3 at 6)

$$\mathbf{1} \xrightarrow[\sim 100\%]{I_2,\ K_2CO_3} \mathbf{2}$$

(**2**: 4,6-dimethyl-3-cyano-2-pyridyl disulfide, $(\text{Ar}S-)_2$)

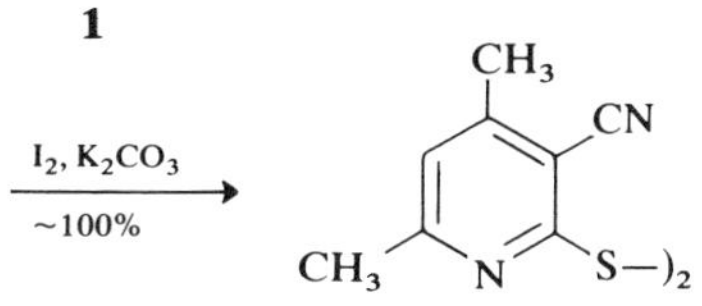

Esterification and lactonization.[2] The disulfide **2**, obtained by oxidation of **1**, reacts rapidly with a carboxylic acid and triphenylphosphine to form a thiolester (**3**) and the sparingly soluble **1**. Reaction of **3** with an alcohol to form **4** proceeds in almost quantitative yield without need for a silver catalyst (**6**, 246). Thus **2** and **3** are considerably more reactive than the corresponding unsubstituted analogs.

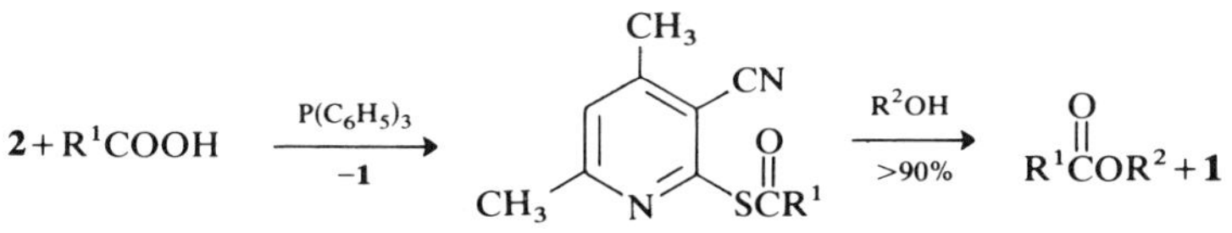

$$\mathbf{2} + R^1COOH \xrightarrow[-\mathbf{1}]{P(C_6H_5)_3} \mathbf{3} \xrightarrow[>90\%]{R^2OH} R^1COR^2 + \mathbf{1}$$

(**3**: 4,6-dimethyl-3-cyano-2-pyridyl thiolester, $ArSC(=O)R^1$; **4**: $R^1C(=O)OR^2$)

The disulfide was used to effect lactonization of 15-hydroxypentadecanoic acid. Under conditions of high dilution the yield of pentadecanolide was 88%.[3]

[1] U. Schmidt and H. Kubitzek, *Ber.*, **93**, 1559 (1960).
[2] U. Schmidt and G. Geisselmann, *ibid.*, **93**, 1590 (1960).
[3] U. Schmidt and D. Heermann, *Angew. Chem., Int. Ed.*, **18**, 308 (1979).

1,3-Dihydroisothianaphthene-2,2-dioxide, 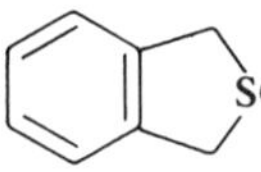 SO_2 (**1**). Mol. wt. 168.21. m.p. 150–151°. Preparation.[1]

Synthesis of estra-1,3,5(10)-triene-17-one. Nicolaou and Barnette[2] report that sulfone **1** is metalated with KH in DME to afford a clear yellow solution of the anion **2**. Alkylation of **2** with tosylate **3** followed by acid hydrolysis affords **4** as a mixture of diastereomers in 77% yield. Thermolysis of **4** affords estra-1,3,5(10)-triene-17-one (**5**) directly in 85% yield, probably by way of an intermediate *o*-quinone methide **a**. This sequence constitutes an exceedingly short and efficient synthesis of the steroid nucleus.

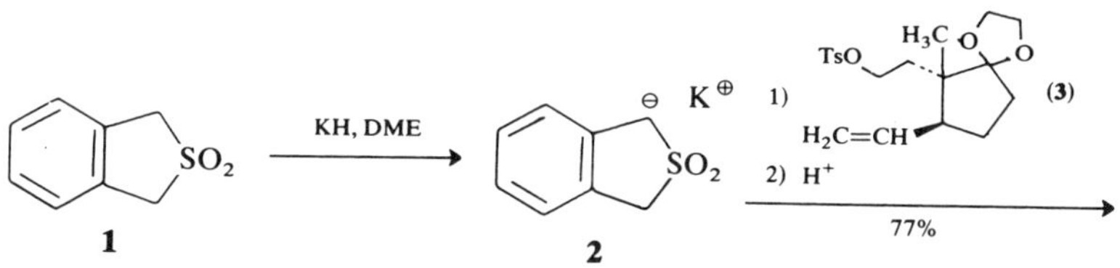

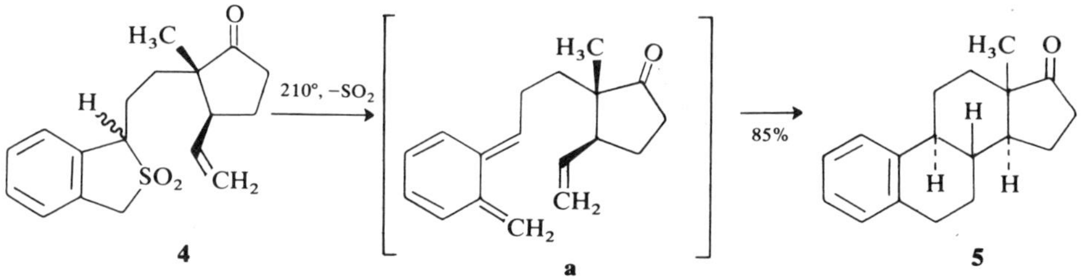

A separate study of the alkylations of **1** has been reported by Oppolzer and co-workers.[3] The group has employed LDA, LiTMP, and n-C_4H_9Li to metalate **1**. The resulting anion condenses with a variety of electrophiles (primary alkyl bromides, iodides, tosylates, disulfides, and acyl derivatives) to afford the corresponding monosubstituted derivatives in 30–95% yield. If the alkyl group contains a double bond in the 4- or 5-position, polycyclic compounds are prepared by thermal extrusion of SO_2 followed by intramolecular Diels–Alder cyclization.

Examples:

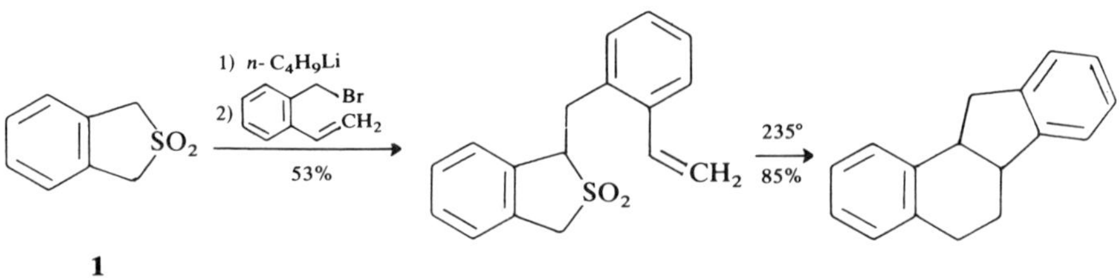

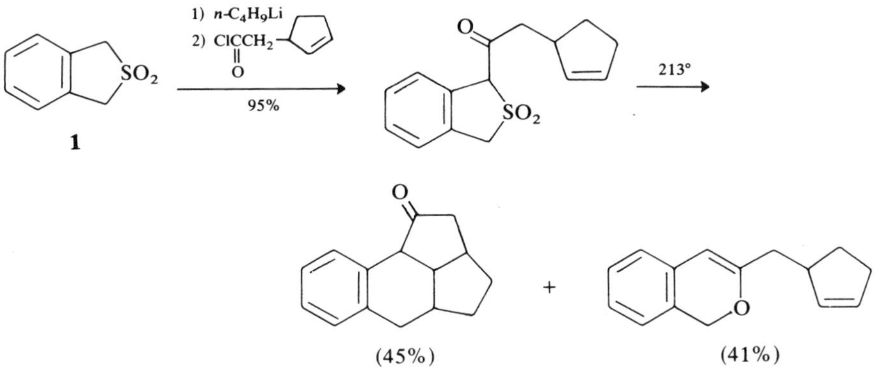

1) n-C_4H_9Li
2) $ClCCH_2$–cyclopentenyl (C=O)
95%

SO_2

1

213°

(45%) + (41%)

[1] M. P. Cava and A. A. Deana, *Am. Soc.*, **81**, 4266 (1959); J. A. Oliver and P. A. Ongler, *Chem. Ind.*, 1024 (1965).
[2] K. C. Nicolaou and W. E. Barnette, *J.C.S. Chem. Comm.*, 1119 (1979).
[3] W. Oppolzer, D. A. Roberts, and T. G. C. Bird, *Helv.*, **62**, 2017 (1979).

2,2-Dihydroxy-1,1-binaphthyl (1)–Lithium aluminum hydride. Supplier of **1**: Aldrich.

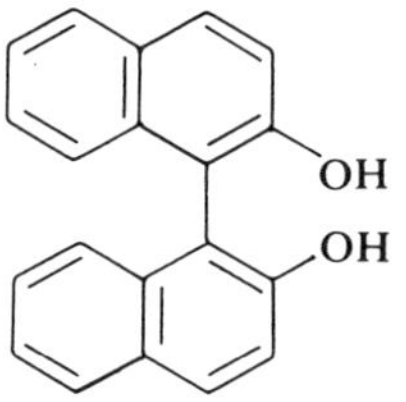

OH
OH

1

Asymmetric reduction of ketones. A complex hydride reagent (**2**) is prepared in THF by reaction of equimolar amounts of $LiAlH_4$, optically active (S)-**1**, and ethanol. This reagent effects asymmetric reduction of aryl and alkenyl ketones with high enantioselectivity.[1] Reductions of dialkyl ketones are less selective (*e.g.*, 13% ee for benzyl methyl ketone). This method has also been applied to ketones that are intermediates in prostaglandin synthesis.[2]

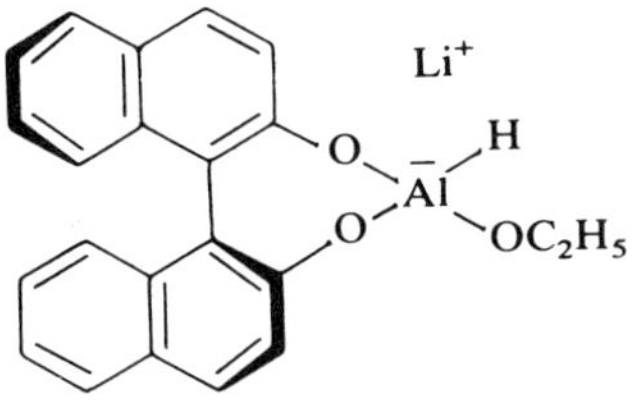

Li^+
H
Al
OC_2H_5

(S)-**2**

Examples:

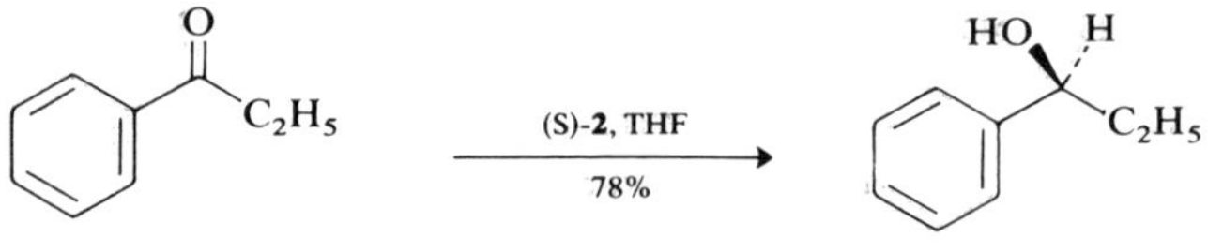

(S)-**2**, THF
78%

(100% ee)

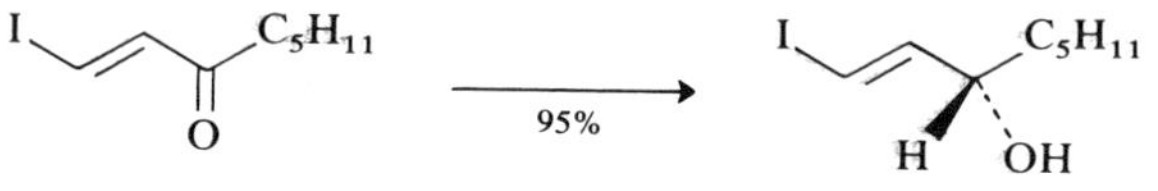

95%

(97% ee)

95%

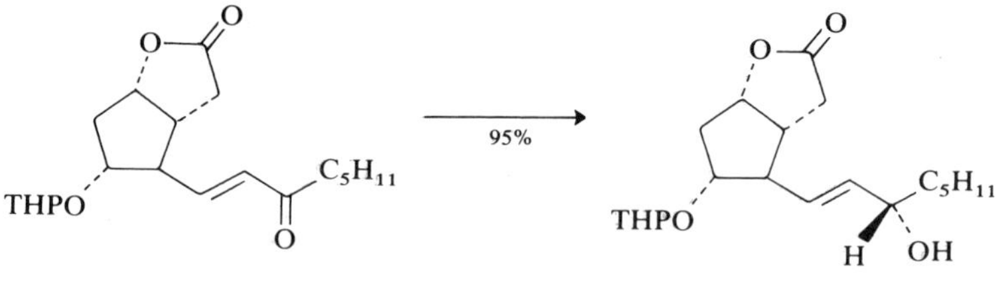

(99.5% ee)

[1] R. Noyori, I. Tomino, and Y. Tanimoto, *Am. Soc.*, **101**, 3129 (1979).
[2] R. Noyori, I. Tomino, and M. Nishizawa, *ibid.*, **101**, 5843 (1979).

Diiododimethylsilane, $(CH_3)_2SiI_2$. Mol. wt. 311.98, b.p. 170°. Preparation.[1]

Reduction of α-aryl alkanols.[2] This reaction can be accomplished in CH_2Cl_2 at room temperature with 1 equiv. of diiododimethylsilane (equation I).

$$\text{(I)}\quad \mathrm{ArC(R^1)(R^2)-OH} + (CH_3)_2SiI_2 \xrightarrow[45\text{–}94\%]{CH_2Cl_2} \mathrm{ArC(R^1)(R^2)-H} + [(CH_3)_2SiO]_n + I_2$$

[1] H. H. Anderson, *Am. Soc.*, **73**, 2351 (1951).
[2] W. Ando and M. Ikeno, *Tetrahedron Letters*, 4941 (1979).

Diiodomethane (Methylene iodide), CH_2I_2. Mol. wt. 267.84, b.p. 181°, m.p. 6°.

Cyclopropanation. Irradiation (350 nm.) of CH_2I_2 in the presence of alkenes yields the corresponding cyclopropane adducts with complete retention of stereochemistry. The reaction is facilitated by increasing alkyl substitution; in contrast to the Simmons-Smith reagent, this reaction is not sensitive to steric factors. Presumably methylene (:CH_2) is the reactive species.[1]

Examples:

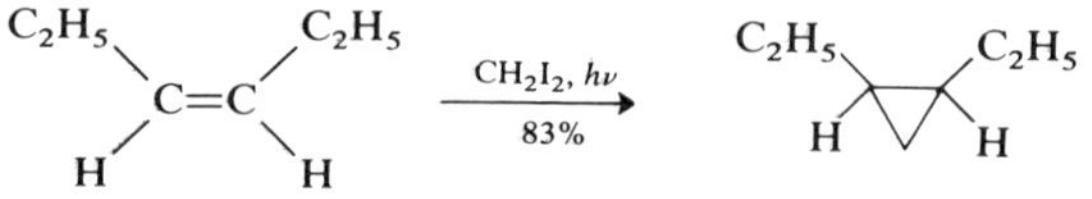

CH_2I_2, $h\nu$
83%

$C_2H_5(H)C{=}C(H)C_2H_5 \xrightarrow{84\%}$ cyclopropane (C_2H_5, H / H, C_2H_5)

$(CH_3)_3CCH{=}CH_2 \xrightarrow{80\%}$ $(CH_3)_3C$(H)cyclopropane

[1] N. J. Pienta and P. J. Kropp, *Am. Soc.*, **100**, 655 (1978).

Diisobutylaluminum 2,6-di-*t*-butyl-4-methylphenoxide, CH_3–C_6H_2[$C(CH_3)_3$]$_2$–$OAl[CH_2CH(CH_3)_2]_2$ **(1).** Mol. wt. 387.54; stable at 0° for several weeks. The reagent is prepared by reaction of diisobutylaluminum hydride with 2,6-di-*t*-butyl-4-methylphenol in toluene at 0°.

Selective reduction. One important problem in prostaglandin syntheses is the stereoselective reduction of the C_{15}-keto group to the desired (15S)-alcohol (*see* **4**, 104). This new reducing agent is highly effective for this purpose. Thus **2** is reduced to **3** with 92% stereoselectivity; and PGE_2 methyl ester is reduced to PGF_2 methyl ester with 100% stereoselectivity. The acetate of **2**, however, is reduced without any stereoselectivity. The effectiveness of **1** therefore results in part from coordination of the bulky aluminum reagent with the 11-hydroxy group.[1]

2 $\xrightarrow[95\%]{\mathbf{1},\ C_6H_5CH_3,\ -78°}$ **3**

[1] S. Iguchi, H. Nakai, M. Hayashi, and H. Yamamoto, *J. Org.*, **44**, 1363 (1979).

Diisobutylaluminum hydride (DIBAH), 1, 260–262; **2**, 140–142; **3**, 101–102; **4**, 158–161; **5**, 224–225; **6**, 198–201; **7**, 111–113; **8**, 173–174.

Aldol cyclization. Although the keto aldehyde **1** is resistant to aldol cyclization under normal alkaline conditions, this reaction can be accomplished by a method originally developed by Raphael *et al.*[1,2] Reduction of the enol lactone (**2**), derived from **1**, with DIBAH produces a bridged ketol, which is oxidized to **3** with chromic acid.[3] This diketone was employed as an intermediate in a synthesis of the sesquiterpene gymnomitrol (**4**).

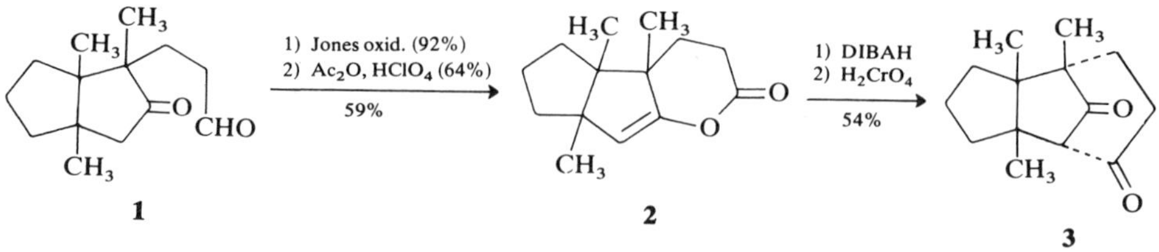

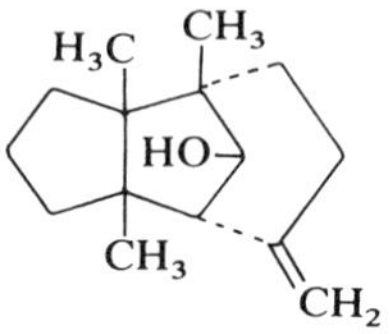

$\xrightarrow[\text{59\%}]{\text{1) Jones oxid. (92\%); 2) } Ac_2O, HClO_4 \text{ (64\%)}}$ 2 $\xrightarrow[\text{54\%}]{\text{1) DIBAH; 2) } H_2CrO_4}$ 3

4

Reduction of propargylamines.[4] Tertiary propargylamines are reduced by this hydride to (E)-allylamines (equation I). The stereoselectivity is the same as that

(I) $$R^1R^2NCH_2C{\equiv}CR^3 \xrightarrow[\text{85–96\%}]{\text{DIBAH}} R^1R^2NCH_2CH{=}CHR^3 \ (E)$$

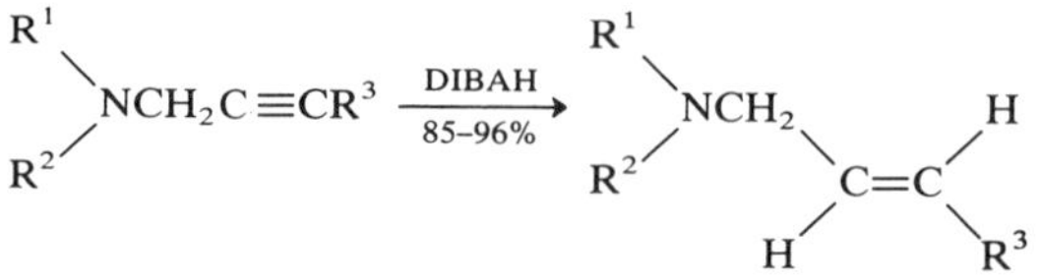

observed on reduction of propargylic alcohols to (E)-allylic alcohols. A conjugated diacetylenic *t*-amine is reduced to an (E, Z)-1,3-diene as the main product (equation II).

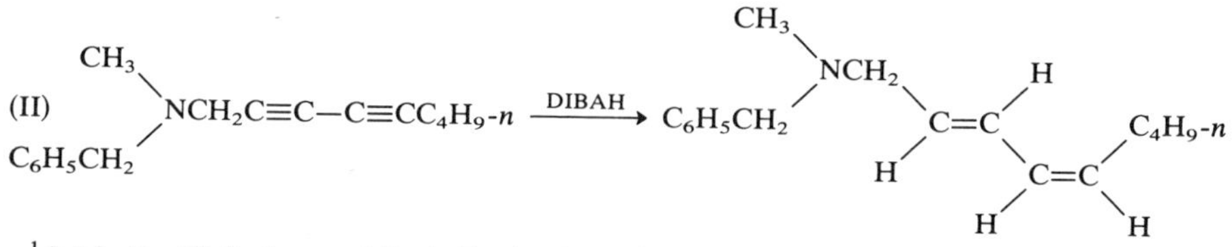

(II) $$CH_3(C_6H_5CH_2)NCH_2C{\equiv}C{-}C{\equiv}CC_4H_9\text{-}n \xrightarrow{\text{DIBAH}} CH_3(C_6H_5CH_2)NCH_2CH{=}CH{-}CH{=}CHC_4H_9\text{-}n$$

[1] J. Martin, W. Parker, and R. A. Raphael, *J. Chem. Soc.*, 289 (1964).
[2] E. W. Colvin, S. Malchenko, R. A. Raphael, and J. S. Roberts, *J.C.S. Perkin I*, 1989 (1973).
[3] R. M. Coates, S. K. Shah, and R. W. Mason, *Am. Soc.*, **101**, 6765 (1979).
[4] W. Granitzer and A. Stütz, *Tetrahedron Letters*, 3145 (1979).

Diisobutylaluminum phenoxide, $[(CH_3)_2CHCH_2]_2AlOC_6H_5$ (**1**). Mol. wt. 234.30. The reagent is prepared *in situ* from diisobutylaluminum hydride and phenol in THF.

Aldol condensation of methyl ketones. This reagent in combination with pyridine has been used to effect aldol condensation of methyl ketones by way of an organoaluminum enolate, as formulated for the aldol condensation of 2-octanone (**2**) to give a mixture of the isomeric C_{16}-ketones **3**.

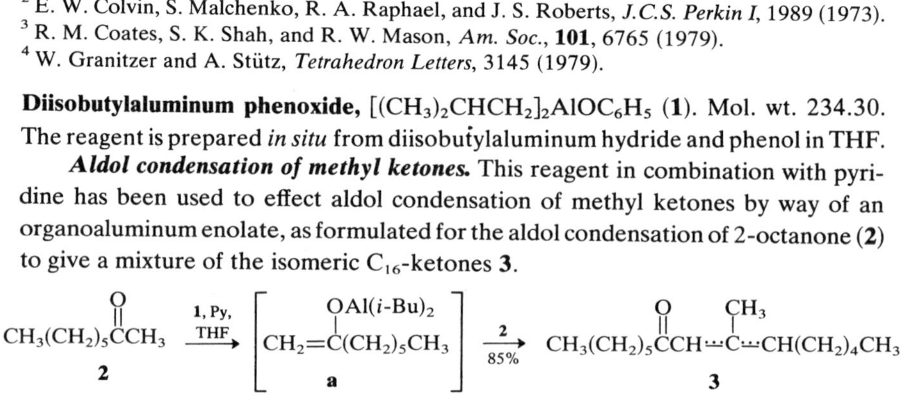

$$\underset{\mathbf{2}}{CH_3(CH_2)_5COCH_3} \xrightarrow[\text{THF}]{\mathbf{1}\text{, Py,}} \left[\underset{\mathbf{a}}{CH_2{=}C(OAl(i\text{-}Bu)_2)(CH_2)_5CH_3}\right] \xrightarrow[\text{85\%}]{\mathbf{2}} \underset{\mathbf{3}}{CH_3(CH_2)_5COCH{\cdots}C(CH_3){\cdots}CH(CH_2)_4CH_3}$$

The same reaction was used to effect intramolecular aldol condensation of 2,15-hexadecanedione (**4**) to a mixture of two dehydromuscones (**5**). In this case a large excess of **1** was necessary for satisfactory results.[1]

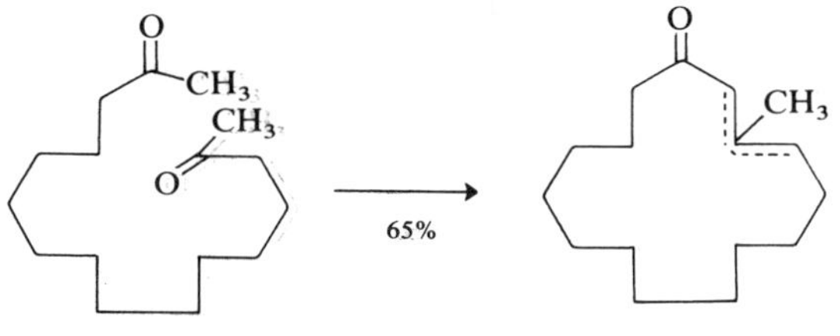

4 → 5 (65%)

[1] J. Tsuji, T. Yamada, M. Kaito, and T. Mandai, *Tetrahedron Letters*, 2257 (1979).

Diisopropylaminomagnesium bromide, $(i\text{-}C_3H_7)_2NMgBr$. Mol. wt. 204.41. This base is prepared by treatment of diisopropylamine in THF with C_2H_5MgBr at 80° and is kept at 50° to prevent precipitation.

Aldol condensation. The aldol reaction of **1** with **2** was achieved by Kishi *et al.*[1] using this base in their total synthesis of monensin. The ratio of diastereomeric aldol products **3** and **4** is sensitive to the reaction temperature. The following ratios of **3**

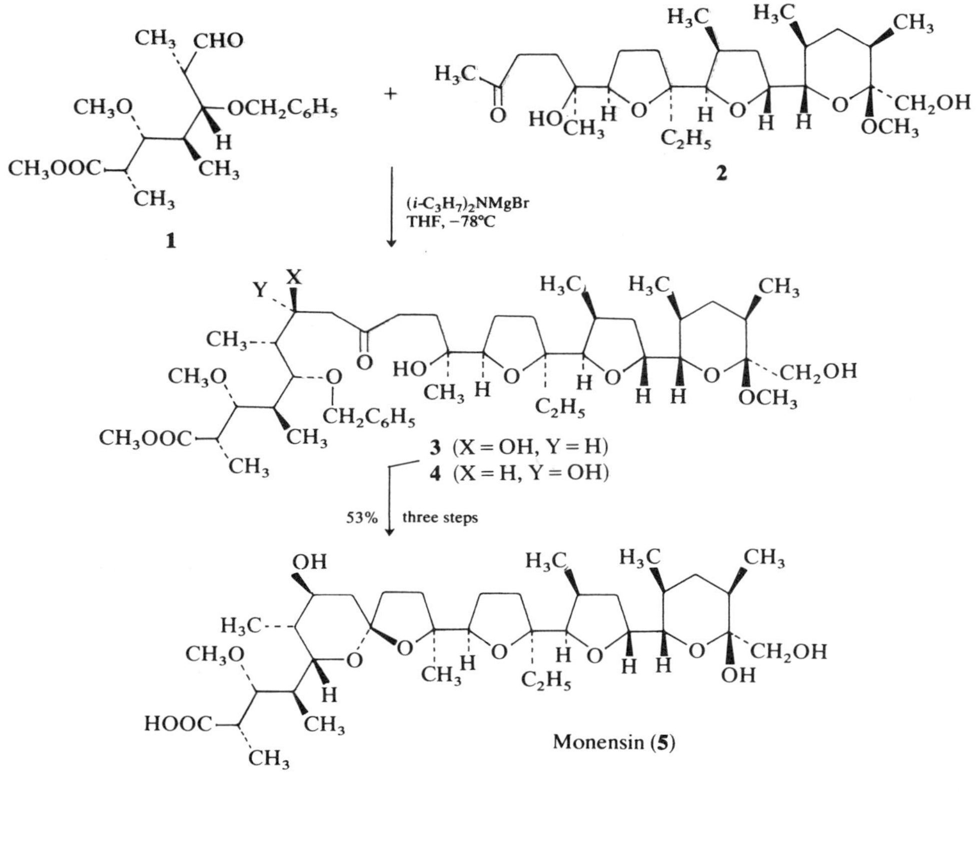

1 + 2 → ($(i\text{-}C_3H_7)_2NMgBr$, THF, −78°C) → **3** (X = OH, Y = H), **4** (X = H, Y = OH) → (53%, three steps) → Monensin (**5**)

and **4** are observed: ~1 : 1 at 0° (71% yield; 90% yield based on consumed **2**), ~2 : 1 at −20° (60%; 91%), ~5 : 1 at −50° (30%; 90%); and >8 : 1 at −78° (21%; 92%). The use of lower reaction temperatures thus improves the stereoselectivity of the reaction, but also decreases the total yield of aldol product.

[1] T. Fukuyama, K. Akasaka, D. S. Karanewsky, C.-L. J. Wang, G. Schmid, and Y. Kishi, *Am. Soc.*, **101**, 262 (1979).

Dilithium tetrachloropalladate, 5, 413; **7**, 114–115; **8**, 176–178.

1,4-Dienes. Larock *et al.*[1] have prepared 1,4-dienes by reaction of allylic halides with vinylmercurials promoted by palladium chloride and lithium chloride (essential). Sometimes only catalytic amounts of Pd(II) are required; other reactions require a full equivalent. Unfortunately an excess of the allylic halide is necessary for high yields.

Examples:

$$(CH_3)_3C(H)C{=}C(H)HgCl + CH_2{=}CHCH_2Cl \xrightarrow[96\%]{Li_2PdCl_4} (CH_3)_3C(H)C{=}C(H)CH_2CH{=}CH_2$$

$$C_6H_5(H)C{=}C(H)HgCl + CH_2{=}CHCH_2Cl \xrightarrow[80\%]{} C_6H_5(H)C{=}C(H)CH_2CH{=}CH_2$$

$$(CH_3)_3C(H)C{=}C(CH_3)HgCl + CH_2{=}CHCH_2Cl \xrightarrow[99\%]{} (CH_3)_3C(H)C{=}C(CH_3)CH_2CH{=}CH_2$$

[1] R. C. Larock, J. C. Bernhardt, and R. J. Driggs, *J. Organometal. Chem.*, **156**, 45 (1978).

4,4′-Dimethoxybenzhydrylamine, $(p\text{-}CH_3OC_6H_4)_2CHNH_2$ (**1**). Mol. wt. 243.30, m.p. 188° dec. The amine can be prepared by reduction of the oxime ether, $(Ar)_2C{=}NOCH_3$, with diborane (70% yield).[1]

Primary allylic amines. Unlike the benzyl group, the 4,4′-dimethoxybenzhydryl (DMB) group is removed by mild acid treatment. Because of this property, it was chosen as the protecting group in a recent synthesis of primary allylic amines by Trost and Keinan.[2] A typical example is shown in equation (I). Treatment of an allylic acetate (**2**) with **1** in the presence of tetrakis(triphenylphosphine)palladium(0) results in the (E)-allylic amine **3** in 81% yield. The protecting group is removed readily to give **4** in 82% yield.

This method has been used for synthesis (equation II) of (±)-gabaculine (**5**), a naturally occurring amino acid involved in metabolism of γ-aminobutyric acid. An interesting point is that alkylation of a cyclohexenyl acetate proceeds without elimination.

(I) $CH_3(CH_2)_4CH(OAc)CH{=}CH_2$ (**2**) $\xrightarrow[81\%]{1,\ Pd[P(C_6H_5)_3]_4}$ $CH_3(CH_2)_4CH{=}CHCH_2NHDMB$ (**3**) $\xrightarrow[82\%]{HCOOH,\ 80^\circ}$ $CH_3(CH_2)_4CH{=}CHCH_2NH_2$ (**4**)

(II) methyl 5-acetoxy-3-cyclohexenecarboxylate (OAc, CO_2CH_3) $\xrightarrow[81\%]{1,\ Pd(0)}$ CO_2CH_3, NHDMB cyclohexene $\xrightarrow{1)\ LDA,\ -78^\circ;\ 2)\ I_2;\ 3)\ DABCO}$ [CO_2CH_3, NHDMB cyclohexadiene] $\xrightarrow[65\text{–}90\%]{OH^-}$ COOH, NHDMB cyclohexadiene $\xrightarrow[80\%]{HCOOH,\ 60^\circ}$ COOH, NH_2 cyclohexadiene (**5**)

[1] H. Feuer and D. M. Braunstein, *J. Org.*, **34**, 1817 (1969).
[2] B. M. Trost and E. Keinan, *ibid.*, **44**, 3451 (1979).

4,6-Dimethoxy-2-pyrone,

(**1**). Mol. wt. 156.14.

Preparation:

$O{=}C(CH_2COOH)_2 \xrightarrow[80\%]{Ac_2O}$ 4-hydroxy-2,6-pyrandione (OH) $\xrightarrow[70\%]{CH_2N_2}$ **1**

Diels–Alder reactions.[1] This pyrone does not react with moderately reactive dienophiles (ethyl acrylate), but does react with reactive symmetrical alkenes to give adducts resulting from addition of 2 equiv. of the dienophile with loss of CO_2 from the primary cycloadduct. It also reacts with acetylenic dienophiles to form *m*-dimethoxyarenes.

Examples:

1 + N-phenylmaleimide $\xrightarrow[34\%]{C_6H_5CH_3,\ 140°}$ bis-adduct (OCH_3, OCH_3, N—C_6H_5)

1 + $HC \equiv CCOOCH_3$ $\xrightarrow[70\%]{140°}$ dimethoxy aromatic ester (OCH_3, $COOCH_3$, CH_3O)

[1] A. P. Kozikowski and R. Schmiesing, *Tetrahedron Letters*, 4241 (1978).

Dimethyl 1,3-acetonedicarboxylate (Dimethyl 3-oxoglutarate), 1, 270; **2**, 144; **4**, 165.

9-Methyl-cis-decalins. The reaction of 3-methyl-4-methylene-2-cyclohexenone (**2**) with **1** in DMSO with KF as catalyst (55–60°, 2 days) leads to a mixture of the stereoisomeric diketo diesters **3** and **4**. The reaction involves a double Michael addition. Both products are converted to the same diketone (**5**) on decarbomethoxylation. 9-Methyl-*cis*-decalin (**6**) was obtained from **5** by standard reactions. 1,9- and 4,9-Dimethyl-*cis*-decalin were also prepared in the same way.[1]

2 + $O{=}C(CH_2COOCH_3)_2$ (**1**) $\xrightarrow[\sim 40\%]{KF,\ DMSO}$ **3** and **4** $\xrightarrow{NaCl,\ DMSO,\ H_2O}$ **5** $\rightarrow$ **6**

[1] H. Irie, J. Katakawa, Y. Mizuno, S. Udaka, T. Taga, and K. Osaki, *J.C.S. Chem. Comm.*, 717 (1978).

Dimethyl acetylenedicarboxylate (DMAD), **1**, 272–273; **2**, 145–146; **3**, 103–104; **4**, 168–172; **5**, 227–230; **6**, 206; **7**, 117–119.

Reaction with enamines (**4**, 170–171). DMAD generally reacts with enamines by the expected [2 + 2] cycloaddition followed by rearrangement to a 1,3-dienamine. Reinhoudt *et al.*[1] have recently found that the reaction can take a different course in a polar solvent (CH_3OH or CH_3CN). Thus the reaction of DMAD with the enamine **1** of cyclopentanone in methanol results in the pyrrolizine **3** in moderate yield. The new reaction is believed to involve a 1,5-dipolar intermediate.

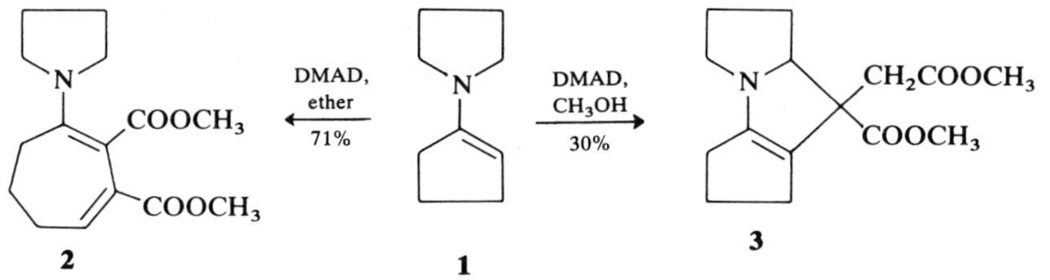

This reaction has been examined as a possible route to mitomycin antibiotics such as **4**,[2] which contain a pyrrolizine unit fused to a benzoquinone ring. Thus DMAD

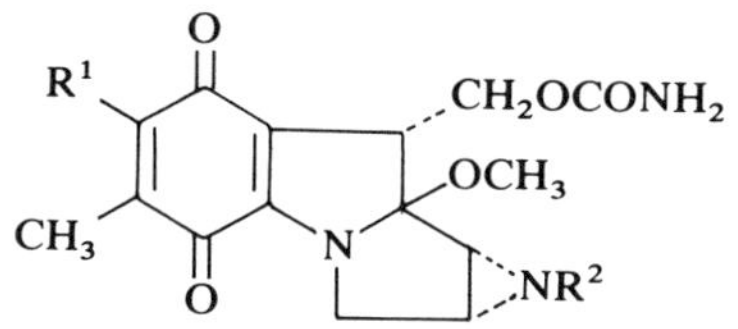

reacts with the pyrrolidinylnaphthoquinone **5** in *n*-butanol at 70° to give **6** in 50% yield.

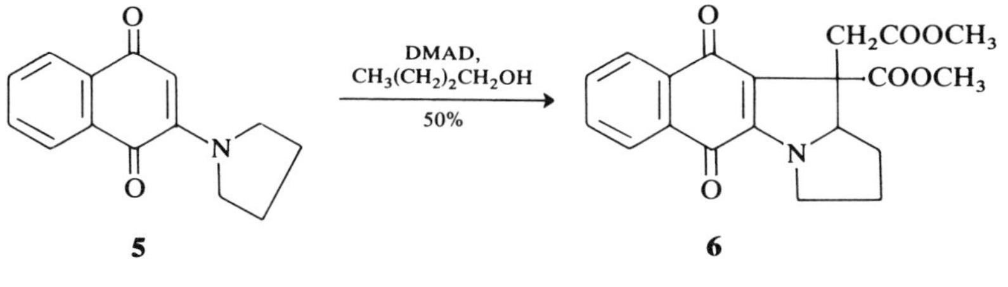

[1] D. N. Reinhoudt, J. Geevers, and W. P. Trompenaars, *Tetrahedron Letters*, 1351 (1978).
[2] J. Geevers, G. W. Visser, and D. N. Reinhoudt, *Rec. trav.*, **98**, 251 (1979).

Dimethylaluminum amides, 8, 182.[1]

Esters → nitriles.[2] This reaction can be accomplished as formulated in equation (I).

$$\text{(I)} \qquad R{-}COOR + (CH_3)_2AlNH_2 \xrightarrow[70\text{–}90\%]{\text{Xylene, }\Delta} R{-}CN$$

[1] M. F. Lipton, A. Basha, and S. M. Weinreb, *Org. Syn.*, **59**, 49 (1979).
[2] J. L. Word, N. A. Khatri, and S. M. Weinreb, *Tetrahedron Letters*, 4907 (1979).

4-Dimethylamino-3-butyne-2-one, $(CH_3)_2NC{\equiv}CCOCH_3$ (**1**), **8**, 183–184. Light yellow oil that polymerizes slowly at room temperature; stable at −70° for 1 month.

Selective activation of carboxylic acids.[1] Carboxylic acids add exclusively to the triple bond of **1** to form an adduct (**a**) that rearranges to the (Z)-enol ester **2**. Hydroxyl, phenolic, amino, and mercapto groups do not react with **1** under the same conditions. The esters (**2**) react with alkali metal salts of thiols and of selenophenol in THF at 0° to form thiol and selenol esters (**3**).

$$1 + RC(=O)OH \rightarrow \underset{\mathbf{a}}{(CH_3)_2N(RCO_2)C{=}C(H)COCH_3} \xrightarrow[91\text{-}98\%]{THF,\ -50^\circ} \underset{\mathbf{2}}{(CH_3)_2NC(=O)CH{=}C(CH_3)OC(=O)R}$$

$$\mathbf{2} + MXR^1 \xrightarrow[80\text{-}90\%]{THF,\ 0^\circ} \underset{\mathbf{3}}{RC(=O)XR^1} + CH_3C(=O)CH(M)CON(CH_3)_2$$

[1] H.-J. Gais and T. Lied, *Angew. Chem., Int. Ed.*, **17**, 267 (1978).

4-Dimethylaminopyridine (DMAP), **3**, 118–119; **5**, 260. Supplier: Aldrich.

Review.[1] A new method suitable for large-scale preparation of DMAP (**1**) is shown in equation (I). The related reagent 4-pyrrolidinopyridine (**2**; **4**, 416) can be obtained in this way by use of N-formylpyrrolidine in place of DMF.

(I) pyridine $\xrightarrow{SOCl_2}$ 1-(4-pyridyl)pyridinium chloride hydrochloride (N^+ Cl^-; $\overset{+}{N}H$ Cl^-) $\xrightarrow[65\text{-}70\%]{(CH_3)_2NCHO}$ 4-$N(CH_3)_2$-pyridine (**1**); 4-pyrrolidinopyridine (**2**)

These two compounds are superior to pyridine as catalysts for acylation of alcohols, particularly for tertiary and sterically hindered alcohols. Even axial 11β-hydroxyl groups of steroids can be acetylated in yields as high as 80%. Several useful C-acylations have been reported. The reagents catalyze the transformation of amino acids into α-acylamino ketones (equation II). They are superior to pyridine for reaction of isocyanates and carboxylic acids to form amides (equation III).

$$\text{(II)}\quad BOC{-}NHCH(CH_3)CONHCH(CH_3)COOH \xrightarrow[86\%]{Ac_2O,\ DMAP,\ N(C_2H_5)_3} BOCNHCH(CH_3)CONHCH(CH_3)COCH_3$$

$$\text{(III)}\quad C_6H_5CH_2COOH + C_6H_5NCO \xrightarrow[66\%]{DMAP} C_6H_5CH_2CONHC_6H_5 + CO_2$$

The enhanced activity of **1** and **2** over pyridine is not a result of increased basicity (DMAP, $pK_a = 9.70$; Py, $pK_a = 5.29$), but is due at least in part to the formation of high concentrations of N-acylpyridinium salts, which are very effective acylating reagents.

Related catalysts. Hassner *et al.*[2] have prepared a number of substituted derivatives of pyridine, pyridazine, and quinoline, but only a few are useful acylation catalysts. Of these, 4-pyrrolidinopyridine (**2**) and 1,1,3,3-tetramethyl-4-(4-pyridyl)guanidine (**3**) are most effective.

N(CH3)2
C
N   N(CH3)2
N

3

Tritylation; silylation. This amine is an efficient catalyst for tritylation of hydroxyl groups with trityl chloride and triethylamine. The method permits use of solvents other than pyridine, such as DMF or CH_2Cl_2. The same conditions can be used to prepare *t*-butyldimethylsilyl ethers. Yields are comparable to those obtained with imidazole (**4**, 57–58). A further advantage is that the selectivity for reaction with a primary hydroxyl group is enhanced in this new procedure.[3]

Enol acetylation.[4] This amine catalyzes the enol acetylation of aldehydes by acetic anhydride and triethylamine with improvement of yields. In some cases use of THF as solvent is beneficial. This method, however, is not satisfactory for enol acetylation of ketones.

Depsipeptides. These esters can be prepared in 90–98% yield by coupling of BOC-protected amino acids with benzyl esters of an α-hydroxy acid by means of DCC and catalysis with 4-dimethylaminopyridine. The reaction is successful even with hindered substrates.[5]

$$\text{(I)}\quad \text{BOCNHC}(R^1)(R^2)\text{COOH} + \text{HOC}(R^3)(R^4)\text{COOCH}_2\text{C}_6\text{H}_5 \xrightarrow[90\text{–}98\%]{\text{DCC, cat., CH}_2\text{Cl}_2,\ 25^\circ} \text{BOCNHC}(R^1)(R^2)\text{COOC}(R^3)(R^4)\text{COOCH}_2\text{C}_6\text{H}_5$$

Activation of quinones. Quinones under ordinary conditions react with phenol ethers to give products of C—C and O—C coupling. Acid catalysis favors C—C coupling (equation I).[6]

(I) O, COOCH3, O + OCH3, OCH3 → 1) TFA, 20°; 2) Ag2O; 40% → O, OCH3, O, COOCH3, OCH3

On the other hand, 4-dimethylaminopyridine or 2-methoxypyridine favors O—C coupling. This reaction provides a regiospecific route to xanthones (equation II).[7]

(II) [2-carbomethoxy-1,4-benzoquinone] + [4-methylanisole] $\xrightarrow[\text{80\%}]{\text{Base, } C_6H_6,\ 20^\circ}$ [2-$COOCH_3$-3-(4-CH_3-phenoxy)-1,4-benzoquinone] $\xrightarrow{\text{1) } H_2\text{, Pd/C, HOAc; 2) } (CH_3)_2SO_4\text{; 3) } OH^-}$

[3,6-di-OCH_3-2-(4-CH_3-phenoxy)benzoic acid ($COOH$)] $\xrightarrow[\text{60\%}]{\text{PPA}}$ [1,4-di-OCH_3-7-CH_3-xanthone]

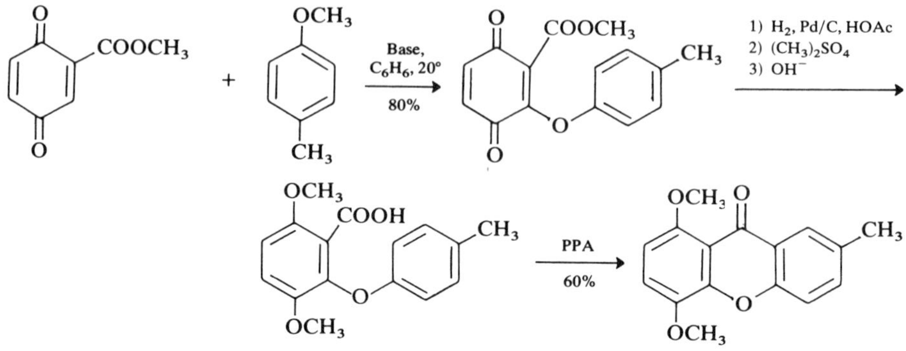

[1] G. Höfle, W. Steglich, and H. Vorbrüggen, *Angew. Chem., Int. Ed.*, **17**, 569 (1978).
[2] A. Hassner, L. R. Krepski, and V. Alexanian, *Tetrahedron*, **34**, 3069 (1978).
[3] S. K. Chaudhary and O. Hernandez, *Tetrahedron Letters*, 95, 99 (1979).
[4] T. J. Cousineau, S. L. Cook, and J. A. Secrist III, *Syn. Comm.*, **9**, 157 (1979).
[5] C. Gilon, Y. Klausner, and A. Hassner, *Tetrahedron Letters*, 3811 (1979).
[6] P. Kuser, M. Inderbitzin, J. Branchli, and C. H. Eugster, *Helv.*, **54**, 980 (1971).
[7] P. Müller, T. Venakis, and C. H. Eugster, *ibid.*, **62**, 2350 (1979).

Dimethyl 3-bromo-2-ethoxypropenylphosphonate (1). Mol. wt. 273.07.

Preparation:

$$CH_3\overset{O}{\overset{\|}{C}}CH_2\overset{O}{\overset{\|}{P}}(OCH_3)_2 \xrightarrow[\text{94\%}]{CH(OC_2H_5)_3,\ FeCl_3} CH_3\overset{C_2H_5O}{\overset{|}{C}}=CH\overset{O}{\overset{\|}{P}}(OCH_3)_2 \xrightarrow[\text{76\%}]{\text{NBS}, h\nu} BrCH_2\overset{C_2H_5O}{\overset{|}{C}}=CH\overset{O}{\overset{\|}{P}}(OCH_3)_2$$

1

Cyclopentenone annelation.[1] A typical example for use of **1** for synthesis of cyclopentenones is illustrated for annelation of cyclohexanone (equation I).

(I) [lithium cyclohexenolate, OLi] $\xrightarrow[\text{quant.}]{\textbf{1}\text{, THF, } -78\rightarrow 20^\circ}$ [2-($CH_2\overset{C_2H_5O}{\overset{|}{C}}=CH\overset{O}{\overset{\|}{P}}(OCH_3)_2$)cyclohexanone] $\xrightarrow[\text{quant.}]{HCl,\ CH_3COCH_3,\ 25^\circ}$

[2-($CH_2\overset{O}{\overset{\|}{C}}CH_2\overset{O}{\overset{\|}{P}}(OCH_3)_2$)cyclohexanone] $\xrightarrow[\text{79\%}]{\text{NaH, DME } 0\rightarrow 60^\circ}$ [bicyclic cyclopentenone, =O]

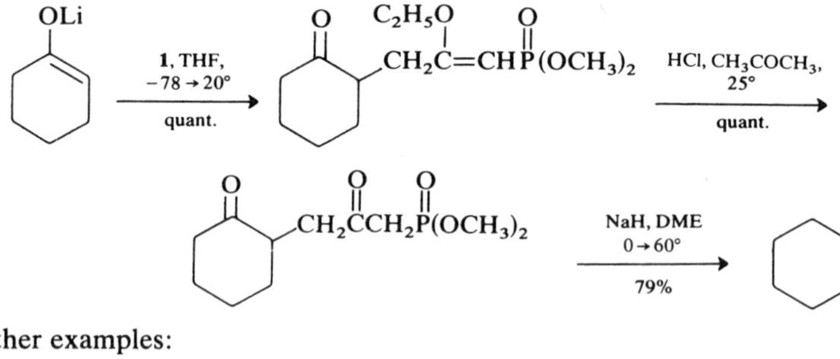

Other examples:

$$CH_3CH_2\overset{O}{\overset{\|}{C}}CH_2CH_3 \xrightarrow[\text{82\%}]{} \text{[3-}C_2H_5\text{-4-}CH_3\text{-cyclopent-2-enone]}$$

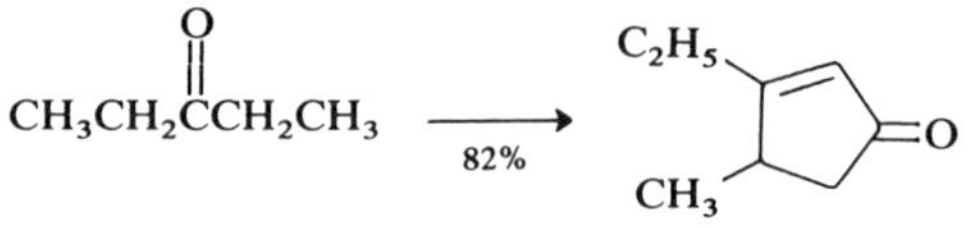

$$\text{(spiro ketal cyclohexanone)} \xrightarrow{52\%} \text{(spiro ketal bicyclic cyclopentenone)}$$

The last example illustrates one advantage of this method: the enol ether function of **1** can be hydrolyzed without effect on a ketal grouping. Another advantage is that the intramolecular Wittig-Horner reaction proceeds without isomerization of the initially formed cyclopentenone.

[1] E. Piers, B. Abeysekera, and J. R. Scheffer, *Tetrahedron Letters*, 3279 (1979).

Dimethyl (diazomethyl)phosphonate, $(CH_3O)_2\overset{O}{\overset{\|}{P}}CHN_2$ (**1**). Mol. wt. 150.09, b.p. 59°/0.1 mm. Preparation.[1,2]

Alkynes from aldehydes or ketones. The reaction of the anion of **1** with diaryl ketones, ArCOAr′, to form alkynes, ArC≡CAr′, was reported first by Colvin and Hamill,[2] but the method was said to fail or give low yield with substrates with enolizable hydrogens. Since then experimental details have been perfected, and the method has proved to be useful.[3] The anion of **1** is prepared with potassium *t*-butoxide, and the reaction with the carbonyl compound is conducted for 12–16 hours at −78° before it is allowed to warm to the ambient temperature. Under these conditions, alkynes can be obtained in 50–80% yield from aldehydes, diaryl ketones, and alkyl aryl ketones, but not from dialkyl ketones. The proposed mechanism is shown in equation (I).

$$\text{(I)} \qquad (CH_3O)_2\overset{O}{\overset{\|}{P}}\bar{C}N_2K^+ + R^1COR^2 \rightarrow \left[N_2C{=}C\begin{matrix} R^1 \\ R^2 \end{matrix} \right] \xrightarrow{-N_2} R^2C{\equiv}CR^1$$

[1] D. Seyferth, R. S. Marmor, and P. Hilbert, *J. Org.*, **36**, 1379 (1971).
[2] E. W. Colvin and B. J. Hamill, *J.C.S. Chem. Comm.*, 151 (1973); *idem*, *J.C.S. Perkin I*, 869 (1977).
[3] J. C. Gilbert and U. Weerasooriya, *J. Org.*, **44**, 4997 (1979).

S,S′-Dimethyl dithiocarbonate, $CO(SCH_3)_2$ (**1**). Mol. wt. 122.21, b.p. 75°/10 mm. The reagent can be obtained by rearrangement of S-methyl xanthate with $AlCl_3$ in CS_2 (50% yield).

α-Methylthiocarbonyl ketones.[1] Ketones react with NaH and **1** preferentially at the least substituted α-carbon to form α-methylthiocarbonyl ketones in 60–100% yield (equation I).

$$\text{(I)} \qquad R^1\overset{O}{\overset{\|}{C}}CHR^2R^3 \xrightarrow[60\text{–}100\%]{\text{NaH, }\mathbf{1}\text{, DME, }\Delta} R^1\overset{O}{\overset{\|}{C}}\underset{COSCH_3}{\underset{|}{C}}R^2R^3$$

[1] H.-J. Liu, S. K. Attah-Poku, and H. K. Lai, *Syn. Comm.*, **9**, 883 (1979).

N,N-Dimethylformamide, 1, 278–281; **2**, 153–154; **3**, 115; **4**, 184; **5**, 247–249; **7**, 124; **8**, 189–190.

Dichlorination of ketones. α-Monochlorination (Cl_2) of alkyl aryl ketones proceeds readily in a variety of solvents. α,α-Dichlorination has been effected with sodium acetate as catalyst in refluxing acetic acid (5 hours, 80–90% yield).[1] Actually this reaction can be effected with DMF as catalyst and solvent at 80–100° in 35–45 minutes in yields of 80–95%.[2]

$$ArC(=O)CH_2R \xrightarrow[80\text{–}95\%]{Cl_2,\ DMF} ArC(=O)-CCl_2-R$$

$$(R = CH_3, C_2H_5, C_3H_7, t\text{-}C_4H_9, C_6H_5)$$

[1] J. G. Aston, J. D. Newkirk, D. M. Jenkins, and J. Dorsky, *Org. Syn. Coll. Vol.*, **3**, 538 (1955).
[2] N. De Kimpe, L. De Bucyk, R. Verhe, F. Wychuyse, and N. Schamp, *Syn. Comm.*, **9**, 575 (1979).

N,N-Dimethylformamide–Dimethyl sulfate adduct, 5, 250.

Acetalization. The adduct (**1**) and an alcohol convert aldehydes and α,β-unsaturated aldehydes, and ketones into acetals and ketals, respectively. For the most part, yields of ketals are only mediocre (10–75%), but reactions with aldehydes proceed satisfactorily.[1]

Carbonyl compounds react with **1** and a 1,2- or 1,3-diol to form 1,3-dioxolanes or 1,3-dioxanes, respectively. The yields are generally about 80%.[2]

$$R^1R^2C{=}O + HOCHR^3(CH_2)_nCHR^4OH\ (n = 0, 1) + [HC(OCH_3){=}N(CH_3)_2]^+\ CH_3OSO_3^-\ (\mathbf{1}) \xrightarrow{\sim 80\%}$$

$$R^1R^2C\langle OCHR^3(CH_2)_nCHR^4O\rangle + CH_3OH + [HC(OH){=}N(CH_3)_2]^+\ CH_3OSO_3^-$$

[1] W. Kantlehner, H.-D. Gutbrod, and P. Gross, *Ann.*, 522 (1979).
[2] *Idem*, *ibid.*, 1362 (1979).

Dimethylformamide diethyl acetal, 1, 281–282; **2**, 154; **3**, 115–116; **4**, 184; **5**, 253–254; **7**, 125.

Enol acetates of oxonucleosides.[1] These derivatives of oxonucleosides are available by reaction of a nucleoside first with a DMF acetal and then with acetic anhydride. This method is limited to compounds with the oxo group in the 4′-position and with a free CO—NH group in the pyrimidine ring. Previous methods of enol

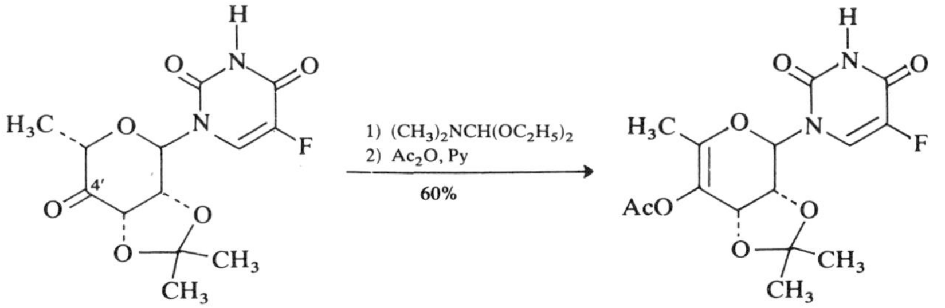

$$\xrightarrow[60\%]{\text{1) }(CH_3)_2NCH(OC_2H_5)_2\text{; 2) }Ac_2O,\ Py}$$

formation fail with oxonucleosides because of instability in alkaline media. This reaction is the first synthesis of oxonucleoside enol acetates by direct enolization.

Enones. A new method for conversion of a ketone to an α,β-unsaturated ketone involves condensation with DMF dimethyl acetal to form an enamino ketone, which is then treated with an alkyllithium.[2]

Examples:

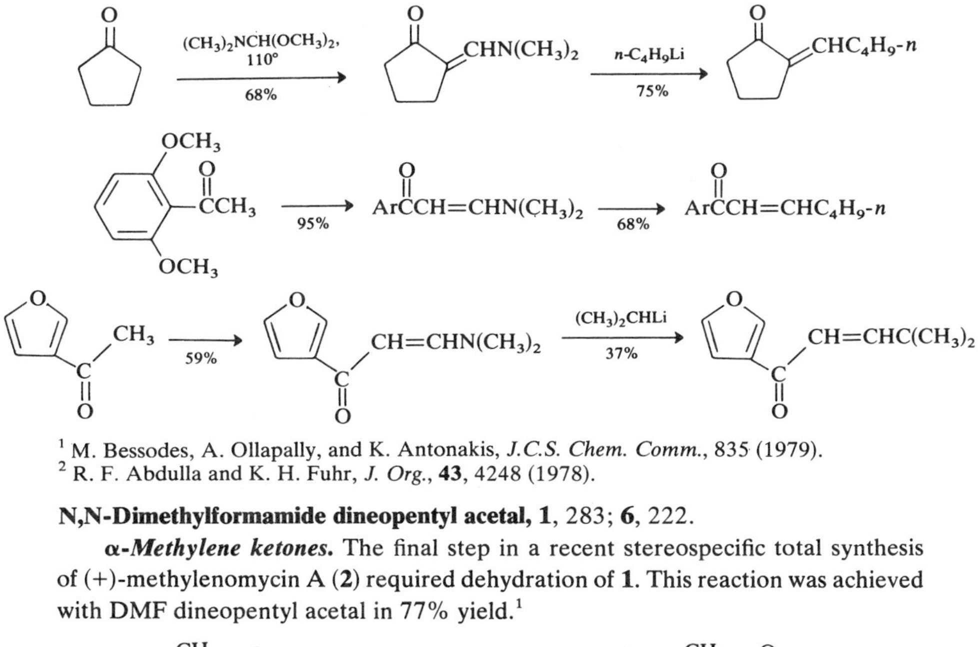

$$\xrightarrow[68\%]{(CH_3)_2NCH(OCH_3)_2,\ 110°} \text{=CHN}(CH_3)_2 \xrightarrow[75\%]{n\text{-}C_4H_9Li} \text{=CHC}_4H_9\text{-}n$$

$$\xrightarrow{95\%} ArCCH{=}CHN(CH_3)_2 \xrightarrow{68\%} ArCCH{=}CHC_4H_9\text{-}n$$

$$\xrightarrow{59\%} CH{=}CHN(CH_3)_2 \xrightarrow[37\%]{(CH_3)_2CHLi} CH{=}CHC(CH_3)_2$$

[1] M. Bessodes, A. Ollapally, and K. Antonakis, *J.C.S. Chem. Comm.*, 835 (1979).
[2] R. F. Abdulla and K. H. Fuhr, *J. Org.*, **43**, 4248 (1978).

N,N-Dimethylformamide dineopentyl acetal, 1, 283; **6,** 222.

α-Methylene ketones. The final step in a recent stereospecific total synthesis of (+)-methylenomycin A (**2**) required dehydration of **1**. This reaction was achieved with DMF dineopentyl acetal in 77% yield.[1]

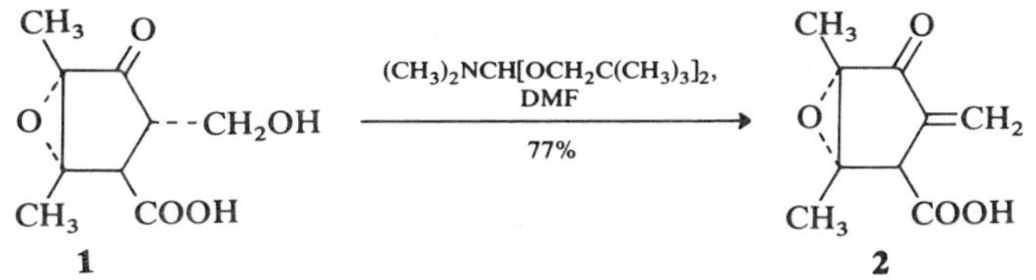

$$\mathbf{1} \xrightarrow[77\%]{(CH_3)_2NCH[OCH_2C(CH_3)_3]_2,\ DMF} \mathbf{2}$$

Dehydration of the related ketone **3** with DCC and CuCl proceeds in low yield to give **4**, originally believed to be methylenomycin B.[2]

CH_3 O O CH_2OH —DCC, CuCl / 15%→ CH_3 O O $=CH_2$ CH_3 CH_3

3 **4**

[1] J. Jernow, W. Tautz, P. Rosen, and J. F. Blount, *J. Org.*, **44**, 4210 (1979).
[2] *Idem, ibid.*, **44**, 4212 (1979).

N,N-Dimethylhydrazine, 1, 289–290; **2**, 154–155; **3**, 117; **5**, 254; **6**, 223; **7**, 126–130; **8**, 192–193.

1,3-Diketones.[1] A new method for synthesis of 1,3-diketones involves regioselective acylation of N,N-dimethylhydrazones of ketones.

Examples:

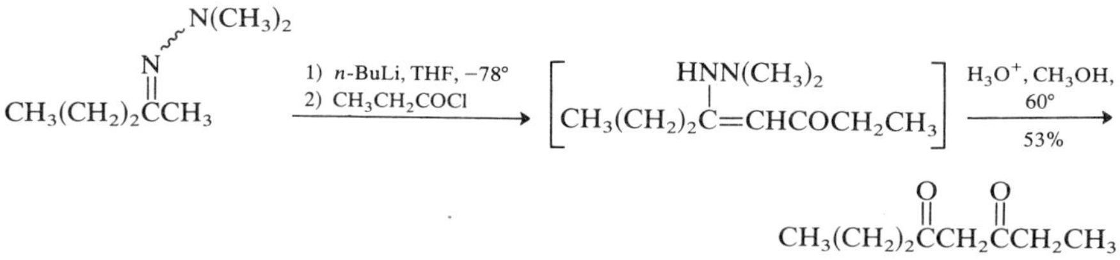

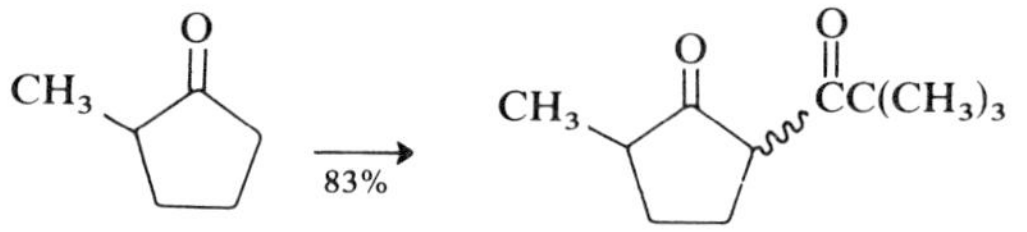

$$CH_3(CH_2)_2C(=NN(CH_3)_2)CH_3 \xrightarrow[2)\ CH_3CH_2COCl]{1)\ n\text{-BuLi, THF, }-78^\circ} \left[CH_3(CH_2)_2C(HNN(CH_3)_2)=CHCOCH_2CH_3\right] \xrightarrow[53\%]{H_3O^+,\ CH_3OH,\ 60^\circ} CH_3(CH_2)_2\overset{O}{\overset{\|}{C}}CH_2\overset{O}{\overset{\|}{C}}CH_2CH_3$$

2-methylcyclopentanone (CH_3, O) —83%→ CH_3, O, $\overset{O}{\overset{\|}{C}}C(CH_3)_3$

Annelation.[2] The Gilman ate reagents derived from acetone N,N-dimethylhydrazone, either the homocuprate derivative (**7**, 149) or the more recently developed phenylthio heterocuprate derivative, add in a conjugate fashion to ethyl 1-cyclohexene-1-carboxylate (**1**) to afford the keto ester (**2**, 60–70% yield) after hydrolysis of the N,N-dimethylhydrazone group. The keto ester **2** serves as a versatile intermediate for the preparation of several complementary annelation products **4–8**.

[1] D. Enders and P. Weuster, *Tetrahedron Letters*, 2853 (1978).
[2] E. J. Corey and D. L. Boger, *ibid.*, 4597 (1978).

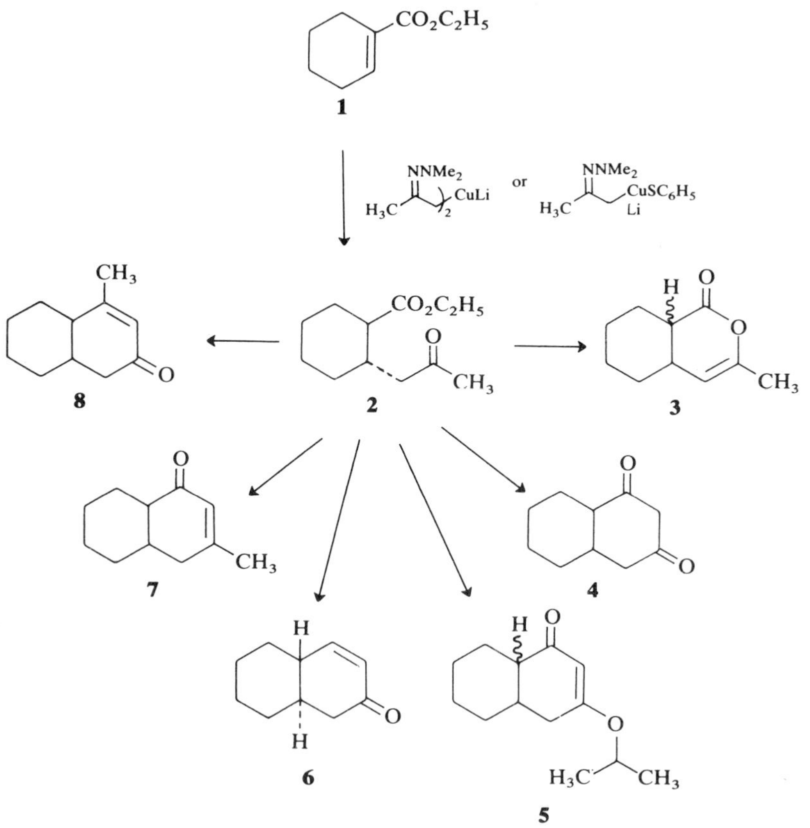

Dimethylketene, 1, 290–292; **6**, 223.

***[2+2] Cycloaddition to iminolactones* (1).** Ketene itself does not undergo this reaction, but dialkylketenes react to form spiro-β-lactams (**2**). A one-pot procedure starting with maleic anhydride, an amine, and a dialkylketene can also be used.[1]

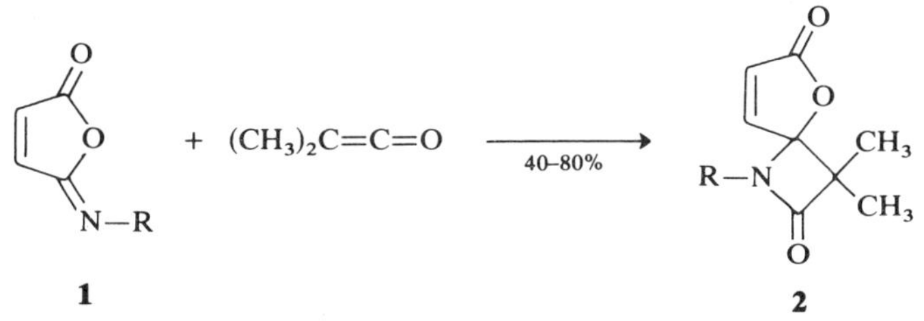

$$\mathbf{1} + (CH_3)_2C{=}C{=}O \xrightarrow{40\text{–}80\%} \mathbf{2}$$

[1] M. Roth, *Helv.*, **62**, 1966 (1979).

4,4-Dimethyl-1,3-oxathiolane-3,3-dioxide (1). Mol. wt. 150.20, m.p. 46–50°.

Preparation:

$$(CH_3)_2CHCHO \xrightarrow{S_2Cl_2} (CH_3)_2C(CHO)S{-}SC(CHO)(CH_3)_2 \xrightarrow{LiAlH_4} (CH_3)_2C(SH)CH_2OH \xrightarrow{H^+,\,HCHO}$$

$$\text{4,4-dimethyl-1,3-oxathiolane} \xrightarrow[22\%\ \text{overall}]{KMnO_4} \mathbf{1}$$

Aldehydes. The reagent can serve as a formyl anion equivalent, which is unmasked by thermolysis. An example of the use of **1** for conversion of an alkyl halide into the corresponding homologous aldehyde is shown in equation (I).

(I) $$\mathbf{1} \xrightarrow[57\%]{1)\ n\text{-BuLi, THF, }-80^\circ;\ 2)\ CH_3(CH_2)_5I} \text{2-}(CH_2)_5CH_3\text{-4,4-dimethyl-1,3-oxathiolane-3,3-dioxide} \xrightarrow[97\%]{\Delta} CH_3(CH_2)_5CHO + SO_2 + CH_2{=}C(CH_3)_2$$

Use of **1** for preparation of an acylsilane is shown in equation (II).[1]

(II) $$\mathbf{1} \xrightarrow[100\%]{1)\ n\text{-BuLi};\ 2)\ ClSi(CH_3)_3} \text{2-}Si(CH_3)_3\text{-derivative} \xrightarrow{1)\ n\text{-BuLi};\ 2)\ CH_3I} \text{2-}CH_3\text{-2-}Si(CH_3)_3\text{-derivative} \xrightarrow[30\%]{\Delta} CH_3COSi(CH_3)_3$$

[1] G. W. Gokel, H. M. Gerdes, D. E. Miles, J. M. Hufnal, and G. A. Zerby, *Tetrahedron Letters*, 3375 (1979); G. W. Gokel and H. M. Gerdes, *ibid.*, 3379 (1979).

Dimethyloxosulfonium methylide, 1, 315–318; **2**, 171–173; **3**, 125–127; **4**, 197–199; **5**, 254–257; **7**, 133; **8**, 194–196.

Asymmetric synthesis of cyclopropanes.[1] The reaction of dimethyloxosulfonium methylide with (E)-(2R,3S)-6-alkylidene-3,4-dimethyl-2-phenylperhydro-1,4-oxazepine-5,7-diones (**1**)[2] yields cyclopropane derivatives (**2**) and dihydrofuranes (**3**). The ratio of the products depends on the solvent and temperature. Use of THF at 25° favors formation of **2**, whereas formation of **3** is favored by use of DMF at −61°. The products (**2** and **3**) can be converted into optically pure cyclopropane-1,1-dicarboxylic acids (**4**) and 3-substituted γ-butyrolactones (**5**), respectively.

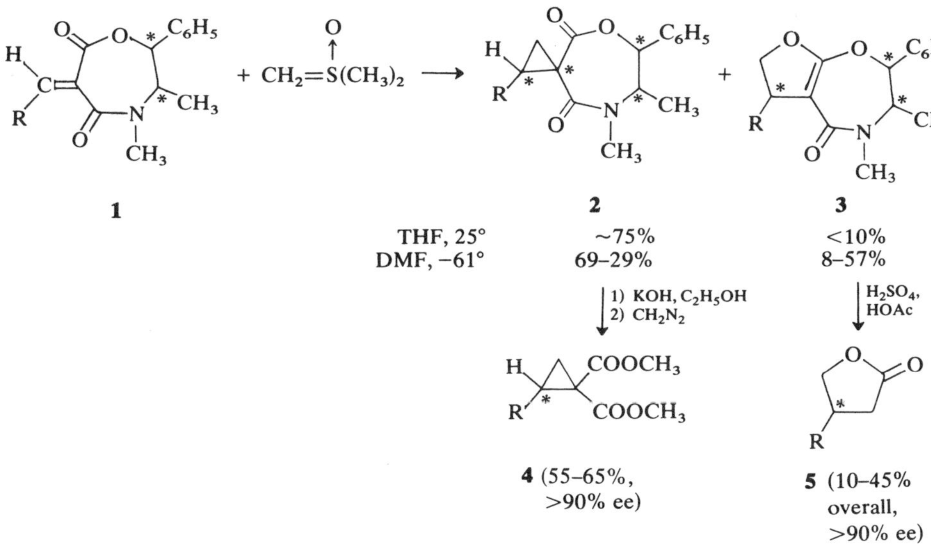

1 + $CH_2{=}S(O)(CH_3)_2$ ⟶ **2** + **3**

	2	3
THF, 25°	~75%	<10%
DMF, −61°	69–29%	8–57%

2 → 1) KOH, C_2H_5OH; 2) CH_2N_2 → **4** (55–65%, >90% ee)

3 → H_2SO_4, HOAc → **5** (10–45% overall, >90% ee)

[1] T. Mukaiyama, K. Fujimoto, and T. Takeda, *Chem. Letters*, 1207 (1979).
[2] T. Mukaiyama, T. Takeda, and M. Osaki, *ibid.*, 1165 (1977); T. Mukaiyama, T. Takeda, and K. Fujimoto, *Bull. Chem. Soc. Japan*, **51**, 3368 (1978).

N,N-Dimethylphosphoramidic dichloride [(Dimethylamido)phosphoryl dichloride], $(CH_3)_2NPOCl_2$. Mol. wt. 161.96, b.p. 194°, 90°/22 mm. The reagent is obtained in 96% yield by the reaction of dimethylamine and phosphorus(V) trichloride oxide, $POCl_3$.[1]

Esterification.[2] Esters can be prepared in 70–95% yield by the reaction of an acid and an alcohol in the presence of pyridine (3 equiv.) and $(CH_3)_2NPOCl_2$ (1.5 equiv.) at room temperature. Phenyl dichlorophosphate, $C_6H_5OPOCl_2$ (**1**, 847), can also be used as the activating agent, often with somewhat improved yields. However, this variation results in much lower yields in the case of penicillins. This esterification is applicable even to tertiary alcohols.

Reaction with alcohols.[3] Primary alcohols are converted in high yield into alkyl chlorides by reaction with **1** at 80° (4–8 hours). Tertiary alcohols are dehydrated by **1** to olefins, again in high yield. Secondary alcohols on reaction with **1** give a mixture of alkyl chlorides and olefins. Displacement is favored in the case of less sterically hindered alcohols. The product of dehydration is usually the more thermodynamically stable isomer.

[1] E. N. Walsh and A. D. F. Toy, *Inorg. Syn.*, **7**, 69 (1963).
[2] H.-J. Liu, W. H. Chan, and S. P. Lee, *Tetrahedron Letters*, 4461 (1978).
[3] *Idem, Chem. Letters*, 923 (1978).

N,N-Dimethylsulfamoyl chloride, $(CH_3)_2NSO_2Cl$ (**1**). Mol. wt. 143.59, b.p. 114°/75 mm. Suppliers: Aldrich, Eastman, Fluka.

Deoxygenation. The 3-hydroxyl group of a number of methyl α-D-glucopyranosides has been replaced by hydrogen via N,N-dimethylsulfamoyl derivatives, prepared with NaH in DMF and **1** or by reaction with sulfuryl chloride, pyridine, and methylamine. The latter method avoids use of a strong base. The reducing agent is sodium in liquid ammonia, which can eliminate some other groups as well: benzylidene, O-acetyl, N-BOC, N-Ts.[1]

Examples:

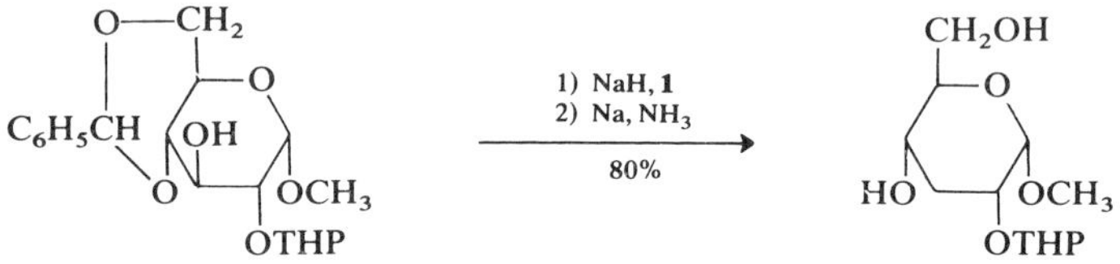

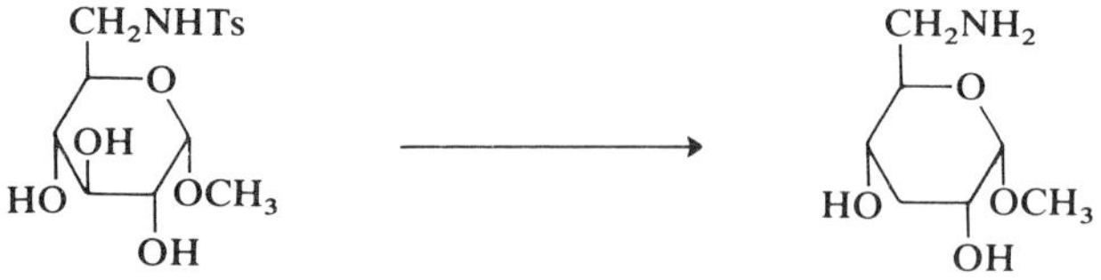

O—CH_2 / C_6H_5CH / O / OH / OCH_3 / OTHP $\xrightarrow[80\%]{1)\ NaH,\ \mathbf{1};\ 2)\ Na,\ NH_3}$ CH_2OH / O / HO / OCH_3 / OTHP

CH_2NHTs / O / OH / HO / OCH_3 / OH $\longrightarrow$ CH_2NH_2 / O / HO / OCH_3 / OH

[1] T. Tsuchiya, I. Watanabe, M. Yoshida, F. Nakamura, T. Usui, M. Kitamura, and S. Umezawa, *Tetrahedron Letters*, 3365 (1978).

Dimethyl sulfide–Trifluoroacetic anhydride.

Reduction of sulfoxides. Sulfoxides are rapidly reduced in almost quantitative yields to sulfides by dimethyl sulfide and TFAA.[1]

$$R\overset{O}{\overset{\uparrow}{S}}R^1 + \underset{(excess)}{CH_3SCH_3} \xrightarrow{TFAA} RSR^1 + CH_3SCH_2OCOCF_3$$

[1] R. Tanikaga, K. Nakayama, K. Tanaka, and A. Kaji, *Chem. Letters*, 395 (1977).

Dimethylsulfonium methylide, 1, 314–315; **2,** 169–171; **3,** 124–125; **4,** 196–197, **8,** 198.

Oxiranes. Ketones can be converted into oxiranes by dimethylsulfonium methylide generated *in situ* as shown (equation I). The method is particularly useful for large-scale preparations.[1]

$$\text{(I)}\quad (CH_3)_2S + (CH_3)_2SO_4 \xrightarrow{CH_3CN} (CH_3)_3S^+CH_3SO_4^- \xrightarrow{NaOCH_3}$$

$$[(CH_3)_2S{=}CH_2] \xrightarrow[60\text{–}90\%]{R^1COR^2} \underset{R^2}{\overset{R^1}{}}C\overset{O}{—}CH_2 + S(CH_3)_2$$

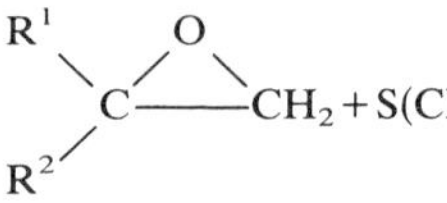

[1] T. Kutsuma, I. Nagayama, T. Okazaki, T. Sakamoto, and S. Akaboshi, *Heterocycles*, **8**, 397 (1977).

Dimethyl sulfoxide, 1, 296–310; **2**, 157–158; **3**, 119–123; **4**, 192–194; **5**, 263–266; **6**, 225–229; **7**, 133–135; **8**, 198–199.

Solvent effects.[1] The final step of a total synthesis of β-vetivone (**2**) involves the intramolecular γ-alkylation of the α,β-unsaturated ketone **1**. Under a variety of base–solvent combinations **1** is converted into **4**, the product of kinetically favored α-alkylation (70% using KO-*t*-Bu, *t*-BuOH). However, alkylation takes the desired course with NaOH in DMSO–H_2O mixtures, the ratio **4/2** being dependent on the amount of water present. Best results are obtained with 25% aqueous DMSO (**4/2** = 7:93). Actually, the β-vetivone as obtained contains 7% of the 10-epimer (**3**).

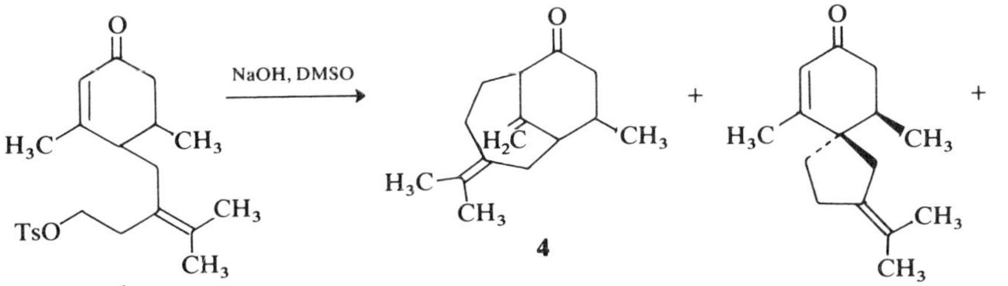

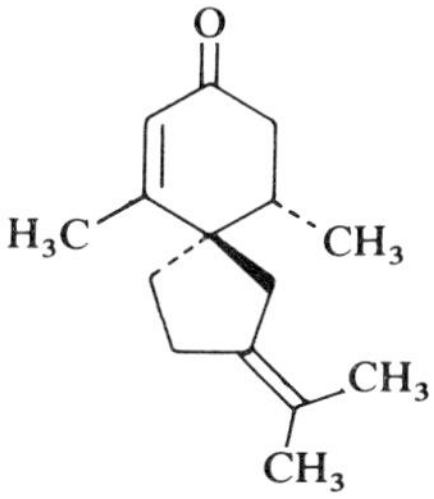

	4/2+3	
<0·5% H_2O	100:0	
5% H_2O	85:15	
10% H_2O	76:24	
15% H_2O	16:84	
20% H_2O	7.2:92.8	
25% H_2O	6.5:93.5	(**2/3** = 92.6:7.4)

Oxidation.[2] Reaction of tropylium tetrafluoroborate (**1**) with DMSO at 55° (4 hours) results in a product (**a**) that is converted to tropone (**2**) on addition of water and chloroform.

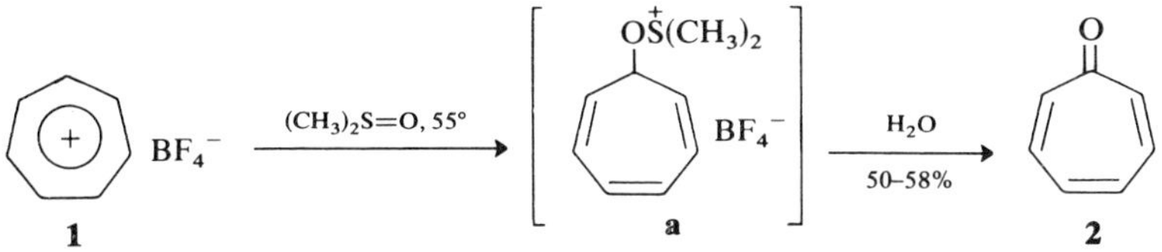

[1] A. P. Johnson and V. Vajs, *J.C.S. Chem. Comm.*, 817 (1979).
[2] E. Garfunkel and I. D. Reingold, *J. Org.*, **44**, 3725 (1979).

Dimethyl sulfoxide–Acetic anhydride, 1, 305; **2**, 163–165; **3**, 121–122; **4**, 199; **7**, 135.

Methylthiomethyl ethers (**6**, 109–110).[1] Alcohols can be converted into these ethers in 40–95% yield by reaction with DMSO, acetic anhydride, and acetic acid at room temperature. (Higher temperatures favor oxidation of the alcohol to the ketone.) This reaction is applicable to primary, secondary, tertiary, and hindered alcohols.

The methylthiomethyl group is removed in nearly quantitative yield by reaction with methyl iodide in moist acetone.

These ethers can also be converted into methyl ethers in high yield by nickel boride (**5**, 472) in boric acid buffer. Ester and acetal groups are stable under these conditions. Raney nickel is less useful for this desulfurization because of erratic results.

[1] P. M. Pojer and S. J. Angyal, *Aust. J. Chem.*, **31**, 1031 (1978).

Dimethyl sulfoxide–*t*-Butyl bromide.

Methylthiomethyl esters. Carboxylic acids are converted into their methylthiomethyl esters by reaction with *t*-butyl bromide–dimethyl sulfoxide in the presence of sodium bicarbonate. Methylthiomethyl esters of N-protected amino acids can be obtained in high yield without racemization by this method.[1]

$$(CH_3)_2SO + (CH_3)_3CBr \rightarrow [(CH_3)_2\overset{+}{S}OC(CH_3)_3Br^-] \xrightarrow[90-99\%]{RCOOH,\ NaHCO_3} RCOOCH_2SCH_3$$

[1] A. Dossena, R. Marchelli, and G. Casnati, *J.C.S. Chem. Comm.*, 370 (1979).

Dimethyl sulfoxide–Chlorotrimethylsilane.

Formaldehyde acetals. The combination of DMSO and $ClSi(CH_3)_3$ in benzene converts an alcohol by an exothermic reaction into the formaldehyde acetal (equation I). The mechanism is not known.[1]

$$\text{(I)} \qquad (CH_3)_2S{=}O + ClSi(CH_3)_3 \xrightarrow{C_6H_6,\ 0^\circ} \xrightarrow[80-95\%]{ROH,\ 0\rightarrow 75^\circ} CH_2(OR)_2$$

[1] B. S. Bal and H. W. Pinnick, *J. Org.*, **44**, 3727 (1979).

Dimethyl sulfoxide–Iodine, 5, 347; **6**, 295; **8**, 200.

Oxidation. DMSO in combination with catalytic amounts of iodine and a strong acid oxidizes methylene groups activated by an adjacent benzoyl group in good yield. The method is only suitable for preparation of diaryl di- and triketones.[1]

Examples:

$$C_6H_5COCH_2C_6H_5 \xrightarrow[69\%]{DMSO,\ I_2,\ H^+} C_6H_5COCOC_6H_5$$

$$C_6H_5COCH_2COC_6H_5 \xrightarrow[89\%]{} C_6H_5CO\overset{OH}{\underset{OH}{C}}COC_6H_5$$

The system also effects dehydrogenation of flavanones (**1**) to flavones (**2**) in high yield. Presumably an intermediate 3-iodoflavanone is formed, which undergoes dehydrohalogenation.[2]

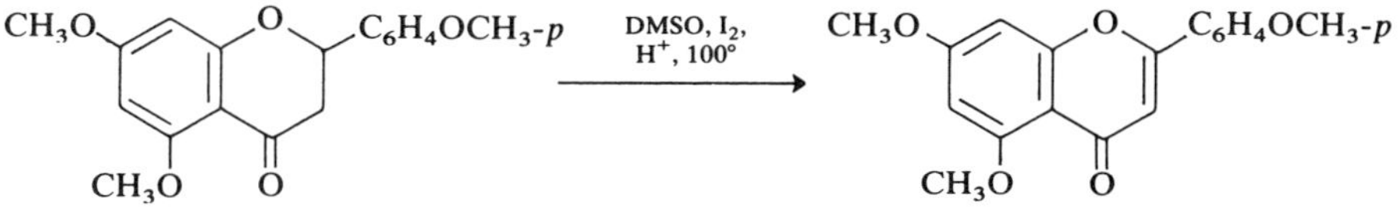

CH_3O ... $C_6H_4OCH_3$-*p* $\xrightarrow{\text{DMSO, } I_2,\ H^+,\ 100°}$ CH_3O ... $C_6H_4OCH_3$-*p*

1 **2**

[1] N. Furukawa, T. Akasaka, T. Aida, and S. Oae, *J.C.S. Perkin I*, 372 (1977).
[2] W. Fatma, J. Igbal, H. Ismail, K. Ishratullah, W. A. Shaida, and W. Rahman, *Chem. Ind.*, 315 (1979).

Dimethyl sulfoxide–Methanesulfonic anhydride, 5, 266.

Oxidative lactonization. A recently discovered group of mycotoxins known as tryptoquivalines has a novel spirolactone unit derived from N-acyltryptophans. A typical member of the group, tryptoquivaline G (**1**), has now been synthesized by Büchi and co-workers[1] and has been shown to be derived from D-tryptophan. A key step involves oxidation of a derivative (**2**) of L-tryptophan with DMSO and methanesulfonic anhydride (2 equiv.) to **3** and **4**. NBS has been used for such a reaction, but it can also effect undesired nuclear bromination.[2] The main product **3** was converted in several steps into tryptoquivaline L, the C_{19} epimer of **1**. Contra-thermodynamic epimerization (KH, then protonation) converted this isomer into **1** and thus proved the absolute configuration.

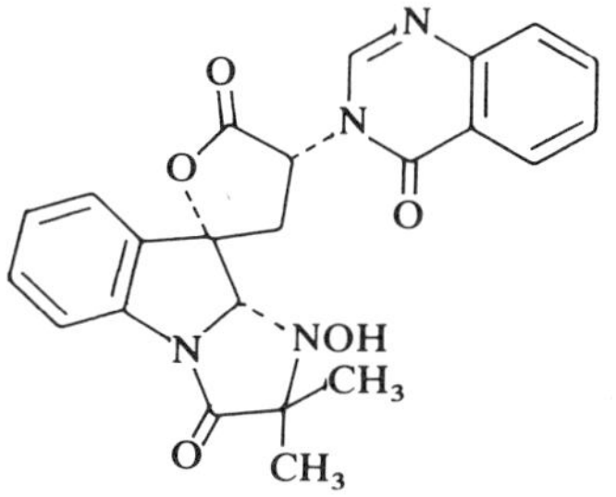

1 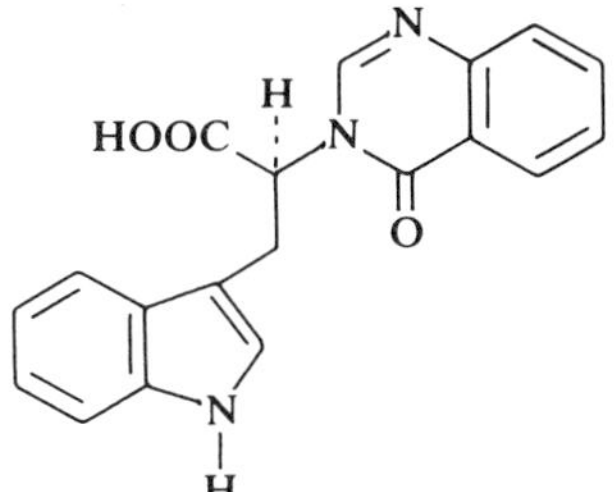 **2**

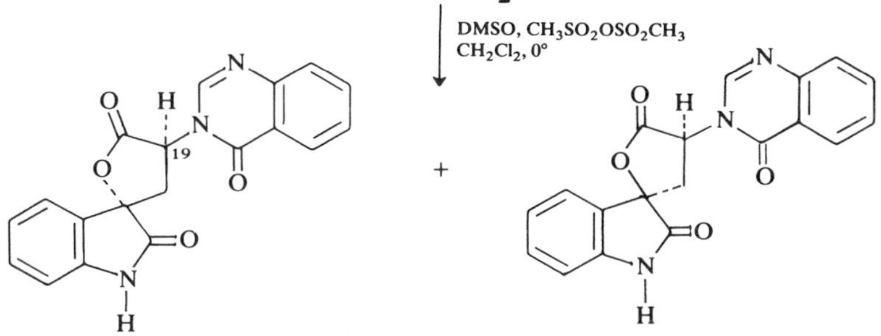

DMSO, $CH_3SO_2OSO_2CH_3$, CH_2Cl_2, 0°

3 (56–66%) **4** (<10%)

[1] G. Büchi, P. R. DeShong, S. Katsumura, and Y. Sugimura, *Am. Soc.*, **101**, 5084 (1979).
[2] A. Patchornik, W. B. Lawson, and B. Witkop, *ibid.*, **80**, 4748 (1958).

Dimethyl sulfoxide–Oxalyl chloride, 8, 200. Swern and co-workers[1] have suggested that the acutal oxidizing agent from DMSO and oxalyl chloride is **1**, formed by almost instantaneous loss of CO_2 and CO. This is also the reagent obtained by Corey and Kim from dimethyl sulfide and chlorine (**4**, 191).

$$\text{(I)}\quad (CH_3)_2SO + (COCl)_2 \xrightarrow[60^\circ]{CH_2Cl_2,} \left[(CH_3)_2\overset{+}{S}O\overset{O}{\overset{\|}{C}}-\overset{O}{\overset{\|}{C}}Cl\,Cl^-\right]_{\mathbf{a}} \xrightarrow{-CO_2,\ CO} \underset{\mathbf{1}}{(CH_3)_2\overset{+}{S}Cl\,Cl^-}$$

The paper lists examples of further useful oxidation of ketals, diols, and ketols, as well as heteroaryl alcohols. It is not useful for phenethyl alcohols (39% yield of ketone).

Iminosulfuranes (sulfilimines).[2] Arenesulfonamides, $ArS(O_2)NH_2$, are converted into S,S-dimethylsulfilimines, $(CH_3)_2\overset{+}{S}\text{-}\overset{-}{N}SO_2Ar$, by DMSO–oxalyl chloride (75–95% yields).

[1] A. J. Mancuso, D. S. Brownfain, and D. Swern, *J. Org.*, **44**, 4148 (1979).
[2] S. L. Huang and D. Swern, *ibid.*, **43**, 4537 (1978).

Dimethyl sulfoxide-3-Sulfopropanoic anhydride complex (1). The two reagents form the insoluble zwitterionic complex **1** when mixed in CH_2Cl_2 at 0°.

$$(CH_3)_2S{=}O + \text{(cyclic 3-sulfopropanoic anhydride, } SO_2\text{)} \longrightarrow \underset{\mathbf{1}}{(CH_3)_2\overset{+}{S}{\cdots}O\overset{\overset{\|}{O}}{C}CH_2CH_2SO_3^-}$$

Sulfuranes.[1] These ylides can be obtained by reaction of complex **1** with active methylene compounds (50–70% yield, 4 examples).

Example:

$$\text{Meldrum's acid (2,2-dimethyl-1,3-dioxane-4,6-dione)} + \mathbf{1} \longrightarrow \underset{(51\%)}{\text{(2,2-dimethyl-4,6-dioxo-1,3-dioxan-5-ide)}-\overset{+}{S}(CH_3)_2} + HOOCCH_2CH_2SO_3H$$

[1] K. Schank and C. Schuhknecht, *Synthesis*, 678 (1978).

Dimethyl sulfoxide–Trichloroacetic anhydride.

Oxidation of alcohols. Oxidation of **1** without glycol cleavage has been achieved with a modified reagent of Swern, DMSO–$(CCl_3CO)_2O$ (*see* **5**, 266–267; **7**, 136).[1]

Trifluoroacetylation of the secondary alcohol complicates oxidations with the original reagent.

$$\mathbf{1} \xrightarrow[75\text{–}80\%]{1)\ DMSO\text{–}(CCl_3CO)_2O;\ 2)\ Et_3N} \mathbf{2}$$

[1] E. J. Corey, R. L. Danheiser, S. Chandrasekaran, P. Siret, G. E. Keck, and J.-L. Gras, *Am. Soc.*, **100**, 8031 (1978).

S,S-Dimethyl-N-tosylsulfilimine, $(CH_3)_2S{=}NSO_2C_6H_4CH_3(p)$. Mol. wt. 231.32, m.p. 154–155°. The sulfilimine is prepared in 92% yield by the reaction of dimethyl sulfide with chloramine T.[1]

o-Methylthiomethylation of phenols.[2] N-Arylsulfonyl-S,S-dimethylsulfilimines react regiospecifically to form *o*-methylthiomethylated phenols in 60–95% yields, based on the phenol that has reacted (equation I). Somewhat higher yields are obtained with S,S-dimethyl-N-(2,4-dinitrophenyl)sulfilimine, but this reagent is less accessible. The products are useful intermediates to *o*-methylphenols (**5**, 131–132).

(I) $$\text{ArOH} + (CH_3)_2S{=}NTs \xrightarrow[60\text{–}95\%]{120\text{–}130^\circ} o\text{-}CH_2SCH_3\text{-ArOH}$$

$(R = H, CH_3)$

[1] C. W. Todd, J. H. Fletcher, and D. S. Tarbell, *Am. Soc.*, **65**, 350 (1943).
[2] T. Yamamoto and M. Okawara, *Bull. Chem. Soc. Japan*, **51**, 2443 (1978).

Dimethyl N-(*p*-tosyl)sulfoximine, $CH_3S(=O)(=NTs)CH_3$ (**1**). Mol. wt. 247.33, m.p. 167–169°.

Supplier: Columbia.

Oxetanes. The sodium anion (**2**) of this sulfoximine reacts with ketones to form epoxides (**3**) in good yield.[1] Welch and Prakasa Rao[2] now report that in the reaction of ketones with an excess of **2** at elevated temperatures intermediate epoxides undergo ring expansion, with formation of the corresponding oxetane derivative.

$$\mathbf{1} \xrightarrow{NaH,\ DMSO} \underset{\mathbf{2}}{CH_3S(=O)(=NTs)CH_2Na} \xrightarrow[25^\circ]{R^1COR^2} \mathbf{3} \xrightarrow[46\text{–}96\%]{\mathbf{2},\ 40^\circ} \mathbf{4}$$

Examples:

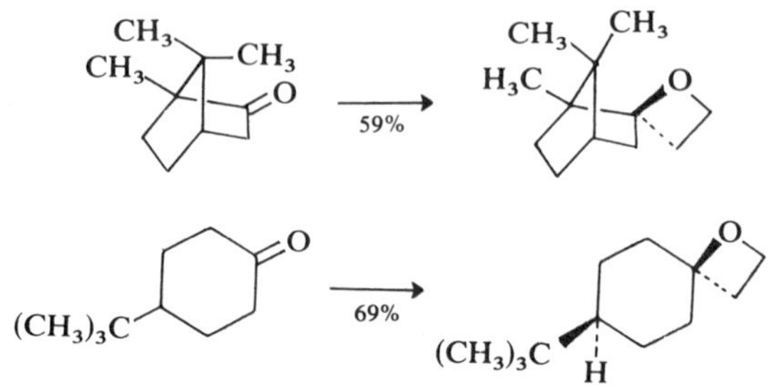

59%

69%

[1] C. R. Johnson, R. A. Kirchoff, R. J. Reischer, and G. F. Katekar, *Am. Soc.*, **95**, 4287 (1973).
[2] S. C. Welch and A. S. C. P. Rao, *ibid.*, **101**, 6135 (1979).

1,3-Dimethyluracil (1). Mol. wt. 140.41, m.p. 122–124°. Suppliers: K and K, Pfaltz and Bauer, Sigma, and Fluka.

[2+2] Cycloadditions. This reagent serves as a formyl acetic ester equivalent in [2+2] cycloadditions with alkenes.[1,2] The addition provides a new method for vicinal attachment of carboxaldehyde and acetic ester appendages to double bonds (e.g., **2** → **5**).[2] An intramolecular version of this reaction was used in a total synthesis of reserpine.[3]

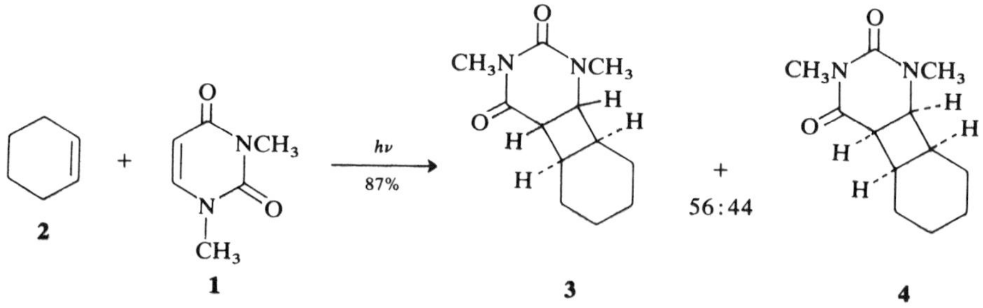

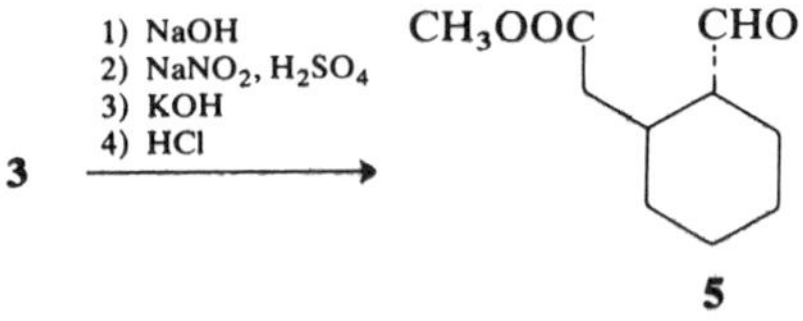

$h\nu$, 87%; **3** + **4** (56:44)

3 → 1) NaOH 2) $NaNO_2$, H_2SO_4 3) KOH 4) HCl → **5**

[1] J. S. Swenton, J. A. Hyatt, J. M. Lisy, and J. Clardy, *Am. Soc.*, **96**, 4885 (1974).
[2] B. A. Pearlman, *ibid.*, **101**, 6398 (1979).
[3] *Idem, ibid.*, **101**, 6404 (1979).

2,4-Dinitrobenzenesulfenyl chloride, 1, 319–320; **6,** 231–232

1,4-Dehydration of allylic alcohols. Treatment of the allylic alcohol **1** with the sulfenyl chloride (2 equiv.) and triethylamine (3 equiv.) in dichloromethane leads to the 1,3-diene **2** in 79% yield. The dehydration evidently involves a [2,3]sigmatropic rearrangement of a sulfenate ester (**a**) to the allyl sulfoxide (**b**), which fragments to the diene **2**.

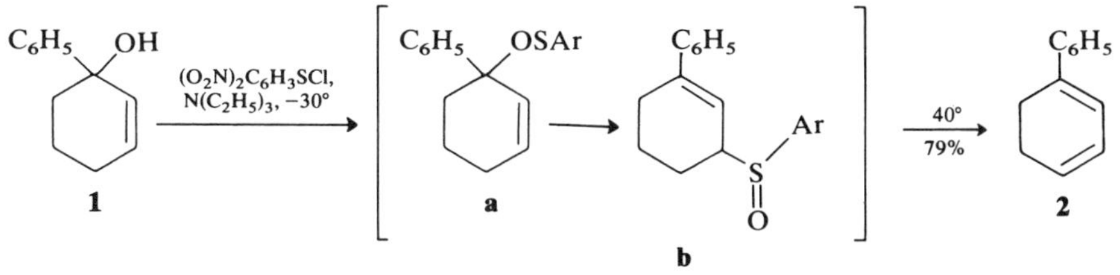

C_6H_5 OH; $(O_2N)_2C_6H_3SCl$, $N(C_2H_5)_3$, −30°; C_6H_5 OSAr; C_6H_5; Ar; S; O; 40°; 79%; C_6H_5

1 **a** **b** **2**

Other examples:

C_6H_5 OH H_3C CH_3 → (75%) C_6H_5 CH_3 CH_3

CH_3 C_6H_5 OH CH_3 → (54%) CH_3 C_6H_5 CH_3 (E, E/E, Z = 60 : 40)

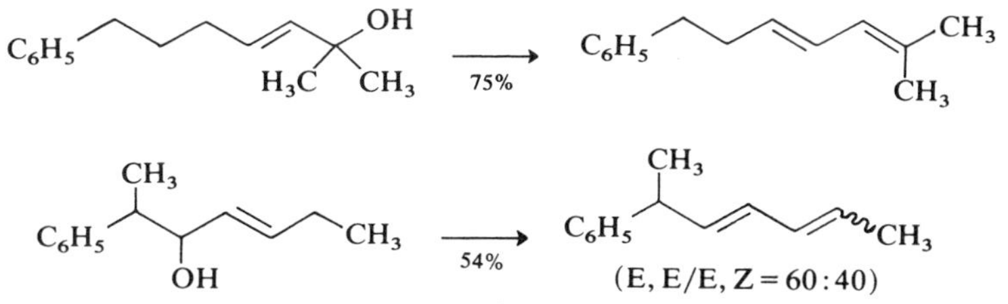

This reaction was used to prepare the oxide of 2,5,2′,5′-tetrachlorobiphenyl (**3**), presumably responsible for the carcinogenic effects of the tetrachlorobiphenyl (equation I). The oxide shows surprising chemical stability, with a half-life in CH_3OH of 20 days at 25°.[1]

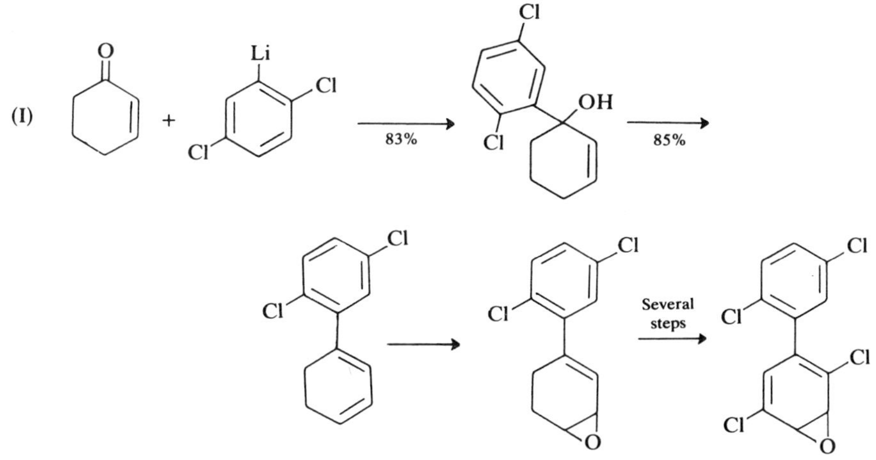

(I) O + Li Cl Cl → (83%) Cl Cl OH → (85%)

Cl Cl → Cl Cl O → (Several steps) Cl Cl Cl Cl O

3

[1] H. J. Reich, I. L. Reich, and S. Wollowitz, *Am. Soc.*, **100**, 5981 (1978).

3,5-Dinitroperoxybenzoic acid, $(NO_2)_2C_6H_3COOOH$. Mol. wt. 228.12, m.p. 113–115°. *Caution*: All peroxyacids are potentially explosive when heated. Prepared by reaction of the carboxylic acid in methanesulfonic acid with 90% H_2O_2 at 50° (*caution*: The reaction is exothermic at temperatures >53°); yield 93.5%. The reagent is stable for about a year at −10°.

Epoxidations and Baeyer–Villiger oxidations.[1] This peroxy acid resembles peroxytrifluoroacetic acid in activity, but buffers are not necessary. The report includes two successful epoxidations with the new peroxy acid (and also *p*-nitroperoxybenzoic acid) for which peroxytrifluoroacetic acid was of no value.

The reagent and sodium carbonate in 1,2-dichloroethane at 54° in the presence of 4,4′-thiobis(6-*tert*-butyl-3-methylphenol) (**4**, 85–86) effect the Baeyer–Villiger oxidation of **1**.[2] This oxidation was unsuccessful with other peracids. The tricyclic compound (**2**) is an intermediate in Corey's total synthesis of gibberellic acid.

1 (H; $C(=O)CH_3$; OH) —71%→ **2** (H; OAc; OH)

[1] W. H. Rastetter, T. J. Richard, and M. D. Lewis, *J. Org.*, **43**, 3163 (1978).
[2] E. J. Corey and J. G. Smith, *Am. Soc.*, **101**, 1038 (1979); J. G. Smith, Ph.D. Thesis, Harvard University, Cambridge, Mass. (1978).

2-(1,3-Dioxan-2-yl)ethylidenetriphenylphosphorane (1). Mol. wt. 376.43.
Preparation:

$CH_2{=}CHCHO \rightarrow [BrCH_2CH_2CHO] \xrightarrow[80\%]{HO(CH_2)_3OH,\ TsOH} BrCH_2CH_2CH(OCH_2CH_2CH_2O) \xrightarrow[96\%]{P(C_6H_5)_3}$

$(C_6H_5)_3\overset{+}{P}{-}CH_2CH_2CH(OCH_2CH_2CH_2O)\ Br^- \xrightarrow{KOC(CH_3)_3} (C_6H_5)_3P{=}CHCH_2CH(OCH_2CH_2CH_2O)$ **1**

cis-β,γ-*Unsaturated aldehydes.* The phosphorane **1**, when generated with potassium *t*-butoxide, reacts with aldehydes to form acetals of (Z)-3-alkenals (**2**) in 40–70% yield. The reagent generated with *n*-butyllithium reacts to form a mixture of (Z)- and (E)-3-alkenals in somewhat higher yield. The free aldehydes are obtained by transacetalization with methanol, followed by hydrolysis with aqueous acetic acid. The free aldehydes are readily isomerized to α,β-unsaturated aldehydes when heated.[1]

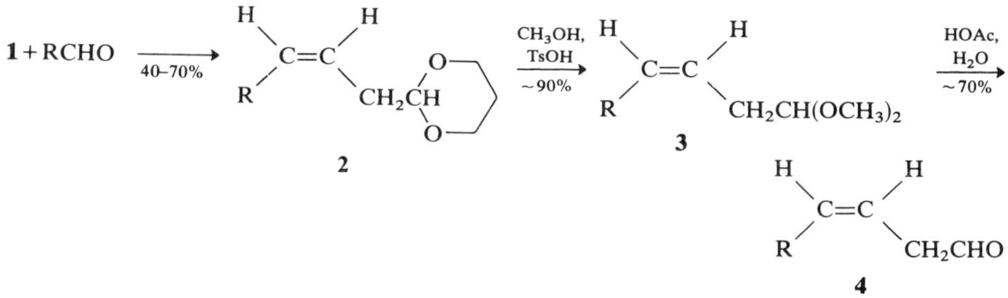

$$1 + RCHO \xrightarrow{40-70\%} \text{(H)(R)C=C(H)}CH_2CH\text{(1,3-dioxane)} \; (2) \xrightarrow[\sim 90\%]{CH_3OH,\ TsOH} \text{(H)(R)C=C(H)}CH_2CH(OCH_3)_2 \; (3) \xrightarrow[\sim 70\%]{HOAc,\ H_2O} \text{(H)(R)C=C(H)}CH_2CHO \; (4)$$

[1] J. C. Stowell and D. R. Keith, *Synthesis*, 132 (1979).

Diperoxo-oxohexamethylphosphoramidomolybdenum(VI), 4, 203–204; **5**, 269–270; **7**, 136; **8**, 206–208.

Enolate hydroxylation (**5**, 269–270; **8**, 207–208). Vedejs hydroxylation was used twice in a biomimetic synthesis of the sesquiterpene phytuberin (**1**) from α-cyperone (**2**).[1] In the first step, a hydroxyl group was introduced at C_2 to give **3**. At

1 2

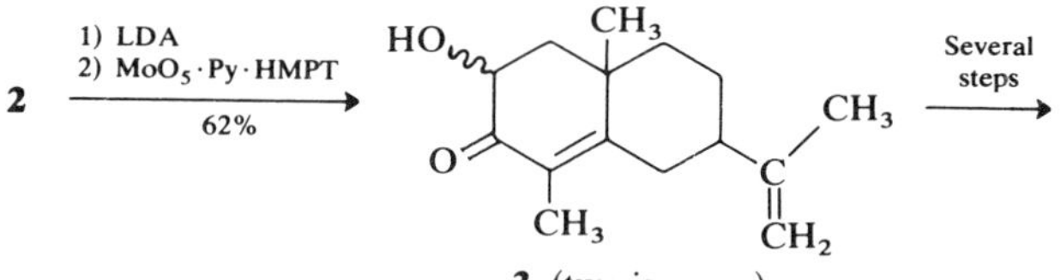

$$\mathbf{2} \xrightarrow[62\%]{1)\ LDA\ \ 2)\ MoO_5 \cdot Py \cdot HMPT} \mathbf{3}\text{ (two isomers)} \xrightarrow{\text{Several steps}}$$

3 (two isomers)

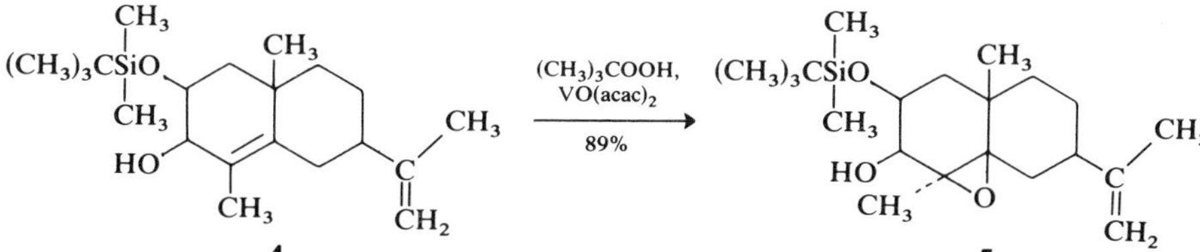

$$\mathbf{4} \xrightarrow[89\%]{(CH_3)_3COOH,\ VO(acac)_2} \mathbf{5}$$

4 5

$$\xrightarrow{\text{Several steps}} \mathbf{6} \xrightarrow[57\%]{1)\ LDA\ \ 2)\ MoO_5 \cdot Py \cdot HMPT} \mathbf{7}$$

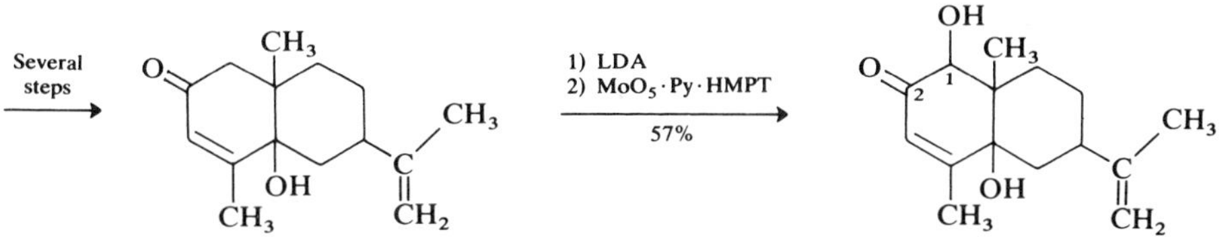

6 7

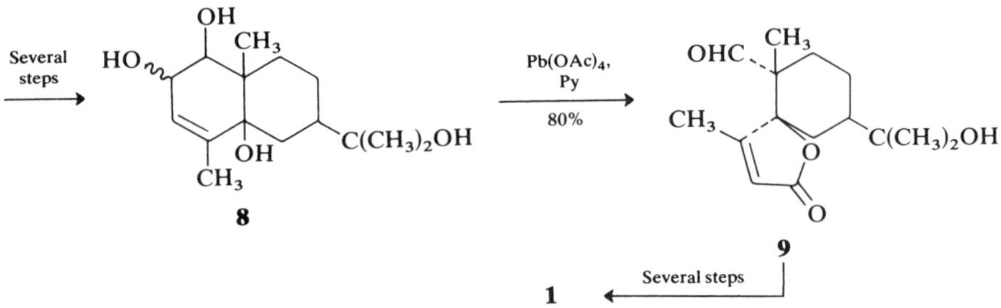

Several steps → 8 → Pb(OAc)$_4$, Py, 80% → 9 → Several steps → 1

a later stage, oxygen was introduced at C_5 by Sharpless allylic epoxidation (**5**, 75–76) to give **5**. A second Vedejs hydroxylation was used to introduce oxygen at C_1 to give **7**. This product was converted into the tetraol **8**, which was oxidatively cleaved by lead tetraacetate in pyridine (**1**, 550) to give **9**. This formyl γ-lactone was then converted by known reactions into **1**. The absolute configuration (R) at C_7 is also established by this synthesis.

Aldehyde-enolate hydroxylation. Enolate hydroxylation has been extended to the aldehyde **1** by Tanis and Nakanishi[2] in a total synthesis of the drimanic sesquiterpene (±)-warburganal (**3**). Thus treatment of **1** with LDA in THF at −78°C followed by $MoO_5 \cdot HMPT$ (1.5 equiv.) affords **2** in 85% yield.

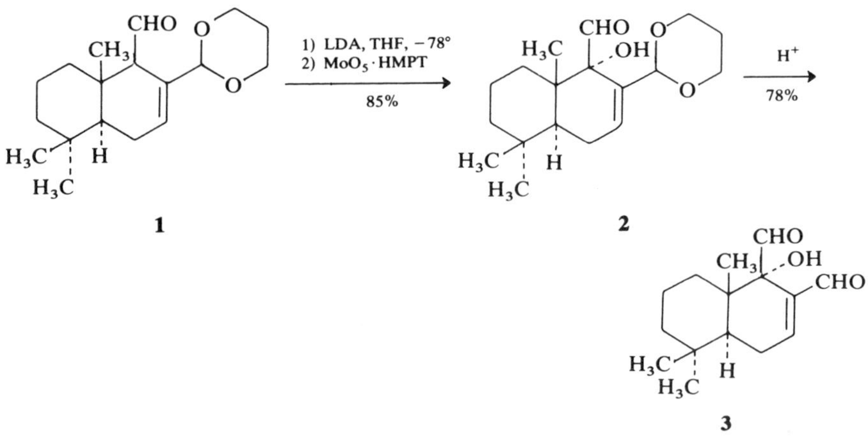

1 → 1) LDA, THF, −78°; 2) $MoO_5 \cdot HMPT$, 85% → **2** → H^+, 78% → **3**

[1] A. Murai, M. Ono, A. Abiko, and T. Masamune, *Am. Soc.*, **100**, 7751 (1978).
[2] S. P. Tanis and K. Nakanishi, *ibid.*, **101**, 4398 (1979).

Diphenylborane, $(C_6H_5)_2BH$ **(1).** The borane has been prepared *in situ* by two methods (equations I and II).[1]

$$\text{(I)}\quad 2C_6H_5MgBr + B(OCH_3)_3 \rightarrow (C_6H_5)_2BOH \xrightarrow[80\%]{CH_3OH} (C_6H_5)_2BOCH_3 \xrightarrow{LiAlH_4} \mathbf{1}$$

$$\text{(II)}\quad (C_6H_5)_2BOCH_3 + Py \xrightarrow{AlH_3} (C_6H_5)_2BH \cdot Py \xrightarrow{BF_3 \cdot O(C_2H_5)_2} \mathbf{1}$$

Conjugate addition of alkyldiphenylboranes. One useful reaction of trialkylboranes is the facile conjugate addition to α,β-unsaturated aldehydes and ketones.[2] One limitation, however, is that only one alkyl group is utilized. Since triphenylborane does not undergo this reaction, the problem is circumvented by use of alkyldiphenylboranes, which are available by hydroboration of an alkene. This hydroboration reaction is highly regioselective; the boron becomes attached almost exclusively to the terminal position of a 1-alkene. The products undergo facile conjugate addition.[1]

Examples:

$$n\text{-}C_6H_{13}CH{=}CH_2 \xrightarrow[95\%]{1)\ HB(C_6H_5)_2\quad 2)\ CH_2{=}CHCOCH_3} n\text{-}C_6H_{13}CH_2CH_2CH_2CH_2COCH_3$$

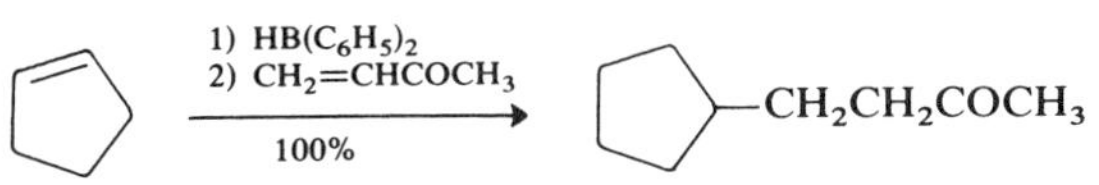

$$\text{cyclopentene} \xrightarrow[100\%]{1)\ HB(C_6H_5)_2\quad 2)\ CH_2{=}CHCOCH_3} \text{cyclopentyl}{-}CH_2CH_2COCH_3$$

[1] P. Jacob III, *J. Organometal. Chem.*, **156**, 101 (1978).
[2] H. C. Brown, *Organic Synthesis via Boranes*, Wiley, New York, 1975, pp. 146–148.

Diphenyl diselenide, 5, 272–276; **6**, 235; **7**, 136–137.

cis–trans *Isomerization.* The isomerization of vitamin D_2 (**1**) to 5,6-*trans*-vitamin D_2 (**2**) was originally carried out by iodine under irradiation in 20–40% yield.[1] It has recently been carried out in higher yield (57%) by irradiation in the presence of diphenyl diselenide.[2]

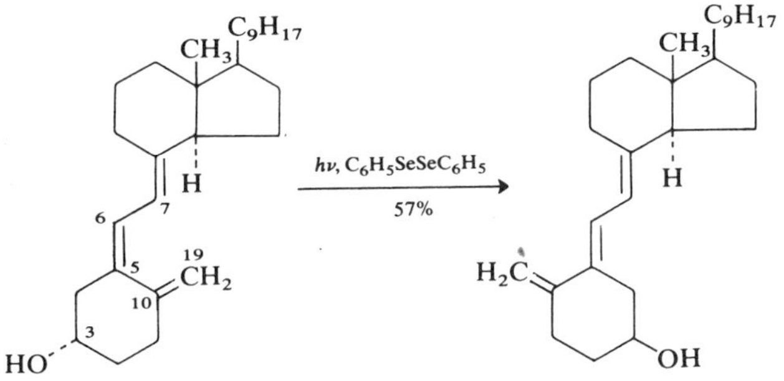

$$\mathbf{1} \xrightarrow[57\%]{h\nu,\ C_6H_5SeSeC_6H_5} \mathbf{2}$$

[1] L. F. Fieser and M. Fieser, *Steroids*, Reinhold, N.Y., 1959, pp. 146–150.
[2] A. G. M. Barrett, D. H. R. Barton, and G. Johnson, *Synthesis*, 741 (1978).

Diphenyl diselenide–Copper(II) acetate.

Acetoxyselenation. Diphenyl diselenide in combination with either $Cu(OAc)_2$ or $Pb(OAc)_4$ effects *trans*-acetoxyselenation of alkenes in about 70% yield. Cupric acetate can be used in catalytic amounts when oxygen is introduced into the reaction.[1] The actual reagent is probably C_6H_5SeOAc (compare Benzeneselenenyl trifluoroacetate, **5**, 522–523).

[1] N. Miyoshi, Y. Ohno, K. Kondo, S. Murai, and N. Sonoda, *Chem. Letters*, 1309 (1979).

Diphenyl disulfide–Tri-*n*-butylphosphine.

Thioacetalization.[1] Aldehydes are converted into diphenyl thioacetals by reaction with diphenyl disulfide and tri-*n*-butylphosphine at room temperature for a few minutes (equation I). Yields of thioacetals are much lower when triphenylphosphine is used and also when diphenyl disulfide is replaced by dialkyl disulfides. Ketones are not reactive to this system even under forcing conditions.

$$\text{(I)}\quad RCHO + C_6H_5SSC_6H_5 + (n\text{-}C_4H_9)_3P \xrightarrow[70\text{–}85\%]{20^\circ} RCH(SC_6H_5)_2 + (n\text{-}C_4H_9)_3P{=}O$$

Epoxides are converted by this method into *vic*-dithioethers (equation II).

$$\text{(II)}\quad \text{R-epoxide} + C_6H_5SSC_6H_5 + (n\text{-}C_4H_9)_3P \xrightarrow[50\text{–}70\%]{20^\circ} R(C_6H_5S)CH{-}CH_2SC_6H_5$$

[1] M. Tazaki and M. Takagi, *Chem. Letters*, 767 (1979).

(S)-α-(R)-2-Diphenylphosphinoferrocenyl ethyldimethylamine [(S)-(R)-PPFA] **(1).** M.p. 139°, α_D +361°.

[Structure **1**: ferrocene bearing $P(C_6H_5)_2$ and $C(CH_3)(H)N(CH_3)_2$]

1

Kumada *et al.*[1] have examined a number of chiral ferrocenylphosphines as ligands for asymmetric reactions catalyzed by transition metals. They are of interest because they contain a planar element of chirality as well as an asymmetric carbon atom. They were first used in combination with rhodium catalysts for asymmetric hydrosilylation of ketones with di- and trialkylsilanes in moderate optical yields (5–50%). High stereoselectivity was observed in the hydrogenation of α-acetamidoacrylic acids (equation I) with rhodium catalysts and ferrocenylphosphines.[2]

$$\text{(I)}\quad RCH{=}C(NHCOCH_3){-}COOH \xrightarrow{H_2,\ \text{Rh cat.}} RCH_2CH(NHCOCH_3)COOH$$

(50–90% ee)

$NiCl_2$ complexed with (S)-(R)-**1** or (R)-(S)-**1** also induces asymmetric coupling of 1-phenylethylmagnesium chloride with vinyl bromide to give (S)- or (R)-3-phenyl-1-

butene (equation II). A related ligand, but lacking an asymmetric carbon atom, was almost as efficient, an indication of the importance of planar chirality.[3]

(II) $$C_6H_5CH(CH_3)MgCl + CH_2{=}CHBr \xrightarrow[83\%]{[Ni]} C_6H_5\overset{*}{C}H(CH_3)CH{=}CH_2$$

(~60% ee)

This reaction has been extended to a synthesis of a terpene, (R)-α-curcumene (**4**), in optical purity of 66% in 34% overall yield. The key step is cross-coupling of **2** with vinyl bromide catalyzed by $NiCl_2$ and **1** to yield optically active **3**.[4]

2 + $CH_2{=}CHBr$ $\xrightarrow[90\%]{NiCl_2,\ 1,\ ether\text{-}C_6H_6}$ **3** (α_D −5.38°) $\xrightarrow[\sim 60\%]{\text{Several steps}}$ bromide (α_D −45.92°) $\xrightarrow[69\%]{(CH_3)_2C{=}CHBr,\ NiCl_2L_2}$ **4** (66% ee)

[1] T. Hayashi, K. Yamamoto, and M. Kumada, *Tetrahedron Letters*, 4405 (1974).
[2] T. Hayashi, T. Mise, S. Mitachi, K. Yamamoto, and M. Kumada, *ibid.*, 1133 (1976).
[3] T. Hayashi, M. Tajika, K. Tamao, and M. Kumada, *Am. Soc.*, **98**, 3718 (1976).
[4] K. Tamao, T. Hayashi, H. Matsumoto, H. Yamamoto, and M. Kumada, *Tetrahedron Letters*, 2155 (1979).

Diphenyl selenoxide, 5, 280–281.

Oxidation of catechols. In accordance with earlier reports, diphenyl selenoxide selectively oxidizes *o*- and *p*-hydroquinones and has no effect on simple phenols. An example is the oxidation of 3,5-di-*t*-butylcatechol in methanol at 0° to the corresponding *o*-quinone, possibly via a selenurane (**a**) (equation I).

(I) 3,5-di-*t*-butylcatechol $\xrightarrow{(C_6H_5)_2SeO,\ CH_3OH,\ 0^\circ}$ [**a**, $Se(C_6H_5)_2$ selenurane] $\xrightarrow{\text{quant.}}$ 3,5-di-*t*-butyl-*o*-quinone

Marino and Schwartz[1] have exploited the selectivity of this selenoxide in an intramolecular phenolic coupling in a synthesis of aporphines and homoaporphins.

An example is the synthesis of N-trifluoroacetylwilsonirine **3** from the benzyltetrahydroisoquinoline **1** in 80% overall yield.[2]

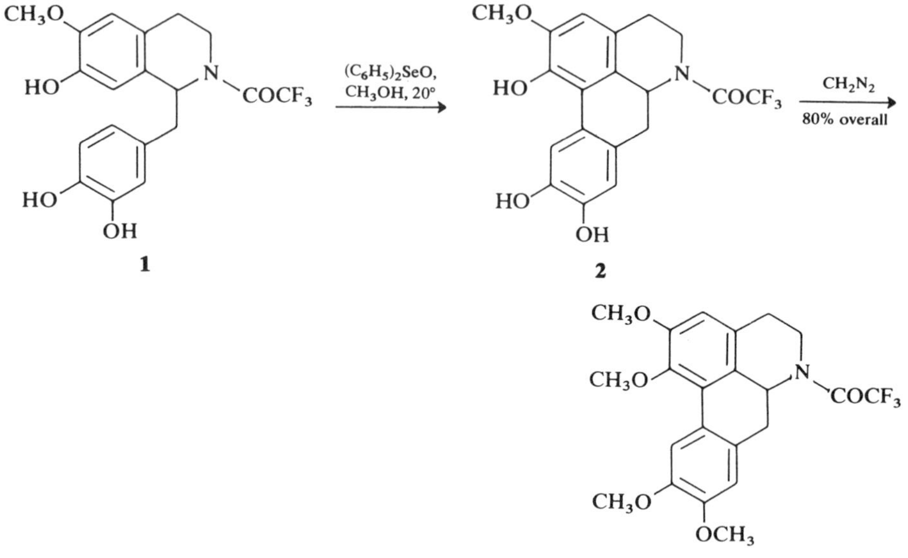

CH_3O, HO, HO, OH, N–$COCF_3$ **1** $\xrightarrow{(C_6H_5)_2SeO,\ CH_3OH,\ 20°}$ CH_3O, HO, HO, OH, N–$COCF_3$ **2** $\xrightarrow[80\%\ overall]{CH_2N_2}$ CH_3O, CH_3O, CH_3O, OCH_3, N–$COCF_3$ **3**

[1] J. P. Marino and A. Schwartz, *Tetrahedron Letters*, 3253 (1979).
[2] For a similar synthesis using VOF_3 as oxidant *see* S. M. Kupchan, O. P. Dhingra, and C.-K. Kim, *J. Org.*, **43**, 4076 (1978).

4,6-Diphenylthieno[3,4-*d*][1,3]dioxol-2-one 5,5-dioxide (2), 7, 140–141.
Preparation[1]:

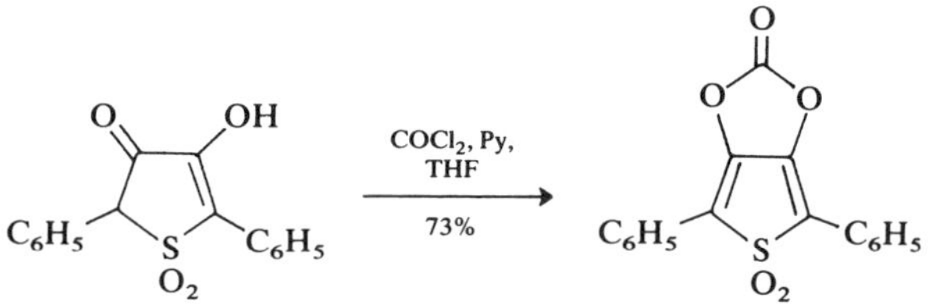

O, OH, C_6H_5, S, O_2, C_6H_5 **1** $\xrightarrow[73\%]{COCl_2,\ Py,\ THF}$ O, O, O, C_6H_5, S, O_2, C_6H_5 **2**

Active BOC-amino acid esters. Steglich *et al.*[1] originally used reagent **2** to prepare active esters **5** of BOC-amino acids (**4**). Actually **2** can also be used to prepare BOC-amino acids by way of **3**. In fact the two reactions can be combined for a one-pot preparation of activated BOC-amino acid esters (**5**). Thus **2** on reaction with *t*-butanol is converted into the activated *t*-butyl carbonate **3**. This compound in combination with 1 equiv. of tetramethylguanidine as base (**6**, 246) converts an amino acid into the BOC-derivative with formation of **1**. After removal of the solvent and neutralization of the base with citric acid, the mixture of **4** and **1** is treated with **2** and pyridine in dry CH_2Cl_2. After removal of the by-product **1** with $NaHCO_3$, the desired activated BOC-amino acid ester **5** is obtained in 70–90% overall yield. These esters are useful for coupling with amino acid esters to form peptides.[2]

$$\mathbf{2} \xrightarrow[90\%]{(CH_3)_3COH,\ Py,\ CH_2Cl_2} \mathbf{3} \xrightarrow{NH_2CH(R)COOH,\ [(CH_3)_2N]_2C{=}NH} \mathbf{1} + (CH_3)_3COC(O)NHCH(R)COOH\ (\mathbf{4})$$

(**3**: 2,5-diphenyl-4-oxo-3-[$OCO_2C(CH_3)_3$]-thiophene S,S-dioxide ring, C_6H_5 groups flanking SO_2)

$$\xrightarrow[70\text{–}90\%\ \text{overall}]{1)\ \mathbf{2},\ Py;\ 2)\ NaHCO_3} CO_2 + \mathbf{5}$$

(**5**: same ring bearing $OC(O){-}NHCH(R)COOC(CH_3)_3$, with C_6H_5, S, O_2)

[1] O. Hollitzer, A. Seewald, and W. Steglich, *Angew. Chem., Int. Ed.*, **15**, 444 (1976).
[2] G. Schnorrenberg and W. Steglich, *ibid.*, **18**, 307 (1979).

Diphosphorus tetraiodide, 1, 349–350.

Deoxygenation of sulfoxides, selenoxides, and nitro alkanes.[1] P_2I_4 is a useful reagent for deoxygenation of sulfoxides, selenoxides, and primary nitroalkanes to sulfides, selenides, and nitriles, respectively. The reactions occur in less than 2 hours at 20°.

Examples:

$$C_9H_{19}CH_2NO_2 \xrightarrow[78\%]{P_2I_4,\ N(C_2H_5)_3,\ 20^\circ} C_9H_{19}C{\equiv}N$$

$$R^1S(O)R^2 \xrightarrow[\sim 80\%]{P_2I_4,\ 20^\circ} R^1SR^2$$

$$R^1Se(O)R^2 \xrightarrow[94\text{–}97\%]{} R^1SeR^2$$

Alkene synthesis.[2] Diphosphorus tetraiodide [or phosphorus(III) iodide] converts di- and trisubstituted β-hydroxy sulfides into alkenes. The reaction is conducted for 2–6 hours in refluxing methylene chloride. The reaction is regiospecific, and involves *anti*-elimination of the OH and RS groups. Yields are 80–90%. Thionyl chloride is the preferred reagent for a similar reaction with tetrasubstituted β-hydroxy sulfides.

Examples:

$$CH_3S-\underset{H}{\overset{CH_3}{C}}-\underset{H}{\overset{C_{10}H_{21}}{C}}-OH \xrightarrow[88\%]{P_2I_4,\ CH_2Cl_2,\ 2\ hr.,\ \Delta} (CH_3)HC=CH(C_{10}H_{21})$$

$$CH_3S-\underset{H}{\overset{C_{10}H_{21}}{C}}-C(CH_3)_2OH \xrightarrow{80\%} C_{10}H_{21}CH=C(CH_3)_2$$

$$(CH_3)_2C(SCH_3)-C(OH)(C_6H_5)(CH_2C_6H_5) \xrightarrow{1)\ P_2I_4;\ 2)\ SiO_2} \underset{(41\%)}{(CH_3)_2C=C(C_6H_5)(CH_2C_6H_5)} + \underset{(41\%)}{CH_2=C(CH_3)-CH(C_6H_5)(CH_2C_6H_5)}$$

$$CH_3S-\underset{CH_3}{\overset{CH_3}{C}}-\underset{CH_2C_6H_5}{\overset{C_6H_5}{C}}-OH \xrightarrow[78\%]{SOCl_2,\ CH_2Cl_2,\ \Delta} (CH_3)_2C=C(C_6H_5)(CH_2C_6H_5)$$

Alkyl iodides.[3] The reagent reacts rapidly with alcohols in CS_2 at 0° to form an intermediate that decomposes more slowly to the corresponding alkyl iodide. The reaction is rapid for tertiary and benzylic alcohols, but much slower for primary and secondary alcohols. Yields are in the range 65–90%. Secondary alcohols react with inversion. An improved preparation of P_2I_4 is also described.

[1] J. N. Denis and A. Krief, *Tetrahedron Letters*, 3995 (1979).
[2] J. N. Denis, W. Dumont, and A. Krief, *ibid.*, 4111 (1979).
[3] M. Lauwers, B. Regnier, M. Van Eenoo, J. N. Denis, and A. Krief, *ibid.*, 1801 (1979).

(2R,3R)-Dipivaloyltartaric acid, (1). Mol. wt. 318.33,

$$\begin{array}{c} COOH \\ H-C-OC(=O)C(CH_3)_3 \\ (CH_3)_3CC(=O)O-C-H \\ COOH \end{array}$$

$\alpha_D - 24.1°$. The reagent is prepared by reaction of (2R,3R)-tartaric acid with excess pivaloyl chloride to give the diacylated anhydride; the free diacyl acid is liberated by treatment with aqueous acetone.[1]

Enantioselective protonation. French chemists[2] have reported a new method for obtaining optically active α-amino acids: A racemic α-amino acid ester is converted into the racemic Schiff base **2**.[3] This is deprotonated with LDA to the anion **3**, which is then protonated with an optically active acid such as **1**. The optically active Schiff base **2** is formed with an enantiomeric ratio as high as 80 : 20. The enantiomeric ratios are dependent largely on the structure of the diacyltartaric acid.

RCHCOOCH$_3$ (N=CH–C$_6$H$_5$) **2** (±) $\xrightarrow[-70^\circ]{\text{LDA}}$ RC=COCH$_3$ (OLi; N=CH–C$_6$H$_5$) **3** $\xrightarrow[\sim 85\%]{\mathbf{1}}$ R$\overset{*}{\text{C}}$HCOOCH$_3$ (N=CH–C$_6$H$_5$) ***2**

This method of "deracemization" by enantioselective protonation has been applied previously to carbonyl compounds. Thus hydrolysis of (E)- and (Z)-enamines of aldehydes or ketones in the presence of chiral acids results in preferential formation of one enantiomer. Reported optical yields are slight to moderate.[4]

[1] L. Vrba and M. Semonský, *Coll. Czech. Chem. Comm.*, **27**, 1732 (1962).
[2] L. Duhamel, J.-C. Plaquevent, *Am. Soc.*, **100**, 7415 (1978).
[3] H. Matsushita, Y. Tsujino, M. Noguchi, M. Sabieri, and S. Yoshidawa, *Bull. Chem. Soc. Japan*, **51**, 862 (1978) and references cited therein.
[4] L. Duhamel and J.-C. Plaquevent, *Tetrahedron Letters*, 2285 (1977).

Disodium tetracarbonylferrate, 3, 267–268; **4,** 461–465; **5,** 624–625; **6,** 550–552; **7,** 341; **8,** 216–217.

Cycloalkanones. Mérour *et al.*[1] have reported that Collman's reagent converts a γ-ethylenic bromide or tosylate into a cyclohexanone and a β-allenic bromide into a cyclopentenone in moderate yield. This carbonylation reaction represents a special

CH$_2$=CH(CH$_2$)$_3$Br $\xrightarrow[40\%]{Na_2Fe(CO)_4}$ cyclohexanone; CH$_2$ … OTs $\xrightarrow[65\%]{}$ O

$Br(CH_2)_2CH{=}C{=}CH_2 \xrightarrow[30\%]{}$ O, CH$_3$

case and is limited to formation of five- and six-membered rings.[2] In any case, it served a very useful role in one of the key steps, **1** → **2**, in a synthesis[3] of the diterpene aphidicolin (**3**), a promising antiviral agent.

CH$_2$OTs, H$_3$C, CH$_3$, O, H$_3$C, H, CH$_3$, O — **1** $\xrightarrow[30\%]{Na_2Fe(CO)_4}$ O, H$_3$C, CH$_3$, O, H$_3$C, H, CH$_3$, O — **2**

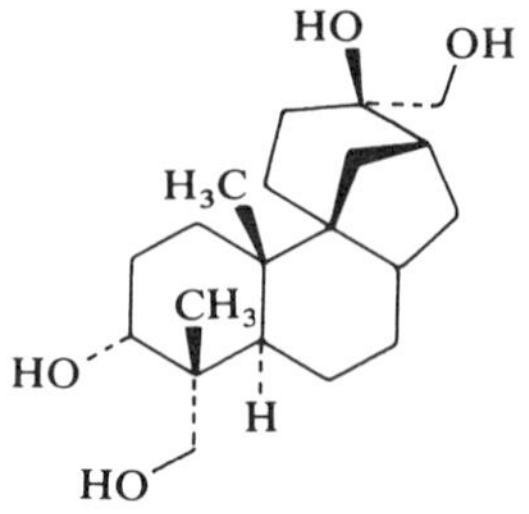

3

α,β-*Unsaturated ketones.*[4] A new route to these ketones involves reaction of terminal epoxides with acyltetracarbonylferrates (equation I).

$$\text{(I)}\quad R^1COCl + Na_2Fe(CO)_4 \xrightarrow{THF} R^1COFe(CO)_4^-Na^+ \xrightarrow[70\text{–}80\%]{R^2\text{-epoxide}} R^1COCH{=}CHR^2$$

Selective reduction of unsaturated aldehydes. Collman's reagent is effective for the selective reduction of the α,β-unsaturation of aldehyde **1**.[5] The dihydro aldehyde **2** is an intermediate in the synthesis of 9,10-dihydroretinals (*e.g.*, **4**).

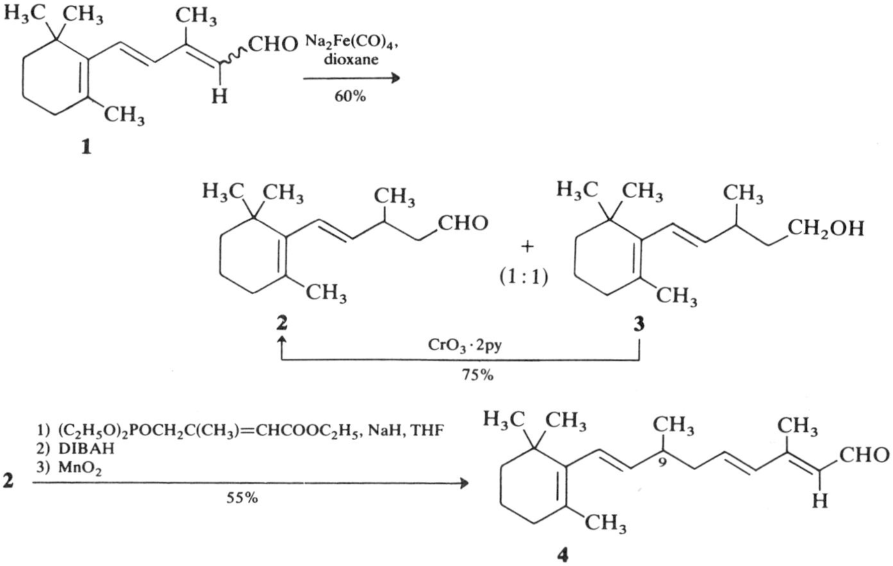

1 $\xrightarrow[60\%]{Na_2Fe(CO)_4,\ \text{dioxane}}$ 2 + 3 (1:1)

3 $\xrightarrow[75\%]{CrO_3\cdot 2py}$ 2

2 $\xrightarrow[55\%]{}$ 4

1) $(C_2H_5O)_2POCH_2C(CH_3){=}CHCOOC_2H_5$, NaH, THF
2) DIBAH
3) MnO_2

Reaction with thioketones.[6] The reaction of this reagent with thiocamphor (**1**) in THF at $-50°$ followed by addition of an acid chloride results in unsaturated thioesters (**2**). Evidently thiocamphor is deprotonated to a thioenolate anion (**a**), which then reacts with the acid chloride to form **2**.

Under these conditions, nonenolizable thioketones are converted into saturated thioesters in modest yields.

$$\text{1} \xrightarrow{Na_2Fe(CO)_4} [\text{a (SNa)}] \xrightarrow[50\text{–}60\%]{RCOCl} \text{2 (SCOR)}$$

[1] J. Y. Mérour, J. L. Roustan, C. Charrier, J. Collin, and J. Benaim, *J. Organomet. Chem.*, **51**, C24 (1973).
[2] J. P. Collman, *Accts. Chem. Res.*, **8**, 342 (1975).
[3] J. E. McMurray, A. Andrus, G. M. Ksander, J. M. Musser, and M. A. Johnson, *Am. Soc.*, **101**, 1330 (1979).
[4] M. Yamashita, S. Yamamura, M. Kurimoto, and R. Suemitsu, *Chem. Letters*, 1067 (1979).
[5] M. Arnaboldi, M. G. Motto, K. Tsujimoto, V. Balogh-Nair, and K. Nakanishi, *Am. Soc.*, **101**, 7083 (1979).
[6] H. Alper, B. Marchand, and M. Tanaka, *Can. J. Chem.*, **57**, 598 (1979).

Disodium tetrachloropalladate, Na_2PdCl_4, **3**, 134.

Decomposition of cyclic hydroperoxides.[1] Cyclohexyl hydroperoxide (**1**) is decomposed mainly to (E)-2-hexenal by this salt and a trace of $FeSO_4 \cdot 7H_2O$ in a two-phase system of water and an organic solvent. Highest yields (58%) are obtained with nitrobenzene. Na_2PdCl_4 is the most satisfactory water-soluble complex of PdX_2 for this purpose. The reaction is general, but yields are low from C_8–C_{12} cyclic hydroperoxides. Cyclopentyl hydroperoxide is converted into the corresponding enal in 73.6% yield.

$$\text{1 (cyclohexyl-OOH)} \xrightarrow[C_6H_5NO_2\text{–}H_2O,\ 80°]{Na_2PdCl_4,\ Fe^{2+}} \underset{(58.4\%)}{CH_3(CH_2)_2CH{=}CHCHO} + \underset{(16.6\%)}{\text{cyclohexanone}} + \underset{(10.2\%)}{\text{cyclohexanol}}$$

[1] K. Formanek, J. P. Aune, M. Jouffret, and J. Metzger, *Nouv. J. Chem.*, 311 (1979).

Disodium tetracyanonickelate(II), $Na_2Ni(CN)_4$ (**1**). Mol. wt. 208.76. Preparation.[1]

Catalytic hydrocyanation of acetylenes.[2] Acetylenes are hydrocyanated to saturated secondary nitriles by **1** with excess cyanide ion and either $NaBH_4$ (in ethylene glycol) or Zn (in water) as reducing agent. Terminal acetylenes RC≡CH afford $RCH(CN)CH_3$ in high yield, whereas internal acetylenes afford a mixture of regioisomers.

Examples:

$$C_6H_5CH_2CH_2C{\equiv}CH \xrightarrow[92\%]{1,\ KCN,\ NaBH_4} C_6H_5CH_2CH_2CH(CN)CH_3$$

$$C_6H_5C{\equiv}C{-}C_2H_5 \xrightarrow[92\%]{} C_6H_5CH(CN){-}CH_2C_2H_5 + C_6H_5CH_2{-}CH(CN)C_2H_5 \quad 51:49$$

[1] W. C. Fernelius and J. J. Burbage, *Inorg. Syn.*, **2**, 227 (1946); *see also* R. L. McCullough, L. H. Jones, and G. A. Crosby, *Spectro. Chim. Acta*, **16**, 929 (1960).
[2] T. Funabiki and Y. Yamazaki, *J.C.S. Chem. Comm.*, 1110 (1979).

N,N'-Disuccinimidyl carbonate (1). Mol. wt. 256.17, m.p. 211–215° dec., stable under refrigeration.

Preparation:

N-Hydroxysuccinimide (N—OH) $\xrightarrow[62\%]{ClCOOCCl_3}$ N—OCO—N (**1**) $\xleftarrow[80\%]{COCl_2}$ N—OSi$(CH_3)_3$

1

Active esters (**1**, 487).[1] Active esters of CBZ- or BOC-amino acids can be obtained by use of **1** without use of DCC as condensing reagent; yields are 80–100%.

[1] H. Ogura, T. Kobayashi, K. Shimizu, K. Kawabe, and K. Takeda, *Tetrahedron Letters*, 4745 (1979).

E

Ephedrin, 5, 289–290.

Asymmetric synthesis of ketones and acids.[1] A new synthesis of chiral ketones and acids starts with the reaction of an anhydride or an acid chloride with either *l*- or *d*-ephedrin to form a chiral N,N-disubstituted amide (**1**) in almost quantitative yield. The amide (**1**) is then alkylated via the anion to give **2**, which contains three asymmetric centers. Acid hydrolysis of **2** gives a carboxylic acid (**3**) with an optical yield of ~75% (two examples). Cleavage of **2** with methyllithium gives a methyl ketone (**4**), in optical yields of 45–75%.

$$(R^1CH_2CO)_2O + C_6H_5\overset{*}{C}H(OH)\overset{*}{C}H(CH_3)NHCH_3 \ (d\text{- or } l\text{-}) \xrightarrow{\Delta} R^1CH_2C(=O)N(CH_3)\overset{*}{C}H(CH_3)\overset{*}{C}H(OH)C_6H_5 \ (\mathbf{1}) \xrightarrow{1)\ LDA;\ 2)\ R^2X}$$

$$R^1R^2\overset{*}{C}HC(=O)N(CH_3)\overset{*}{C}H(CH_3)\overset{*}{C}H(OH)C_6H_5 \ (\mathbf{2}) \xrightarrow[\sim 75\%]{HCl} R^1R^2\overset{*}{C}HCOOH \ (\mathbf{3})$$

$$\mathbf{2} \xrightarrow[50\text{–}70\%]{CH_3Li} R^1R^2\overset{*}{C}HCOCH_3 \ (\mathbf{4})$$

[1] M. Larchevêque, E. Ignatova, and T. Cuvigny, *Tetrahedron Letters*, 3961 (1978).

N-(2,3-Epoxypropylidene)cyclohexylamine N-oxide (1). Mol. wt. 169.22, m.p. 97°.
Preparation:

$$H_2C{=}CHCHO \xrightarrow[64\%]{H_2O_2,\ OH^-} H_2C\overset{O}{—}CHCHO \xrightarrow[89\%]{C_6H_{11}NHOH} \mathbf{1}$$

(**1**: epoxide–CH=N$^+$(O$^-$)–C$_6$H$_{11}$)

α-Methylene-γ-lactones.[1] Treatment of **1** with a potent electrophilic reagent such as the trimethylsilyl ester of trifluoromethanesulfonic acid at −78° results in

O-silylation to give **a**, identified by the NMR spectrum. At $-30°$ the epoxide ring is opened to give **b**. The ion **b** resembles the active intermediate from Eschenmoser's

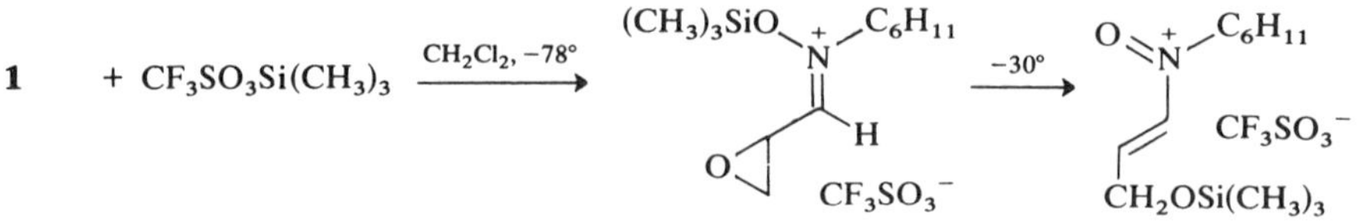

$$\mathbf{1} + CF_3SO_3Si(CH_3)_3 \xrightarrow{CH_2Cl_2,\ -78°} \mathbf{a} \xrightarrow{-30°} \mathbf{b}$$

α-chloro nitrone reagent (**4**, 80–82; **5**, 110–113), which undergoes Diels–Alder reactions with olefins. As expected, **b** also gives [2+4] cycloadducts (**2**) with olefins; these can be isolated as the stable cyano derivatives **3**, as formulated for the reaction with cyclohexene. Three diastereoisomeric adducts of structure **3** are obtained and after hydrolysis of the silyl ether group are separable by chromatography to give **4**.

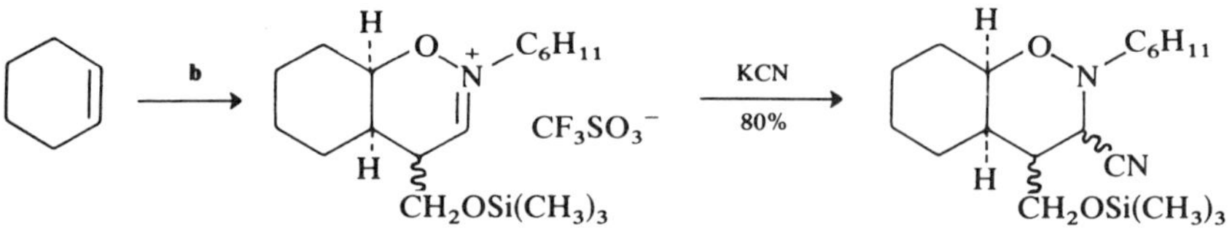

$$\text{cyclohexene} \xrightarrow{\mathbf{b}} \mathbf{2} \xrightarrow[80\%]{KCN} \mathbf{3}$$

$$\mathbf{4} \xrightarrow[98\%]{} \mathbf{5} \xrightarrow[65-75\%]{KOC(CH_3)_3} \mathbf{6} \xrightarrow[80\%]{H_2SO_4,\ CHCl_3,\ 60°} \mathbf{7}$$

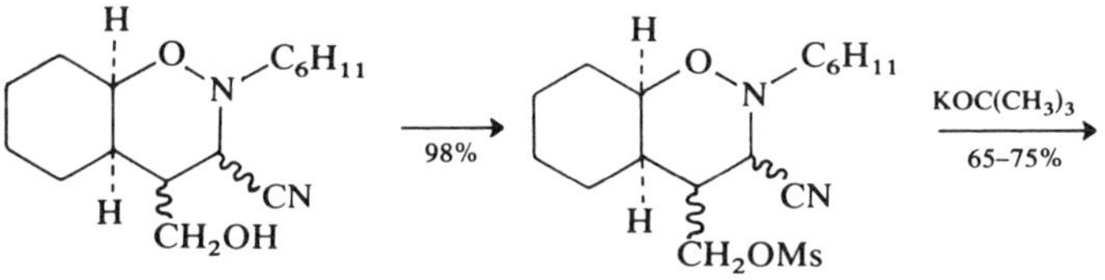

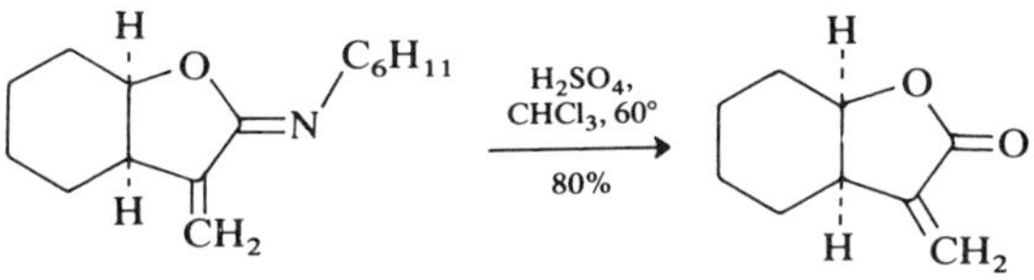

The adducts (**4**) can be converted into α-methylene-γ-lactones (**7**) by the method used earlier by Eschenmoser (**5**, 111) for related adducts.

In the reaction of **b** with 1-methylcyclohexene substitution at C_2 competes with cycloaddition.

[1] M. Riediker and W. Graf, *Helv.*, **62**, 205 (1979).

Erbium chloride, $ErCl_3$. Mol. wt. 273.64. Commercially available (Alfa) as the hexahydrate, $ErCl_3 \cdot 6H_2O$.

Acetals.[1] Various lanthanide chlorides are efficient catalysts for acetalization of aldehydes by methanol. Lanthanum chloride and cerium chloride are satisfactory for aliphatic aldehydes, but erbium chloride and ytterbium chloride are generally superior, particularly for aromatic and bicyclic aldehydes. Addition of trimethyl orthoformate as a water scavenger allows use of the commerically available hydrated forms of the salts. Acetals can be obtained in 80–100% yield from reactions conducted for 10 minutes at 25°.

$$RCHO + CH_3OH \xrightarrow[80\text{–}100\%]{ErCl_3,\ HC(OCH_3)_3} RCH(OCH_3)_2$$

[1] J.-L. Luche and A. L. Gemal, *J.C.S. Chem. Comm.*, 976 (1978).

Ethenesulfonyl chloride–Trimethylamine, $CH_2{=}CHSO_2Cl{-}N(CH_3)_3$.

Betylates. Ammonioalkanesulfonate esters (**1**) have been given the trivial name betylates because they form betaines on substitution or elimination. They are prepared by reaction of alcohols with ethenesulfonyl chloride and trimethylamine and then with dimethylamine and CH_3X (equation I).[1]

$$\text{(I)}\quad ROH \xrightarrow[80\text{–}90\%]{1)\ CH_2{=}CHSO_2Cl,\ N(CH_3)_3;\ 2)\ HN(CH_3)_2,\ CH_3X} ROSO_2CH_2CH_2\overset{+}{N}(CH_3)_3X^- \quad \mathbf{1}$$

Betylates are highly reactive to nucleophiles, particularly in an aqueous medium. Betylates possibly act as phase-transfer catalysts in the displacement reaction and are of particular use in effecting reactions with hydrophilic nucleophiles.[2]

Examples:

$$(CH_3)_3\overset{+}{N}CH_2CH_2SO_2O(CH_2)_3CH_3\ FSO_3^- \xrightarrow[91\%]{KSCN, H_2O,\ 25°} CH_3(CH_2)_3SCN$$

$$(CH_3)_2\overset{+}{N}HCH_2CH_2SO_2OCH_2C(CH_3)_3\ I^- \xrightarrow[68\%]{DMF,\ 120°} (CH_3)_3CCH_2I$$

The second example shows that displacement reactions can be effected with neopentyl alcohol via a betylate. Of even greater interest, S_N2 displacements of chiral secondary alcohols can be effected via a betylate with high stereoselectivity: 100% inversion is commonly observed.[3]

Examples:

$$(CH_3)_2\overset{+}{N}H(CH_2)_2SO_2O{-}CH(CH_3)(CH_2)_5CH_3\ Cl^- \xrightarrow[85\%]{C_6H_6,\ 80°} HC(CH_3)(Cl)(CH_2)_5CH_3$$

(94.6% ee) → (97–101% ee)

$$(CH_3)_3\overset{+}{N}(CH_2)_2SO_2O-\underset{(CH_2)_5CH_3}{\overset{CH_3}{CH}} \; FSO_3^- \xrightarrow[60\%]{NaSCN,\ H_2O,CH_2Cl_2} H\underset{(CH_2)_5CH_3}{\overset{CH_3}{C}}-SCN$$

(101% ee)

[1] J. F. King and S. M. Loosmore, *J.C.S. Chem. Comm.*, 1011 (1976).
[2] J. F. King, S. M. Loosmore, J. D. Lock, and M. Aslam, *Am. Soc.*, **100**, 1637 (1978).
[3] J. F. King, M. Aslam, and J. D. Lock, *Tetrahedron Letters*, 3615 (1979).

N-(Ethoxycarbonyl)phthalimide (1), 1, 111–112.

Phthaloylamino acids.[1] The original method of Nefkens for preparation of these derivatives of amino acids gives only low yields in the case of the branched-chain amino acids valine (**2**) and isoleucine, probably because of steric hindrance. When the reaction was conducted in aqueous Na_2CO_3 for a prolonged period, the main product was the hydrolysis product (**5**) of **1**. When ethyl acetate was added to the reaction as solvent to decrease decomposition to **5**, the main product was **3**, which can be converted to the desired phthaloyl-L-valine (**4**) when heated. The best procedure to obtain **4** directly is treatment of valine first with the aqueous base followed by addition of **1** in ethyl acetate. This method gives optically pure **4** in 70–75% yield. The procedure is also satisfactory for isoleucine.

1 + $NH_2-CH(CH(CH_3)_2)-COOH$ (2) $\xrightarrow[50\text{–}60\%]{Na_2CO_3,\ H_2O,\ CH_3COOC_2H_5}$ 3 ($C_6H_4(CONHCOOC_2H_5)(CONH-CH(CH(CH_3)_2)COOH)$)

1: N-$COOC_2H_5$ phthalimide

1 $\xrightarrow{Na_2CO_3,\ H_2O,\ 75\ min.}$ 5 ($C_6H_4(CONHCOOC_2H_5)(COOH)$)

3 $\xrightarrow[78\%]{C_6H_5CH_3,\ \Delta}$ 4

2 $\xrightarrow[70\text{–}75\%]{1)\ Na_2CO_3,\ H_2O;\ 2)\ \mathbf{1}\ in\ CH_3COOC_2H_5}$ 4 (phthalimido N$-CH(CH(CH_3)_2)COOH$)

[1] P. Adriaens, B. Meesschaert, G. Janssen, L. Dumon, and H. Eyssen, *Rec. trav.*, **97**, 260 (1978).

Ethyl 3-bromo-2-hydroxyiminopropanoate, $BrCH_2C(=NOH)CO_2C_2H_5$ (**1**). Mol. wt. 210.03, m.p. 76–78°. The reagent is prepared from ethyl pyruvate by bromination followed by reaction of the bromo ester with hydroxylamine sulfate ($CHCl_3$, H_2O, ~50% overall yield).[1]

α-Amino acids.[1] The hydroxyimino ester **1** reacts with a variety of nucleophiles in the presence of sodium carbonate. Indole, pyrrole, thiols, and dialkyl malonates give products of substitution (for example, equation I), whereas furane, 3-methylindole, cyclopentadiene, and nucleophilic alkenes give cycloadducts (equation II). The nitroso olefin **a** is a probable intermediate in these reactions. Adducts such as **2** are reduced with aluminum amalgam to esters of α-amino acids (52–96% yield).

(I) $BrCH_2C(=NOH)CO_2C_2H_5$ (**1**) $\xrightarrow{Na_2CO_3}$ [$CH_2=C(CO_2C_2H_5)N=O$] (**a**) $\xrightarrow[82\%]{Indole}$ **2** $\xrightarrow[96\%]{Al/Hg}$ **3**

(II) **4** $\xrightarrow[79\%]{1,\ Na_2CO_3}$ **5**

Thiophene and 1,3-dimethoxybenzene fail to give adducts of **1** in synthetically useful yields. However, the related reagent 3-chloro-2-hydroxyiminopropanal[2] (**6**) reacts with 1,3-dimethoxybenzene as formulated in equation (III). The adduct **7** is oxidized to the corresponding acid (silver oxide) and esterified (CH_2N_2) to provide an alternate route to a α-hydroxyimino ester.

(III) $ClCH_2C(=NOH)-CHO$ (**6**) + 1,3-$(CH_3O)_2C_6H_4$ $\xrightarrow[41\%]{Na_2CO_3}$ **7**

γ-Hydroxybutyronitriles.[3] Carboxylic acids **4**, which are available either by addition of **1** to an olefin followed by alkaline hydrolysis or by the analogous reaction of **2** followed by Ag_2O oxidation (equation I), are decarboxylated when heated briefly at 150°. The overall process is highly stereoselective, the geometry of the original oxazine being retained in the γ-hydroxynitriles. Thus reagents **1** and **2** give

(I)

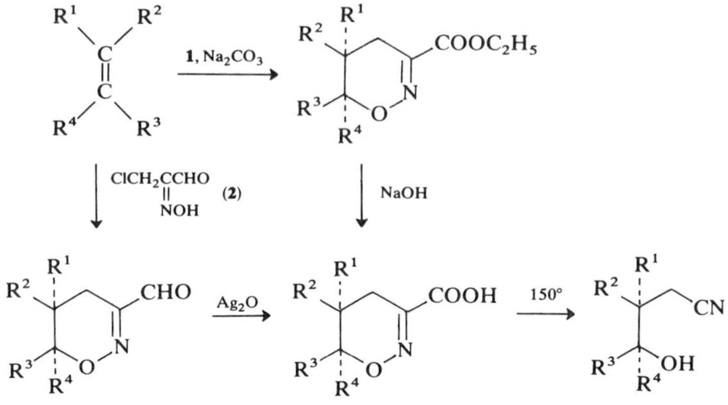

cis-adducts of —OH and —CH_2CN to olefins. Since γ-hydroxynitriles are converted into γ-lactones in acidic media, **1** and **2** are also useful for transformation of an olefin into a lactone.

Examples:

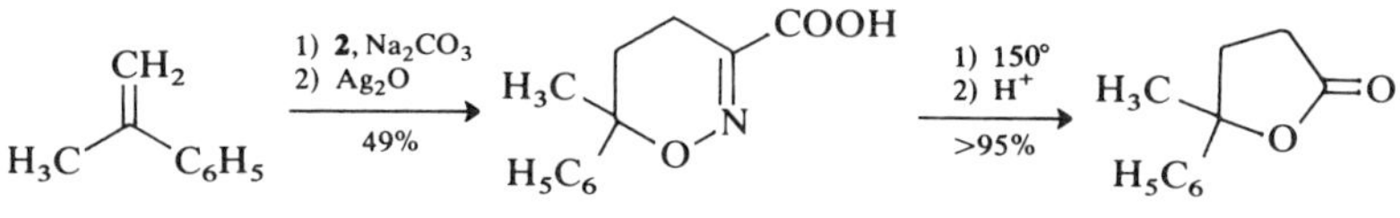

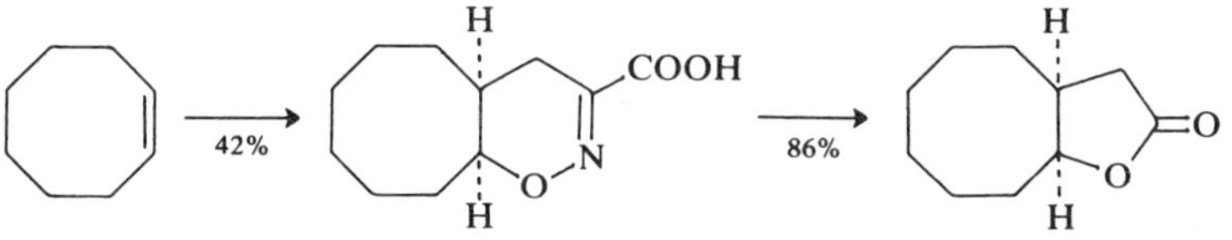

[1] T. L. Gilchrist, D. A. Lingham, and T. G. Roberts, *J.C.S. Chem. Comm.*, 1089 (1979).
[2] Prepared by the addition of nitrosyl chloride to acrolein: K. A. Ogloblin and A. A. Potekhin, *J. Org. Chem. USSR*, **1**, 1370 (1965).
[3] T. L. Gilchrist and T. G. Roberts, *J.C.S. Chem. Comm.*, 1090 (1979).

Ethyl chloroformate, 1, 364–367; **2**, 193; **4**, 228–230; **5**, 294–295; **7**, 147.

Regioselective C–N cleavage. When heated in ethyl chloroformate, tetrahydroberberine (**1**) is cleaved to **2** as the only product (81% yield). The product was converted by known reactions into (±)-canadaline (**3**), representative of a new group of isoquinoline alkaloids, the secoberberines.[1]

The same cleavage has been used in the transformation of the acetate of (±)-ophiocarpine (**4**) into (±)-α-hydrastine (**6**).[2]

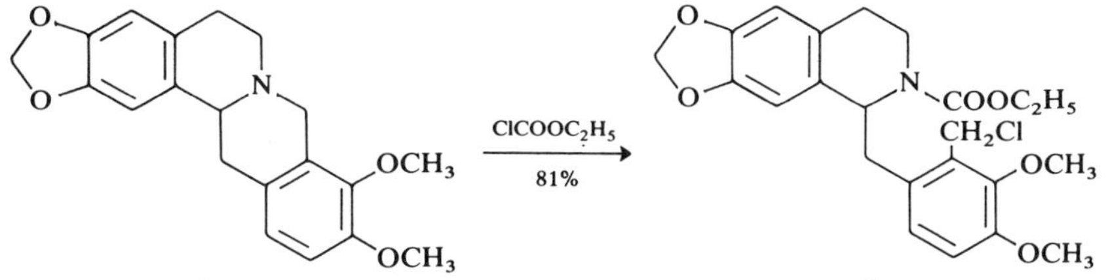

[1] M. Hanaoka, K. Nagami, and T. Imanishi, *Heterocycles*, **12**, 497 (1979).
[2] *Idem*, *Chem. Pharm. Bull. Japan*, **27**, 1947 (1979).

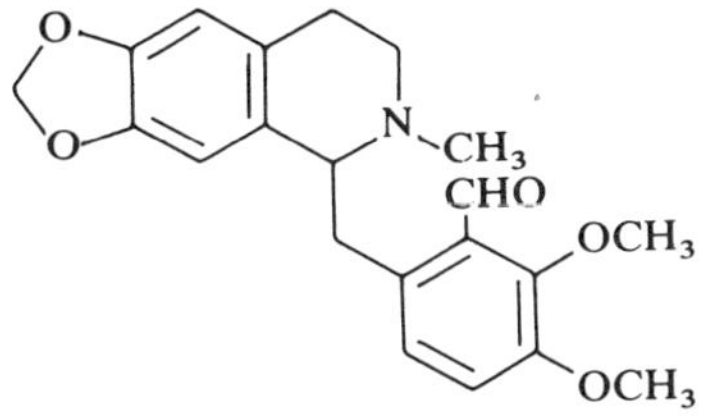

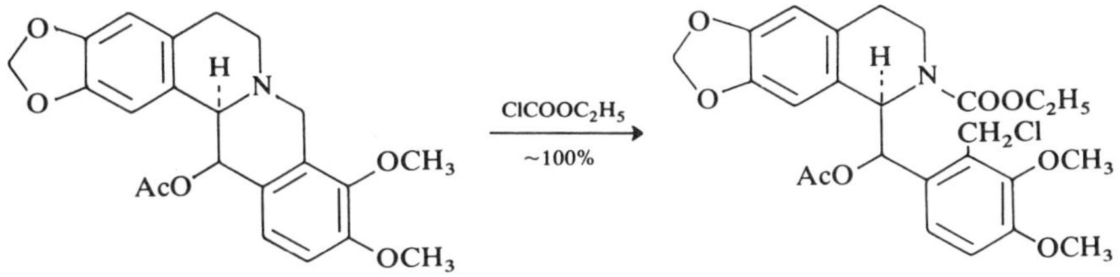

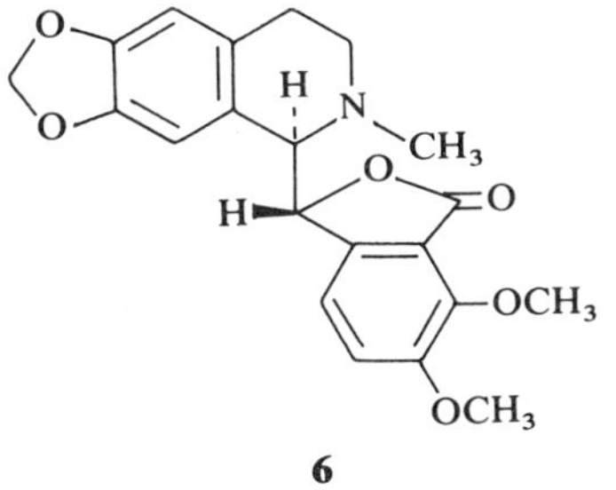

O-Ethyl S-cyanomethyl dithiocarbonate, $C_2H_5OC(=S)-SCH_2CN$ (**1**). Mol. wt. 161.25, yellow oil. Prepared by reaction of $C_2H_5OC(S)SK$ with $ClCH_2CN$.[1]

Vinylic nitriles.[2] The reagent (**1**) reacts with aldehydes or ketones under phase-transfer conditions to form vinylic nitriles (**2**) (equation I).

$$R^1R^2C{=}O + \mathbf{1} \xrightarrow[\text{NaOH, } H_2O, CH_3CN]{[(C_8H_{17})_3NCH_3]^+Cl^-,} \left[\text{episulfide: } R^1R^2C\text{–S–}C(H)(CN)\right]\ (\mathbf{a}) \rightarrow R^1R^2C{=}C(H)CN\ (\mathbf{2}) + S$$

[1] J. Troger and F. Volkmer, *J. pr. Chem.*, **70**, 442 (1904).
[2] K. Tanaka, N. Ono, Y. Kubo, and A. Kaji, *Synthesis*, 890 (1979).

N-Ethyl(diethylphosphono)methylketenimine (1). Mol. wt. 219.22, b.p. 86–90°/1 mm.

Preparation:

$$(C_2H_5O)_2PONa + ClCH(CH_3)CONHC_2H_5 \xrightarrow[83\%]{\text{THF}}$$

$$(C_2H_5O)_2P(O)CH(CH_3)CONHC_2H_5 \xrightarrow[82\%]{(C_6H_5)_3P,\ Br_2,\ (C_2H_5)_3N} (C_2H_5O)_2P(O)C(CH_3){=}C{=}NC_2H_5\ (\mathbf{1})$$

Annelation reactions. The ketenimine **1** acts chemically as a vinyl phosphonate activated by the cumulated imino group and is useful as an annelation reagent for the synthesis of heterocyclic compounds. As such it condenses with the sodium salt of salicylaldehyde (**2**) to afford the benzopyrane in **3** in 52% yield. Similarly, the pyrrolizine **5** is available from the sodium salt of 2-formylpyrrole in 51% yield.[1]

$$\text{2 (2-CHO-C}_6H_4\text{ONa)} \xrightarrow[52\%]{\mathbf{1},\ \text{DMF, } 80^\circ} \text{3 (3-CH}_3\text{-2-NC}_2H_5\text{-benzopyran)}$$

$$\text{4 (2-CHO-pyrrole N-Na)} \xrightarrow{51\%} \text{5 (CH}_3\text{, NC}_2H_5\text{ pyrrolizine)}$$

[1] J. Motoyoshiya, J. Enda, Y. Ohshiro, and T. Agawa, *J.C.S. Chem. Comm.*, 900 (1979).

Ethylenebis(triphenylphosphine)platinum(0), $[(C_6H_5)_3P]_2Pt\text{—}\| \begin{smallmatrix}CH_2\\CH_2\end{smallmatrix}$ (**1**). Mol. wt. 747.72, m.p. 124–125° dec. The complex is prepared by reaction of $[(C_6H_5)_3P]_2PtO_2$ with ethylene and then with $NaBH_4$.[1]

Isomerization of anti-Bredt alkenes.[2] The reagent **1** forms a complex with **2** that is stable for some months at −20°, but that rearranges at 60° largely to a mixture of **2**, **3**, and **4**, in the ratio 2:10:2 with a trace of **5**.

$$\mathbf{2} \xrightarrow[-C_2H_4]{\mathbf{1},\ C_6H_5CH_3,\ -20^\circ} \mathbf{a}\ (\text{—Pt}[P(C_6H_5)_3]_2) \xrightarrow{60^\circ,\ 1\ \text{hr.}} \mathbf{2} + \mathbf{3} + \mathbf{4} + \mathbf{5}$$

2:10:2

[1] R. Ugo, F. Cariati, G. LaMonica, and J. J. Mrowca, *Inorg. Syn.*, **11**, 105 (1968); C. D. Cook and G. S. Jauhal, *Am. Soc.*, **90**, 1464 (1968).
[2] E. Stamm, K. B. Becker, P. Engel, and R. Keese, *Helv.*, **62**, 2181 (1979); E. Stamm, H.-R. Leu, and R. Keese, *ibid.*, **62**, 2174 (1979).

Ethylene glycol, 1, 375–377; **5**, 296.

Mono ethylene ketals of α-diketones. Yields are generally low for monoketalization of α-diketones. A recent expedient is outlined in equation (I).[1]

$$\text{(I)}\quad C_2H_5OOC\text{—}COOC_2H_5 \xrightarrow[88\%]{HN(C_2H_5)_2} C_2H_5OOC\text{—}CON(C_2H_5)_2 \xrightarrow[75\text{–}95\%]{R^1MgX}$$

$$R^1CO\text{—}CON(C_2H_5)_2 \xrightarrow[60\text{–}70\%]{HOCH_2CH_2OH,\ TsOH} \text{(ethylene ketal of } R^1, CON(C_2H_5)_2) \xrightarrow[75\text{–}90\%]{R^2Li \text{ or } R^2MgBr} \text{(ethylene ketal of } R^1, COR^2)$$

[1] T. Cuvigny, M. Larchevêque, and H. Normant, *Synthesis*, 875 (1978).

Ethyl lithiopropiolate, $LiC{\equiv}C\text{—}CO_2C_2H_5$ (**1**). Mol. wt. 104.03. The preparation involves dropwise addition of a solution of LDA (1.0 equiv.) in THF to a solution of ethyl propiolate in THF at −70°.

γ-Substituted α,β-butenolides. Treatment of aldehydes or ketones with the reagent (**1**) affords products of 1,2-addition in good yield. The resulting adducts are hydrolyzed and then reduced by 5% Pd on $BaSO_4$ in methanol–quinoline to afford butenolides.[1]

$$R^1COR^2 + LiC{\equiv}C\text{—}CO_2C_2H_5 \xrightarrow{-70^\circ} R^1R^2C(OH)\text{—}C{\equiv}C\text{—}CO_2C_2H_5 \xrightarrow[45\text{–}65\%\ \text{overall}]{1)\ KOH,\ CH_3OH;\ 2)\ H_2,\ Pd/BaSO_4,\ CH_3OH,\ \text{quinoline};\ 3)\ H^+} \text{butenolide } (R^1, R^2)$$

[1] J. L. Herrmann, M. H. Berger, and R. H. Schlessinger, *Am. Soc.*, **101**, 1544 (1979).

Ethylmagnesium bromide, 1, 415–424; **2,** 205.

Intramolecular phenol alkylation. Phenols of type **1**, **3**, and **4** when treated with ethylmagnesium bromide in benzene at reflux undergo intramolecular alkylation. These reactions are thought to proceed via *o*-quinone methide intermediates such as **b** formulated for the case of **1**. These cyclizations fail with NaH or *n*-butyllithium as

C_2H_5MgBr, C_6H_6, Δ

1 **a** **b** **2** (71%) (11%)

3 (n = 1) 43% 10:90
4 (n = 2) 90% 7:93

base; and no cyclization is observed if either phenolic oxygen is blocked as a methyl ether. The high *ortho*-specificity of **1** may be a result of magnesium chelation, whereas the *para*-specificity of **3** and **4** is probably a result of steric hindrance of the *ortho*-position (chelation is not possible in these cases). *para*-Specificity for **1** is observed when the cyclization is conducted with $SnCl_4$. The dimethyl ether of **4** cyclizes with an *ortho* to *para* ratio of 10:90 (71%).

1 $\xrightarrow[72\%]{SnCl_4,\ CH_2Cl_2}$ + **2** 95:5

[1] W. S. Murphy and S. Wattansin, *J.C.S. Chem. Comm.*, 788 (1979).

Ethylmagnesium bromide–Copper(I) iodide.

Reduction of vinylic phenyl sulfoxides. The combination of ethylmagnesium bromide and 10% copper(I) iodide reduces α,β-unsaturated phenyl sulfoxides in ether at 0° to the corresponding sulfides in 60–93% yield with retention of the original configuration of the vinyl group. Phenyl sulfoxides are reduced more easily than benzylic sulfoxides and alkyl sulfoxides. Methylmagnesium bromide is less effective than the ethyl Grignard reagent.

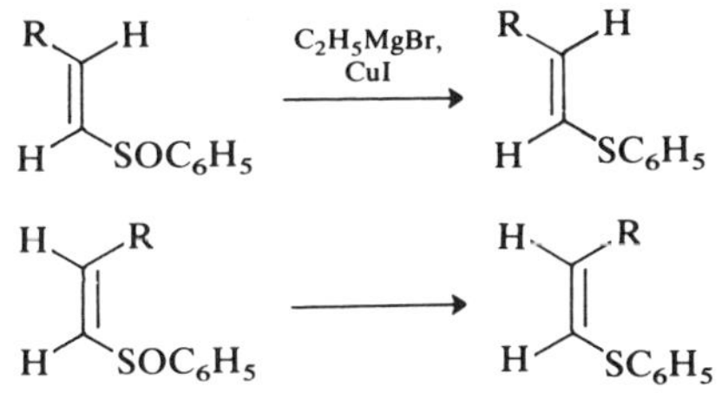

R(H)C=C(H)SOC_6H_5 —(C_2H_5MgBr, CuI)→ R(H)C=C(H)SC_6H_5

H(R)C=C(H)SOC_6H_5 ——→ H(R)C=C(H)SC_6H_5

[1] G. H. Posner and P.-W. Tang, *J. Org.*, **43**, 4131 (1978).

Ethyl (E)-3-methylthio-2-methyl-2-butenoate (1). Mol. wt. 174.26.

Preparation:

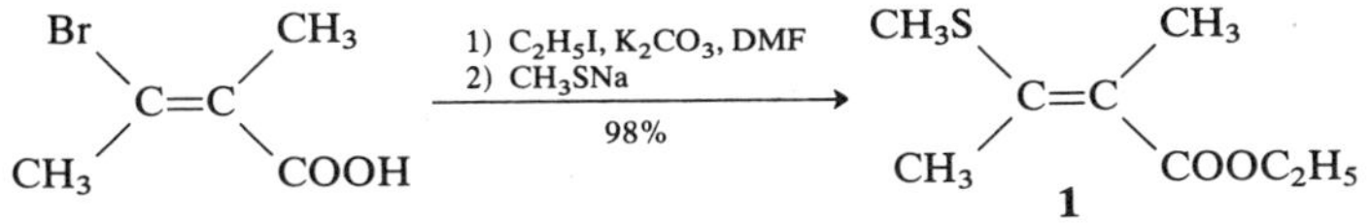

$(Br)(CH_3)C{=}C(CH_3)(COOH)$ —1) C_2H_5I, K_2CO_3, DMF; 2) CH_3SNa; 98%→ $(CH_3S)(CH_3)C{=}C(CH_3)(COOC_2H_5)$ **1**

2-Methyl-(2Z,4E)-dienoic acids.[1] (2E,4E)-Dienoic acids are available by Wittig-type reactions. A stereospecific synthesis of the less stable (2Z,4E)-isomers from aldehydes uses **1** as the key reagent. Deprotonation of **1** (LDA, −95°) leads to the lithio derivative **2**, which reacts with the aldehyde **3**, to give, in 72% yield, the δ-lactone **4**, which is desulfurized quantitatively to **5**. Reaction of **5** with potassium t-butoxide is accompanied by elimination to the desired acid **6**.

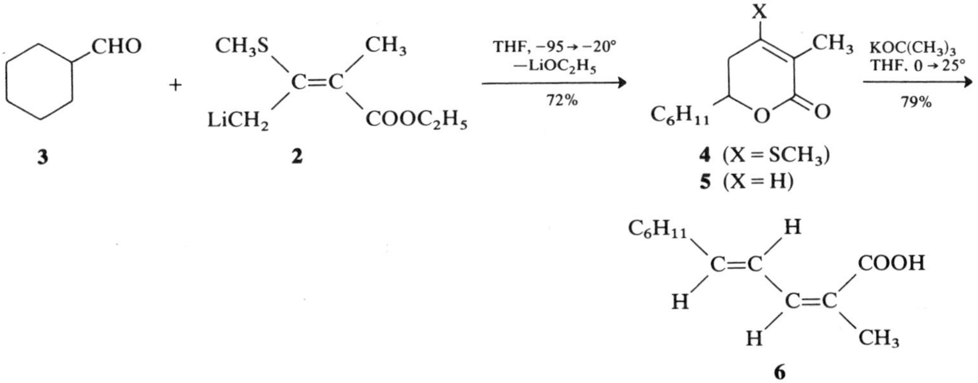

Cyclohexyl-CHO (**3**) + $(CH_3S)(LiCH_2)C{=}C(CH_3)(COOC_2H_5)$ (**2**) —THF, −95→−20°; $-LiOC_2H_5$; 72%→ C_6H_{11}-substituted δ-lactone with X and CH_3 (**4** (X = SCH_3); **5** (X = H)) —$KOC(CH_3)_3$, THF, 0→25°; 79%→ $C_6H_{11}(H)C{=}C(H)$–$C(H){=}C(COOH)(CH_3)$ (**6**)

The same sequence was used to convert the aldehyde **7** into **8**. This methodology was developed in connection with a projected synthesis of rifamycin S, which contains a sequence similar to that of **8**.

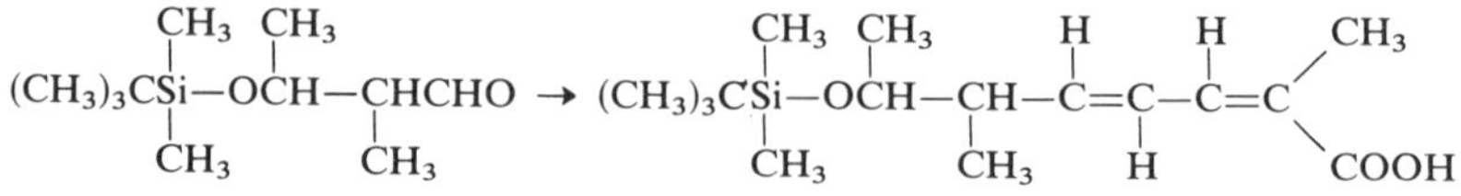

$$\underset{\mathbf{7}}{(CH_3)_3C\overset{CH_3}{\underset{CH_3}{Si}}-O\overset{CH_3}{CH}-\underset{CH_3}{CH}CHO} \rightarrow \underset{\mathbf{8}}{(CH_3)_3C\overset{CH_3}{\underset{CH_3}{Si}}-O\overset{CH_3}{CH}-\underset{CH_3}{CH}-\overset{H}{C}=\underset{H}{C}-\overset{H}{C}=C(CH_3)COOH}$$

[1] E. J. Corey and G. Schmidt, *Tetrahedron Letters*, 2317 (1979).

O-Ethylperoxycarbonic acid, $C_2H_5OCOO_2H$. Mol. wt. 106.08. The reagent is prepared *in situ* by reaction of ethyl chloroformate and H_2O_2 (equation I).

$$\text{(I)}\quad C_2H_5OCOCl+H_2O_2 \rightarrow C_2H_5O\overset{O}{\overset{\|}{C}}OOH+HCl$$

Epoxidation of alkenes.[1] The peracid is generated at the interface of the biphasic solvent system CH_2Cl_2 and H_2O, buffered to control the pH either to 6.8–4.5 or to 9.5–8.8, depending on the substrate. This reagent epoxidizes alkenes in yields generally between 50 and 85% (isolated). It does not effect Baeyer–Villiger oxidation. An advantage is the use of either acidic or basic conditions.

[1] R. D. Bach, M. W. Klein, R. A. Ryntz, and J. W. Holubka, *J. Org.*, **44**, 2569 (1979).

N-Ethyl-2-pyrrolidone (NEP), (N-ethyl-2-pyrrolidone ring, N–C_2H_5, C=O). Mol. wt. 113.16, b.p. 78°/90 mm.

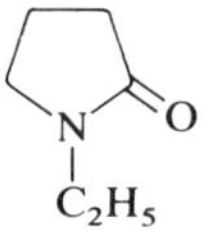

Solvent effects. Sowinski and Whitesides[1] have examined several polar, aprotic solvents as substitutes for HMPT, a possible carcinogen. Of these, N,N-diethylacetamide, $CH_3CON(C_2H_5)_2$, and N-ethyl-2-pyrrolidone appear to be the most useful. S_N2 displacement of neopentyl tosylates with lithium halides proceeds satisfactorily in either solvent. Neither one is as useful as HMPT for dissolving-metal reductions of alkenes to alkanes, even though blue solutions are formed with sodium with these solvents. However, a new reaction was developed that occurs in NEP but not in HMPT: reductive coupling of ketones and terminal alkenes to give tertiary alcohols in fair to good yield (equation I). An intramolecular version of the reaction to give five- and six-membered tertiary alcohols is also possible (equation II). In both reactions, yields decrease as the scale is increased.

$$\text{(I)}\quad (CH_3)_2C=O+CH_2=CH_2 \xrightarrow[55\%]{Na,\ (CH_3)_3COH,\ NEP} CH_3-\overset{OH}{\underset{CH_3}{C}}-CH_2CH_3$$

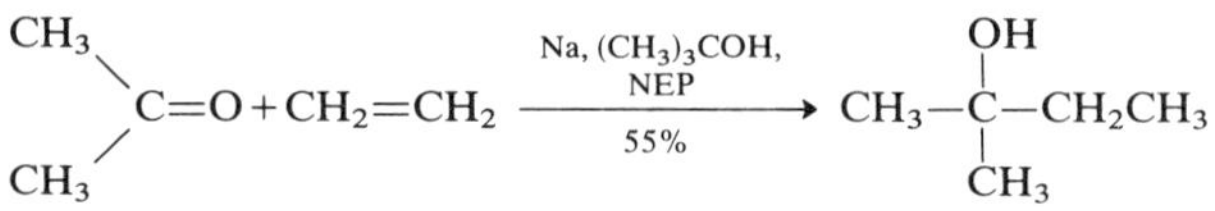

(II) $CH_2{=}CH\ \overset{O}{\overset{\|}{C}}CH_3$ (cyclopentane ring) $\xrightarrow[65\%]{}$ 1-CH_3-2-OH-2-CH_3 cyclopentane

[1] A. F. Sowinski and G. M. Whitesides, *J. Org.*, **44**, 2369 (1979).

1-Ethylsulfinyl-3-pentanone (1). Mol. wt. 162.25, m.p. 60–61°.

Preparation:

$$CH_2{=}CHCHO \xrightarrow[\sim 90\%]{1)\ C_2H_5SH;\ 2)\ HS(CH_2)_3SH} \text{2-}(C_2H_5SCH_2CH_2)\text{-2-H-1,3-dithiane} \xrightarrow{1)\ n\text{-BuLi};\ 2)\ C_2H_5I}$$

$$\text{2-}(C_2H_5SCH_2CH_2)\text{-2-}(CH_2CH_3)\text{-1,3-dithiane} \xrightarrow{1)\ Tl(ONO_2)_3;\ 2)\ NaIO_4} C_2H_5\overset{O}{\overset{\uparrow}{S}}CH_2CH_2\overset{O}{\overset{\|}{C}}C_2H_5 \quad \mathbf{1}$$

Vinyl ketone equivalent (*cf.*, **5**, 525–526). γ-Keto sulfoxides such as **1** can function as equivalents to vinyl ketones. Thus **1** reacts with dimedone in the presence of triethylamine (1.5 equiv.) in refluxing CH_3OH to give the Michael adduct **2** in 79% yield.[1]

$$\text{dimedone} + \mathbf{1} \xrightarrow[79\%]{N(C_2H_5)_3,\ CH_3OH,\ \Delta} \text{2-}(CH_2CH_2C(=O)C_2H_5)\text{-5,5-}(CH_3)_2\text{-cyclohexane-1,3-dione} \quad \mathbf{2}$$

[1] Y. Nagao, K. Seno, and E. Fujita, *Tetrahedron Letters*, 3167 (1979).

F

Ferric chloride, 1, 390–392; **2**, 199; **3**, 145; **4**, 236; **5**, 307–308; **6**, 260; **7**, 153–155; **8**, 228.

Coupling of benzylic halides. Treatment of the Grignard reagents derived from **1**[1] and **4**[2] with anhydrous ferric chloride affords useful intermediates for the synthesis of cyclophanes.

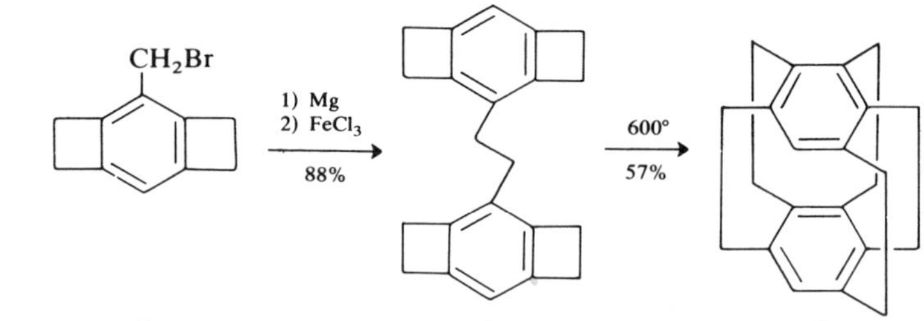

CH_2Br

1) Mg
2) $FeCl_3$
88%

600°
57%

1 **2** **3**

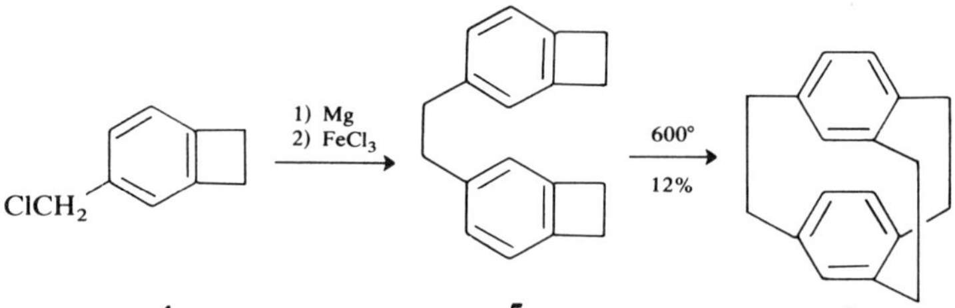

1) Mg
2) $FeCl_3$

600°
12%

$ClCH_2$

4 **5** **6**

γ-Lactones. Treatment of the dimethylethylenetricarboxylates **1a** and **1b** with $FeCl_3$ (25°) gives the chloro lactones **2a** and **2b**, respectively, with stereospecific *trans*-addition to the double bond.[3]

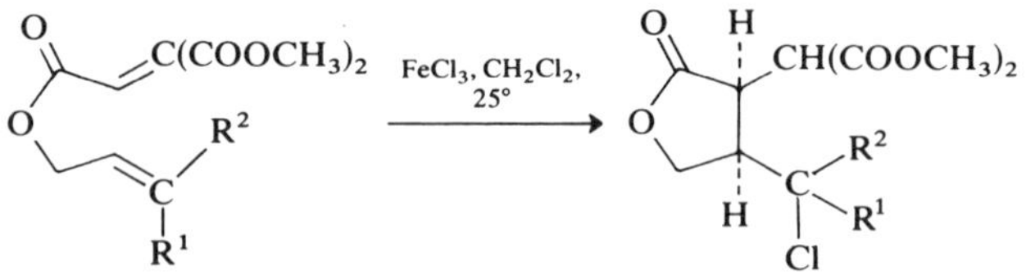

$C(COOCH_3)_2$ R^2 R^1

$FeCl_3, CH_2Cl_2,$ 25°

$CH(COOCH_3)_2$ R^2 R^1 Cl

1a ($R^1 = H, R^2 = CH_3$) **2a** (85%)
1b ($R^1 = CH_3, R^2 = H$) **2b** (75%)

[1] P. F. T. Schirch and V. Boekelheide, *Am. Soc.*, **101**, 3125 (1979).
[2] G. D. Ewing and V. Boekelheide, *J.C.S. Chem. Comm.*, 207 (1979).
[3] B. B. Snider and D. M. Roush, *J. Org.*, **44**, 4229 (1979).

Florisil, 1, 393–394.

Diels–Alder catalyst. The furane **1** undergoes an intramolecular Diels–Alder cycloaddition to give **2** in 93% yield in florisil–CH_2Cl_2 at 20° in 6 days. In the absence of florisil, **2** is formed to the extent of 55% in the same time. The acidic nature of florisil may be involved in this catalysis in addition to the known effect of materials with high surface area.[1]

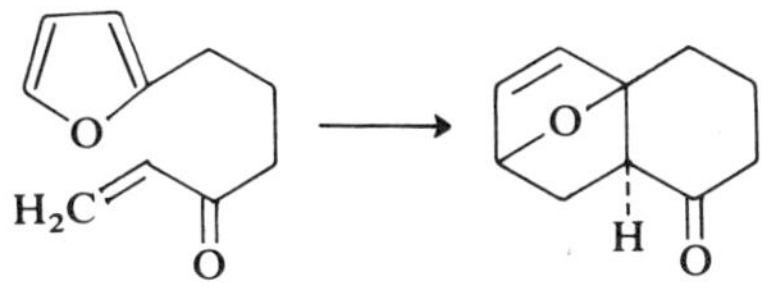

H₂C O O H O
1 **2**

[1] P. J. DeClercq and L. A. Van Royen, *Syn. Comm.*, **9**, 771 (1979).

Fluorine–Sodium trifluoroacetate.

α-Fluoro ketones.[1] Passage of fluorine gas through a suspension of sodium trifluoroacetate in Freon at −75° produces a solution containing CF_3CF_2OF and other perfluoroxy compounds. Reaction of steroidal enol acetates with this oxidative solution at −75° affords α-fluoro ketones in good yield.[1]

Examples:

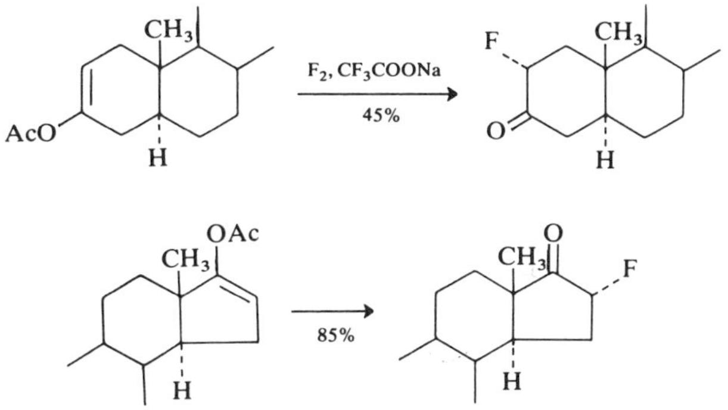

CH_3 AcO H $\xrightarrow[45\%]{F_2,\ CF_3COONa}$ F CH_3 O H

OAc CH_3 H $\xrightarrow[85\%]{}$ O CH_3 F H

[1] S. Rozen and Y. Menachem, *J.C.S. Chem. Comm.*, 479 (1979); S. Rozen and O. Lerman, *Am. Soc.*, **101**, 2782 (1979).

Fluorotri-*n*-butylmethylphosphorane (1). Mol. wt. 236.3, m.p. 11–13°, b.p. 74–76°/0.1 mm.

Preparation:[1]

$$(n\text{-}C_4H_9)_3P{=}CH_2 + HF \xrightarrow[75\%]{-70^\circ} (n\text{-}C_4H_9)_3\overset{+}{P}CH_3\ \overset{-}{F}$$

1

Fluorination. Primary and secondary alkyl halides and tosylates are converted into alkyl fluorides by reaction with **1** in THF. Alkenes are formed as by-products.

The reaction proceeds with inversion of configuration.[2] The reagent has been compared with other fluorinating reagents for the preparation of 2-fluorooctane. In this case, the highest yield (40%) is obtained with KF or diethyl(2-chloro-1,1,2-trifluoroethyl)amine (**5**, 26), but the highest optical yields are obtained with **1** (100%) and (diethylamino)sulfur trifluoride (97.6%, **6**, 183–184).[3]

[1] H. Schmidbaur, K. H. Mitschke, W. Buchner, H. Stühler, and J. Weindlein, *Ber.*, **106**, 1226 (1973).
[2] J. Bensoam, J. Leroy, F. Mathey, and C. Wakselman, *Tetrahedron Letters*, 353 (1979).
[3] J. Leroy, E. Hebert, and C. Wakselman, *J. Org.*, **44**, 3406 (1979).

Fluoroxytrifluoromethane, 2, 200; **3**, 146–147; **4**, 237–238; **5**, 312; **6**, 263–264; **7**, 156–158.

Addition–elimination to an arene. CF_3OF reacts with 4-acetoxypyrene (**1**) at low temperatures to give the adduct **2**, which loses CF_3OH at room temperature to give **3**. Extended reaction of **1** or of **3** with CF_3OF results in the difluoro ketone **4**.[1]

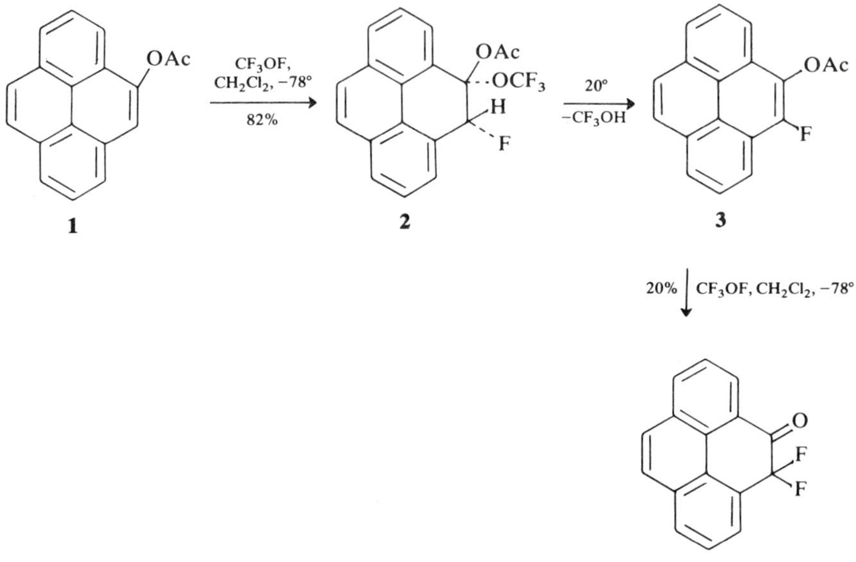

1 **2** **3** **4**

[1] T. B. Patrick, G. L. Cantrell, and C.-Y. Chang, *Am. Soc.*, **101**, 7434 (1979).

Formaldehyde, 1, 397–402; **2**, 200–201; **4**, 238–239; **6**, 264–267; **7**, 158–160; **8**, 231–232.

Vinyl ketones. Methyl ketones (**1**) undergo a Mannich-type condensation with paraformaldehyde (**2**), when refluxed in THF or dioxane in the presence of N-methylanilinium trifluoroacetate (this volume) to give vinyl ketones. The only limitation is that a linear methyl ketone such as **4** reacts at both the α- and α'-position, but reaction at the methyl group is greatly favored.[1]

$$\underset{\mathbf{1}}{RCOCH_3} + \underset{\mathbf{2}}{(HCHO)_n} \xrightarrow[70\text{–}90\%]{CF_3COO^- \overset{+}{H_2}N(CH_3)(C_6H_5)} \underset{\mathbf{3}}{RC(=O)\text{—}CH{=}CH_2}$$

$$\underset{\mathbf{4}}{n\text{-}C_8H_{17}CH_2COCH_3} \rightarrow \underset{\mathbf{5}}{n\text{-}C_8H_{17}CH_2COCH{=}CH_2} + \underset{\mathbf{6}}{n\text{-}C_8H_{17}C(=CH_2)COCH_3}$$

85:15

***Deacylative condensation*; *α-methylenation*.**[2] Reaction of the carbanion of α-alkylated β-keto esters with paraformaldehyde in THF at $-78°$ initially and finally at reflux leads to α-methylenated esters (equation I).

$$\text{(I)} \qquad CH_3COCH(R)COOC_2H_5 \xrightarrow{\text{1) LDA, THF; 2) }(CH_2O)_n} \underset{(50\text{–}95\%)}{RC(=CH_2)COOC_2H_5} + CH_3COOH$$

The reaction is also possible with lactones.

Examples:

$$C_6H_5CH_2CH(COCH_3)COOC_2H_5 \xrightarrow{96\%} C_6H_5CH_2C(=CH_2)COOC_2H_5$$

$$\alpha\text{-acetyl-}\gamma\text{-butyrolactone (COCH}_3\text{)} \xrightarrow{74\%} \alpha\text{-methylene-}\gamma\text{-butyrolactone (=CH}_2\text{)}$$

$$C_6H_5CH_2CH(COCH_3)_2 \xrightarrow{85\%} C_6H_5CH_2C(=CH_2)COCH_3$$

[1] J.-L. Gras, *Tetrahedron Letters*, 2955 (1978); *Org. Syn.*, submitted (1979).
[2] Y. Ueno, H. Setoi, and M. Okawara, *Tetrahedron Letters*, 3753 (1978).

Formaldehyde–Diisopropylamine–Copper(I) bromide.

Homologation of acetylenes to allenes.[1] A one-pot homologation of acetylenes to allenes involves reaction of an acetylenic compound with formaldehyde (1.6 equiv.), diisopropyl amine (1.2 equiv.), and copper(I) bromide at reflux in dioxane or THF (equation I). Other amines and metal salts can be employed, but diisopropylamine and copper(I) bromide give the most satisfactory results (26–97% yields). The reaction is general only for terminal acetylenes and is successful with propargylic alcohols, ethers, and acetates. Racemization is not observed with optically active

$$\text{(I)}\quad R-C\equiv C-H \xrightarrow{HCHO,\ HN(i\text{-}C_3H_7)_2} \underset{\mathbf{a}}{[R-C\equiv C-CH_2-N(i\text{-}C_3H_7)_2]} \xrightarrow{CuBr} \begin{matrix} H \\ R \end{matrix}\!\!>C=C=CH_2$$

substrates. The Mannich base **a** is an intermediate in the reaction, but need not be isolated. The origin of the hydrogen atom that becomes bonded to the allene is uncertain.

Examples:

$$\underset{OH}{CH_3(CH_2)_4CH}-C\equiv CH \xrightarrow[97\%]{} \underset{OH}{CH_3(CH_2)_4CH}-\overset{H}{C}=C=CH_2$$

$$\underset{\alpha_D\ -69°}{\underset{OAc}{CH_3(CH_2)_4CH}-C\equiv CH} \xrightarrow[41\%]{} \underset{\alpha_D\ -104°}{\underset{OAc}{CH_3(CH_2)_4CH}-\overset{H}{C}=C=CH_2}$$

This reaction was discovered during preparation of the Mannich base from the ethynyl A-nor steroid **1**. Allene **3** is obtained from this reaction as a side product in 31% yield. Allene **3** is also available from **2** by quaternization and $LiAlH_4$ reduction.[2]

H_2CO, CuBr, $HN(i\text{-}C_3H_7)_2$: **1** (HO, HC≡C, CH₃, OH, H) → **2** ($H_2C-C\equiv C$, $N(i\text{-}C_3H_7)_2$, HO, H) + **3** ($H_2C=C=C$, HO, H)

2 → **3**: 1) CH_3I; 2) $LiAlH_4$

[1] P. Crabbé, H. Fillion, D. André, and J.-L. Luche, *J.C.S. Chem. Comm.*, 859 (1979).
[2] P. Crabbé, D. André, and H. Fillion, *Tetrahedron Letters*, 893 (1979).

Formic acid, 1, 404–407; **2**, 202–203; **3**, 147; **4**, 239–240; **5**, 316–316; **7**, 160; **8**, 232.

Catalytic hydrogen transfer. Two laboratories[1,2] have recently reported that formic acid is a particularly useful hydrogen donor for catalytic hydrogenolysis of benzyl-type protecting groups from peptides. Benzyl and benzyloxycarbonyl groups are generally removed in less than 10 minutes at room temperature. *t*-Butyloxycarbonyl groups are removed only slowly. Formic acid is superior to the conventional donors, cyclohexene and cyclohexadiene, in that it is an excellent solvent for most peptides.

[1] B. ElAmin, G. M. Anantharamaiah, G. P. Royer, and G. E. Means, *J. Org.*, **44**, 3442 (1979).
[2] K. M. Sivanandaiah and S. Gurusiddappa, *J. Chem. Res.* (*S*), 108 (1979).

Furane, 7, 161.

Modification of allenes. α-Allenic ketones are excellent dienophiles in the Diels–Alder reaction. This fact coupled with the reversibility of the cycloaddition provides a method for modification of the allenic ketones. An example is cited in equation (I).[1]

(I) furane + $CH_2{=}C{=}CHCOCH_3$ $\xrightarrow{>90\%}$ Diels–Alder adduct ($=CH_2$, $C(=O)CH_3$) (two isomers) $\longrightarrow$

oxime adduct ($=CH_2$, $C(CH_3){=}N{-}OH$) $\xrightarrow[\text{53\% overall}]{450^\circ}$ $CH_2{=}C{=}CH{-}C(CH_3){=}N{-}OH$

[1] M. Bertrand, J.-L. Gras, and B. S. Galledou, *Tetrahedron Letters*, 2873 (1978).

G

Glyoxylic acid, 5, 320; **7**, 162.

Marschalk alkylation (**8**, 456). The intramolecular Marschalk alkylation is useful in syntheses of the tetracyclic ring system of the cytostatic anthracycline antibiotics. A bimolecular version of the reaction involves condensation of **1** with glyoxylic acid and alcoholic KOH. The product of this reaction is reduced and esterified to afford **2** in 33% overall yield. Less than 5% of the isomeric product **3** is obtained. 9-Deoxy-ε-rhodomycinone **4** is available from **2** in two steps (details not provided).[1]

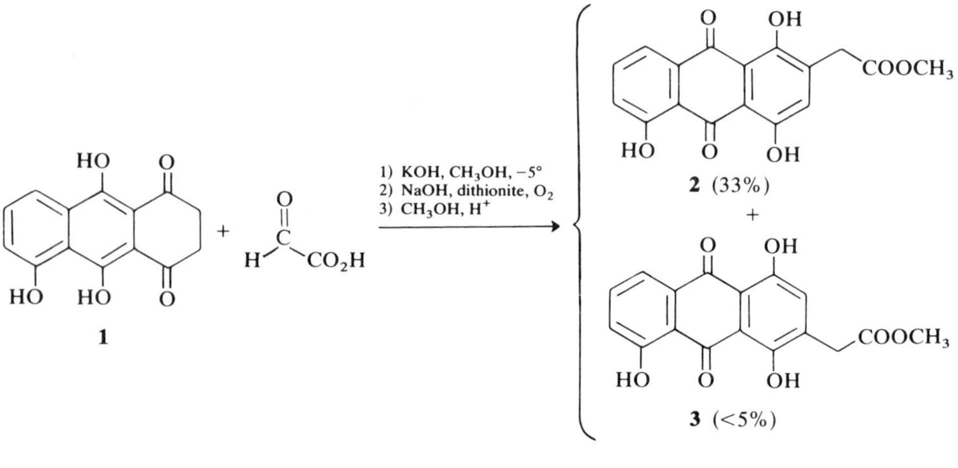

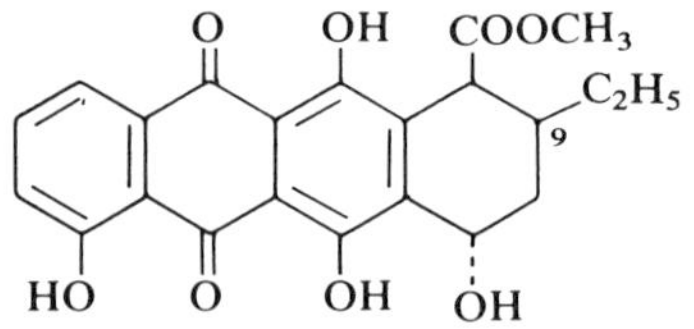

4

[1] K. Krohn, *Angew. Chem., Int. Ed.*, **18**, 621 (1979).

Glyoxylic acid dithioethyl acetal, $(C_2H_5S)_2CHCOOH$ (**1**). Mol. wt. 180.29. This reagent is conveniently prepared in quantitative yield by addition of dichloroacetic acid to 3 equiv. of NaH in THF followed by reaction overnight with an excess of ethanethiol.[1]

Glyoxylic acid dianion equivalent.[1] Treatment of **1** with 2 equiv. of potassium bis(trimethylsilyl)amide (**6**, 482) in THF at 0° generates the dianion (**2**), which reacts in good yield with a variety of electrophiles.

$$\underset{\mathbf{1}}{(C_2H_5S)_2CHCOOH} \xrightarrow[\text{THF}]{KN[Si(CH_3)_3]_2} \underset{\mathbf{2}}{(C_2H_5S)_2C{=}CO_2^{-2}\cdot 2K^+}$$

$$\mathbf{2} \xrightarrow[100\%]{CH_3I} \underset{\mathbf{3}}{(C_2H_5S)_2C(CH_3){-}COOH}$$

$$\mathbf{2} \xrightarrow[93\%]{\text{PhCH}_2\text{-aziridine (N-Ts)}} \mathbf{4}\ \text{(Ph-CH}_2\text{-CH(NHTs)-CH}_2\text{-C(SC}_2\text{H}_5)_2\text{COOH)}$$

[1] G. S. Bates, *J.C.S. Chem. Comm.*, 161 (1979).

Grignard reagents, 1, 415–424; **2**, 205; **5**, 321; **6**, 269–270; **7**, 163–164; **8**, 235–238.

Alkene synthesis. A new synthesis of alkenes, with the bond in a predetermined position, involves reaction of an α-chloro ketone with a Grignard reagent in ether at −60°; the adduct is then treated with lithium powder and THF at −60° for 2 hours. When the temperature is raised to 20°, LiOMgBr is eliminated with formation of an alkene. Yields are moderate to good (35–99%).[1]

Examples:

$$CH_3C(=O){-}CHClCH_3 + C_6H_5CH_2MgBr \xrightarrow{-60^\circ} \left[C_6H_5CH_2{-}C(OMgBr)(CH_3){-}CHClCH_3\right] \xrightarrow{Li,\ -60^\circ}$$

$$\left[C_6H_5CH_2C(OMgBr)(CH_3){-}CHLiCH_3\right] \xrightarrow[74\%]{20^\circ} (C_6H_5CH_2)(CH_3)C{=}CHCH_3$$

(E and Z)

$$\text{2-chlorocyclohexanone} \xrightarrow[87\%]{1)\ C_2H_5MgBr;\ 2)\ Li} \text{1-ethylcyclohexene}$$

$$\text{2-chlorocyclopentanone} \xrightarrow[99\%]{1)\ CH_2{=}CHCH_2MgBr;\ 2)\ Li} \text{1-}(CH_2CH{=}CH_2)\text{cyclopentene}$$

Ketone synthesis. It has been an article of faith that Grignard reagents are too reactive to be useful for synthesis of ketones from acid chlorides. Actually the reaction affords ketones in almost quantitative yields when conducted in THF at −30 to −78°.[2]

Examples:

$$C_6H_5MgBr + C_6H_5COCl \xrightarrow[89\%]{THF,\ -78^\circ} C_6H_5COC_6H_5$$

$$CH_3CH_2CH_2MgBr + (CH_3)_3CCOCl \xrightarrow[88\%]{} CH_3CH_2CH_2COC(CH_3)_3$$

Reaction with α-chloroacylsilanes. Methylmagnesium iodide reacts with α-chloroacyltrimethylsilanes (**2**), readily available from silyl enol ethers (**1**), to give β-ketoalkyltrimethylsilanes (**3**). The reaction involves rearrangement of the silyl group to the adjacent carbon atom.[3]

$$\underset{\mathbf{1}}{RCH{=}\overset{OSi(CH_3)_3}{\overset{|}{C}}Si(CH_3)_3} \xrightarrow{Cl_2,\ CCl_4} \underset{\mathbf{2}}{R\underset{Cl}{\underset{|}{C}}HCOSi(CH_3)_3} \xrightarrow{CH_3MgI}$$

$$\underset{\mathbf{a}}{\left[RCH(Cl)\text{—}C(OMgI)(Si(CH_3)_3)\text{—}CH_3\right]} \xrightarrow[90\text{–}98\%]{} \underset{\mathbf{3}}{RCH(Si(CH_3)_3)COCH_3}$$

Ketone enolates undergo a similar reaction with **2** to give 1,3-dicarbonyl compounds.[4]

Example:

$$C_2H_5\overset{O}{\overset{\|}{C}}C_2H_5 \xrightarrow{1)\ LDA\ \ 2)\ \mathbf{2}} \left[C_2H_5\overset{O}{\overset{\|}{C}}\text{—}\underset{CH_3}{CH}\text{—}\overset{OLi}{\underset{Si(CH_3)_3}{C}}\text{—}CHClR \rightarrow\right.$$

$$\left.C_2H_5\overset{O}{\overset{\|}{C}}\underset{CH_3}{CH}\text{—}\overset{O}{\overset{\|}{C}}\underset{Si(CH_3)_3}{CHR}\right] \xrightarrow[93\%]{H_3O^+} C_2H_5\overset{O}{\overset{\|}{C}}\underset{CH_3}{CH}\overset{O}{\overset{\|}{C}}CH_2R$$

$(R = CH_2C_6H_5)$

Cyclopropyl phenyl sulfones.[5] 3-Bromo-1-(phenylsulfonyl)-1-propene (**1**), prepared as shown, reacts with allyl, propargyl, aryl, and benzyl Grignard reagents to give (E)-2-substituted cyclopropyl phenyl sulfones (**2**) in moderate to good yield. This reaction is not observed with alkyl Grignard reagents.

$$CH_2{=}CHCH_2SO_2C_6H_5 \xrightarrow{1)\ Br_2\ \ 2)\ N(C_2H_5)_3} \underset{\mathbf{1}\ (E/Z = 75:25)}{BrCH_2CH{=}C(SO_2C_6H_5)(H)} \xrightarrow{RMgX}$$

$$\underset{\mathbf{a}}{\left[BrCH_2\overset{R}{\overset{|}{C}}H\text{—}\underset{MgX}{\underset{|}{C}}HSO_2C_6H_5\right]} \xrightarrow[40\text{–}75\%]{} \underset{\mathbf{2}}{\text{(E)-cyclopropane: R, H / H, }SO_2C_6H_5} + MgBrX$$

Regioselective anthraquinone synthesis. The methoxy-substituted phthalic anhydride **1** reacts with the Grignard reagent **2** to give essentially only the pseudoacid **3**. The product is cyclized (conc. H_2SO_4, 25°) and demethylated (HBr–HOAc) to digitopurpone (**4**) in 51% overall yield.

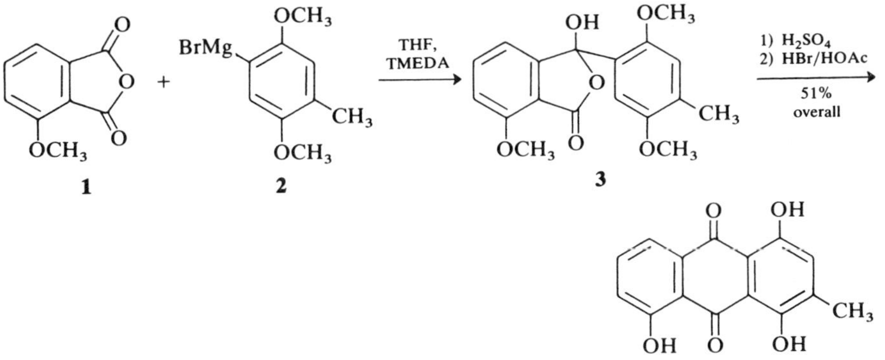

1 + 2 → (THF, TMEDA) → 3 → (1) H_2SO_4 2) HBr/HOAc; 51% overall) → 4

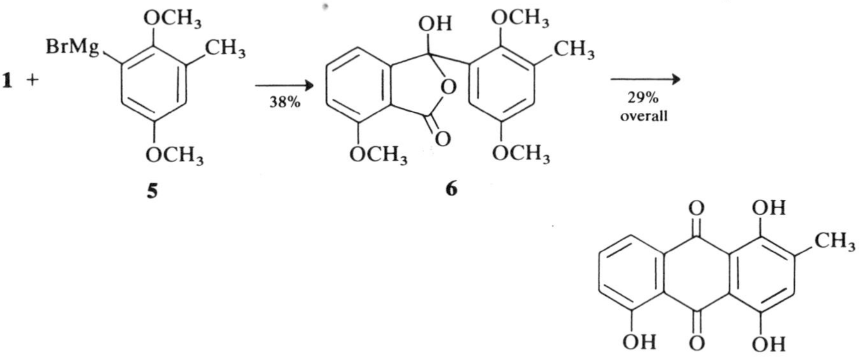

1 + 5 → (38%) → 6 → (29% overall) → 7

Similar reactions of **1** with the isomeric Grignard reagent **5** result in islandicin (**7**) in a somewhat lower overall yield. In both cases, only 3–5% of the other possible isomer is formed. Surprisingly, the aryllithium reagent corresponding to **2** shows no regioselectivity in the reaction with **1**.[6]

Similar reactions starting with 3,5-dimethoxyphthalic anhydride and the Grignard reagent **5** were used to synthesize catenarin (**8**) in 53% overall yield.[7]

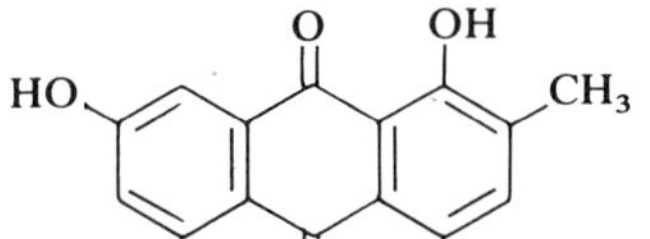

8

The classical phthalic anhydride synthesis is not useful for the synthesis of islandicin. Thus both the benzoylbenzoic acids **9** and **10** on cyclization give exclusively the same anthraquinone **11**.

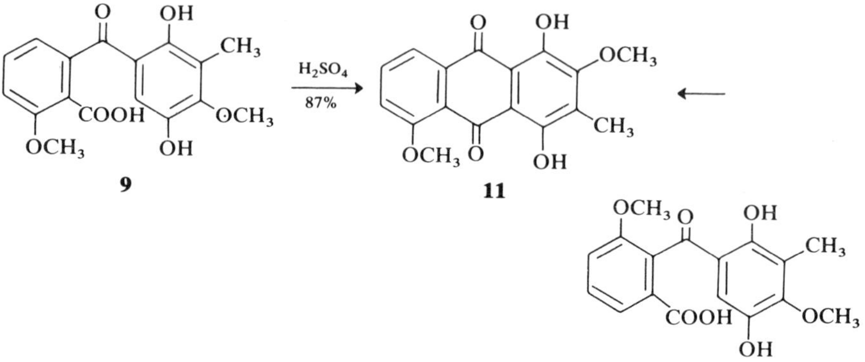

O OH CH_3 COOH OCH_3 OCH_3 OH

9

H_2SO_4 87%

O OH OCH_3 CH_3 OCH_3 O OH

11

OCH_3 O OH CH_3 COOH OCH_3 OH

10

Another regioselective synthesis of islandicin monomethyl ether is a variation of the succinic acid synthesis (equation I).[8]

(I) $OCOC_2H_5$ OCH_3 1) $BF_3 \cdot (C_2H_5)_2O$ 2) $(CH_3)_2SO_4$, K_2CO_3 80% OCH_3 O C CH_2 CH_3 OCH_3 Br_2 79%

OCH_3 O CH_3 Br OCH_3 1) $CH_2(COOC_2H_5)_2$, NaH, DMF, HMPT 90% OCH_3 O CH_3 $CH(COOC_2H_5)_2$ OCH_3 1) $AlCl_3$, CH_2Cl_2 (86%) 2) KOH, H_2O (85%) 3) Py, 110° (98%)

OH O CH_3 COOH OCH_3 $K_2S_2O_8$ 48% OH O CH_3 COOH OCH_3 OH H_2SO_4

[OH O CH_3 OCH_3 OH O] Oxid. 30% O OH CH_3 OCH_3 O OH

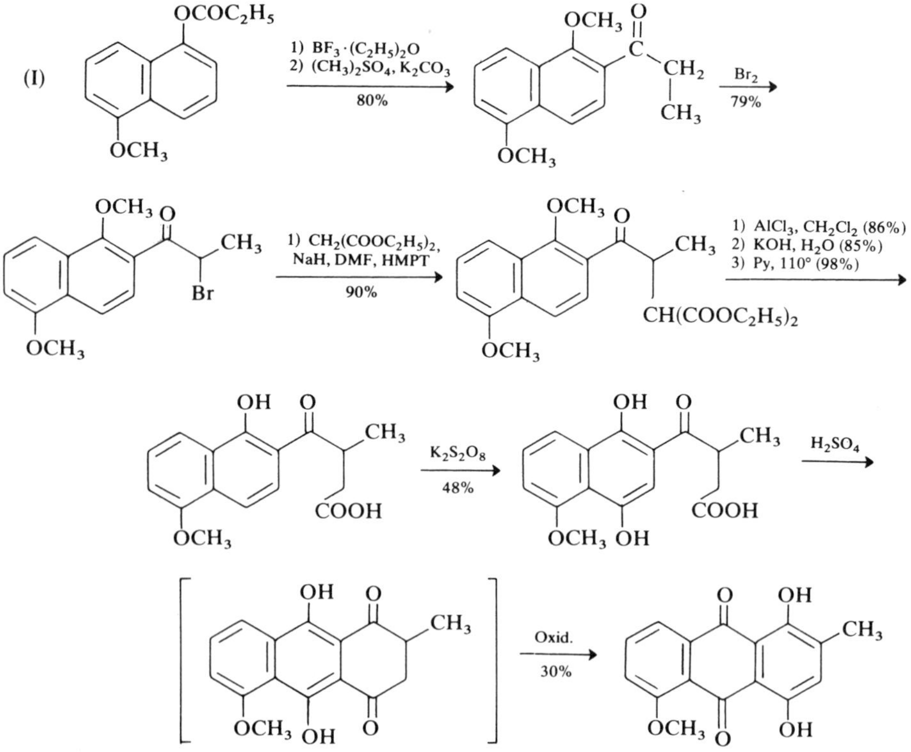

[1] J. Barluenga, M. Yus, and P. Bernad, *J.C.S. Chem. Comm.*, 847 (1978).
[2] F. Sato, M. Inone, K. Oguro, and M. Sato, *Tetrahedron Letters*, 4303 (1979).
[3] T. Sato, T. Abe, and I. Kuwajima, *ibid.*, 259 (1976).
[4] I. Kuwajima and K. K. Matsumoto, *ibid.*, 4095 (1979).
[5] J. J. Eisch and J. E. Galle, *J. Org.*, **44**, 3277 (1979).
[6] M. Braun, *Angew. Chem., Int. Ed.*, **17**, 945 (1978).
[7] *Idem, Tetrahedron Letters*, 2885 (1979).
[8] R. D. Gleim, S. Trenbeath, F. Suzuki, and C. J. Sih, *J.C.S. Chem. Comm.*, 242 (1978).

H

2-Halopyridinium salts.

Review.[1] New synthetic reactions based on the onium salts of aza-arenes have been reviewed (75 references). The reactions discussed involve activation of carboxylic acids or alcohols with 2-halopyridinium, benzoxazolium, benzothiazolium, and pyridinium salts to afford 2-acyloxy or 2-alkoxy intermediates, which can be transformed into esters, amides, thiol esters, (macrocyclic) lactones, acid fluorides, olefins, allenes, carbodiimides, isocyanates, isothiocyanates, and nitriles under appropriate conditions.

[1] T. Mukaiyama, *Angew. Chem., Int. Ed.*, **18**, 707 (1979).

Hexamethyldisilazane, 1, 427; **2**, 207–208; **5**, 323; **6**, 273; **7**, 167.

Lactams.[1] Lactams, γ-, δ, and ϵ-, are obtained in high yield by treatment of amino acids with this silylating agent [or N,O-bis(trimethylsilyl)acetamide] at reflux for 4–48 hours. Work-up includes dilution with methanol or ethanol and evaporation. No racemization is observed. Trimethylsilyl esters are intermediates.

Examples:

(R)-(−)-$H_2NCH_2CH(OH)CH_2COOH$ —1) $[(CH_3)_3Si]_2NH, \Delta$; 2) C_2H_5OH; 89%→ HO-substituted pyrrolidinone (N–H, C=O)

$H_2N(CH_2)_4COOH$ —95%→ 2-piperidinone (N–H, C=O)

S-(+)-$H_2N(CH_2)_4CH(NH_2)COOH$ —82%→ 3-amino-2-azepanone (NH_2, N–H, C=O)

[1] R. Pellegata, M. Pinza, and G. Pifferi, *Synthesis*, 614 (1978).

Hexamethylenetetramine, 1, 427; **2**, 208; **4**, 243.

Review.[1] Use of hexamethylenetetramine for introduction of amino and formyl groups has been reviewed. The tetramine has also been used for synthesis of nitrogen-containing heterocycles.

[1] N. Blažević, D. Kolbah, B. Belin, V. Šunjić, and F. Kajfež, *Synthesis*, 161 (1979).

Hexamethylphosphoric triamide (HMPT), **1**, 430–431; **2**, 208–210; **3**, 149–153; **4**, 244–247; **5**, 323–325; **6**, 273–274; **7**, 168–170; **8**, 240–245.

Carcinogenicity. HMPT is now thought to be a potent chemical carcinogen and should be handled with extreme care.[1]

Solvent Effects

Conjugate addition.[2,3] The ability of HMPT to promote conjugate addition of lithium carbanions to α,β-enones has been noted in several reactions. Reagents of the type $(R'S)_2CRLi$ and $(R'Se)_2CRLi$ generally react with α,β-enones in THF at $-78°$ to give mainly or exclusively 1,2-adduct, but react in THF–HMPT at $-78°$ to give mainly or exclusively 1,4-adducts. Both reactions generally proceed under kinetic control; once formed in THF, the 1,2-adducts usually do not rearrange to the 1,4-adducts on addition of HMPT. However, kinetically controlled 1,4-addition is dependent on the structure of both the carbanion and the enone. Some carbanions require HMPT for 1,4-addition to one enone but not to a closely related one, and some enones undergo 1,4-addition under thermodynamic control with one carbanion but not with a closely related carbanion.

Example[3]:

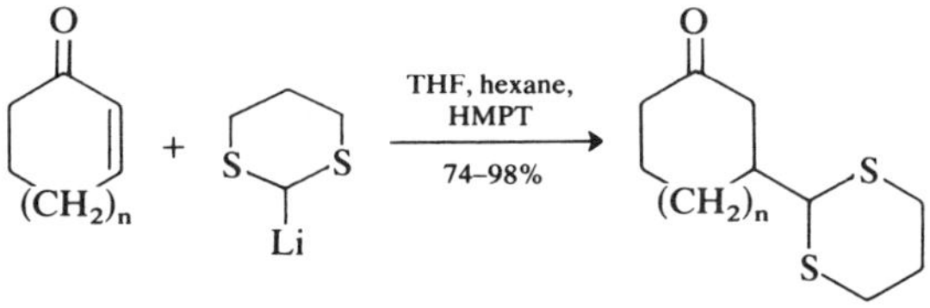

O, $(CH_2)_n$ + S, S, Li —THF, hexane, HMPT; 74–98%→ O, $(CH_2)_n$, S, S

Conjugate addition of lithium enolates to α,β-unsaturated thioamides is also promoted by the addition of 1–2 equiv. of HMPT.[4]

[1] H. Spencer, *Chem. Ind.*, 728 (1979).
[2] L. Wartski, M. El Bouz, J. Seyden-Penne, W. Dumont and A. Krief, *Tetrahedron Letters*, 1543 (1979); J. Luchetti, W. Dumont, and A. Krief, *ibid.*, 2695 (1979).
[3] C. A. Brown and A. Yamaichi, *J.C.S. Chem. Comm.*, 100 (1979).
[4] Y. Tamaru, T. Harada and Z. Yoshida, *Am. Soc.*, **101**, 1316 (1979).

Hexamethylphosphorous triamide, 1, 425; **2**, 207; **3**, 148–149; **6**, 279–280.

Desulfurization (**4**, 242; **6**, 280). Harpp and co-workers[1] have prepared polymeric reagents, **1** and **2**, related to hexamethylphosphorous triamide and also a new aminophosphine (**3**). In general, **1** and **2** are somewhat less reactive for

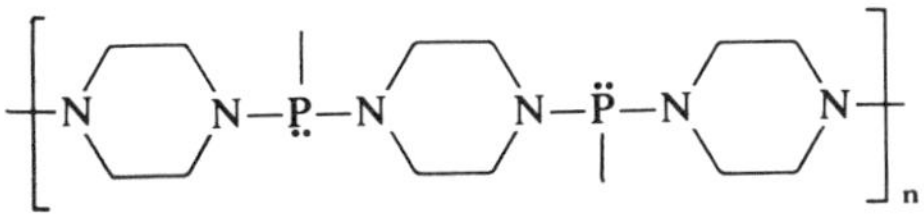

N, N—P—N, N—P—N, N, n

1

N, N—P—N, N—P, n

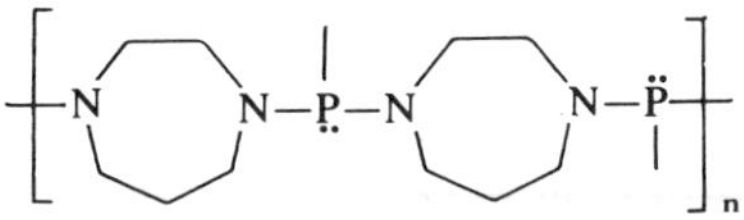

2

desulfurization than **3**, but offer the advantages of polymeric reagents.

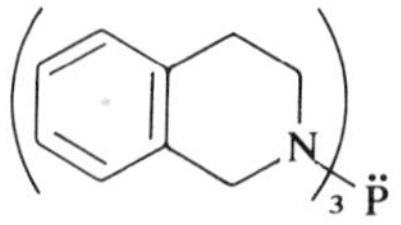

3

Examples:

$$(C_6H_5CH_2S)_2 \xrightarrow[88\%]{\mathbf{2}} C_6H_5CH_2SCH_2C_6H_5$$

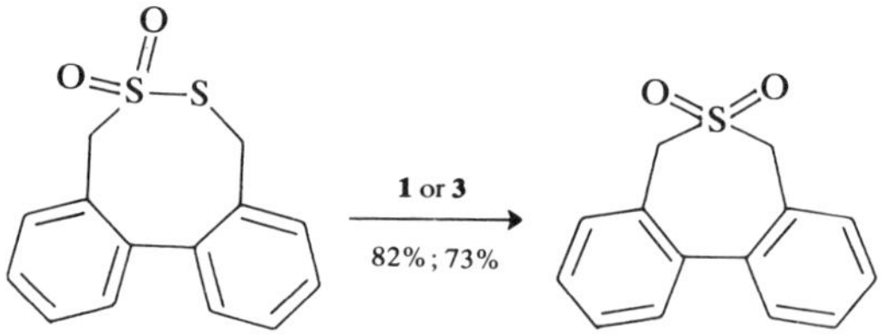

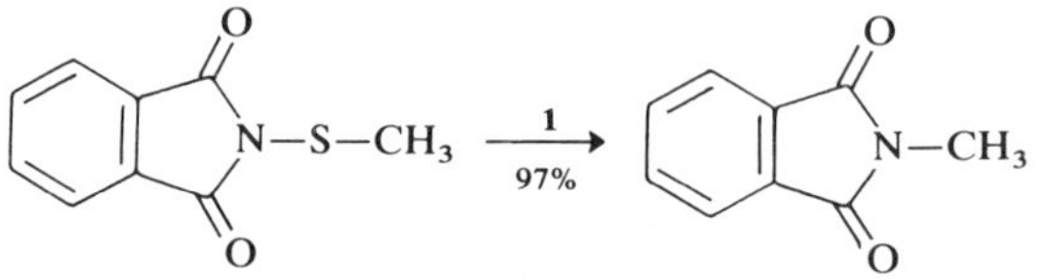

1 or **3**; 82%; 73%

1; 97%

[1] D. N. Harpp, J. Adams, J. G. Gleason, D. Mullins, and K. Steliou, *Tetrahedron Letters*, 3989 (1978).

Hexamethylphosphorous triamide–Iodine.

Deoxygenation of sulfoxides and azoxyarenes.[1] This combination is superior to triphenylphosphine–iodine[2] for deoxygenation of sulfoxides to sulfides (70–95% yield) and of azoxy benzenes to azobenzenes (~90% yield, two examples). The reaction can be promoted by addition of sodium iodide. One advantage is that the by-product, HMPT, is soluble in water and easily removed.

$$C_6H_5\overset{O}{\overset{\|}{S}}C_6H_5 + \underset{\mathbf{1}}{[(CH_3)_2N]_3P} \xrightarrow[95\%]{I_2,\ CH_3CN,\ 12\ hr.,\ 23^\circ} C_6H_5SC_6H_5 + [(CH_3)_2N]_3P{=}O$$

Substrate/**1**/I_2/NaI = 1 : 2 : 1 : 4

[1] G. A. Olah, B. G. B. Gupta, and S. C. Narang, *J. Org.*, **43**, 4503 (1978).
[2] *Idem, Synthesis*, 137 (1978).

Hydrazine, 1, 434–445; **2,** 211; **3,** 153; **4,** 248; **5,** 327–329; **6,** 280–281; **7,** 170–171; **8,** 245.

Synthesis of pyridazines. Treatment of 1,4-dicarbonyl compounds with hydrazine in EtOH, THF, or aqueous THF followed by oxidation with PtO_2 affords pyridazines in 80–90% yield.[1]

Examples:

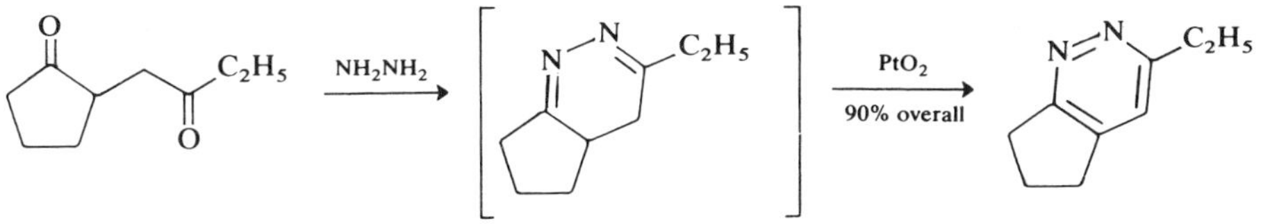

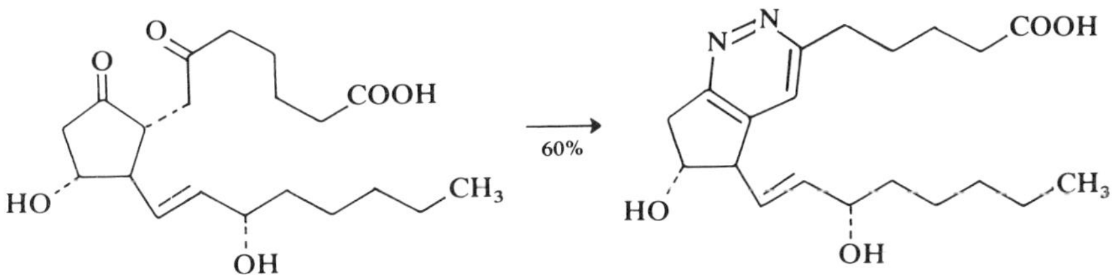

$\xrightarrow[\text{}]{NH_2NH_2}$ $\xrightarrow[\text{90\% overall}]{PtO_2}$

$\xrightarrow[60\%]{}$

Wolff–Kishner reduction. One step in a study of the absolute configuration of **2** involved a Wolff–Kishner reduction of **1** by the procedure indicated in the formulation.[2]

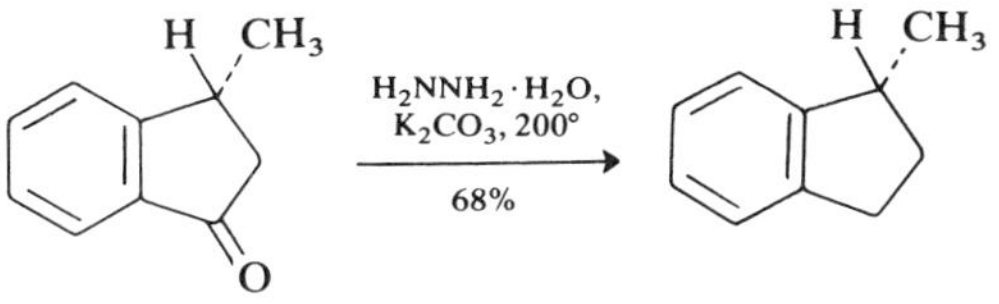

$\xrightarrow[68\%]{H_2NNH_2 \cdot H_2O,\ K_2CO_3,\ 200°}$

(R)-**1** (R)-**2**

Paquette and Han[3] used these conditions in the final step in a synthesis of (±)-isocomene (**3**).

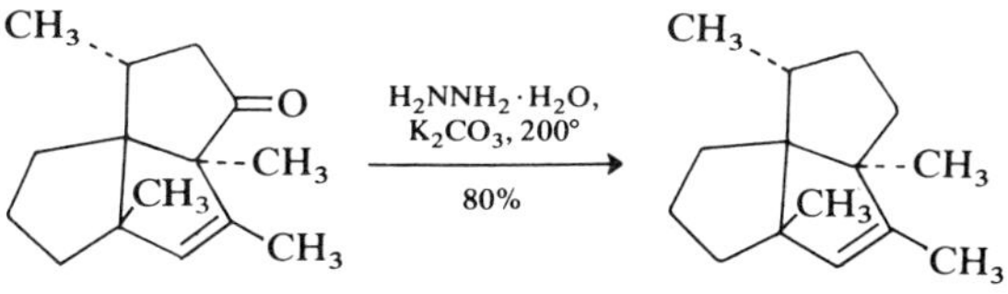

$\xrightarrow[80\%]{H_2NNH_2 \cdot H_2O,\ K_2CO_3,\ 200°}$

3

Reduction of nitroarenes (*cf.*, **1**, 440).[4] Nitroarenes are reduced to amines in generally good yields by hydrazine hydrate with hydrated β-iron(III) oxide, β-$Fe_2O_3 \cdot H_2O$[5] as catalyst.

[1] K. C. Nicolaou, W. E. Barnette, and R. L. Magolda, *Am. Soc.*, **101**, 766 (1979).
[2] H.-J. Hansen, H.-R. Sliwka, and W. Hug, *Helv.*, **62**, 1120 (1979).
[3] L. A. Paquette and Y. K. Han, *J. Org.*, **44**, 4014 (1979).
[4] T. Miyata, Y. Ishino, and T. Hirashima, *Synthesis*, 834 (1978).
[5] H. B. Weiser, W. O. Milligan, and E. L. Cook, *Inorg. Syn.*, 215 (1946).

Hydridotris(triisopropylphosphine)rhodium(I), $RhH[P(i\text{-}C_3H_7)_3]_3$ (**1**). Mol. wt. 584.64. This rhodium hydride is prepared by reduction of $RhCl \cdot 3H_2O$ with Na/Hg in the presence of triisopropylphosphine in THF.[1]

Homogeneous hydrogenation of nitriles.[2] Saturated nitriles are hydrogenated to primary amines in 67–100% yield with **1** as catalyst in THF. The catalytic activity of **1** for aromatic nitriles is not so high as for hydrogenation of aliphatic nitriles. The C=C bond of unsaturated nitriles is reduced more readily than the C≡N bond. This catalyst is also effective for dehydrogenation of amines (benzylamine → benzonitrile, 27% yield).

Examples:

$$C_6H_5CH_2CN \xrightarrow[96\%]{} C_6H_5CH_2CH_2NH_2$$

$$C_6H_5CN \xrightarrow[45\%]{} C_6H_5CH_2NH_2$$

$$CH_3CH{=}CHCN \rightarrow \underset{72\%}{CH_3(CH_2)_3NH_2} + \underset{28\%}{\begin{array}{c}CH_3CHCH_2CH_2CN\\ |\\ CH_2CH_2CH_3\end{array}}$$

[1] T. Yoshida, T. Okano, and S. Otsuka, *J.C.S. Chem. Comm.*, 855 (1978).
[2] *Idem, ibid.*, 870 (1979).

Hydriodic acid, 1, 449–450; **2,** 213–214; **8,** 246.

Reductive cyclization. 2-(9′-Phenanthroyl)benzoic acid (**1**) is converted into dibenz[*a,c*]anthracene (**2**) in high yield when refluxed in acetic acid with 50% HI. Experiments with compounds related to **1** suggest that reduction to **a** precedes cyclization to a dibenzanthrone, which is then reduced by HI to **2.**[1]

Hydriodic acid is recommended for reduction of polycyclic quinones and phenols to polycyclic arenes.[2]

[1] R. G. Harvey, C. Leyba, M. Konieczny, P. P. Fu, and K. B. Sukumaran, *J. Org.*, **43,** 3423 (1978).
[2] M. Konieczny and R. G. Harvey, *ibid.*, **44,** 4813 (1979).

Hydrofluoric acid, 6, 284.

Cleavage of* t*-butyldimethylsilyl ethers.[1] Aqueous hydrofluoric acid (5–30% of a 40% aqueous solution of HF) is an excellent reagent for removal of *t*-butyldimethylsilyl and trimethylsilyl groups. The method is particularly useful for regeneration of β-ketols, which are prone to dehydration. Acetonitrile, nitromethane, or ethanol can be used as solvent. Typical yields of the regenerated alcohols are indicated in the formulas [$R = Si(CH_3)_2C(CH_3)_3$].

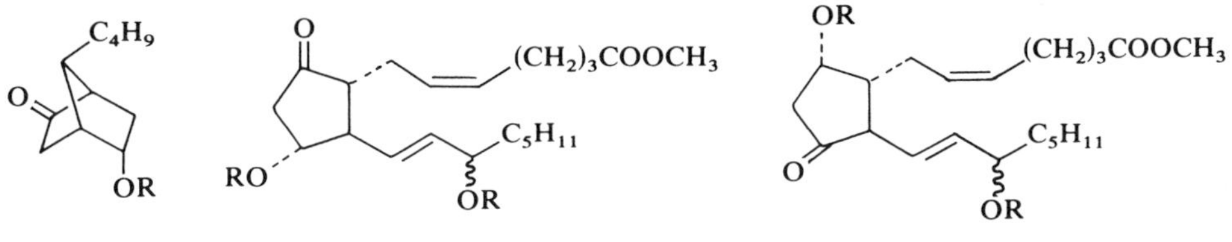

(100%) (100%) (70%)

[1] R. F. Newton, D. P. Reynolds, M. A. W. Finch, D. R. Kelly, and S. M. Roberts, *Tetrahedron Letters*, 3981 (1979).

Hydrogen chloride, 2, 215; **4,** 252; **5,** 335–336; **6,** 285; **7,** 172–173; **8,** 246.

Protodesilylation.[1] A new route to 4-alkylidenecyclohexenes involves protodesilylation with aqueous HCl in THF–methanol (equation I). The method provides a facile synthesis of terpinolene (**1**) as formulated in equation (II).

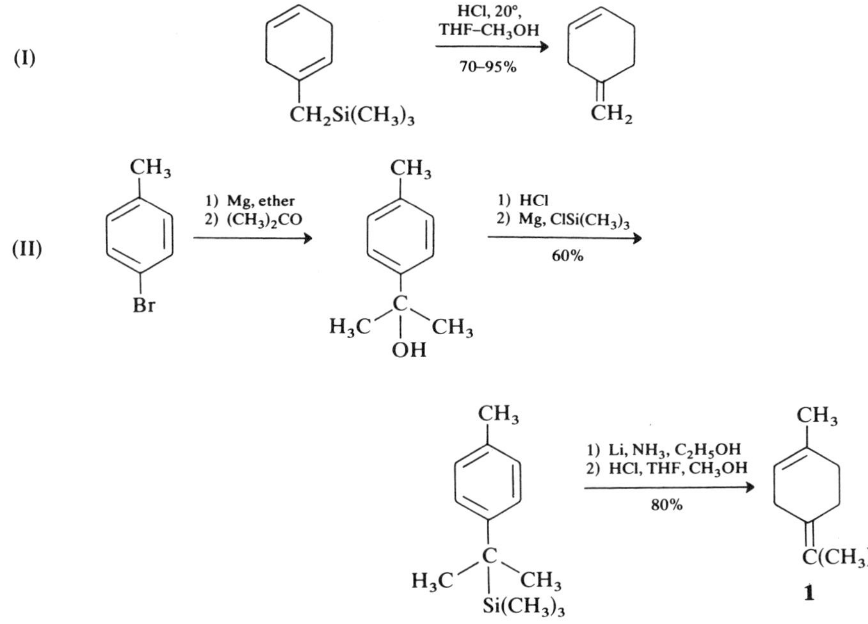

(I) $CH_2Si(CH_3)_3$ —[HCl, 20°, THF–CH_3OH; 70–95%]→ CH_2

(II) CH_3 / Br —[1) Mg, ether 2) $(CH_3)_2CO$]→ CH_3 / H_3C–C(OH)–CH_3 —[1) HCl 2) Mg, $ClSi(CH_3)_3$; 60%]→

CH_3 / H_3C–C($Si(CH_3)_3$)–CH_3 —[1) Li, NH_3, C_2H_5OH 2) HCl, THF, CH_3OH; 80%]→ CH_3 / $C(CH_3)_2$ **1**

[1] D. J. Coughlin and R. G. Salomon, *J. Org.*, **44**, 3784 (1979).

Hydrogen cyanide, 1, 454–455; **4,** 252.

Asymmetric addition of HCN to alkenes. [(+)-DIOP]Pd (**6,** 309) catalyzes the asymmetric addition of HCN to norbornene derivatives in 17–33% optical yield.[1]

Since HCN does not react with 7,7-dimethylnorbornene under these conditions, the reaction is apparently very susceptible to steric hindrance.

Example:

HCN, [(+)-DIOP]Pd, 53% → CN, H (29% ee)

[1] P. S. Elmes and W. R. Jackson, *Am. Soc.*, **101**, 6128 (1979).

Hydrogen fluoride, 1, 455–456; **2**, 215–216; **4**, 252; **5**, 336–337; **6**, 285; **7**, 173.

Friedel–Crafts acylation. Posner *et al.*[1] have developed a remarkably efficient route to the methyl ether of the steroid 11-oxoequilenin (**5**) from 2-methyl-2-cyclopentenone (**1**). β-Addition of the organocoppermagnesium reagent **2** to **1** followed by α-alkylation with ethyl iodoacetate proceeds stereospecifically to give the secosteroid **3** in 94% yield. The final step requires an intramolecular Friedel–Crafts acylation, a reaction that has proved troublesome in previous syntheses of steroids via 9,11-secosteroids. And indeed attempts to cyclize the free acid corresponding to **3** with HF proceeded in yields of 10%. However, cyclization of the ketal acid **4** gives stereochemically pure **5** in 75% yield based on recovered secosteroid. The overall yield from 2-bromo-6-methoxynaphthalene is 52%.

CH_3O ... $(CuC{\equiv}CC_3H_7\text{-}n)MgBr$ (**2**) → 1) **1** (CH_3, O); 2) $ICH_2COOC_2H_5$, HMPT; 94% → **3** ($C_2H_5OOCCH_2$, CH_3, O, H, CH_3O)

4 (HOOC, CH_3, O, O, H, CH_3O) →(HF) **5** (65%) (O, CH_3, O, H, CH_3O) + acid corresponding to **3** (10%)

Protodesilylation. A new preparation of β,γ-unsaturated carboxylic acids involves the amide acetal Claisen rearrangement of a 3-(trimethylsilyl)allyl alcohol such as **1** to allylsilanes (**2**). Desilylation of **2** by usual methods results in formation of stereoisomers. However, use of liquid HF at low temperatures results only in (E)-**3** in 88% yield.[2] The HF–pyridine complex is not so stereoselective. Amides such as **3** can be converted into carboxylic acid esters by Meerwein's trialkyloxonium salts (~70% yield).

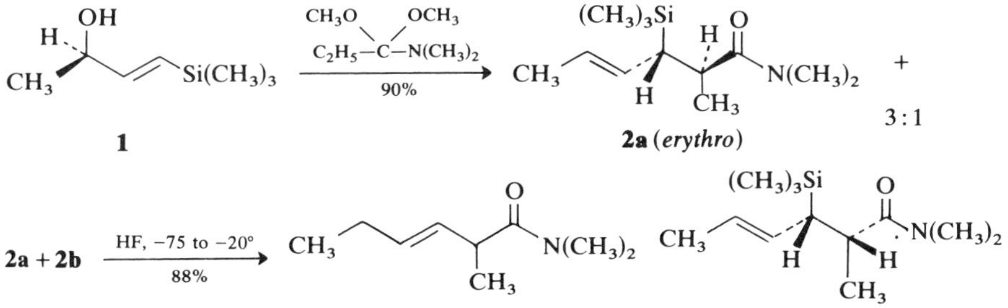

CH_3O OCH_3 / $C_2H_5-C-N(CH_3)_2$ — 90% → **2a** (*erythro*) + **2b** (*threo*), 3:1

1

2a + **2b** —HF, −75 to −20°, 88%→ **3**

[1] G. H. Posner, M. J. Chapdelaine, and C. M. Lentz, *J. Org.*, **44**, 3661 (1979); *see also*, G. H. Posner and C. M. Lentz, *Tetrahedron Letters*, 3769 (1978).
[2] P. R. Jenkins, R. Gut, H. Wetter, and A. Eschenmoser, *Helv.*, **62**, 1922 (1979).

Hydrogen peroxide, 1, 457–471; **2**, 216–217; **3**, 154–155; **4**, 253–255; **5**, 337–339; **6**, 286; **7**, 174; **8**, 247–248.

Saponification. Corey *et al.*[1] employed 2,2′-dihydroxydiethyl sulfide to suppress cleavage of the methylthiomethyl ether group during the hydrolysis of the lactone **1** with $LiOH-H_2O_2$.[1] Treatment of the saponification product with KOH (to effect benzoic ester cleavage) and then diazomethane gives **2**, a key intermediate in the synthesis of erythronolide A (**3**).

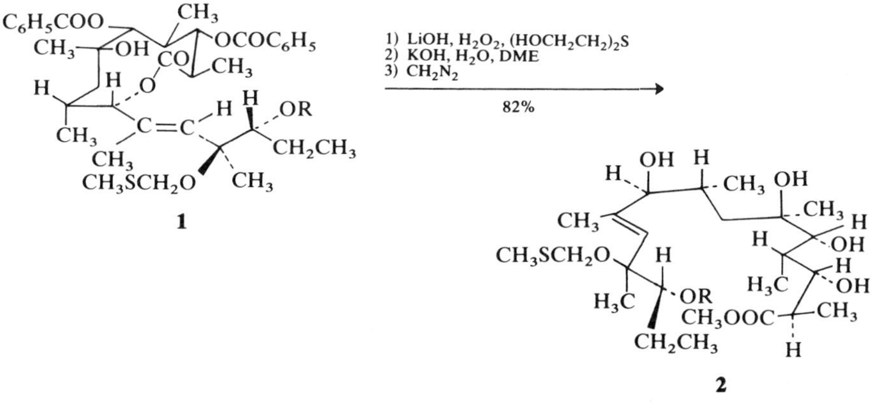

1 —1) LiOH, H_2O_2, $(HOCH_2CH_2)_2S$; 2) KOH, H_2O, DME; 3) CH_2N_2; 82%→ **2**

$R = Si(CH_3)_2C(CH_3)_3$

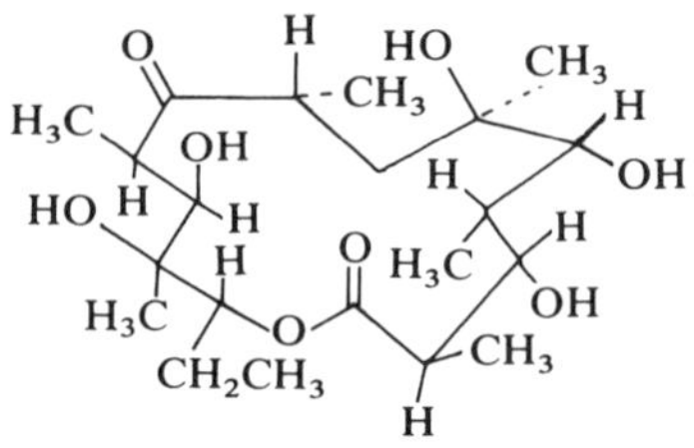

3

Oxidation of codeinone.[2] Codeinone (**1**) and codeine are oxidized by peracids quantitatively to the N-oxide. However, codeinone can be oxidized in part to the 7,8-oxide (**2**) by H_2O_2 (3%) in the presence of NaOH at 0°. The product can be reduced to the 7,8-oxide of codeine (**3**), which may be a metabolite of codeine.

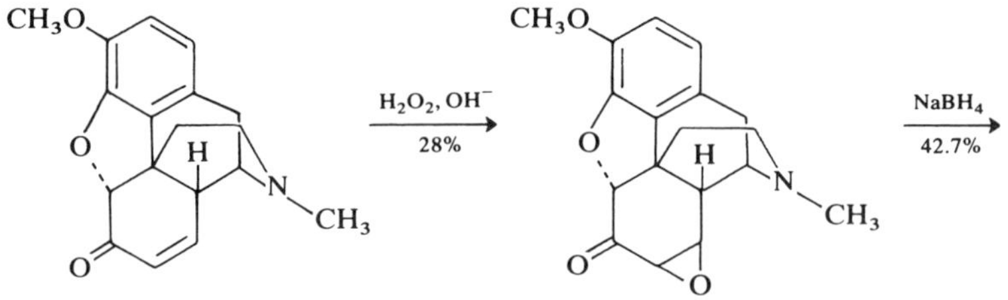

H_2O_2, OH^- 28% ; $NaBH_4$ 42.7%

1 2 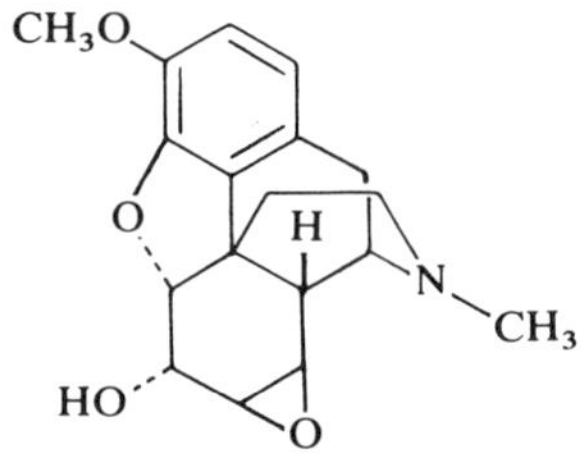 3

[1] E. J. Corey, P. B. Hopkins, S. Kim, S. Yoo, K. P. Nambiar, and J. R. Falck, *Am. Soc.*, **101**, 7131 (1979).

[2] K. Uba, N. Miyata, K. Watanabe, and M. Hirobe, *Chem. Pharm. Bull. Japan*, **27**, 2257 (1979).

Hydrogen peroxide–N,N′-Carbonyldiimidazole.

***Epoxidation of arachidonic acid* (1).**[1] Chemical oxidations of arachidonic acid are of current interest because enzymic oxidation of this acid is involved in the biogenesis of prostaglandins and related natural products. Direct epoxidation of **1** results in a mixture of all possible epoxides. Usual methods for conversion of an acid into a peroxy acid fail with arachidonic acid, but this reaction can be accomplished by treatment with N,N′-carbonyldiimidazole (**1**, 114–115) to form an imidazolide (**a**), which on treatment with anhydrous hydrogen peroxide is converted into peroxyarachidonic acid (**2**). This peracid is unstable and on standing is slowly converted into 14,15-epoxyarachidonic acid, isolated as the methyl ester **3** in >98% yield.

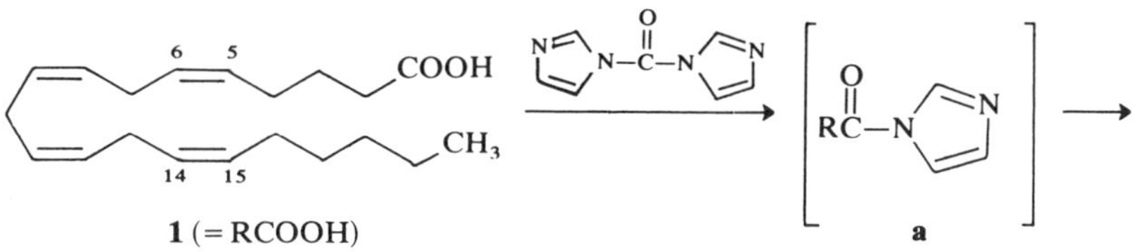

1 (= RCOOH) → **a** → RCOOOH **2** $\xrightarrow[>98\%\ \text{overall}]{1)\ 20^\circ;\ 2)\ CH_2N_2}$ **3**

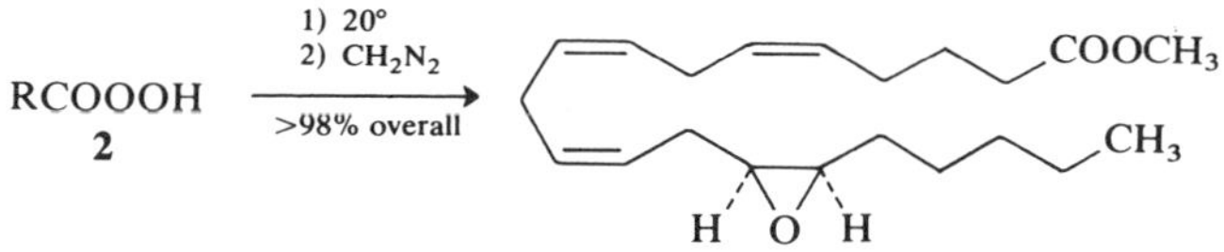

Inspection of models shows that this internal oxygen transfer is energetically favorable in the case of the 14,15-double bond by way of a 15-membered strain-free cyclic structure. The same sequence of reactions was also reported for the 5,6-dihydro derivative of **1**. Internal epoxidation does not appear to be favorable in the case of a double bond in the 9,10-position of a peroxy acid (peroxyoleic acid).

The same paper reports selective epoxidation of the 5,6-double bond of **1** by conversion to an unstable iodo-δ-lactone followed by hydrolysis and treatment with base to convert the intermediate iodohydrin into an epoxide.

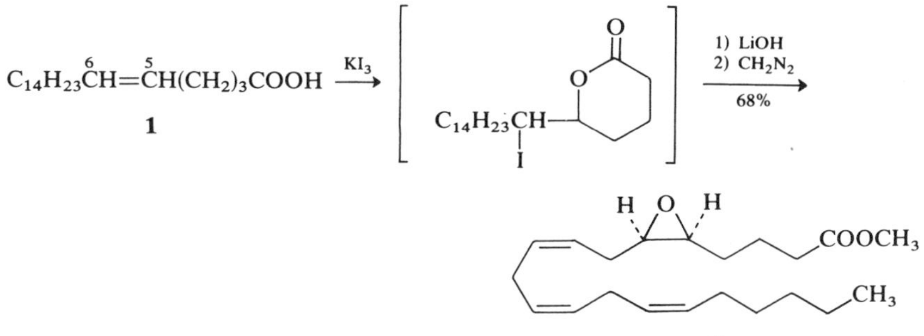

$$C_{14}H_{23}\overset{6}{C}H{=}\overset{5}{C}H(CH_2)_3COOH \xrightarrow{KI_3} \left[\ \right] \xrightarrow[68\%]{1)\ LiOH;\ 2)\ CH_2N_2} \mathbf{4}$$

[1] E. J. Corey, H. Niwa, and J. F. Falck, *Am. Soc.*, **101**, 1586 (1979).

Hydrogen peroxide–Potassium carbonate.

Oxidative cleavage of arenesulfonylhydrazones. The combination of hydrogen peroxide (30–50%) and potassium carbonate in methanol or dioxane is recommended for regeneration of both aldehydes and ketones from arenesulfonylhydrazones. Most of the known methods for this reaction are suitable only for regeneration of ketones. Yields are in the range of 70–95% for the crude carbonyl compounds.[1]

[1] J. Jiricny, D. M. Orere, and C. B. Reese, *Synthesis*, 919 (1978).

Hydrogen peroxide–Pyridinium polyhydrogen fluoride.

Hydroxylation of arenes. This reaction can be carried out conveniently with 30% H_2O_2 in pyridinium polyhydrogen fluoride at 0° (20–60% yield). In the case of alkylbenzenes predominate *ortho–para* substitution obtains. 1,2-Methyl shifts are sometimes observed. Loss of an alkyl group can occur to a slight extent by *ipso* hydroxyl attack.[1]

[1] G. A. Olah, T. Keumi, and A. P. Fung, *Synthesis*, 536 (1979).

Hydrogen peroxide-Selenium dioxide.

Sulfides → sulfoxides.[1] Sulfides are oxidized to sulfoxides rapidly and in 80–95% yield by H_2O_2–SeO_2. This method is particularly useful for oxidation of α-phosphoryl sulfides (equation I), but is not suitable for oxidation of α-halo sulfides. The actual oxidant is probably perseleninic acid, HOOSe(O)OH.

$$(R^1O)_2\overset{O}{\overset{\|}{P}}CH_2SR^2 \xrightarrow[95\text{–}100\%]{H_2O_2,\ SeO_2} (R^1O)_2\overset{O}{\overset{\|}{P}}CH_2\overset{O}{\overset{\|}{S}}R^2 \qquad \text{(I)}$$

[1] J. Drabowicz and M. Mikołajczyk, *Synthesis*, 758 (1978).

2-Hydroperoxyhexafluoro-2-propanol (1). Mol. wt. 162.05, stable at −5° for about 2 months.

Preparation[1,2]:

$$CF_3\overset{O}{\overset{\|}{C}}CF_3 \underset{-H_2O}{\overset{H_2O_2}{\rightleftarrows}} CF_3-\underset{OOH}{\overset{OH}{C}}-CF_3$$

1

Epoxidation.[3,4] The reagent effects epoxidation of unhindered alkenes in CH_2Cl_2 or 1,2-dichloroethane at room temperature or at reflux. Oxidations can be performed using stoichiometric quantities of **1** or with reagent prepared by the catalytic cycle shown in equation I. In the latter case, a two-phase mixture of substrate, solvent, excess H_2O_2, and 10–15 mole % of **1** or **2** is refluxed in 1,2-dichloroethane. Although the paper recommends use of 90% H_2O_2, 30% H_2O_2 can be substituted with only a minor effect on reaction rate. High *syn*-stereoselectivity is observed in epoxidation of allylic alcohols and acetates. Baeyer–Villager

$$\underset{\mathbf{1}}{CF_3-\underset{OOH}{\overset{OH}{C}}-CF_3} + \;>C=C<\; \rightarrow \underset{\mathbf{2}}{CF_3-\underset{OH}{\overset{OH}{C}}-CF_3} + \;>C\overset{O}{\frown}C< \qquad \text{(I)}$$

($\mathbf{2} \xrightarrow{+H_2O_2,\ -H_2O} \mathbf{1}$)

oxidation of ketones with **1** is slow at room temperature. A useful reaction of **1** is the oxidation of aldehydes to carboxylic acids in high yield.

Examples:

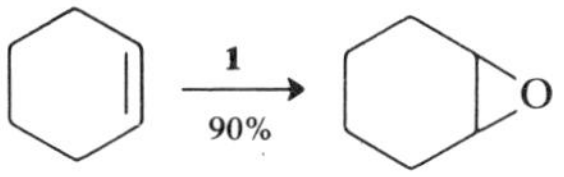

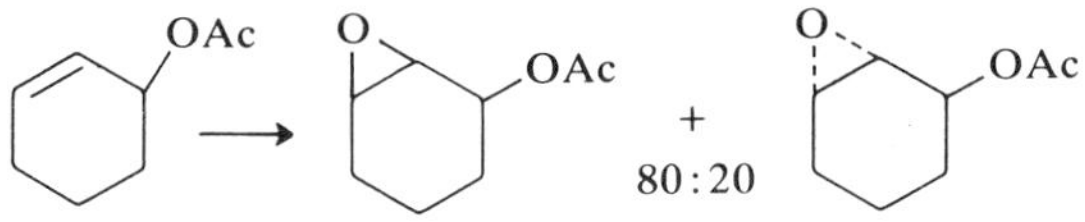

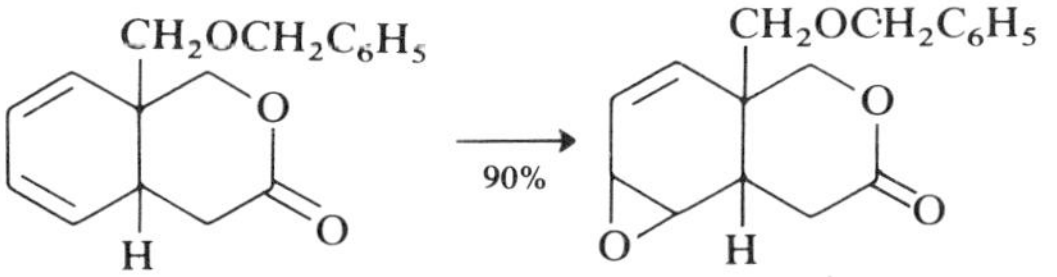

1, 90%

OAc

OAc + OAc

80:20

$CH_2OCH_2C_6H_5$ $CH_2OCH_2C_6H_5$

90%

[1] R. D. Chambers and M. Clark, *Tetrahedron Letters*, 2741 (1970).
[2] C. T. Ratcliffe, C. V. Hardin, L. R. Anderson, and W. B. Fox, *J.C.S. Chem. Comm.*, 784 (1971).
[3] R. P. Heggs and B. Ganem, *Am. Soc.*, **101**, 2484 (1979); B. Ganem, A. J. Biloski, and R. P. Heggs, *Org. Syn.*, submitted (1979).
[4] This epoxidation was developed independently by L. Kim, U.S. Patent Appl. 171,734 [*C.A.* **78**, 159400n (1973)].

Hydroxylamine hydrochloride, 1, 478–481; **7,** 176–177.

Nitriles. Aldehydes are converted in one step into nitriles when heated with hydroxylamine hydrochloride in formic acid. An aldoxime formate is a probable intermediate.[1] The method is a variation of that of van Es.[2]

[1] G. A. Olah and T. Keumi, *Synthesis*, 112 (1979).
[2] T. van Es, *J. Chem. Soc.*, 1564 (1965).

Hydroxylamine-O-sulfonic acid, 1, 481–484; **2,** 217–219; **3,** 156–157; **4,** 256; **5,** 343–344; **6,** 290; **8,** 250–251.

Nitriles. This reagent converts aliphatic and aromatic aldehydes into nitriles in high yield via intermediate oxime-O-sulfonic acids, which are isolable in the case of formylpyridines (equation I). The reaction is usually conducted in water; for aldehydes having slight solubility addition of acetonitrile accelerates the first step.[1]

$$\text{(I)} \quad RCHO + NH_2OSO_3H \longrightarrow RCH{=}NOSO_3H \xrightarrow[75\text{–}79\%]{} RCN + H_2SO_4$$

Nitriles have also been obtained recently from active methylene compounds of type **1** by a two-step process.[2] Treatment with the aminal ester **2** (**5**, 71–73) leads to an enamine (**3**), which reacts with hydroxylamine-O-sulfonic acid in an aqueous medium to give a nitrile (**4**). This reaction does not proceed through an aldehyde, although the morpholine enamine of phenylacetaldehyde reacts with hydroxylamine-O-sulfonic acid to give phenylacetonitrile.

$$\underset{\mathbf{1}}{ArCH_2R} \xrightarrow{[(CH_3)_2N]_2CHOC(CH_3)_3\ (\mathbf{2})} \underset{\mathbf{3}}{ArC(R){=}CHN(CH_3)_2} \xrightarrow[50\text{–}80\%]{NH_2OSO_3H} \underset{\mathbf{4}}{ArCH(R)CN}$$

($R = H, CH_3, COOC_2H_5$)

[1] J. Streith, C. Fizet, and H. Fritz, *Helv.*, **59**, 2786 (1976).
[2] H. Biere and R. Russe, *Tetrahedron Letters*, 1361 (1979).

(2S,2'S)-2-Hydroxymethyl-1-[(1-methylpyrrolidine-2-yl)methyl]pyrrolidine,

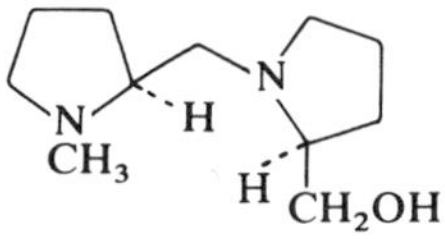

(**1**). Mol. wt. 182.31, b.p. 112°/4.5 mm, $\alpha_D - 130°$. The chiral alcohol is prepared from (S)-proline.

Asymmetric addition of alkyllithium to aldehydes.[1] This ligand promotes asymmetric addition of an alkyllithium or a dialkylmagnesium to aldehydes (equation I). The optical yields are 45–95%, being dependent on the bulk of R^1 and R^2, the temperature, and the solvent.

(I) $$R^1Li + R^2CHO \xrightarrow{1} R^1R^2\overset{*}{C}HOH$$

Enantioselective ethynylation of benzaldehyde.[2] The reaction of benzaldehyde with lithium trimethylsilylacetylide in the presence of **1** in DME affords the corresponding (S)-alkynyl alcohol **2** in 92% optical yield. The optical yield is ~75% with lithium acetylide itself.

$$C_6H_5CHO + (CH_3)_3SiC{\equiv}CH \xrightarrow{\mathbf{1},\ n\text{-BuLi}} C_6H_5\overset{*}{C}H(OH)C{\equiv}CSi(CH_3)_3 \xrightarrow[87\%]{NaOH} C_6H_5\overset{*}{C}H(OH)C{\equiv}CH$$

2, 92% ee

[1] T. Mukaiyama, K. Soai, T. Sato, H. Shimizu, and K. Suzuki, *Am. Soc.*, **101**, 1455 (1979).
[2] T. Mukaiyama, K. Suzuki, K. Soai, and T. Sato, *Chem. Letters*, 447 (1979).

N-Hydroxysuccinimide, 1, 487.

Peptide synthesis.[1] Amides can be prepared from an acid and an amine in the presence of an isocyanide, which serves as a dehydrating agent (equation I).

(I) $$R^1COOH + H_2NR^2 + R^3N{=}C \rightarrow R^1CONHR^2 + R^3NHCHO$$

When this method is applied to synthesis of a dipeptide, however, considerable racemization of the acid component is observed. This racemization is suppressed when N-hydroxysuccinimide (1.1 equiv.) is also present to form an active ester (equation II), which then reacts with the amino acid in the presence of base to form the dipeptide (20–80% yield).

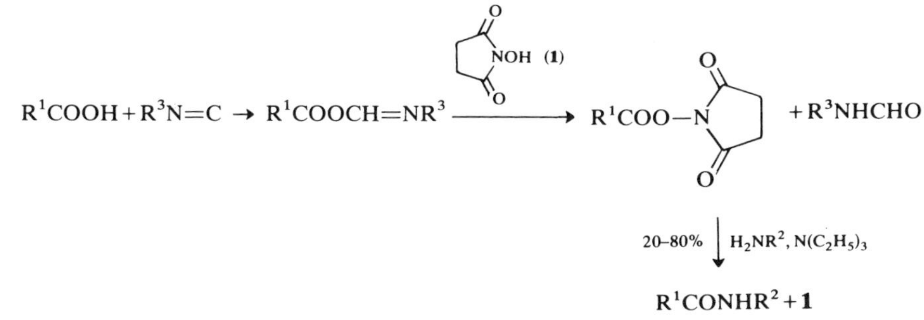

$$\text{(II)}\quad R^1COOH + R^3N{=}C \rightarrow R^1COOCH{=}NR^3 \xrightarrow{\text{N-hydroxysuccinimide } (\mathbf{1})} R^1COO{-}N(\text{succinimide}) + R^3NHCHO$$

$$\xrightarrow[20\text{–}80\%]{H_2NR^2,\ N(C_2H_5)_3} R^1CONHR^2 + \mathbf{1}$$

[1] L. Wackerle, *Synthesis*, 197 (1979).

4-Hydroxy-2,2,6,6-tetramethylpiperidinooxy radical (1). Mol. wt. 172.25, m.p. 69–71°. Supplier: Aldrich.

Antioxidant. This free radical is an outstanding antioxidant for protecting readily oxidized polyunsaturated fatty acid derivatives.[1]

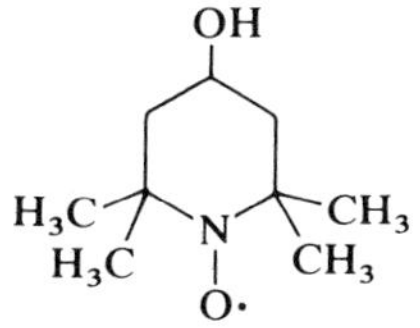

OH, H_3C, H_3C, N, O·, CH_3, CH_3

1

[1] E. J. Corey, Y. Arai, and C. Mioskowski, *Am. Soc.*, **101**, 6748 (1979).

I

Iodine, 1, 495–500; **2**, 220–222; **3**, 159–160; **4**, 258–260; **5**, 346–347; **6**, 293–295; **7**, 179–181; **8**, 256–260.

Iodolactonization. A highly stereoselective iodolactonization is a key step in a synthesis of α-multistriatin (**6**), an aggregation pheromone of the elm bark beetle. Reaction of the unsaturated acid **1** with iodine in CH_3CN at 0° gives the iodolactone **2** in 85–90% yield in purity of >95%. When NIS is used about 20% of the isomeric iodolactone is formed. The former reaction presumably is under thermodynamic control, since **2** has the all-equatorial configuration. Alkaline methanolysis of **2** leads to the epoxy ester **3**, which was converted into **6** by steps that also proceed with high selectivity.[1]

1 $\xrightarrow[85\text{–}90\%]{I_2}$ **2** $\xrightarrow[92\%]{CH_3OH,\ OH^-}$ **3** $\xrightarrow{CH_3\overset{Li}{C}HSO_2C_6H_5}$ **4** $\xrightarrow{Al/Hg}$ **5** $\xrightarrow[\text{50\% overall from 3}]{SnCl_4}$ **6**

(Z)-Vinylic iodides. These iodides can be prepared by reaction of iodine with (Z)-vinylic cuprates, prepared by addition of lithium dialkylcuprates to acetylene (equation I).[2]

$$\text{(I)}\quad (n\text{-}C_5H_{11})_2CuLi + 2HC\equiv CH \rightarrow \left(\begin{matrix} H & & H \\ & C=C & \\ n\text{-}C_5H_{11} & & \end{matrix}\right)_2 CuLi \xrightarrow[\text{67–75\% overall}]{I_2} \begin{matrix} H & & H \\ & C=C & \\ n\text{-}C_5H_{11} & & I \end{matrix}$$

Chrysenes. In the presence of an oxidizing agent (I_2) styrylnaphthalenes (**1**) can be cyclized to chrysenes (**2**) by irradiation.[3]

$h\nu$, I_2 → 50–95%

1 (X = CN, COOH, $COCH_3$) **2**

Prostacyclin (**8**, 256–257). The Upjohn synthesis of prostacyclin has been described in detail. The key step, iodination of prostaglandin $F_{2\alpha}$ methyl ester, is greatly improved by use of I_2 and KI rather than I_2 alone. The paper also reports synthesis of the related prostaglandin I_1 (5,6-dihydroprostacyclin), prostaglandin I_3 (derived from prostaglandin $F_{3\alpha}$), and 6-ketoprostaglandin $F_{1\alpha}$, the hydrolysis product of prostacyclin.[4]

[1] P. A. Bartlett and J. Myerson, *J. Org.*, **44**, 1625 (1979).
[2] A. Alexakis, G. Cahiez, and J. F. Normant, *J. Organomet. Chem.*, **177**, 293 (1979); *idem, Org. Syn.*, submitted (1979).
[3] P. H. Gore and F. S. Kamonah, *Synthesis*, 773 (1978), and references cited therein.
[4] R. A. Johnson, F. H. Lincoln, E. G. Nidy, W. P. Schneider, J. L. Thompson, and U. Axen, *Am. Soc.*, **100**, 7690 (1978).

Iodine–Silver acetate.

α-Iodo ketones. Reaction of enol silyl ethers with silver acetate and iodine in CH_2Cl_2 at 25° followed by desilylation gives α-iodo carbonyl compounds regioselectively in 65–95% yield.[1] Enol acetates also have been used for this reaction, but yields are lower.[2]

[1] G. M. Rubottom and R. C. Mott, *J. Org.*, **44**, 1731 (1979).
[2] R. C. Cambie, R. C. Hayward, J. L. Jurlina, P. S. Rutledge, and P. D. Woodgate, *J.C.S. Perkin I*, 126 (1978).

Iodobenzene dichloride (IBD), **1**, 505–506; **2**, 225–226; **3**, 164–165; **4**, 264–265; **5**, 352–353; **6**, 298–300.

Chlorination of alkanes.[1] Photoinduced chlorination of alkanes with IBD is a well established reaction (**1**, 506) with the observed selectivity of tertiary > secondary ≫ primary positions. The reaction can also be induced with trialkylboranes. The reactivities of both reagents are similar, but yields are somewhat higher in the trialkylborane catalyzed reaction.

Examples:

$$(CH_3)_2CHCH(CH_3)_2 \xrightarrow{IBD} (CH_3)_2CHC(Cl)(CH_3)_2$$

$B(C_6H_{13})_3$	99%
$h\nu$	78%

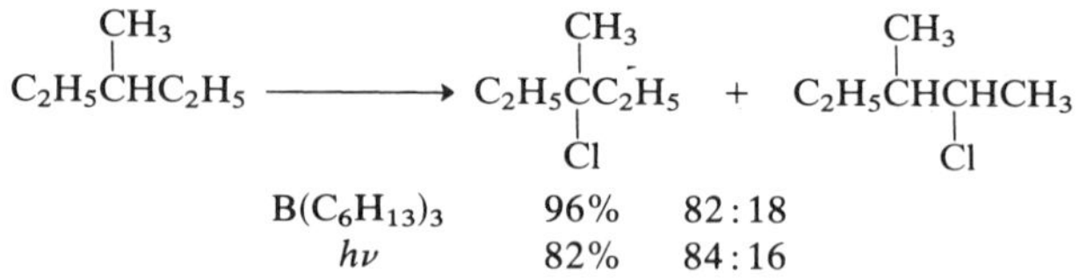

$$C_2H_5CH(CH_3)C_2H_5 \longrightarrow C_2H_5C(CH_3)(Cl)C_2H_5 + C_2H_5CH(CH_3)CH(Cl)CH_3$$

$B(C_6H_{13})_3$	96%	82:18
$h\nu$	82%	84:16

[1] A. Arase, M. Hoshi, and Y. Masuda, *Chem. Letters*, 961 (1979).

Iodomethyl methyl ether, ICH_2OCH_3. A new synthesis of this ether has the virtue that bis(iodomethyl)ether, which is probably carcinogenic, is not formed as a by-product.

$$CH_3OCH_2OCH_3 + ISi(CH_3)_3 \xrightarrow[75\text{–}93\%]{} ICH_2OCH_3 + (CH_3)_3SiOCH_3$$

$$ICH_2OCH_3 \xrightarrow[78\%]{P(C_6H_5)_3,\ 20^\circ} (C_6H_5)_3\overset{+}{P}CH_2OCH_3I^-$$

Iodomethyl methyl ether reacts at room temperature with triphenylphosphine to form methoxymethyltriphenylphosphonium iodide.[1]

[1] M. E. Jung, M. A. Mazurek, and R. M. Lim, *Synthesis*, 588 (1978).

Iodomethyltri-*n*-butyltin, $(C_4H_9)_3SnCH_2I$ (**1**). Mol. wt. 430.96, b.p. 100–110°/0.01 mm. This reagent is prepared[1] in high yield by the reaction of $(C_4H_9)_3SnH$ and ICH_2ZnI.[2]

(Z)-Homoallylic alcohols. The tributylstannylmethyl ether of an allylic alcohol on treatment with *n*-butyllithium at −78° undergoes tin–lithium exchange followed by [2,3] sigmatropic rearrangement to a homoallylic alcohol. A valuable feature is that the (Z)-isomer is formed. An example is shown in equation (I).[1]

$$\text{(I)}\quad CH_3(CH_2)_3CH(OH)C(CH_3){=}CH_2 \xrightarrow[\text{quant.}]{1)\ KH;\ 2)\ \mathbf{1}} CH_3(CH_2)_3CH(OCH_2Sn(C_4H_9)_3)C(CH_3){=}CH_2 \xrightarrow[>95\%]{n\text{-BuLi}} (Z)\text{-}CH_3(CH_2)_3CH{=}C(CH_3)CH_2CH_2OH$$

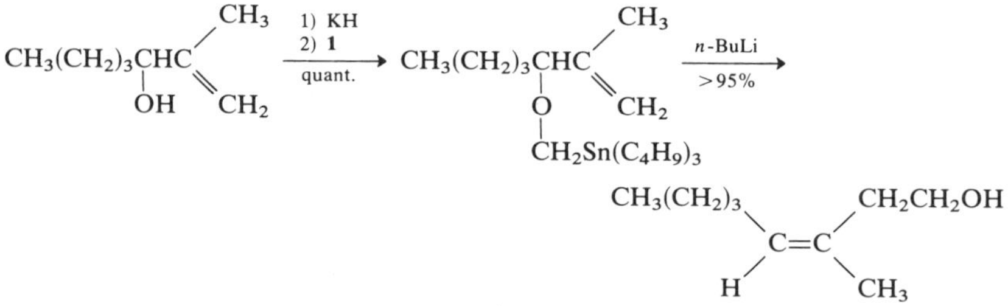

This rearrangement was a key step in a synthesis of the homoallylic acetate **2**, a sex pheromone of the California red scale.[3]

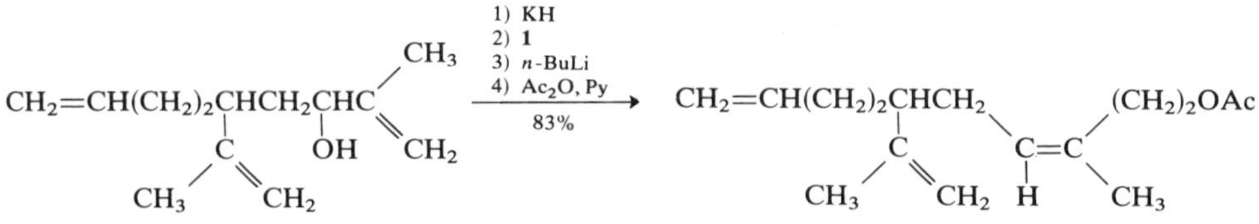

$$CH_2{=}CH(CH_2)_2CH(C(CH_3){=}CH_2)CH_2CH(OH)C(CH_3){=}CH_2 \xrightarrow[83\%]{1)\ KH;\ 2)\ \mathbf{1};\ 3)\ n\text{-BuLi};\ 4)\ Ac_2O,\ Py} CH_2{=}CH(CH_2)_2CH(C(CH_3){=}CH_2)CH_2CH{=}C(CH_3)(CH_2)_2OAc\ (\mathbf{2})$$

It has also been used in a convenient synthesis of bishomofarnesol (**3**),[4] a precursor in one synthesis of the C_{18}-Cecropia juvenile hormone (**4**).[5]

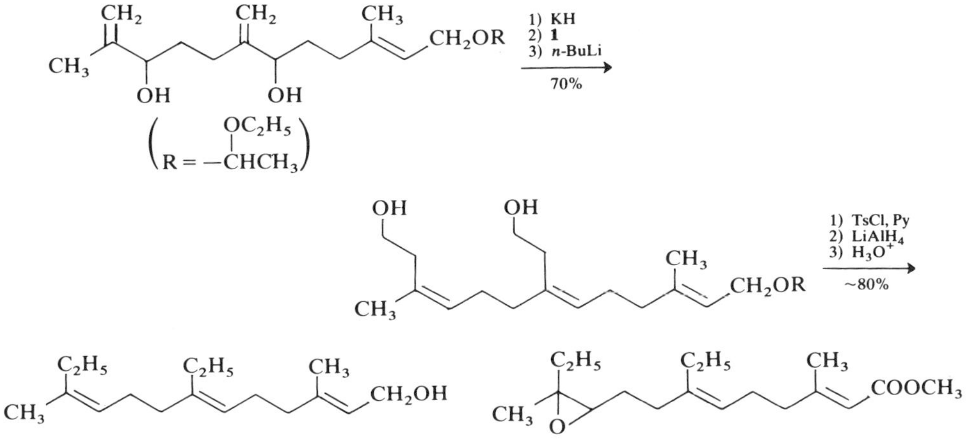

CH_3, CH_2, OH, CH_2, OH, CH_3, CH_2OR $\xrightarrow[70\%]{1)\ KH;\ 2)\ \mathbf{1};\ 3)\ n\text{-BuLi}}$

$\left(R = -\overset{OC_2H_5}{CH}CH_3\right)$

OH, OH, CH_3, CH_3, CH_2OR $\xrightarrow[\sim 80\%]{1)\ TsCl,\ Py;\ 2)\ LiAlH_4;\ 3)\ H_3O^+}$

C_2H_5, C_2H_5, CH_3, CH_3, CH_2OH — **3**

C_2H_5, C_2H_5, CH_3, CH_3, O, $COOCH_3$ — **4**

[1] W. C. Still, *Am. Soc.*, **100**, 1481 (1978).
[2] D. Seyferth and S. B. Andrews, *J. Organometal. Chem.*, **30**, 151 (1971).
[3] W. C. Still and A. Mitra, *Am. Soc.*, **100**, 1927 (1978).
[4] W. C. Still, J. H. McDonald III, D. B. Collum, and A. Mitra, *Tetrahedron Letters*, 593 (1979).
[5] E. J. Corey, J. A. Katzenellenbogen, N. W. Gilman, S. A. Roman, and B. W. Erickson, *Am. Soc.*, **90**, 5618 (1968).

Iodotrimethylsilane, 8, 261–263.

In situ *generation.* Two new methods for this purpose are formulated in equations (I) and (II). Both methods can be conducted at room temperature. Reagent generated in this way can be used in the same way as pure isolated material.[1]

(I) $C_6H_6 \xrightarrow[55\%]{1)\ Li,\ THF;\ 2)\ ClSi(CH_3)_3} (CH_3)_3Si\text{-}C_6H_6\text{-}Si(CH_3)_3 \xrightarrow{I_2} 2\ ISi(CH_3)_3 + C_6H_6$

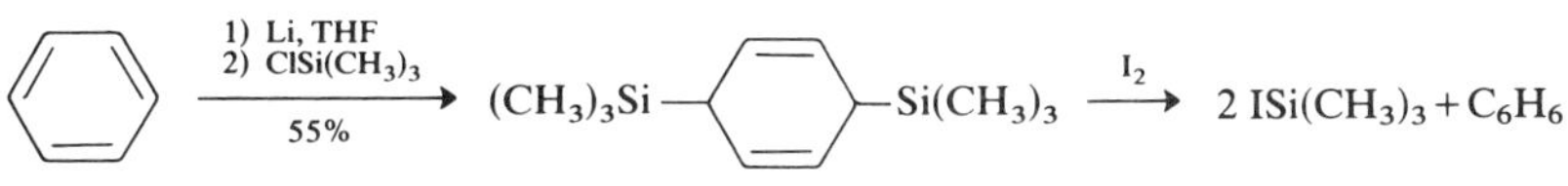

(II) $(CH_3)_3SiCH_2CH{=}CH_2 \xrightarrow{I_2} ISi(CH_3)_3 + ICH_2CH{=}CH_2$

Another convenient *in situ* preparation involves the reaction of hexamethyldisilane[2] with iodine at 50° (equation I).[2–5]

(III) $2(CH_3)_3SiCl \xrightarrow[97\%]{\text{Li dispersion}} (CH_3)_3SiSi(CH_3)_3 \xrightarrow{I_2,\ 50°} 2\ (CH_3)_3SiI$

Iodotrimethylsilane equivalent. Two laboratories[6–10] have reported that the combination of chlorotrimethylsilane and sodium iodide (1 : 1 ratio) in acetonitrile can function as a convenient equivalent to the rather sensitive iodotrimethylsilane. Thus the *in situ* reagent is satisfactory for deoxygenation of sulfoxides[6] and for cleavage of esters, lactones, carbamates, and ethers.[7,8] It is particularly useful for conversion of alcohols to iodides.[9] Successful cleavage of phosphonate esters is possible with the *in situ* reagent.[10]

Trimethylsilyl enol ethers.[11] This reagent in conjunction with hexamethyldisilazane converts ketones and aldehydes into trimethylsilyl enol ethers rapidly and in high yield at room temperature or below. The thermodynamically more stable products predominate. A useful feature is that ketones can be selectively silylated in the presence of an ester group.

Examples:

2-Methylcyclohexanone $\xrightarrow[90\%]{ISi(CH_3)_3,\ HN[Si(CH_3)_3]_2,\ 10°,\ C_6H_{12}}$ 1-$OSi(CH_3)_3$-2-CH_3-cyclohexene + 6-$OSi(CH_3)_3$-1-CH_3-cyclohexene (9 : 1)

$$CH_3(CH_2)_6CHO \xrightarrow[89\%]{} (C_6H_{13})HC{=}CH(OSi(CH_3)_3) + (C_6H_{13})HC{=}CH(OSi(CH_3)_3) \quad 23:77$$

$$C_6H_5\overset{O}{\overset{\|}{C}}(CH_2)_3COOCH_3 \xrightarrow[96\%]{} (C_6H_5)((CH_3)_3SiO)C{=}CH(CH_2)_2COOCH_3$$

Trimethylsiloxy esters.[12] The reaction of iodotrimethylsilane with esters to form trimethylsiloxy esters is markedly catalyzed by iodine, possibly by formation of triiodotrimethylsilane, $(CH_3)_3SiI_3$. This observation explains the fact that trimethylphenylsilane in combination with excess iodine can be more effective than iodotrimethylsilane. Aryl trimethylsiloxy esters, unlike ordinary esters, can be reduced to methyl groups by the method of Benkeser (**4**, 525–526). The reaction can be conducted without isolation of intermediates (equation I).

$$\text{(I)}\quad Ar\overset{O}{\overset{\|}{C}}OR \xrightarrow{ISi(CH_3)_3,\ I_2} \left[Ar\overset{O}{\overset{\|}{C}}OSi(CH_3)_3 \xrightarrow{HSiCl_3,\ N(Pr\text{-}n)_3,\ CH_3CN} ArCH_2SiCl_3\right] \xrightarrow[40\text{–}70\%]{KOH,\ CH_3OH} ArCH_3$$

$[R = CH_3, C_2H_5, C(CH_3)_3]$

Reduction of α-ketols.[13] α-Ketols are deoxygenated to the corresponding ketones in 75–95% yield by reaction with 2.5 equiv. of iodotrimethylsilane in $CHCl_3$ or CH_2Cl_2 at room temperature.

$$RCOCHR'(OH) \xrightarrow[-HI]{ISi(CH_3)_3} \left[RCOCHR'(OSi(CH_3)_3) \xrightarrow[-[(CH_3)_3Si]_2O]{ISi(CH_3)_3} RCOCHR'(I)\right] \xrightarrow[75\text{–}95\%]{HI} RCOCH_2R'$$

β- and γ-Iodo ketones.[14] The reagent reacts with α,β-enones in CH_2Cl_2 at $-40°$ to give, after aqueous work-up, β-iodo ketones in 75–90% yield. The reagent also reacts with α,β-cyclopropyl ketones to form γ-iodo ketones. The mechanism of this reaction is not certain.

Examples:

$$C_2H_5\overset{O}{\overset{\|}{C}}CH{=}CH_2 \xrightarrow[85\%]{ISi(CH_3)_3,\ -40°} C_2H_5\overset{O}{\overset{\|}{C}}CH_2CH_2I$$

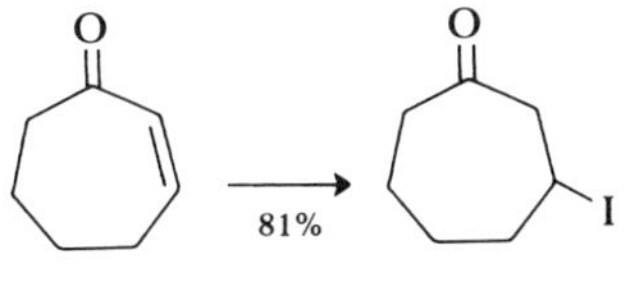

$$\xrightarrow[81\%]{}$$

$$C_6H_5\overset{O}{\overset{\|}{C}}-\underset{CH_3}{\underset{|}{C}}{=}CH_2 \xrightarrow[93\%]{} C_6H_5\overset{O}{\overset{\|}{C}}-\underset{CH_3}{\underset{|}{C}}HCH_2I$$

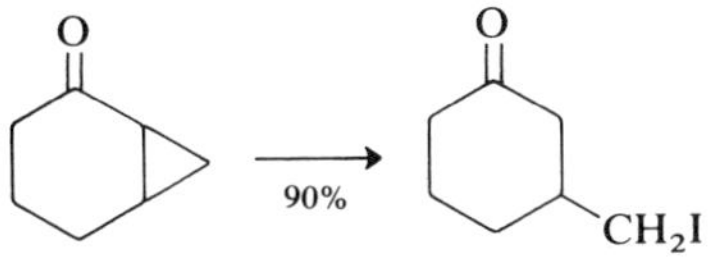

$$\xrightarrow[90\%]{}$$

CH$_2$I

Aldehyde–iodotrimethylsilane adducts. Benzaldehyde and $ISi(CH_3)_3$ react to form an iodohydrin trimethylsilyl ether (**1**), which is only stable in solution. This substance is converted into α,α-diiodotoluene (**2**) on further treatment with the reagent.

$$C_6H_5CHO + (CH_3)_3SiI \xrightarrow{CHCl_3,\ 25°} \underset{\mathbf{1}}{C_6H_5\underset{I}{\underset{|}{C}}HOSi(CH_3)_3} \xrightarrow{(CH_3)_3SiI} \underset{\mathbf{2}\ (51\%)}{C_6H_5CHI_2} + [(CH_3)_3Si]_2O$$

Extension of this reaction to phenylacetaldehyde leads to the unexpected formation of **3** in yields as high as 50% (equation I). Probable mechanisms for formation of **3** are discussed in the report.[15]

(I)

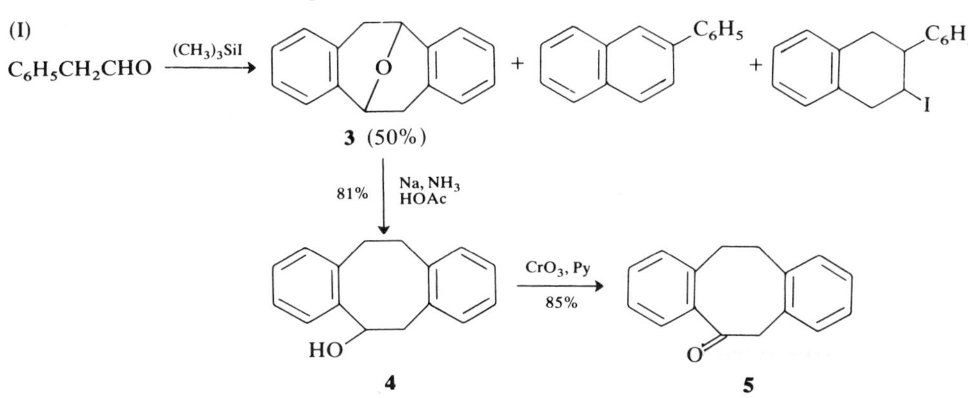

$$C_6H_5CH_2CHO \xrightarrow{(CH_3)_3SiI}$$ **3** (50%) + (C$_6$H$_5$) + (C$_6$H$_5$, I)

3 $\xrightarrow[81\%]{Na,\ NH_3,\ HOAc}$ **4** (HO) $\xrightarrow[85\%]{CrO_3,\ Py}$ **5** (O)

α-Methylene-γ-butyrolactones.[16] A novel route to this unit utilizes dibromocarbene adducts (**1**), which are converted by known reactions into the ester **3**. The

$$\mathbf{1}\ \xrightarrow[60\text{–}70\%]{1)\ n\text{-BuLi},\ -75^\circ;\ 2)\ CH_2{=}N^+(CH_3)_2I^-}\ \mathbf{2}\ \xrightarrow[60\text{–}75\%]{1)\ n\text{-BuLi};\ 2)\ (CH_3O)_2C{=}O}$$

$$\mathbf{3}\ \xrightarrow{ISi(CH_3)_3}\ [\mathbf{a}]\ \xrightarrow[60\text{–}65\%]{\Delta}\ \mathbf{4}$$

ester is treated with iodotrimethylsilane in C_6H_6 or $ClCH_2CH_2Cl$ at 90–95° for several hours, the solvent is evaporated, and the residue is distilled under reduced pressure at 170°. The lactone **4** is obtained in 60–65% yield. Probably **a** is the intermediate in the conversion of **3** to **4**; however, thermolysis of **a**, prepared by methylation of **3** and KOH hydrolysis, gives **4** in somewhat lower yield than in the original reaction. The reaction is regio- and stereoselective. It was used for synthesis of tulipalin (**6**).

$$\mathbf{5}\ \xrightarrow{43\%}\ \mathbf{6}$$

Cleavage of carbamates.[17] N-Benzyloxy- and N-*t*-butoxycarbonyl protected peptide derivatives can be selectively cleaved by reaction with $ISi(CH_3)_3$ at 25–50° in the presence of side-chain blocking groups such as methyl esters or benzyl ethers.

Decarboalkoxylation.[18] *gem*-Diesters can be hydrolyzed by treatment with iodotrimethylsilane at 100°; if the reaction is prolonged, decarboxylation can be effected to give a monocarboxylic acid. Similarly, *β*-keto esters are hydrolyzed and then decarboxylated to ketones.

Examples:

$$C_6H_5\overset{O}{\overset{\|}{C}}CH_2COOC_2H_5\ \xrightarrow[82\%]{ISi(CH_3)_3,\ 100^\circ}\ C_6H_5\overset{O}{\overset{\|}{C}}CH_3$$

$$n\text{-}C_4H_9CH(COOCH_3)_2\ \xrightarrow[68\%]{}\ n\text{-}C_4H_9CH_2COOH$$

Selective O-demethylation.[19] O-Methyl aryl ethers can be selectively O-demethylated in the presence of a methylenedioxy group by $ISi(CH_3)_3$ in refluxing quinoline, but not in the standard solvents (CH_2Cl_2, $HCCl_3$). Under these conditions **1** and **3** are converted into **2** and **4**, respectively, in about 70% yield.

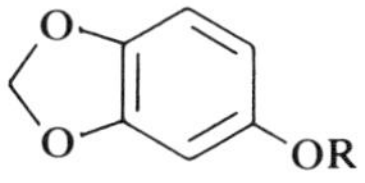

1 ($R = CH_3$) **3** ($R = CH_3$)
2 ($R = H$) **4** ($R = H$)

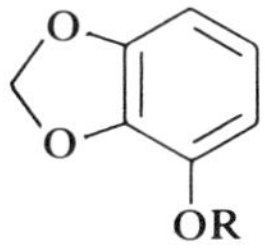

Eisenbraun and co-workers[20] have compared BBr_3 and $ISi(CH_3)_3$ for demethylation of aryl ethers. The former reagent is more reactive, but is less effective for regiospecific attack of hindered methoxyl groups. For example, **5** is cleaved to **6** in 94% yield by $ISi(CH_3)_3$ (1.1 equiv.). BBr_3 is less useful for this reaction.

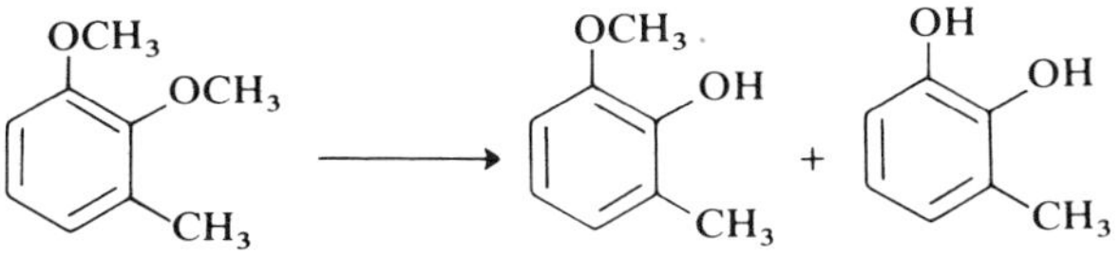

	5 → **6**	+ **7**
$ISi(CH_3)_3$	94%	0%
BBr_3	42%	18%

Dealkylation of phosphonates.[21] Vigorous acid treatment is usually required for this reaction. To circumvent this difficulty, both bromo-[22] and chlorotrimethylsilane[23] have been used with limited success. Iodotrimethylsilane is more selective; the parent phosphonic acids can be obtained under very mild conditions without modification of a variety of other functional groups.[24] However, monodealkylation is not practical with this reagent. Aryl esters are not cleaved; hence selective dealkylation of alkyl aryl esters of phosphonic acids may be possible.

This cleavage can also be carried out in high yield with equimolar amounts of chlorotrimethylsilane and sodium iodide in acetonitrile. Sodium iodide cannot be replaced by potassium bromide.[25]

Cyclization of a δ-amino alcohol to a pyrrolidine.[26] This reaction was used in the final steps of a synthesis of isoretronecanol (**4**) from the δ-amino alcohol (**2**), prepared as shown in equation (I).

[1] M. E. Jung and T. A. Blumenkopf, *Tetrahedron Letters*, 3657 (1978).
[2] Preparation of $(CH_3)_3Si{-}Si(CH_3)_3$: H. Sakurai, and F. Kondo, *J. Organomet. Chem.*, **117**, 149 (1976).
[3] D. E. Seitz and L. Ferreira, *Syn. Comm.*, **9**, 931 (1979).
[4] G. A. Olah, S. C. Narang, B. G. B. Gupta, and R. Malhotra, *Angew. Chem., Int. Ed.*, **18**, 612 (1979).
[5] H. Sakurai, A. Shirahata, K. Sasaki, and A. Hosomi, *Synthesis*, 740 (1979).
[6] G. A. Olah, S. C. Narang, B. G. B. Gupta, and R. Malhotra, *Synthesis*, 61 (1979).
[7] T. Morita, Y. Okamoto, and H. Sakurai, *J.C.S. Chem. Comm.*, 874 (1978).

(I) → 91% → **1** → $LiAlH_4$, quant. → **2** → $(C_2H_5)_2NSi(CH_3)_3$ → **3** → $ISi(CH_3)_3$ → [**a**] → $C_6H_5CH_2\overset{+}{N}(CH_3)_3F^-$, 58% → **4**

[8] G. A. Olah, S. C. Narang, B. G. B. Gupta, and R. Malhotra, *J. Org.*, **44**, 1247 (1979).
[9] T. Morita, S. Yoshida, Y. Okamoto, and H. Sakurai, *Synthesis*, 379 (1979).
[10] T. Morita, Y. Okamoto, and H. Sakurai, *Tetrahedron Letters*, 2523 (1978).
[11] R. D. Miller and D. R. McKean, *Synthesis*, 730 (1979).
[12] R. A. Benkeser, E. C. Mozdzen, and C. L. Muth, *J. Org.*, **44**, 2185 (1979).
[13] T.-L. Ho, *Syn. Comm.*, **9**, 665 (1979).
[14] R. D. Miller and D. R. McKean, *Tetrahedron Letters*, 2305 (1979).
[15] M. E. Jung, A. B. Mossman, and M. A. Lyster, *J. Org.*, **43**, 3698 (1978).
[16] T. Hiyama, H. Saimoto, K. Nishio, M. Shinoda, H. Yamamoto, and H. Nozaki, *Tetrahedron Letters*, 2043 (1979).
[17] R. S. Lott, V. S. Chauhan, and C. H. Stammer, *J.C.S. Chem. Comm.*, 495 (1979).
[18] T.-L. Ho, *Syn. Comm.*, **9**, 233 (1979).
[19] J. Minamikawa and A. Brossi, *Tetrahedron Letters*, 3085 (1978).
[20] E. H. Vickery, L. F. Pahler, and E. I. Eisenbraun, *J. Org.*, **44**, 4444 (1979).
[21] R. Engel, *Chem. Rev.*, **77**, 349 (1977).
[22] C. E. McKenna, M. T. Higa, N. H. Cheung, and M.-C. McKenna, *Tetrahedron Letters*, 155 (1977).
[23] R. Rabinowitz, *J. Org.*, **28**, 2975 (1963).
[24] G. M. Blackburn and D. Ingleson, *J.C.S. Chem. Comm.*, 870 (1978).
[25] T. Morita, Y. Okamoto, and H. Sakurai, *Tetrahedron Letters*, 2523 (1978).
[26] T. Iwashita, T. Kusumi, and H. Kakisawa, *Chem. Letters*, 1337 (1979).

Ion-exchange resins, 1, 511–517; **2**, 227–228; **4**, 266–267; **5**, 355–356; **6**, 302–304; **7**, 182; **8**, 263–264.

Tetrahydropyranylation. Alcohols and phenols are converted into tetrahydropyranyl ethers by reaction with 3,4-dihydro-2*H*-pyrane catalyzed by a strongly acidic ion-exchange resin such as Amberlyst H-15. Yields are >90%. Deprotection of the group can be effected by methanolysis catalyzed by the same resin[1] or by treatment with acid-washed Dowex-50-W-X8.[2]

Ketalization. Ethylene ketals are conveniently prepared by running a solution of the ketone and ethylene glycol through a column of an acidic ion-exchange resin such as Amberlyst at a rate dependent on the reactivity of the ketone (in general a 5–10 minute pass is sufficient). The same technique can be used for preparation of trimethylene ketals and for acetonides.[3]

[1] A. Bongini, G. Cardillo, M. Orena, and S. Sandri, *Synthesis*, 618 (1979).
[2] R. Beier and B. P. Mundy, *Syn. Comm.*, **9**, 271 (1979).
[3] A. E. Dann, J. B. Davis, and M. J. Nagler, *J.C.S. Perkin I*, 158 (1979).

Iridium black. Mol. wt. 192.20. Supplier: Alfa.

Hydrogenation catalyst. Iridium black is a highly selective hydrogenation catalyst.[1] Yamamoto and Sham[2] used this catalyst to effect the remarkably stereoselective (>98%) hydrogenation of **1** to **2**. Mixtures of stereoisomers are obtained when **1** is hydrogenated with palladium on charcoal or Raney nickel.

CH_3, H_3C, CH_2, O, CH_3 — **1** $\xrightarrow[C_2H_5OH,\ 25°C]{Ir,\ H_2\ (1\ atm.)}$ CH_3, O, CH_3, $HC(CH_3)_2$ — **2**

[1] S. Nishimura, F. Mochizuki, and S. Kobayakawa, *Bull. Chem. Soc. Japan*, **43**, 1919 (1970); S. Nishimura, H. Sakamoto, and T. Ozawa, *Chem. Letters*, 855 (1973).
[2] H. Yamamoto and H. L. Sham, *Am. Soc.*, **101**, 1609 (1979).

Iron, 1, 519; **2**, 229; **3**, 167; **5**, 357.

Catalyst for Diels–Alder reaction of ynamines.[1] A zero-valent iron species prepared by reduction of iron(III) chloride with isopropylmagnesium chloride[2] serves as a unique catalyst for cycloaddition of butadiene and ynamines (**1**) to form 1,4-cyclohexadienamines (**2**).[3] These products are hydrolyzed by mild acid treatment to β,γ-cyclohexenones (**3**), which are isomerized to either **4** or **5** by catalytic amounts of rhodium catalysts.

$CH_2{=}CH{-}CH{=}CH_2$ + $N(C_2H_5)_2{-}C{\equiv}C{-}R(Ar)$ (**1**) $\xrightarrow[40\text{–}80\%]{Fe(0),\ 25°}$ **2** ($N(C_2H_5)_2$, R) $\xrightarrow[quant.]{HOAc}$ **3** (O, R)

4 (O, R) $\xleftarrow[quant.]{RhCl[P(C_6H_5)_3]_3}$ **3** $\xrightarrow{RhCl_3}$ **5** (O, R) + **4** (~10%).

[1] J. P. Genet and J. Ficini, *Tetrahedron Letters*, 1499 (1979).
[2] A. Carbonaro, A. Greco, and D. Dall'Asta, *J. Org.*, **33**, 3945 (1968).
[3] Birch reduction of *ortho*-substituted anilines results in conjugated 1,3-dienamines.

Iron(II) perchlorate, $Fe(ClO_4)_2$, **6**, 260–261.

Hydroxylation of C—H bonds.[1] The reaction of cholestanyl xanthate (**1**) with $Fe(ClO_4)_2$ in acetic acid at 120° and in the presence of Fe(III) affords the diacetate **2** in 45% yield. In the absence of Fe(III), the yield of **2** is 34%. No oxidation of the tertiary hydrogen at C_5 is observed. The xanthate probably serves as a ligand for an active iron–oxygen intermediate, which specifically oxidizes the axial hydrogen at C_1.

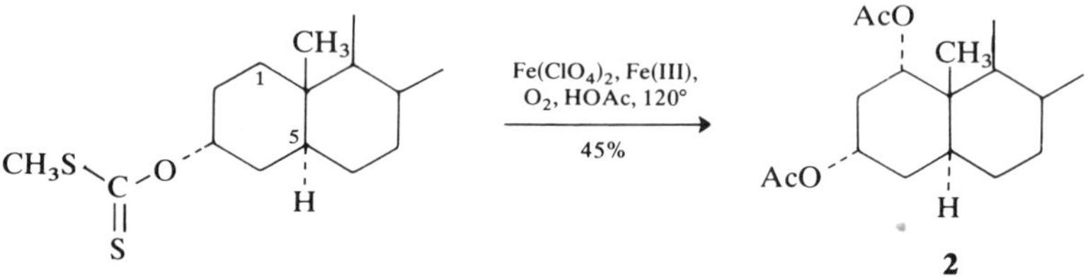

$Fe(ClO_4)_2$, Fe(III), O_2, HOAc, 120°; 45%

[1] H. Patin and G. Mignani, *J.C.S. Chem. Comm.*, 685 (1979).

Isoamyl nitrite, 1, 40–41; **4**, 270; **5**, 358–359; **6**, 307.

Dethioacetalization.[1] Dithioacetals are cleaved to the parent carbonyl compounds by reaction with this nitrite in CH_2Cl_2 for 0.5–1.5 hours. An intermediate red color develops that probably corresponds to an intermediate $S^+(NO)R$ compound. Yields range from 65 to 90%. Oxathiolanes can be cleaved in the same way.

[1] K. Fuji, K. Ichikawa, and E. Fujita, *Tetrahedron Letters*, 3561 (1978).

Isopropylidene diazomalonate (1). Mol. wt. 170.13, m.p. 93–95°. This compound is prepared by diazo transfer from tosyl azide to Meldrum's acid.[1]

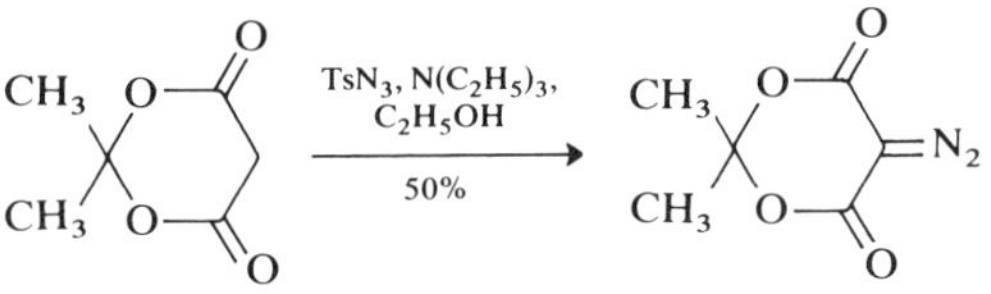

TsN_3, $N(C_2H_5)_3$, C_2H_5OH; 50%

Cyclobutanones. Irradiation of **1** in the presence of cyclohexene produces the cyclobutanone **2**[2] rather than the cyclopropane **3**, as originally believed.[3] This reaction involves a Wolff rearrangement of the initially formed carbene (**a**) to a ketene (**b**), which undergoes regio- and stereoselective cycloaddition to the olefin.

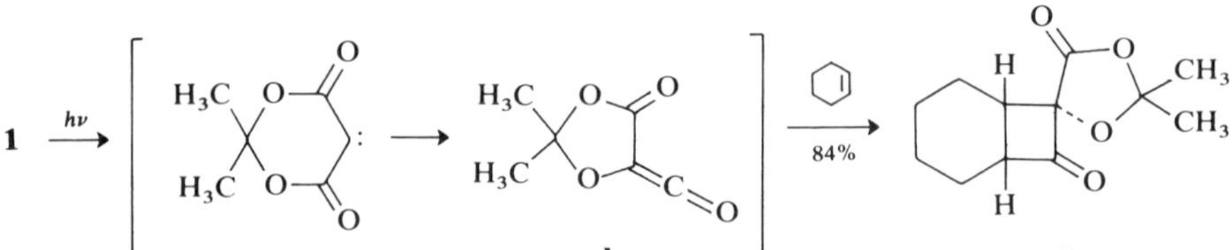

1 —hν→ [a → b] —cyclohexene, 84%→ 2

3

[1] B. Eistert and F. Geiss, *Ber.*, **94**, 929 (1961).
[2] R. V. Stevens, G. S. Bisacchi, L. Goldsmith, and C. E. Strouse, *J. Org.*, **45**, 2708 (1980).
[3] T. Livinghouse and R. V. Stevens, *Am. Soc.*, **100**, 6479 (1978).

2,3-O-Isopropylidene-2,3-dihydroxy-1,4-bis(diphenylphosphino)butane (DIOP), 4, 273; 5, 360–361.

Enantioselective catalytic hydrogenation.[1] Hoffmann-LaRoche chemists[2] have examined the efficiency of DIOP (**1**) and a number of related phosphines as ligands for asymmetric hydrogenation of acrylic acids for synthesis of (R)-6-methyl-tryptophan (**6**), of potential value as a non-nutritive sweetening agent. Of these, **2**[3] and the less available **3** showed the most promise. Thus catalytic hydrogenation of **4**

$P(C_6H_4CH_3\text{-}m)_2$ $P(C_6H_4CH_3\text{-}m)_2$ **2**

C_6H_5 $CH_2\text{+}_2$ OCH_3 **3**

with a Rh(I) catalyst complexed with **2** gave **5** in 82% optical yield. Deacetylation of **5** gave enantiomerically pure **6**.

4 $\xrightarrow{\text{Rh cat., 2}}$

5 (R = $COCH_3$, 82% ee)
6 (R = H, 91% ee)

Asymmetric hydroformylation.[4] Optically active aldehydes can be obtained by hydroformylation ($CO/H_2 = 1:1$) of conjugated dienes with $HRh(CO)[P(C_6H_5)_3]_3$-(−)-DIOP as catalyst. The highest optical yield (32%) was obtained in the hydroformylation of isoprene to give 3-methylpentanal.

[1] Review: R. E. Merrill, *Asymmetric synthesis using chiral Phosphine ligands*, Reaction Design Corp., Hillside, N.J., 1979. This review covers the literature to mid-1979 (234 references). It discusses mechanisms and applications to asymmetric hydrogenation, hydrosilylation, hydroformylation and alkylation.

[2] U. Hengartner, D. Valentine, Jr., K. K. Johnson, M. E. Larscheid, F. Pigott, F. Scheidl, J. W. Scott, R. C. Sun, J. M. Townsend, and T. H. Williams, *J. Org.*, **44**, 3741 (1979).
[3] T.-P. Dang, J.-C. Poulin, and H. B. Kagan, *J. Organometal. Chem.*, **91**, 105 (1975).
[4] C. Botteghi, M. Branca, and A. Saba, *ibid.*, **184**, C17 (1980).

K

Ketene bis(methylthio)ketal monoxide, $H_2C{=}C(S(O)CH_3)(SCH_3)$ (**1**), **5**, 302.

Regiospecific Michael addition. A new total synthesis of (±)-prostaglandin $F_{2\alpha}$ (**4**) utilizes this Michael acceptor in a key step to obtain the epimeric alcohols **2** and **3**.[1] Remaining steps to **4** included a Wittig condensation and a sulfenate–sulfoxide rearrangement.[2]

1) LiCu(cis-$CH{=}CHCH(C_5H_{11}$-$n)OC(CH_3)_2OCH_3)_2$
2) **1**

2 (15β-OH, 45%)
3 (15α-OH, 8%)

4

[1] R. Davis and K. G. Untch, *J. Org.*, **44**, 3755 (1979).
[2] J. G. Miller, W. Kurz, K. G. Untch, and G. Stork, *Am. Soc.*, **96**, 6774 (1974).

Ketene bis(phenylthio)ketal, $CH_2{=}C(SC_6H_5)_2$ (**1**). Mol. wt. 244.37, b.p. 140°. Two recent syntheses are shown in equations (I)[1] and (II).[2]

(I) $CH_3COOH + C_6H_5SH \xrightarrow[50\%]{Al(CH_3)_3} CH_3C(SC_6H_5)_3 \xrightarrow[98.5\%]{CuOTf,\ lutidine}$ **1**

(II) $(C_6H_5S)_2CHB(OC(CH_3)_2C(CH_3)_2O) \xrightarrow[73\%]{1)\ LDA;\ 2)\ HCHO}$ **1**

Cyclopropanone phenyl thioketals. A typical new one-step synthesis is shown in equation (III).[1]

$$\text{(III)}\quad (CH_3)_2C{=}CHCH_2SC_6H_5 \xrightarrow[71\%]{1)\ n\text{-BuLi, TMEDA};\ 2)\ \mathbf{1},\ 25^\circ} (CH_3)_2C{=}C(H){-}\text{cyclopropane}(H)(SC_6H_5)_2$$

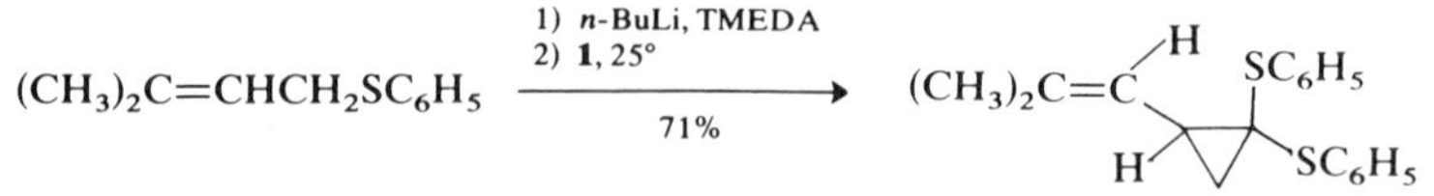

[1] T. Cohen, R. B. Weisenfeld, and R. E. Gapinski, *J. Org.*, **44**, 4744 (1979).
[2] A. Mendoza and D. S. Matteson, *ibid.*, **44**, 1352 (1979).

Ketene diethyl acetal (1,1-Diethoxyethylene), $CH_2{=}C(OC_2H_5)_2$ (**1**). Mol. wt. 116.16, b.p. 86°/200 mm.

Preparation:[1]

$$BrCH_2CH(OC_2H_5)_2 \xrightarrow[67\text{–}75\%]{KOC(CH_3)_3} \mathbf{1}$$

Reactions with quinones; anthraquinone synthesis. Early studies of McElvain[2] indicated that the reagent reacts in two different ways with quinones. In the case of benzoquinones the products are 1 : 1 adducts, which are ethoxyfurane derivatives. 2-Halonaphthoquinones were found to react at least in part with 2 equiv. of **1** to form 1,3-diethoxyanthraquinones (with loss of the halogen substituent). A more recent study established that the reaction of halonaphthazarins with **1** is stereospecific, *i.e.*, only one anthraquinone is formed under suitable conditions; yield of 1 : 2 adducts can be as high as 60%. A typical reaction is formulated in equation (I). The 1 : 1 adducts are major products, however, in the absence of a *free* peri-hydroxyl group.[3]

(I) 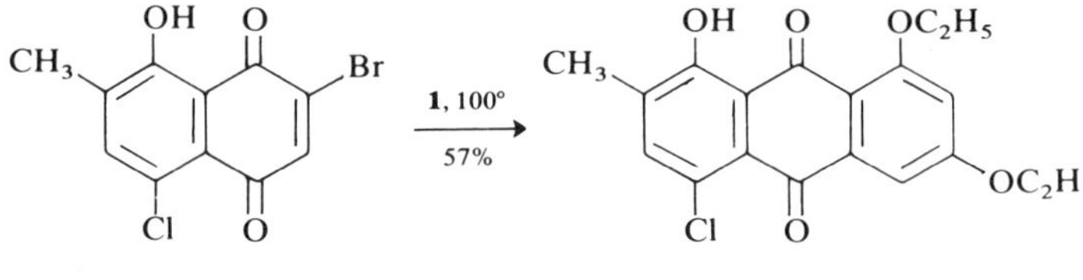 [2-bromo-5-chloro-8-hydroxy-7-methyl-1,4-naphthoquinone] $\xrightarrow[57\%]{\mathbf{1},\ 100^\circ}$ [5-chloro-8-hydroxy-7-methyl-1,3-diethoxyanthraquinone]

Labels: OH, O, CH_3, Br, Cl, O → OH, O, OC_2H_5, CH_3, OC_2H_5, Cl, O

Use of DMSO as solvent favors formation of the 1 : 2 adducts, and even permits use of naphthoquinones themselves as substrates.[4] However, a 2-halo substituent is useful for control of the regioselectivity (equations II and III).[5]

(II) 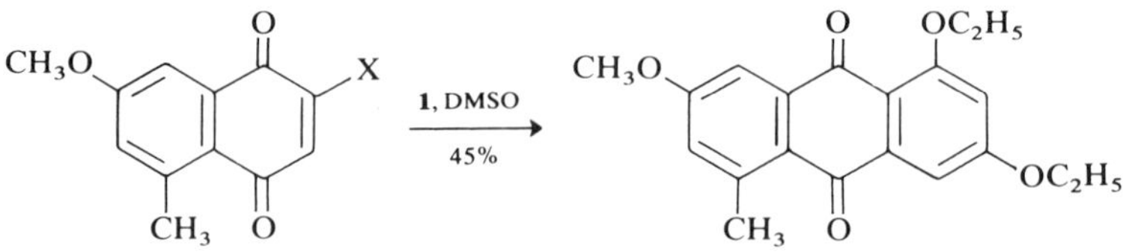 [6-methoxy-8-methyl-1,4-naphthoquinone, 2-X] $\xrightarrow[45\%]{\mathbf{1},\ DMSO}$ [1,3-diethoxy-6-methoxy-8-methylanthraquinone]

Labels: O, CH_3O, X, CH_3, O → O, OC_2H_5, CH_3O, OC_2H_5, CH_3, O

(X = H or Br)

(III)

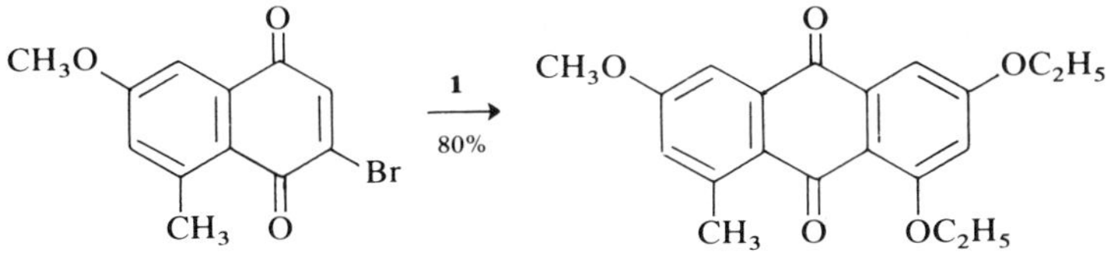

CH_3O, O, Br, CH_3, O $\xrightarrow[80\%]{\mathbf{1}}$ CH_3O, O, OC_2H_5, CH_3, O, OC_2H_5

This reaction has been used for synthesis of various natural anthraquinones; a recent example is the synthesis of the insect pigment kermisic acid (**2**).[6]

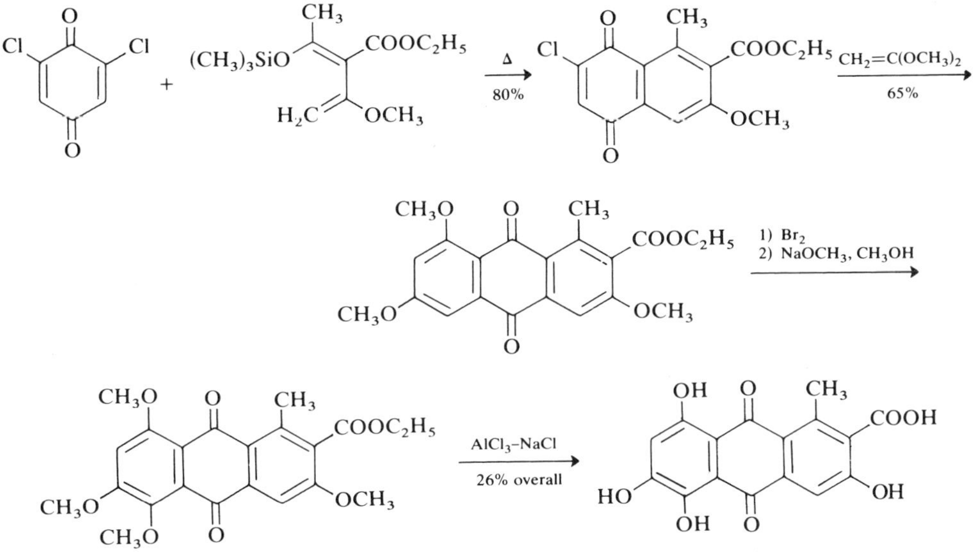

O, Cl, Cl, O + $(CH_3)_3SiO$, CH_3, $COOC_2H_5$, H_2C, OCH_3 $\xrightarrow[80\%]{\Delta}$ Cl, O, CH_3, $COOC_2H_5$, OCH_3, O $\xrightarrow[65\%]{CH_2{=}C(OCH_3)_2}$

CH_3O, O, CH_3, $COOC_2H_5$, CH_3O, OCH_3, O $\xrightarrow{1)\ Br_2;\ 2)\ NaOCH_3,\ CH_3OH}$

CH_3O, O, CH_3, $COOC_2H_5$, CH_3O, CH_3O, O, OCH_3 $\xrightarrow[26\%\ overall]{AlCl_3\text{–}NaCl}$ OH, O, CH_3, COOH, HO, OH, OH, O

2

The exact mechanism for the reaction of **1** with quinones to form anthraquinones is not known, but the intermediate **3** has been isolated in the reaction with 2,3-

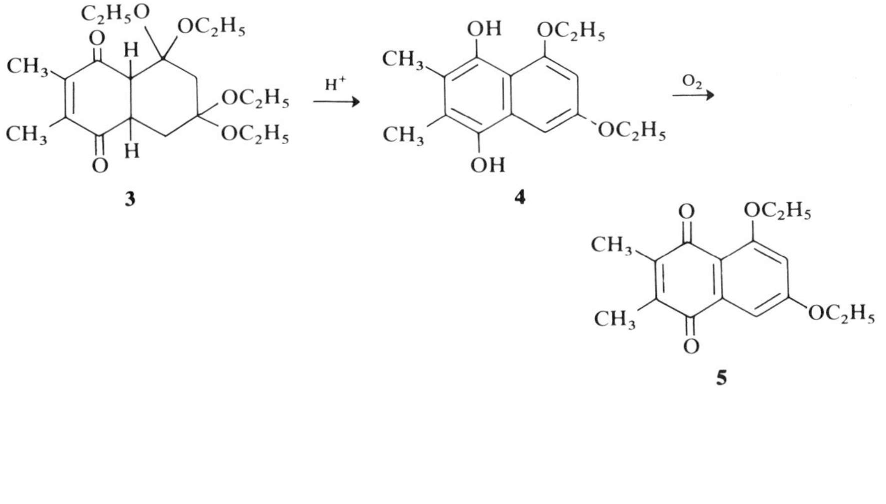

C_2H_5O, OC_2H_5, O, H, CH_3, CH_3, OC_2H_5, OC_2H_5, H, O

3

$\xrightarrow{H^+}$

OH, OC_2H_5, CH_3, CH_3, OC_2H_5, OH

4

$\xrightarrow{O_2}$

O, OC_2H_5, CH_3, CH_3, OC_2H_5, O

5

dimethylbenzoquinone; it is unstable and rearranges in the presence of HCl and absence of air to **4**, which gives **5** quantitatively in the presence of air.

[1] S. M. McElvain and D. Kundiger, *Org. Syn. Coll. Vol.*, **3**, 506 (1955).
[2] S. M. McElvain and E. L. Englehardt, *Am. Soc.*, **66**, 1077 (1944).
[3] J. Banville, J.-T. Grandmaison, G. Lang, and P. Brassard, *Can. J. Chem.*, **52**, 80 (1974).
[4] D. W. Cameron, M. J. Crossley, and G. I. Fentrill, *J.C.S. Chem. Comm.*, 275 (1976).
[5] D. W. Cameron, M. J. Crossley, G. I. Fentrill, and P. G. Griffiths, *ibid.*, 297 (1977).
[6] D. W. Cameron, G. I. Fentrill, P. G. Griffiths, and D. J. Hodder, *ibid.*, 688 (1978).

L

Lead carbonate, $(PbCO_3)_2 \cdot Pb(OH)_2$. Mol. wt. 775.67.

Cyclopentenes.[1] Vinylcyclopropanes (**1**) when pyrolyzed at 500° through a glass column coated with lead carbonate rearrange to annulated cyclopentenes (**2**). In the absence of the salt, the yields are lowered to 15–20%.

$(CH_2)_n$ ... $CH{=}CH_2$ $\xrightarrow[65-72\%]{500°}$ $(CH_2)_n$

1 (n = 1, 2, 3) **2**

[1] T. Hudlicky, J. P. Sheth, V. Gee, and D. Barnvos, *Tetrahedron Letters*, 4889 (1979).

Lead tetraacetate (LTA), **1**, 537–563; **2**, 234–238; **3**, 168–171; **4**, 278–282; **5**, 365–370; **6**, 313–317; **7**, 185–188; **8**, 269–272.

Oxidative lactonization. Oxidation of γ,δ-unsaturated carboxylic acids with LTA produces γ-lactones[1] in satisfactory yield (*see* **2**, 236). Corey and Pearce[2] found this reagent uniquely effective for the key double lactonization of **1** in their total synthesis of picrotoxinin. The double bond of **1** is extremely hindered; as a consequence the usual oxidants are ineffective.

$BrCH_2$, CH_3, COOH, O, $OCOC_6H_5$, CH_3, COOH $\xrightarrow[99\%]{LTA,\ CH_3CN}$ $BrCH_2$, CH_3, O, CO, $OCOC_6H_5$, O, CH_3, H, O—C=O

1 **2**

Conversion of —Si(CH₃)₃ to —OH. Searle chemists[3] have used this transformation in a synthesis of 4-demethoxydaunomycinone (**6**). Thus aryltrimethylsilanes were known to be converted into aryl trifluoroacetates by lead tetrakistrifluoroacetate (**4**, 282–283); the Searle chemists found that this trifluoroacetoxylation was also applicable to benzyltrimethylsilane. With this information and knowing that 1,3-butadienes substituted by trimethylsilyl groups show selectivity in Diels–Alder reactions, they then devised the route shown in the formulation. The first step involved a Diels–Alder reaction to give **3**. This product was converted into **4** by known reactions. The desired conversion, **4** into **5**, proceeded in 79% yield with lead tetraacetate catalysed by KF. Remaining steps to **6** involved hydration of the triple bond and hydrolysis of the acetyl groups.

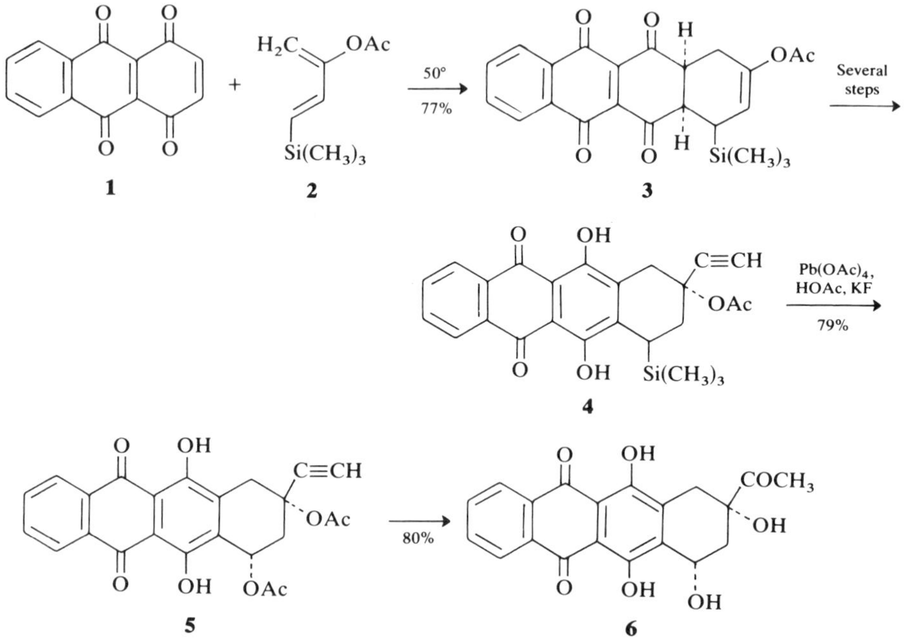

H_2C OAc; $Si(CH_3)_3$; 50°, 77%; OAc; H; H; $Si(CH_3)_3$; Several steps

1 **2** **3**

OH; $C{\equiv}CH$; OAc; OH; $Si(CH_3)_3$; $Pb(OAc)_4$, HOAc, KF; 79%

4

OH; $C{\equiv}CH$; OAc; OH; OAc; 80%; OH; $COCH_3$; OH; OH; OH

5 **6**

Favorski-type rearrangement.[4] Reaction of lead tetraacetate and BF_3 in benzene with the ketone (**1**) results in a mixture of α-acetoxylated ketones, **2a** and **2b**. If

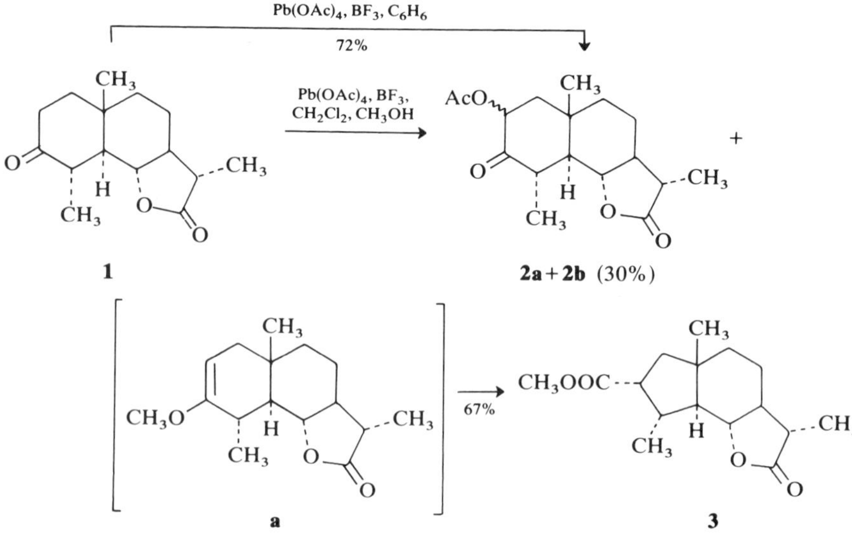

$Pb(OAc)_4$, BF_3, C_6H_6; 72%

CH_3; H; CH_3; CH_3; $Pb(OAc)_4$, BF_3, CH_2Cl_2, CH_3OH; AcO; CH_3; H; CH_3; CH_3; +

1 **2a + 2b** (30%)

CH_3; CH_3O; H; CH_3; CH_3; 67%; CH_3OOC; CH_3; H; CH_3; CH_3

a **3**

the reaction is conducted in the presence of methanol, the major product is the ring-contracted **3**, considered to be formed via the enol ether **a**.

Ketene thioacetals.[5] The oxidation of α-hydroxy thioacetals with 1 equiv. of lead tetraacetate in some cases results in clean fragmentation to a ketene thioacetal. The reaction proceeds particularly well when the reacting center is part of a strained ring system, as in examples formulated in equations (I) and (II). The oxidation of one open-chain substrate also proceeded cleanly. In general, the reaction tends to be slow and capricious (**7**, 186–187), but can be a useful approach to ketene acetals with an unprotected carbonyl group.

(I) OH, CH_3, S, S $\xrightarrow[100\%]{\text{LTA, } CH_3COCH_3\text{, 20°}}$ CH_3, CHO, S, S

(II) OH, S, S $\xrightarrow[92\%]{\text{LTA, } C_6H_6\text{, 80°}}$ CHO, S, S

(III) S, S, $C_6H_5CH_2$, $CHCH_2CH(CH_3)_2$, OH $\xrightarrow{82\%}$ $C_6H_5CH=$ S, S

Aziridines.[6] N-S-Aziridines (**1**) are available by oxidation of 2,4-dinitrobenzenesulfenamide in the presence of olefins (38–64% yield). These heterocycles are also obtained from 2-nitrobenzenesulfenamide, but not from 4-nitro- or 4-chlorobenzenesulfenamides. The N—S bond of these sulfenylated aziridines is cleaved with

CH_3 $\xrightarrow[64\%]{ArSNH_2\text{, LTA}}$ SAr, H, N, CH_3, H **1** $\xrightarrow[56\%]{NaBH_4}$ H, NH, CH_3, H **2**

CH_3 $\longrightarrow$ **1** + H, NSAr, H, CH_3 (1:1)

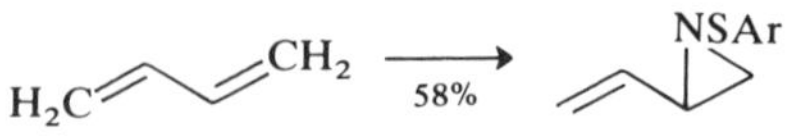

$H_2C{=}CH{-}CH{=}CH_2$ $\xrightarrow{58\%}$ (vinyl aziridine, NSAr)

$NaBH_4$ in ethanol-dichloromethane, as illustrated for **1** → **2**. Aryl sulfenylnitrenes ($Ar\ddot{N}$:) are postulated as intermediates in these reactions.[7]

1,4-Dithiepanes.[8] On treatment with $Pb(OAc)_4$ (benzene, 55°) 2-alkylidene-1,3-dithianes undergo ring expansion to 3-alkyl-1,4-dithiepane-2-ones in moderate yield (equation I).

(I) 2-(=CHR)-1,3-dithiane $\xrightarrow[35\text{–}75\%]{Pb(OAc)_4}$ 3-R-1,4-dithiepan-2-one

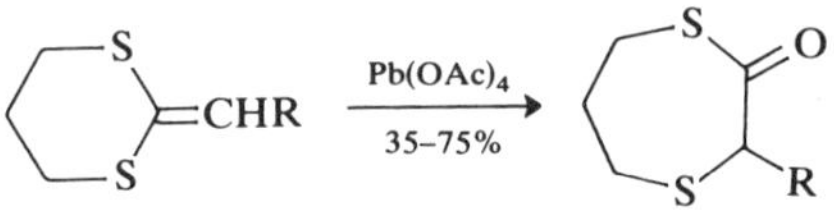

Acetoxylation of enol thioethers.[9] The reaction of enol thioethers with lead tetraacetate (1.1 equiv.) in THF for 1 hour followed by addition of BF_3 etherate (or 5 *N* KOH–ether) results in allylic acetoxylation. The reaction is considered to involve bisacetoxylation of the double bond. Oxidation of sulfur is not observed. Examples:

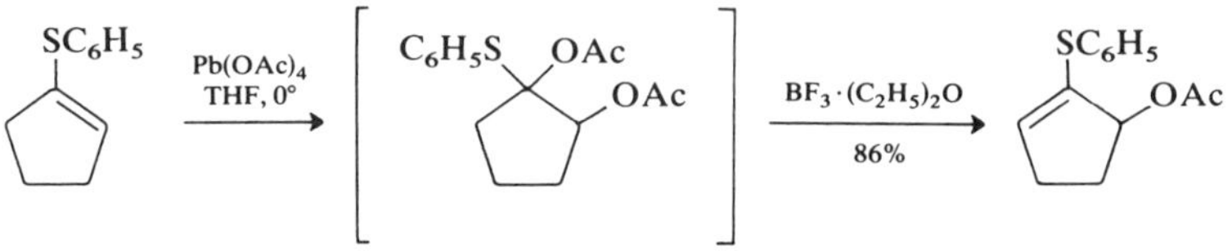

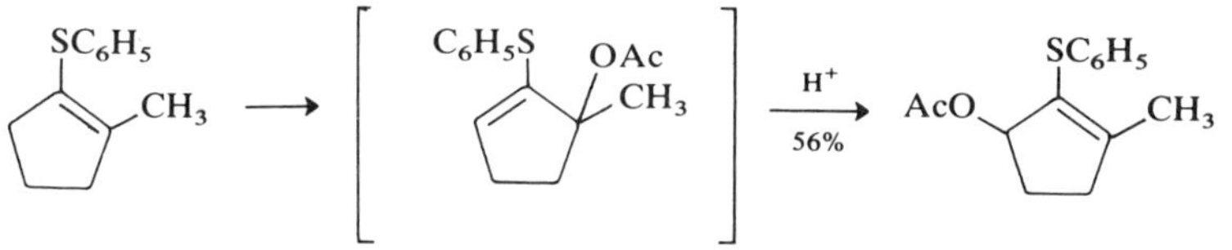

1-SC_6H_5-cyclopentene $\xrightarrow[THF,\ 0°]{Pb(OAc)_4}$ [C_6H_5S, OAc, OAc cyclopentane] $\xrightarrow[86\%]{BF_3\cdot(C_2H_5)_2O}$ 1-SC_6H_5-5-OAc-cyclopentene

1-SC_6H_5-2-CH_3-cyclopentene $\longrightarrow$ [C_6H_5S, OAc, CH_3 cyclopentene] $\xrightarrow[56\%]{H^+}$ AcO, SC_6H_5, CH_3 cyclopentene

Two interesting reactions of these products have been reported. One is conversion to sulfenylated enones (equation I). The other is coupling with cuprates with complete regioselectivity and inversion (equation II).

(I) SC_6H_5, OAc cyclopentene $\xrightarrow{K_2CO_3}$ SC_6H_5, OH cyclopentene $\xrightarrow{MnO_2}$ SC_6H_5 cyclopentenone (=O)

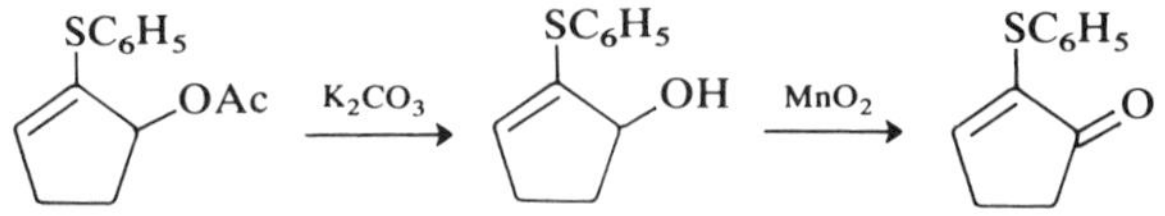

(II) SC_6H_5, ..OAc, $C(CH_3)_3$ cyclohexene $\xrightarrow[71\%]{(C_6H_5)_2CuLi}$ SC_6H_5, C_6H_5, $C(CH_3)_3$ cyclohexene

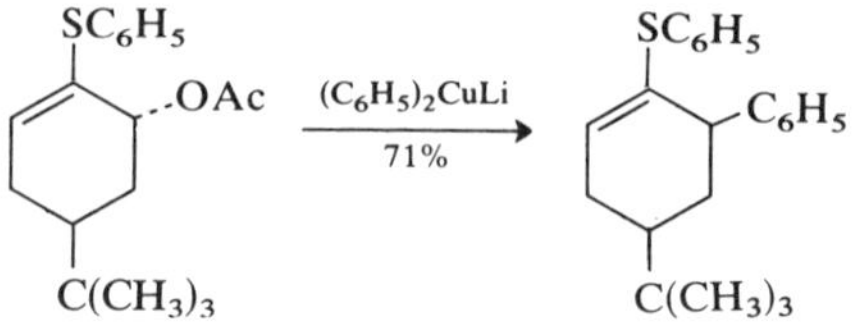

Cleavage of the methylenedioxy group.[10] The methylenedioxy group is cleaved to the corresponding catechol in 20–40% yield by lead tetraacetate in dry benzene (but not in acetic acid).

Example:

CH_3, CH_3 — $\xrightarrow[40\%]{Pb(OAc)_4,\ C_6H_6}$ — CH_3, CH_3, OH, OH

Oxidative decarboxylation. Oxidation of [n.2.2]propellanecarboxylic acids (**1**) with lead tetraacetate in pyridine at 80° gives bicyclic acetates **2** and/or tricyclic acetates **3**. The latter products are converted into **2** on vapor-phase thermolysis.[11]

HOOC, $(CH_2)_n$ (**1**) $\xrightarrow{Pb(OAc)_4}$ $(CH_2)_n$, OAc (**2**) + $(CH_2)_n$, OAc (**3**); **3** → **2**: 250°, quant.

	2	**3**
n = 4	0	68%
n = 5	13%	60%
n = 6	81%	0

[1] K. Alder and S. Schneider, *Ann.*, **524**, 189 (1936).
[2] E. J. Corey and H. L. Pearce, *Am. Soc.*, **101**, 5841 (1979).
[3] R. B. Garland, J. R. Palmer, J. A. Schulz, P. B. Sollman, and R. Pappo, *Tetrahedron Letters*, 3669 (1978).
[4] H. Miura, Y. Fumimoto, and T. Tatsuno, *Synthesis*, 898 (1979).
[5] W. Lottenbach and W. Graf, *Helv.*, **61**, 3087 (1978).
[6] R. S. Atkinson and B. D. Judkins, *J.C.S. Chem. Comm.*, 832 (1979).
[7] *Idem, ibid.*, 833 (1979).
[8] K. Hiroi and S. Sato, *Chem. Letters*, 923 (1979).
[9] B. M. Trost and Y. Tanigawa, *Am. Soc.*, **101**, 4413 (1979).
[10] Y. Ikeya, H. Taguchi, and I. Yosioka, *Chem. Pharm. Bull. Japan*, **27**, 1383 (1979).
[11] Y. Sakai, S. Toyotani, Y. Tobe, and Y. Odaira, *Tetrahedron Letters*, 3855 (1979).

Lead(IV) trifluoroacetate–Diphenyl disulfide.

α-Hydroxysulfenylation.[1] Lead tetraacetate and trifluoroacetic acid form lead(IV) trifluoroacetate *in situ.*[2] When an olefin and diphenyl disulfide are added to this reagent at 0 or −40°, a β-trifluoroacetoxy sulfide is formed, which on basic work-up generates a thiohydrin (equation I). The reaction involves *trans*-addition.

$$\text{(I)}\quad >C=C< + Pb(OAc)_4 + CF_3COOH + (C_6H_5S)_2 \rightarrow >C(OCOCF_3)-C(SC_6H_5)< \xrightarrow{OH^-} >C(OH)-C(SC_6H_5)<$$

The regiochemistry is governed by the temperature, the olefin, and the disulfide.

Many of these α-hydroxy sulfides can be cleaved by a reagent composed of lead tetraacetate, pyridine, and acetic acid in a molar ratio of about 2:4:3 per mole of substrate. A *trans*-diaxial arrangement is required and highly hindered α-hydroxy sulfides are not cleaved.

Examples:

LTA, TFA, $(C_6H_5S)_2$, 61%; OH, SC_6H_5; LTA, Py, HOAc, 88%; AcO, $CH(CH_2)_3CHO$, C_6H_5S

$CH_3(CH_2)_4CH{=}CH_2 \xrightarrow{87\%} C_5H_{11}CH(OH)CH_2SC_6H_5 + C_5H_{11}CH(SC_6H_5)CH_2OH$ (2:1)

CH_3, H; −40°, 82%; HO, CH_3, C_6H_5S, H; LTA, Py, HOAc, >40%; OHC, CH_3, AcO, SC_6H_5, H

[1] B. M. Trost, M. Ochiai, and P. G. McDougal, *Am. Soc.*, **100**, 7103 (1978).
[2] S. R. Jones and J. M. Meller, *J.C.S. Perkin II*, 511 (1977).

Lindlar catalyst, 1, 566–567; **3**, 171–172; **4**, 283; **6**, 319.

(Z,E,E)-Trienes.[1] Lindlar hydrogenation of substituted diene acetylenic esters of type **1–3** affords the corresponding (Z,E,E)-triene esters in high yield. The site selectivity of this reduction is excellent for all substrates examined except **3**; evidently the steric bulk of the isopropyl or methyl substituents of **1** and **2** suppresses the rate of butadiene reduction in these systems. Only *cis* C_2–C_3 double bonds are obtained.

[1] W. R. Roush, H. R. Gillis, and S. E. Hall, *Tetrahedron Letters*, **21**, 1023 (1980).

$(CH_3)_2CH$, $COOCH_3$, O, O, **1**; H_2, Lindlar cat., 88%; $(CH_3)_2CH$, CH_3OOC, O, O

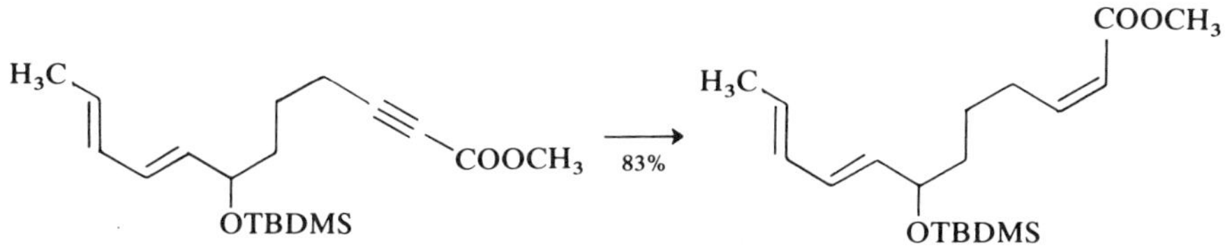

H_3C … $COOCH_3$ (OTBDMS) $\xrightarrow{83\%}$ H_3C … $COOCH_3$ (OTBDMS)

2

H_2C … $COOCH_3$ (OTHP) $\xrightarrow[74\%]{H_2,\ cat.,\ quinoline}$ CH_2 … CH_3OOC (OTHP) + CH_3 … $COOCH_3$ (OTHP) ~3 : 1

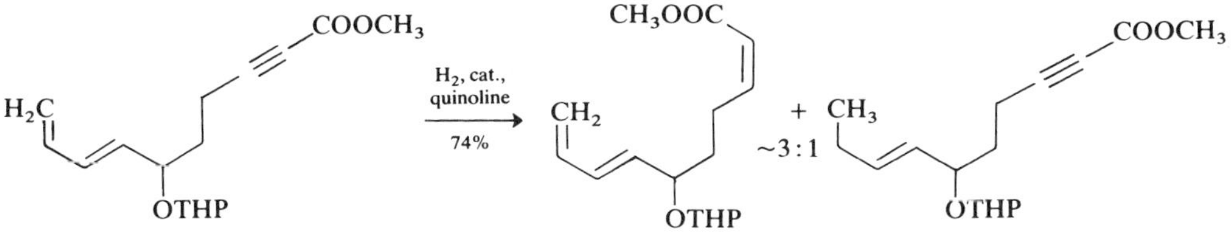

3

Lithiochloromethyl phenyl sulfoxide, $C_6H_5S(O)CHClLi$. The anion is generated from chloromethyl phenyl sulfoxide by reaction with LDA in THF at −78°.

Vinyl chlorides.[1] The anion can be alkylated by alkyl iodides (HMPT) or benzylic bromides in high yield. On pyrolysis at 160°, the products are converted into vinyl chlorides (*cis*- and *trans*-isomers).

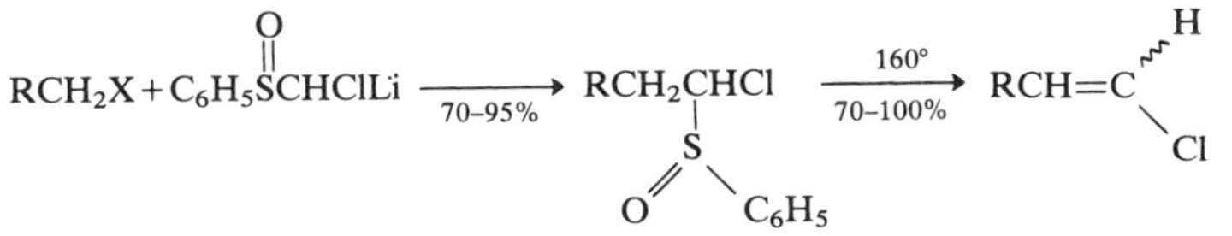

$$RCH_2X + C_6H_5S(O)CHClLi \xrightarrow{70-95\%} RCH_2CHCl(S(O)C_6H_5) \xrightarrow[70-100\%]{160^\circ} RCH{=}C(H)Cl$$

[1] V. Reutrakul and P. Thamnusan, *Tetrahedron Letters*, 617 (1979).

1-Lithiocyclopropyl phenyl sulfide, 5, 372–373; **6**, 319–320; **7**, 190.

γ-Keto sulfides. Another rearrangement of the adducts of this anion with carbonyl compounds has been reported. When treated with $ZnCl_2$ and anhydrous HCl in CH_2Cl_2, the cyclopropylcarbinyl alcohols are converted into ring opened γ-keto sulfides with a 1,2-carbonyl transposition (equation I).[1] The products can be desulfurized or converted into enones by sulfoxide elimination.

$$\text{(I)}\quad R_2C{=}O + \text{1-lithio-1-}(SC_6H_5)\text{cyclopropane} \longrightarrow R_2C(OH)\text{–}C(SC_6H_5)\text{(cyclopropyl)} \xrightarrow[45-70\%]{ZnCl_2,\ HCl} R_2CHC(O)CH_2CH_2SC_6H_5$$

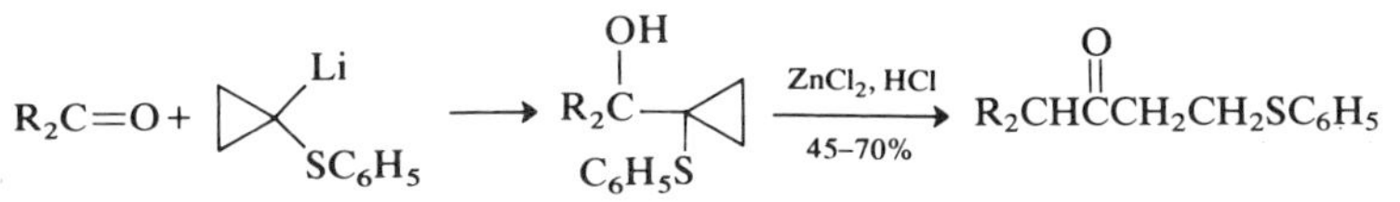

The adducts obtained from the anion and enones under similar conditions can be converted into β-haloethyl vinyl sulfides (equation II).[2] The products can be selectively hydrogenated to give homoallylic halides or dehydrohalogenated (DBN) to conjugated trienes.

(II) 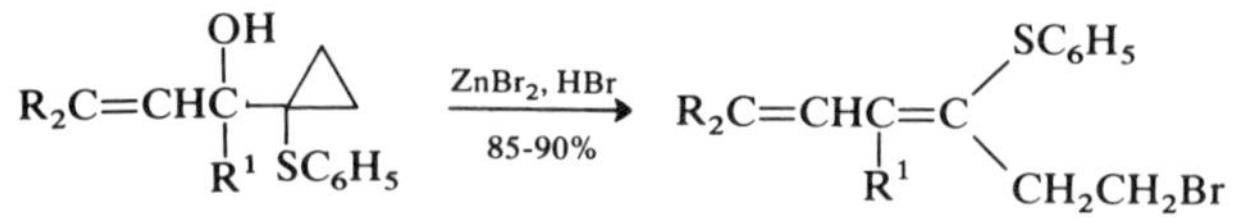 $R_2C{=}CHC(OH)(R^1)$–[1-(SC_6H_5)cyclopropyl] $\xrightarrow[85\text{-}90\%]{ZnBr_2,\ HBr}$ $R_2C{=}CHC(R^1){=}C(SC_6H_5)CH_2CH_2Br$

The rearrangement shown in equation (I) proceeds through an intermediate cyclobutanone, previously obtained from the adducts by treatment with fluoroboric acid (**5**, 372–373). Thus the cyclobutanone (**1**) rearranges to the γ-keto sulfide (**2**) on treatment with $ZnCl_2$, HCl and thiophenol. This reaction is also applicable to

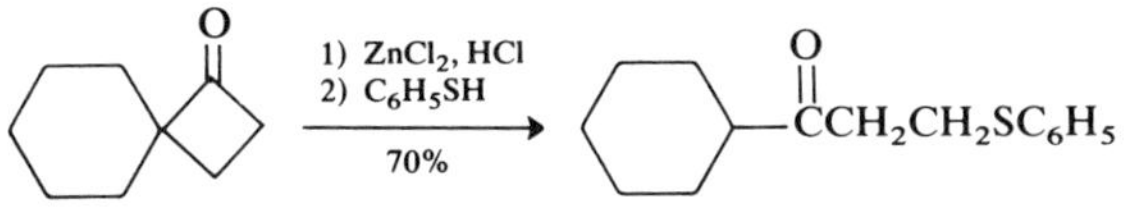

1 $\xrightarrow[70\%]{1)\ ZnCl_2,\ HCl;\ 2)\ C_6H_5SH}$ C_6H_{11}–$\overset{O}{\overset{||}{C}}CH_2CH_2SC_6H_5$ **2**

fused cyclobutanones such as **3**, prepared by addition of dichloroketene to a cyclic olefin followed by dechlorination (**4**, 134–135). In this case the reaction results in a two-carbon ring expansion of the original cycloalkene.[3]

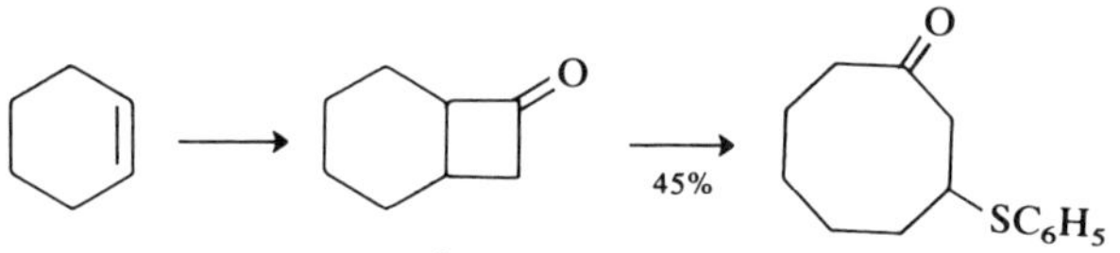

Cyclohexene → **3** $\xrightarrow{45\%}$ **4** (SC_6H_5)

[1] R. D. Miller and D. R. McKean, *Tetrahedron Letters*, 583 (1979).
[2] R. D. Miller, D. R. McKean, and D. Kaufmann, *ibid.*, 587 (1979).
[3] R. D. Miller and D. R. McKean, *ibid.*, 1003 (1979).

α-Lithio-α-methoxyallene, $(Li)(CH_3O)C{=}C{=}CH_2$ (**1**). Mol. wt. 76.02. The reagent is prepared by deprotonation of methoxyallene (**7**, 225–226) with *n*-butyllithium in THF at −78°.

Dihydrofurane-3(2H)-ones.[1] This anion reacts with cyclohexenone to give the 1,2-adduct (**2**). On treatment of **2** with KH and dicyclohexyl-18-crown-6 in THF, cyclization to **3** occurs. Mild acid hydrolysis of the enol ethers gives the spirodihydrofurane-3(2*H*)-one (**4**). This reaction has also been carried out successfully with saturated five- and six-membered cyclic ketones.

[1] D. Gange and P. Magnus, *Am. Soc.*, **100**, 7746 (1978).

1-Lithio-2-vinylcyclopropane, *cis-* and *trans-*, 7, 192–193.

Cycloheptane derivatives (**7**, 192–193). Wender *et al.*[1] have applied their cycloheptane annelation procedure to total syntheses of the pseudoguaianolides damsinic acid (**1**) and confertin (**2**). The paper describes the synthesis of 1-lithio-1-methyl-2-vinylcyclopropane, which is the annelation reagent utilized in the pseudoguaianolide syntheses.

[1] P. A. Wender, M. A. Eissenstat, and M. P. Filosa, *Am. Soc.*, **101**, 2196 (1979).

Lithium–Ammonia, 1, 601–603; **2,** 205; **3,** 179–182; **4,** 288–290; **5,** 379–381; **6,** 322–323; **7,** 195; **8,** 282–284.

Cleavage of esters. Lithium or sodium in ethereal ammonia cleaves esters to the alcohol in 70–95% yield. The method is not useful for highly hindered esters and for esters of allylic alcohols (which are converted to the corresponding hydrocarbons by alkyl–oxygen cleavage).[1]

Reduction of thebaine. Thebaine (**1**) is reduced by potassium (2.3 equiv.) in liquid ammmonia to a 1:1 mixture of **2** and **3**.[2] Reduction under similar conditions but with calcium, lithium, or sodium gives only the nonconjugated diene **3**. This diene is not isomerized by the usual basic reagents,[3] but can be converted in part into **2** by K–NH_3 in the presence of $Fe(NO_3)_3 \cdot 9H_2O$ (**1**, 907).

1 H. W. Pinnick and E. Fernandez, *J. Org.*, **44**, 2810 (1979).
2 R. K. Razdan, D. E. Portlock, H. C. Dalzell, and C. Malmberg, *ibid.*, **43**, 3604 (1978).
3 A. J. Birch and M. Fitton, *Aust. J. Chem.*, **22**, 971 (1969).

Lithium–Ethylenediamine, 1, 601–603; **2**, 205; **3**, 179–182; **4**, 288–290; **5**, 379–381; **6**, 322–323; **7**, 195.

Reduction of methylenecyclohexanes.[1] Reduction of 4-*t*-butylmethylenecyclohexane (**1**) with lithium in ethylenediamine is highly stereoselective and independent of the temperature. The result implies that the stereochemistry is controlled by the relative stability of the possible intermediate carbanions. In this case the stable one has an equatorial C—C bond. The stereochemistry of reduction of **4**, however, is dependent on the temperature, with the less stable product (**6**) favored by a rise in temperature. Apparently the rate of protonation of the equatorial carbanion becomes a significant factor at higher temperatures. In even more hindered systems equatorial protonation can be significant, even at 35°.

Li, $H_2N(CH_2)_2NH_2$, dioxane, 35–100°

1 → **2** (90–95%) + **3** (10–5%)

4 → **5** + **6**

	5 : 6
35°	95:5
100°	60:40

1 F. Ficini, J. Francillette, and A. M. Touzin, *J. Chem. Res. (M)*, 1820 (1979).

Lithium aluminum hydride, 1, 581–595; **2**, 242; **3**, 176–177; **4**, 291–293; **5**, 382–389; **6**, 325–326; **7**, 196; **8**, 286–289.

(E)-2-Alkenenitriles. 2-Alkynenitriles, RC≡C—CN, are reduced by lithium aluminum hydride (0.5 equiv.) in ether to (E)-2-alkenenitriles in 70–98% yield.[1]

Stereospecific reduction of an enone. Reduction of the hydroazulenone **1** with lithium aluminum hydride (or $NaBH_4$, DIBAH) gives only the allylic alcohol **2** in quantitative yield.[2] The result is surprising since epoxidation of **1** followed by reduction with sodium borohydride gives only the epoxy alcohol **3**.[3] These stereospecific reactions permitted conversion of **1** into *dl*-bigelovin (**4**) and into *dl*-helenalin (**5**), two members of the pseudoguaianolide sesquiterpene lactones.

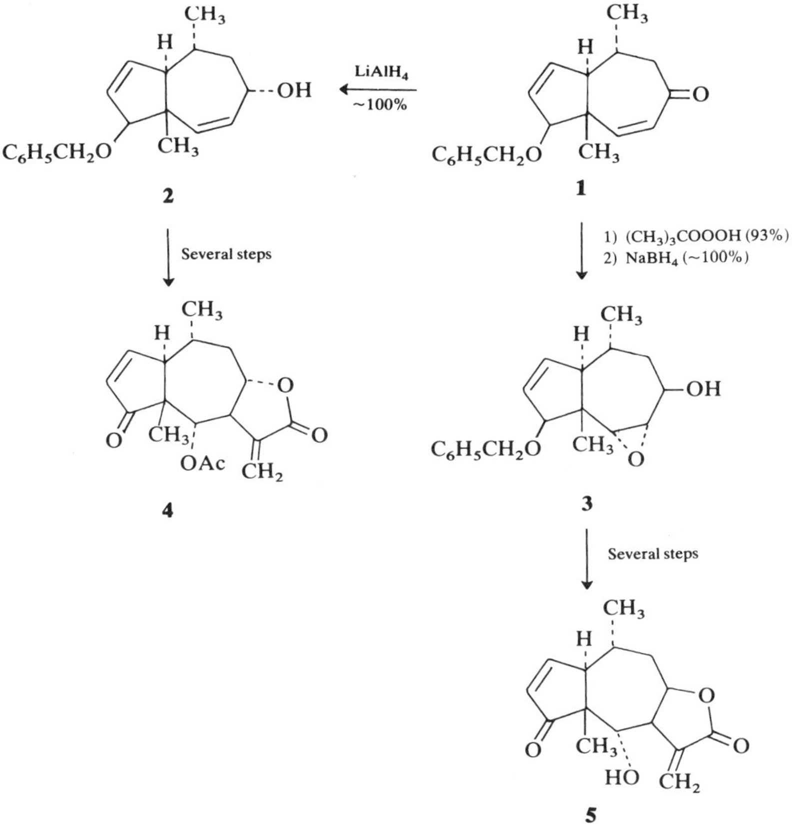
LiAlH4
~100%
1) (CH3)3COOOH (93%)
2) NaBH4 (~100%)
Several steps
Several steps
2
1
4
3
5

Reductive Grob-like fragmentation.[4] Reaction of the mesyloxy alcohol **1** with $LiAlH_4$ in refluxing DME gives **2** in almost quantitative yield. The corresponding keto mesylate is also converted into **2** in high yield under the same conditions. This fragmentation was used for a total synthesis of racemic γ-elemene **3**.

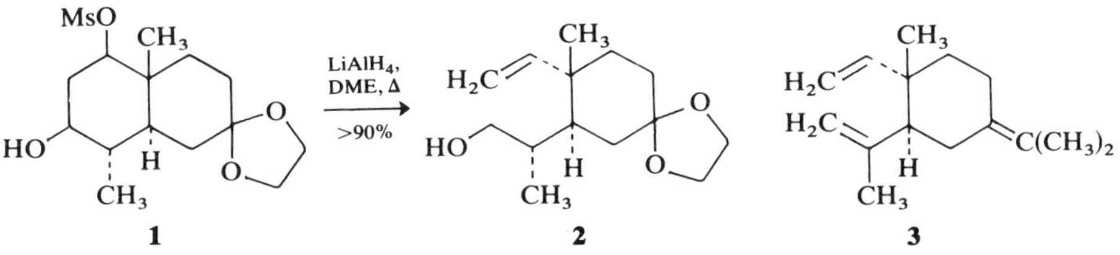
LiAlH4,
DME, Δ
>90%
1
2
3

Hydroalumination of alkenes. Hydroalumination of alkynes is a well-known reaction, but hydroalumination of alkenes has been achieved only recently under catalysis by $TiCl_4$ or $ZrCl_4$, (**8**, 288). As expected hydroalumination affords a convenient, high-yield route to primary alkanes (by hydrolysis), terminal primary alcohols (by oxygenation), and primary alkyl halides (reaction with halogens, N-halosuccinimides, or CuX_2).[5]

$$\text{(I)}\quad 4RCH{=}CH_2 \xrightarrow{LiAlH_4,\ TiCl_4} LiAl(CH_2CH_2R)_4 \xrightarrow[\sim 90\%]{O_2} 4RCH_2CH_2OH$$

$$LiAl(CH_2CH_2R)_4 \xrightarrow[\sim 90\%]{X_2} 4RCH_2CH_2X$$

The conversion to primary acetates is carried out in higher yield if the hydroalumination is conducted with only 2 equiv. of the alkene followed by reaction with lead tetraacetate (equation II).[5]

$$\text{(II)}\quad 2RCH{=}CH_2 \xrightarrow{LiAlH_4} LiAl(CH_2CH_2R)_2H_2 \xrightarrow[50\text{–}70\%]{Pb(OAc)_4} 2RCH_2CH_2OAc$$

A number of coupling reactions are possible with the organoaluminum complexes, as shown in equations (III)[6], (IV),[7] and (V).[8]

$$\text{(III)}\quad LiAl(CH_2CH_2R)_4 \xrightarrow[65\text{–}80\%]{4CH_2{=}CHCH_2X,\ CuCl} 4RCH_2CH_2CH_2CH{=}CH_2$$

$$\text{(IV)}\quad LiAl(CH_2CH_2R)_4 \xrightarrow[45\text{–}55\%]{4CH_2{=}C{=}CHBr,\ CuCl} 4RCH_2CH_2CH_2C{\equiv}CH$$

$$LiAl(CH_2CH_2R)_4 \xrightarrow{4BrCH_2C{\equiv}CH,\ CuCl} 4RCH_2CH_2CH{=}C{=}CH_2$$

$$\text{(V)}\quad LiAl(CH_2CH_2R)_4 \xrightarrow[70\%]{4Cu(OAc)_2} 2RCH_2CH_2CH_2CH_2R$$

$$LiAl(CH_2CH_2R)_4 \xrightarrow[20\text{–}30\%]{CO,\ Cu(OAc)_2} 2RCH_2CH_2C(=O)CH_2CH_2R$$

Hydroalumination of 1-chloro-1-alkynes.[9] Lithium aluminum hydride adds to 1-chloro-1-alkynes (**1**) regio- and stereoselectively to form the α-chlorovinyl alanates **2**, which are moderately stable at 0°C. On methanolysis they are converted into (E)-1-chloro-1-alkenes (**3**). They can also be converted into (Z)-1-bromo-1-chloro-1-alkenes (**5**) and into (Z)-1-chloro-1-iodo-1-alkenes (**6**).

$$\text{(I)}\quad RC{\equiv}CH \xrightarrow{1)\ n\text{-BuLi, THF, }-78^\circ;\ 2)\ NCS,\ -25\rightarrow 20^\circ} \underset{\mathbf{1}}{RC{\equiv}CCl} \xrightarrow{LiAlH_4,\ THF,\ -30\rightarrow 0^\circ}$$

$$\underset{\mathbf{2}}{\left[\begin{matrix} R & & AlH_3 \\ & C{=}C & \\ H & & Cl \end{matrix}\right]^- Li^+} \xrightarrow[75\text{–}95\%]{CH_3OH} \underset{\mathbf{3}}{\begin{matrix} R & & H \\ & C{=}C & \\ H & & Cl \end{matrix}}$$

(II) **2** $\xrightarrow{3(CH_3)_2C{=}O}$ (R)(H)C=C($\bar{A}l(O\text{-}i\text{-}Pr)_3$)(Cl) Li^+ **4** $\xrightarrow[60-90\%]{Br_2,\ -78°}$ (R)(H)C=C(Br)(Cl) **5**

4 $\xrightarrow[85-90\%]{ICl,\ -30 \rightarrow 25°}$ (R)(H)C=C(I)(Cl) **6**

Reduction of ω-alkynols to (E)-alkenols.[10] This reaction is often one step in the synthesis of insect pheromones. Sodium in liquid ammonia is generally preferred for this reduction (**5**, 590). An alternative, general method is reduction with a large excess of lithium aluminum hydride in diglyme–THF.

$$RC{\equiv}C{-}(CH_2)_n{-}OH \xrightarrow[85-95\%]{LiAlH_4,\ 140°,\ \text{diglyme, THF}} RCH{=}CH{-}(CH_2)_n{-}OH$$

(n = 2–7)

[1] H. Westmijze, H. Kleijn, and P. Vermeer, *Synthesis*, 430 (1979).
[2] P. A. Grieco, Y. Ohfune, and G. Majetich, *J. Org.*, **44**, 3092 (1979).
[3] Y. Ohfune, P. A. Grieco, C.-L. J. Wang, and G. Majetich, *Am. Soc.*, **100**, 5946 (1978).
[4] M. Kato, H. Kurihara, and A. Yoshikoshi, *J.C.S. Perkin I*, 2740 (1979).
[5] F. Sato, Y. Mori, and M. Sato, *Tetrahedron Letters*, 1405 (1979).
[6] F. Sato, H. Kodama, and M. Sato, *J. Organometal. Chem.*, **157**, C30 (1978).
[7] (a) *Idem*, *Chem. Letters*, 789 (1978); (b) F. Sato, K. Ogura, and M. Sato, *ibid.*, 805 (1978).
[8] F. Sato, Y. Mori, and M. Sato, *ibid.*, 1337 (1978).
[9] G. Zweifel, W. Lewis, and H. P. On, *Am. Soc.*, **101**, 5101 (1979).
[10] R. Rossi and A. Carpita, *Synthesis*, 561 (1977).

Lithium aluminum hydride–Boron trifluoride.

Reaction with sapogenins.[1] Smilagenin (**1**) reacts with a reagent prepared from lithium aluminum hydride and BF_3 etherate in THF to give dihydrosmilagenin (**2**) and a mixture of the epimeric tetraols **3a** and **3b**.

[1] G. R. Pettit, J. J. Einck, and J. C. Knight, *Am. Soc.*, **100**, 7781 (1978).

1 (structure: HO, H, CH_3, CH_3, CH_3, O, O, CH_3) $\xrightarrow{LiAlH_4,\ BF_3 \cdot (C_2H_5)_2O}$

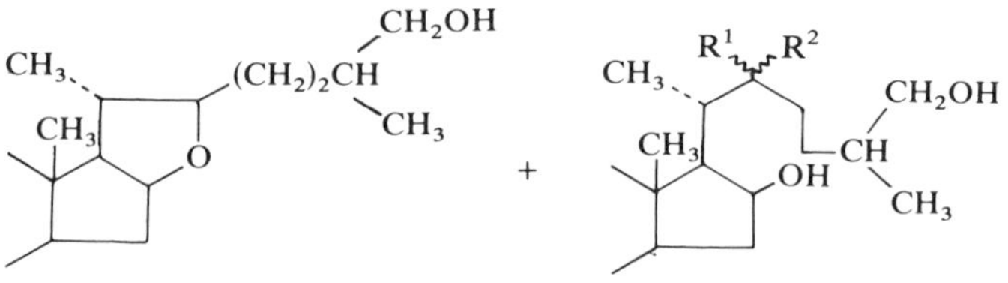

2 (~50%) + **3a** ($R^1 = OH, R^2 = H$)
3b ($R^1 = H, R^2 = OH$)

Lithium aluminum hydride–Di-μ-carbonyldecacarbonyltri-*triangulo*-iron, $[Fe_3(CO)_{12}]$.

Reductive dimerization of ketones.[1] Aromatic α,β-unsaturated ketones are reduced and dimerized to 1,5-hexadienes by sequential reaction with $LiAlH_4$, $Fe_3(CO)_{12}$, and HCl. The effectiveness of iron carbonyls shows the order: $Fe_3(CO)_{12} > Fe_2(CO)_9 > Fe(CO)_5$.

Examples:

$$C_6H_5CH{=}CHCOC_6H_5 \xrightarrow[3)\ HCl]{1)\ LiAlH_4;\ 2)\ Fe_3(CO)_{12}} \begin{matrix} C_6H_5CH{=}CHCHC_6H_5 \\ | \\ C_6H_5CH{=}CHCHC_6H_5 \end{matrix} + C_6H_5COCH_2CH_2C_6H_5$$

(78%) (7%)

$$\text{2-thienyl-}CH{=}CHCOC_6H_5 \xrightarrow{59\%} \begin{matrix} \text{2-thienyl-}CH{=}CHCHC_6H_5 \\ | \\ \text{2-thienyl-}CH{=}CHCHC_6H_5 \end{matrix}$$

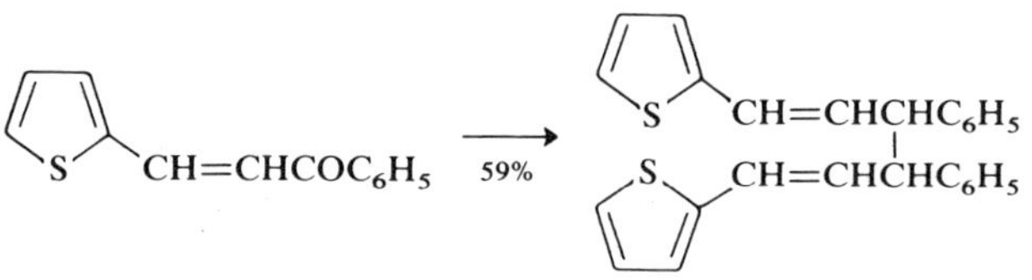

[1] S. Nakanishi, T. Shundo, T. Nishibuchi, and Y. Otsuji, *Chem. Letters*, 955 (1979).

Lithium aluminum hydride–Triethylenediamine.

Lithium trialkylborohydrides.[1] These hydrides, even those with bulky alkyl groups, can be prepared in very high yield by reaction of trialkylboranes in ether with lithium aluminum hydride (1 equiv.) and triethylenediamine (1 equiv.) at 0°. The function of the diamine is to form an insoluble complex with the AlH_3 formed (equation I).

$$\text{(I)}\quad R_3B + LiAlH_4 \xrightarrow{THF,\ -78^\circ} LiR_3BH + AlH_3 \xrightarrow{DABCO} AlH_3 \cdot N(CH_2CH_2)_3N$$

[1] H. C. Brown, J. L. Hubbard, and B. Singaram, *J. Org.*, **44**, 5004 (1979).

Lithium amide–Ammonia.

ω-Acetylenic alcohols.[1] 1-Alkyne-ω-ols (**1**) can be alkylated by primary or secondary alkyl halides in the presence of lithium amide in liquid ammonia in

50–80%. Yields are poor in reactions with tertiary halides. It is not necessary to protect the hydroxyl group.

$$HC{\equiv}C(CH_2)_nCH_2OH \xrightarrow[50-80\%]{1)\ LiNH_2,NH_3\ \ 2)\ RX} RC{\equiv}C(CH_2)_nCH_2OH$$

1 (n = 1, 2, 3) **2**

[1] J. Flahant and P. Miginiac, *Helv.*, **61**, 2275 (1978).

Lithium bis(N,N-diethylcarbamoyl)cuprate (1). Mol. wt. 270.76, thermally stable at 80°.

Preparation:

$$CuCl + LiN(C_2H_5)_2 \xrightarrow{THF,\ HMPT} CuN(C_2H_5)_2 \xrightarrow{LiN(C_2H_5)_2}$$

$$[(C_2H_5)_2N]_2CuLi \xrightarrow{CO} [(C_2H_5)_2NCO]_2CuLi$$

1

Carbamoylation.[1] The reagent does not react with benzaldehyde or ethyl benzoate. It reacts with alkyl iodides at 80°. A more useful reaction is that with acid halides, particularly acid bromides, which provides a useful route to α-keto amides. The reagent does not react with cyclohexenone, but does react with methyl vinyl ketone in THF–HMPT (4 : 1) at −78° to give an adduct in 78% yield.

Examples:

$$CH_3I \xrightarrow[10\%]{\mathbf{1},\ 80°} CH_3CON(C_2H_5)_2$$

$$CH_3COBr \xrightarrow[70\%]{\mathbf{1},\ -78\rightarrow 20°} CH_3COCON(C_2H_5)_2$$

$$C_6H_5COCl \xrightarrow[61\%]{\mathbf{1},\ 80°} C_6H_5COCON(C_2H_5)_2$$

$$CH_2{=}CHCOCH_3 \xrightarrow{\mathbf{1},\ -78\rightarrow 20°,\ THF\text{–}HMPT} \xrightarrow[78\%]{H_3O^+} (C_2H_5)_2NCOCH_2CH_2COCH_3$$

[1] T. Tsuda, M. Miwa, and T. Saegusa, *J. Org.*, **44**, 3734 (1979).

Lithium chloropropargylide, 5, 397.

Homopropargylic alcohols; α-allenic alcohols.[1] The synthesis of alkylallenes (**5**, 397) by the reaction of trialkylboranes and lithium chloropropargylide, $ClCH_2C{\equiv}CLi$ (**1**), has been extended to these two systems. Thus addition of acrolein to the organoborane **a** can lead to either **2** or **3**, depending on the temperature at which **a** is kept before reaction with the aldehyde (equation I). If the aldehyde is added to **a** at −78°, the product, after oxidative work-up is **2**. If **a** is first allowed to warm to 25°, then reaction with an aldehyde gives **3**, after usual work-up. The method is applicable to a wide variety of aldehydes and trialkylboranes.

$$\text{(I)}\quad R_3B + \mathbf{1} \xrightarrow{-90^\circ} \underset{\mathbf{a}}{Li(R_3BC{\equiv}CCH_2Cl)} \xrightarrow[84\%]{\text{1) } -90\to25^\circ;\ \text{2) } CH_2{=}CHCHO,\ -78^\circ;\ \text{3) [O]}} \underset{\mathbf{3}}{CH_2{=}CH\overset{OH}{C}H\underset{R}{C}{=}C{=}CH_2}$$

$$\mathbf{a} \xrightarrow[75\%]{\text{1) } CH_2{=}CHCHO,\ -78^\circ;\ \text{2) [O]}} \underset{\mathbf{2}}{RC{\equiv}CCH_2\underset{OH}{C}HCH{=}CH_2}$$

[1] G. Zweifel, S. J. Backlund, and T. Leung, *Am. Soc.*, **100**, 5561 (1978).

Lithium diisopropylamide (LDA), 1, 611; **2**, 249; **3**, 184–185; **4**, 298–302; **5**, 400–406; **6**, 334–339; **7**, 204–207; **8**, 292.

Stereoselective aldol condensation. Heathcock and Buse have previously employed 2-methyl-2-trimethylsiloxy-3-pentanone (**1**) in a highly stereoselective route to 3-hydroxy-2-methylcarboxylic acids (**8**, 295). Aldol condensation of the lithium enolate derived from **1** with a chiral aldehyde yields *erythro*-aldols, which are cleaved with periodic acid to *β*-hydroxy carboxylic acids. However, when **1** is condensed with a chiral aldehyde such as **2**, two *erythro*-products (**3** and **4**) are produced. Heathcock and co-workers now report that the 1,2-diastereoselectivity of these aldol condensations can be enhanced by use of the ketone **5**.[1] Reaction of racemic **5** with racemic aldehyde **2** furnishes a single (racemic) adduct **6**.

$$\underset{\mathbf{1}}{CH_3CH_2\overset{O}{\overset{\|}{C}}{-}\overset{OSi(CH_3)_3}{C}(CH_3)_2} \xrightarrow{\text{1) LDA, THF, } -70^\circ;\ \text{2) } \mathbf{2}} \mathbf{3}\ (75\%) + \mathbf{4}\ (25\%)$$

(**2**: CH₃CH₂, H, CH₃, CHO-substituted 2,2-dimethyl-1,3-dioxolane; **3** and **4**: CH₃CH₂, H, CH₃, OH, O, CH₃, CH₃, CH₃, OSi(CH₃)₃ aldol adducts bearing the dioxolane with CH₃, CH₃)

3 (75%)

4 (25%)

$$CH_3CH_2\overset{O}{\overset{\|}{C}}-\overset{OSi(CH_3)_3}{\overset{|}{C}}HC(CH_3)_3 \xrightarrow[2)\ \mathbf{2}]{1)\ LDA,\ THF,\ -70^\circ} \mathbf{6}$$

5 **6**

α-Sulfenylation of dialkyl sulfoxides. The deprotonation of unsymmetrical sulfoxides with LDA is highly regioselective. The resulting α-sulfinyl carbanion is sulfenylated with dimethyl disulfide in 65–70% yield (two examples reported).[2]

$$(CH_3)_2HCCH_2S(O)CH_3 \xrightarrow[65\%]{1)\ LDA,\ THF,\ -78^\circ;\ 2)\ CH_3S-SCH_3} (CH_3)_2HCCH_2S(O)CH_2SCH_3$$

An alternate route to α-thiomethyl methyl sulfoxides involves substitution of an α-chloromethyl sulfoxide with sodium thiomethoxide.[3]

$$R-S(O)-CH_2Cl \xrightarrow[64\%]{CH_3SNa,\ CH_3OH,\ 40^\circ} R-S(O)-CH_2SCH_3$$

Allenes.[4] 1-Alkenyl aryl sulfoxides (**1**) are lithiated at the α-position by LDA at -78° in THF (equation I). Both (Z)- and (E)-**1** give almost entirely (E)-**a**.

$$\text{(I)}\quad RCH_2CH{=}CH{-}S(O)Ar\ (\mathbf{1}) \xrightarrow{LDA} \left[RCH_2CH{=}C(Li){-}S(O)Ar\right]\ (\mathbf{a}) \xrightarrow[76\text{–}94\%]{R'X} RCH_2CH{=}C(R'){-}S(O)Ar\ (\mathbf{2})$$

Treatment of (E)-2-alkenyl sulfoxides (**2**) with lithium 2,2,6,6-tetramethylpiperidide induces elimination to allenes. An example is the conversion of **3** to **4**. This transformation permits conversion of aldehydes and ketones into terminal allenes.

$$\mathbf{3} \xrightarrow[60\%]{LiTMP,\ -100^\circ} \mathbf{4}$$

3 **4**

Aromatic rings by base-induced cyclizations.[5] A novel cyclization occurs when **1** or **3** is treated with 4–5 equiv. of LDA in THF at -78°. These reactions may involve intermediates such as A, which undergo a pericyclic ring closure under mild conditions. The protecting groups of **4** are removed with HBr in phenol at 100° to give pretetramide (**5**), a biosynthetic precursor of the tetracyclines.[6]

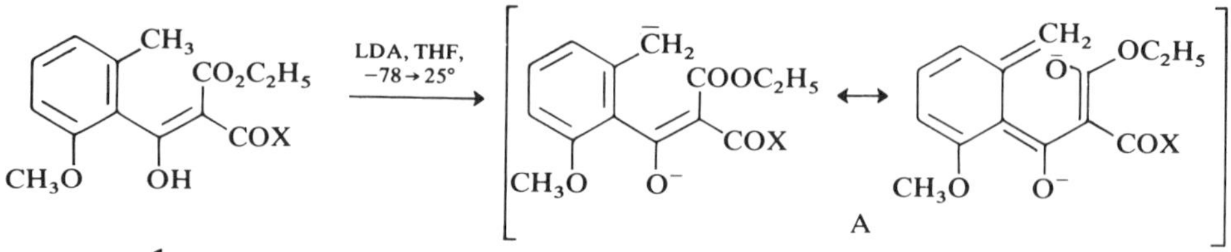

LDA, THF, −78→25°

A

1
a ($X = OCH_3$)
b [$X = NHC(CH_3)_3$]

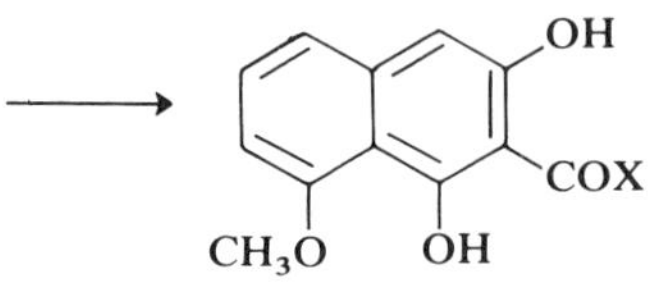

2a (68%)
2b (87%)

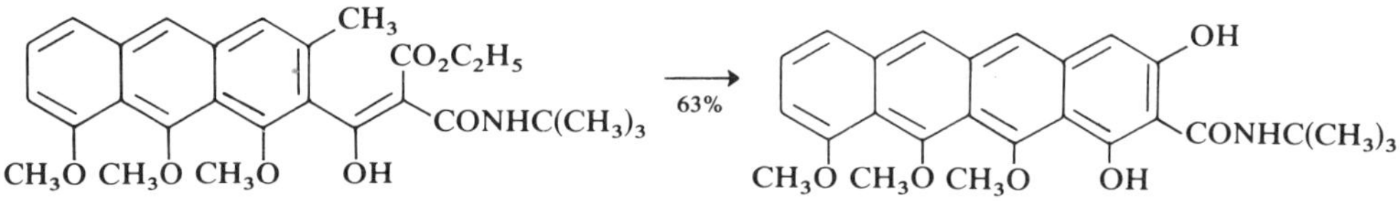

3

63%

4

HBr, phenol, 100°

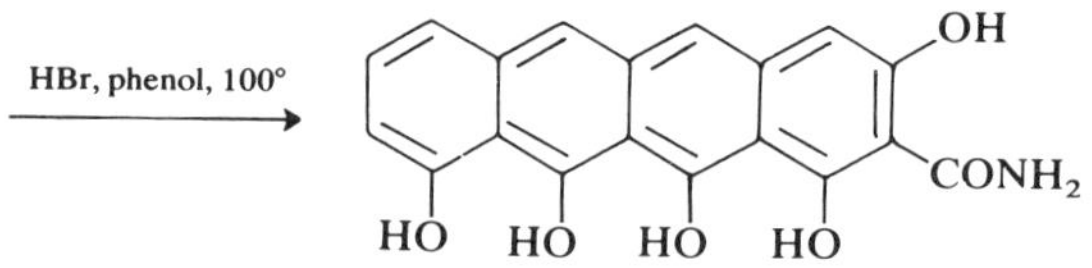

5

α-Lithio selenides and selenoxides. Full details are available[7,8] for deprotonation of alkyl phenyl selenides and selenoxides, and the application of the resulting anions for synthesis of olefins, dienes, and allylic alcohols (**6**, 335).

Reaction with phenyl vinyl selenide **(1).** Substitution as well as metalation is observed on reaction of **1** with alkyllithium reagents. LDA is the most effective reagent for metalation, but elimination can occur as well. Addition of HMPT exerts a marked effect, as shown in equation (I).[9]

$$\text{(I)}\quad \underset{\mathbf{1}}{CH_2{=}CHSeC_6H_5} \xrightarrow[]{\text{1) LDA; 2) } CH_3(CH_2)_9Br} \underset{\mathbf{2}}{CH_2{=}C(-SeC_6H_5)(CH_2)_9CH_3} + \underset{\mathbf{3}}{C_6H_5Se(CH_2)_9CH_3}$$

	2	3
HMPT–THF (1:20)	70%	23%
HMPT–THF (1:3)		65%

[1] C. H. Heathcock, M. C. Pirrung, C. T. Buse, J. P. Hagen, S. D. Young, and J. E. Sohn, *Am. Soc.*, **101**, 7077 (1979).
[2] P. Helquist and M. S. Shekhani, *ibid.*, **101**, 1057 (1979).
[3] H. C. J. Ottenheijm and R. M. J. Liskamp, *Tetrahedron Letters*, 2437 (1978).
[4] G. H. Posner, P.-W. Tang, and J. P. Mallamo, *ibid.*, 3995 (1978).
[5] J. A. Murphy and J. Staunton, *J.C.S. Chem. Comm.*, 1165, 1166 (1979).
[6] J. R. D. McCormick, S. Johnson, and N. O. Sjolander, *Am. Soc.*, **85**, 1692 (1963).
[7] H. J. Reich, F. Chow, and S. K. Shah, *ibid.*, **101**, 6638 (1979).
[8] H. J. Reich, S. K. Shah, and F. Chow, *ibid.*, **101**, 6648 (1979).
[9] M. Sevrin, J. N. Denis, and A. Krief, *Angew. Chem., Int. Ed.*, **17**, 526 (1978).

Lithium iodide, 5, 410–411; **7**, 208.

Decarboethoxylation. Attempted decarboethoxylation of the O-benzoyltartronate **1** with wet DMSO and NaCl as catalyst to form **2** proceeds in low yield (1–6%). The conversion is greatly improved by heating **1** with LiI in pyridine at reflux (12 hours); acidification gives the free acid **a**, which spontaneously loses CO_2 to give **2** in 82% yield.[1]

$C_6H_5CH_2$, CN, C_6H_5OCO, $COOC_2H_5$ (**1**) —DMSO, NaCl, 175–200°, 1–6%→ $C_6H_5CH_2$, CN, C_6H_5OCO, H (**2**)

1 —1) LiI, Py, Δ; 2) HCl→ [$C_6H_5CH_2$, CN, C_6H_5OCO, COOH] (**a**) —82%, $-CO_2$→ **2**

[1] J. Kristensen and S.-O. Lawesson, *Tetrahedron*, **35**, 2075 (1979).

Lithium isopropoxide, $(CH_3)_2CHOLi$ (**1**).

Phosphonoacetate cyclization. Intramolecular cyclization of keto phosphonates can be used for construction of macrocyclic α,β-unsaturated lactones. Stork's laboratory[1] found that use of lithium isopropoxide or lithium hexamethyldisilazide in THF containing 1% HMPT minimized formation of cyclic dilactones. Use of sodium or potassium counterions was much less satisfactory. An example is shown in equation (I).

(I) O–C(=O)–$CH_2P(O)(OC_2H_5)_2$ … CH_2CHO —**1**, THF, C_6H_6, 66%→ macrocyclic α,β-unsaturated lactone

Nicolaou *et al.*[2] used NaH in DME under conditions of high dilution to carry out the cyclization. Yields were typically 45–70%.

[1] G. Stork and E. Nakamura, *J. Org.*, **44**, 4010 (1979).
[2] K. C. Nicolaou, S. P. Seitz, M. R. Pavia, and N. A. Petasis, *ibid.*, **44**, 4011 (1979).

Lithium methoxy(trimethylsilyl)methylide $(CH_3)_3SiCH(Li)OCH_3$ (**1**). Mol. wt. 124.19.

This carbanion is prepared[1] by deprotonation of methoxymethyltrimethylsilane[2] with *sec*-butyllithium in THF (−70 → −25°).

Reductive nucleophilic acylation.[1] The reagent **1** condenses with aldehydes and ketones to give products (**2**) of carbonyl addition, which do not undergo *syn*-elimination of $—OSi(CH_3)_3$ *in situ*. However, treatment of **2** with KH in THF affords enol ethers (**3**) in excellent yields. The latter products are readily converted into aldehydes (**4**) on hydrolysis with 90% aqueous formic acid (~90% yield, equation I). The adducts (**2**) are desilylated to **5** with CsF in DMSO (equation II). In this case, **1** functions as an equivalent of the $—CH_2OCH_3$ anion.

(I) $R^1R^2C{=}O \xrightarrow{1} R^1R^2C(OH)—CH(Si(CH_3)_3)OCH_3$ (**2**) $\xrightarrow{KH} R^1R^2C{=}CH(OCH_3)$ (**3**) $\xrightarrow[\sim 90\%]{H_3O^+} R^1R^2CH—CHO$ (**4**)

(II) **2** $\xrightarrow{CsF,\ DMSO} R^1R^2C(OH)—CH_2OCH_3$ (**5**)

[1] P. Magnus and G. Roy, *J.C.S. Chem. Comm.*, 822 (1979).
[2] J. L. Spier, *Am. Soc.*, **70**, 4142 (1948).

Lithium naphthalenide, 2, 288–289; **3,** 208; **4,** 348–349; **5,** 468; **6,** 415; **8,** 305–306.

Reductive lithiation of dithioketals. Cyclopropanone dithioketals (**1**) are reduced by lithium naphthalenide to the sulfur-stabilized cyclopropyl anions **2** in quantitative yield. Lithium in HMPT is less effective.[1]

$$\underset{\mathbf{1}}{R^1R^2\text{-cyclopropane-}C(SC_6H_5)_2} + 2[C_{10}H_8]^{\dot{-}}Li^+ \xrightarrow[100\%]{THF,\ -70^\circ} \underset{\mathbf{2}}{R^1R^2\text{-cyclopropane-}C(SC_6H_5)Li} + LiSC_6H_5 + 2C_{10}H_8$$

$$\underset{\mathbf{3}}{R^1R^2C{=}C(SC_6H_5)_2} \xrightarrow{[C_{10}H_8]^{\dot{-}}Li^+} \underset{\mathbf{4}}{R^1R^2C{=}C(SC_6H_5)Li} \xrightarrow[56\%]{\text{cyclohexanone}} \underset{\mathbf{5}}{R^1R^2C{=}C(SC_6H_5)\text{-}C_6H_{10}OH}$$

$$\mathbf{4} \xrightarrow[49\%]{CO_2} \underset{\mathbf{6}}{R^1R^2C{=}C(SC_6H_5)COOH}$$

The same reaction can be used for conversion of ketene bis(phenylthio)acetals (**3**) to the vinyllithium reagents **4**.[2] This method may be more useful in some cases than deprotonation of vinyl sulfides. Thus **4** generated in this way reacts as expected with cyclohexanone and carbon dioxide. Both these reactions fail when **4** is generated by deprotonation of vinyl sulfides.

Lithium naphthalenide has also been used to convert tertiary benzylic sulfides, $(C_6H_5)_2C(R)SC_6H_5$, into lithium reagents, $(C_6H_5)_2C(R)Li$.[3]

[1] T. Cohen, W. M. Daniewski, and R. B. Weisenfeld, *Tetrahedron Letters*, 4665 (1978).
[2] T. Cohen and R. B. Weisenfeld, *J. Org.*, **44**, 3601 (1979).
[3] C. G. Screttas and M. Micha-Screttas, *ibid.*, **44**, 713 (1979).

Lithium phthalide, 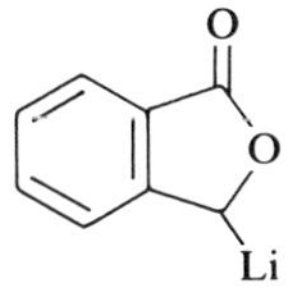 (**1**). This reagent is prepared by deprotonation of phthalide with LDA in THF at −60°.

Anthraquinones. Lithium phthalide reacts with benzynes to afford anthraquinones after acidification and air oxidation.[1] The benzynes can be generated *in situ* from substituted bromobenzenes.

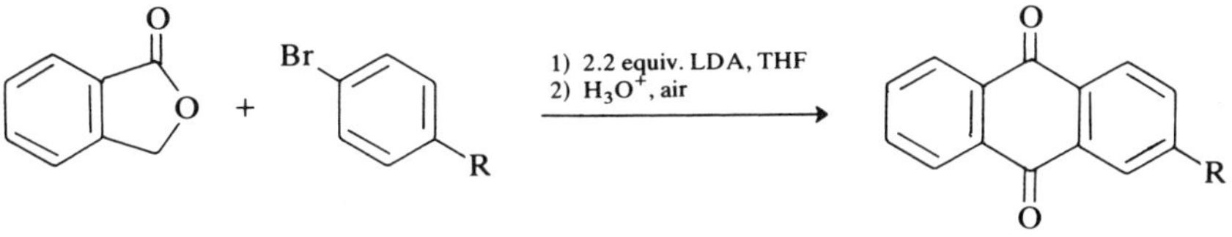

1) 2.2 equiv. LDA, THF
2) H_3O^+, air

(R = H, 74%)
(R = OCH_3, 45%)
(R = CH_3, 40%)

[1] P. G. Sammes and D. J. Dodsworth, *J.C.S. Chem. Comm.*, 33 (1979).

Lithium 2,2,6,6-tetramethylpiperidide (LiTMP), 4, 310–311; **5,** 417; **6,** 345–348; **7,** 213–215; **8,** 307–308.

Pentannelation[1] (**6,** 208–209). The deprotonation of vinyl sulfides such as **1** was originally thought to afford allyl anions (**2**).[2] Actually, the vinyl hydrogen is abstracted instead. The resulting anions **3** undergo cycloaddition reactions with unsaturated

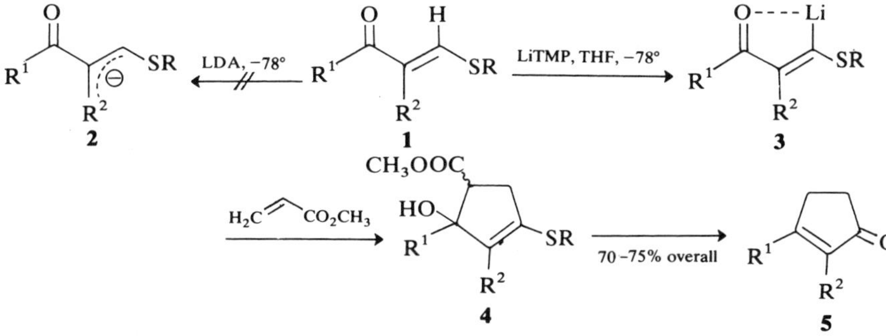

LDA, −78° · LiTMP, THF, −78° · $H_2C{=}CHCO_2CH_3$ · 70–75% overall

2 **1** **3** **4** **5**

esters to adducts **4**, which are hydrolyzed and decarboxylated to cyclopentenones **5**. The overall sequence proceeds in 70–75% yield. The vinyllithium derivative **3** serves as an equivalent for **6** in this process.

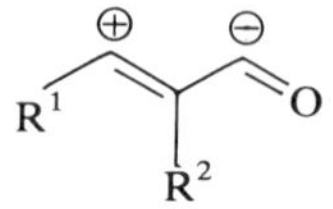

R^1 ⊕ ⊖ O R^2 **6**

Example:

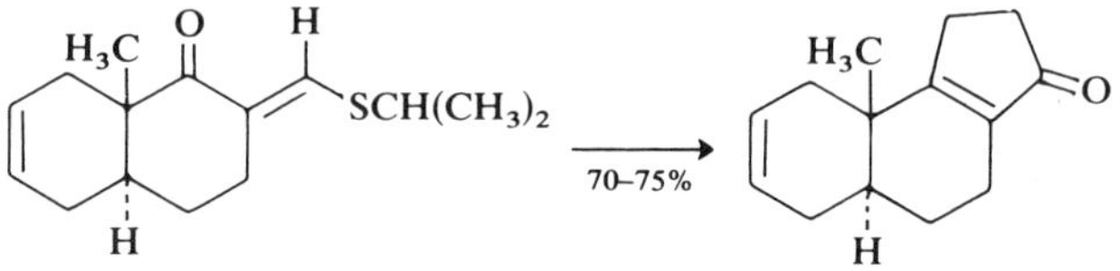

H_3C O H $SCH(CH_3)_2$ H $\xrightarrow{70–75\%}$ H_3C O H

[1] J. P. Marino and L. C. Katterman, *J.C.S. Chem. Comm.*, 946 (1979).
[2] J. P. Marino and W. B. Mesberger, *Am. Soc.*, **96**, 4051 (1974).

Lithium triethylborohydride, 4, 313–314; **6**, 348–349; **7**, 215–216; **8**, 309–310.

Reactions with organometallic substrates.[1] One useful reaction of this and other trialkylborohydrides is cleavage of metal carbonyl dimers to metal carbonyl anions (equation I). The by-products are H_2 and $B(C_2H_5)_3$. Another useful reaction is the generation of anionic formyl complexes (equation II). Sulfur (S_8) can be cleaved to give either Li_2S or Li_2S_2 (equations III and IV). Disulfides are cleaved by this reaction to lithium mercaptides.

(I) $$(CO)_4Co{-}Co(CO)_4 \xrightarrow{2Li(C_2H_5)_3BH} 2Li[Co(CO)_4] + 2B(C_2H_5)_3 + H_2$$

(II) $$L_nMC{=}O \xrightarrow{Li(C_2H_5)_3BH} Li^+L_nM^-CHO + B(C_2H_5)_3$$

(III) $$S \xrightarrow{2Li(C_2H_5)_3BH} Li_2S + 2B(C_2H_5)_3 + H_2$$

(IV) $$2S \xrightarrow{2Li(C_2H_5)_3BH} Li_2S_2 + 2B(C_2H_5)_3 + H_2$$

Reduction of methoxyethoxymethyl (MEM) esters. MEM esters are more easily reduced by lithium triethylborohydride than other esters. Selective reduction of a less hindered MEM ester is also possible.[2]

Example:

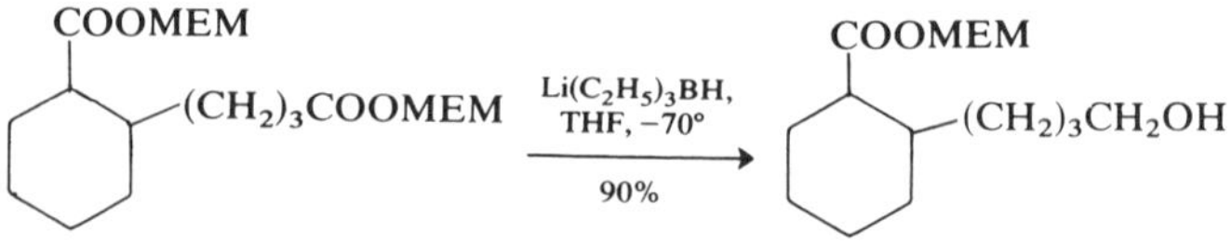

COOMEM $(CH_2)_3COOMEM$ $\xrightarrow[90\%]{Li(C_2H_5)_3BH,\ THF,\ -70°}$ COOMEM $(CH_2)_3CH_2OH$

[1] J. A. Gladysz, *Aldrichim. Acta*, **12**, 13 (1979); *idem*, *Tetrahedron*, **35**, 2329 (1979).
[2] R. E. Ireland and W. J. Thompson, *Tetrahedron Letters*, 4705 (1979).

Lithium triethylborohydride–Aluminum *t*-butoxide complex, $LiB(C_2H_5)_3H-Al-[OC(CH_3)_3]_3$ (**1**). The complex is formed rapidly in THF by reaction of lithium tri-*t*-butoxyaluminohydride with triethylborane.

Reductive cleavage of cyclic ethers.[1,2] This complex is effective for reductive cleavage of cyclic ethers. The order of reactivity is epoxide > oxetane > tetrahydrofurane > tetrahydropyrane > oxepane. It is less effective for cleavage of acyclic ethers, except for methyl ethers. The reaction involves formation of a complex of the ethereal oxygen with aluminum *t*-butoxide followed by S_N2 displacement with lithium triethylborohydride. Steric and electronic factors are involved, but yields are > 90% in favorable cases.

[1] H. C. Brown, S. Krishnamurthy, J. L. Hubbard, and R. A. Coleman, *J. Organometal. Chem.*, **166**, 281 (1979).
[2] S. Krishnamurthy and H. C. Brown, *J. Org.*, **44**, 3678 (1979).

M

Magnesium bromide etherate, 1, 629–630; **7**, 218–220.

Wagner–Meerwein rearrangement of β-lactones.[1] β-Lactones, which are readily available from aldehydes and carboxylic acids via 3-hydroxy carboxylic acids,[2] are rearranged to γ-lactones by $MgBr_2$ in ether. The stereochemistry of the 3,4-bond is preserved in the process and migration occurs from a position *anti* to the C—O bond that is broken. In unsymmetrically substituted systems the group that migrates depends on the configuration of the system. The isopropyl group in **2** has nearly unrestricted conformational mobility; hence a H shift predominates (78%). In the *cis*-isomer **3**, steric hindrance between the phenyl and isopropyl groups forces the system to adopt the conformation formulated, and CH_3-migration predominates. Systems that are incapable of ring expansion are decarboxylated under these conditions.

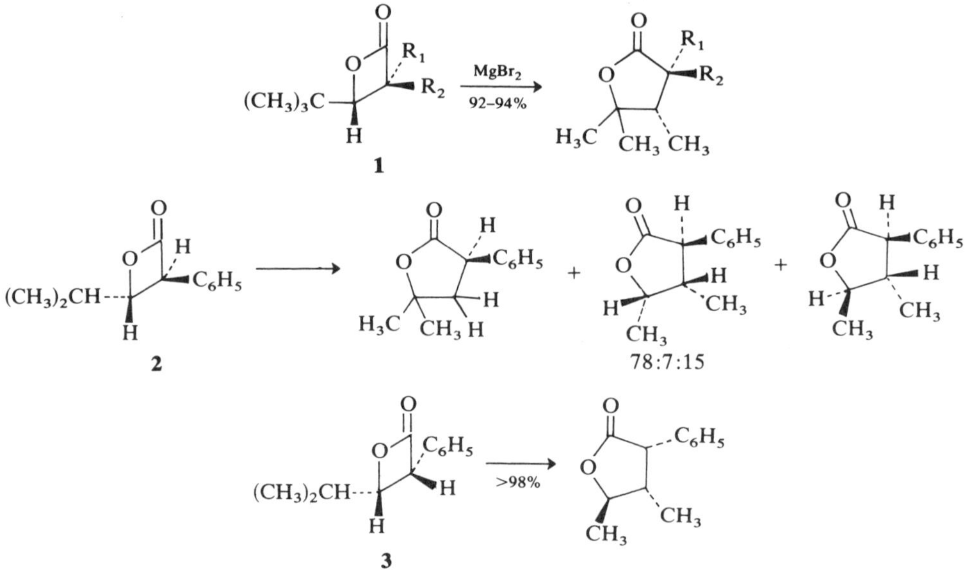

$(CH_3)_3C$ (**1**, R_1, R_2) $\xrightarrow[92\text{–}94\%]{MgBr_2}$ γ-lactone (R_1, R_2, H_3C, CH_3, CH_3)

$(CH_3)_2CH$ (**2**, H, C_6H_5) ⟶ γ-lactones, 78:7:15

$(CH_3)_2CH$ (**3**, C_6H_5, H) $\xrightarrow{>98\%}$ γ-lactone (C_6H_5, CH_3, CH_3)

[1] J. Mulzer and C. Brüntrup, *Angew. Chem., Int. Ed.*, **18**, 793 (1979).
[2] J. Mulzer, A. Pointner, A. Chucholowski and G. Brüntrup, *J.C.S. Chem. Comm.*, 52 (1979).

Manganese(II) chloride, 7, 222.

Organomanganese(II) chlorides; ketone synthesis.[1] The synthesis of ketones by RMnI reagents (**8**, 312–313) is limited by the requirement of ether as solvent. However, RMnCl reagents can be prepared from Grignard reagents, RLi, and

$MnCl_2$. The chloromanganese reagents can be used in THF, a more satisfactory solvent for the synthesis of ketones from mixed carbonic–carboxylic anhydrides (prepared by the reaction of carboxylic acids with ethyl chloroformate).

Examples:

$$C_2H_5MnCl + C_6H_5COOCOOC_2H_5 \xrightarrow[89\%]{THF,\ -20\rightarrow 20^\circ} C_2H_5\overset{O}{\overset{\|}{C}}C_6H_5$$

$$C_2H_5\overset{CH_3}{\overset{|}{C}}HCH_2MnCl + (CH_3)_2C{=}CHCOOCOOC_2H_5 \xrightarrow[88\%]{} C_2H_5\overset{CH_3}{\overset{|}{C}}HCH_2\overset{O}{\overset{\|}{C}}CH{=}C(CH_3)_2$$

$$(CH_3)_2C{=}CHCH_2MnCl + C_6H_5COOCOOC_2H_5 \xrightarrow[92\%]{} C_6H_5\overset{O}{\overset{\|}{C}}{-}\underset{CH_3}{\underset{|}{\overset{CH_3}{\overset{|}{C}}}}{-}CH{=}CH_2$$

[1] G. Cahiez, A. Alexakis, J. F. Normant, *Syn. Comm.*, **9**, 639 (1979).

Manganese(II) iodide, 7, 222; **8**, 312–313.

α-Acetylenic ketones (**8**, 313). Details are available for use of an RMnI reagent for synthesis of 2,2-dimethyl-4-nonyne-3-one (equation I).[1]

$$\text{(I)}\quad n\text{-}C_4H_9C{\equiv}CH \xrightarrow{1)\ CH_3Li,\ ether;\ 2)\ MnI_2} n\text{-}C_4H_9C{\equiv}CMnI \xrightarrow[78\%\ overall]{(CH_3)_3CCOCl} n\text{-}C_4H_9C{\equiv}C\overset{O}{\overset{\|}{C}}C(CH_3)_3$$

[1] G. Cahiez, A. Alexakis, and J. F. Normant, *Org. Syn.*, submitted (1979).

Manganese(III) sulfate, $Mn_2(SO_4)_3$. This reasonably stable compound is prepared by oxidation of $MnSO_4 \cdot 4H_2O$ with $KMnO_4$ in dilute H_2SO_4.[1]

Oxidation of arenes.[2] This oxidant is somewhat superior to CrO_3 for oxidation of arenes to quinones but less efficient than ceric ammonium sulfate (**8**, 80–81). An example is the oxidation of naphthalene to 1,4-naphthoquinone in 75% yield.

[1] A. R. J. P. Ubhelohde, *J. Chem. Soc.*, 1605 (1935).
[2] M. Periasamy and M. V. Bhatt, *Tetrahedron Letters*, 4561 (1978).

Menthoxyaluminum dichloride (1). Mol. wt. 253.15. This reagent is prepared[1] by reaction of *l*-menthol and ethylaluminum dichloride in heptane at −78°.

1 (menthyl–$OAlCl_2$; substituents CH_3, $OAlCl_2$, H_3C, CH_3)   **2** (stereoisomer; CH_3, $OAlCl_2$, H_3C, CH_3)

Asymmetric Diels–Alder reactions. The Diels–Alder adduct **3** is produced in 72% optical purity by reaction of methacrolein and cyclopentadiene in the presence of this catalyst.[2] Substitution of the chiral alkoxyaluminum dichloride **2** for **1** leads to

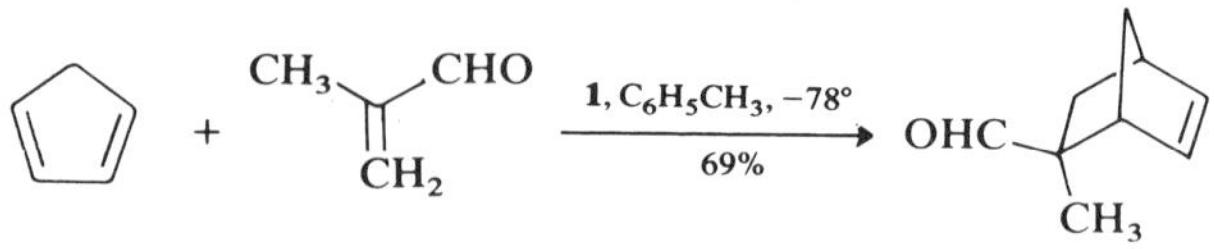

CH3 CHO; CH2; + ; 1, $C_6H_5CH_3$, −78°; 69%; OHC; CH3; **3**

the formation of the enantiomeric aldehyde (66% optical purity). Cycloadditions of acrolein and methyl acrylate, however, proceed with little asymmetric induction in reactions with these catalysts.

[1] Y. Hayakawa, T. Fueno, and J. Furukawa, *J. Polymer. Sci.*, *A*-1, **5**, 2099 (1967).
[2] S.-I. Hashimoto, N. Komeshima, and K. Koga, *J.C.S. Chem. Comm.*, 437 (1979).

(−)-10-Mercaptoisoborneol, (structure: CH3, CH3, OH, CH_2SH) **(1).** Mol. wt. 186.32, m.p. 76–78°, α_D −55.4°. The reagent is prepared by reduction of (+)-10-camphorsulfonyl chloride with $LiAlH_4$.

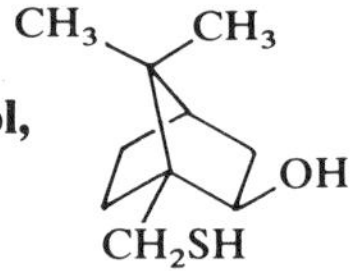

Diastereospecific reaction of a Grignard reagent. This reagent has been used as the chiral adjunct in place of 4,6,6-trimethyl-1,3-oxathiane (**8**, 508–509) in a synthesis of atrolactic acid methyl ether (**6**) in 97±2% enantiomeric excess (equation I).[1]

(I) **1** —HCHO→ **2** —1) *sec*-BuLi; 2) C_6H_5CHO; 3) Oxid.; 65%→ **3** (α_D −103.2°) (S–COC_6H_5) —CH_3MgBr, 92–100%→

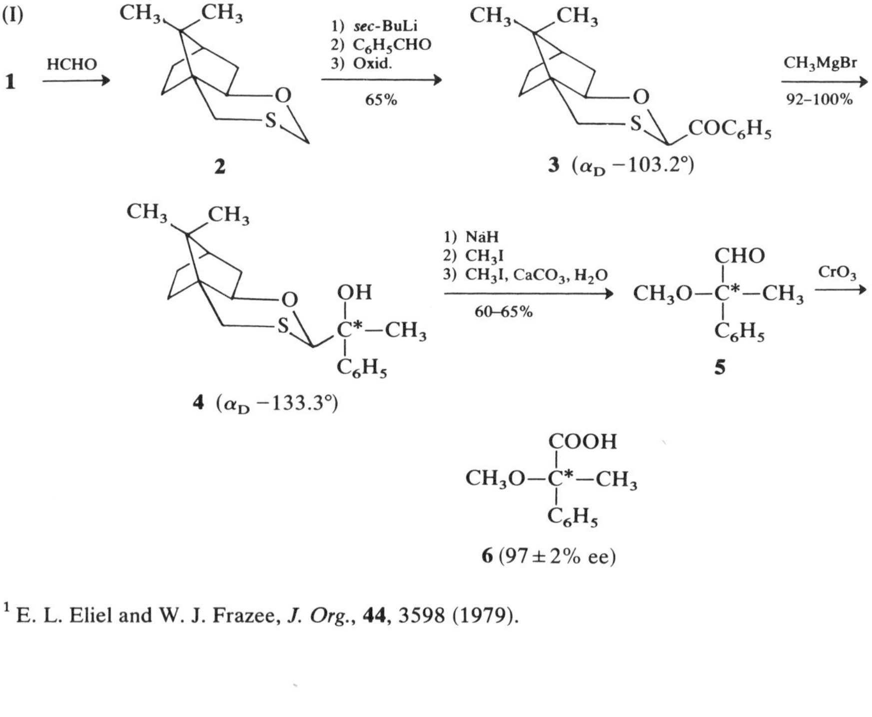

4 (α_D −133.3°) (S–C*(OH)(CH3)C_6H_5) —1) NaH; 2) CH_3I; 3) CH_3I, $CaCO_3$, H_2O; 60–65%→ CH_3O—C*(CHO)(C_6H_5)—CH_3 **5** —CrO_3→

CH_3O—C*(COOH)(C_6H_5)—CH_3

6 (97±2% ee)

[1] E. L. Eliel and W. J. Frazee, *J. Org.*, **44**, 3598 (1979).

Mercury (Hg). At. wt. 200.59, m.p. −38.87, b.p. 356.9°.

Reduction of α,α′-dibromo ketones. Reduction of 2,4-dibromo-2,4-dimethylpentane-3-one (**1**) with ultrasonically dispersed mercury in the presence of a ketone such as acetone leads to a 4-methylene-1,3-dioxolane (**2**) in about 50% yield. A similar reaction has been observed in the reduction of dibromo ketones with zinc in dimethylformamide (**5**, 221).[1] 2-Oxyallyl cations such as **a** have been invoked as intermediates in reductions of dihalo ketones with various metals.

Hg(0); CH_3COCH_3, 49%

[1] A. J. Fry, G. S. Ginsburg, and R. A. Parente, *J.C.S. Chem. Comm.*, 1040 (1978).

Mercury(II) acetate, 1, 644–652; **2,** 264–267; **3,** 194–196; **4,** 319–323; **5,** 424–427; **6,** 358–359; **7,** 222–223; **8,** 315–316.

Carbopenams. The β-lactam **1** is cyclized to a bicyclic β-lactam (**2**) by reaction with a number of electrophiles (I_2, C_6H_5SBr), but highest yields are obtained with mercury(II) acetate. Reduction of **2** leads to **3**, which has the ring skeleton of the thienamycin antibiotics.[1]

$Hg(OAc)_2$, ~100%; $NaBH_4$, 75%

[1] T. Aida, R. Legault, D. Dugat, and T. Durst, *Tetrahedron Letters*, 4993 (1979).

Mercury(II) chloride, 1, 652–654; **5,** 427–428; **6,** 359.

Polyene cyclization. Cyclization of dienes by oxymercuration[1] ordinarily is not stereoselective. However, the presence of an allylic hydroxyl group[2] can increase the stereoselectivity. Thus oxymercuration–demercuration of linalool (**1**) under the best conditions leads mainly to the iridanols **2** and **3**, which constitute 85% of the

1) $HgCl_2$, $(CH_3)_3COH$
2) $NaBH_4$
~55%

HO CH3 / CH2 / C(CH3)2 / CH3 → (~20%) OH CH3 / CH3 / C(CH3)2 / CH3 OH

4 **5**

isolable products. The reaction was used to prepare the fungal metabolite cyclonerodiol (**5**) from nerolidol (**4**).[3]

[1] Review: R. C. Larock, *Angew. Chem., Int. Ed.*, **17**, 27 (1978).

[2] H. B. Henbest and B. Nicholls, *J. Chem. Soc.*, 227 (1959); H. B. Henbest and R. S. McElkinney, *ibid.*, 1834 (1959).

[3] Y. Matsuki, M. Kodama, and S. Itô, *Tetrahedron Letters*, 2901 (1979).

Mercury(II) nitrite, $Hg(NO_2)_2$. Mol. wt. 292.60.

Cyclic nitro olefins. Some years ago Bachman and Whitehouse[1] reported the Markownikoff addition of mercury(II) nitrite to olefins. The salt is generated *in situ* from mercury(II) chloride and sodium nitrite. Because of the requirement for an aqueous medium the rate is slow; the reaction is also subject to steric effects.

This reaction has now been shown to provide a general route to nitro olefins, which had previously not been readily available. Thus on treatment with sodium hydroxide (or a tertiary amine) the adducts are converted into nitro olefins in high yield. The overall process is formulated in equation (I).[2]

(I) cyclohexene $\xrightarrow[80\%]{Hg(NO_2)_2,\ H_2O}$ **1** (NO_2, HgCl) $\xrightarrow[>98\%]{NaOH,\ H_2O}$ **2** (NO_2) + Hg + NaCl

In general, this demercuration step proceeds in almost quantitative yield; the nitromercuration step is variable with respect to yield and rate. Even so a wide range of nitro olefins is available by this two-step process. A few products with the yields are shown in the formulas. One point of interest is the regioselectivity of nitromercuration.

NO_2 (77%) NO_2, CH_3 (80%) NO_2, CN (78%) $n\text{-}C_4H_9C(NO_2){=}CH_2$ (71%)

The nitro alkenes are useful for further transformations, 12 of which were reported in the case of **2**. Thus **2** is converted into cyclohexenone in 70% yield by application of the Nef reaction (treatment with potassium *t*-butoxide and then with sulfuric acid). The nitro group can be replaced by a cyano group by addition of HCN followed by treatment with DBU (85% overall yield). Other particularly useful reactions of nitro olefins are conjugate addition and Diels–Alder cycloaddition.

[1] G. B. Bachman and M. L. Whitehouse, *J. Org.*, **32**, 2303 (1967).
[2] E. J. Corey and H. Estreicher, *Am. Soc.*, **100**, 6294 (1978).

Mercury(II) oxide, 1, 655–658; **2**, 267–268; **4**, 323–324; **5**, 428; **6**, 360; **7**, 224; **8**, 316.

Hunsdiecker reaction. The Cristol–Firth modification (**1**, 657; **6**, 428) of this reaction generally proceeds in improved yield if irradiated with a 100-W bulb.[1]

[1] A. I. Meyers and M. P. Fleming, *J. Org.*, **44**, 3405 (1979).

Mercury(II) oxide–Tetrafluoroboric acid, $HgO{-}HBF_4$, white, hygroscopic solid.

Diamination of alkenes.[1] This reaction can be effected with a primary or secondary aromatic amine (2 equiv.) and 1 equiv. of **1** in refluxing THF (4 hours).

$$R^1R^2C{=}CR^3R^4 + \underset{\mathbf{1}}{HgO\cdot HBF_4} + \underset{\mathbf{2}}{HN(R^5)Ar} \xrightarrow{\text{THF, }66^\circ} \left[R^1R^2C(NR^5Ar){-}CR^3R^4(HgBF_4)\right] \xrightarrow[60\text{–}90\%]{HN(R^5)Ar} \underset{\mathbf{3}}{R^1R^2C(NR^5Ar){-}CR^3R^4(NR^5Ar)} + Hg(0)$$

[1] J. Barluenga, L. Alonso-Cires, and G. Asensio, *Synthesis*, 962 (1979).

Mercury(II) thiocyanate, $Hg(SCN)_2$. Mol. wt. 316.78. Supplier: Alfa.

Addition of HSCN to C≡C. Thiocyanic acid (HSCN), formed *in situ* from $(n\text{-}C_4H_9)_4N^+SCN^-$ and a strong acid (H_2SO_4, HCl, or HBF_4), adds to acetylenes in the presence of Lewis acids such as $Hg(SCN)_2$ or $HgSO_4$. $Hg(SCN)_2$ is the preferred catalyst for this reaction, which affords 2-thiocyanatoalkenes in high yield. The corresponding methyl ketone is a side reaction product.

$$C_7H_{15}C{\equiv}CH \xrightarrow[\sim 100\%]{(n\text{-}C_4H_9)_4N^+SCN^-,\ H_2SO_4,\ HgSO_4} C_7H_{15}C(SCN){=}CH_2 + C_7H_{15}\overset{O}{\overset{\|}{C}}CH_3 \quad 28:72$$

$$\xrightarrow{Hg(SCN)_2} \quad 86:14$$

[1] M. Gifford and J. Cousseau, *J.C.S. Chem. Comm.*, 1026 (1979).

Mercury(II) trifluoroacetate, 3, 195; **4,** 325; **6,** 360; **8,** 316–317.

Cyclization. Treatment of the unsaturated enol phosphate ester **1** with 1 equiv. of mercury(II) trifluoroacetate in nitromethane at 0° followed by aqueous sodium chloride produced the mercurated bicyclic keto ester **2** in 60% yield along with 20% of the monocarbocyclic product **3**. Keto ester **2** is an intermediate in a total synthesis of aphidicolin (**4**). This reaction is the first example of the use of a mercury(II) salt for simultaneous construction of two carbocyclic rings.[1]

1) $Hg(OCOCF_3)_2$, CH_3NO_2 2) aq. NaCl

1 → **2** (60%) + **3** (20%)

4

Hydrolysis of vinyl chlorides. One step in the Wichterle annelation with 1,3-dichloro-*cis*-2-butene (**1,** 214–215; **2,** 111–112) involves hydrolysis of an intermediate vinyl chloride to a ketone. This reaction has been conducted with conc. H_2SO_4. A new method involves reaction with mercury(II) trifluoroacetate, which can result in either a methyl or an ethyl ketone depending on the solvent. For example, hydrolysis of **1** with the mercury salt in CH_3NO_2, CH_2Cl_2, or HOAc gives only the 1,5-diketone **2** in 90–97% yield; hydrolysis in CH_3OH gives **3** as the major product in 83% yield.[2]

(I) **1** —$Hg(OCOCF_3)_2$→ **2** and/or **3**

	2	**3**
CH_3NO_2	~95%	
CH_3OH	6%	83%

Brominative cyclization of dienes.[3] Reaction of homogeranic acid (**1**) with this mercury salt in CH_3NO_2 results in a mercury-substituted lactone **2**, which can be converted into either **4** or **5** via **3**. Lower yields are obtained when the mercury salt is replaced by silver tetrafluoroborate (15%).[2]

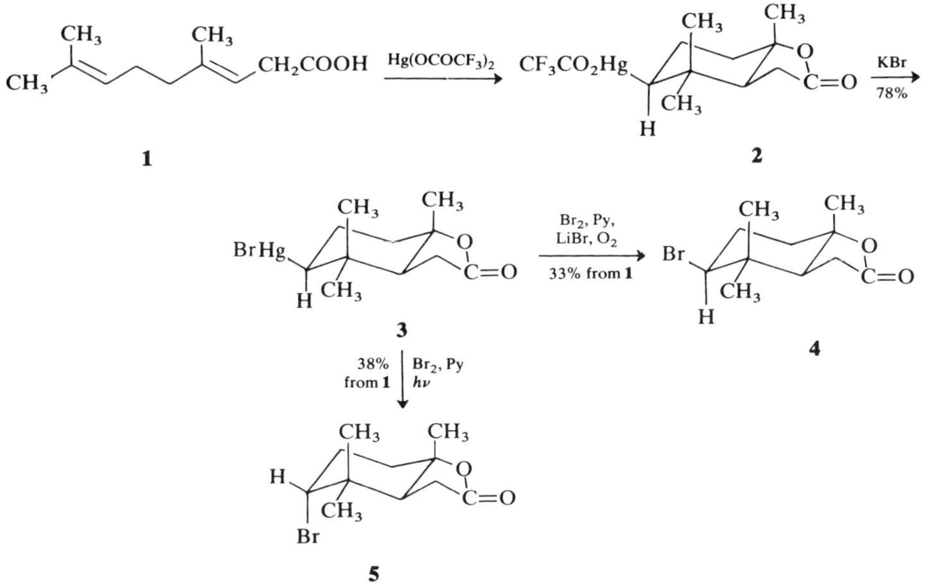

CH_3 CH_3 CH_3 CH_2COOH $\xrightarrow{Hg(OCOCF_3)_2}$ CF_3CO_2Hg CH_3 CH_3 O C=O CH_3 H $\xrightarrow[78\%]{KBr}$

1 **2**

BrHg CH_3 CH_3 O C=O H CH_3 $\xrightarrow[33\% \text{ from } \mathbf{1}]{Br_2, Py, LiBr, O_2}$ Br CH_3 CH_3 O C=O CH_3 H

3 **4**

38% from **1** | Br_2, Py, $h\nu$

H CH_3 CH_3 O C=O CH_3 Br

5

The method has been used to synthesize *dl*-3β-bromo-8-epicaparrapi oxide (**6**), a natural product in a marine organism, from a derivative of geranylacetone (scheme I).[4]

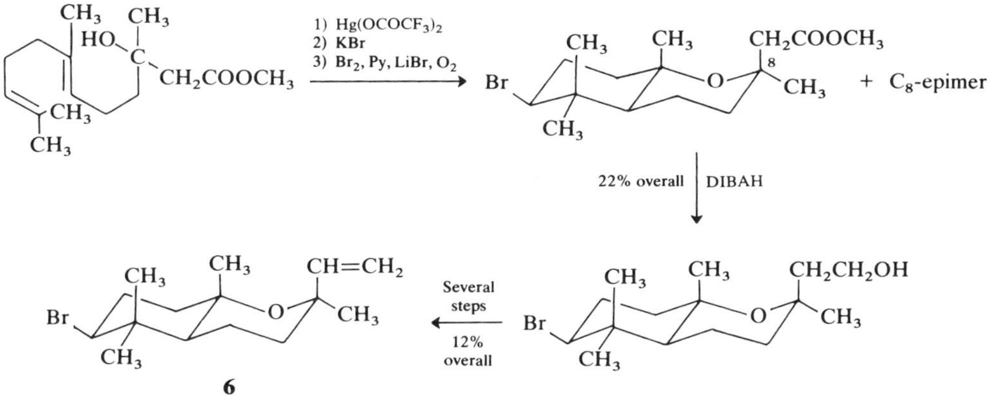

CH_3 CH_3 HO C CH_2COOCH_3 CH_3 CH_3 → 1) $Hg(OCOCF_3)_2$ 2) KBr 3) Br_2, Py, LiBr, O_2 → Br CH_3 CH_3 CH_2COOCH_3 8 O CH_3 CH_3 + C_8-epimer

22% overall | DIBAH

Br CH_3 CH_3 CH_2CH_2OH O CH_3 CH_3 → Several steps, 12% overall → Br CH_3 CH_3 CH=CH_2 O CH_3 CH_3

6

Scheme (I)

Azidomercuration. Azidomercuration of the unsaturated sugar **1** with NaN_3 and mercury(II) acetate proceeds in low yield (20%). However, use of NaN_3 and mercury(II) trifluoroacetate (1 : 3 molar ratio) in THF–H_2O (1 : 1) at 60° for 48 hours results in quantitative mercuration to give **2** and **3**. The reaction affords a high yield route to the amino sugar **4**.[5]

$CH_2{=}CH$ (furanose, OBzl, O–C(CH_3)$_2$–O) **1** $\xrightarrow[\sim 100\%]{1)\ NaN_3,\ Hg(OCOCF_3)_2;\ 2)\ KBH_4}$ CH_2N_3–CH_2 (furanose, OBzl, O–C(CH_3)$_2$–O) **2** + N_3–CH(CH_3) (furanose, OBzl, O–C(CH_3)$_2$–O) **3** (20 : 80)

3 $\xrightarrow{H_2,\ Pd/C}$ H_2N–C(H)(CH_3) (furanose, OH, O–C(CH_3)$_2$–O) **4**

[1] E. J. Corey, M. A. Tius, and J. Das, *Am. Soc.*, **102**, 1742 (1980).
[2] H. Yoshioka, K. Takasaki, M. Kobayashi, and T. Matsumoto, *Tetrahedron Letters*, 3489 (1979).
[3] T. R. Hoye and M. J. Kurth, *J. Org.*, **44**, 3461 (1979).
[4] *Idem, ibid.*, **43**, 3693 (1978).
[5] S. Czernecki, C. Georgoulis, and C. Provelenghiou, *Tetrahedron Letters*, 4841 (1979).

Mesityloxytris(dimethylamino)phosphonium azide, 2,4,6-$(CH_3)_3C_6H_2$–$O\overset{+}{P}[N(CH_3)_2]_3\,N_3^-$ (**1**). Mol. wt. 340.42, m.p. 110–112°, soluble in organic solvents. The salt is prepared by reaction of 2,4,6-trimethylphenol with hexamethylphosphorous triamide, $P[N(CH_3)_2]_3$, and then with sodium azide; yield (pure) 68%.[1]

Glycosyl azides. This reagent has been used in place of the hazardous silver azide[2] for a synthesis of glycosyl azides that permits isolation of the kinetic products. The free anomeric hydroxyl group is converted into the alkoxytris(dimethylamino)phosphonium chloride by reaction with hexamethyl-phosphorous triamide and carbon tetrachloride (**5**, 259).[3] The anomeric configuration of these salts is *trans* with respect to the substituent at C_2.[4] The salt is then allowed to react with **1** with inversion of configuration to give the glycosyl azide (equation I).

(I) [Structure: sugar with —O and OR substituents, ∼OH → 2, $\overset{+}{O}P[N(CH_3)_2]_3$ Cl^- → (1, 65–80% overall) → 3, with N_3 and OR]

2 3

Deoxy sugars. The reagent can also be used for reductive elimination of a hydroxyl group. Thus the alkoxytris(dimethylamino)phosphonium salts are reduced in ~90% yield by lithium triethylborohydride (equation I).[5]

$$\text{(I)} \qquad RO\overset{+}{P}[N(CH_3)_2]_3Cl^- \xrightarrow[\sim 90\%]{LiBH(C_2H_5)_3,\ THF, 66°} RH + O{=}P[N(CH_3)_2]_3 + LiCl$$

[1] F. Chretien, B. Castro, and B. Gross, *Synthesis*, 937 (1979).
[2] B. Paul and W. Korytnyk, *Carbohydrate Res.*, **67**, 457 (1978).
[3] B. Castro, Y. Chapleur, B. Gross, and C. Selve, *Tetrahedron Letters*, 5001 (1972).
[4] R. A. Boigegrain, F. Chretien, B. Castro, and B. Gross, *J. Chem. Res.* (*M*), 1929 (1978).
[5] P. Simon, J.-C. Ziegler, and B. Gross, *Synthesis*, 951 (1979).

p-Methoxybenzenesulphonyl chloride, $CH_3O-C_6H_4-SO_2Cl$ **(1).** Mol. wt. 206.65, m.p. 40–43°. Supplier: Aldrich.

Protection of imidazole NH.[1] The imidazole NH group of the histidine derivative **2** is protected as the *p*-methoxysulfonamide **3** by reaction with **1** and Et_3N. This sulfonamide derivative is stable to the acidic conditions required to remove the

[Scheme: **2** (imidazole N–H, $-CH_2CH(HNCO_2C(CH_3)_3)-COOH$) → $(C_2H_5)_3N$, **1** → **3** (imidazole N–$SO_2-C_6H_4-OCH_3$, $-CH_2CH(HN-CO_2C(CH_3)_3)-COOH$); **3** → TFA, $(CH_3)_2S$, ~100% → **2**]

N-*t*-butoxycarbonyl group (*e.g.*, 25% HBr–HOAc, TFA–anisole), but is removed by TFA in the presence of dimethyl sulfide within 40–60 minutes at 23°. The N^{im}-tosyl protecting group[2] is also removed under these conditions, but at a slower rate. The *p*-methoxysulfonyl group is also removed by N-hydroxybenzotriazole or by 1 *N* NaOH within 1 hour, but only partially by 80% hydrazine hydrate after 24 hours.

[1] K. Kitagawa, K. Kitade, Y. Kiso, T. Akita, S. Funakoshi, N. Fujii, and H. Yajima, *J.C.S. Chem. Comm.*, 955 (1979).
[2] T. Fuji and S. Sakakibara, *Bull. Chem. Soc. Japan*, **47**, 3146 (1974).

Methoxybutenynylcopper, $CuC{\equiv}C{-}\overset{H}{C}{=}\overset{H}{C}OCH_3$ **(1).** Mol. wt. 144.65, air sensitive, yellow solid. The cuprous acetylide is prepared by treatment of methoxybutenyne (Aldrich) with N-ethylpiperidine and cuprous iodide in DMF (75% yield).[1]

Coupling with aryl halides.[2] The reagent undergoes coupling with aryl iodides in pyridine at 85–125° (10–12 hours). Air must be rigorously excluded. The adducts (**3**) resist mild acid hydrolysis, but can be converted to a mixture of isomeric enol ethers (**4**) by KOH in refluxing CH_3OH.

$$\mathbf{2} + \mathbf{1} \xrightarrow[60\text{–}95\%]{Py} \mathbf{3} \xrightarrow{KOH,\ CH_3OH} \mathbf{4}$$

2 ($R^1 = H, CH_3, OH, NO_2$; $R^2 = H, OH, NO_2$) (aryl iodide, I, R^1, R^2)

3 ($C{\equiv}CCH{=}CHOCH_3$, R^1, R^2)

4 (OCH_3, $CH_2CH(OCH_3)_2$, R^1, R^2)

One use of the adducts is for synthesis of indoles. For example, **5** can be converted into **7**.

$$\mathbf{5} \xrightarrow{H^+,\ CH_3OH} \mathbf{6} \xrightarrow[75\%]{Al(Hg)} \mathbf{7}$$

5 (OCH_3, $CH_2CH(OCH_3)_2$, NO_2)

6 (O, $CH_2CH(OCH_3)_2$, NO_2)

7 (N, H, $CH_2CH(OCH_3)_2$)

Some other uses of **1** are formulated in the examples.

Examples:

$$C_6H_5COCl \xrightarrow[45\%]{\mathbf{1}} C_6H_5COC{\equiv}CCH{=}CHOCH_3$$

$$CH_2{=}CHCH_2Cl \xrightarrow[63\%]{\mathbf{1}} CH_2{=}CHCH_2C{\equiv}CCH{=}CHOCH_3$$

[1] C. E. Castro, E. J. Gaughan, and D. C. Owsley, *J. Org.*, **31**, 4071 (1966).
[2] G. A. Kraus and K. Frazier, *Tetrahedron Letters*, 3195 (1978).

2-Methoxy-1,3-dioxolane (1). Mol. wt. 120.11, b.p. 129°.

Preparation[1]:

$$HOCH_2CH_2OH + (CH_3O)_3CH \xrightarrow[70\%]{H_2SO_4} \mathbf{1}$$

1 (H, CH_3O, O, O)

Ethylene ketals. The exchange reaction between dialkyl acetals or dialkyl ketals and 1,2-glycols as a route to cyclic acetals and ketals was first described by Delépine.[2] This method in combination with Claisen's orthoester ketalization procedure (**1**, 1206) constitutes a convenient route to ethylene ketals (**1**, 376). The mixed orthoester **1** is probably the actual reagent involved in the Delépine method. It is convenient to prepare and use for ketalization of carbonyl compounds under mild conditions. A striking example is the ready conversion of the acid-sensitive **2** into **3**,[3] a reaction that proceeds by other known methods in yields of only ~30%.

CH_3O $COOCH_3$ CH_3O O **2** —**1**, 87%→ CH_3O $COOCH_3$ CH_3O O O **3**

[1] H. Baganz and L. Domaschke, *Ber.*, **91**, 650 (1958).
[2] M. Delépine, *Bull. soc.*, **25**, 580 (1901); *Ann. Chim.* [7], **23**, 482 (1901).
[3] B. Glatz, G. Helmchen, H. Muxfeldt, H. Porcher, R. Prewo, J. Senn, J. J. Stezowski, R. J. Stojda, and D. R. White, *Am. Soc.*, **101**, 2171 (1979).

β-Methoxyethoxymethyl chloride, 7, 227–229.

Protection of hydroxyl groups.[1] A recent synthesis of norchorismic acid (**7**), of interest because of its relationship to shikimic acid, depended on use of the β-methoxyethoxymethyl (MEM) group for protection of one hydroxyl group of a *vic*-diol. Thus the dienol **2**, prepared as shown from **1**, after esterification of the carboxyl group, was successfully alkylated under mild conditions to give **3**. The MEM group of **3** was completely stable to $ZnBr_2$ and $TiCl_4$, but could be cleaved by a catalytic quantity of HBF_4 at 0°. The resulting diester (**4**) on basic hydrolysis gave the desired **7** in low yield. Hence a less direct method was used: cyclization by

Br H O O $OCH_2OCH_2CH_2OCH_3$ **1** —KOH, THF, H_2O, 84%→ COOH OH OMEM **2** —1) CH_3I, HMPT (85%) 2) $N_2CHCOOC_2H_5$, HBF_4 (70%)→

$COOCH_3$ $OCH_2COOC_2H_5$ OMEM **3** —HBF_4, CH_2Cl_2, 0°, 50–60%→ $COOCH_3$ $OCH_2COOC_2H_5$ OH **4** —TsOH, C_6H_6, 50%→

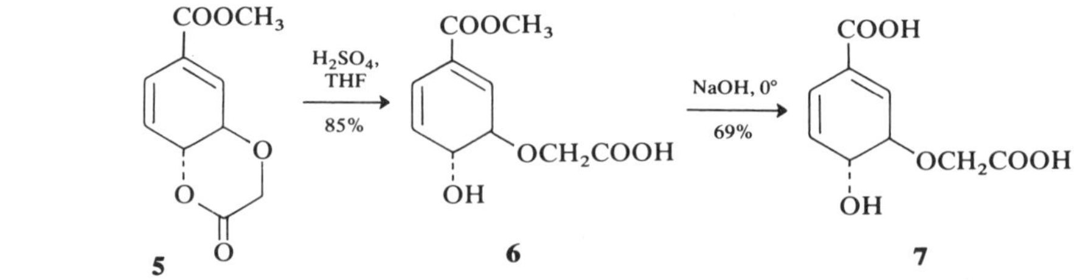

p-toluenesulfonic acid to the δ-lactone (**5**). This was hydrolyzed by acid to the monoester **6**, which gave **7** on mild treatment with base.

Cleavage of MEM ethers of an alkaloid. The cleavage of MEM ethers of alkaloids can present problems. Thus attempted cleavage of the bis-β-methoxyethoxymethyl ether (**1**) of abresoline failed with the usual reagents ($ZnBr_2$ and $TiCl_4$) for this purpose. However, if the basic nitrogen of **1** is converted to the hydrochloride, cleavage with $TiCl_4$ proceeds in 70% yield. Actually, **1** can be cleaved directly by TFA in CH_2Cl_2 at 0° in 75% yield.[2]

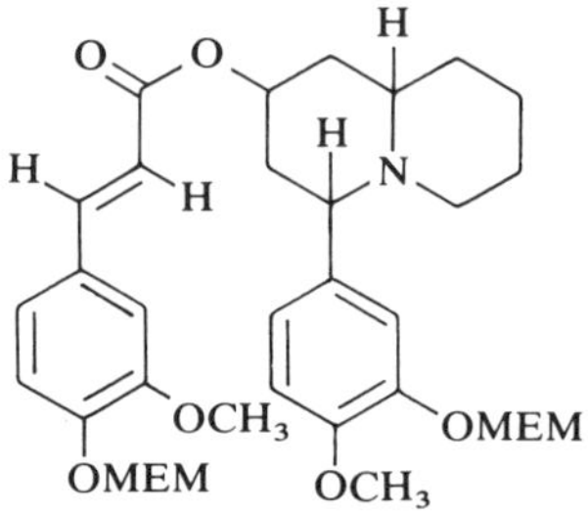

[1] N. Ikota and B. Ganem, *J.C.S. Chem. Comm.*, 869 (1978).
[2] J. Quick and R. Ramachandra, *Syn. Comm.*, **8**, 511 (1978).

3-Methoxy-3-methylbutynylcopper, $CH_3O-C(CH_3)_2-C\equiv C-Cu$ (**1**), **8**, 324–325.

Mixed cuprates. Corey *et al.*[1] allowed the mixed cuprate **2** to react with the 2-pyridinethiol ester **3** to produce the α,β-unsaturated ketone **4**, a key intermediate in the synthesis of erythronolide A (**5**)[1]

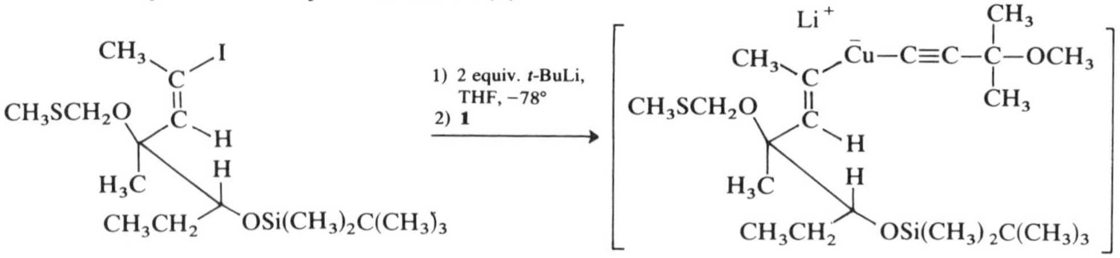

1) 2 equiv. *t*-BuLi, THF, −78°
2) **1**

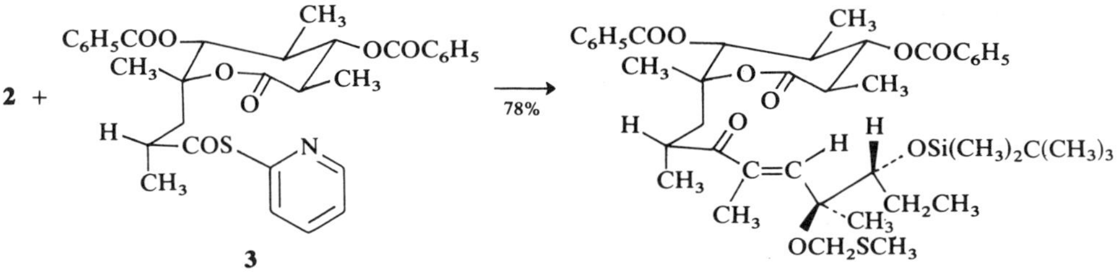

$$\mathbf{2} + \mathbf{3} \xrightarrow{78\%} \mathbf{4}$$

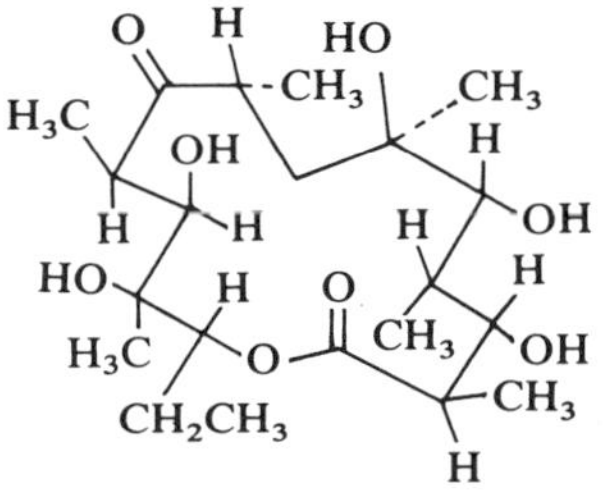

5

[1] E. J. Corey, P. B. Hopkins, S. Kim, S. Yoo, K. P. Nambiar, and J. R. Falck, *Am. Soc.*, **101**, 7131 (1979).

Methoxymethyl(diphenyl)phosphine oxide, $(C_6H_5)_2P(=O)CH_2OCH_3$ (**1**). Mol. wt. 246.23, m.p. 114–116°.

Preparation:

$$P(C_6H_5)_3 + CH_3OCH_2Cl \rightarrow (C_6H_5)_3\overset{+}{P}CH_2OCH_3Cl^- \xrightarrow[90\%]{NaOH,\ H_2O,\ \Delta} \mathbf{1}$$

Vinyl ethers. The anion **2** (which must be generated with LDA for satisfactory results) has been used in a Wittig-Horner reaction for preparation of vinyl ethers from both aldehydes and ketones. Usually a 1 : 1 mixture of isomeric adducts (**3**) is formed, which can be separated by crystallization or chromatography. Treatment of each diastereoisomeric **3** with base gives a single (E)- or (Z)-vinyl ether (**4**).[1]

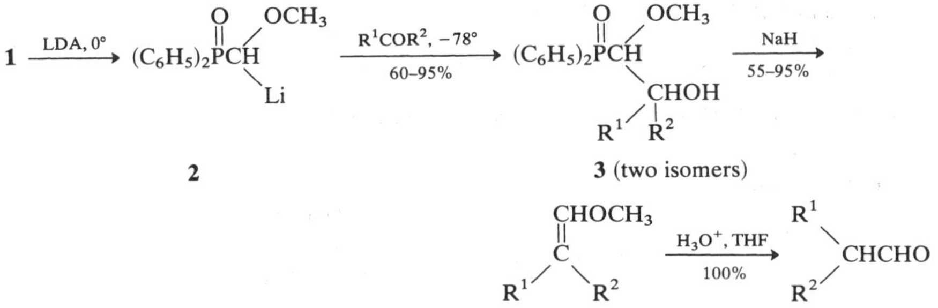

$$\mathbf{1} \xrightarrow{LDA,\ 0°} (C_6H_5)_2P(=O)CH(OCH_3)Li \ (\mathbf{2}) \xrightarrow[60\text{–}95\%]{R^1COR^2,\ -78°} (C_6H_5)_2P(=O)CH(OCH_3)CHOH(R^1)(R^2) \ (\mathbf{3}\text{ (two isomers)}) \xrightarrow[55\text{–}95\%]{NaH}$$

$$R^1R^2C{=}CHOCH_3 \ (\mathbf{4}) \xrightarrow[100\%]{H_3O^+,\ THF} R^1R^2CHCHO \ (\mathbf{5})$$

[1] C. Earnshaw, C. J. Wallis, and S. Warren, *J.C.S. Perkin I*, 3099 (1979).

4-Methoxy-6-methyl-2-pyranone (1). Mol. wt. 140.13, m.p. 83–86°. Supplier: Aldrich. This ether can be prepared by methylation of triacetic acid lactone (**2**) with dimethyl sulfate.[1] Exposure of **2** to methyl fluorosulfonate produces the isomeric ether (**3**).[2]

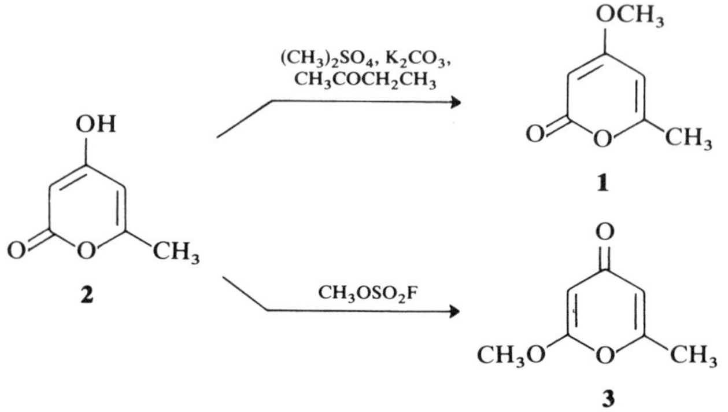

$(CH_3)_2SO_4$, K_2CO_3, $CH_3COCH_2CH_3$

CH_3OSO_2F

Polyketides. These pyranones combine with carbanions obtained from orsellinic acid derivatives to afford polyketide-type compounds.[3,4]

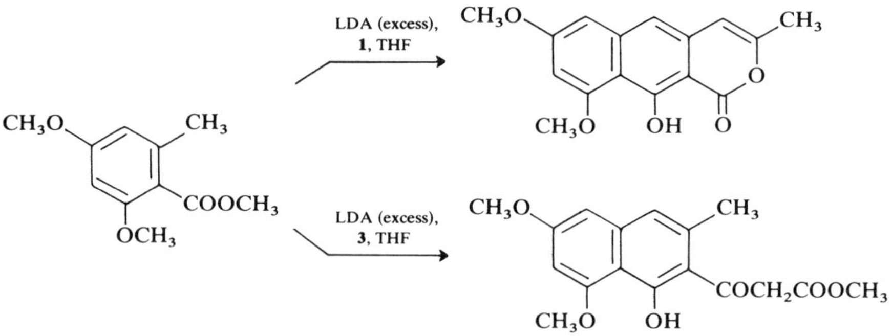

LDA (excess), **1**, THF

LDA (excess), **3**, THF

[1] J. D. Bu'Lock and H. G. Smith, *J. Chem. Soc.*, 502 (1960).
[2] P. Beak and J. K. Lee, *J. Org.*, **43**, 1367 (1978).
[3] G. E. Evans, F. J. Leeper, J. A. Murphy, and J. Staunton, *J.C.S. Chem. Comm.*, 205 (1979).
[4] F. J. Leeper and J. Staunton, *ibid.*, 206 (1979).

1-Methoxy-3-methyl-1-trimethylsilyloxy-1,3-butadiene (1). Mol. wt. 174.31, b.p. 80–82°/16 mm.

Preparation:

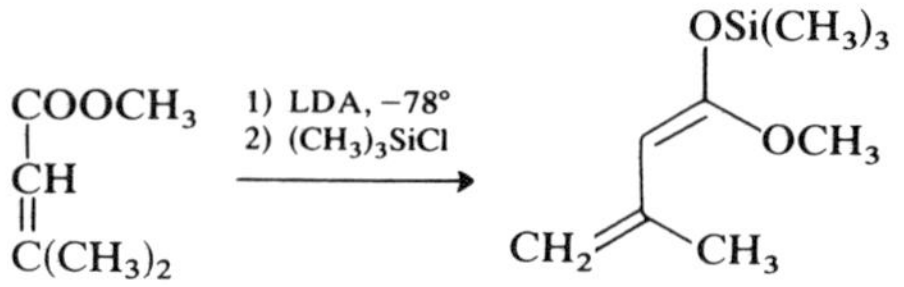

$COOCH_3$–CH=$C(CH_3)_2$ → 1) LDA, −78° 2) $(CH_3)_3SiCl$ → **1**

Annelation of quinones.[1] This mixed acetal (and related acetals) reacts regioselectively with 2-chloro-1,4-quinones to form annelated quinones.

Examples:

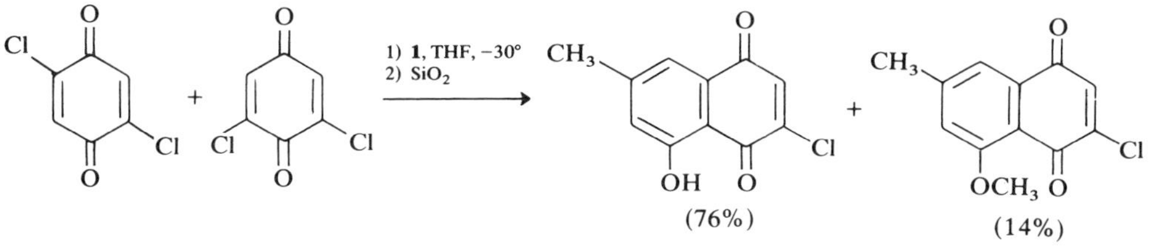

1) **1**, THF, −30° 2) SiO_2 (76%) (14%)

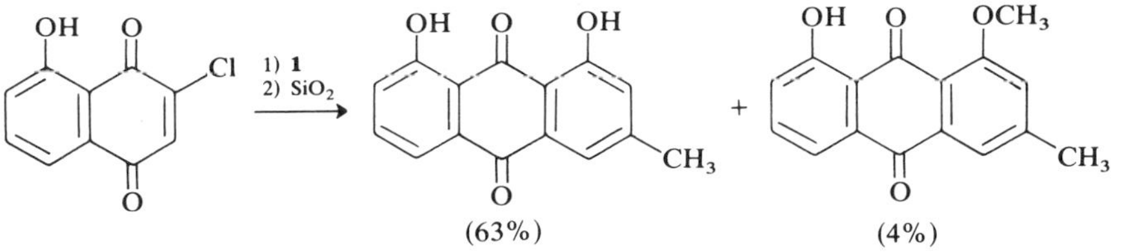

1) **1** 2) SiO_2 (63%) (4%)

[1] J. Savard and P. Brassard, *Tetrahedron Letters*, 4911 (1979).

***trans*-1-Methoxy-3-trimethylsilyloxy-1,3-butadiene (1), 6,** 370.

Diels–Alder reactions. A key step in the synthesis[1] of the sesquiterpene *dl*-pentalenolactone (**5**) was a Diels–Alder reaction of **2**, available from cyclopentadiene by several steps in 46% overall yield, with **1**. This reaction proceeds in

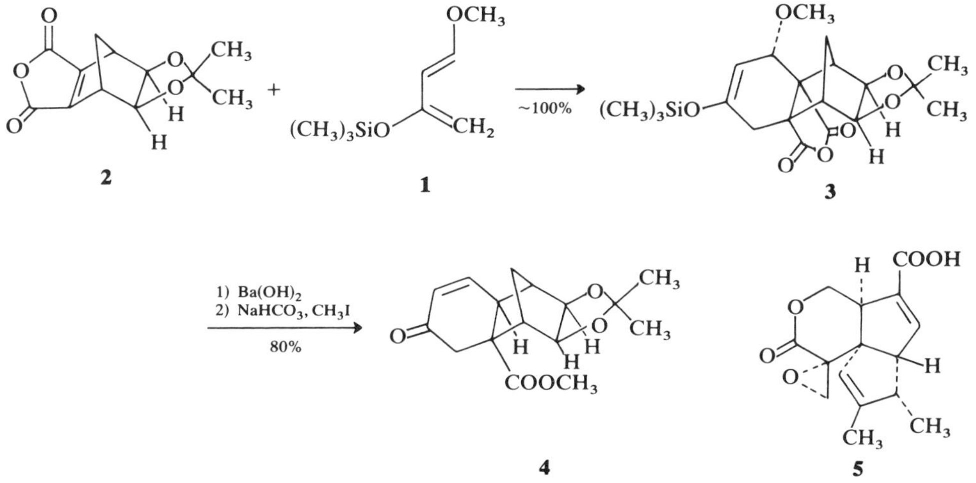

2 + **1** → (~100%) **3**

1) $Ba(OH)_2$ 2) $NaHCO_3$, CH_3I (80%) → **4** **5**

almost quantitative yield in benzene under reflux or at 20° for 20 hours. Treatment of **3** with barium hydroxide liberates the enone system, hydrolyzes the anhydride group, and effects decarboxylation of the vinylogous β-keto ester[2] to give **4**.

Danishefsky and Walker[3] have employed the cycloaddition of **1** with **6** in a new synthesis of the antifungal agent griseofulvin (**8**).

$$\mathbf{1} + \mathbf{6} \xrightarrow[54\%]{C_6H_5CH_3,\ \Delta} \mathbf{7} \xrightarrow[58\%]{H_2,\ Pd/C} \mathbf{8}$$

[1] S. Danishefsky, M. Hirama, K. Gombatz, T. Harayama, E. Berman, and P. Schuda, *Am. Soc.*, **100**, 6536 (1978).
[2] R. B. Miller and R. D. Nash, *Tetrahedron*, **30**, 2961 (1974).
[3] S. Danishefsky and F. J. Walker, *Am. Soc.*, **101**, 7108 (1979).

α-Methoxyvinyllithium, 6, 372–373; **7**, 331; **8**, 331.

Reaction with trialkylboranes (**7**, 234). α-Methoxyvinyllithium reacts with a trialkylborane at −80° in THF to form an ate complex (**1**), which rearranges when warmed to room temperature to another complex (**2**). This latter complex can be converted into methyl ketones, dialkylmethylcarbinols, and hindered tertiary alcohols. These reactions all involve electrophilic attack at the olefinic carbon β to boron.[1]

$$R_3B + LiC(OCH_3){=}CH_2 \xrightarrow{-80^\circ} Li^+[R_3B{-}C(OCH_3){=}CH_2]^- \ (\mathbf{1}) \xrightarrow{20^\circ} Li^+[R_2B{-}C(OCH_3)(R){=}CH_2]^- \ (\mathbf{2})$$

$$R_2C(OH)CH_3 \xleftarrow[85\text{–}100\%]{1)\ HCl;\ 2)\ H_2O_2,\ OH^-} \mathbf{2} \xrightarrow[75\text{–}95\%]{H_2O_2,\ OH^-,\ C_2H_5OH} RC(O)CH_3$$

$$\mathbf{2} \xrightarrow[60\text{–}95\%]{1)\ R'X;\ 2)\ H_2O_2,\ OH^-} R_2C(OH)CH_2R'$$

[1] A. B. Levy, S. J. Schwartz, N. Wilson, and B. Christie, *J. Organometal. Chem.*, **156**, 123 (1978).

o-(Methylamino)phenyl disulfide, $[2\text{-}(CH_3NH)C_6H_4S]_2$ **(1).** Mol. wt. 252.36, m.p. 64–65°. Preparation.[1]

Medium-sized cyclic ketones. A method for synthesis of 8–12 membered cyclic ketones is formulated in equation I.[2]

(I) 1 + $COCl(CH_2)_{n-2}CH_2Br$ (n = 8–12) —[1) $N(C_2H_5)_3$; 2) $NaBH_4$, NaH; 60–85%]→ N-methyl benzothiazine lactam with $CH_2(CH_2)_{n-2}$ ring —[$ClC_6H_4CO_3H$; 85–95%]→ sulfone lactam with $(CH_2)_{n-1}$ —[$KOC(CH_3)_3$, THF, DMSO; quant.]→ NHCH$_3$-aryl sulfonyl cyclic ketone, $O{=}C\text{—}CH\text{—}(CH_2)_{n-2}$ —[Al/Hg; 70–85%]→ cyclic ketone $C{=}O$, $(CH_2)_{n-1}$

[1] A. I. Kiprianov and Z. N. Pazenko, *Zhur. Obschei. Khim.*, **19**, 1523 (1949). [C. A., **44**, 3487g (1950).]
[2] Y. Ohtsuka and T. Oishi, *Tetrahedron Letters*, 4487 (1979).

Methyl α-benzalaminocrotonate (1). Mol. wt. 203.34. This Schiff base is obtained as a 60:40 (E)/(Z) mixture from DL-β-chloro-α-aminobutyric acid methyl ester hydrochloride. It is stable at 0°, but isomerizes slowly to the pure (E)-isomer.

$$CH_3CHClCH(NH_2\cdot HCl)COOCH_3 \xrightarrow{C_6H_5CHO} CH_3CHClCH(N{=}CHC_6H_5)COOCH_3 \xrightarrow[80\%]{DBU} CH_3CH{=}C(N{=}CHC_6H_5)COOCH_3$$

1 (E/Z = 60:40)

α-Vinyl-α-amino acids.[1] The anion derived from **1** is alkylated exclusively at the α-position; the products, such as **2**, can be converted into α-vinyl-α-amino esters (**3, 5**) or hydrochlorides of the acid (**4**).

Examples:

$$\mathbf{1} \xrightarrow{1)\ LiN[Si(CH_3)_3]_2;\ 2)\ ArCH_2Br} CH_2{=}CHC(CH_2Ar)(N{=}CHC_6H_5)COOCH_3\ (\mathbf{2}) \xrightarrow[65\text{–}84\%]{SiO_2} CH_2{=}CHC(CH_2Ar)(NH_2)COOCH_3\ (\mathbf{3}) \xrightarrow[80\text{–}90\%]{HCl} CH_2{=}CHC(CH_2Ar)(NH_2\cdot HCl)COOH\ (\mathbf{4})$$

$$\mathbf{1} \xrightarrow[76\%]{1)\ LiN[Si(CH_3)_3]_2;\ 2)\ CH_2{=}CHCH_2Br;\ 3)\ SiO_2} CH_2{=}CHC(CH_2CH{=}CH_2)(NH_2)COOCH_3\ (\mathbf{5})$$

[1] W. J. Greenlee, D. Taub, and A. A. Patchett, *Tetrahedron Letters*, 3999 (1978).

Methyl chloroformate, 6, 376; **7**, 236.

Carbomethoxylation of terminal alkynes.[1] Grignard reagents of 1-alkynes react with methyl chloroformate to form methyl 2-alkyne-1-carboxylates. The reaction proceeds more readily than carboxylation of the Grignard reagent, which is usually carried out in an autoclave.

$$THPOCH_2C{\equiv}CH \xrightarrow{C_2H_5MgBr} \left[THPOCH_2C{\equiv}CMgBr \xrightarrow[THF,\ -15^\circ]{ClC(O)OCH_3,} THPOCH_2C{\equiv}CCOOCH_3\right] \xrightarrow[60\text{–}65\%]{H^+\text{-resins},\ CH_3OH} HOCH_2C{\equiv}CCOOCH_3$$

[1] R. A. Earl and L. B. Townsend, *Org. Syn.*, submitted (1979).

Methyl cyclobutenecarboxylate, 8, 335–336. Wilson *et al.*[1] recommend a sulfenylation–dehydrosulfenylation approach to the synthesis of this ester (equation I).

$$\text{(I)}\quad \text{cyclobutane-}COOCH_3 \xrightarrow[58\%]{1)\ LDA,\ THF,\ C_6H_5SSC_6H_5;\ 2)\ NaIO_4} \text{1-}SOC_6H_5\text{-1-}COOCH_3\text{-cyclobutane} \xrightarrow[75\%]{160^\circ} \text{cyclobutene-}COOCH_3$$

[1] S. R. Wilson, L. R. Phillips, Y. Pelister, and J. C. Huffman, *Am. Soc.*, **101**, 7373 (1979).

Methyl 3-(N,N-dimethylamino)propionate, $(CH_3)_2NCH_2CH_2COOCH_3$ (**1**). Mol. wt. 131.18, b.p. 45°/20 mm. The reagent is prepared by reaction of methyl acrylate with dimethylamine.[1]

Methyl acrylates.[2] The reagent is converted by LDA into the stable ester enolate **a**, which is alkylated by primary alkyl iodides and allylic bromides in the presence of HMPT to give the useful Mannich bases **2**. These products can be converted into methyl acrylates (**4**) by quaternization (quantitative) followed by treatment with diazabicyclo[4.3.0]nonene-5 (DBN) in refluxing benzene.

$$\mathbf{1} \xrightarrow[-78^\circ]{LDA,\ THF,} \left[(CH_3)_2NCH_2CH{=}C(OLi)(OCH_3)\right]_{\mathbf{a}} \xrightarrow[60\text{–}90\%]{RX,\ HMPT,\ -78\rightarrow25^\circ} \underset{\mathbf{2}}{(CH_3)_2NCH_2CH(R)COOCH_3} \xrightarrow[100\%]{CH_3I}$$

$$\underset{\mathbf{3}}{I^-(CH_3)_3\overset{+}{N}CH_2CH(R)COOCH_3} \xrightarrow[75\text{–}85\%]{DBN,\ C_6H_6,\ \Delta} \underset{\mathbf{4}}{H_2C{=}C(R)COOCH_3}$$

[1] E. Rouvier, J.-C. Giacomoni, and A. Cambon, *Bull. soc.*, 1717 (1971).
[2] L.-C. Yu and P. Helquist, *Tetrahedron Letters*, 3423 (1978).

Methyl fluorosulfonate, 3, 202; **4,** 339–340; **5,** 445–446; **6,** 381–383; **7,** 240–241; **8,** 340–341.

Alkenes from selenides. Preparation of alkylidenecyclopropanes by selenoxide elimination proceeds in rather low yields. Yields are improved considerably if the selenide is first methylated (CH_3OSO_2F or CH_3I–$AgBF_4$) and the methyl selenonium salt is treated with potassium *t*-butoxide in THF or DMSO (equation I).

$$\text{(I)}\quad \text{cyclopropane}(SeCH_3)(CH_2C_8H_{17}) \xrightarrow{CH_3OSO_2F} \text{cyclopropane}(\overset{+}{Se}(CH_3)_2\ FSO_2O^-)(CH_2C_8H_{17}) \xrightarrow[\text{60\% overall}]{KOC(CH_3)_3,\ DMSO} \text{cyclopropylidene}{=}CHC_8H_{17}$$

Actually this elimination is general for primary and secondary alkyl selenides, and even potassium hydroxide can be used as base.[1]

Examples:

$$C_8H_{17}CH_2Br \xrightarrow[80\%]{CH_3SeCH_2Li,\ THF,\ HMPT} CH_3SeCH_2CH_2C_8H_{17} \xrightarrow[88\%]{1)\ CH_3OSO_2F\quad 2)\ KOC(CH_3)_3} C_8H_{17}CH{=}CH_2$$

$$C_6H_5SeCH(CH_3)CH_2C_{10}H_{21} \xrightarrow[88\%]{} CH_2{=}CHCH_2C_{10}H_{21} + (CH_3)HC{=}CH(C_{10}H_{21})\quad (65:35)$$

[1] S. Halazy and A. Krief, *Tetrahedron Letters*, 4233 (1979).

Methylenetriphenylphosphorane, 1, 678; **6,** 380–381; **8,** 339–340.

Isopropenyl terpenes. The most convenient synthesis of the natural sesquiterpene (±)-iso-α-curcumene (**4**)[1] employs the general method of van der Gen (**6,** 380–381).

$$p\text{-}CH_3C_6H_4COCH_3\ (\mathbf{1}) \xrightarrow[60\%]{O{=}P(OC_2H_5)_2CH_2CH{=}CHCOOCH_3} p\text{-}CH_3C_6H_4C(CH_3){=}CHCH{=}CHCOOCH_3\ (\mathbf{2}) \xrightarrow[80\%]{H_2,\ Pd/C}$$

$$p\text{-}CH_3C_6H_4CH(CH_3)CH_2CH_2CH_2COOCH_3\ (\mathbf{3}) \xrightarrow{(C_6H_5)_3P{=}CH_2} p\text{-}CH_3C_6H_4CH(CH_3)CH_2CH_2CH_2C(CH_3){=}CH_2\ (\mathbf{4},\ 55\%)$$

[1] P. N. Chaudhari, *Bull. soc.*, II–429 (1979).

N-Methylephedrine, $H—\overset{\overset{CH_3}{\vdots}}{C}—N(CH_3)_2$ (**1**).
$H—\underset{\underset{C_6H_5}{\vdots}}{C}—OH$

Asymmetric reduction. The complex of $LiAlH_4$, N-methylephedrine, and 3,5-dimethylphenol (1:1:2) reduces aryl alkyl ketones and α-acetylenic ketones to optically active alcohols with a purity of 75–90%. The optical yields are comparable to those observed with the structurally related darvon alcohol (**8**, 184–186). The reduction of benzoylacetylene, $C_6H_5COC{\equiv}CH$, results in a racemic alcohol.[1]

[1] J.-P. Vigneron and V. Bloy, *Tetrahedron Letters*, 2065 (1974); *idem, ibid.*, 2683 (1979).

Methyl iodide, 1, 682–685; **2**, 274; **3**, 206; **4**, 341; **5**, 448–454; **6**, 384–385; **7**, 242–243; **8**, 342.

Hydrolysis of thioacetals.[1] Exposure of the thioacetal **1** to excess methyl iodide in aqueous acetonitrile at reflux results in formation of the lactam **2**, an intermediate in the synthesis of quebrachamine.[1] This transformation involves initial hydrolysis of **1** to an aldehyde intermediate (**4**, 341), followed by an HI-catalyzed Pictet–Spengler cyclization.[2]

CH_3I, H_2O, CH_3CN; 83%

1 → **2** (β-ethyl/α-ethyl = 6:1)

[1] S. Takano, S. Hatakeyama, and K. Ogasawara, *Am. Soc.*, **101**, 6414 (1979).
[2] W. M. Whaley and T. R. Govindachari, *Org. React.*, **6**, 151 (1951).

Methyl iodide–1,3-Propanedithiol.

Ketene S,S-acetals. Thioamides (**1**) can be converted into ketene S,S-acetals (**3**) by methylation to form the onium salt **2**, which is then converted into **3** by reaction with 1,3-propanedithiol in the presence of base. The intermediate onium salt need not be isolated. The presence of at least one hydrogen atom α to the thiocarbonyl group is required; the method is compatible with olefin, ketone, ester, and amide

$$\underset{\mathbf{1}}{RCH_2\overset{\overset{S}{\|}}{C}N(CH_3)_2} \xrightarrow{CH_3I,\ CH_2Cl_2} \underset{\mathbf{2}}{RCH_2\overset{\overset{^+SCH_3}{\|}}{C}N(CH_3)_2I^-} \xrightarrow[\text{80–90\% from 1}]{HS(CH_2)_3SH,\ K_2CO_3}$$

$$\underset{\mathbf{3}}{RCH{=}C\langle S(CH_2)_3S\rangle} \xrightarrow[\text{65–85\%}]{(C_2H_5)_3SiH,\ CF_3COOH} \underset{\mathbf{4}}{RCH_2CH\langle S(CH_2)_3S\rangle}$$

groups.[1] The ketene S,S-acetals can be reduced to 1,3-dithianes (**4**) with triethylsilane and TFA.[2]

[1] T. Harada, Y. Tamaru, and Z. Yoshida, *Tetrahedron Letters*, 3525 (1979).
[2] F. A. Carey and A. S. Court, *J. Org.*, **37**, 1926 (1972).

5-Methylisoxazole, (**1**). Mol. wt. 83.09, b.p. 122°. Supplier: Aldrich.

***Acetoacetonitrile dianion* (2).**[1] This dianion is prepared most conveniently by treatment of **1** with 2 equiv. of LDA in THF at −10°. Reactions with a number of electrophiles have been reported.

Examples:

$$\underset{\mathbf{2}}{Li^+\,^-CH_2C(OLi^+)=CHCN^-} \xrightarrow{C_6H_5CH_2Br} C_6H_5CH_2CH_2COCH_2CN \xrightarrow[50\%]{NaBH_4} C_6H_5CH_2CH_2CH(OH)CH_2CN$$

$$\mathbf{2} + p\text{-}ClC_6H_4CHO \xrightarrow[82\%]{NaBH_4} p\text{-}ClC_6H_4CH(OH)CH_2CH(OH)CH_2CN \text{ (two isomers)}$$

$$\mathbf{2} + C_6H_5CN \xrightarrow[24\%]{30\ min} C_6H_5C(NH_2)=CHCOCH_2CN$$

$$\mathbf{2} + p\text{-}CH_3OC_6H_4CN \xrightarrow[62\%]{18\ hrs.} p\text{-}CH_3OC_6H_4\text{-(2-amino-4-pyridone, NH)}$$

[1] F. J. Vinick, Y. Pan, and H. W. Gschwend, *Tetrahedron Letters*, 4221 (1978).

3-Methyl-5(4*H*)-isoxazolone, (**1**). Mol. wt. 99.09, b.p. 65–66°/0.3 mm. The reagent is obtained in about 50% yield by the reaction of ethyl acetoacetate and hydroxylamine.[1]

Arylacetylenes.[2] Aryl- and heteroarylacetylenes can be prepared in useful yields by condensation of aryl aldehydes with **1** to form 4-arylmethylene-5(4*H*)-isoxazoles (**2**). On flash pyrolysis (700–800°) these compounds lose CO_2 and CH_3CN with formation of acetylenes **3**.

$$\mathbf{1} + ArCHO \xrightarrow{40\text{–}95\%} \underset{\mathbf{2}}{\text{3-}CH_3\text{-4-}(=CHAr)\text{-isoxazol-5-one}} \xrightarrow[-CO_2,\ -CH_3CN]{700\text{–}800^\circ} \underset{\mathbf{a}}{[ArCH=C:]} \xrightarrow{65\text{–}95\%} \underset{\mathbf{3}}{ArC\equiv CH}$$

[1] A. R. Katritzky, S. Oksme, and A. J. Boulton, *Tetrahedron*, **18**, 777 (1962).
[2] C. Wentrup and W. Reichen, *Helv.*, **59**, 2615 (1976); C. Wentrup and H.-W. Winter, *Angew. Chem., Int. Ed.*, **17**, 609 (1978).

Methylketene methyl trimethylsilyl acetal (1). Mol. wt. 158.27, b.p. 46.5°/23 mm. The acetal is prepared by a modification of the method of Ainsworth *et al.*[1] (equation I).

$$\text{(I)}\quad CH_3CH_2COOCH_3 \xrightarrow[57\%]{1)\ LDA,\ -78°;\ 2)\ ClSi(CH_3)_3} (CH_3)(H)C{=}C(OCH_3)(OSi(CH_3)_3)\quad \textbf{(1)}$$

Silylation.[2] The reagent silylates alcohols, phenols, mercaptans, carboxylic acids, and amides at 15–55° in the absence of a base, acid, or catalyst. Only methyl propionate (b.p. 80°) is formed as by-product.

Examples:

$$C_6H_5CH_2CH_2OH \xrightarrow[95\%]{\mathbf{1},\ CH_3CN,\ 50°} C_6H_5CH_2CH_2OSi(CH_3)_3 + CH_3CH_2COOCH_3$$

$$C_6H_5CH_2COOH \xrightarrow[99\%]{\mathbf{1},\ CH_2Cl_2} C_6H_5CH_2COOSi(CH_3)_3$$

$$C_6H_5OH \xrightarrow[99\%]{} C_6H_5OSi(CH_3)_3$$

$$CH_3C_6H_4NHCOCH_3 \xrightarrow[95\%]{} CH_3C_6H_4N(Si(CH_3)_3)COCH_3$$

[1] C. Ainsworth, F. Chen, and Y.-N. Kuo, *J. Organometal. Chem.*, **46**, 59 (1972).
[2] Y. Kita, J. Haruta, J. Segawa, and Y. Tamura, *Tetrahedron Letters*, 4311 (1979).

Methyllithium, 1, 686–689; **2**, 274–278; **3**, 202–204; **5**, 448–459; **6**, 384–385; **7**, 242–243; **8**, 342–344.

Lithiation of vinylic sulfones. Phenyl vinyl sulfones (**1**), prepared as indicated,[1] react with methyllithium regiospecifically at −95° at the α-vinyl position to give the lithium derivatives **2**. As expected, **2** can be alkylated to give **3**. The reaction of **2** with enolizable carbonyl compounds proceeds more satisfactorily by prior conversion to the vinylic Grignard reagent **a**. This sequence constitutes a route to disubstituted alkenes, since a sulfone group is reductively cleaved by sodium amalgam (**7**, 326).[2]

$$\underset{(R = CH_3,\ C_6H_5)}{RCH{=}CH_2} \xrightarrow[85\text{–}95\%]{C_6H_5SO_2Cl,\ CuCl} RCH(Cl){-}CH_2SO_2C_6H_5 \xrightarrow[70\text{–}90\%]{N(C_2H_5)_3} \underset{\mathbf{1}}{(H)(R)C{=}C(SO_2C_6H_5)(H)} \xrightarrow{CH_3Li,\ THF,\ -95°}$$

$$\underset{\mathbf{2}}{(H)(R)C{=}C(SO_2C_6H_5)(Li)} \xrightarrow[70\text{–}95\%]{R^1I} \underset{\mathbf{3}}{(H)(R)C{=}C(SO_2C_6H_5)(R^1)}$$

$$\mathbf{2} \xrightarrow{MgBr_2} \underset{\mathbf{a}}{\left[(H)(R)C{=}C(SO_2C_6H_5)(MgBr)\right]} \xrightarrow[\sim 75\%]{R^1COR^2} \underset{\mathbf{4}}{(H)(R)C{=}C(SO_2C_6H_5)C(OH)R^1R^2}$$

Heteroconjugate addition. The addition of methyllithium to the heteroalkene **1** is accompanied by high asymmetric induction (>99%) to give **2** in 95% yield, in which C_1 and C_2 are in the *threo* orientation. The MEM group is undoubtedly responsible for the stereochemical control by coordination of the oxygen atoms with the lithium reagent.[3]

$SO_2C_6H_5$; $(CH_3)_3Si$; OMEM; $CH_2C_6H_5$ — **1** (E/Z = 1:4) → 1) CH_3Li, −78°; 2) H_3O^+; 95% → **2**

[1] M. Asscher and D. Vofsi, *J. Chem. Soc.*, 4962 (1964); J. Sinnreich and M. Asscher, *J.C.S. Perkin I*, 1543 (1972).
[2] J. J. Eisch and J. E. Galle, *J. Org.*, **44**, 3279 (1979).
[3] M. Isobe, M. Kitamura, and T. Goto, *Tetrahedron Letters*, 3465 (1979).

Methyllithium–Lithium bromide.

Selective desilylation of bis(trimethylsilyl)acetylenes. The methyllithium–lithium bromide complex reacts with **1** or **2** to afford the corresponding monolithium derivatives in nearly quantitative yield. These readily condense with aldehydes and ketones to afford alcohols in 46–90% yield. Note that different solvents are used for **1** and **2**. Selective desilylation of **1** or **2** can also be effected with 1–10 mole % KF–18-crown-6 in the presence of a carbonyl compound. In this case, the yields of product are usually somewhat lower.[1]

$$\underset{\mathbf{1}}{(CH_3)_3SiC\equiv CSi(CH_3)_3} \xrightarrow{CH_3Li\text{-}LiBr,\ THF} [(CH_3)_3SiC\equiv CLi] \xrightarrow[88\%]{(CH_3)_2CHCHO} (CH_3)_3SiC\equiv C\overset{OH}{C}HCH(CH_3)_2$$

$$\underset{\mathbf{2}}{(CH_3)_3SiC\equiv C-C\equiv CSi(CH_3)_3} \xrightarrow{CH_3Li\text{-}LiBr,\ (C_2H_5)_2O} [(CH_3)_3Si(C\equiv C)_2Li] \xrightarrow[46\%]{\text{cyclohexanone}} (CH_3)_3Si(C\equiv C)_2\text{-}C_6H_{10}OH$$

[1] A. B. Holmes, C. L. D. Jennings-White, A. H. Schulthess, B. Akinde, and D. R. M. Walton, *J.C.S. Chem. Comm.*, 840 (1979).

Methyllithium–Tetramethylethylenediamine (TMEDA).

Cyclopropanone dithioketals.[1] Methyllithium (1 equiv.) in combination with TMEDA (2 equiv.) is an excellent reagent for elimination of thiophenoxide from a 1,1,3-tris(phenylthio)alkane to form cyclopropanone dithioketals (equation I).

$$\text{(I)}\quad R^1-\overset{C_6H_5S}{\underset{R^2}{C}}CH_2CH(SC_6H_5)_2 \xrightarrow[-C_6H_5SLi]{CH_3Li,\ TMEDA} \left[\begin{array}{c} C_6H_5\overset{+}{S}-\overset{-}{C}SC_6H_5 \\ R^1-\underset{R^2}{C}-CH_2 \end{array}\right] \longrightarrow \text{1,1-bis}(C_6H_5S)\text{-2-}R^1\text{-2-}R^2\text{-cyclopropane}$$

The substrates can be prepared by several routes, two of which are shown in the examples:

Examples:

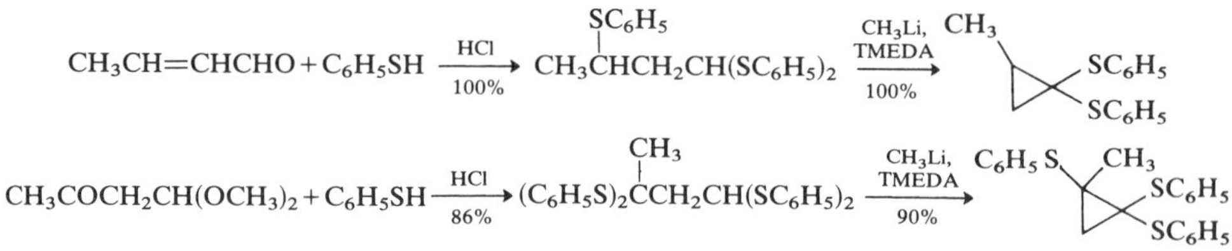

$$CH_3CH{=}CHCHO + C_6H_5SH \xrightarrow[100\%]{HCl} CH_3CH(SC_6H_5)CH_2CH(SC_6H_5)_2 \xrightarrow[100\%]{CH_3Li,\ TMEDA} \text{1-methyl-2,2-bis(phenylthio)cyclopropane}$$

$$CH_3COCH_2CH(OCH_3)_2 + C_6H_5SH \xrightarrow[86\%]{HCl} (C_6H_5S)_2C(CH_3)CH_2CH(SC_6H_5)_2 \xrightarrow[90\%]{CH_3Li,\ TMEDA} \text{1-methyl-1,2,2-tris(phenylthio)cyclopropane}$$

Ring closure to cyclobutanes and cyclopentanes.[2] The synthesis of cyclopropanones by elimination of lithium thiophenoxide with this base has been extended to similar syntheses of functionalized cyclobutanes and cyclopentanes, as shown in equations (I) and (II). When extended to synthesis of a cyclohexane derivative, this method resulted in a very low yield. One possible mechanism is formation of a dianion followed by loss of thiophenoxide ion to give a carbene-anion, which cyclizes with loss of the second thiophenoxide ion.

(I) $$\text{2,5-dimethoxytetrahydrofuran} \xrightarrow[100\%]{C_6H_5SH,\ HCl} (C_6H_5S)_2CH\text{-}CH_2CH_2\text{-}CH(SC_6H_5)_2 \xrightarrow[47\%]{CH_3Li,\ TMEDA} \text{1,2-bis(phenylthio)cyclobutene} + 2LiSC_6H_5$$

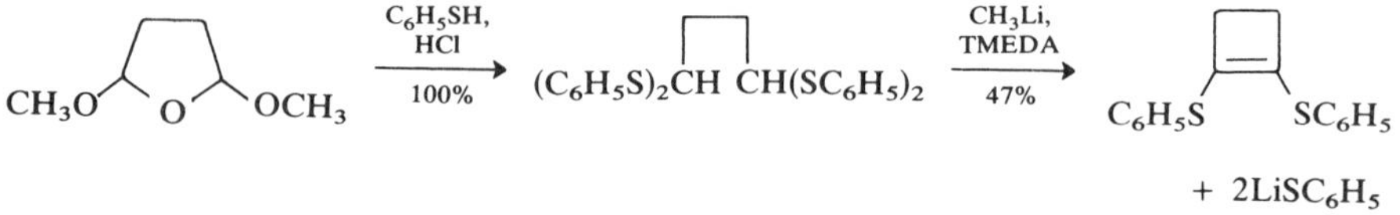

(II) $$\text{4-methyl-2-ethoxy-3,4-dihydro-2H-pyran} \xrightarrow{100\%} (C_6H_5S)_2CH\text{-}CH_2CH(CH_3)CH_2\text{-}CH(SC_6H_5)_2 \xrightarrow{52\%} \text{4-methyl-1,2-bis(phenylthio)cyclopentene} + 2LiSC_6H_5$$

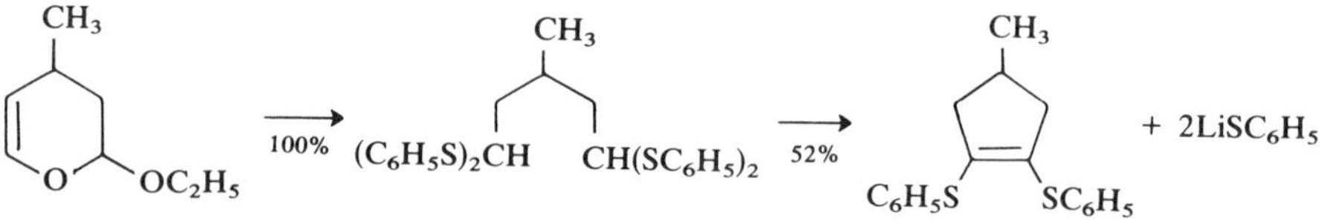

[1] T. Cohen and W. M. Daniewski, *Tetrahedron Letters*, 2991 (1978).
[2] T. Cohen, D. Ouellette, and W. M. Daniewski, *ibid.*, 5063 (1978).

(4S,5S)-2-Methyl-4-methoxymethyl-5-phenyl-2-oxazoline (1), 6, 386–389.

3-*Alkylalkanoic acids* (**6**, 386–387). Full details including improved procedures have been published for the synthesis of these acids in high optical purity from **1**. The oxazoline is converted into **3** by Wittig-Horner olefination via **2**; only the (E)-isomer is formed. Remaining steps are addition of an alkyllithium to **3**, followed by hydrolysis to **5**. One limitation is that use of stabilized carbanions ($LiCH_2COOC_2H_5$, $LiCH_2CN$, etc.) results in essentially racemic products.

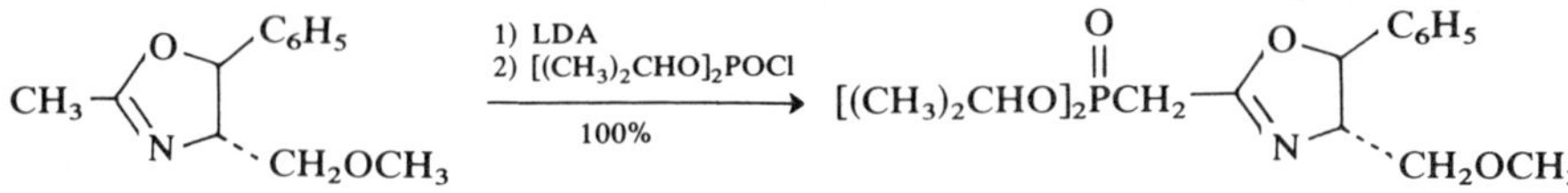

$$\mathbf{1} \xrightarrow[100\%]{1)\ LDA;\ 2)\ [(CH_3)_2CHO]_2POCl} [(CH_3)_2CHO]_2P(O)CH_2\text{-oxazoline}\ (\mathbf{2})$$

(oxazoline ring: 5-C_6H_5, 4-CH_2OCH_3)

2 → (RCHO, $KOC(CH_3)_3$; 80–90%) → **3** (E, 100%) [R—CH=CH— oxazoline with C_6H_5 and CH_2OCH_3] → (R^1Li, −78°; 40–85%) → **4** [R, R^1, H—C—CH_2— oxazoline with C_6H_5 and CH_2OCH_3] → (H_3O^+, Δ; 40–85% from **3**) → **5** [R, H, R^1—C—CH_2COOH] (ee, 91–99%)

The method is also applicable to the preparation of β-substituted valerolactones in almost 100% enantiomeric excess (equation I). In this case, 3-methoxypropionaldehyde is used to obtain the methoxy substituted acids **6**, which are converted into the chiral lactones **7** on treatment with BBr_3.[1]

(I) **6** [R, H, CH_3O, HOOC] → (BBr_3, 20°; 45–50%) → **7** [H, R; O, O] (S, 95% ee)

The sequence has been used for synthesis of natural (+)-ar-turmerone (**10**), an aromatic sesquiterpene. Reaction of **3** ($R = CH_3$) with p-tolyllithium followed by hydrolysis resulted in **8** in 98.8% enantiomeric purity, although in rather low yield (~25%). The acid was converted into **10** in essentially 100% ee.[2]

(+)-**8** [CH_3-C$_6$H$_4$, H, CH_3—C—CH_2COOH] → ($LiCH_2\overset{O}{\overset{\|}{P}}[OCH(CH_3)_2]_2$; 75%) → (+)-**9** [$CH_3$-C$_6H_4$, H, CH_3—C—$CH_2\overset{O}{\overset{\|}{C}}CH_2\overset{O}{\overset{\|}{P}}[OCH(CH_3)_2]_2$] → ($KOC(CH_3)_3$, CH_3COCH_3; 71%) → (+)-**10** [CH_3-C$_6$H$_4$, H, CH_3—C—$CH_2\overset{O}{\overset{\|}{C}}CH{=}C(CH_3)_2$]

[1] A. I. Meyers, R. K. Smith, and C. E. Whitten, *J. Org.*, **44**, 2250 (1979).
[2] A. I. Meyers and R. K. Smith, *Tetrahedron Letters*, 2749 (1979).

Methyl 5-methoxy-3-oxopentanoate (Nazarov's reagent) (1). Mol. wt. 160.2, b.p. 75–76°/0.5 mm.

Preparation[1]:

$$ClCH_2CH_2COCl + CH_2{=}CCl_2 \xrightarrow[73\%]{AlCl_3} ClCH_2CH_2\overset{O}{\overset{\|}{C}}CH_2CCl_3 \xrightarrow[50\%]{1)\ CH_3OH,\ CH_3ONa;\ 2)\ H_3O^+}$$

$$\underset{\mathbf{1}}{CH_3OCH_2CH_2\overset{O}{\overset{\|}{C}}CH_2COOCH_3} \xrightarrow{TsOH} \underset{\mathbf{2}}{CH_2{=}CH\overset{O}{\overset{\|}{C}}CH_2COOCH_3}$$

The paper refers to other, less convenient methods.

Annelation. The reagent **1** is converted by acid treatment into **2**, which has been used for annelations; an example is shown in equation (I).[2]

(I) 2-methylcyclohexane-1,3-dione (CH3, O, O) + **1** $\xrightarrow[75\%]{TsOH}$ bicyclic enedione (CH3, O, O, COOCH3)

[1] C. Wakselman and H. Molines, *Synthesis*, 622 (1979).
[2] J. E. Ellis, J. S. Dutcher, and C. H. Heathcock, *Syn. Comm.*, **4**, 71 (1974).

Methyl methylthiomethyl sulfoxide, 4, 341–342; **5**, 456–457; **6**, 390–392; **8**, 344–345.

α-Arylpropionic acids. Aryl methyl ketones such as **1** can be converted to esters (**5**) of these acids by the method shown in equation (I).[1]

$$\text{(I)}\quad \underset{\mathbf{1}}{(C_6H_5)(CH_3)C{=}O} \xrightarrow{LiCH(SCH_3)(SOCH_3)\ \mathbf{2}} \underset{\mathbf{a}}{\left[LiO{-}C(C_6H_5)(CH_3){-}CH(SCH_3)(SOCH_3)\right]} \xrightarrow{Ac_2O,\ -75^\circ}$$

$$\underset{\mathbf{3}}{AcO{-}C(C_6H_5)(CH_3){-}CH(SCH_3)(SOCH_3)} \xrightarrow[69\%\ \text{from}\ \mathbf{2}]{KOC(CH_3)_3,\ HOC(CH_3)_3} \underset{\mathbf{4}}{(C_6H_5)(CH_3)C{=}C(SCH_3)(SOCH_3)} \xrightarrow[69\%]{HCl,\ CH_3OH} \underset{\mathbf{5}}{(C_6H_5)(CH_3)CHCOOCH_3}$$

[1] K. Ogura, S. Mitamura, K. Kishi, and G. Tsuchihashi, *Synthesis*, 880 (1979).

Methyl 2-nitrophenyl disulfide, $2-NO_2C_6H_4SSCH_3$ **(1).** The disulfide is obtained in 83% yield by reaction of 2-nitrophenylsulfenyl chloride in THF at 0° with methyl mercaptan.

Bissulfenylation of α-oxomethylene groups; α-diketones.[1] Treatment of an α-oxomethylene group with NaH (2.4 equiv.) and **1** (2.2–2.4 equiv.) results in bissulfenylation with formation of the dimethyl monothioketal of an α-diketone. The thioketal group can be converted into a dimethyl ketal group by reaction with thallium(III) nitrate (**7**, 364) or into a keto group by reaction with mercury(II) perchlorate.[2]

(I) 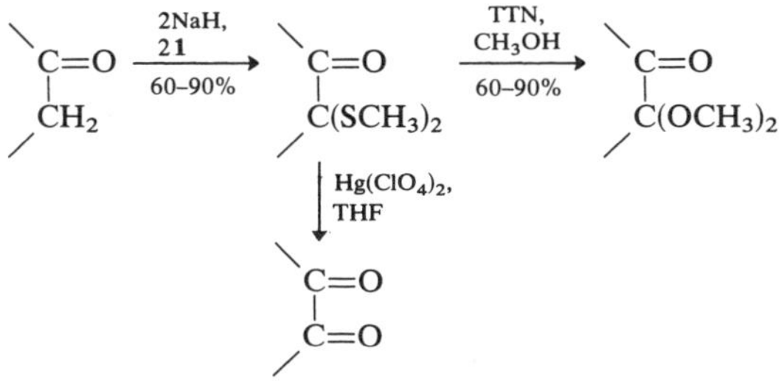 $>C{=}O\text{-}CH_2<$ $\xrightarrow[60\text{–}90\%]{2NaH,\ \mathbf{21}}$ $>C{=}O\text{-}C(SCH_3)_2<$ $\xrightarrow[60\text{–}90\%]{TTN,\ CH_3OH}$ $>C{=}O\text{-}C(OCH_3)_2<$

$>C{=}O\text{-}C(SCH_3)_2<$ $\xrightarrow{Hg(ClO_4)_2,\ THF}$ $>C{=}O\text{-}C{=}O<$

The sequence also has been used to prepare an α-methylene ketone (equation II).

(II) 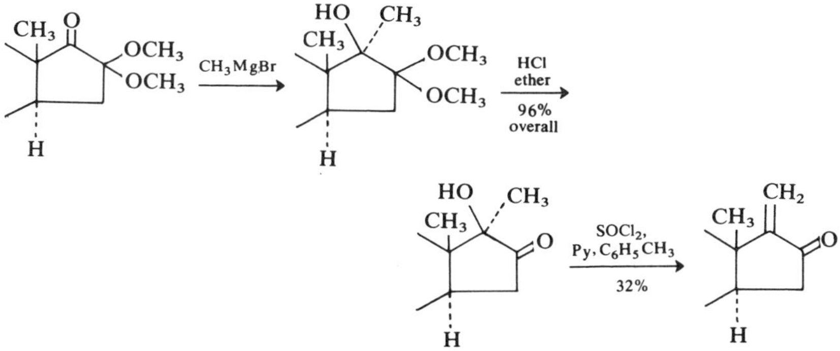 $\xrightarrow{CH_3MgBr}$ $\xrightarrow[96\%\ overall]{HCl,\ ether}$ $\xrightarrow[32\%]{SOCl_2,\ Py,\ C_6H_5CH_3}$

[1] Y. Nagao, K. Kaneko, K. Kawabata, and E. Fujita, *Tetrahedron Letters*, 5021 (1978).
[2] E. Fujita, Y. Nagao, and K. Kaneko, *Chem. Pharm. Bull. Japan*, **24**, 1115 (1976); *ibid.*, **26**, 3743 (1978).

Methyl(phenylthio)ketene, $(C_6H_5S)(CH_3)C{=}C{=}O$ **(1).** Mol. wt. 164.22. The ketene is generated *in situ* by dehydrochlorination of α-(phenylthio)propanoyl chloride with triethylamine.

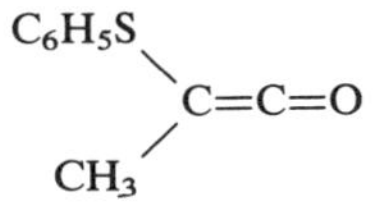

Methyleneketene equivalent.[1] This ketene can be used as the equivalent of the unstable methyleneketene. When it is generated in the presence of cyclopentadiene,

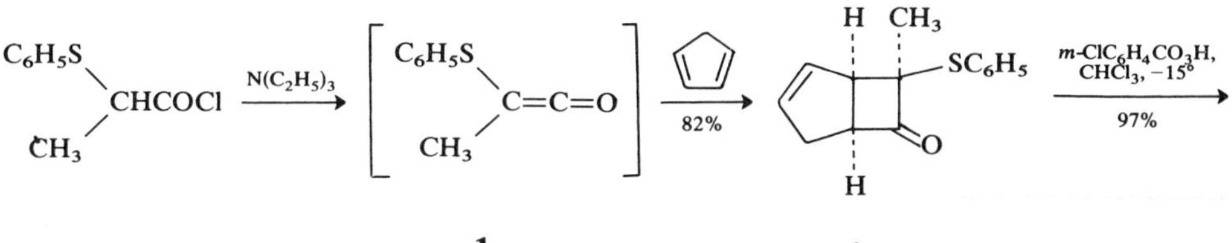

$(C_6H_5S)(CH_3)CHCOCl \xrightarrow{N(C_2H_5)_3} [(C_6H_5S)(CH_3)C{=}C{=}O]\ (\mathbf{1}) \xrightarrow[82\%]{\text{cyclopentadiene}} \mathbf{2} \xrightarrow[97\%]{m\text{-}ClC_6H_4CO_3H,\ CHCl_3,\ -15^\circ}$

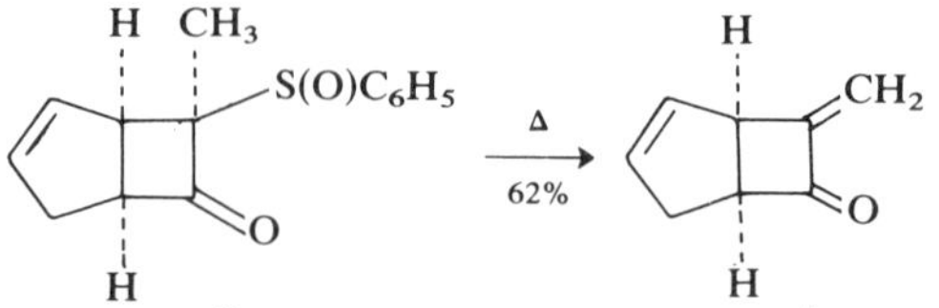

H CH₃ S(O)C₆H₅ → (Δ, 62%) H CH₂

3 4

the cyclobutanone **2** is obtained in 82% yield. This product is convertible into the methylenecyclobutanone **4**.

[1] T. Minami, M. Ishida, and T. Agawa, *J.C.S. Chem. Comm.*, 12 (1978); *J. Org.*, **44**, 2067 (1979).

N-Methyl-2-pyrrolidone, (**1**), **1**, 696; **2**, 281.

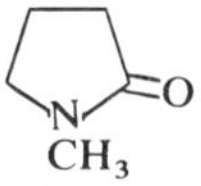

Oxy-Cope rearrangement. Fujita *et al.*[1] have examined the effect of various solvents on the course of the oxy-Cope rearrangement of 2-propynols such as **2**. N-Methylpyrrolidone markedly favors formation of **4**. HMPT also has the same effect, but in this case it is a result in part of isomerization of **3** to **4**. This solvent effect was useful in a synthesis of polyprenyl ketones via an oxy-Cope rearrangement. ε-Caprolactam shows a similar solvent effect.

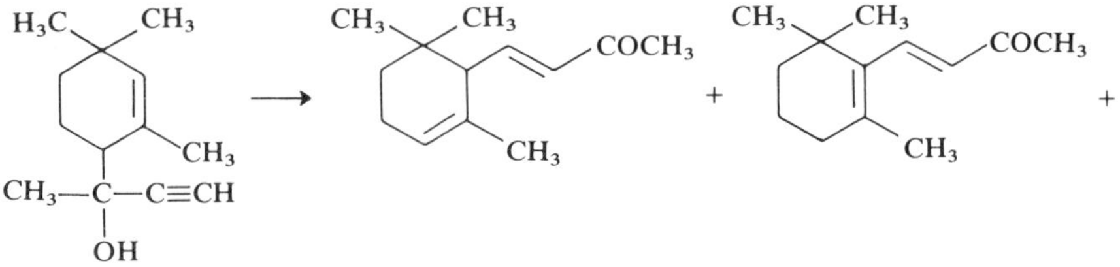

2 → 3 + 4 + 5

160°, neat ⟶ **3** + **4** + **5** = 61 : 11 : 28
1, Δ ⟶ **3** + **4** = 42 : 58
HMPT, Δ ⟶ **3** + **4** = 36 : 64
n-Decane, Δ ⟶ **3** + **5** = 61 : 39

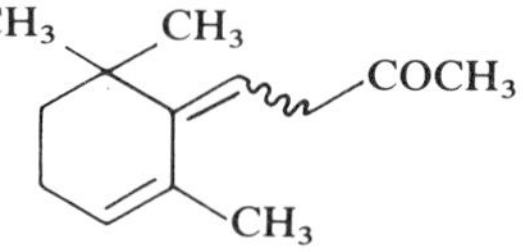

[1] T. Onishi, Y. Fujita, and T. Nishida, *J.C.S. Chem. Comm.*, 651 (1978); Y. Fujita, T. Onishi, and T. Nishida, *Synthesis*, 934 (1978).

Molecular sieves, 1, 703–705; **2**, 286–287; **3**, 306; **4**, 345; **5**, 465; **6**, 411–412.

Acetylation of tertiary hydroxyl groups. Midecamycin (**1**), a macrolide antibiotic from *S. mycarofacience*, is readily converted into the 9,2′-diacetyl derivative by Ac_2O–Py; forcing conditions required for acetylation of the 3″-hydroxyl group result in decomposition. However, reaction of the diacetate with acetyl chloride in ethyl acetate in the presence of a 4 Å molecular sieve to scavenge the hydrogen chloride formed (**6**, 412) gives the 9,2′,3″-triacetate of **1** in satisfactory yield. This triacetate undergoes partial hydrolysis to the 9,3″-diacetate on treatment with aqueous methanol. The overall yield is 51%.[1]

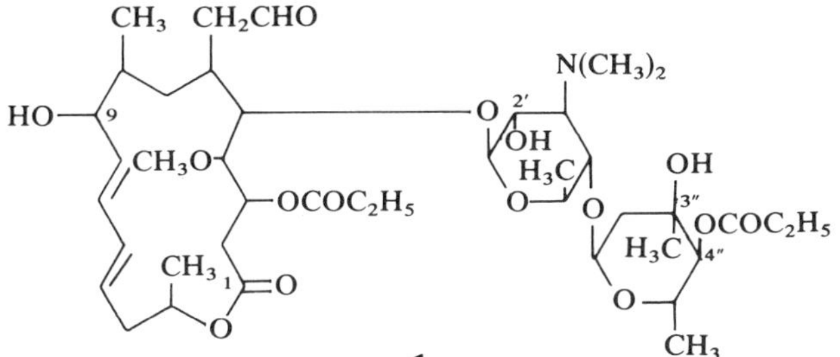

1

[1] T. Nakamura, S. Fukatsu, S. Seki, and T. Niida, *Chem. Letters*, 1293 (1978).

Molybdenum carbonyl, 2, 287; **3**, 206–207; **4**, 346; **7**, 247–248.

Dehalogenation (**7**, 247). The reduction of α-halo ketones to the corresponding ketone by $Mo(CO)_6$ is facilitated by use of this metal carbonyl adsorbed on alumina. Yields are considerably higher because products of condensation are not formed, and the rate of reduction is markedly enhanced with the supported reagent.[1]

[1] H. Alper and L. Patter, *J. Org.*, **44**, 2568 (1979).

Monoisopinocampheylborane [(IPC)BH_2], *see* diisopinocampheylborane, **1**, 266–268; **4**, 161–162. The laboratories of Pelter and of Brown[1] have succeeded in the preparation of monoisopinocampheylborane by hydroboration of α-pinene. In one method, equimolar proportions of (+)-α-pinene and BH_3·THF are mixed in THF and the solution is allowed to stand for 96 hours, after which time it contains about 90% (IPC)BH_2, 4.75% $(IPC)_2BH$, and 4.75% BH_3·THF. The last material is precipitated as TMEDA·$2BH_3$, and the solution is used for hydroboration. Alternatively, addition of TMEDA precipitates TMEDA·2(IPC)BH_2 as a white solid, m.p. 139–140°. This is dissolved in THF, and then BF_3 etherate is added to generate (IPC)BH_2.

Either material when used for hydroboration of 1-methylcyclopentene gives, after oxidation, *trans*-2-methylcyclopentanol in 70–75% optical purity.

[1] A. Pelter, D. J. Ryder, J. H. Sheppard, C. Subrahmanyam, H. C. Brown, and A. K. Mandal, *Tetrahedron Letters*, 4777 (1979).

Morpholine–Camphoric acid,

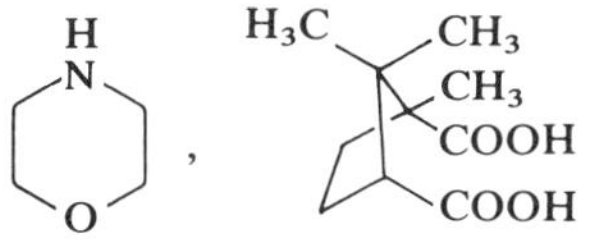

Intramolecular aldol reaction. The regioselectivity of the intramolecular aldol condensation of the dialdehyde (**1**) is markedly affected by the base used. Thus use of·

L-proline followed by reaction of the product with diethyl cyanomethylphosphonate (**1**, 250; **2**, 130–131) gives **2** and **3** in a 1:1 ratio (18.3% total yield).[1] Use of piperidine acetate[2] gives **2** and **3** in a 1:22 ratio (50.3% total yield). Dibenzylammonium trifluoroacetate[3] shows similar selectivity: **2**/**3** = 1:19 (76.4% total yield). However, use of morpholine–camphoric acid in ether–HMPT favors formation of **2**. With this catalyst **2** and **3** are obtained in 37.5% and 1.5% yield, respectively.[4]

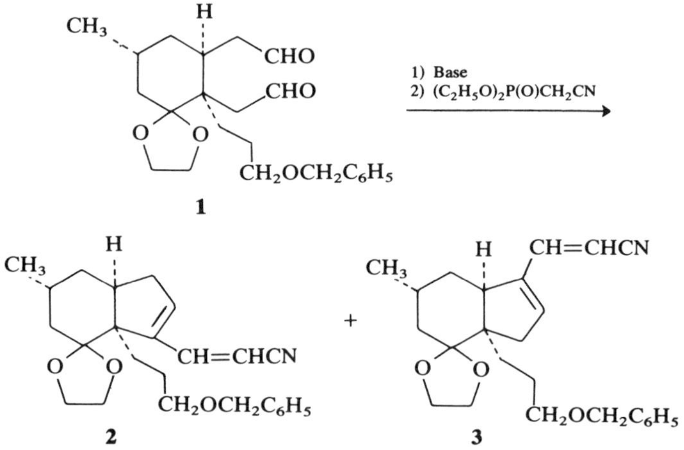

This reaction was used in a synthesis of the lycopodium alkaloid 8-deoxyserratinine (**4**).[4]

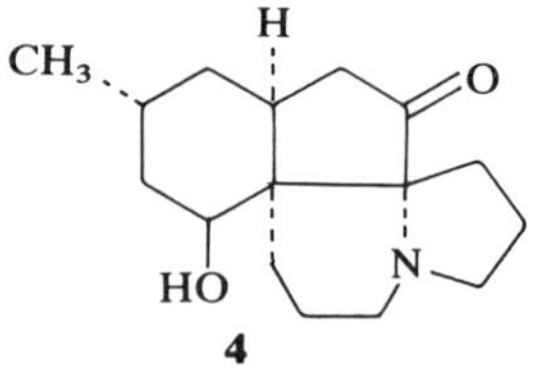

4

[1] T. Harayama, M. Ohtani, M. Oki, and Y. Inubushi, *Chem. Pharm. Bull. Japan*, **21**, 1061 (1973).
[2] R. B. Woodward, F. Sondheimer, D. Taub, K. Heusler, and W. M. McLamore, *Am. Soc.*, **74**, 4223 (1952).
[3] E. J. Corey, R. L. Danheiser, S. Chandrasekaran, P. Siret, G. E. Keck, and J.-L. Gras, *ibid.*, **100**, 8031 (1978).
[4] T. Harayama, M. Takatani, and Y. Inubushi, *Tetrahedron Letters*, 4307 (1979).

N-Morpholinomethyldiphenylphosphine oxide (1). Mol. wt. 301.32, m.p. 159°.

Preparation:

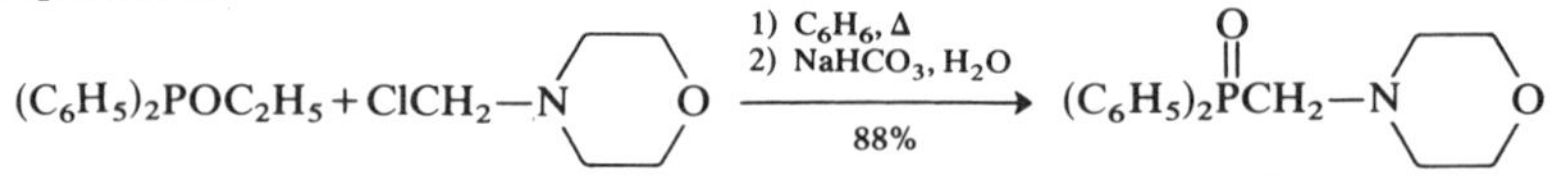

1

Morpholino enamines.[1] The lithium anion **2** of the reagent reacts with aldehydes to form the adducts **3** in high yield. These are converted into (E)-enamines (**4**) by treatment with potassium hydride. The method is applicable to ketones, but yields are lower because of formation of the enolate.

$$RCHO + (C_6H_5)_2\overset{O}{\overset{\|}{P}}-\overset{Li}{\overset{|}{C}}H-N(CH_2CH_2)_2O \ (\mathbf{2}) \xrightarrow{THF} (C_6H_5)_2\overset{O}{\overset{\|}{P}}-\underset{RCHOH}{\underset{|}{C}H}-N(CH_2CH_2)_2O \ (\mathbf{3}) \xrightarrow[65\text{–}99\%]{KH,\ THF} \underset{H}{\overset{R}{>}}C{=}C\underset{N(CH_2CH_2)_2O}{\overset{H}{<}} \ (\mathbf{4})$$

Since enamines are readily hydrolyzed by dilute acid, this sequence can be used for sẏnthesis of homologous aldehydes.

[1] N. L. J. M. Broeckhof, F. L. Jonkers, and A. van der Gen, *Tetrahedron Letters*, 2433 (1979).

N

Nafion-H. Nafion is the trade name of du Pont Co. for a perfluorinated resin acid commercially available as the potassium salt. The H-form is generated from the salt by treatment with nitric acid, followed by washing with water and drying at 105°.

Olah and co-workers have used this solid superacid as a catalyst in various aromatic substitution reactions: alkylation,[1] aroylation,[2] and nitration.[3]

Nafion-H is a convenient acid catalyst for pinacolone rearrangements.[4] Hydration of acetylenes can be conducted with Nafion-H impregnated with mercury(II) ions.[5]

Diels–Alder catalyst. This protic resin catalyzes Diels–Alder reactions, but longer reaction times are needed than in reactions catalyzed by Lewis acids. The reactions are generally conducted in refluxing benzene or chloroform. In the case of dienes that polymerize readily, the reaction is conducted at room temperature for 1–2 days.[6]

[1] G. A. Olah, J. Kaspi, and J. Bukala, *J. Org.*, **42**, 4187 (1977).
[2] G. A. Olah, R. Malhotra, S. C. Narang, and J. A. Olah, *Synthesis*, 672 (1978).
[3] G. A. Olah, R. Malhotra, and S. C. Narang, *J. Org.*, **43**, 4628 (1978).
[4] G. A. Olah and D. Meidar, *Synthesis*, 358 (1978).
[5] *Idem, ibid.*, 671 (1978).
[6] G. Olah, D. Meidar, and A. P. Fung, *ibid.*, 270 (1979).

(S)-1-α-Naphthylethylamine, 8, 355–356.

Chiral synthesis of a prostanoid. Trost *et al.*[1] have devised an enantioconvergent prostanoid synthesis from **1**, in which a 1,3-hydroxy shift interconverts the two enantiomers. The interconversion was possible by equilibration of the urethane of **1** with (S)-1-α-naphthylethylamine catalyzed with mercury(II) trifluoroacetate (**7**, 382). At equilibrium the ratio of **2a/2b** is about 1:3. By separation and repeated

CH_3OOC / HO — **1a**; $COOCH_3$ / OH — **1b**

CH_3OOC / OCR(=O) — **2b** $\underset{}{\overset{(CF_3CO_2)_2Hg}{\rightleftharpoons}}$ RCO(=O) / $COOCH_3$ — **2a** $\xrightarrow[76\%]{Cl_3SiH,\ N(C_2H_5)_3}$ **1a** (1R, 5R)

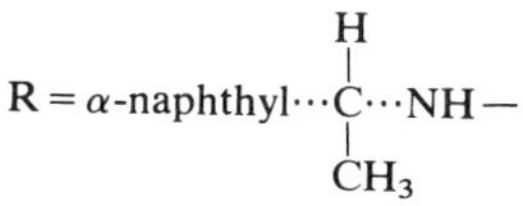

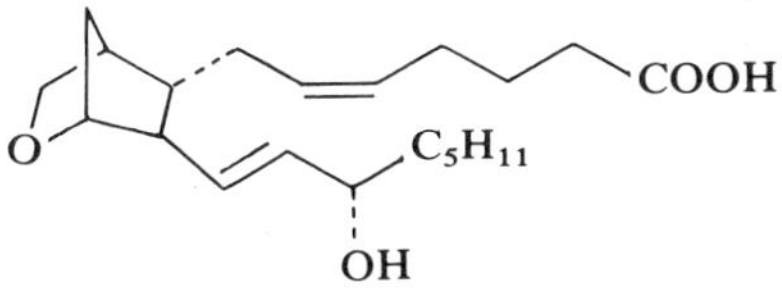

R = α-naphthyl···C(H)(CH₃)···NH—

3

equilibration, the racemate can be converted into the single enantiomer **1a**. This sequence was used in a synthesis of the prostanoid **3**.

[1] B. M. Trost, J. M. Timko, and J. L. Stanton, *J.C.S. Chem. Comm.*, 436 (1978).

(R)-1-(1-Naphthyl)ethyl isocyanate (1), 8, 356–357.

Chiral epoxides. A new method for resolution of racemic epoxides involves cleavage with sodium thiophenoxide to give racemic β-hydroxy sulfides (**2a** and **2b**); these are converted by reaction with **1** into diastereomeric carbamates, which can be resolved by chromatography. Cleavage of either carbamate with trichlorosilane affords optically pure β-hydroxy sulfides (**2**) in yields of 80–90%. These are converted into optically active epoxides (**3**) by trimethyloxonium tetrafluoroborate followed by treatment with base. The epoxides are obtained in optical purities of 85–100%.[1]

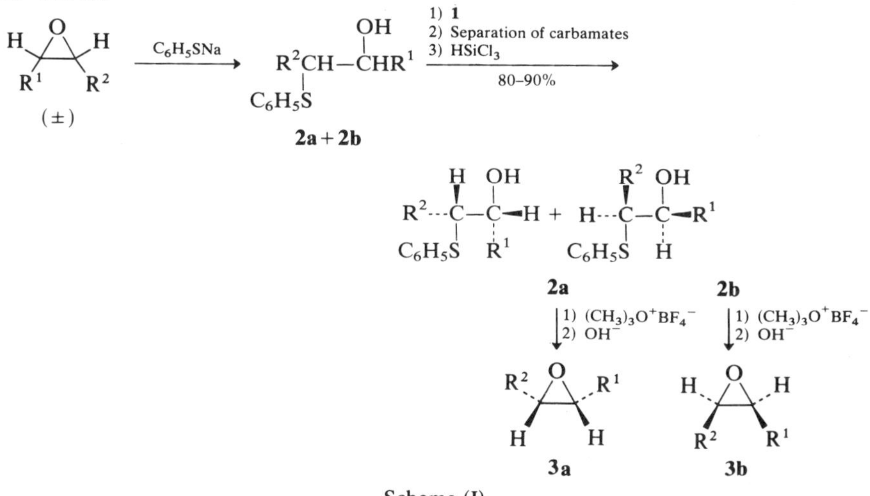

(±) $\xrightarrow{C_6H_5SNa}$ $R^2CH(SC_6H_5)-CH(OH)R^1$ **2a + 2b** $\xrightarrow[80-90\%]{1)\ \mathbf{1};\ 2)\ \text{Separation of carbamates};\ 3)\ HSiCl_3}$ **2a** + **2b**

2a $\xrightarrow{1)\ (CH_3)_3O^+BF_4^-;\ 2)\ OH^-}$ **3a**

2b $\xrightarrow{1)\ (CH_3)_3O^+BF_4^-;\ 2)\ OH^-}$ **3b**

Scheme (I)

[1] W. H. Pirkle and P. L. Rinaldi, *J. Org.*, **43**, 3803 (1978).

Nickel–Alumina.

Catalytic hydrogenolysis of functional groups. A variety of functional groups (—Br, $—NH_2$, —OH, $—OSi(CH_3)_3$, $—CH_2OH$, —COOH, $—COOCH_3$, —COCl, —CN, $—CH_3$, $—CH_2Br$, C=O) are hydrogenolyzed in the gas phase using 30% nickel on alumina and hydrogen. The yield of hydrocarbon product is high (70–99%). Some selectivity for various functional groups is observed.[1]

Examples:

CN (H_2, Ni, Al_2O_3, 180°; 99%) → CH_3 (280°; 80%) →

OH, CH_3 (150°; 99%) → CH_3 (235°; 99%) →

[1] W. F. Maier, P. Grubmüller, I. Thies, P. M. Stein, M. A. McKervey, and P. v. R. Schleyer, *Angew. Chem., Int. Ed.*, **18**, 939 (1979).

Nickel(II) bromide–Zinc.

Ullmann reaction (**4**, 33; **6**, 654). The nickel species generated from $NiBr_2$ and zinc in HMPT is an effective catalyst for the preparation of biaryls from aryl iodides.[1] An excess of zinc is generally helpful; potassium iodide exerts a favorable effect. The reaction is conducted at 40–70° for 3–30 hours. Yields are >90% for iodobenzene itself and are also similar for *para*- and *meta*-substituted derivatives. Yields are lowered by the presence of *ortho*-substituents.

[1] K. Takagi, N. Hayama, and S. Inokawa, *Chem. Letters*, 917 (1979).

Nickel peroxide, 1, 731–732; **5**, 474; **7**, 250–251; **8**, 357–358. Review (90 references).[1]

α-Allenic aldehydes, ketones, and amides.[2] α-Allenic alcohols with two γ-alkyl substituents are oxidized in satisfactory yield to α-allenic carbonyl compounds by excess nickel peroxide. Amides can also be obtained by oxidation in the presence of ammonia.

Examples:

$$(C_2H_5)(CH_3)C{=}C{=}CHCH_2OH \xrightarrow[88\%]{NiO_2} (C_2H_5)(CH_3)C{=}C{=}CHCHO$$

$$(CH_3)_2C{=}C{=}CHCH(OH)CH_3 \xrightarrow[89\%]{} (CH_3)_2C{=}C{=}CHCOCH_3$$

$$(CH_3)_2C{=}C{=}CHCH_2OH \xrightarrow[60\%]{NiO_2,\ NH_3} (CH_3)_2C{=}C{=}CHCONH_2$$

Dehydrogenation of heterocycles (**8**, 357–358). Hecht, Meyers, and coworkers[3] have examined the dehydrogenation of various partially reduced heterocycles containing oxygen, sulfur, and nitrogen with NiO_2. This oxidant is particularly useful for dehydrogenation of thiazolines to thiazoles even in the presence of various functional groups. It can also effect oxidation of oxazolines to 1,3-oxazoles, a conversion that had not been observed previously. The paper also reports several other related dehydrogenations.

[1] M. V. George and K. S. Balachandran, *Chem. Rev.*, **75**, 491 (1975).
[2] M. Bertrand, G. Gil, and J. Viola, *Tetrahedron Letters*, 1595 (1979).
[3] D. L. Evans, D. K. Minster, U. Jordis, S. M. Hecht, A. L. Mazzu, Jr., and A. I. Meyers, *J. Org.*, **44**, 497 (1979).

Nitromethane, 1, 739; **4**, 357.

Nitromethylation of aldehydes. A few years ago Bachman and Maleski[1] reported the preparation of 1-nitroalkanes (**3**) by condensation of nitromethane with an aldehyde catalyzed by base, acetylation of the product (**1**), and finally reduction with $NaBH_4$. Reported overall yields were 34–70%.

The reaction has been considerably improved by several modifications. The initial condensation is accelerated by use of KF and 18-crown-6; isopropanol is a suitable solvent. Acetylation is effected with acetic anhydride at 25° with 4-dimethylaminopyridine as catalyst (**3**, 118). The reduction is carried out with $NaBH_4$ in ethanol. These mild conditions are compatible with various functional groups often present in the synthesis of natural products.[2]

$$RCHO + CH_3NO_2 \xrightarrow[(CH_3)_2CHOH]{KF,} \underset{\mathbf{1}}{RCH(OH)CH_2NO_2} \xrightarrow{Ac_2O} \underset{\mathbf{2}}{RCH(OAc)CH_2NO_2} \xrightarrow[\text{(65–90\% isolated, overall)}]{NaBH_4,\ C_2H_5OH} \underset{\mathbf{3}}{RCH_2CH_2NO_2}$$

Cyclopropanation. Chemists at Schering AG (Berlin)[3] discovered a novel cyclopropanation in the reaction of the derivative (**1**) of estrone with nitromethane and sodium methoxide in methanol. The product is **2**, evidently formed by Michael addition of $^-CH_2NO_2$ and ring closure with elimination of the nitro group. This

NC COOCH₃ / C / H₃C — CH₃O — **1** $\xrightarrow[85\%]{CH_3NO_2,\ NaOCH_3,\ HOCH_3}$ NC COOCH₃ / H₃C — **2**

3 (cyclohexylidene)C(CN)₂ $\xrightarrow{33\%}$ **4** (spirocyclopropane)(CN)₂

simple cyclopropanation is limited to alkenes of the type $R^1R^2C{=}CXY$, where X and Y are electron-attracting groups (CN or COOR). Unfortunately yields are rather low in simple systems such as cyclohexylidenemalononitrile (**3**).

[1] G. B. Bachman and R. J. Maleski, *J. Org.*, **37**, 2810 (1972).
[2] R. H. Wollenberg and S. J. Miller, *Tetrahedron Letters*, 3219 (1978).
[3] K. Annen, H. Hofmeister, H. Laurent, A. Seeger, and R. Wiechert, *Ber.*, **111**, 3094 (1978).

Nitronium tetrafluoroborate, 1, 742–743; **4**, 358; **5**, 476–477; **8**, 361.

Dethioacetalization.[1] Ethylene dithioacetals are cleaved to carbonyl compounds by nitronium tetrafluoroborate in excellent yield. Other somewhat less effective reagents are $NO^+HSO_4^-$ and $NaNO_2$ or $NaNO_3$ in TFA.

$$R^1R^2C(SCH_2CH_2S) \xrightarrow[80\text{–}95\%]{NO_2^+BF_4^-,\ CH_2Cl_2,\ -15^\circ} R^1R^2C{=}O$$

Acetamides.[2] Alkyl halides and ethers can be converted into acetamides by reaction with this reagent in acetonitrile. The reactivity is RI > RBr ~ ROR ≫ RH ~ RCl > RF.

$$RX(OR^1) \xrightarrow{NO_2^+BF_4^-} \left[R^+ \xrightarrow{CH_3C\equiv N} R\overset{+}{N}\equiv CCH_3\right] \xrightarrow[50\text{–}90\%]{H_2O} RNHCOCH_3$$

[1] G. A. Olah, S. C. Narang, G. F. Salem, and B. G. B. Gupta, *Synthesis*, 273 (1979).
[2] R. D. Bach, J. W. Holubka, and T. A. Taaffee, *J. Org.*, **44**, 1739 (1979).

***p*-Nitroperbenzoic acid, 1**, 743.

Hydroxylation.[1] Tertiary C—H bonds of alicyclic hydrocarbons are oxidized to the corresponding alcohols by *p*-nitroperbenzoic acid. Unlike the oxidation with trifluoroperacetic acid (**7**, 281–282), the reaction occurs regioselectively and with practically complete retention of configuration, and thus is comparable to dry ozonation (**6**, 440).

$$\text{1,2-dimethylcyclohexane} \xrightarrow[70\%]{p\text{-}NO_2C_6H_4CO_3H,\ CHCl_3,\ \Delta} \text{1,2-dimethylcyclohexan-1-ol (OH, CH}_3\text{, CH}_3\text{)}$$

The peracid oxidizes secondary hydroxy groups to ketones, which then undergo the Baeyer–Villiger reaction.

[1] W. Muller and H.-J. Schneider, *Angew. Chem., Int. Ed.*, **18**, 407 (1979).

p-Nitrophenyl selenocyanate, $NO_2-C_6H_4-SeCN$ **(1).** Mol. wt. 227.08, m.p. 142°. Preparation.[1]

Isomerization of primary allylic alcohols.[2] Reaction of 2-alkenols (**2**) in pyridine with this reagent (1.2 equiv.) and tri-*n*-butylphosphine (1.2 equiv.) leads to terminal selenides (**3**) usually in 90–95% yield. The selenides (**3**) are converted into the rearranged alcohol (**4**) by oxidation with 15% H_2O_2 at room temperature. Two processes are involved: a [2,3]sigmatropic rearrangement of **a** to give **b**, and hydrolysis. (Compare a similar transposition using allylic sulfoxides, **5**, 400–402, **6**, 30–31.)

$$\underset{\mathbf{2}}{RCH{=}CHCH_2OH} + \mathbf{1} \xrightarrow[90\text{–}95\%]{(C_4H_9)_3P,\ Py.,\ 20°} \underset{\mathbf{3}}{RCH{=}CHCH_2SeAr} \xrightarrow{H_2O_2,\ Py.}$$

$$\left[\underset{\mathbf{a}}{RCH{=}CHCH_2\overset{O}{\overset{\|}{Se}}Ar} \rightarrow \underset{\mathbf{b}}{R\overset{OSeAr}{\overset{|}{C}}HCH{=}CH_2}\right] \xrightarrow[75\text{–}85\%]{H_2O} \underset{\mathbf{4}}{R\overset{OH}{\overset{|}{C}}HCH{=}CH_2}$$

[1] H. Bauer, *Ber.*, **46**, 92 (1913).
[2] D. L. J. Clive, G. Chittattu, N. J. Curtis, and S. M. Menchen, *J.C.S. Chem. Comm.*, 770 (1978).

o-Nitrophenyl thiocyanate, $o\text{-}NO_2C_6H_4SCN$ **(1).** Mol. wt. 180.1, m.p. 136°. Preparation.[1]

Amides.[2] Carboxylic acids can be converted directly into amides by reaction with this thiocyanate, tri-*n*-butylphosphine, and an amine in THF (20°, 7 hours). Yields are 95–100%.

$$RCOOH + R^1R^2NH \xrightarrow[95\text{–}100\%]{\mathbf{1},\ Bu_3P} RCONR^1R^2$$

The reaction can be used to prepare five-, six-, and seven-membered lactams from γ, δ, and ω-aminocarboxylic acids.

[1] T. Wagner-Jauregg and E. Helmut, *Ber.*, **75**, 935 (1942).
[2] P. A. Grieco, D. S. Clark, and G. P. Withers, *J. Org.*, **44**, 2945 (1979).

3-Nitro-2-pyridinesulfenyl chloride, (3-nitro-2-pyridyl)-SCl **(1).** Mol. wt. 190.59, m.p. 217–222° dec., yellow needles, stable for at least 6 months in a refrigerator. The sulfenyl chloride is obtained in about quantitative yield by reaction of chlorine with bis(3-nitro-2-pyridyl) disulfide. The unusual stability is associated with the presence of the nitro group.

Esterification of cephalosporanic acids.[1] Usual esterification methods of these acids lead to mixtures of Δ^3- and Δ^2-esters. This undesired isomerization is prevented by conversion of the acid into the 3-nitro-2-pyridinethiol ester (**2**) by reaction with **1** and triphenylphosphine. This stable intermediate (obtainable in about 90% yield) reacts with various alcohols to form the desired esters in 60–80% yield (equation I).

(I) [cephalosporanate salt, R^1CONH–, CH_2R^2, $COO^-\overset{+}{H}NR_3^3$] $\xrightarrow[\sim 90\%]{1,\ P(C_6H_5)_3}$ **2** [3-nitro-2-pyridinethiol ester, O_2N] $\xrightarrow[60\text{–}80\%]{R^4OH}$ **3** [ester, $COOR^4$, CH_2R^2]

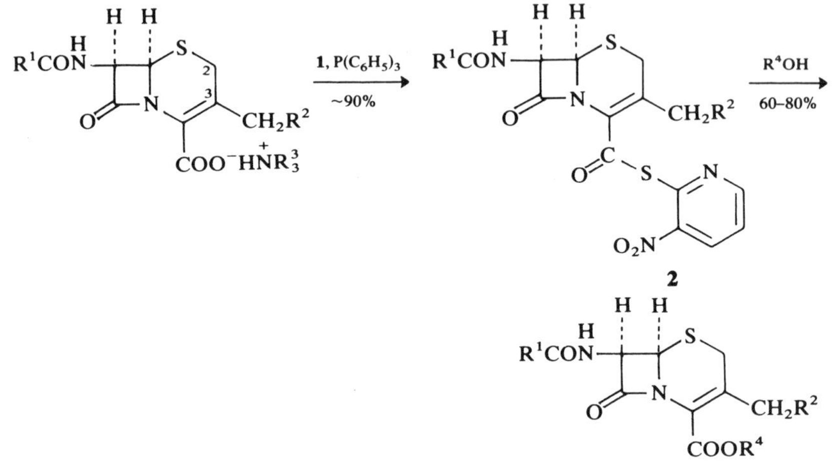

[1] R. Matsuda and K. Aiba, *Chem. Letters*, 951 (1978); R. Matsuda, *ibid.*, 979 (1978).

Nitrosocarbonylmethane, CH_3CONO (**1**).

Allylic amidation.[1] Nitrosocarbonylmethane undergoes an ene reaction with various olefins to form allylic N-hydroxy amides, often in useful yields.

Examples:

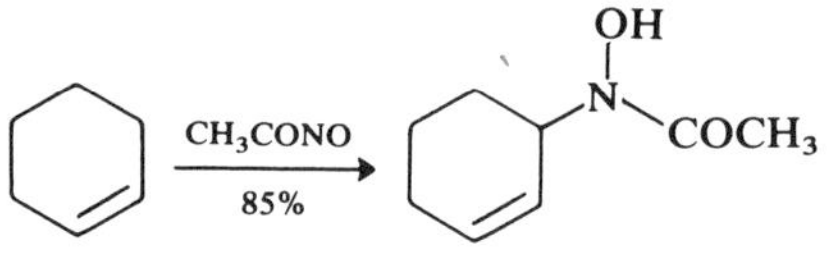

Cyclohexene $\xrightarrow[85\%]{CH_3CONO}$ 3-[N(OH)COCH$_3$]cyclohexene

$$CH_3(CH_2)_4CH_2CH{=}CH_2 \xrightarrow[83\%]{} CH_3(CH_2)_4CH{=}CHCH_2N(OH)COCH_3$$

(*cis*/*trans* = 1 : 1)

[1] G. E. Keck and J. B. Yates, *Tetrahedron Letters*, 4627 (1979).

Nitrosonium hexafluorophosphate, 1, 747.

Amides.[1] Amides can be prepared by reaction of alkyl halides with nitriles in the presence of nitrosonium hexafluorophosphate at −15 to 25°.

$$R^1X + R^2CN \xrightarrow[-NOX]{NO^+PF_6^-} [R^1\overset{+}{N}{\equiv}CR^2PF_6^-] \xrightarrow[30\text{–}90\%]{H_2O} R^1NHCOR^2$$

[1] G. A. Olah, B. G. B. Gupta, and S. C. Narang, *Synthesis*, 274 (1979).

O

1,7-Octadiene-3-one, $CH_2{=}CH(CH_2)_3C(=O)CH{=}CH_2$ (**1**). Mol. wt. 124.18, b.p. 31°/4 mm; semicarbazone, m.p. 180–182°.

Preparation:

$$CH_2{=}CHCH{=}CH_2 \xrightarrow{Pd(OAc)_2,\ P(C_6H_5)_3} CH_2{=}CH(CH_2)_3CH(OAc)CH{=}CH_2 \xrightarrow[\text{"high"}]{1)\ H_2O;\ 2)\ Cu\text{–}Zn,\ 360°\ \text{or DMSO–ClCOCOCl}} \mathbf{1}$$

Bisannelation (*see also* **6**, 409). Tsuji *et al.*[1] have used this reagent to effect bisannelation. The product of addition of **1** to the pyrrolidine enamine of cyclohexanone gives, after hydrolysis and aldol condensation, **2** in 64% yield. The terminal double bond of **2** is oxidized with $PdCl_2$ and CuCl in aqueous DMF under O_2 (**7**, 278), and the resulting methyl ketone **3** is cyclized with base to the tricycle **4** in 60% yield from **2**.

1) **1**; 2) H^+ (64%) → **2** ($H_2C{=}CH$); $PdCl_2$, CuCl, O_2 (83%) → **3** (H_3C); $KO{-}C(CH_3)_2C_2H_5$ (73%) → **4**

5 (CH_3, $OC(CH_3)_3$, H, $COOCH_3$) → 1) NaH, **1**; 2) NaI, HMPT (68%) → **6** (CH_3, $OC(CH_3)_3$, H, $CH_2{=}CH$)

→ 1) NaOH, C_2H_5OH (90%); 2) $PdCl_2$, CuCl, O_2 (78%); 3) Pd/C (95%) → **7** (CH_3, $OC(CH_3)_3$, H, H_3C) → HCl, CH_3OH, Δ (76%) → **8** (CH_3, OH, H)

The reagent **1** also reacts with stabilized carbanions, as in the synthesis of (+)-19-nortestosterone (**8**) from optically active **5**.

[1] J. Tsuji, I. Shimizu, H. Suzuki, and Y. Naito, *Am. Soc.*, **101**, 5070 (1979).

Organoaluminum cyanides.

Hydrocyanation of conjugated carbonyl compounds. This reaction has been reviewed (305 references).[1] Organoaluminum cyanides offer some advantages over use of hydrogen cyanide or an alkali metal cyanide. They are more reactive and more selective for enones. The two more commonly used reagents are hydrogen cyanide in combination with triethylaluminum (**2**, 427; **4**, 526; **5**, 688–689) and diethylaluminum cyanide (**1**, 244; **2**, 127–128; **4**, 146–147). The latter reagent is more reactive. Another difference is that the addition with the former reagent is irreversible and thus under kinetic control, whereas with the latter reagent the reaction is reversible and under thermodynamic control. Thus if the kinetic and thermodynamic products are different, one or the other product can often be obtained selectively. An example is the hydrocyanation of cholestenone under different conditions as shown.

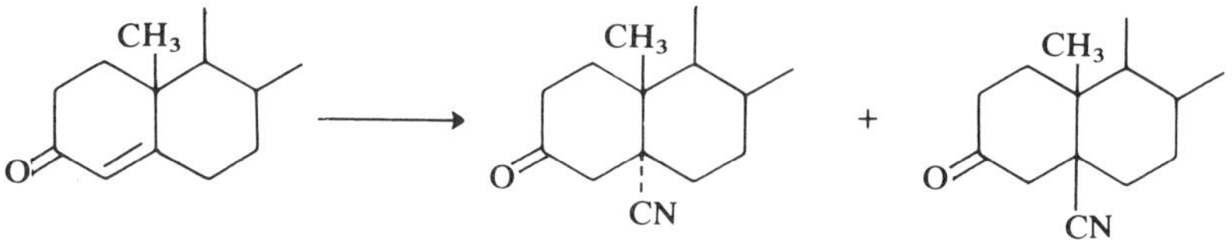

Reagent	Ratio
$HCN-Al(C_2H_5)_3$	49:42
$(C_2H_5)_2AlCN$ (4 hr.)	45:42
$(C_2H_5)_2AlCN$ (10 hr.)	10:90
$KCN-NH_4Cl$ (8 hr.)	33:51

When the 1,2-adduct is unusually stable or reversal of 1,4-addition is favorable, hydrogen cyanide with diethylaluminum chloride is the preferred reagent.

[1] W. Nagata and M. Yoshioka, *Org. React.*, **25**, 255 (1977).

Organocopper reagents.

Trisubstituted alkenes. Complete details have been published for preparation of these alkenes by addition of alkylcopper complexes to acetylenes. The paper includes numerous references to work of other laboratories as well. The paper also presents convincing evidence that the reaction involves *syn*-addition with a high degree of stereoselectivity.[1]

Cycloheptane annelation (**7**, 212). The mixed cuprate **1** reacts with acid chlorides to afford vinylcyclopropyl ketones. Previously these ketones were prepared from aldehydes by condensation with 1-lithio-2-vinylcyclopropane followed by oxidation (**7**, 192–193). These compounds are rearranged to 4-cycloheptenones on conversion to trimethylsilyl enol ethers, thermolysis, and hydrolysis.[2]

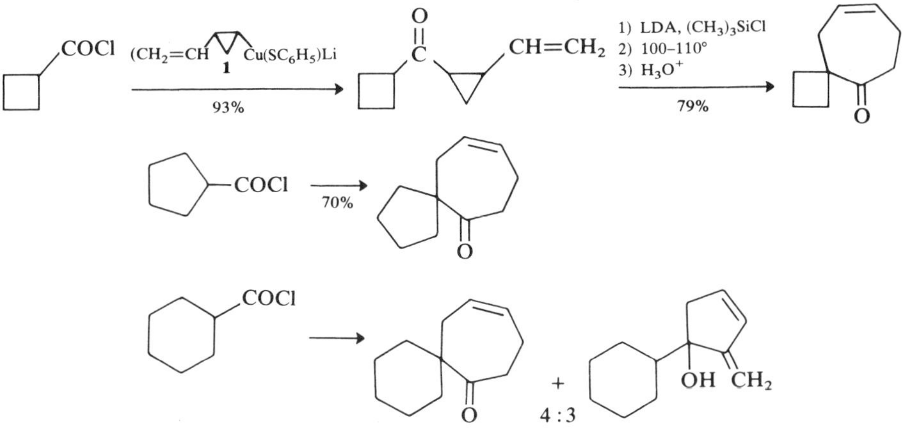

COCl (CH$_2$=CH Cu(SC$_6$H$_5$)Li 1 93% C=O CH=CH$_2$ 1) LDA, (CH$_3$)$_3$SiCl 2) 100–110° 3) H$_3$O$^+$ 79% O

COCl 70% O

COCl O + 4:3 OH CH$_2$

Mixed cyanocuprates, [RCuCN]Li. These cuprates (**1**) are prepared by addition of 1 equiv. of copper(I) cyanide to an alkyllithium in ether at −40°. [CH_3CuCN]Li and [C_6H_5CuCN]Li add stereo- and regiospecifically to 1,3-cyclohexadiene monoepoxide (**2**) to give **3**, which can be epoxidized by *m*-chloroperbenzoic acid to give **4** in high yield. The products (**4**) are valuable for stereocontrolled synthesis of trisubstituted cyclohexenols (scheme I).[3]

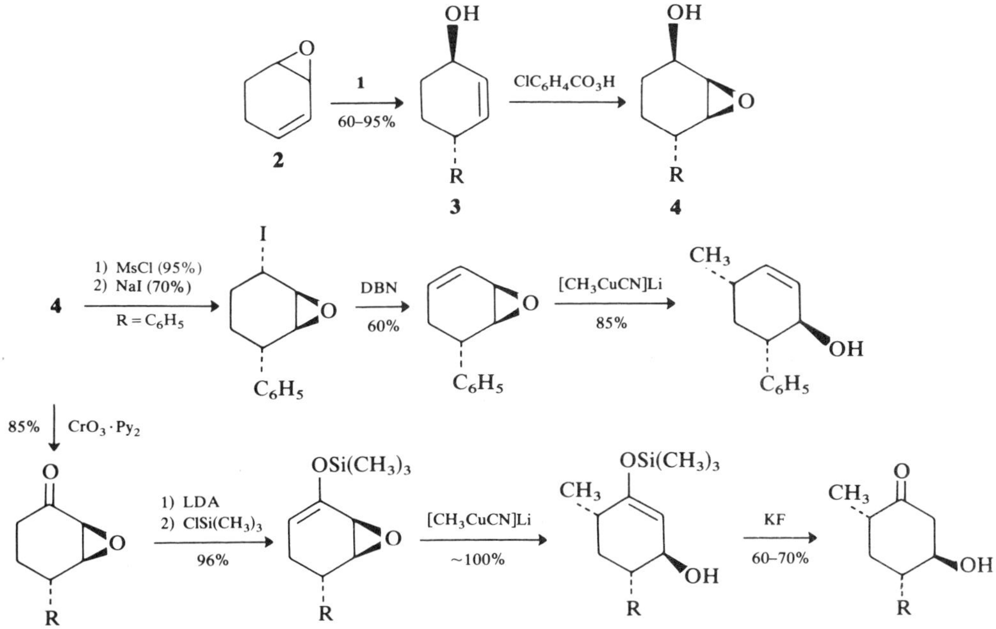

O 2 1 60–95% OH R 3 ClC$_6$H$_4$CO$_3$H OH O R 4

4 1) MsCl (95%) 2) NaI (70%) R = C$_6$H$_5$ I O C$_6$H$_5$ DBN 60% O C$_6$H$_5$ [CH$_3$CuCN]Li 85% CH$_3$ OH C$_6$H$_5$

85% CrO$_3$·Py$_2$

O O R 1) LDA 2) ClSi(CH$_3$)$_3$ 96% OSi(CH$_3$)$_3$ O R [CH$_3$CuCN]Li ~100% OSi(CH$_3$)$_3$ CH$_3$ OH R KF 60–70% O CH$_3$ OH R

Scheme (I)

The reaction of the *n*-butylcyanocuprate is more stereoselective than that of the cyanomethylcuprate (equation I).[4]

(I) + $(CH_3CuCN)^-Li^+$ → (Ether, −78°; 95%) 9:1

$[n\text{-}C_4H_9CuCN]^-Li^+$

OH

C_4H_9-*n*

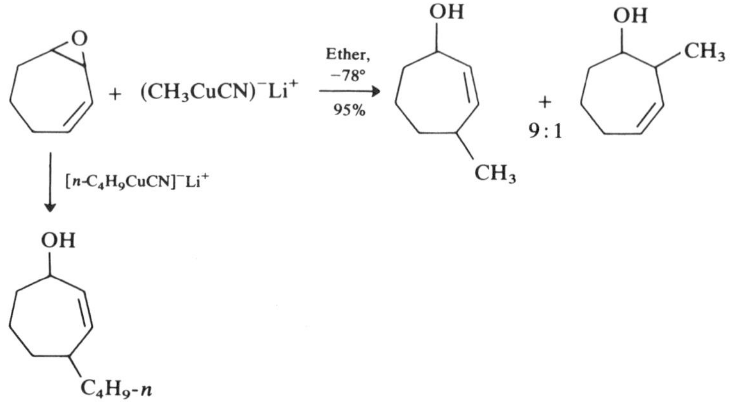

Lithium phenylthio(trimethylstannyl)cuprate, $C_6H_5S[(CH_3)_3Sn]CuLi$ (**1**).[5] The deep red cuprate is prepared by reaction of C_6H_5SCu with $(CH_3)_3SnLi$ in THF at −20°. It converts *β*-iodo enones efficiently into *β*-trimethylstannyl-*α*,*β*-enones. It is somewhat less efficient than $(CH_3)_3SnLi$ for conjugate addition to enones.[6]

Examples:

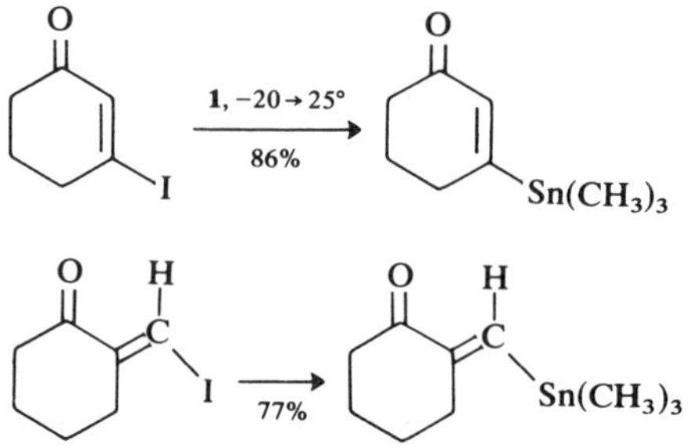

1, −20→25°; 86% → $Sn(CH_3)_3$

77% → $Sn(CH_3)_3$

Hydrindanediones. Conjugate addition of an organocuprate to a 4,4-disubstituted cyclohexenone is not stereospecific, but steric control can be achieved indirectly by conjugate addition to enone lactones such as **1** for construction of either *cis*- or *trans*-hydrindanediones such as **6**.[7]

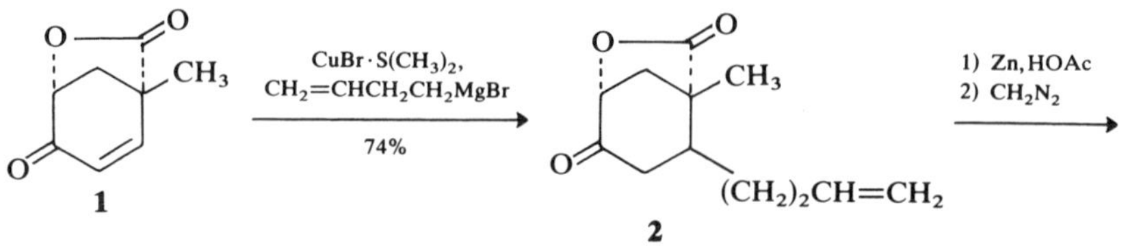

1 → ($CuBr \cdot S(CH_3)_2$, $CH_2{=}CHCH_2CH_2MgBr$; 74%) → **2** ($(CH_2)_2CH{=}CH_2$) → 1) Zn, HOAc 2) CH_2N_2 →

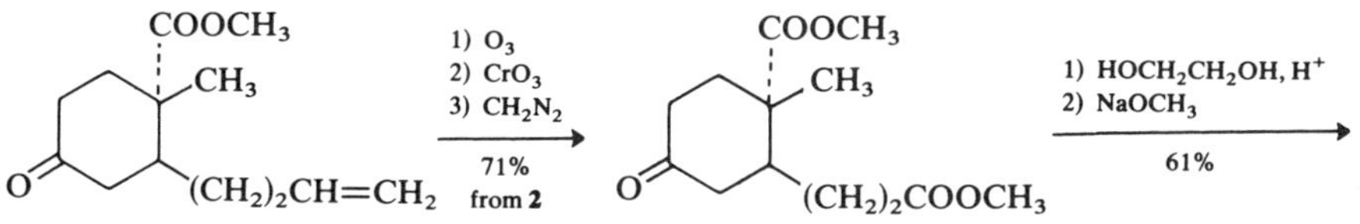

Coupling with allylic carbamates.[8] Cyclic allylic carbamates **1** and **2** (R = NHC_6H_5) react with $LiCu(CH_3)_2$ to afford products of syn γ-substitution with greater than 98% regio- and stereospecificity. All other acyl derivatives examined [R = H, CH_3, C_6H_5, OC_2H_5, and $N(CH_3)C_6H_5$] afford nearly 1 : 1 mixtures derived from anti α- and γ-substitution. The high stereo- and regioselectivity of this process are associated with the presence of the acidic hydrogen in the N-phenylcarbamate group. Since 2 equiv. of $LiCu(CH_3)_2$ are required for the reaction to proceed, an intermediate lithium carbamate may be involved.

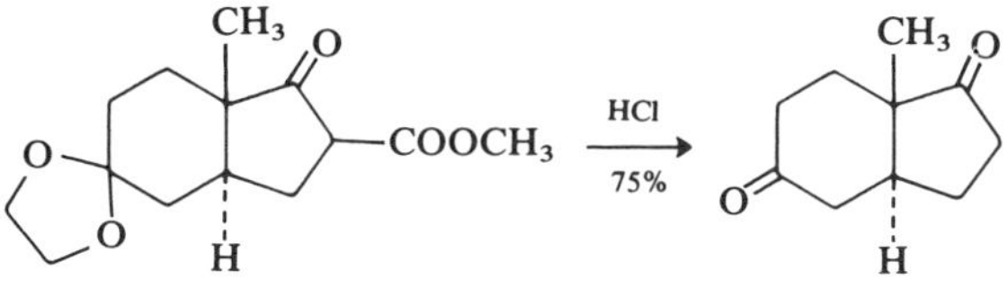

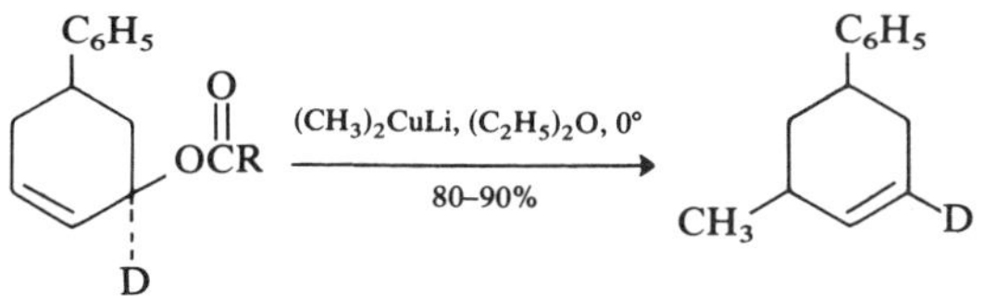

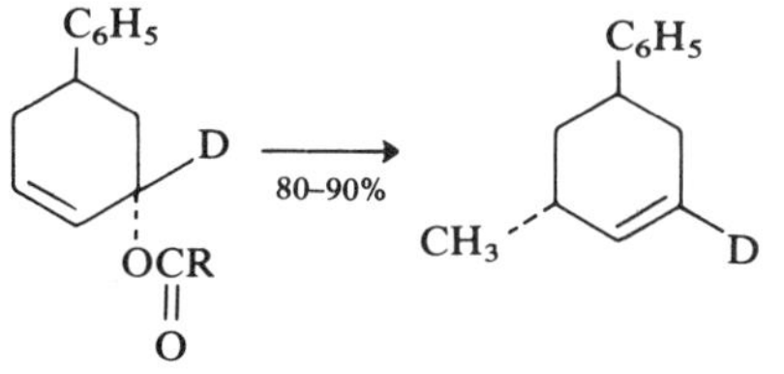

Coupling with the enol phosphate of β-keto esters.[9] The coupling of lithium dialkylcuprates with enol phosphates of ketones (**7**, 93) is also possible with enolates of β-keto esters. The reaction is useful for stereoselective synthesis of β-substituted α,β-unsaturated esters.

Examples:

$$n\text{-}C_3H_7\overset{O}{\overset{\|}{C}}CH_2COOCH_3 \xrightarrow[]{\text{NaH, ClPO(OC}_2\text{H}_5)_2} n\text{-}C_3H_7\overset{OPO(OC_2H_5)_2}{\overset{|}{C}}{=}CHCOOCH_3 \xrightarrow[\text{83\% overall}]{(CH_3)_2CuLi} \begin{matrix} CH_3 & & COOCH_3 \\ & C{=}C & \\ n\text{-}C_3H_7 & & H \end{matrix}$$

(>94% E)

$$\text{2-(methoxycarbonyl)cyclohexanone} \xrightarrow[79\%]{(C_2H_5)_2CuLi} \text{methyl 2-ethyl-1-cyclohexenecarboxylate}$$

Reactions with 2-propynylic acetates.[10] R_2CuLi reacts with 2-propynylic acetates to give allenes as major products (equation I).

$$CH_3C{\equiv}CCH_2OAc \xrightarrow{R_2CuLi} \begin{matrix} CH_3 \\ R \end{matrix}\!\!>C{=}C{=}CH_2$$

On the other hand, R_2CuLi reacts with **1** mainly by direct displacement of the acetate group (equation II). $(CH_3)_2CuLi$ reacts with the enyne acetate **2** to give mainly a coupled product **3** (equation III).

$$\text{(II)} \quad \underset{\mathbf{1}}{(CH_3)_3SiC{\equiv}C{-}\overset{OAc}{\overset{|}{C}}HC_5H_{11}} \xrightarrow[65\text{–}85\%]{R_2CuLi} (CH_3)_3SiC{\equiv}C\overset{R}{\overset{|}{C}}HC_5H_{11}$$

$$\text{(III)} \quad \underset{\mathbf{2}}{2CH_2{=}CH\overset{OAc}{\overset{|}{C}}HC{\equiv}CSi(CH_3)_3} \xrightarrow[62\%]{(CH_3)_2CuLi} \underset{\mathbf{3}}{CH_2{=}CH\underset{\underset{CSi(CH_3)_3}{\overset{|||}{C}}}{\underset{|}{C}}H{-}CH_2{-}\overset{H}{\overset{|}{C}}{=}\overset{H}{\overset{|}{C}}C{\equiv}CSi(CH_3)_3}$$

Dithioesters. These esters have been prepared by reaction of Grignard reagents with carbon disulfide in THF and then with an alkyl halide. However, yields were low with t-alkylmagnesium halides.[11] The yields are greatly improved by addition of catalytic amounts of copper(I) bromide, and a wider range of Grignard reagents can be used (equation I). The actual reagents are undoubtedly cuprates, and indeed preformed cuprates can be used equally well.[12]

$$\text{(I)}\quad (CH_3)_3CMgBr + CS_2 \xrightarrow[\text{THF}]{\text{CuBr,}} \left[(CH_3)_3C{-}C{\overset{S}{\underset{SMgBr}{\lessgtr}}}\right] \xrightarrow[95\%]{CH_3I} (CH_3)_3C{-}C{\overset{S}{\underset{SCH_3}{\lessgtr}}}$$

1,4-Addition to α,β-acetylenic esters and ketones (**3**, 108; **6**, 163–164). The addition of organocopper reagents to conjugated acetylenic carbonyl compounds is usually not stereospecific, although *cis*-addition predominates. Japanese chemists[13] have found that the stereoselectivity in the reaction with alkylcopper reagents is markedly enhanced by use of the complex $RCu{\cdot}BR'_3$. Thus the reaction of dimethyl acetylenedicarboxylate with *n*-butylcopper complexed with tri-*n*-butylboron (or triethylboron) results in exclusive formation of the *cis*-adduct. In the absence of a trialkylborane the *cis*- and *trans*-adducts are formed in the ratio 85:15. Stereoselectivity was similarly increased for additions to α,β-acetylenic acids and, to a lesser extent, to α,β-acetylenic ketones.

$$CH_3OOCC{\equiv}CCOOCH_3 + n\text{-}BuCu\cdot B(n\text{-}Bu)_3 \xrightarrow[93\%]{\text{Ether, }-70^\circ}$$

$$\underset{>99\%}{(CH_3OOC)(n\text{-}Bu)C{=}C(COOCH_3)(H)} + \underset{<1\%}{(CH_3OOC)(n\text{-}Bu)C{=}C(H)(COOCH_3)}$$

α,β-Acetylenic sulfoxides undergo stereospecific *cis*-addition with CH_3Cu, *n*-BuCu, and C_6H_5Cu without an additive.[14]

Alkylation of allylic alcohols.[15] The alkyl substitution of allylic halides with allylic rearrangement by reagents of the type $RCu{\cdot}BF_3$ (**8**, 334–335) can also be effected with allylic alcohols. However, in this case the alkylation is somewhat less regioselective and some α-substitution is also observed. On the other hand, allylic alcohols are more readily available than allylic halides.

Example:

$$C_6H_5\overset{\gamma}{C}H{=}CH\overset{\alpha}{C}H_2OH + n\text{-}C_4H_9Cu\cdot BF_3 \xrightarrow[97\%]{}$$

$$(C_6H_5)(n\text{-}C_4H_9)CHCH{=}CH_2 + C_6H_5CH{=}CHCH_2C_4H_9\text{-}n \qquad 91:9$$

[1] A. Marfat, P. R. McGuirk, and P. Helquist, *J. Org.*, **44**, 3888 (1979).
[2] E. Piers and H.-U. Reissig, *Angew. Chem., Int. Ed.*, **18**, 791 (1979).
[3] J. P. Marino and N. Hatanako, *J. Org.*, **44**, 4467 (1979).
[4] J. P. Marino and D. M. Floyd, *Tetrahedron Letters*, 675 (1979).

[5] E. Piers and H. E. Morton, *J.C.S. Chem. Comm.*, 1033 (1978).
[6] W. C. Still, *Am. Soc.*, **99**, 4836 (1977).
[7] G. Stork and E. W. Logusch, *Tetrahedron Letters*, 3361 (1979); *see also* E. W. Logusch, *ibid.*, 3365 (1979).
[8] C. Gallina and P. G. Ciattini, *Am. Soc.*, **101**, 1035 (1979).
[9] F. W. Sum and L. Weiler, *Canad. J. Chem.*, **57**, 1431 (1979).
[10] R. S. Brinkmeyer and T. L. Macdonald, *J.C.S. Chem. Comm.*, 876 (1978).
[11] J. Meijer, P. Vermeer, and L. Brandsma, *Rec. trav.*, **92**, 601 (1973).
[12] H. Westmijze, H. Kleijn, J. Meijer, and P. Vermeer, *Synthesis*, 432 (1979).
[13] Y. Yamamoto, H. Yatagai, and K. Maruyama, *J. Org.*, **44**, 1744 (1979).
[14] W. E. Truce and M. J. Lusch, *ibid.*, **43**, 2252 (1978).
[15] Y. Yamamoto and K. Maruyama, *J. Organometal. Chem.*, **156**, C9 (1978).

Osmium tetroxide–N-Methylmorpholine oxide, 7, 256–257.

cis-*Hydroxylation.*[1] One step in a total synthesis of lacinilene C methyl ether (**4**) required oxidation of **1** to the α-ketol **3**. This oxidation is complicated by the ease of cleavage of the olefinic bond, but was accomplished successfully with the $OsO_4 \cdot NMMO \cdot H_2O$ reagent in acetone to give a mixture of diols (**2**). The mixture was oxidized with CrO_3 to a mixture of **3** and **4**; treatment of the mixture with DDQ transformed **3** into **4**, obtained in this way in 27% overall yield from **1**.

[1] J. P. McCormick, J. P. Pachlatko, and T. R. Schafer, *Tetrahedron Letters*, 3993 (1978).

Osmium tetroxide–Potassium chlorate.

Oxidation of acetylenic ethers. 1-Alkoxy-1-alkynes (**1**) are oxidized by OsO_4 (the function of $KClO_3$ is to regenerate OsO_4) to α-keto esters (**2**) in H_2O–ether in 50–80% yield. The paper also reports use of this system for the novel oxidation of **3** to the *erythro*-isomer (**4**) of 3,4-dihydroxyhexane-2,5-dione.[1]

$$\underset{\mathbf{1}}{RC{\equiv}COR'} \xrightarrow[50\text{–}80\%]{OsO_4,\ KClO_3} \underset{\mathbf{2}}{RC(=O)\text{–}C(=O)OR'}$$

CH_3 O CH_3 —61%→ CH_3 HO H O H O H $COCH_3$

3 4

[1] L. Bassignani, A. Brandt, V. Caciagli, and L. Re, *J. Org.*, **43**, 4245 (1978).

Oxalyl chloride–Dimethylformamide.

Carboxylic acid chlorides. Conventional methods for conversion of an acid into the acid chloride usually require acidic reagents. Lederle chemists[1] have accomplished this transformation under neutral conditions by conversion of the acid into the *t*-butyldimethylsilyl ester. The esters react with oxalyl chloride in CH_2Cl_2 in the presence of catalytic quantities of DMF[2] to form the acid chloride with evolution of gas. Since DMF is essential, dimethylformiminium chloride (Vilsmeier reagent) is probably the reactive species.[3] The acid chloride is formed in 85–95% yield, as shown by conversion to esters with ethanol–pyridine.

[1] A. Wissner and C. V. Grudzinskas, *J. Org.*, **43**, 3972 (1978).
[2] A. W. Burgstahler, L. O. Weigel, and C. G. Shaefer, *Synthesis*, 767 (1976).
[3] H. H. Bosshard, R. Mory, M. Schmid, and H. Zollinger, *Helv.*, **42**, 1653 (1959).

Oxalyl chloride–Sodium iodide.

Deoxygenation of sulfoxides. This combination [$(COCl)_2/NaI = 1:2$] reduces sulfoxides to sulfides in 75–95% yield. Oxalyl chloride can be replaced by $SOCl_2$, $POCl_3$, and PCl_5.[1]

[1] G. A. Olah, R. Malhotra, and S. C. Narang, *Synthesis*, 58 (1979).

μ-Oxobis(chlorotriphenylbismuth), $[(C_6H_5)_3BiCl]_2O$. Mol. wt. 976.52, m.p. 147°, soluble in CH_2Cl_2 and C_6H_6. This reagent is prepared by treatment of $(C_6H_5)_3BiCl_2$ with NaOH.[1]

Oxidation of hydroxyl groups.[2] This Bi(V) compound can be used, in the presence of K_2CO_3 or $NaHCO_3$, for oxidation of primary and secondary alcohols to carbonyl compounds. Allylic alcohols are oxidized in 75–95% yield at room temperature. α-Glycols are cleaved readily in good yield.

[1] R. G. Goel and H. S. Prasad, *J. Organometal. Chem.*, **36**, 323 (1972).
[2] D. H. R. Barton, J. P. Kitchin, and W. B. Motherwell, *J.C.S. Chem. Comm.*, 1099 (1978).

Oxygen, 4, 362; **5,** 482–486; **6,** 426–430; **7,** 258–260; **8,** 366–367.

Hydroxylation of* 1,3*-dicarbonyl compounds. Hydroxylation of **1**[1] and **2**[2] with oxygen in the presence of either NaH or potassium *t*-butoxide and triethyl phosphite in DMF affords the corresponding alcohols **3** and **4** in 50–70% yields. In the case of **1**, an isomeric product (**5**) is obtained in 14% yield. When NaH is employed as base, a trace of H_2O is required to initiate the reaction.

Previously reported methods for hydroxylation failed or gave yields of less than 10% (O_2–Pt; O_2–$NaNO_2$; O_2–$CeCl_3$).

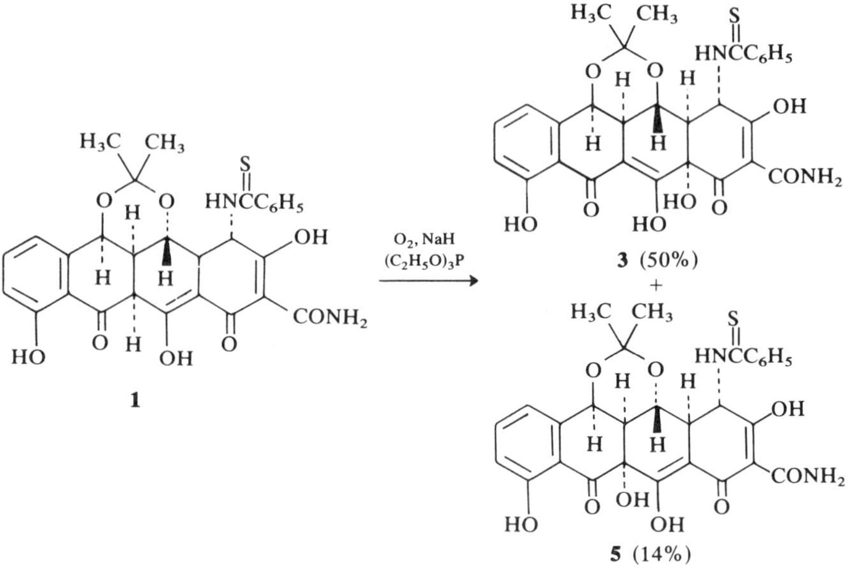

O₂, NaH
(C₂H₅O)₃P

1

3 (50%)

\+

5 (14%)

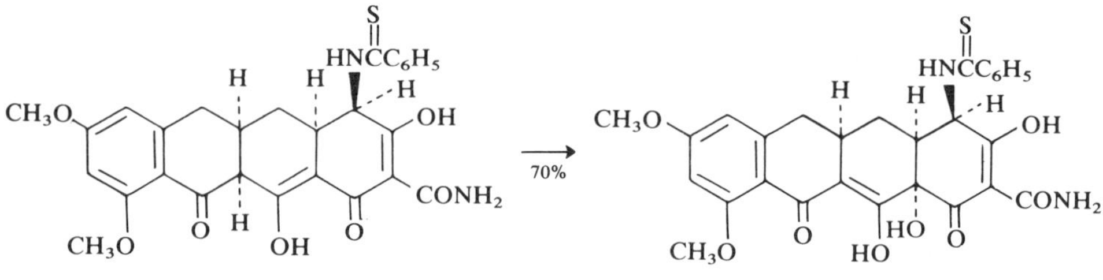

70%

2

4

***Oxygenation of* $>C{=}C<$ *and* $-C{\equiv}C-$.** 9,10-Dicyanoanthracene (DCA) sensitizes photooxygenation of some alkenes in CH_3CN. The reaction does not involve singlet oxygen and is believed to involve electron transfer. Thus tetraphenylethylene is oxidized in this way mainly to benzophenone (equation I). Under the same conditions *trans*-stilbene is oxidized to benzaldehyde. Sulfides are oxidized to sulfoxides and/or sulfones.[3]

$$\text{(I)}\quad (C_6H_5)_2C{=}C(C_6H_5)_2 \xrightarrow[h\nu]{\text{DCA, }O_2,} \left[\begin{array}{c} O{-}O \\ |\qquad | \\ (C_6H_5)_2C{-}C(C_6H_5)_2 \end{array}\right] \xrightarrow[57\%]{} 2(C_6H_5)_2C{=}O$$

Some alkynes are also subject to electron-transfer photooxidation. Thus diphenylacetylene is converted to benzil and benzoic acid, the product of further oxidation (equation II).[4]

$$\text{(II)}\quad C_6H_5C{\equiv}CC_6H_5 \xrightarrow[h\nu]{\text{DCA, }O_2,} \begin{array}{c} C_6H_5 \\ | \\ C{=}O \\ | \\ C{=}O \\ | \\ C_6H_5 \end{array} \xrightarrow{O_2,\ h\nu} 2C_6H_5COOH$$

Oxidative decyanation (**6**, 430). Two laboratories[5,6] have reported regiospecific syntheses of anthracyclines based on conjugate addition of the enolate of a nitrile to an α,β-unsaturated ester. For the synthesis of daunomycinone (**7**), the reaction involved addition of the enolate of **1** to **2** to give **3** in 94% yield. The ester group of **3** was hydrolyzed and the resulting acid was cyclized to **4**. Oxygenation of the enolate of **4** by the procedure of Watt results in the quinone **5**, which is oxidatively demethylated; deketalization of the product led to **6**, which had been converted previously into **7**.

1 + **2** —(NaH, THF, 0°; 94%)→ **3**

3 —(1) KOH, C_2H_5OH; 2) CF_3COOH, $(CF_3CO)_2O$; 75%)→ **4**

4 —(1) LDA, O_2; 2) Na_2SO_3; 70%)→ **5**

5 —(1) AgO; 2) H^+; 70%)→ **6**

7

[1] H. Muxfeldt, G. Haas, G. Hardtmann, F. Kathawala, J. B. Mooberry, and E. Vedejs, *Am. Soc.*, **101**, 689 (1979).

[2] B. Glatz, G. Helmchen, H. Muxfeldt, H. Porcher, R. Prewo, J. Senn, J. Stezowski, R. J. Stojda, and D. R. White, *ibid.*, **101**, 2171 (1979).

[3] J. Erickson, C. S. Foote, and T. L. Parker, *ibid.*, **99**, 6455 (1977).

[4] N. Berenjian, P. de Mayo, F. H. Phoenix, and A. C. Weedon, *Tetrahedron Letters*, 4179 (1979).

[5] K. A. Parker and J. L. Kallmerten, *ibid.*, 1197 (1979).

[6] A. S. Kende, J. Rizzi, and J. Riemer, *ibid.*, 1201 (1979).

Oxygen, singlet, 4, 362–363; **5**, 486–491; **6**, 431–436; **7**, 261–269; **8**, 367–374. Review.[1]

Oxidation of vitamin D_2. Vitamin D_2 (**1**) is photooxidized to two isomeric 6,19-epidioxides (**2**), in 35% yield.[2]

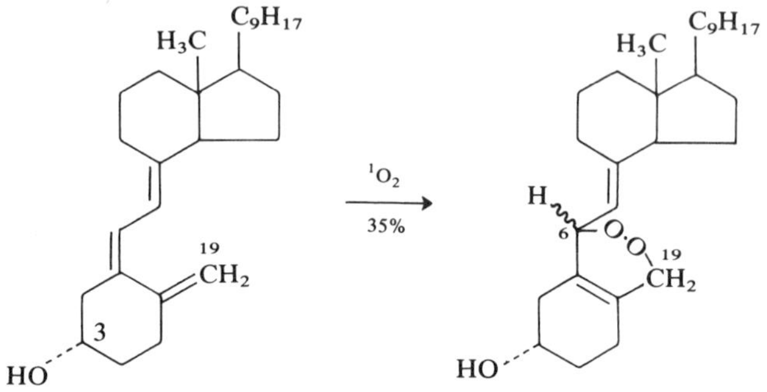

C_9H_{17} H_3C 19 CH_2 3 HO **1** $\xrightarrow[35\%]{^1O_2}$ C_9H_{17} H_3C H 6 O O 19 CH_2 HO **2**

Reaction with cycloheptatriene.[3] Reaction of cycloheptatriene with singlet oxygen leads to a mixture (~1 : 1) of **2** and **3**. Formation of **2** results from the unusual [2+6] addition. Some transformations of **2** and **3** have been reported.

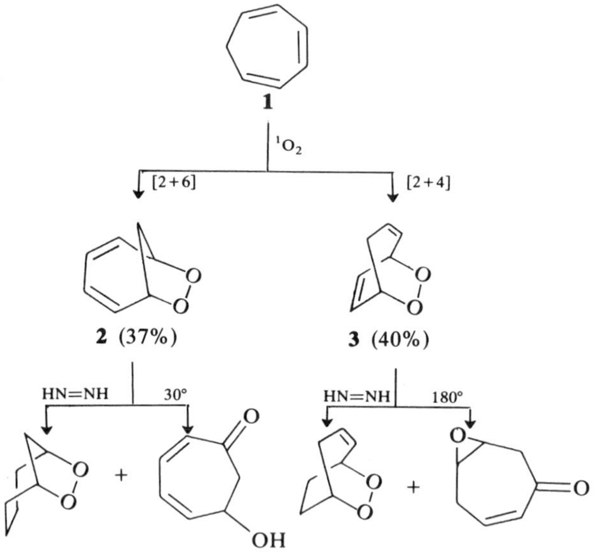

1 $\xrightarrow{^1O_2}$ [2+6] **2** (37%); [2+4] **3** (40%)

2 $\xrightarrow{HN{=}NH}$ **4**; **2** $\xrightarrow{30°}$ **5** (O, OH)

3 $\xrightarrow{HN{=}NH}$ **6**; **3** $\xrightarrow{180°}$ **7** (O, =O)

A thiaozonide.[4] One of the few known thiaozonides (**2**) has been prepared as shown in equation (I) by photooxygenation (tetraphenylporphyrin) of the thiophene **1**, followed by reduction with diimide (**8**, 172–173). The compound is thermally unstable and decomposes at 25° to **3**.

(I)

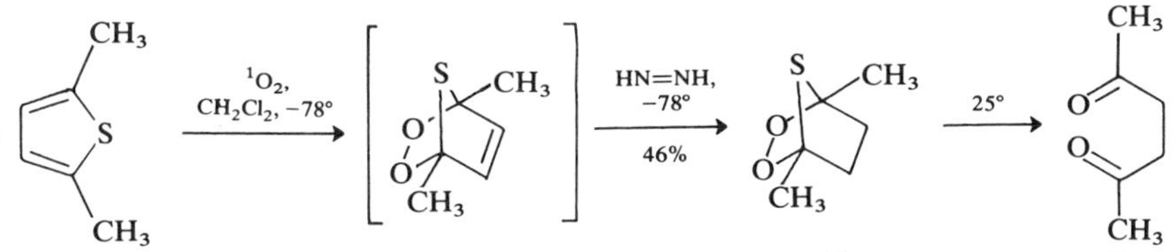

CH_3, S, CH_3 — **1** → ($^1O_2$, CH_2Cl_2, −78°) → [S, CH_3, O, O, CH_3 — **a**] → (HN=NH, −78°; 46%) → S, CH_3, O, O, CH_3 — **2** → (25°) → CH_3, O, O, CH_3 — **3**

Protopine alkaloids.[5] A notably efficient route to protopanes is the photooxygenation (Rose Bengal or Methylene Blue) of an enamine such as **1** to give an amido ketone (**2**), which can be converted into **3** by reduction with lithium aluminum hydride in THF followed by oxidation with activated manganese dioxide. The overall yield is about 65%.

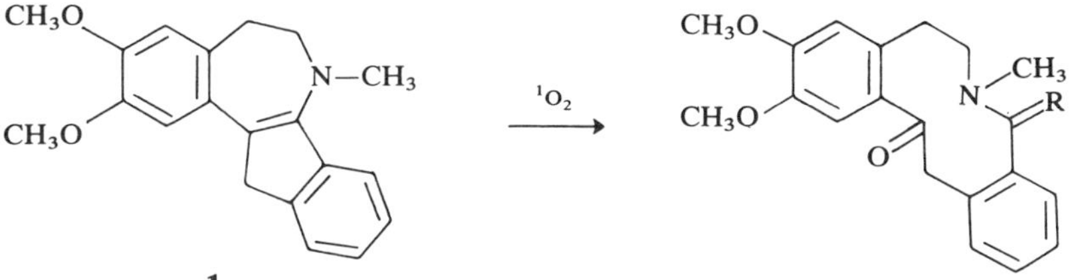

CH_3O, CH_3O, N—CH_3 — **1** → ($^1O_2$) → CH_3O, CH_3O, N, CH_3, R, O — **2** (R = O), **3** (R = H_2)

Homobenzoquinones.[6] A new route to these compounds (**4**) is shown in equation (I). The yield of **4** from **2** is low (23%) when X = CHO (*exo*).

(I)

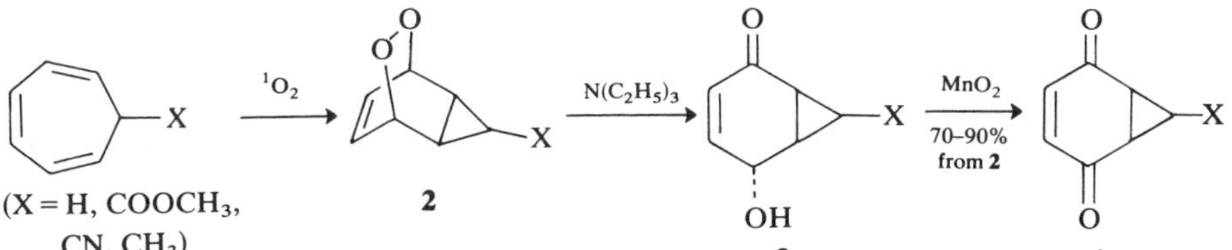

X — **1** (X = H, $COOCH_3$, CN, CH_3) → ($^1O_2$) → O, O, X — **2** → ($N(C_2H_5)_3$) → O, X, OH — **3** → (MnO_2; 70–90% from **2**) → O, X, O — **4**

Photooxygenation of β-methoxystyrene. Foote *et al.*[7] have reported that singlet oxygen undergoes 1,4-cycloaddition to 1,1-diphenyl-2-methoxyethylene to give an unstable endoperoxide, which decomposes to a variety of products (equation I).

(I)

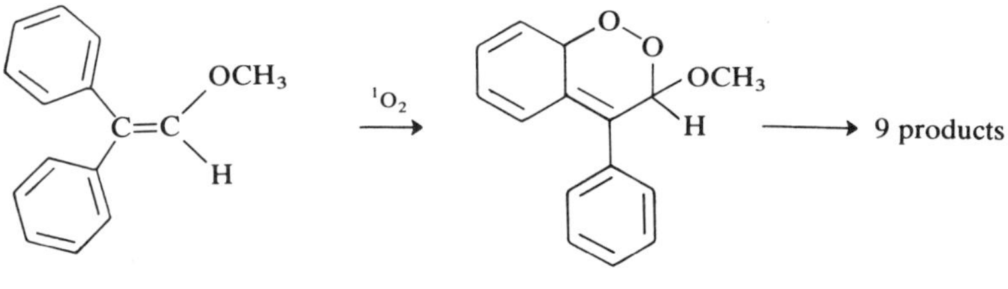

C=C, OCH_3, H → ($^1O_2$) → O, O, OCH_3, H → 9 products

The simpler system β-methoxystyrene (**1**) has now been examined (equation II).[8] Both the *cis*- and *trans*-isomers of **1** also give the unstable endoperoxides **2**, which lose methanol to form the *o*-formylmethide quinone **3**.

(II) $C_6H_5CH{=}CH{\sim}OCH_3$ → (endoperoxide, OCH_3, H) → (quinone methide, H, CHO, O)

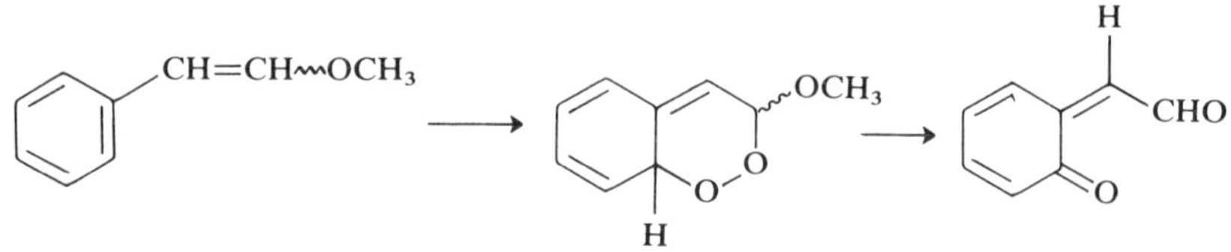

1 **2** **3**

Addition to 3,5-cycloheptadiene-1-ol **(1).** Photogenerated singlet oxygen (hematoporphyrin hydrochloride, sensitizer) adds to **1** in a stereoselective manner (*syn* to OH) to give mainly the endoperoxide **2**. The reaction of the acetate of **1** is less selective because the acetyl derivative of **3** rearranges readily under both photolytic and thermal conditions to a diepoxide. Under the same conditions the acetyl derivative of **2** is relatively stable.[9]

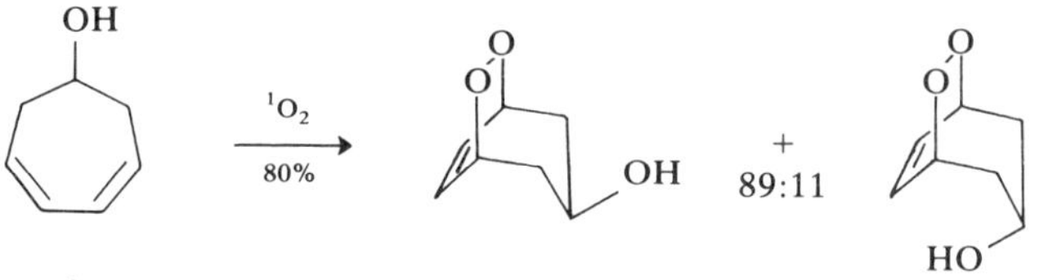

1 (OH) $\xrightarrow[80\%]{^1O_2}$ **2** (OH) + **3** (HO), 89:11

Oxygenation of vinylsilanes; allylic alcohols.[10] Singlet oxygen undergoes the ene reaction with vinylsilanes to give, after reduction with sodium borohydride, β-silylated allylic alcohols. These products are desilylated by tetra-*n*-butylammonium fluoride with preservation of the geometry of the double bond.

Examples:

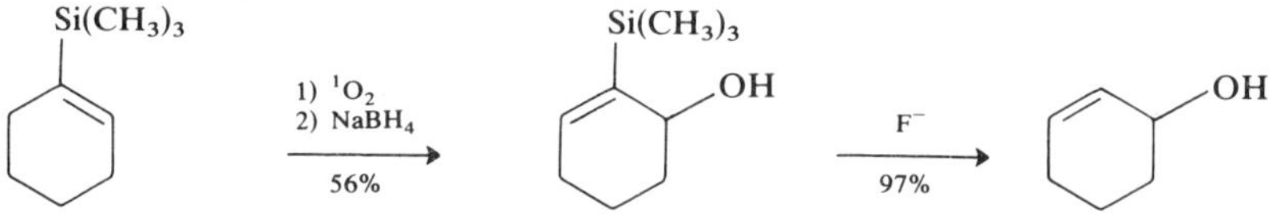

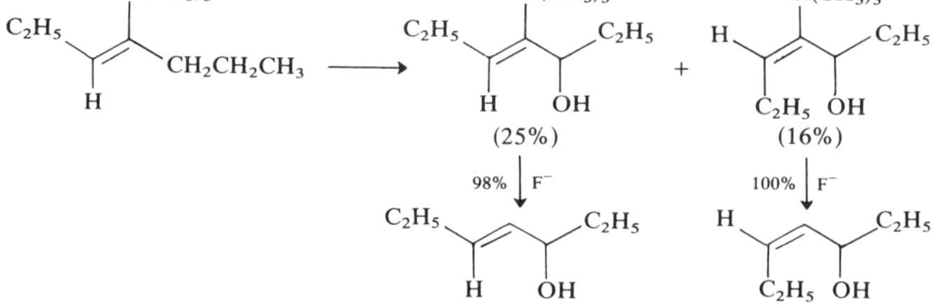

1-$Si(CH_3)_3$-cyclohexene $\xrightarrow[56\%]{1)\ ^1O_2\ 2)\ NaBH_4}$ $Si(CH_3)_3$, OH product $\xrightarrow[97\%]{F^-}$ 2-cyclohexen-1-ol (OH)

C_2H_5(H)C=C($Si(CH_3)_3$)$CH_2CH_2CH_3$ → C_2H_5, H, $Si(CH_3)_3$, C_2H_5, OH (25%) + H, C_2H_5, $Si(CH_3)_3$, C_2H_5, OH (16%)

(25%) $\xrightarrow[98\%]{F^-}$ C_2H_5, H, C_2H_5, OH

(16%) $\xrightarrow[100\%]{F^-}$ H, C_2H_5, C_2H_5, OH

α-Keto lactones.[11] Lactones can be converted into α-keto lactones by conversion to an enamino lactone by reaction with tris(dimethylamino)methane[12] followed by photooxidation in CH_2Cl_2 at $-78°$. The α-keto lactones exist mainly or exclusively in the enol form and do not undergo normal Wittig condensation.

Examples:

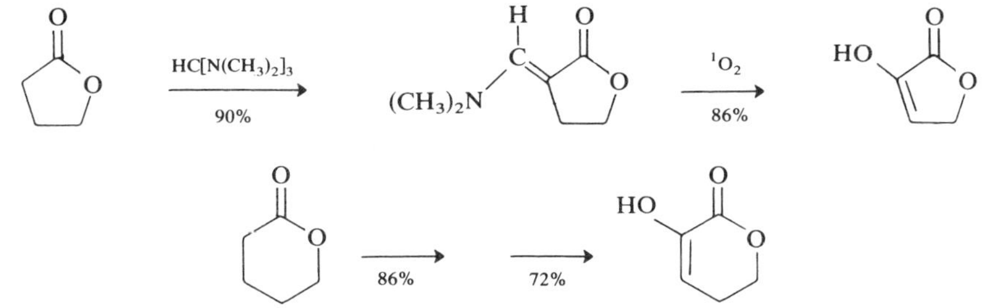

$$\xrightarrow[90\%]{HC[N(CH_3)_2]_3} \quad (CH_3)_2N\text{–}CH= \quad \xrightarrow[86\%]{^1O_2} \quad HO$$

$$\xrightarrow[86\%]{} \quad \xrightarrow[72\%]{} \quad HO$$

[1] *Singlet Oxygen* (Organic Chem. Series), H. H. Wasserman and R. W. Murray, Eds., Academic Press, N.Y. 1979.

[2] S. Yamada, K. Nakayama, and H. Takayama, *Tetrahedron Letters*, 4895 (1978).

[3] W. Adam and M. Balchi, *Angew. Chem., Int. Ed.*, **17**, 954 (1978).

[4] W. Adam and H. J. Eggette, *ibid.*, **17**, 765 (1978).

[5] K. Orito and M. Itoh, *J.C.S. Chem. Comm.*, 812 (1978).

[6] W. Adam, M. Balchi, and J. Rivera, *Synthesis*, 807 (1979).

[7] C. S. Foote, P. A. Burns, S. Mazur, and D. Lerdal, *Am. Soc.*, **95**, 586 (1973); D. S. Steichen and C. S. Foote, *Tetrahedron Letters*, 4363 (1979).

[8] M. Matsumoto and K. Kuroda, *ibid.*, 1607 (1979).

[9] D. M. Floyd and C. M. Cimarusti, *ibid.*, 4129 (1979).

[10] W. E. Fristad, T. R. Bailey, L. A. Paquette, R. Gleiter, and M. C. Böhm, *Am. Soc.*, **101**, 4420 (1979).

[11] H. H. Wasserman and J. L. Ives, *J. Org.*, **43**, 3238 (1978).

[12] S. F. Martin and D. R. Moore, *Tetrahedron Letters*, 4459 (1976).

Ozone, 1, 773–777; **4**, 363–364; **5**, 491–495; **6**, 436–441; **7**, 269–271; **8**, 374–377.

Cleavage of alkenes to esters.[1] 1,2-Disubstituted alkenes can be cleaved directly to carboxylic esters by ozonation in methanol containing hydrochloric acid. The reaction probably involves initial cleavage to an aldehyde and/or a methoxy hydroperoxide, which on further reaction with methanol are converted into esters.

$$C_6H_5CH{=}CH_2 \xrightarrow[72\%]{O_3,\ CH_3OH,\ HCl} C_6H_5COOCH_3$$

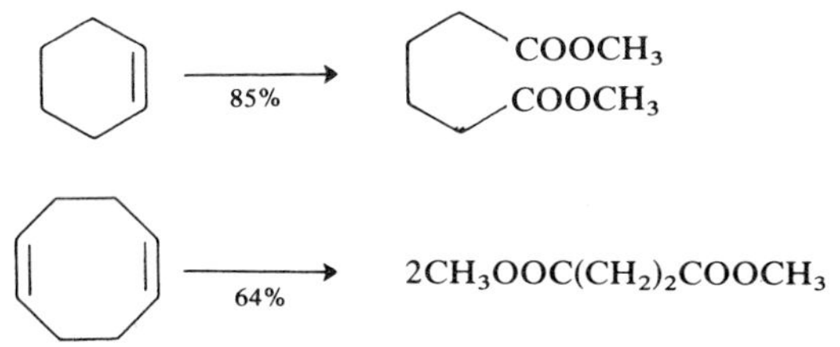

$$\xrightarrow[85\%]{} \quad \begin{matrix} COOCH_3 \\ COOCH_3 \end{matrix}$$

$$\xrightarrow[64\%]{} \quad 2CH_3OOC(CH_2)_2COOCH_3$$

A stable ozonide of an arene.[2] Ozonolysis of the cyclopentophenanthrene **1** leads to an ozonide (**2**) that is stable in the dark at 20° for several weeks. The ozonide is also unusual in that it is converted into **3** and **4** in high yield; products of this type are not usually formed from ozonides.

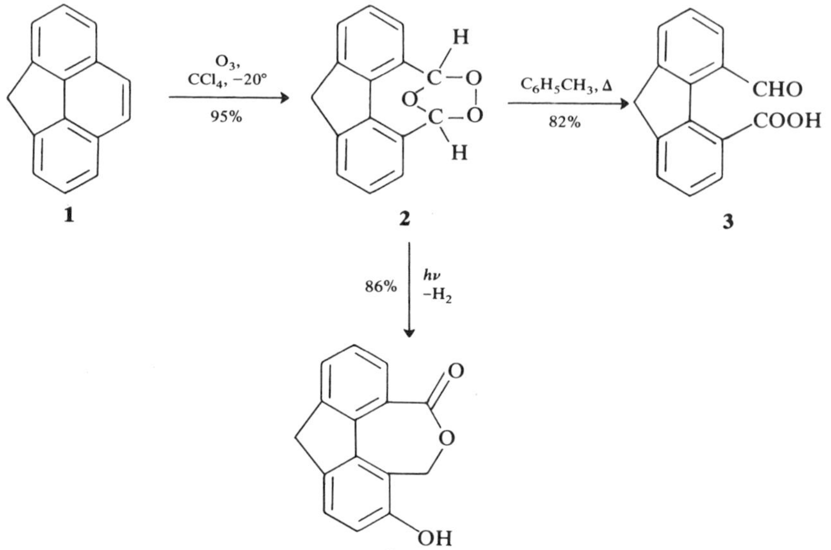

O_3, CCl_4, −20°; 95%
$C_6H_5CH_3$, Δ; 82%
hν, $-H_2$; 86%

1 **2** **3** **4**

Selective ozonation of a trisubstituted double bond. 1-Methyl-(1Z,5Z)-cyclooctadiene (**1**) can be selectively cleaved to the aldehyde **2** by ozone. The product was used for synthesis of sex pheromones of the (Z)-alkenyl acetate series (**3**).[3]

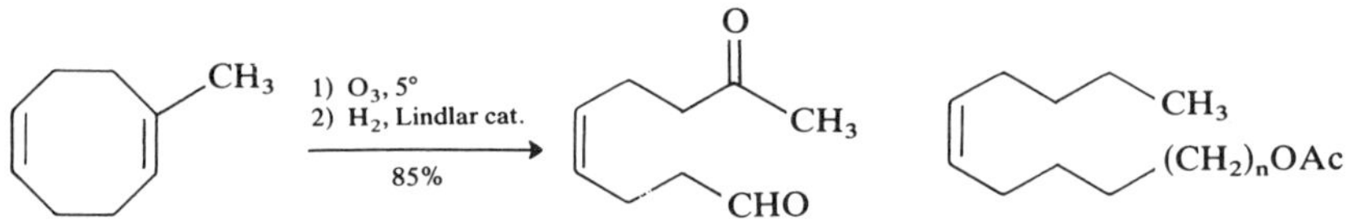

1) O_3, 5°
2) H_2, Lindlar cat.
85%

1 **2** **3** (n = 3, 5, 7)

Ozonation of the endoperoxide* (1) *of 7-dehydrocholesterol. The two main products of this reaction are the 5,8-diketone **2** and the hemiacetal **3**.[4]

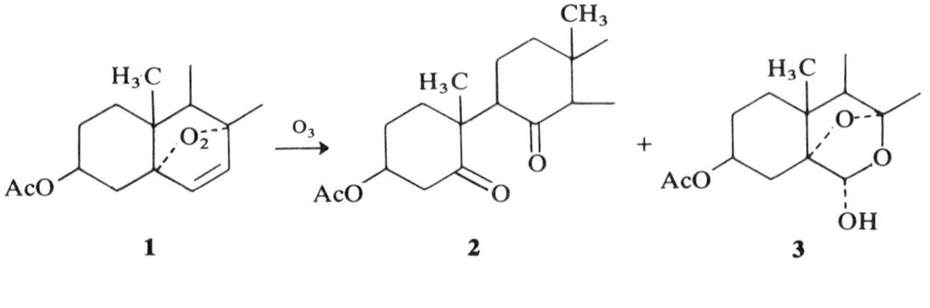

O_3

1 **2** **3**

[1] J. Neumeister, H. Keul, M. P. Saxena, and K. Griesbaum, *Angew. Chem., Int. Ed.*, **17**, 939 (1978).
[2] A. Kadokura, M. Yoshida, M. Minabe, and K. Suzuki, *Chem. Ind.*, 734 (1978).
[3] G. A. Tolstikov, V. N. Odinokov, R. I. Galeeva, R. S. Bakeeva, and V. R. Akhunova, *Tetrahedron Letters*, 4851 (1979).
[4] J. Gumulka, W. J. Szczepek, and Z. Wielogorski, *ibid.*, 4847 (1979).

Ozone–Silica gel, 6, 440–441; **7**, 271–273; **8**, 375–377.

Regioselective oxidation of methylene groups. Ozonation of cyclododecyl acetate adsorbed on silica gel results in formation of only three keto acetates, 5-keto-, 6-keto-, and 7-ketocyclododecyl acetate in yields of 15, 27, and 9%, respectively. This same regioselectivity is observed in microbiological hydroxylation of the substrate by various fungi.[1]

Hydroxylation of tertiary carbon atoms (**6**, 440; **7**, 271–273).[2] The hydrogen succinates of alcohols bind strongly to silica gel. Unactivated tertiary C—H bonds in these bound substrates are oxidized efficiently with a solution of O_3 in Freon 11. This method is preferable to the oxygenation of acetates adsorbed on silica gel.

Examples:

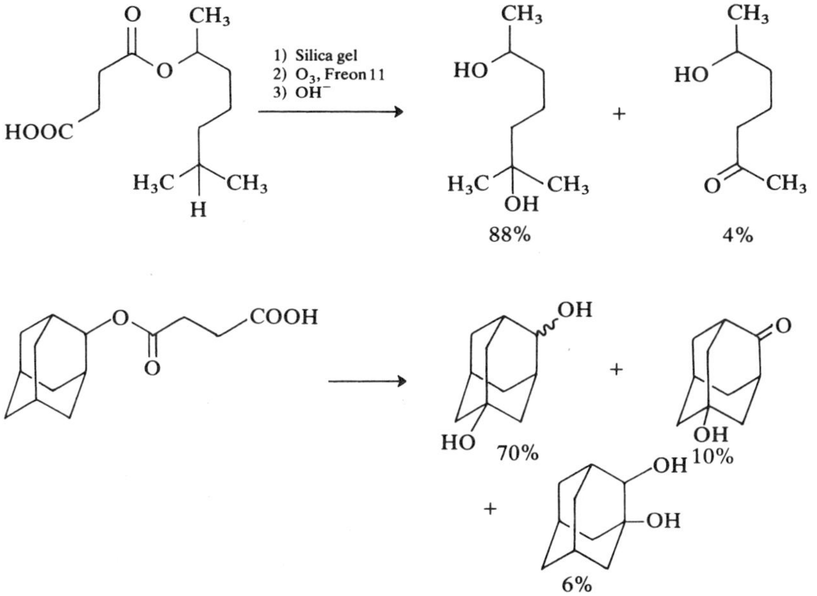

O CH₃ HOOC O H₃C CH₃ H — 1) Silica gel 2) O_3, Freon 11 3) OH^- → HO CH₃ H₃C CH₃ OH 88% + HO CH₃ O CH₃ 4%

O COOH O → OH HO 70% + O OH 10% + OH OH 6%

[1] A. L. J. Beckwith and T. Duong, *J.C.S. Chem. Comm.*, 413 (1978).
[2] *Idem, ibid.*, 690 (1979).

P

Palladium(0)–Phosphines.

γ-Lactones.[1] Palladium(0)–phosphine complexes[2] catalyze the high pressure reaction of CO_2 and a methylenecyclopropane with opening of the three-membered ring to give five-membered lactones. The product obtained depends on the catalyst used, as formulated in the examples.

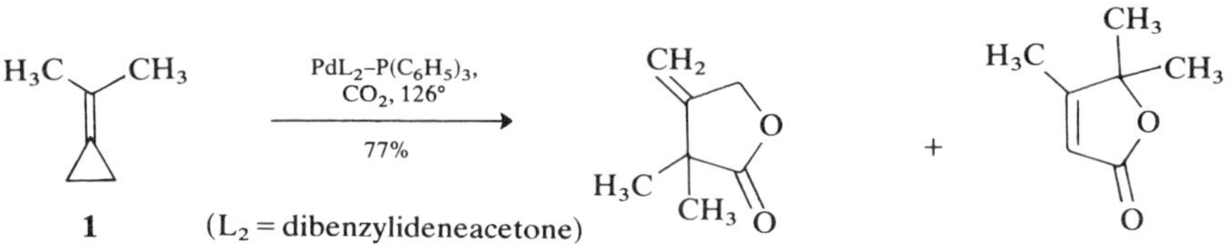

H₃C CH₃ — PdL_2–$P(C_6H_5)_3$, CO_2, 126°, 77% → 2 + 3

1 (L_2 = dibenzylideneacetone)

1 —Pd(diphos)₂, 48%→ **3** + trace of **2**

$Pd[P(C_6H_5)_3]_4$, 67%

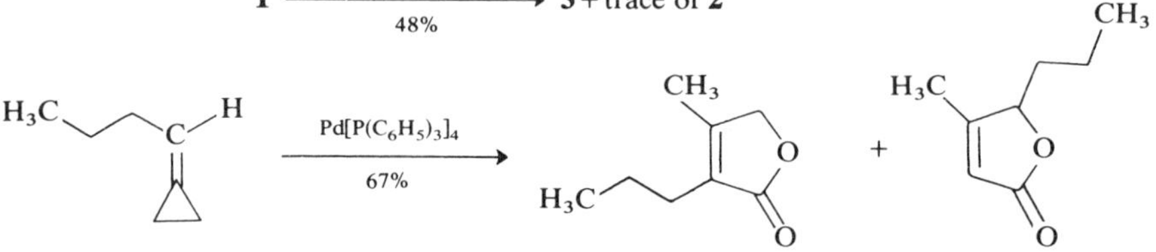

[1] Y. Inoue, T. Hibi, M. Satake, and H. Hashimoto, *J.C.S. Chem. Comm.*, 982 (1979).
[2] Y. Takahashi, T. Ito, S. Sakai, and Y. Ishii, *ibid.*, 1065 (1970).

Palladium(II) acetate, 1, 778; **2,** 303; **4,** 365; **5,** 496–497; **6,** 442–443; **7,** 274–277; **8,** 378–382.

2,2,7,7,12,13,17,18-Octamethylisobacteriochlorin. The isobacteriochlorins **1a** and **1b** are important biosynthetic precursors of vitamin B_{12}. A synthetic approach to the isobacteriochlorin macrocycle has been reported from the laboratories of Eschenmoser and Battersby.[1] The critical step involves coupling of the corrinoid A/B component (**2**) and the porphinoid C/D partner (**4**).

Preliminary experiments with a related A/B component established that the coupling reaction could not be accomplished directly with base or with a mild coordinating metal ion such as Zn(II) or Ni(II). However, the coupling was achieved by conversion of the ion of **2** into **3** by the $(DBU)_2$-adduct of $Pd(OAc)_2$ followed by treatment of **4** with equimolar amounts of DBU and excess **3**. Palladium(II) was removed from the resulting **5** (40% yield) by KCN. Potassium *t*-butoxide in the presence of Zn(II) then cyclized the free seco-ligand to the Zn complex **6**. Acid treatment removed Zn(II) and also the *t*-butylcarbonyl group to afford **7**.

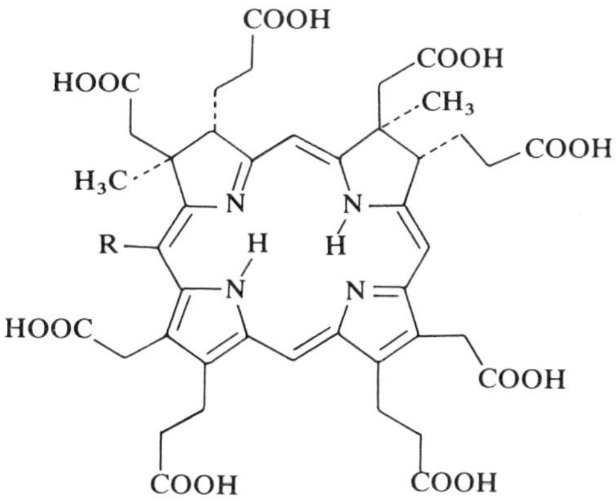

Sirohydrochlorin (**1a**, R = H)
20-Methylsirohydrochlorin (**1b**, R = CH_3)

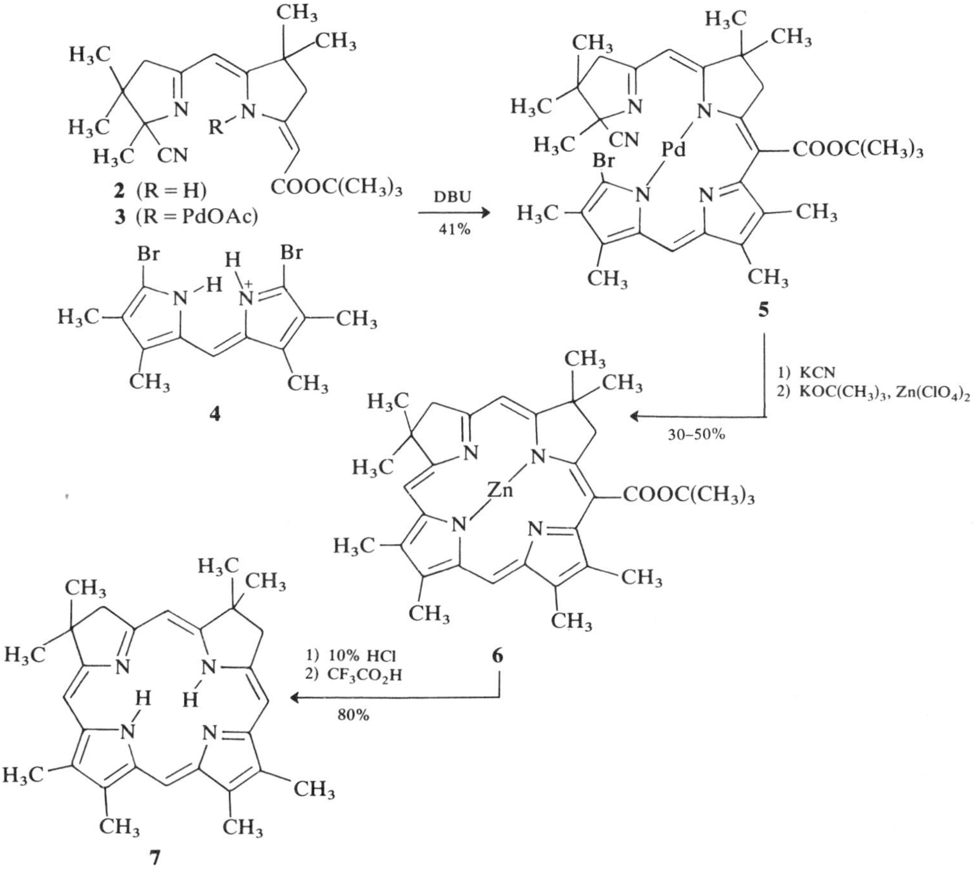

2 (R = H)
3 (R = PdOAc)

4

DBU
41%

5

1) KCN
2) $KOC(CH_3)_3$, $Zn(ClO_4)_2$
30–50%

6

1) 10% HCl
2) CF_3CO_2H
80%

7

α,β-*Enones* (**8**, 378). The method of Saegusa *et al.* for introduction of unsaturation α,β to a keto group has been used by Schlessinger and co-workers[2] in the synthesis of the pseudoguaianolides *dl*-helenalin, *dl*-confertin, and *dl*-damsin. An example is the conversion of **1** to **3**, a precursor to helenalin (**4**).

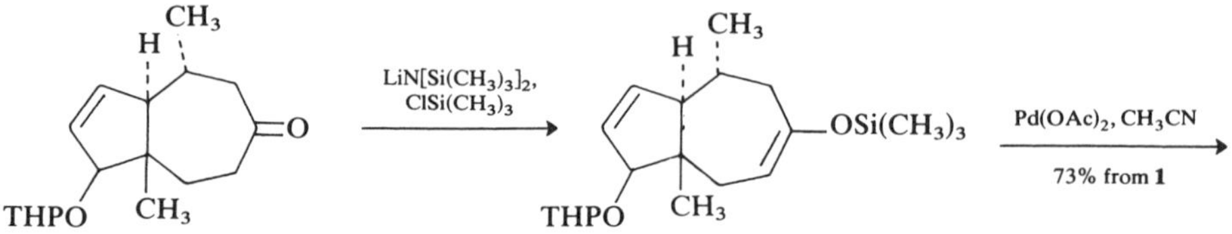

$LiN[Si(CH_3)_3]_2$, $ClSi(CH_3)_3$; $Pd(OAc)_2$, CH_3CN, 73% from **1**

1, **2**, 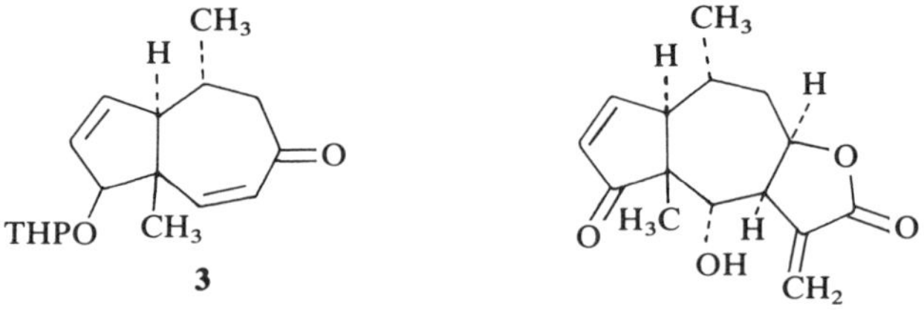 **3**, **4**

Benzohydrofuranes. 2-Allylphenols are cyclized oxidatively by $Pd(OAc)_2$ (1 equiv.) to isomeric benzofuranes. In the presence of oxygen, the Pd(II) salt can be used in catalytic amounts. An example is formulated in equation (I).[3]

(I) 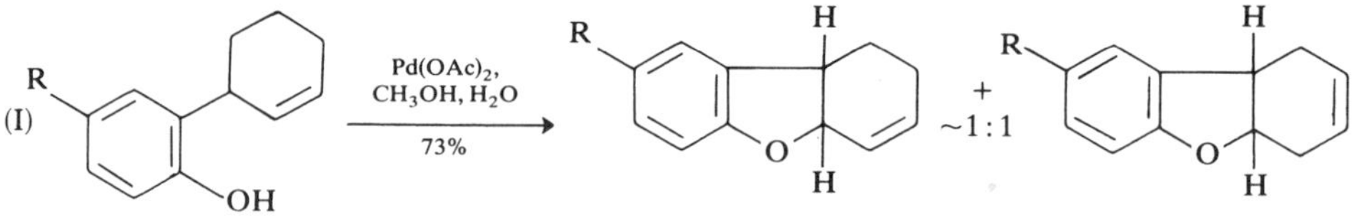 ($R = H, OCH_3$) $\xrightarrow{Pd(OAc)_2,\ CH_3OH,\ H_2O;\ 73\%}$ + ~1:1

If this cyclization is carried out in the presence of catalytic amounts of (−)-β-pinene, optically active products can be formed. Thus if the cyclization of the allylphenol **1** is conducted in the presence of $Pd(OAc)_2$ (1 mmole), $Cu(OAc)_2 \cdot H_2O$ (10 mmole), and (−)-β-pinene (1 mmole), an excess of (S)-2,3-dihydro-2-vinylbenzofurane (**2**) is formed in 12% optical yield. Surprisingly, if β-pinene is present in larger amounts (2 mmoles), the cyclization does not proceed.[4]

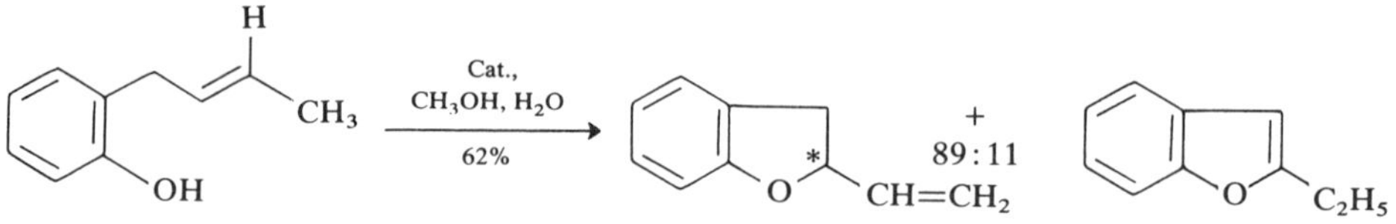

1 $\xrightarrow{Cat.,\ CH_3OH,\ H_2O;\ 62\%}$ **2** ($\alpha_D + 2.79°$) + **3** (89:11)

***Transannular reactions of grayanotoxin-II* (1).** The diterpenoids isolated from *Leucothoe grayana* belong to three groups: grayanotoxins, leucothols, and grayanols. Japanese chemists have converted grayanotoxin-II (**1**) to leucothol D (**2**) in one step by treatment with palladium acetate in methanol at 20°.[5] They also have effected conversion of **1** into **3** using thallium(III) nitrate. This compound can be regarded as a derivative of grayanol B (4).[6]

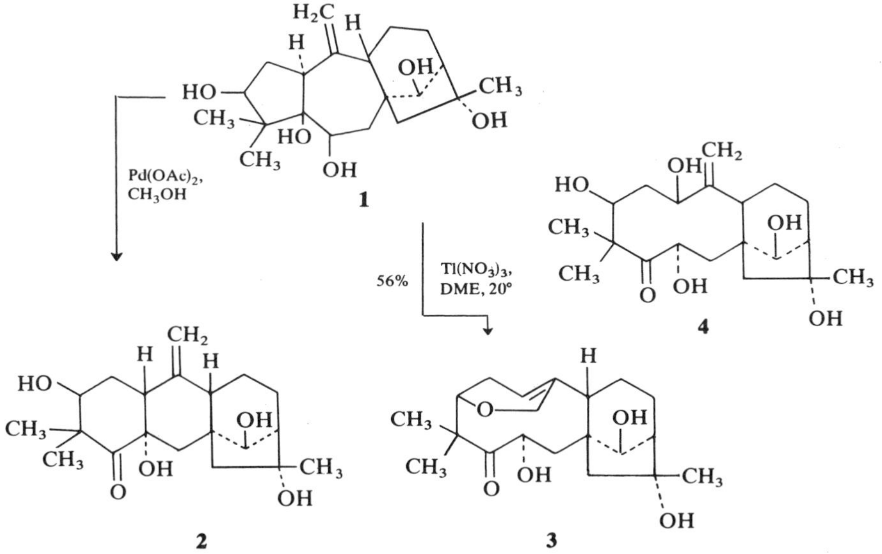

$Pd(OAc)_2$, CH_3OH

$Tl(NO_3)_3$, DME, 20° 56%

1 **2** **3** **4**

Cyclopentenes. Palladium(II) acetate catalyzes the cyclization of 1,6-dienes to substituted cyclopentenes.[7] However, several dienes (**1**, R = CN or $NHCOCH_3$; **3**, R = $(CH_2)_3CH{=}CH_2$; and **5**) fail to cyclize under these conditions.

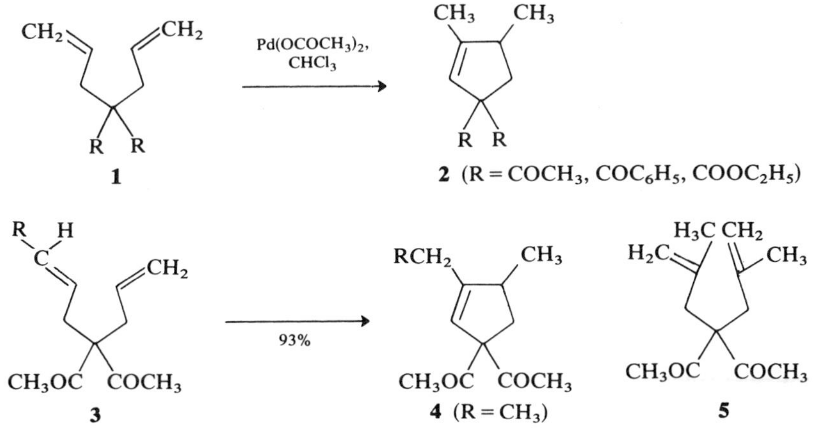

$Pd(OCOCH_3)_2$, $CHCl_3$

1 **2** (R = $COCH_3$, COC_6H_5, $COOC_2H_5$)

93%

3 **4** (R = CH_3) **5**

3-Methyl-2-cyclopentenones.[8] Enol silyl ethers of 3-butenyl methyl ketones such as **1** and **3** are cyclized exclusively to 2-cyclopentenones (**2, 4**) in the presence of palladium(II) acetate (1 equiv.) at room temperature with deposition of Pd(0).

$CH{=}CH_2$, CH_2, $OSi(CH_3)_3$ **1** → ($Pd(OAc)_2$, CH_3CN, 25°; 87%) → CH_3, O **2**

CH_2, CH_2, $OSi(CH_3)_3$ **3** → (99%) → CH_3, O **4**

The cyclization can be extended to preparation of 3-methyl-2-cyclohexenones and 3-methyl-2-cycloheptenones, but yields are low.

Polycyclic indoles. The treatment of the 3-benzoylindoles **1** and **3** with palladium acetate (0.5 equiv.) in acetic acid leads to **2** and **4**, respectively, by intramolecular oxidative ring closure.[9]

O, C, N, CH_3 **1** → ($Pd(OAc)_2$, HOAc; 60%) → O, N, CH_3 **2**

O, C, CH_3, N, CH_3 **3** → (31%) → O, CH_3, N, CH_3 **4**

β-Lactams. Carbon monoxide inserts into 2-bromo-3-aminopropene derivatives in the presence of a catalytic amount of $Pd(OAc)_2$ and triphenylphosphine to give α-methylene-β-lactams in 38–89% yield. $Pd(acac)_2$ can be used instead of $Pd(OAc)_2$ with similar results.[10]

Examples:

$$H_2C{=}C(Br)CH_2NHCH_2C_6H_5 \xrightarrow[67\%]{CO,\ Pd(OAc)_2,\ (n\text{-}C_4H_9)_3N,\ (C_6H_5)_3P}$$ 3-methylene-1-benzyl-2-azetidinone

$$H_2C{=}C(Br)CH_2NHCH_2CH_2CO_2CH_3 \xrightarrow{38\%}$$ 3-methylene-1-($CH_2CH_2COOCH_3$)-2-azetidinone

$$C_6H_5CH{=}C(Br)CH_2NHCH_2C_6H_5 \xrightarrow{89\%}$$ 3-benzylidene-1-benzyl-2-azetidinone

[1] F.-P. Montforts, S. Ofner, V. Rasetti, A. Eschenmoser, W.-D. Woggon, K. Jones, and A. R. Battersby, *Angew. Chem.*, *Int. Ed.*, **18**, 675 (1979).
[2] M. R. Roberts and R. H. Schlessinger, *Am. Soc.*, **101**, 7626, 7627 (1979).
[3] T. Hosokawa, S. Miyagi, S.-I. Murahashi, and A. Sonada, *J. Org.*, **43**, 2752 (1978).
[4] *Idem*, *J.C.S. Chem. Comm.*, 687 (1978).
[5] T. Kaiya, N. Shirai, and J. Sakakibara, *ibid*, 431 (1979).
[6] T. Kaiya, N. Shirai, J. Sakakibara, and Y. Iitaka, *Tetrahedron Letters*, 4297 (1979).
[7] R. Grigg, T. R. B. Mitchell, and A. Ramasubbu, *J.C.S. Chem. Comm.*, 669 (1979).
[8] Y. Ito, H. Aoyama, T. Hirao, A. Moehizuki, and T. Saegusa, *Am. Soc.*, **101**, 494 (1979).
[9] T. Itahara and T. Sakakibara, *Synthesis*, 607 (1978).
[10] M. Mori, K. Chiba, M. Okita, and Y. Ban, *J.C.S. Chem. Comm.*, 498 (1979).

Palladium(II) acetate–Triphenylphosphine, 5, 497–498.

Terminal 1,3-dienes. Palladium(II) acetate combined with $P(C_6H_5)_3$ is an effective catalyst for conversion of allylic acetates or allylic phenyl ethers into terminal dienes. Allylic methyl ethers or alcohols do not undergo this elimination.[1]

Example:

$$\left.\begin{array}{c} CH_3\overset{O}{\overset{\|}{C}}(CH_2)_3CH{=}CHCH_2OAc \\ \text{or} \\ CH_3\overset{O}{\overset{\|}{C}}(CH_2)_3CH(OAc)CH{=}CH_2 \end{array}\right\} \xrightarrow[70\text{–}80\%]{Pd(OAc)_2,\ P(C_6H_5)_3,\ C_6H_5CH_3,\ \Delta} CH_3\overset{O}{\overset{\|}{C}}(CH_2)_2CH{=}CHCH{=}CH_2$$

This reaction was used in a synthesis of pyrethrolone (**1**) from butadiene (scheme I).[2]

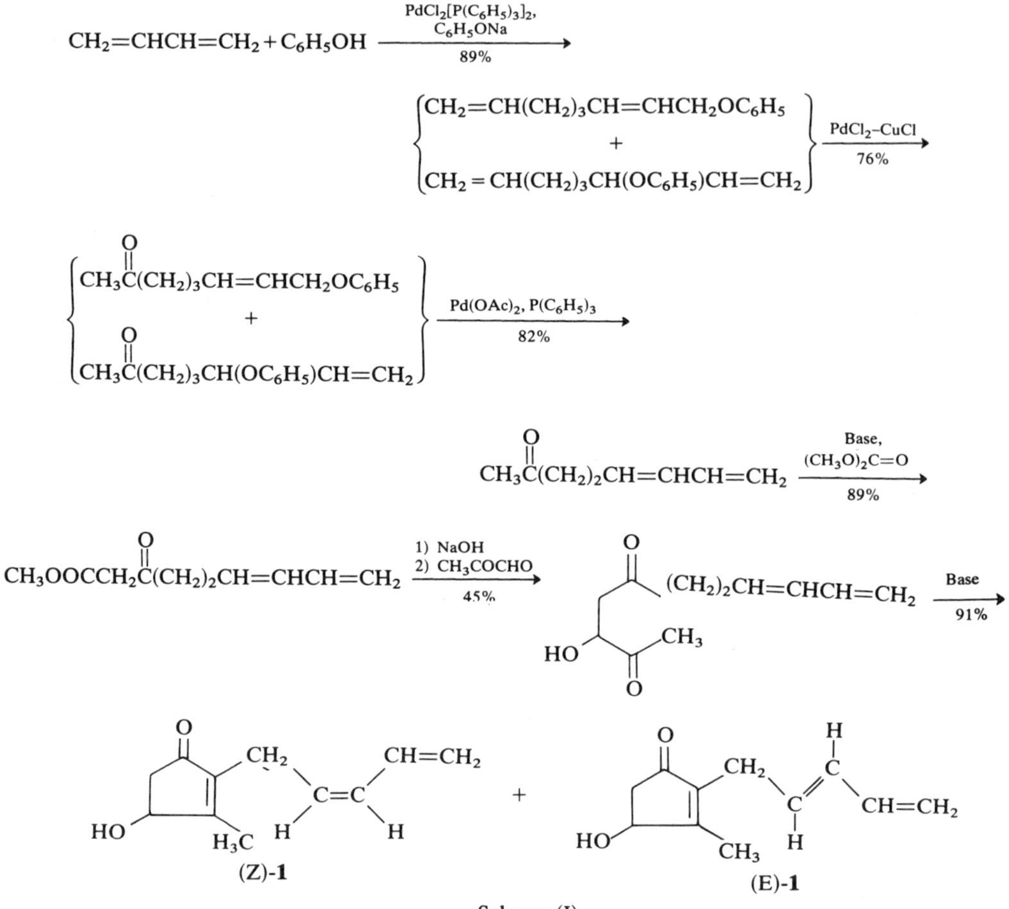
CH2=CHCH=CH2+C6H5OH
PdCl2[P(C6H5)3]2, C6H5ONa
89%
CH2=CH(CH2)3CH=CHCH2OC6H5
+
CH2=CH(CH2)3CH(OC6H5)CH=CH2
PdCl2–CuCl
76%
CH3C(CH2)3CH=CHCH2OC6H5
CH3C(CH2)3CH(OC6H5)CH=CH2
Pd(OAc)2, P(C6H5)3
82%
CH3C(CH2)2CH=CHCH=CH2
Base, (CH3O)2C=O
89%
CH3OOCCH2C(CH2)2CH=CHCH=CH2
1) NaOH
2) CH3COCHO
45%
(CH2)2CH=CHCH=CH2
CH3
HO
Base
91%
(Z)-1
(E)-1

Scheme (I)

[1] J. Tsuji, T. Yamakawa, M. Kaito, and T. Mandai, *Tetrahedron Letters*, 2075 (1979).
[2] J. Tsuji, T. Yamakawa, and T. Mandai, *ibid.*, 3741 (1979).

Palladium(II) acetate–Tri-*o*-tolylphosphine, $Pd(OAc)_2$–P 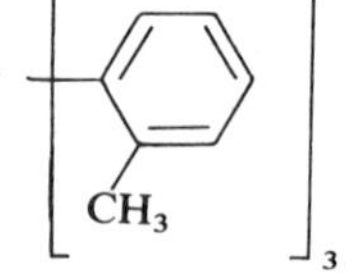 (**1**). Preparation of the phosphine;[1] supplier: Strem.

Vinylic substitution with aryl halides.[2] This reaction can be effected with catalysis by $Pd(OAc)_2$ and a phosphine; yields are generally improved when tri-*o*-

tolylphosphine is used rather than the more usual triphenylphosphine. An example is the preparation of diethyl (Z)-phenylmaleate (equation I).

$$\text{(I)}\quad C_6H_5I + \underset{H}{\overset{C_2H_5OOC}{}}\!\!C{=}C\!\!\underset{COOC_2H_5}{\overset{H}{}} + N(C_2H_5)_3 \xrightarrow[73\%]{\mathbf{1},\ CH_3CN,\ 100^\circ} \underset{H}{\overset{C_2H_5OOC}{}}\!\!C{=}C\!\!\underset{C_6H_5}{\overset{COOC_2H_5}{}} + (C_2H_5)_3\overset{+}{N}HI^-$$

[1] C. B. Ziegler, Jr., and R. F. Heck, *J. Org.*, **43**, 2941 (1978).
[2] N. A. Cortese, C. B. Ziegler, B. J. Hrnjez, and R. F. Heck, *ibid.*, **43**, 2952 (1978).

Palladium catalysts, 1, 778–782; **2,** 203; **4,** 368–369; **5,** 499; **6,** 445–446; **7,** 275–277; **8,** 382–383.

Acylsilanes. A new route to acylsilanes from allyloxysilanes is shown in equation (I). A key step is the palladium-promoted isomerization of the double bond under neutral conditions.[1]

$$\text{(I)}\quad \underset{R^2}{\overset{R^1}{}}\!\!C{=}CHCH_2OSi(R^3)_3 \xrightarrow[50\text{–}90\%]{\substack{1)\ \textit{sec}\text{-BuLi, THF, TMEDA}\\ 2)\ (CH_3)_3SiCl}} \underset{R^2}{\overset{R^1}{}}\!\!C{=}CH\underset{OSi(CH_3)_3}{\underset{|}{C}}HSi(R^3)_3 \xrightarrow[50\text{–}80\%]{Pd/C,\ CH_3OH,\ \Delta} R^2{-}\underset{H}{\overset{R^1}{C}}{-}CH_2\underset{O}{\underset{\|}{C}}Si(R^3)_3$$

Transfer hydrogenation.[2] Hydrazine is apparently superior to cyclohexene for transfer hydrogenation with palladium black as catalyst for hydrogenolysis of various protective groups of peptides. It can be used for cleavage of CBZ groups, benzyl esters, and benzyl ethers; it is particularly useful for removal of nitro groups.

Hydrogenation of a sulfur-containing substrate. Usually sulfur-compounds are not satisfactory substrates for catalytic hydrogenations. However, the diacetyl derivative (**2**) of dehydrobiotin (**1**) can be hydrogenated successfully with Pd/C as catalyst to enantiomerically pure N,N-diacetylbiotin methyl ester (**3**). The yield can be made quantitative by using hydrogen pressures of about 3000 psi and increased amounts of catalyst.[3]

O, R^1N, NR′, S, COOR

1 ($R = H, R^1 = H$)
2 ($R = CH_3, R^1 = COCH_3$)

O, AcN, NAc, H, H, S, $COOCH_3$

3

Review. Rylander[4] has provided a brief guide to selection of noble metals for catalytic hydrogenation. In general, palladium is more effective than platinum for hydrogenation of various functional groups. For example, platinum is relatively active only for hydrogenation of double bonds, whereas palladium is effective for hydrogenolysis of conjugated cyclopropyl systems, for isomerization of double bonds, for hydrogenation of alkynes to *cis*-alkenes, and for reduction of acid chlorides to aldehydes.

[1] A. Hosomi, H. Hashimoto, and H. Sakurai, *J. Organometal. Chem.*, **175**, C1 (1979).
[2] M. K. Anwer, S. A. Kahn, and K. M. Sivandaiah, *Synthesis*, 751 (1978).
[3] J. Vasilevskis, J. A. Gualtieri, S. D. Hutchings, R. C. West, J. W. Scott, D. R. Parrish, F. T. Bizzarro, and G. F. Field, *Am. Soc.*, **100**, 7423 (1978).
[4] P. N. Rylander, *Aldrichim. Acta*, **12**, 53 (1979); *Catalytic Hydrogenation in Organic Synthesis*, Academic Press, New York, 1979.

Palladium(II) chloride, 1, 782; **3,** 303–305; **4,** 367–370; **5,** 500–503; **6,** 447–450; **7,** 277; **8,** 384–385.

Phenylation of enones. This reaction can be conducted with C_6H_5HgCl or $Sn(C_6H_5)_4$ and with $PdCl_2$ and tetra-*n*-butylammonium chloride (TBA^+Cl^-) as catalysts in acidic CH_2Cl_2–H_2O. The active catalyst is probably $TBA^+PdCl_3^-$. An acidic medium is essential.[1]

Example:

$$CH_3\overset{O}{\overset{\|}{C}}CH{=}CHC_6H_5 + C_6H_5HgCl \xrightarrow[85\%]{PdCl_2,\ TBA^+Cl^-,\ CH_2Cl_2\text{–}H_2O,\ HCl} CH_3\overset{O}{\overset{\|}{C}}CH_2CH(C_6H_5)_2$$

Polymer-bound reagent.[2] A polymer-bound form of $PdCl_2$ (**1**) has been prepared by anchoring anthranilic acid to chloromethylated polystyrene followed by complexation with $PdCl_2$ in CH_3CN.

Ⓟ–C₆H₄–CH₂NH–C₆H₄(COOH) $\xrightarrow{PdCl_2}$ **1**

This catalyst (**1**) is a selective catalyst for hydrogenation of 4-octyne to *cis*-4-octene (90% yield). If the product is left in contact with **1**, it isomerizes to the *trans*-isomer. Since monoalkenes are hydrogenated slowly with this catalyst, cyclooctene can be obtained from either 1,3- or 1,5-cyclooctadiene. Of more interest, benzene can be hydrogenated to cyclohexane (83% yield). One drawback is that the catalyst loses about 90% of its activity after one cycle.

[1] S. Cacchi, F. La Torre, and D. Misiti, *Tetrahedron Letters*, 4591 (1979).
[2] N. L. Holy, *J. Org.*, **43**, 4686 (1978).

Palladium(II) chloride–Copper(II) chloride.

Oxidative cyclization of dienes. Dienes are oxidized with formation of a new carbon–carbon bond by catalytic amounts of $PdCl_2$ and $CuCl_2$ [to regenerate Pd(II)] in NaOAc-buffered acetic acid.[1]

CH_2 $\xrightarrow[75\%]{PdCl_2, CuCl_2, HOAc}$ Cl, Cl

H_2C $\xrightarrow[50\text{–}70\%]{}$ AcO, Cl

(*exo*/*endo* = 7 : 3)

CH_2, CH_2 $\xrightarrow[77\%]{}$ H, OAc, CH_3

[1] A. Heumann, M. Reglier, and B. Waegell, *Angew. Chem., Int. Ed.*, **18**, 866 (1979).

Palladium hydroxide, 2, 305; **7**, 385–386.

β-Amino acids. Chiral enamines (**1**) are hydrogenated by 10% palladium hydroxide catalyst on charcoal to β-amino acids (**2**) in optical purity of 7–28%.[1] Optical yields are lower if the enamines are reduced with sodium cyanoborohydride. The two methods lead to opposite configurations of **2**.

$$\overset{*}{R}{}^1NH_2 + R^2COCH_2COOC_2H_5 \rightarrow \underset{\textbf{1}}{R^2\underset{\displaystyle NH\overset{*}{R}{}^1}{C}{=}CHCOOC_2H_5} \xrightarrow{H_2,\ Pd(OH)_2/C} \underset{\textbf{2}\ (7\text{–}28\%\ ee)}{R^2\overset{*}{\underset{\displaystyle NH_2}{C}}HCH_2COOH}$$

[1] M. Furukawa, T. Okawara, Y. Noguchi, and Y. Terawaki, *Chem. Pharm. Bull. Japan*, **27**, 2223 (1979).

Pentamethoxyphosphorane, $P(OCH_3)_5$. Mol. wt. 186.15, b.p. 37°. The material can be prepared in about 55% yield by reaction of trimethyl phosphite with methyl benzenesulfenate (which is converted during the reaction into diphenyl disulfide). It is stable to 80°. It reacts with acids to form methyl esters (methyl benzoate, 90% yield). It converts phenols into methyl ethers (anisole, 90%; 2,4-dimethylanisole, 77%; thioanisole, 87%).[1]

[1] D. B. Denney, R. Melis, and A. D. Pendse, *J. Org.*, **43**, 4672 (1978).

3,3,6,9,9-Pentamethyl-2,10-diazabicyclo[4,4,0]-1-decene, 1, Mol. wt. 208.35, m.p. 15°, b.p. 65°/0.07 mm.

Preparation:

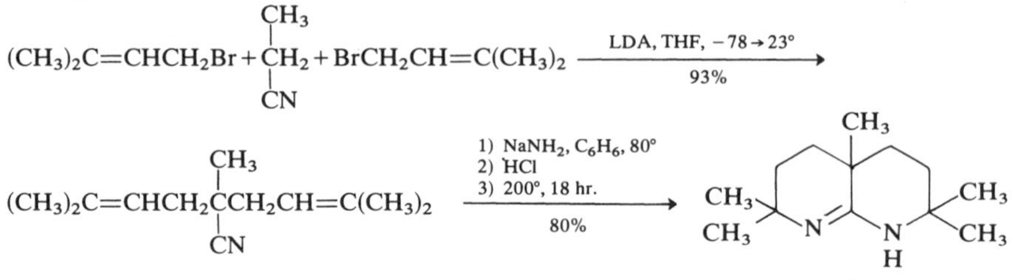

$$(CH_3)_2C{=}CHCH_2Br + CH_3CH_2CN + BrCH_2CH{=}C(CH_3)_2 \xrightarrow[93\%]{\text{LDA, THF, } -78 \rightarrow 23°}$$

$$(CH_3)_2C{=}CHCH_2C(CH_3)(CN)CH_2CH{=}C(CH_3)_2 \xrightarrow[80\%]{\text{1) NaNH}_2\text{, C}_6\text{H}_6\text{, 80°; 2) HCl; 3) 200°, 18 hr.}} \mathbf{1}$$

1

Salts.[1] The amidine **1** is a useful reagent for formation of salts of carboxylic acids and related proton complexes of bidentate ligands. The salts of **1** have favorable solubility characteristics.

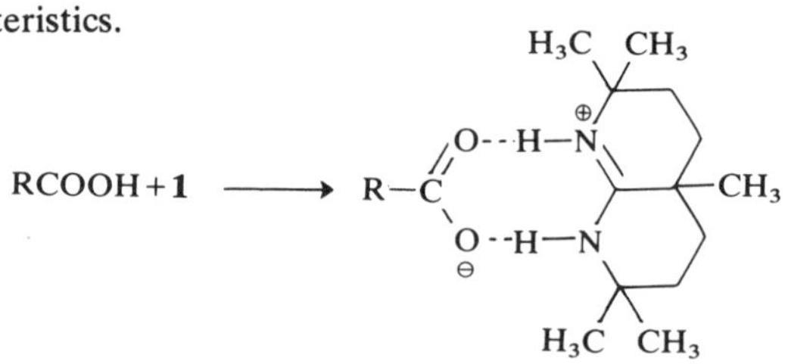

RCOOH + **1** ⟶ R—C(O⁻)(=O)···H—N salt of **1**

Fragmentation reactions.[2,3] Heating salts **2–5** to the melting point for 1–3 minutes effects decarboxylative double fragmentation reactions, as formulated. The simplicity of this procedure and the high yields are consequences of the use of amidine **1**. However, the ability of tosyloxycarboxylic acids corresponding to **2** and **3** to undergo fragmentation is independent of **1**.

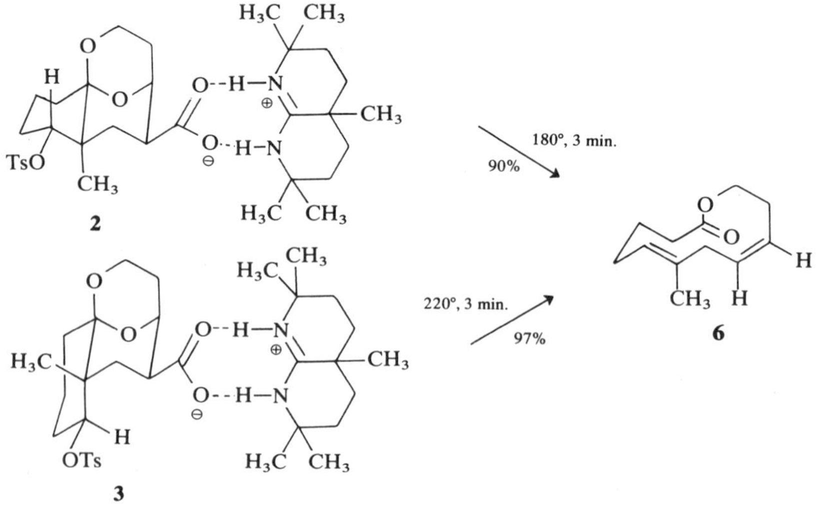

2 $\xrightarrow[90\%]{180°, \text{ 3 min.}}$ **6**

3 $\xrightarrow[97\%]{220°, \text{ 3 min.}}$ **6**

2

3

6

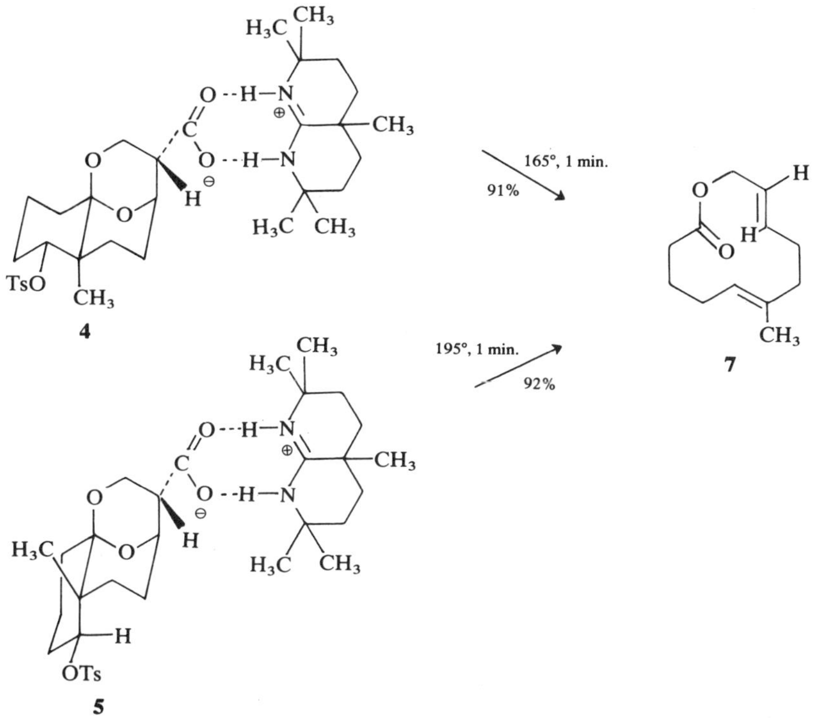

$$\mathbf{4} \xrightarrow[91\%]{165°,\ 1\ \text{min.}} \mathbf{7} \xleftarrow[92\%]{195°,\ 1\ \text{min.}} \mathbf{5}$$

[1] F. Heinzer, M. Soukup, and A. Eschenmoser, *Helv.*, **61**, 2851 (1978).
[2] D. Sternbach, M. Shibuya, F. Jaisli, M. Bonetti, and A. Eschenmoser, *Angew. Chem., Int. Ed.*, **18**, 634 (1979).
[3] M. Shibuya, F. Jaisli, and A. Eschenmoser, *ibid.*, **18**, 636 (1979).

Pentane-1,5-di(magnesium bromide), $BrMg(CH_2)_5MgBr$. Mol. wt. 278.58.

Grignard reagents from trialkylboranes. Trialkylboranes are converted into Grignard reagents by reaction with this di-Grignard reagent in benzene; the efficiency of this reaction is probably a result of formation of the stable bicyclic borate (equation I).

$$\text{(I)}\quad R_3B + 2BrMg(CH_2)_5MgBr \rightarrow 3RMgBr + \text{(spiro-borate)}^{-}\ \overset{+}{M}gBr$$

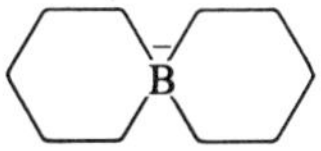

Some representative reactions conducted with Grignard reagents prepared in this way are formulated.[1]

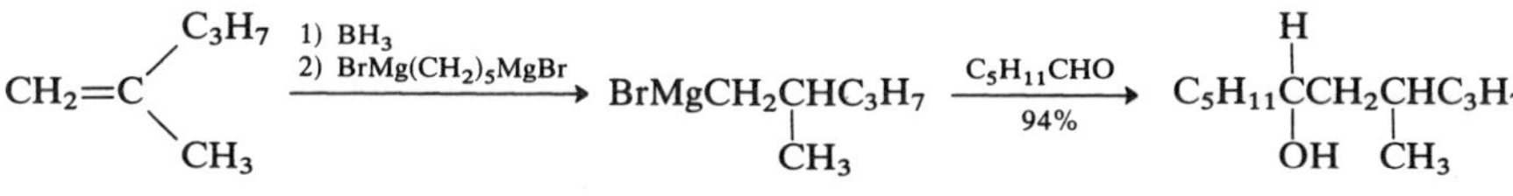

$$CH_2{=}C(C_3H_7)CH_3 \xrightarrow[2)\ BrMg(CH_2)_5MgBr]{1)\ BH_3} BrMgCH_2CH(CH_3)C_3H_7 \xrightarrow[94\%]{C_5H_{11}CHO} C_5H_{11}CH(OH)CH_2CH(CH_3)C_3H_7$$

$$CH_2{=}CHCH_2CH_3 \rightarrow CH_3CH_2CH_2CH_2MgBr + \underset{C_6H_5}{\overset{H}{}}\!\!>\!C{=}C\!<\!\!\underset{H}{\overset{Br}{}} \xrightarrow[77\%]{Pd[P(C_6H_5)_3]_4} \underset{C_6H_5}{\overset{H}{}}\!\!>\!C{=}C\!<\!\!\underset{H}{\overset{CH_2CH_2CH_2CH_3}{}}$$

$$CH_2{=}CH(CH_2)_9Cl \rightarrow Cl(CH_2)_9CH_2CH_2MgBr + CH_2{=}CHCH_2Br \xrightarrow[88\%]{Li_2CuCl_4} CH_2{=}CH(CH_2)_{12}Cl$$

[1] K. Kondo and S.-I. Murahashi, *Tetrahedron Letters*, 1237 (1979).

Phase-transfer catalysts, 8, 387–391.

Reviews. Gokel and Weber[1] have reviewed the principles involved in phase-transfer catalysis and the applications to synthesis (128 references). The review includes crown ethers and cryptates as well as quaternary ammonium and phosphonium salts.

Keller's review[2] is organized according to the type of compound that has been prepared by PTC. It contains 691 literature references. It also includes a section by G. Simchen on his work with quaternary ammonium salts in homogeneous phase.

Starks and Liotta[3] have reviewed this field in somewhat more detail.

Hydrolysis of esters. Dehmlow and Naranjo[4] have examined the effect of various catalysts on the hydrolysis of esters in an organic solvent and 50% aqueous NaOH. In general, only slight acceleration is noted. Of various catalysts, tetra-*n*-butylammonium hydrogen sulfate is effective and also superior to the corresponding chloride. For no apparent reason, crown ethers are not effective. Anionic and neutral surfactants such as $C_{17}H_{35}COONa$ and n-$C_{16}H_{33}O[CH_2CH_2O]_{29}COCH_3$, respectively, are about as effective as quaternary ammonium salts.

Dehydrohalogenation.[5] Aryl vinyl ethers are prepared conveniently by dehydrohalogenation of aryl 2-haloethyl ethers with aqueous sodium hydroxide with tetra-*n*-butylammonium hydrogen sulfate as phase-transfer catalyst (equation I).

$$\text{(I)} \quad \underset{(X = Cl, Br)}{ArOCH_2CH_2X} \xrightarrow[90\text{–}97\%]{NaOH, H_2O, C_6H_6, (C_4H_9)_4NHSO_4} ArOCH{=}CH_2$$

N-Methyldi-t-butylamine. This amine is the first N-alkyldi-*t*-butylamine to be prepared.[6] It was obtained by reaction of di-*t*-butylamine[7] with chloroform and sodium hydroxide under phase-transfer conditions to give di-*t*-butylformamide, which was then reduced with lithium aluminum hydride (equation I).

$$\text{(I)} \quad HN[C(CH_3)_3]_2 \xrightarrow[55\%]{CHCl_3, NaOH, C_6H_5CH_2N^+(C_2H_5)_3Cl^-} [(CH_3)_3C]_2NCHO \xrightarrow[\text{"high"}]{LiAlH_4} CH_3N[C(CH_3)_3]_2$$

Reductive methylation of di-*t*-butylamine failed.

Alkyl iodides. Alkyl bromides or chlorides can be converted in the gas phase (160°) into alkyl iodides by reaction with solid KI in the presence of tetra-*n*-butylphosphonium iodide as catalyst (equation I).[8]

$$\text{(I)} \qquad RBr(Cl) + KI \xrightarrow[45\text{-}85\%]{SiO_2,\ \text{cat.},\ 160°} RI$$

Esterification. N-Protected (BOC or CBZ) amino acids can be esterified by reaction with 2 equiv. of an alkyl bromide or iodide and 1 equiv. of Adogen 464 (purified[9]) in aqueous $NaHCO_3$–CH_2Cl_2. The yields are usually 70–95%.[10a]

The same laboratory[10b] reports esterification of protected amino acids via the sodium salts in $NaHCO_3$–DMF in comparable yields.

Tribromomethylation of allylic bromides. Exposure of primary allylic bromides to bromoform–aqueous NaOH in the presence of TEBA results in substitution of bromide by tribromomethyl anion.[11]

$$R_1R_2C{=}C(R_3)CH_2Br \xrightarrow[38\text{-}58\%]{CHBr_3,\ \text{TEBA},\ NaOH,\ H_2O} R_1R_2C{=}C(R_3)CH_2CBr_3$$

Acetylation of phenols and anilines. These substrates can be acetylated rapidly and in generally high yield by reaction with acetyl chloride using powdered NaOH and tetrabutylammonium hydrogen sulfate in an organic solvent (CH_2Cl_2, dioxane, THF). Even 2,6-di-*t*-butyl-*p*-cresol can be acetylated in this way in 72% yield. Selective acylation of the phenolic hydroxyl group of estradiol is possible.[12]

Oxidation of alcohols.[13] Primary and secondary alcohols can be oxidized to aldehydes and ketones, respectively, by potassium chromate and sulfuric acid in a two-phase system ($CHCl_3$–H_2O) in the presence of tetra-*n*-butylammonium hydrogen sulfate as catalyst. The method is particularly useful for oxidation of primary

$$3RCH_2OH + 2K_2CrO_4 + 5H_2SO_4 \xrightarrow{(C_4H_9)_4\overset{+}{N}HSO_4^-,\ CHCl_3,\ H_2O} 3RCHO + 2K_2SO_4 + Cr_2(SO_4)_3 + 8H_2O$$

alcohols, since aldehydes are not oxidized further to carboxylic acids under these conditions. However, yields are low with water-soluble alcohols. With other alcohols yields of aldehydes as high as 90% can be achieved. The method is suitable for oxidation of allylic alcohols and for acid-labile alcohols.

Alkylation of aldehydes and β-keto esters.[14] This reaction can be conducted by use of solid–liquid phase-transfer catalysis using powdered sodium hydroxide as base and benzene as the solvent. Under these conditions aldehydes with only one α-hydrogen, such as isobutyraldehyde, are alkylated in reasonable yield even by less reactive halides (equation I).

$$\text{(I)} \qquad (CH_3)_2CHCHO + RCl(Br) \xrightarrow[50\text{-}85\%]{NaOH,\ C_6H_6,\ (n\text{-}Bu)_4N^+I^-} (CH_3)_2C(R)CHO$$

Methyl acetoacetate can also be monoalkylated under these conditions (equation II).

$$\text{(II)}\quad CH_3\overset{O}{\overset{\|}{C}}CH_2COOCH_3 + RCl(Br) \xrightarrow[40\text{–}80\%]{NaOH,\ C_6H_6,\ CH_3(CH_2)_{15}\overset{+}{N}(CH_3)_2CH_2C_6H_5Cl^-} CH_3\overset{O}{\overset{\|}{C}}\underset{R}{\underset{|}{C}}HCOOCH_3$$

Spiroalkylation of cyclopentanone. Näf and Decorzant[15] have reported a novel reaction between cyclopentanone and 1,4-dibromo-2-pentene (**1**) resulting in the spiro[2.4]heptene-4-ones **2**. NaH or $NaNH_2$ can be used as base (THF, 65°), but yields are higher using phase-transfer conditions: Aliquat 336, aqueous KOH, 65°; yield of **2a** and **2b**, 40–55%. Descotes *et al.*[16] had previously reported that **2a** rearranged to **3** when heated, but his yield for the preparation of **2** from cyclopentanone was only about 7.5%. Methods for preparation of methyl jasmonate (**4**) from **3** and for preparation of (Z)-jasmone (**5**) from **3**[17] had already been developed.

Cyclopentanone + $BrCH_2CH{=}CHCHBrCH_3$ (**1** (*cis* and *trans*)) $\xrightarrow{40\text{–}55\%}$ **2a + 2b** $\xrightarrow[89\%]{300\text{–}350°}$ **3**

3 $\xrightarrow{\text{1) Na, } CH_2(COOCH_3)_2;\ \text{2) NaOH};\ \text{3) } CH_3OH, H_2SO_4}$ **4**

3 $\xrightarrow[90\%]{\text{1) } CH_3Li;\ \text{2) } CrO_3, H^+}$ **5**

Activation of alkali metal carbonates.[18] One difficulty with the classical Mąkosza phase-transfer catalytic reaction employing aqueous NaOH as one phase is that carboalkoxy groups in the substrate are hydrolyzed to a considerable extent. Actually anhydrous K_2CO_3 and Na_2CO_3 can be used as bases for generation of carbanions in a solid–liquid two-phase system with tetraalkylammonium salts or crown ethers as catalysts. Probably the carbanions generated on the surface of the carbonate migrate as ion pairs into the organic phase.

Some typical reactions conducted by this technique are shown.

$$H_2C(COOCH_3)(CN) + C_6H_5CH_2Cl \xrightarrow[87\%]{Na_2CO_3,\ 90°,\ C_6H_5CH_2\overset{+}{N}(C_2H_5)_3Cl^-} (C_6H_5CH_2)_2C(COOCH_3)(CN)$$

$$\text{9-fluorenyl}-CH(COOC_2H_5) + C_6H_5COCH_2Cl \xrightarrow[89\%]{K_2CO_3,\ 40°,\ \text{18-crown-6}} \text{9-fluorenyl}-C(COOC_2H_5)(CH_2COC_6H_5)$$

$$(C_6H_5)_2CHCHO + NO_2-C_6H_3(NO_2)-Cl \xrightarrow[77\%]{K_2CO_3,\ \text{crown ether}} (C_6H_5)_2C{=}CH-O-C_6H_3(NO_2)_2$$

$$C_6H_5CH_2CN + ClCH_2COOC_2H_5 \xrightarrow[45\%]{120°} C_6H_5CH(CN)CH_2COOC_2H_5$$

$$(C_6H_5)(C_2H_5)CHCN + C_6H_5COCl \xrightarrow[35\%]{150°} (C_6H_5)(C_2H_5)C(CN)(COC_6H_5)$$

$$C_6H_5CHO + ClCH_2COOC_2H_5 \xrightarrow[72\%]{130°} C_6H_5CH\overset{O}{-}CHCOOC_2H_5$$

$$C_6H_5C(CH_3){=}CH_2 + CHBr_3 \xrightarrow[68\%]{140°} \text{1-}C_6H_5\text{-1-}CH_3\text{-2,2-dibromocyclopropane}$$

Sulfonates.[19] Both methane- and benzenesulfonates can be prepared in good to excellent yield under phase-transfer conditions (benzyltriethylammonium chloride in benzene or CH_2Cl_2).

$$R^1(Ar)SO_2Cl + R^2CH_2OH \xrightarrow[75\text{–}90\%]{} R^1(Ar)SO_2OCH_2R^2$$

Cleavage of ethers.[20] Dialkyl and alkyl aryl ethers are cleaved by 47% hydrobromic acid in an aqueous organic solvent in the presence of an onium salt (hexadecyltributylphosphonium bromide, tetraoctylammonium bromide) that is completely soluble in the organic phase. Dialkyl ethers are converted into alkyl bromides (65–90% yield),whereas a phenol and an alkyl bromide are obtained from alkyl aryl ethers (85–90% yields).

Dichlorocarbene addition to alkenes. Dehmlow and Lissel[21] have examined the reaction variables in the generation of dichlorocarbene by PTC. Optimal conditions include use of 4 molar excess each of $CHCl_3$ and 50% aqueous NaOH, 1 mole % of catalyst, and efficient stirring. The reaction should be conducted initially at 0–5°, then at 20° for 1–2 hours, and finally at 50° for 2–4 hours. Most quaternary ammonium salts are suitable as catalysts; the anions should be chloride or hydrogen sulfate. From the point of cost/efficiency, the most useful are benzyltriethylammonium chloride, tetra-*n*-butylammonium chloride, Aliquat 336, and tri-*n*-propylamine. The reaction rate is strongly dependent on the nucleophilicity of the alkene.

Polyethylene glycol. Cram *et al.*[22] have compared the ability of macrocyclic polyethers and open-chain analogs to complex various salts and have concluded that the former ethers are more efficient. Nevertheless several laboratories have recently reported that the inexpensive polyethylene glycols, $H(OCH_2CH_2)_nOH$, or their ethers can function about as efficiently as crown ethers as phase-transfer catalysts in reactions involving potassium salts. Examples are displacement reactions of benzyl bromide with various potassium salts[23] and oxidations with potassium permanganate.[24] Polyethylene glycol can also serve as substitute for 18-crown-6[25] in Sandmeyer and Gomberg-Bachman reactions of aryldiazonium tetrafluoroborates initiated by potassium acetate according to the newer method of Gokel *et al.*[26]

Further examples of the ability of PEG-400 to serve as a phase-transfer catalyst have been reported. It is also able to solubilize $KMnO_4$ in benzene.[27]

α-Methylene-β-lactams.[28] A new and useful route to these lactams (**3**) involves cyclization of N-arylamides (**1**), readily obtainable from 3-bromo-2-bromomethylpropionic acid,[29] under phase-transfer conditions with triethylpentylammonium bromide. The bromomethyl-*β*-lactam **2** can be isolated as an intermediate.

Unfortunately yields tend to be low with N-alkylamides corresponding to **1**; unsubstituted amides do not undergo this reaction.

$(BrCH_2)_2CHCONHAr$ (**1**) —[40% NaOH, CCl_4, Cat., 20°]→ **2** ($BrCH_2$, O, N, Ar) —[75–95% overall]→ **3** (H_2C, O, N, Ar)

Michael addition to α,β-unsaturated aldehydes.[30] The Michael addition of malonic ester and acetoacetic ester to *α,β*-unsaturated aldehydes can be carried out in moderate yields under phase-transfer conditions with sodium or potassium carbonate as base and benzyltriethylammonium chloride as catalyst (equation I).

$$\text{(I)}\quad CH_3CH{=}CHCHO + CH_2(COOC_2H_5)_2 \xrightarrow[60\%]{\text{Cat., }K_2CO_3,\ C_6H_6,\ H_2O} (C_2H_5OOC)_2CH\overset{\overset{\displaystyle CH_3}{|}}{C}HCH_2CHO$$

Under the same conditions cyanoacetic ester undergoes Knoevenagel condensation (equation II).

$$\text{(II)}\quad C_6H_5CH{=}CHCHO + CNCH_2COOC_2H_5 \xrightarrow[64\%]{} \begin{matrix} CN \\ \quad\diagdown \\ \quad\; C{=}CHCH{=}CHC_6H_5 \\ \quad\diagup \\ C_2H_5OOC \end{matrix}$$

[1] G. W. Gokel and W. P. Weber, *J. Chem. Ed.*, **35**, 350 (1978).
[2] W. E. Keller, *Compendium of Phase-transfer Reactions and Related Synthetic Methods*, Fluka, 1979.
[3] C. M. Starks and C. Liotta, *Phase-transfer Catalysis*, Academic Press, New York, 1978.
[4] E. V. Dehmlow and S. B. Naranjo, *J. Chem. Res.* (*M*), 238 (1979).
[5] K. Mizuno, Y. Kimura, and Y. Otsuji, *Synthesis*, 688 (1979).
[6] C. A. Audeh, S. E. Fuller, R. J. Hutchinson, and J. R. L. Smith, *J. Chem. Res.* (*M*), 2984 (1979).
[7] T. G. Back and D. H. R. Barton, *J.C.S. Perkin I*, 924 (1977).
[8] P. Tundo and P. Venturello, *Synthesis*, 952 (1979).
[9] A. Dossena, V. Rizzo, R. Marchelli, G. Casnati, and P. L. Luisi, *Biochem. Biophys. Acta*, **446**, 493 (1976).
[10] (a) V. Bocchi, G. Casnati, A. Dossena, and R. Marchelli, *Synthesis*, 957 (1979); (b) *idem*, *ibid.*, 961 (1979).
[11] M. S. Baird, A. G. W. Baxter, B. R. J. Devlin, and R. J. G. Searle, *J.C.S. Chem. Comm.*, 210 (1979).
[12] V. O. Illi, *Tetrahedron Letters*, 2431 (1979).
[13] D. Landini, F. Montanari, and F. Rolla, *Synthesis*, 134 (1979).
[14] V. G. Purohit and R. Subramanian, *Chem. Ind.*, 731 (1978).
[15] F. Näf and R. Decorzant, *Helv.*, **61**, 2524 (1978).
[16] Y. Bahurel, L. Cottier, and G. Descotes, *Synthesis*, 118 (1974).
[17] G. Büchi and B. Egger, *J. Org.*, **36**, 2021 (1971).
[18] M. Fedoryński, K. Wojciechowski, Z. Matacz, and M. Mąkosza, *ibid.*, **43**, 4682 (1978).
[19] W. Szeza, *Synthesis*, 822 (1979).
[20] D. Landini, F. Montanari, and F. Rolla, *ibid.*, 771 (1978).
[21] E. V. Dehmlow and M. Lissel, *J. Chem. Res.* (*M*), 4163 (1978).
[22] E. P. Kyba, R. C. Helgeson, K. Madar, G. W. Gokel, T. L. Tarnowski, S. S. Moore, and D. J. Cram, *Am. Soc.*, **99**, 2565 (1977).
[23] H. Lehmkuhl, F. Rabeb, and K. Hauschild, *Synthesis*, 184 (1977).
[24] D. G. Lee and V. S. Chang, *J. Org.*, **43**, 1532 (1978).
[25] R. A. Bartsch and I. W. Yang, *Tetrahedron Letters*, 2503 (1979).
[26] S. H. Korzeniowski, L. Blum, and G. W. Gokel, *ibid.*, 1871 (1977); S. H. Korzeniowski and G. W. Gokel, *ibid.*, 3519 (1977).
[27] D. Balasubramanian, P. Sukumar, and B. Chandani, *ibid.*, 3543 (1979).
[28] S. R. Fletcher and I. T. Kay, *J.C.S. Chem. Comm.*, 903 (1978).
[29] Preparation: A. F. Ferris, *J. Org.*, **20**, 780 (1955).
[30] G. V. Kryshtal, V. V. Kulganek, V. F. Kucherov, and L. A. Yanovskaya, *Synthesis*, 107 (1979).

Phenoxyacetic acid, $C_6H_5OCH_2COOH$ (**1**). Mol. wt. 152.15, m.p. 98–100°. Supplier: Aldrich. Preparation.[1]

$R_3B \rightarrow RCH_2COOH$.[2] Phenoxyacetic acid forms a dianion (**2**) by treatment with 2 equiv. of LDA in THF at 0°. The dianion reacts with trialkylboranes to form an intermediate **a**, which undergoes the usual rearrangement of the R group from boron to carbon to give **b**. Alkaline oxidation gives the carboxylic acids (equation I).

$$\text{(I)}\quad R_3B + C_6H_5O\overset{\text{Li}}{\overset{|}{C}}HCOOLi \rightarrow \underset{\mathbf{2}}{}$$

$$\left[\underset{\mathbf{a}}{C_6H_5O(R_2\bar{B}R)CHCOOLi\ Li^+} \rightarrow \underset{\mathbf{b}}{R_2B(R)CHCOOLi}\right] \xrightarrow[\text{NaOH}]{H_2O_2,} \underset{\mathbf{3}}{RCH_2COOH}$$

Yields are low when R is bulky; however, they can be improved by use of B-alkyl-9-BBN derivatives instead of trialkylboranes.

[1] J. van Alphen, *Rec. trav.*, **46**, 144 (1927).
[2] S. Hara, K. Kishimura, and A. Suzuki, *Tetrahedron Letters*, 2891 (1978).

Phenyl benzenethiosulfonate, 8, 391–392.

1,2-Carbonyl transposition. Yee and Schultz[1] have used a modification of the method of Trost (**8**, 392) for this transposition. The first step also involves bissulfenylation with this reagent, but more satisfactory results were obtained when the lithium enolate was generated with lithium tetramethylpiperidide. The remaining steps are formulated in equation (I). It is also possible to convert the intermediate hydroxy thioketal to the ketol, which is then mesylated and reduced to the trans-

(I) [ketone with H, O, CH₃ substituents, CH₃ on aryl ring] $\xrightarrow[85\%]{1)\ \text{LiTMP}\quad 2)\ C_6H_5SO_2SC_6H_5}$ [2,2-bis(C_6H_5S) ketone, H, CH_3] $\xrightarrow[\text{quant.}]{LiAlH_4}$

[bis(C_6H_5S) alcohol, OH, H, CH_3] $\xrightarrow{\text{quant.}}$ [bis(C_6H_5S) mesylate, OMs, H, CH_3] $\xrightarrow[73\%]{HgO,\ HgCl_2,\ CH_3CN,\ H_2O}$

[keto mesylate, O, OMs, H, CH_3] $\xrightarrow[\text{58\% overall}]{CrCl_2}$ [transposed ketone, O, H, O, CH_3, CH_3]

posed ketone. This methodology was used in a more complex system in a total synthesis of lycoramine (**1**); the transposition in this case was accomplished in 64% overall yield (equation II).

(II) O, O, OCH_3, N, CH_3 —64%→ O, O, OCH_3, N, CH_3 —$LiAlH_4$→ HO, O, OCH_3, N, CH_3 **1**

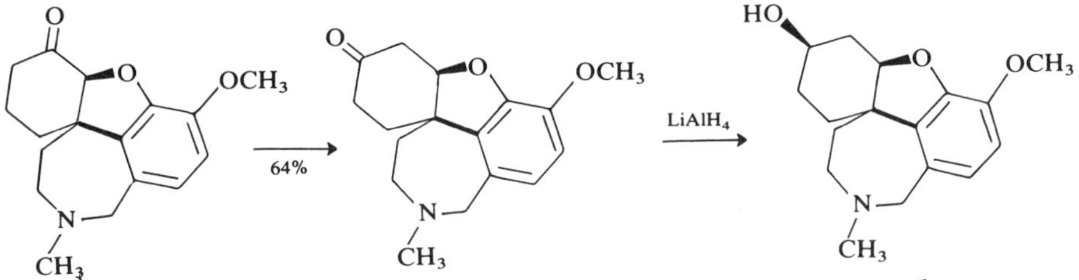

[1] Y. K. Yee and A. G. Schultz, *J. Org.*, **44**, 719 (1979); A. G. Schultz, Y. K. Yee, and M. H. Berger, *Am. Soc.*, **99**, 8065 (1977).

(+)-3-Phenyl-2,3-bornanediol, H_3C, CH_3, OH, OH, CH_3, C_6H_5 (**1**). Mol. wt. 246.34.

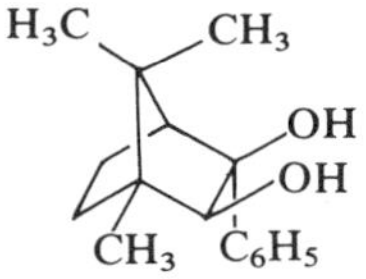

Reagent **1** is prepared by addition of phenyllithium to (−)-2-hydroxy-3-bornanone.[1]

Chiral homoallylic alcohols.[2] The glycol **1** has been used as the chiral matrix in an enantioselective synthesis of homoallylic alcohols (**4**) from aldehydes and allylboranes (equation I).

(I) $1 + B(CH_2C(R^1){=}CH_2)_3$ —THF, 50°, >95%→ **2** (H_3C, CH_3, O, O, $BCH_2C(R^1){=}CH_2$, CH_3, C_6H_5) —R^2CHO→

3 (H_3C, CH_3, O, O, $B{-}OCH(R^2)CH_2C(R^1){=}CH_2$, CH_3, C_6H_5) —$N(CH_2CH_2CH_2CH_3)_3$, 83–93% from **2**→ **1** (80%) + $R^2CH(OH)CH_2C(R^1){=}CH_2$ **4** (45–75% ee)

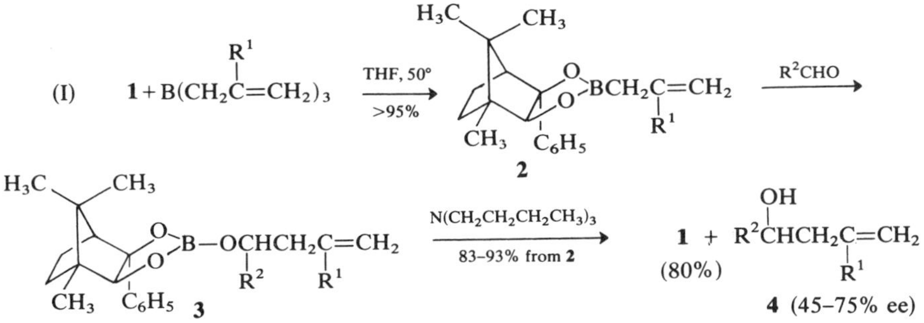

[1] I. Fleming and R. B. Woodward, *J. Chem. Soc. C*, 1289 (1968).
[2] T. Herold and R. W. Hoffmann, *Angew. Chem., Int. Ed.*, **17**, 768 (1978).

α-Phenylethylamine, $CH_3CH(C_6H_5)NH_2$, **1**, 838; **2**, 271–273; **3**, 199–200; **6**, 457. S-(−), $\alpha_D -39.4°$; R-(+), $\alpha_D +39.7°$. Resolution.[1]

Asymmetric alkylation of aldehydes.[2] Aldimines derived from the S-(−)-amine and propionaldehyde undergo alkylation (LDA, 1 equiv. of $MgBr_2$) in a stereoselective manner (equation I).

(I) $\mathrm{CH_3CH_2(H)C{=}N{-}CH(CH_3)C_6H_5} \xrightarrow[70\text{–}88\%]{1)\ \mathrm{LDA, MgBr_2}\quad 2)\ \mathrm{RX}} \mathrm{CH_3CH(R)(H)C{=}N{-}CH(CH_3)C_6H_5}$ (67–70% ee)

[1] A. Ault, *Org. Syn. Coll. Vol.* **5**, 932 (1973).
[2] R. R. Fraser, F. Akiyama, and J. Banville, *Tetrahedron Letters*, 3929 (1979).

(S)-(−)-5-(α-Phenylethyl)semioxamazide (1). Mol. wt. 207.22, m.p. 169°, α_D −103°.

Preparation[1]:

$\mathrm{C_6H_5(CH_3)CHNH_2} \xrightarrow[91\%]{\mathrm{HOOCCOOC_2H_5}} \mathrm{C_6H_5(CH_3)CHNHCOCOOC_2H_5} \xrightarrow[93\%]{\mathrm{H_2NNH_2}} \mathrm{C_6H_5(CH_3)CHNHCOCONHNH_2}$ (S)-**1**

Resolution of aldehydes.[2] The reagent was introduced by Leonard and Boyer[1] for resolution of carbonyl compounds. It is particularly useful for resolution of chiral aromatic aldehydes (**2**) complexed with tricarbonylchromium.

2 (R-substituted benzaldehyde–$\mathrm{Cr(CO)_3}$, CHO) $\xrightarrow[70\text{–}90\%]{\mathbf{1}}$ $\mathrm{C_6H_5(CH_3)CHNHCOCONHN{=}CH{-}C_6H_3(R)Cr(CO)_3}$ $\xrightarrow[\sim 90\%]{\text{Chromatography}}$ (R)- + (S)-derivative $\xrightarrow[\sim 90\%]{60\%\ \mathrm{H_2SO_4},\ \mathrm{C_6H_6}}$ (R)-**2** and (S)-**2**

[1] N. J. Leonard and J. H. Boyer, *J. Org.*, **15**, 42 (1950).
[2] A. Solladie-Cavallo, G. Solladie, and E. Tsamo, *ibid.*, **44**, 4189 (1979).

Phenyllithium, $\mathrm{C_6H_5Li}$.

Reaction with 1-chloro-2-methylcyclohexene **(1).** This vinyl halide reacts with phenyllithium (5 equiv.) to give the bicyclo[4.1.0]heptane (**2**) in 90% yield. The paper presents evidence for the intermediacy of the cyclopropene **a**. Other organolithium reagents react in the same way, but yields are lower, 47% in the case of $\mathrm{CH_3Li}$.

1 (1-chloro-2-methylcyclohexene; $\mathrm{CH_3}$, Cl) $\xrightarrow{\mathrm{C_6H_5Li}}$ [**a**] $\xrightarrow{90\%}$ **2** ($\mathrm{C_6H_5}$, H)

[1] P. G. Gassman, J. T. Valcho, and G. S. Proehl, *Am. Soc.*, **101**, 231 (1979).

Phenyl propargyl selenide, $C_6H_5SeCH_2C{\equiv}CH$ (**1**). Mol. wt. 195.11, b.p. 76°/0.5 mm.

Preparation[1]:

$$C_6H_5SeNa + BrCH_2C{\equiv}CH \rightarrow \mathbf{1} + C_6H_5SeCH{=}C{=}CH_2$$

87% (2–4%)

***epi-7-Hydroxymyoporone* (7).**[2] The dianion (**2**) of **1**, prepared with LDA, was the key reagent in the synthesis of **7**.

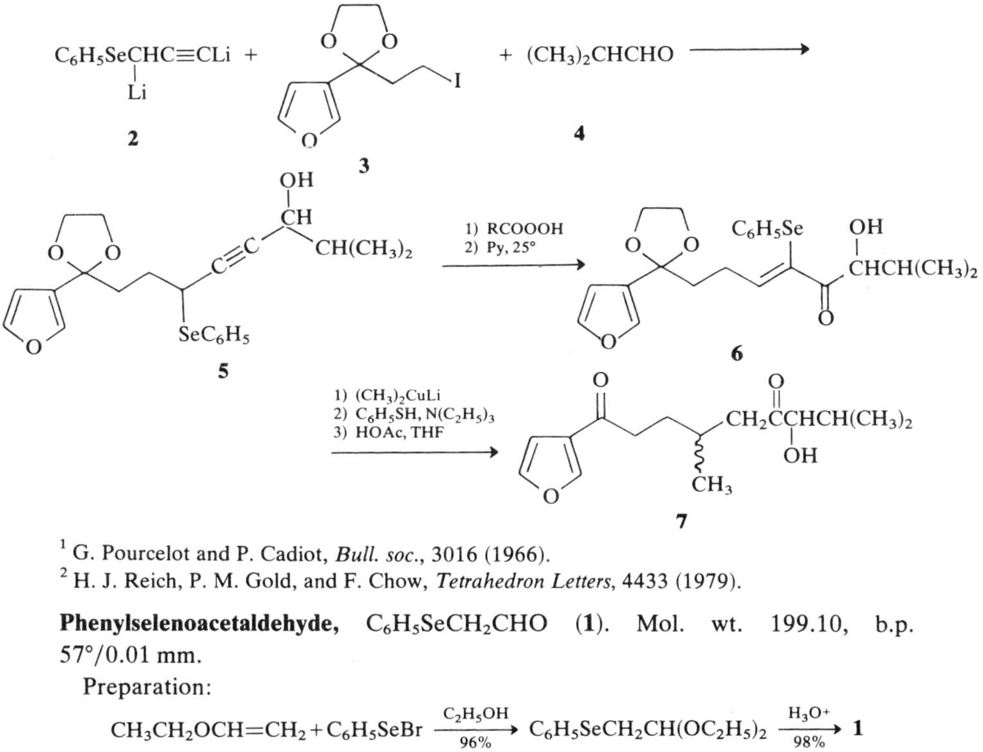

$C_6H_5SeCH(Li)C{\equiv}CLi$ (**2**) + **3** + $(CH_3)_2CHCHO$ (**4**) → **5** $\xrightarrow{\text{1) RCOOOH; 2) Py, 25°}}$ **6** $\xrightarrow{\text{1) }(CH_3)_2CuLi\text{; 2) }C_6H_5SH, N(C_2H_5)_3\text{; 3) HOAc, THF}}$ **7**

[1] G. Pourcelot and P. Cadiot, *Bull. soc.*, 3016 (1966).
[2] H. J. Reich, P. M. Gold, and F. Chow, *Tetrahedron Letters*, 4433 (1979).

Phenylselenoacetaldehyde, $C_6H_5SeCH_2CHO$ (**1**). Mol. wt. 199.10, b.p. 57°/0.01 mm.

Preparation:

$$CH_3CH_2OCH{=}CH_2 + C_6H_5SeBr \xrightarrow[96\%]{C_2H_5OH} C_6H_5SeCH_2CH(OC_2H_5)_2 \xrightarrow[98\%]{H_3O^+} \mathbf{1}$$

Phenylselenomethyl ketones.[1] The reagent reacts with Grignard reagents to form β-hydroxy selenides (**2**), generally in 80–90% yield. Oxidation of the selenides to the desired phenylselenomethyl ketones (**3**) proved to be more difficult than anticipated. In the case of saturated alcohols, the Corey–Kim reagent (**4**, 87–90) is satisfactory. Allylic alcohols are best oxidized with DDQ.

Examples:

$$CH_3MgI \xrightarrow[78\%]{\mathbf{1}} \underset{\mathbf{2}}{CH_3CH(OH)CH_2SeC_6H_5} \xrightarrow[61\%]{NCS\text{-}(CH_3)_2S,\ N(C_2H_5)_3} \underset{\mathbf{3}}{CH_3COCH_2SeC_6H_5}$$

$C_6H_{11}MgCl \xrightarrow{87\%} C_6H_{11}CH(OH)CH_2SeC_6H_5 \xrightarrow{60\%} C_6H_{11}C(=O)CH_2SeC_6H_5$

$CH_2{=}CHMgBr \xrightarrow{74\%} CH_2{=}CHCH(OH)CH_2SeC_6H_5 \xrightarrow[63\%]{DDQ} CH_2{=}CHC(=O)CH_2SeC_6H_5$

[1] R. Baudat and M. Petrzilka, *Helv.*, **62**, 1406 (1979).

N-Phenylselenophthalimide (1).[1] Mol. wt. 302.19, m.p. 171–175°. **N-Phenylselenosuccinimide (2).**[2] Mol. wt. 254.15, m.p. 114–120°.

Preparation:

Potassium phthalimide (N—K) $\xrightarrow[88\%]{C_6H_5SeCl,\ C_6H_{14},\ 25°}$ N—SeC_6H_5 phthalimide **1**

C_6H_5SeNa $\xrightarrow[69\%]{1)\ CH_2{=}CHCH_2Cl,\ EtOH,\ 0°;\ 2)\ NCS,\ CH_2Cl_2}$ N—SeC_6H_5 succinimide **2**

Oxyselenation of alkenes.[1,2] Treatment of olefins with **1** or **2**, water, and an acid catalyst (e.g., p-TsOH) in CH_2Cl_2 affords β-hydroxy selenides in excellent yield. Unsaturated carboxylic acids, phenols, alcohols, thioacetates, and urethanes react with **1** or **2** and an acid catalyst ($-78 \rightarrow 25°$) to afford products of oxidative cyclization. These reagents are superior to benzeneselenenyl halides for selenium-induced ring closures. This reaction is also useful for synthesis of 14- and 16-membered lactones. Benzeneselenenyl halides and benzeneselenenic acid do not promote macrolide formation under similar conditions.

Examples:

1-Methylcyclohexene (CH_3) → 1-methyl-2-(SeC_6H_5)cyclohexanol (OH, CH_3, SeC_6H_5)
(**1**, 89%)
(**2**, 84%)

o-Divinylbenzene (CH_2, CH_2) → 1,3-bis($CH_2SeC_6H_5$)-1,3-dihydroisobenzofuran (O, SeC_6H_5, SeC_6H_5)
(**1**, 71%)
(**2**, 70%)

			+ macrodiolides
n = 11	50%	14%	10%
n = 13	54%	15%	8%

Allylic chlorination.[2,3] β-Pinene reacts with **2** to afford selenide **3** in excellent yield. When **3** is treated with NCS, the rearranged allylic chloride **4** is formed as the principal product. The latter reaction is general for allylic selenides. An analogous reaction of **3** with chloramine-T effects allylic amination to **6** (44% yield). The overall conversion of β-pinene to **4** can be performed with **2** generated *in situ* by use of a catalytic amount of diphenyl diselenide and 1 equiv. of NCS. Olefins less reactive than β-pinene do not react with **2** (in the absence of water). In such cases, allylic chlorination is achieved by use of catalytic amounts of an arylselenenyl chloride or an aryl diselenide with NCS (equation I). The mechanism of this reaction is thought to be different than that of the allylic chlorination of β-pinene.

(I) NCS, $(C_6H_5Se)_2$, pyridine, CH_2Cl_2; 87%; 96:4

[1] K. C. Nicolaou, D. A. Claremon, W. E. Barnette, and S. P. Seitz, *Am. Soc.*, **101**, 3704 (1979).
[2] T. Hori and K. B. Sharpless, *J. Org.*, **44**, 4208 (1979).
[3] *Idem, ibid.*, **44**, 4204 (1979).

3-Phenylsulfonyl-1(*3H*)-isobenzofuranone (γ-Phenylsulfonylphthalide),

$SO_2C_6H_5$ O O

(**1**). Mol. wt. 275.29, m.p. 209–211°. The phthalide **1** is prepared in 87% yield by condensation of phthaldehydic acid with thiophenol followed by oxidation with *m*-chloroperbenzoic acid.

Aromatic annelation. The anion of **1** (LDA, THF, −78°) undergoes Michael addition to α,β-unsaturated esters and ketones to give 2,3-disubstituted-1,4-hydroxynaphthalenes. The phenylsulfonyl group plays a dual role: it stablizes the anion and it provides a leaving group in the cyclization step (equation I). The reaction is regiospecific, and the specificity is not altered by the presence of a methoxyl group in the 7-position of **1**. It thus offers an unambiguous route to naphthohydroquinones.[1]

(I) $SO_2C_6H_5$ O O + R^1 CH CH COR^2 —LDA→ [$SO_2C_6H_5$ R^1 O C O COR^2] —$-C_6H_5SO_2Li$→

OH R^1 COR^2 OH —$(CH_3O)_2SO_2$, K_2CO_3, 30–85%→ OCH_3 R^1 COR^2 OCH_3

Hauser has used repetitive annelations of this type for synthesis of linear aromatic systems. This methodology provides a simple regiospecific route to the tetracyclic system of the antibiotic anthracyclinones.[2] For this purpose 5-ethoxy-2(5*H*)-furanone (**2**) is used as the Michael acceptor in the first annelation of the anion of **1**. After methylation, **3** is obtained in 65% yield. It is then converted into **4**, for a second annelation with cyclohexenone or a derivative (**5**). The reaction results in a tetracyclic hydronaphthacene **6**.

(II) $SO_2C_6H_5$ O O + OC_2H_5 O O (**2**) —LDA→ —$(CH_3O)_2SO_2$, K_2CO_3, 65%→ OCH_3 OC_2H_5 O O OCH_3 (**3**) —1) C_6H_5SH 2) $ClC_6H_4CO_3H$→

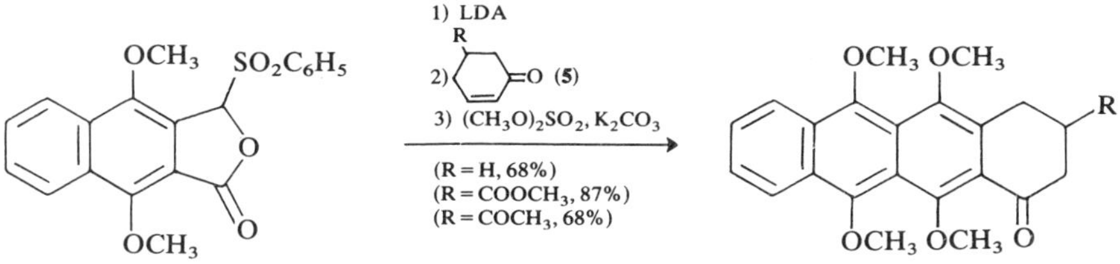

4 **6**

A similar methodology has been used for synthesis of the methyl ether of kidamycinone (**13**), the aglycone of the antibiotic kidamycin.[3] This substance contains a 4*H*-anthra[1,2*b*]pyrane nucleus. The anthracene unit was constructed by Michael addition of **7** with methyl crotonate followed by methylation to give **8**. This

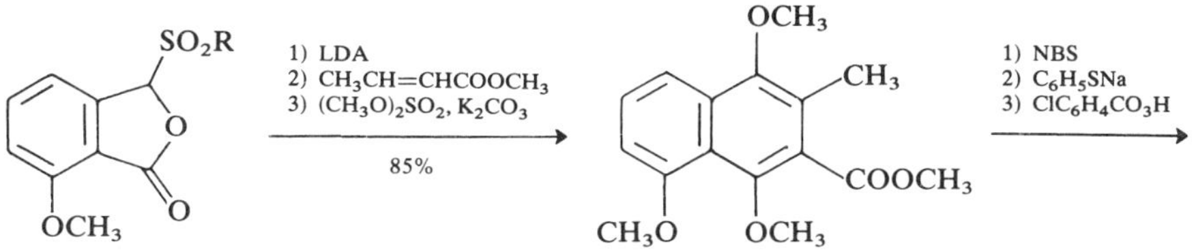

7 **8**

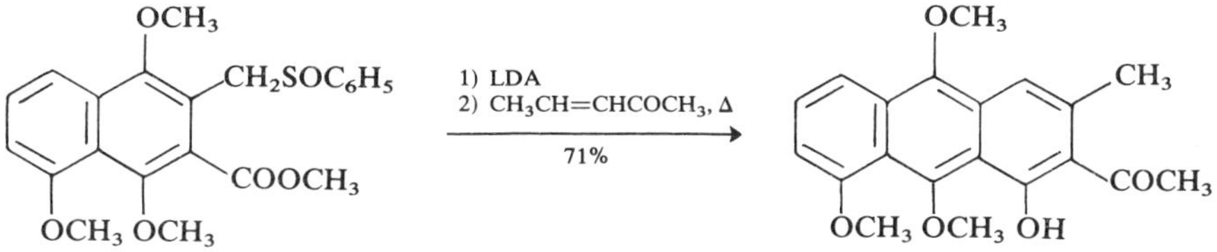

9 **10**

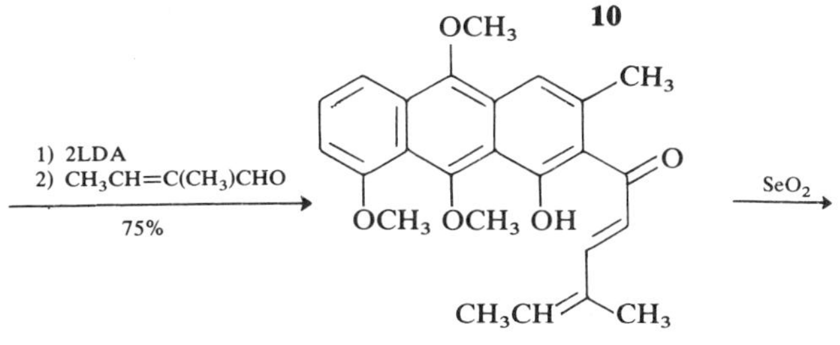

11

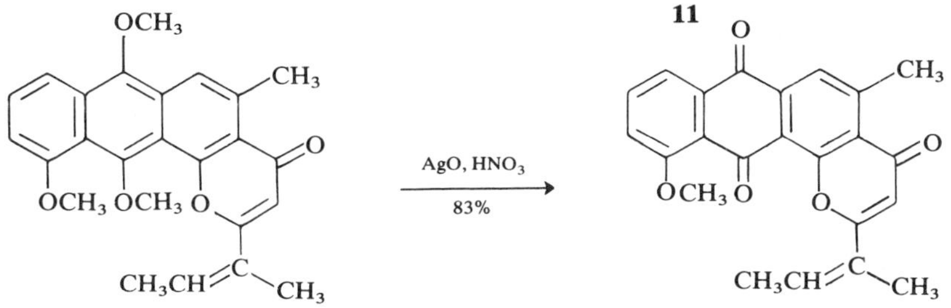

12 **13**

was converted into **9** for a second Michael addition to 3-pentene-2-one to give the anthracene **10**. The final steps involved construction of the pyrone unit. The dianion of **10** was condensed with tiglaldehyde to give the dienone **11**, which underwent cyclization and dehydrogenation to **12** on treatment with selenium dioxide (**5**, 576). The final step was the oxidative cleavage to the anthraquinone (**13**) by silver(II) oxide (**4**, 432).

[1] F. M. Hauser and R. P. Rhee, *J. Org.*, **43**, 178 (1978).
[2] F. M. Hauser and S. Prasanno, *ibid.*, **44**, 2596 (1979).
[3] F. M. Hauser and R. P. Rhee, *Am. Soc.*, **101**, 1628 (1979).

Phenylthioacetyl chloride, $C_6H_5SCH_2COCl$. Phenylthioacetic acid is available commerically.

Lactonization. Japanese chemists[1] have reported a new method for synthesis of medium- and large-membered lactones from the phenylthioacetate of an ω-iodo alcohol. An example is the synthesis of the 12-membered lactone recifeiolide (**4**) shown in equation (I). The high yield (75%) obtained on the cyclization of **2** to **3** may be associated with the presence of a double bond; dihydro-**2** is cyclized to a lactone in 51% yield.

The paper reports two other examples of this method.

(I) **1** (CH₃, OH, I) $\xrightarrow[82\%]{C_6H_5SCH_2COCl,\ CH_2Cl_2}$ **2** (CH₃, O–C=O, $CH_2SC_6H_5$, I) $\xrightarrow[75\%]{KN[Si(CH_3)_3]_2}$ **3** (CH₃, O, O, SC_6H_5) $\xrightarrow[80\%]{\text{Raney Ni}}$ **4** (CH₃, O, O)

[1] T. Takahashi, S. Hashiguchi, K. Kasuga, and J. Tsuji, *Am. Soc.*, **100**, 7424 (1978).

Phenylthioacetylene, $C_6H_5SC{\equiv}CH$ (**1**). Mol. wt. 134.20, b.p. 78–79°/7 mm.

Preparation:

$$C_6H_5SH + BrCH_2CH(OC_2H_5)_2 \xrightarrow[85\%]{NaOC_2H_5} C_6H_5SCH_2CH(OC_2H_5)_2 \xrightarrow[82\%]{LDA} \mathbf{1}$$

Dienals.[1] The sodium (or potassium) derivatives of allylic alcohols such as **2** add to this acetylene derivative (**1**) to form adducts **3**, which on oxidation and pyrolysis undergo Claisen rearrangement and elimination of C_6H_5SOH to yield 2,4-dienals (**5**); see also **8**, 535.

$$\underset{\mathbf{2}}{HOCH_2C(CH_3)=CH_2} \xrightarrow[75\%]{1)\ NaH,\ THF;\ 2)\ \mathbf{1}} \underset{\mathbf{3}}{(CH=CHSC_6H_5)OCH_2C(CH_3)=CH_2} \xrightarrow[\sim 100\%]{ClC_6H_4CO_3H} \mathbf{4} \xrightarrow{ZnCO_3,\ 130^\circ} [\mathbf{a}] \xrightarrow[84\%]{-C_6H_5S(O)H} \underset{\mathbf{5}}{CH_2=C(CH_3)CH=CHCHO}$$

[1] R. C. Cookson and R. Gopalan, *J.C.S. Chem. Comm.*, 924 (1978).

1-Phenylthiovinyllithium, $CH_2=C(Li)SC_6H_5$ (**1**), **8**, 281. The anion can be obtained conveniently by reaction of phenyl vinyl sulfide in THF with *n*-butyllithium and TMEDA at −90°.

α,β-Unsaturated ketones. The anion **1** reacts with aliphatic and aromatic aldehydes to form 2-phenylthioallyl alcohols (**2**) in high yield.[1] These alcohols are rearranged by hydrogen chloride to the isomeric α-phenylthioethyl ketones (**3**). Oxidation of **3** followed by pyrolysis in CCl_4 yields vinyl ketones (**4**).[2]

$$RCHO \xrightarrow[\sim 75\%]{\mathbf{1}} \underset{\mathbf{2}}{RCH(OH)C(SC_6H_5)=CH_2} \xrightarrow[80\%]{HCl,\ CH_3CN} \underset{\mathbf{3}}{RC(=O)CH(SC_6H_5)CH_3} \xrightarrow[70\%]{1)\ ClC_6H_4CO_3H;\ 2)\ CCl_4,\ \Delta} \underset{\mathbf{4}}{RC(=O)CH=CH_2}$$

This rearrangement can also be applied to 2-phenylthiohomoallenyl alcohols (**5**)[3] to give 2,4-dienones (**7**).

$$\underset{\mathbf{5}}{(R^1CH_2)(R^2)C=C=C(SC_6H_5)CHR^3OH} \xrightarrow{HCl} \underset{\mathbf{6}}{(R^1CH_2)(R^2)C=CH-CH(SC_6H_5)C(=O)R^3} \xrightarrow{1)\ Oxid.;\ 2)\ \Delta} \underset{\mathbf{7}}{R^1CH=C(R^2)CH=CHC(=O)R^3}$$

[1] R. C. Cookson and P. J. Parsons, *J.C.S. Chem. Comm.*, 990 (1976).
[2] *Idem, ibid.*, 821 (1978).
[3] *Idem, ibid.*, 822 (1978).

N-Phenyl-1,2,4-triazoline-3,5-dione (PTAD), 1, 849–850; **2,** 324–326; **3,** 223–224; **4,** 381–383; **5,** 528–530; **6,** 467; **7,** 287–288.

Ene reactions with allylsilanes. This reactive enophile adds to allylsilanes to afford vinyl- and alkynylsilanes.[1,2]

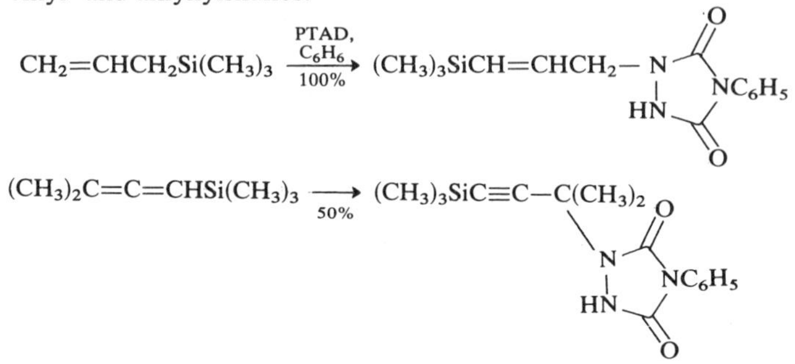

$$CH_2{=}CHCH_2Si(CH_3)_3 \xrightarrow[100\%]{PTAD,\ C_6H_6} (CH_3)_3SiCH{=}CHCH_2{-}N$$

O, NC_6H_5, HN, O

$$(CH_3)_2C{=}C{=}CHSi(CH_3)_3 \xrightarrow[50\%]{} (CH_3)_3SiC{\equiv}C{-}C(CH_3)_2$$

O, N, NC_6H_5, HN, O

The regiochemical outcome of this ene reaction is reversed when the double bond of the allylsilane bears a cyano substituent.[3]

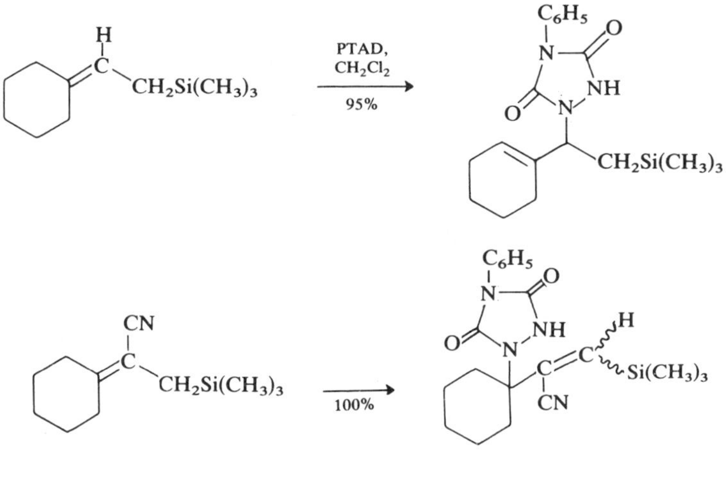

H, C, $CH_2Si(CH_3)_3$ $\xrightarrow[95\%]{PTAD,\ CH_2Cl_2}$ C_6H_5, O, N, NH, O, N, $CH_2Si(CH_3)_3$

CN, C, $CH_2Si(CH_3)_3$ $\xrightarrow[100\%]{}$ C_6H_5, O, N, NH, O, N, H, C=C, $Si(CH_3)_3$, CN

Protection of steroidal 5,7-dienes (**3,** 223–224; **6,** 467).[4] The cycloadducts of these dienes with 4-phenyl-1,2,4-triazoline-3,5-dione revert to the diene in quantitative yield when heated at 120° (7 hours) with anhydrous K_2CO_3 in DMSO. This method is superior to the original method (reduction with lithium aluminum hydride) for substrates containing reducible groups.

[1] A. LaPorterie, J. Dubac, and M. Lesbre, *J. Organometal. Chem.*, **101,** 187 (1975).
[2] A. LaPorterie, J. Dubac, G. Manuel, G. Deleris, J. Kowalski, J. Dunogues, and R. Calas, *Tetrahedron*, **34,** 2669 (1978).
[3] A. Gopalan, R. Moerck, and P. Magnus, *J.C.S. Chem. Comm.*, 548 (1979).
[4] M. Tada and A. Oikawa, *ibid.*, 727 (1978).

Phenyl trimethylsilylpropargyl ether, $(CH_3)_3SiC{\equiv}CCH_2OC_6H_5$ (**1**). Mol. wt. 214.34. The ether is prepared by reaction of 3-phenoxy-1-propyne with *n*-butyllithium followed by chlorotrimethylsilane.

1-Alkynyl(trimethyl)silanes.[1] A recent route to these silanes is shown in equation (I).

$$\text{(I)}\quad \mathbf{1} \xrightarrow[\text{2) } R_3B,\ -78^\circ]{\text{1) } n\text{-BuLi, TMEDA}} \left[(CH_3)_3SiC{\equiv}CCH\begin{matrix} BR_2 \\ R \end{matrix}\right] \xrightarrow{\text{HOAc, HMPT, } -40\rightarrow 0^\circ} \underset{(40\text{–}50\%)}{(CH_3)_3SiC{\equiv}CCH_2R} + \underset{\text{(minor product)}}{(CH_3)_3SiCH{=}C{=}CHR}$$

[1] T. Yogo, J. Koshino, and A. Suzuki, *Syn. Comm.*, **9**, 809 (1979).

Phenyl trimethylsilyl selenide (Phenylselenotrimethylsilane), $C_6H_5SeSi(CH_3)_3$ (**1**). Mol. wt. 229.25, b.p. 70°/1.9 mm.; sensitive to oxygen and moisture.

Several preparations have been reported, two of which are formulated.[1,2]

$$C_6H_5SeH \xrightarrow[91\%]{\text{1) } CH_3Li \text{; 2) } ClSi(CH_3)_3} \mathbf{1} \xleftarrow[60\text{–}70\%]{\text{1) Na; 2) } ClSi(CH_3)_3} C_6H_5SeSeC_6H_5$$

Reactions. The reagent converts aldehydes into hemiselenoacetals under catalysis with $ZnCl_2$ or $P(C_6H_5)_3$. α,β-Unsaturated aldehydes and ketones give predominately products of 1,4-addition. Saturated ketones react very slowly with the reagent.[1]

Examples:

$$CH_3(CH_2)_2CHO \xrightarrow[100\%]{\mathbf{1},\ P(C_6H_5)_3} CH_3(CH_2)_2CH\begin{matrix} OSi(CH_3)_3 \\ SeC_6H_5 \end{matrix}$$

$$CH_3\overset{O}{\overset{\|}{C}}CH{=}C(CH_3)_2 \xrightarrow[70\%]{} CH_3\overset{OSi(CH_3)_3}{\overset{|}{C}}{=}CH\underset{SeC_6H_5}{\underset{|}{C}}(CH_3)_2$$

$$\text{2-cyclohexen-1-one} \xrightarrow[100\%]{} \text{1-}OSi(CH_3)_3\text{-3-}SeC_6H_5\text{-cyclohexene}$$

The reagent is converted by KF (18-crown-6) into potassium benzeneselenolate (**5**, 272–276; **6**, 235).[2]

[1] D. Liotta, P. B. Paty, J. Johnston, and G. Zima, *Tetrahedron Letters*, 5091 (1978).
[2] M. R. Detty, *ibid.*, 5087 (1978).

Phenyl vinyl selenide, $CH_2{=}CHSeC_6H_5$ (**1**). Mol. wt. 183.10.

Preparation:

$$ClCH_2CH_2Cl \xrightarrow{NaSeC_6H_5} C_6H_5SeCH_2CH_2Cl \xrightarrow[95\%]{KOC(CH_3)_3} \mathbf{1}$$

Disubstituted alkenes.[1] Under controlled conditions to prevent α-deprotonation or C—Se cleavage (DME or ether, 0°) alkyllithium reagents (with the exception of methyllithium) add to phenyl vinyl selenide to give α-lithioalkyl phenyl selenides (**2**), which can be trapped with electrophiles to give **3**. On oxidation, these products form alkenes (**4**) by elimination of benzeneselenenic acid. The reagent thus functions as $\overset{+}{C}H{=}\overset{-}{C}H$.

$$\mathbf{1} + n\text{-}C_4H_9Li \rightarrow n\text{-}C_4H_9CH_2\overset{Li}{\overset{|}{C}}HSeC_6H_5 \ (\mathbf{2}) \xrightarrow[71\%]{C_6H_5CHO}$$

$$n\text{-}C_4H_9CH_2\overset{HOCHC_6H_5}{\overset{|}{C}}HSeC_6H_5 \ (\mathbf{3}) \xrightarrow[75\%]{O_3} (H)(n\text{-}C_4H_9)C{=}C(CH(OH)C_6H_5)(H) \ (\mathbf{4})$$

$$\mathbf{1} + (CH_3)_2CHLi \rightarrow (CH_3)_2CHCH_2\overset{Li}{\overset{|}{C}}HSeC_6H_5 \xrightarrow[72\%]{CH_3COCH_3}$$

$$(CH_3)_2CHCH_2\overset{HOC(CH_3)_2}{\overset{|}{C}}HSeC_6H_5 \xrightarrow[81\%]{O_3} (H)((CH_3)_2CH)C{=}C(\overset{OH}{\overset{|}{C}}(CH_3)_2)(H)$$

[1] S. Raucher and G. A. Koolpe, *J. Org.*, **43**, 4252 (1978).

Phosphorus(V) sulfide (Tetraphosphorus decasulfide), **1**, 870–871; **3**, 226–228; **4**, 389; **5**, 534–535; **8**, 401.

Reaction with unsaturated amides. The reaction of α,β-unsaturated amides with P_4S_{10} and $NaHCO_3$ in CH_3CN gives β-mercaptoalkyl thioamides in 31–56% yield (equation I). The reaction involves Michael addition of SH^- and thionation.[1]

$$\text{(I)} \quad R^1NH\overset{O}{\overset{\|}{C}}C(R^2){=}CH_2 \xrightarrow[31\text{–}56\%]{P_4S_{10},\ NaHCO_3,\ CH_3CN} R^1NH\overset{S}{\overset{\|}{C}}{-}\overset{R^2}{\overset{|}{C}}HCH_2SH$$

[1] H. Alper, J. K. Currie, and R. Sachdeva, *Angew. Chem., Int. Ed.*, **17**, 689 (1978).

Phosphoryl chloride, **1**, 876–882; **2**, 330–331; **3**, 228; **4**, 390; **5**, 535–537; **7**, 292–293; **8**, 401–402.

Decarbonylation of α-amino acids (**7**, 292–293). Bates and Rapoport[1] have applied this method for decarbonylation of α-amino acids to the synthesis of the iminium ion **1**, which cyclizes to **2** in 47% overall yield on treatment with refluxing acidic methanol. These transformations constitute the key steps in the formal total synthesis of anatoxin a (**3**).

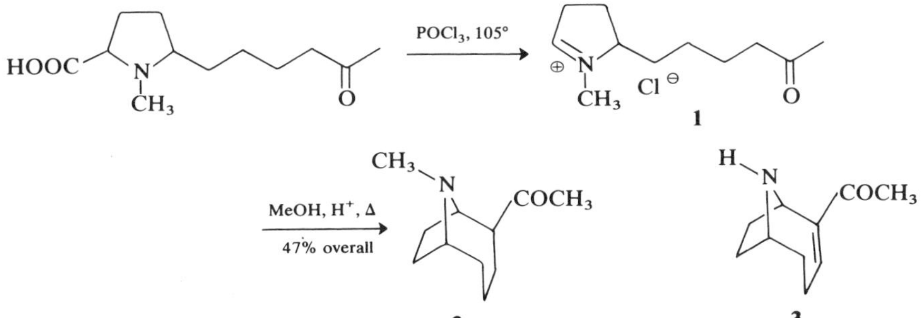

HOOC, N, CH_3, O — $POCl_3$, 105° → ⊕N, CH_3, Cl⊖, O — **1**

MeOH, H^+, Δ; 47% overall → CH_3, N, $COCH_3$ — **2**

H, N, $COCH_3$ — **3**

[1] H. A. Bates and H. Rapoport, *Am. Soc.*, **101**, 1259 (1979).

Phthalimide-N-sulfenyl chloride, N—SCl (**1**). Mol. wt. 213.65, m.p. 115–117°. The reagent is prepared by reaction of potassium phthalimide with S_2Cl_2 to form the disulfide, which is then cleaved by Cl_2 in $CHCl_3$.

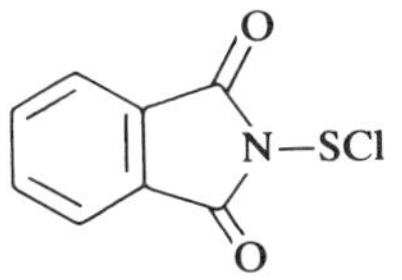

Episulfides.[1] Addition of **1** to alkenes forms an adduct that is reduced by $LiAlH_4$ to the episulfide.

Examples:

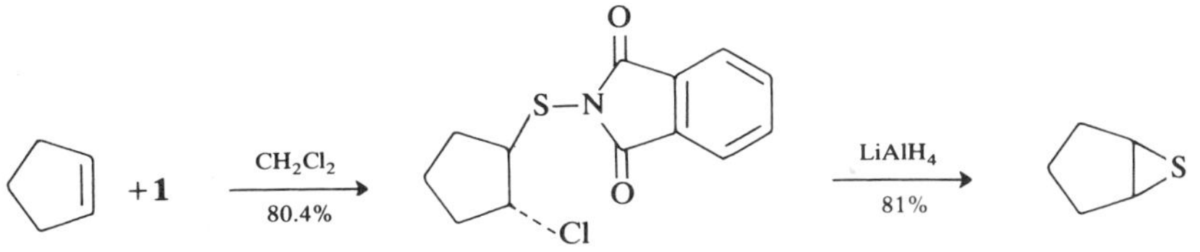

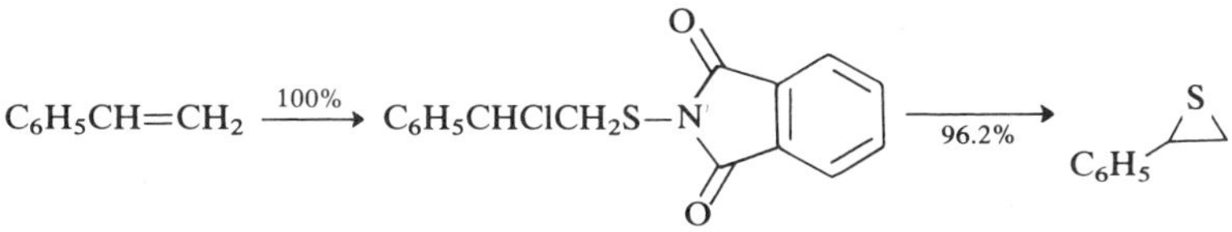

cyclopentene + **1** —CH_2Cl_2, 80.4%→ (S—N phthalimide, Cl adduct) —$LiAlH_4$, 81%→ cyclopentene episulfide (S)

$C_6H_5CH{=}CH_2$ —100%→ $C_6H_5CHClCH_2S{-}N$(phthalimide) —96.2%→ C_6H_5-thiirane (S)

Succinimide-N-sulfenyl chloride can be used in place of **1**, but yields tend to be somewhat lower.

[1] M. U. Bombala and S. V. Ley, *J.C.S. Perkin I*, 3013 (1979).

Piperidinium acetate, $\overset{+}{N}H_2$ (piperidinium ring) OAc^- Mol. wt. 145, m.p. 131–133°.

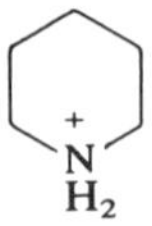

Marschalk reaction. Some years ago Marschalk *et al.*[1] found that leucoquinizarin (**1**), prepared *in situ* by reduction of quinizarin with alkaline dithionite, reacts with aldehydes to form 2-alkylquinizarins (**2**). It is not possible to obtain 2,3-disubstituted quinizarins in this way. Lewis[2] found that pyridinium acetate with isopropyl alcohol as solvent is superior to a base or an acid catalyst for this reaction; yields as high as 90% of **2** can be obtained. In addition, aromatic aldehydes can be used successfully. The leuco forms of **2** can be alkylated in this way to give 2,3-dialkylquinizarins in 40–70% yields.

Because of the current interest in anthracycline antibiotics, Sutherland *et al.*[3] have extended the earlier studies to leuco-5-hydroxyquinizarins (**3**). When **3** is generated *in situ* by alkaline dithionite reduction, it reacts with propionaldehyde to give the

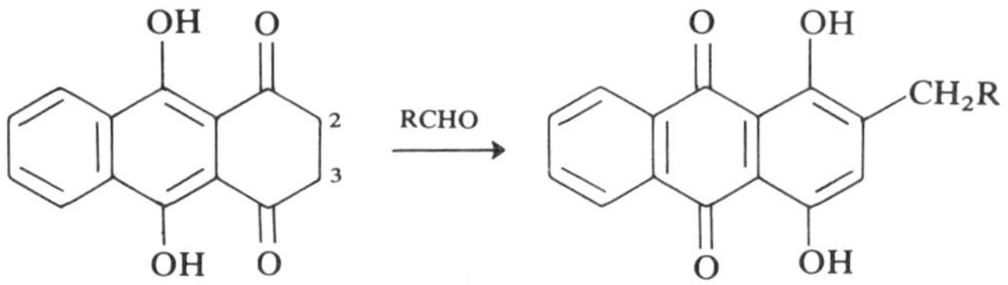

OH O; 2; 3; OH O; **1** —RCHO→ O OH; CH_2R; O OH; **2** OH O; 2; 3; 5; HO OH O; **3**

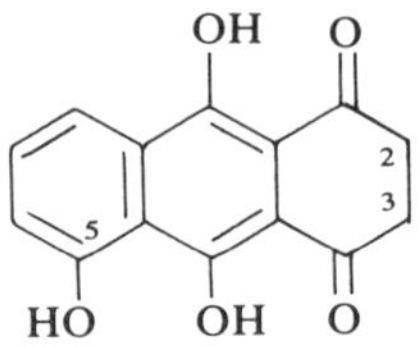

2-propyl substituted derivative in 50% yield. No product of alkylation at C_3 is observed. However, when the reaction is conducted in isopropyl alcohol with piperidinium acetate as catalyst, the 3-propyl derivative is the major product with only traces of the 2-isomer (ratio 9 : 1). The same reversibility in regioselectivity was observed with $CH_3CH(OH)CH_2CH_2CHO$. The paper includes a possible reason for this unusual effect.

[1] C. Marschalk, F. Koenig, and N. Ouroussoff, *Bull. soc.*, 1545 (1936).
[2] C. E. Lewis, *J. Org.*, **35**, 2938 (1970).
[3] L. M. Harwood, L. C. Hodgkinson, and J. K. Sutherland, *J.C.S. Chem. Comm.*, 712 (1978).

Polyethylene glycols (PEG), $HO(CH_2{-}CH_2O)_nH$.

Solvent uses.[1] PEG 400 can dissolve some inorganic salts and most organic substrates. Oxidations with $K_2Cr_2O_7$ in PEG are comparable to those in HMPT or crown ethers. Reductions with $NaBH_4$ proceed readily in this solvent.

[1] E. Santaniello, A. Manzocchi, and P. Sozzani, *Tetrahedron Letters*, 4581 (1979).

Polyphosphate ester, 1, 892–894; **2,** 333–334; **3,** 229–231; **4,** 394–395; **5,** 539–540; **6,** 474.

Aryl esters.[1] Esters can be obtained in 50–90% yield from phenols and aromatic carboxylic acids by reaction with PPE in DMF at 20–85° for 3–150 hours. Sterically hindered acids and hydrogen-bonded phenols react satisfactorily. Chloroform is less useful as solvent.

[1] J. H. Adams, J. R. Lewis, and J. G. Paul, *Synthesis*, 429 (1979).

Potassium–Ammonia.

Reductive cyclization. Potassium in liquid ammonia is an effective reagent for reductive cyclization of 5-alkynyl ketones.[1] This reaction provides an efficient route to the D ring of gibberellic acid (**1** → **2**).[2] Sodium naphthalenide has been employed in a similar fashion for the preparation of 2-methylenecyclopentanols and 2-methylenecyclohexanols (**3** → **4**).[3]

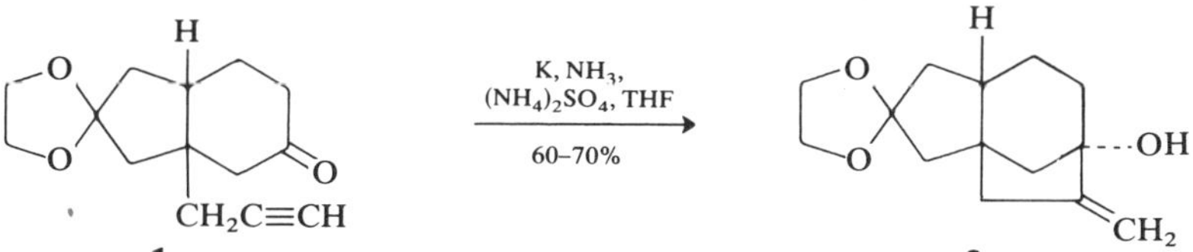

H, O, O, O, $CH_2C{\equiv}CH$, **1** — K, NH_3, $(NH_4)_2SO_4$, THF, 60–70% → H, O, O, ---OH, CH_2, **2**

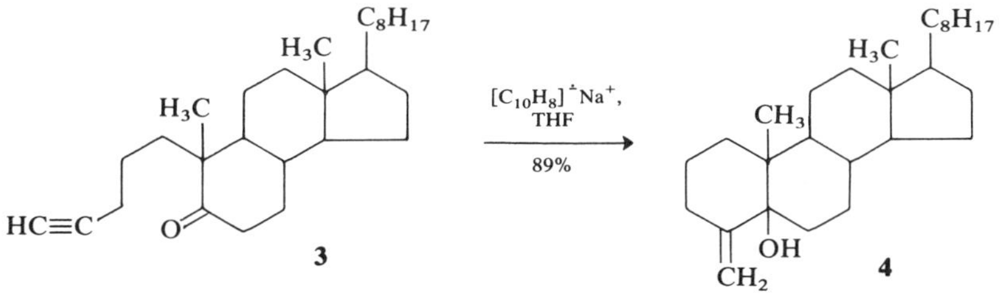

C_8H_{17}, H_3C, H_3C, HC≡C, O, **3** — $[C_{10}H_8]^{\dot{-}}Na^+$, THF, 89% → C_8H_{17}, H_3C, CH_3, OH, CH_2, **4**

[1] G. Stork, S. Malhotra, H. Thompson, and M. Uchibayashi, *Am. Soc.*, **87**, 1148 (1965).
[2] G. Stork, R. K. Boeckmann, D. F. Taber, W. C. Still, and J. Singh, *ibid.*, **101**, 7107 (1979).
[3] S. K. Pradhan, T. V. Radhakrishnan, and R. Subramanian, *J. Org.*, **41**, 1943 (1976).

Potassium–t-Butylamine–18-Crown-6.

Deoxygenation of alcohols. Esters of hindered alcohols are selectively deoxygenated with lithium in ethylamine[1] or by potassium solubilized by 18-crown-6 in *t*-butylamine.[2] Two examples are formulated in equations (I) and (II).[2]

(I) H_3C, AcO, H, OAc, **1** — Li, $C_2H_5NH_2$, 89% → H_3C, HO, H, **2** + H_3C, HO, H, OH, **3** (2:1)

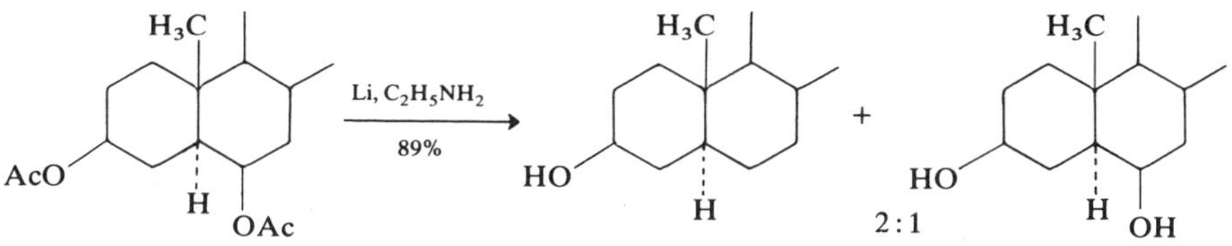

A related procedure involves reduction of thiocarbonates or thiocarbamates of primary or secondary alcohols.[3] The alcohols need not be hindered in this case. This deoxygenation procedure is more convenient than many other existing methods and

(II)

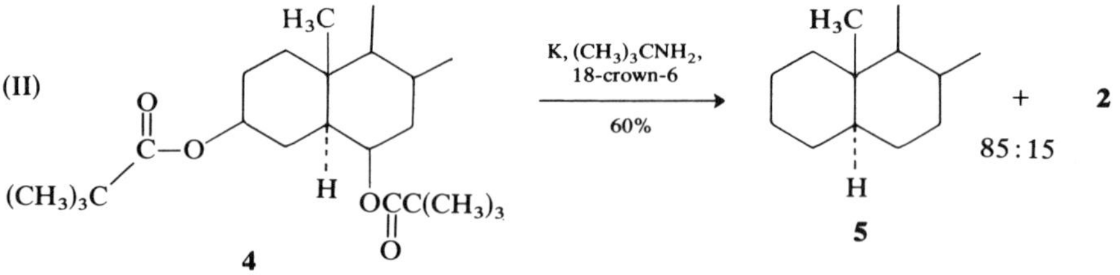

is applicable to carbohydrate derivatives. The thiocarbamates are readily available from the alcohol via the methylthiocarbonyl derivative and diethylamine.[4]

Examples:

$$CH_3(CH_2)_{16}CH_2O{-}\overset{S}{\overset{\|}{C}}{-}N(C_2H_5)_2 \xrightarrow[99\%]{K,\ (CH_3)_3CNH_2,\ 18\text{-crown-6}} CH_3(CH_2)_{16}CH_3 + CH_3(CH_2)_{16}CH_2OH$$

87 : 13

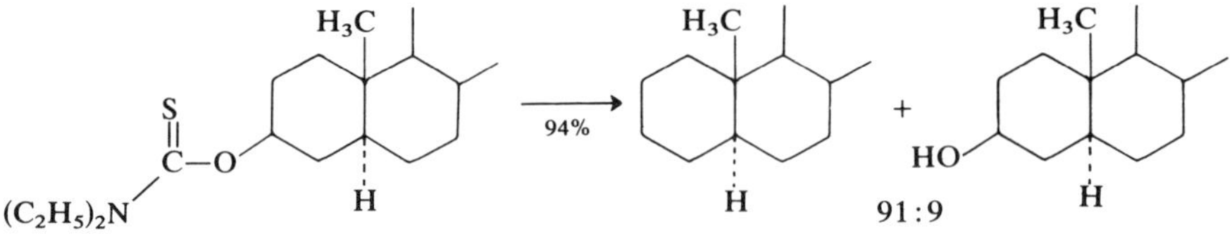

[1] R. B. Boar, L. Joukhadar, J. F. McGhie, S. C. Misra, A. G. M. Barrett, D. H. R. Barton, and P. A. Prokopiou, *J.C.S. Chem. Comm.*, 68 (1978); *see also* H. Deshayes and J.-P. Pete, *ibid.*, 567 (1978).

[2] A. G. M. Barrett, P. A. Prokopiou, D. H. R. Barton, R. B. Boar, and J. F. McGhie, *ibid.*, 1173 (1979).

[3] A. G. M. Barrett, P. A. Prokopiou, and D. H. R. Barton, *ibid.*, 1175 (1979).

[4] D. H. R. Barton, R. V. Stick, and R. Subramanian, *J.C.S. Perkin I*, 2112 (1976).

Potassium–Graphite, 4, 397; **7**, 296; **8**, 405–406.

Reduction of conjugated double bonds. This reagent reduces double bonds conjugated with a keto or carboxyl group in good to excellent yield. It also reduces aryl imines to the corresponding secondary amines.[1]

[1] M. Contento, D. Savoia, C. Trombini, and A. Umani-Ronchi, *Synthesis*, 30 (1979).

Potassium 3-aminopropylamide, 6, 476; **7**, 296; **8**, 406–407.

Ring enlargement of aminolactams. Schmid and co-workers[1] have described a method for enlarging a 13-membered aminolactam to a 25-membered polyaminolactam either in two steps via a 4-membered unit or in one operation. This enlargement is called the "zip" reaction because of the similarity to a zipper. An example is the conversion of **1** to **3**. In the early steps **1** is alkylated with $CH_2{=}CHCN$ (the 4-membered ring enlargement unit) and the product is reduced to the amine $-CH_2CH_2CH_2NH_2$. Repetition of these steps leads to **2** in 69% overall yield. Treatment of **2** with KAPA in 1,3-propanediamine at 20° results in cyclization

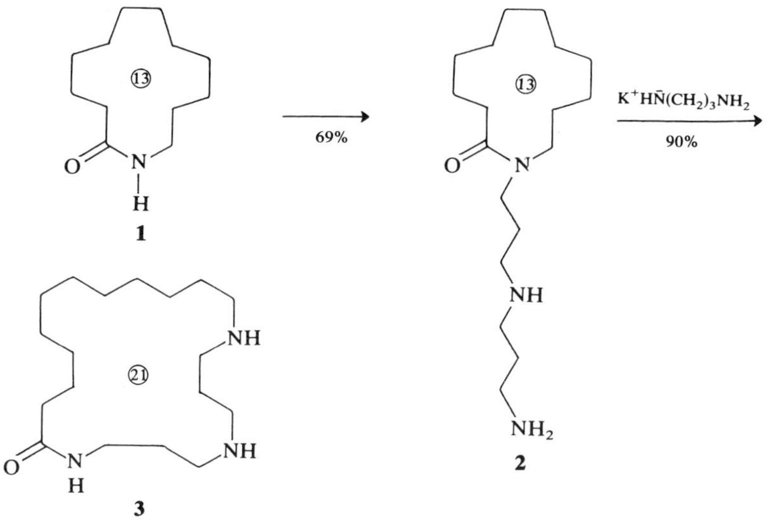

1

69%

2

$K^+H\bar{N}(CH_2)_3NH_2$

90%

3

to **3** in 90% yield. Use of potassium *t*-butoxide in this cyclization requires 6 hours reflux in toluene. The starting lactam (**1**) was also converted by this zipper reaction into a 25-membered polyaminolactam.

[1] U. Kramer, A. Guggisberg, M. Hesse, and H. Schmid, *Helv.*, **61**, 1342 (1978).

Potassium *t*-amyloxide, $C_2H_5C(CH_3)_2OK$ **(1).** Mol. wt. 126.24. The base is generated conveniently in benzene (or toluene) by reaction of *t*-amyl alcohol with potassium hydride in oil. In the absence of moisture, it can be stored at 35° for 1 year.

Wittig reactions. The Wittig reaction with 20-keto steroids usually proceeds in poor yields. However, Schmit *et al.*[1] reported that Wittig reactions of pregnenolones with some phosphoranes results in (E)-20(22)-alkenes in satisfactory yields when sodium *t*-amyloxide is used as base. This reaction has been further improved by use of potassium *t*-amyloxide. Thus (E)-20(22)-dehydrocholesterol can be prepared in this way in 78% yield.[2] It is hydrogenated under the conditions (PtO_2, dioxane–acetic acid) of Hershberg *et al.*[3] without reduction of the 5,6-double bond to give (20R)- and (20S)-cholesterol in the ratio of 4 : 1. The natural configuration is 20R; pure cholesterol is obtained in 66% yield.

α-Alkylation of ketones.[4] Base-catalyzed alkylation of ketones is often unsatisfactory because of polyalkylation and/or formation of isomeric products, particularly in the case of cyclopentanone. A reexamination of this reaction with a variety of bases and solvents indicates that use of potassium *t*-amyloxide as base and DME as solvent can lead to satisfactory yields, although a large excess of the alkylating agent is necessary for satisfactory yields. Thus 2-methyl- and 2-benzylcyclopentanone can be obtained in about 75% yield from reactions conducted at 0 to −20°. For enone alkylations, the reaction is allowed to warm to 20° before quenching.

[1] J. P. Schmit, M. Piraux, and J. F. Pilette, *J. Org.*, **40**, 1585 (1975).
[2] S. R. Schow and T. C. McMorris, *ibid.*, **44**, 3760 (1979).
[3] E. B. Hershberg, E. P. Oliveto, C. Gerold, and L. Johnson, *Am. Soc.*, **73**, 5073 (1951).
[4] H. N. Edwards, A. F. Wycpalek, N. C. Corbin, and J. D. McChesney, *Syn. Comm.*, **8**, 563 (1978).

Potassium t-butoxide, 1, 911–927; **2**, 338–339; **3**, 233–234; **4**, 399–405; **5**, 544–553; **6**, 477–479; **7**, 296–298; **8**, 407–408.

Cleavage of 1,1,1-triphenylmethyl ketones.[1] 1,1,1-Triphenylmethyl ketones are cleaved to carboxylic acids and triphenylmethane by potassium *t*-butoxide (6 equiv.) and H_2O (2 equiv.) in ether (equation I). This combination of reagents had previously been used by Gassman and co-workers[2] to hydrolyze amides. For the preparation of triphenylmethyl ketones, *see* Organolithium compounds, this volumc.

$$\text{(I)}\quad R{-}\overset{O}{\overset{\|}{C}}{-}C(C_6H_5)_3 \xrightarrow[(C_2H_5)_2O]{KOC(CH_3)_3,\ H_2O,} \underset{82\text{–}84\%}{R{-}COOH} + \underset{>96\%}{(C_6H_5)_3CH}$$

Dehydrochlorination. The dehydrochlorination of the sterically hindered *meso*-3,4-dichloro-2,2,5,5-tetramethylhexane (**1**) by potassium *t*-butoxide in DMSO proceeds normally by *anti*-coplanar elimination to give the less stable (E)-olefin **2** as the major product. However, *syn*-elimination to the (Z)-isomer **3** becomes the main reaction in THF or *t*-butyl alcohol. The difference may be related to the differing aggregates of the base in the various solvents.[3]

$$\underset{\mathbf{1}}{(CH_3)_3C{-}CHCl{-}CHCl{-}C(CH_3)_3} \xrightarrow[DMSO]{KOC(CH_3)_3,} \underset{\mathbf{2}}{(CH_3)_3C(H)C{=}C(Cl)C(CH_3)_3} + \underset{\mathbf{3}}{(CH_3)_3C(H)C{=}C(Cl)C(CH_3)_3} \quad 93:7$$

Indoles.[4] The potassium enolates of ketones, prepared with potassium *t*-butoxide, undergo $S_{RN}1$ reactions with *o*-iodoaniline on brief irradiation in NH_3 at −33°. Indoles are obtained in 36–83% yield after acidic work-up.

Examples:

$$o\text{-}IC_6H_4NH_2 + CH_3COCH_3 \xrightarrow[83\%]{1)\ NH_3,\ KOC(CH_3)_3,\ h\nu;\ 2)\ NH_4Cl,\ H_2O} \text{2-methylindole}$$

o-Iodoaniline + cyclohexanone ⟶ tetrahydrocarbazole (N–H) + aniline (NH_2)

36% 59%

Oxiranes.[5] A recent, general synthesis of epoxides involves α-sulfenylation of a ketone, reduction of the C=O group to CHOH, methylation to give a β-hydroxy dimethylsulfonium iodide, and, finally, cyclization with potassium *t*-butoxide (or sodium hydride) in DMSO. *cis*-Epoxides are formed exclusively.

Examples:

$$C_6H_5COCH(CH_3)SCH_3 \xrightarrow{NaBH_4} C_6H_5CH(OH)CH(CH_3)SCH_3 \xrightarrow{CH_3I} C_6H_5CH(OH)CH(CH_3)\overset{+}{S}(CH_3)_2I^-$$

$$\xrightarrow[(70\%)]{KOC(CH_3)_3,\ DMSO} \text{cis-}C_6H_5CH(O)CHCH_3 \text{ (H, O, H; } C_6H_5, CH_3)$$

2-(Methylthio)cyclohexanone (SCH_3) ⟶ 2-hydroxycyclohexyl-$\overset{+}{S}(CH_3)_2I^-$ $\xrightarrow{NaH,\ DMSO}$ cyclohexene oxide

Enamines.[6] A new route to enamines involves alkylation of an α-amino nitrile (**1**) derived from an aldehyde to give an α-amino nitrile of type **2**. When **2** is treated with potassium *t*-butoxide (or KOH), hydrogen cyanide is eliminated with formation of an enamine (**3**). The sequence is also applicable to synthesis of dienamines.

$$\underset{\mathbf{1}}{R^1(H)C(CN)NR^2R^3} \xrightarrow[80\text{–}90\%]{1)\ LDA,\ THF;\ 2)\ R^4R^5CHX} \underset{\mathbf{2}}{R^4R^5CH\text{–}C(R^1)(CN)NR^2R^3} \xrightarrow[50\text{–}90\%]{KOC(CH_3)_3,\ C_6H_6,\ \Delta} \underset{\mathbf{3}}{R^2R^3N(R^1)C{=}CR^4R^5}$$

[1] D. Seebach and R. Locher, *Angew. Chem., Int. Ed.*, **18**, 957 (1979).
[2] P. G. Gassman, P. K. G. Hodgson, and R. J. Balchunis, *Am. Soc.*, **98**, 1275 (1976).
[3] M. Schlosser and T. D. An, *Helv.*, **62**, 1194 (1979).
[4] R. Beugelmans and G. Roussi, *J.C.S. Chem. Comm.*, 950 (1979).
[5] S. Kano, T. Yokomatsu, and S. Shibuya, *ibid.*, 785 (1978).
[6] H. Ahlbrecht, W. Raab, and C. Vonderheid, *Synthesis*, 127 (1979).

Potassium *t*-butoxide–Benzophenone.

Aldol cyclization. Exposure of the hydroxy ketone **1** to excess potassium *t*-butoxide and benzophenone in refluxing benzene provides the α,β-unsaturated ketone **2** directly.[1] This transformation involves a modified Oppenauer oxidation followed by aldol closure of the resulting keto aldehyde. The product serves as an intermediate in the synthesis of *Elaeocarpus* alkaloids.

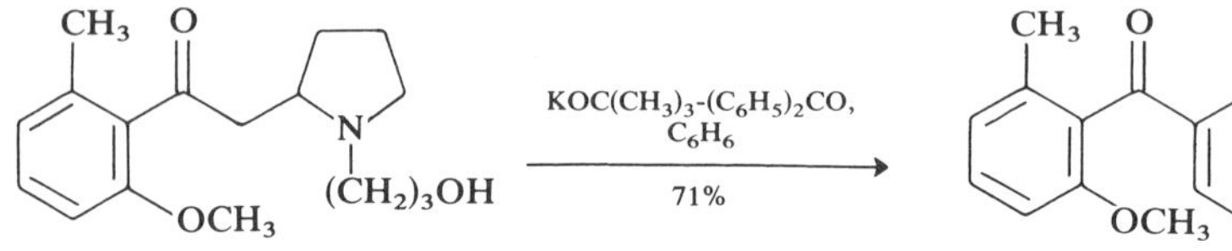

CH_3 O N $(CH_2)_3OH$ OCH_3 **1** $\xrightarrow[71\%]{KOC(CH_3)_3\text{-}(C_6H_5)_2CO,\ C_6H_6}$ CH_3 O N OCH_3 **2**

[1] J. J. Tufariello and S. A. Ali, *Am. Soc.*, **101**, 7114 (1979).

Potassium carbonate, 5, 552–553; **8,** 408.

α,β-Unsaturated ketones and esters. Aldehydes can undergo condensation with α-acetyl ketones, esters, and lactones in the presence of this base with loss of the acetyl group to form α,β-unsaturated ketones and esters. The suggested mechanism is shown in equation (I). The yields are low with simple aldehydes, but are markedly improved by substitution of chlorine in the α-position of the aldehyde.[1]

$$\text{(I)}\quad R^1CHO + CH_3\overset{O}{\overset{\|}{C}}CH_2COR^2 \xrightarrow[THF]{K_2CO_3,} R^1\underset{}{CH}(O\!-\!C(CH_3)(OK))\!-\!CHCOR^2 \xrightarrow[20\text{–}75\%]{-CH_3COOK} R^1CH{=}CHCOR^2$$

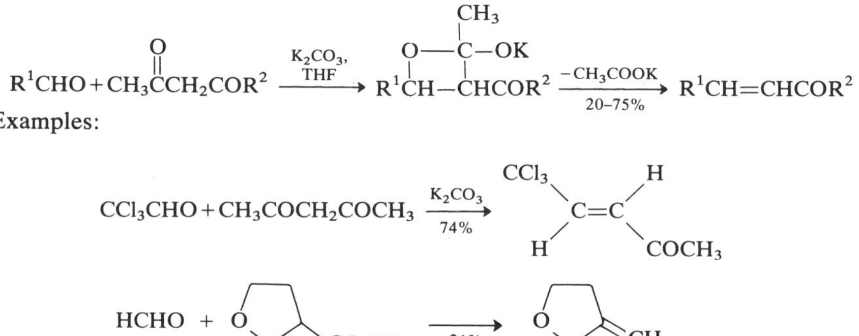

Examples:

$$CCl_3CHO + CH_3COCH_2COCH_3 \xrightarrow[74\%]{K_2CO_3} (CCl_3)(H)C{=}C(H)(COCH_3)$$

HCHO + O (lactone, O) COCH$_3$ $\xrightarrow[21\%]{}$ O (lactone, O) =CH$_2$

CCl$_3$CHO + O (lactone, O) COCH$_3$ $\xrightarrow[74\%]{}$ O (lactone, O) =CHCCl$_3$

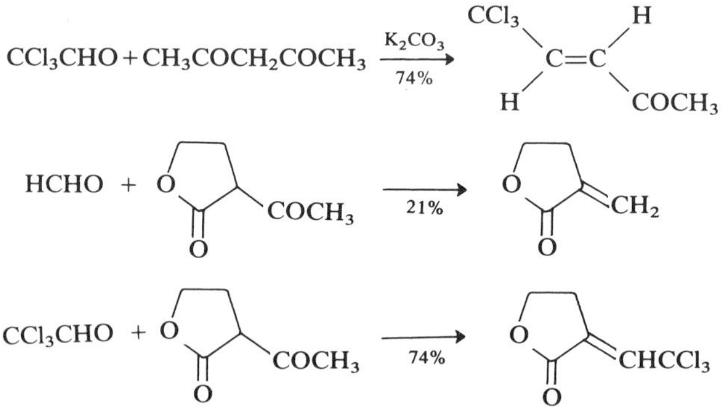

Rearrangement. A key step in a synthesis of the sesquiterpene hirsutene (**3**) is rearrangement of **1** to **2**. This rearrangement involves selective migration of the C_2–C_7 bond, which is *trans*-coplanar to the C_6-tosyloxy bond broken in the reaction.[2]

[1] S. Tsuboi, T. Uno, and A. Takeda, *Chem. Letters*, 1325 (1978).
[2] K. Tatsuta, K. Akimoto, and M. Kinoshita, *Am. Soc.*, **101**, 6116 (1979).

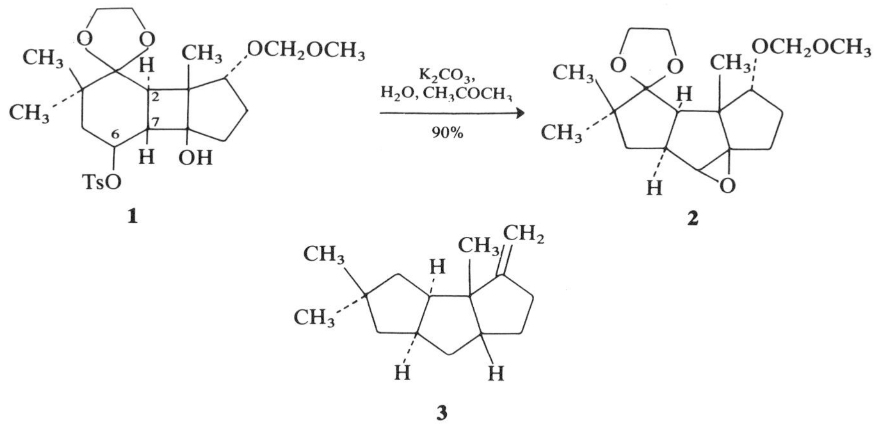

K_2CO_3, H_2O, CH_3COCH_3 → 90%

1 2

3

Potassium dichromate, 8, 410.

3-Chromenes.[1] *o*-Allylphenols are oxidized to 3-chromenes by potassium dichromate dissolved in benzene in the presence of Adogen 464[2] in yields of 45–83%. *o*-Quinone methides are postulated intermediates. These heterocycles are also obtained on oxidation of *o*-allylphenols with DDQ, but in this case ether is preferable to benzene (82–90% yield; *cf.* **2**, 116).

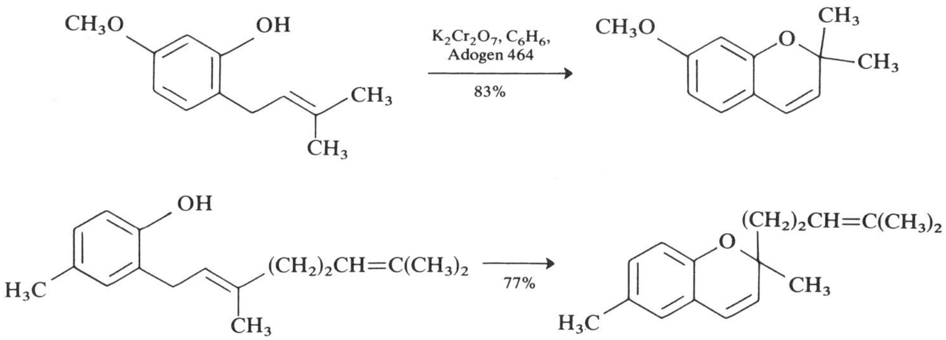

$K_2Cr_2O_7$, C_6H_6, Adogen 464 → 83%

→ 77%

[1] G. Cardillo, M. Orena, G. Porzi, and S. Sundri, *J.C.S. Chem. Comm.*, 836 (1979).
[2] R. O. Hutchins, N. R. Natale, W. J. Cook, and J. Ohr, *Tetrahedron Letters*, 4167 (1977).

Potassium diisopropylamide–Lithium *t*-butoxide (1). This nonnucleophilic strong base combination is stable in hexane at 0° for at least 1 hour. Potassium diisopropylamide alone is much less stable. The mixture (**1**) can be prepared in two ways: addition of diisopropylamine to *n*-BuK–$LiOC(CH_3)_3$ in hexane at 0° or addition of *n*-BuLi to a solution of $KOC(CH_3)_3$ and diisopropylamine in THF at −78°.

The base is able to deprotonate some selenium-containing substrates that are stable to LDA and to $KN[Si(CH_3)_3]_2$ (equations I and II.)[1]

(I)

$$R^1R^2C{=}CH(SeC_6H_5) \xrightarrow[80\text{–}95\%]{1)\ \mathbf{1};\ 2)\ RX} R^1R^2C{=}C(R)(SeC_6H_5)$$

$$R^1R^2C{=}CH(SeC_6H_5) \xrightarrow[80\text{–}95\%]{1)\ \mathbf{1};\ 2)\ R^3COR^4} R^1R^2C{=}C(SeC_6H_5)C(OH)R^3R^4$$

(II)

$$R^2R^3COH\text{–}C(SeC_6H_5)_2R^1 \xleftarrow[65\text{–}90\%]{1)\ \mathbf{1};\ 2)\ R^2COR^3} R^1CH(SeC_6H_5)_2 \xrightarrow[70\text{–}95\%]{1)\ \mathbf{1};\ 2)\ RX} R^1CR(SeC_6H_5)_2$$

[1] S. Raucher and G. A. Koolpe, *J. Org.*, **43**, 3794 (1978).

Potassium diphenyl-4-pyridylmethide, $(C_6H_5)_2(4\text{-}C_5H_4N)CK$ **(1).** Mol. wt. 283.42, stable at low temperatures if protected from oxygen. This soluble base is prepared by reaction of KH with diphenyl-4-pyridylmethane (Aldrich). One advantage of this reagent is the ease of removal of diphenylpyridylmethane with dilute HCl.

Intramolecular alkylation of oximes. Fuchs *et al.*[1] reasoned that C-alkylation of oximes to give nitroso compounds should be possible when O- or N-alkylation would lead to strained imines. Thus treatment of the oxime tosylate (**2**) with **1** in a homogeneous reaction results in ring contraction to **3**, presumably through a nitrosocyclopropane intermediate **a**. The reaction with bases insoluble in THF is much slower.

$$\mathbf{2}\ (\text{NOH},\ CH_3,\ CH_3,\ OSO_2C_7H_7) \xrightarrow{2\ \mathbf{1},\ \text{THF},\ 0°,\ <1\ \text{min.}} [\mathbf{a}\ (H_3C,\ CH_3,\ N{=}O)] \xrightarrow{93\%} \mathbf{3}\ (N\text{–}OH,\ C(CH_3){=}CH_2)$$

Another example:

$$(\text{H, H; OH–N; }CH_3,\ CH_3,\ OSO_2C_6H_5) \xrightarrow[88\%]{\mathbf{1},\ \text{THF},\ -78\rightarrow-40°} (\text{H, H, H; N–OH; }C(CH_3){=}CH_2)$$

In the second reaction the ring contraction is stereospecific. A similar contraction of a five-membered ring is not possible. In all cases the product oxime has the *syn*-configuration.

This reaction can also be used for ring expansion as shown in the conversion of **4** to **5**.

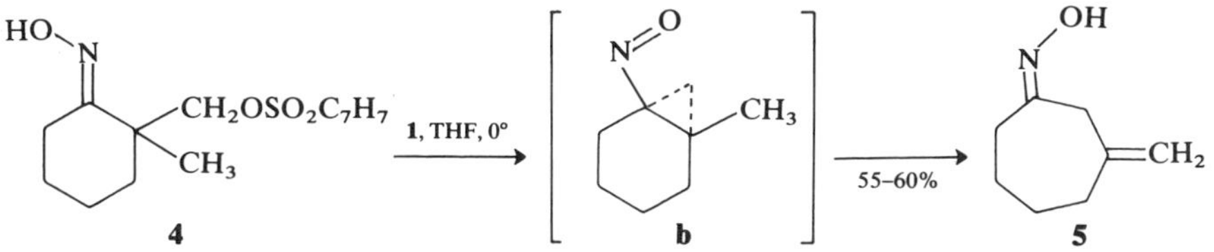

HO–N; $CH_2OSO_2C_7H_7$; CH_3; **4**; **1**, THF, 0°; [N=O; CH_3; **b**]; 55–60%; OH; N; $=CH_2$; **5**

[1] D. A. Clark, C. A. Bunnell, and P. L. Fuchs, *Am. Soc.*, **100**, 7777 (1978).

Potassium ferricyanide, 1, 929–933; **2,** 345; **4,** 406–407; **5,** 554–555; **6,** 480–482; **7,** 300–301; **8,** 410.

A dipheno-2,2'-quinone.[1] Oxidation of sesamol (**1**) with potassium ferricyanide leads to the dark violet, high-melting quinone **3** in 66% yield. A somewhat superior route to **3** involves coupling of the benzyl ether **2** of **1** with vanadyl trifluoride (**5**, 745–746) to give the biphenyl **4**. Hydrogenolysis to **5** followed by oxidation with $K_3Fe(CN)_6$ gives **3** in high yield.

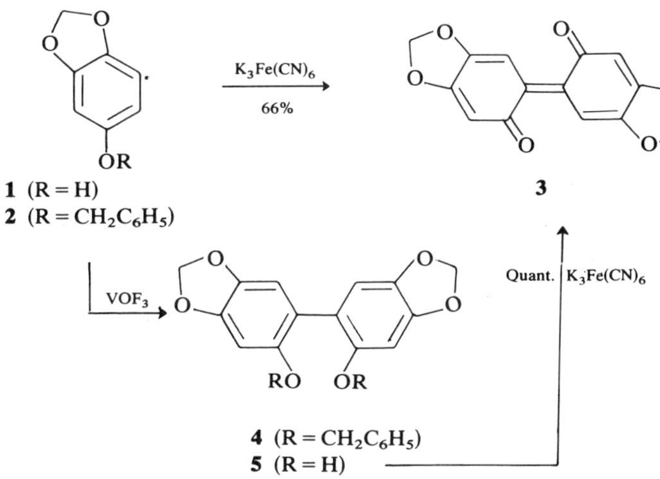

1 (R = H)
2 (R = $CH_2C_6H_5$)

$K_3Fe(CN)_6$, 66% → **3**

VOF_3 →

4 (R = $CH_2C_6H_5$)
5 (R = H)

Quant. $K_3Fe(CN)_6$ → **3**

[1] F. R. Hewgill, *Tetrahedron*, **34**, 1595 (1978).

Potassium hexamethyldisilazide, 4, 407–409.

Macrolides. A new method of lactonization involves intramolecular alkylation of an ω-haloalkyl 2-phenylthiomethylbenzoate.[1] An example is shown in equation (I). The 15-membered lactone corresponding to **3** was also obtained readily in this way, but the method failed in attempted cyclization to a 12-membered lactone.

(I) $\mathbf{1}$ + $CH_3CH(OH)(CH_2)_7CH_2I$ $\xrightarrow{85\%}$ $\mathbf{2}$ $\xrightarrow[75\%]{2KN[Si(CH_3)_3]_2}$ $\mathbf{3}$ $\xrightarrow[\sim 100\%]{1)\ NaIO_4;\ 2)\ \Delta}$ $\mathbf{4}$

(II) $\mathbf{5}$ $\xrightarrow[40\%]{Base}$ $\mathbf{6}$ $\xrightarrow[70\%]{Raney\ Ni}$ $\mathbf{7}$

However, the desired lactone was prepared by the modification shown in equation (II). The final product (**7**) is the dimethyl ether of lasiodiplodin.

[1] T. Takahashi, K. Kasuga, and J. Tsuji, *Tetrahedron Letters*, 4917 (1978).

Potassium hydride, 1, 935; **2**, 346; **4**, 409; **5**, 557; **6**, 482–483; **7**, 302–303; **8**, 412–415.

Potassium enolates of aldehydes. Enolates of aldehydes are somewhat difficult to generate because of competing polymerization by base. They have been obtained recently in high yield by use of potassium hydride in THF at 0° and successfully alkylated,[1] sulfenylated with diphenyl disulfide,[2] and converted into α-iodo aldehydes by iodine. The last two reactions have not been observed previously. Sulfenylation of aldehydes has previously used indirectly generated lithium enolates and a reactive sulfenyl chloride. All three reactions are useful, however, for aldehydes with only one α-proton. Otherwise yields of monosubstituted aldehydes are low and largely by-products are obtained.

Potassium enolates can be converted into the enol acetate by reaction with acetyl chloride and a catalytic amount of 4-dimethylaminopyridine or into the enol trimethylsilyl ether by reaction with chlorotrimethylsilane and triethylamine.[3]

$$RCH_2CHO \xrightarrow{-5°, KH, DME} [RCH{=}CHOK] \xrightarrow[80\text{–}90\%]{AcCl,\ 4\text{-}(CH_3)_2N\text{-pyridine}} RCH{=}CH{\sim}OAc \quad (cis/trans = 1:1)$$

$$[RCH{=}CHOK] \xrightarrow[70\text{–}80\%]{ClSi(CH_3)_3,\ N(C_2H_5)_3} RCH{=}CH{\sim}OSi(CH_3)_3$$

The same laboratory reports a method for transformation of the enol acetate of an aldehyde into an α,β-unsaturated aldehyde, as formulated in equation (I).[4]

$$\text{(I)}\quad RCH_2CH{=}CHOAc \xrightarrow{NBS,\ CCl_4} \left[\underset{\text{(main product)}}{RCHBrCH{=}CHOAc} + \underset{\text{(minor product)}}{RCH{=}CHCHBrOAc}\right] \xrightarrow[85\text{–}90\%]{NaOH,\ H_2O,\ (n\text{-}C_4H_9)_4NHSO_4} \underset{(R,H\ /\ H,CHO)}{RCH{=}CHCHO}$$

[1] P. Groenewegen, H. Kallenberg, and A. van der Gen, *Tetrahedron Letters*, 491 (1978).
[2] *Idem, ibid.*, 2817 (1978).
[3] D. Ladjama and J. J. Riehl, *Synthesis*, 504 (1979).
[4] F. Jung, D. Ladjama, and J. J. Riehl, *ibid.*, 507 (1979).

Potassium hydride–18-Crown-6.

Alkylation of N-substituted trifluoroacetamides. N-alkyl or N-aryl trifluoroacetamides can be alkylated by primary bromides or iodides by use of potassium hydride in combination with 18-crown-6 as base. This reaction is a useful route to secondary amines (equation I).[1]

$$\text{(I)}\quad RNHCCF_3 \,(C{=}O) \xrightarrow[80\text{–}95\%]{1)\ KH,\ crown\ ether;\ 2)\ R^1CH_2X} (R^1CH_2)(R)NCCF_3 \,(C{=}O) \xrightarrow[85\text{–}99\%]{KOH,\ CH_3OH,\ 20°} (R^1CH_2)(R)NH$$

[1] J. E. Nordlander, D. B. Catalane, T. H. Eberlein, L. V. Farkas, R. S. Howe, R. M. Stevens, R. E. Stansfield, N. A. Tripoulas, J. L. Cox, M. J. Payne, and A. Viehbeck, *Tetrahedron Letters*, 4987 (1978).

Potassium hydroxide, 5, 557–560; **6,** 486; **7,** 303–304; **8,** 415–416.

α,β-Unsaturated nitriles (**7,** 303). The complete paper on the preparation of these nitriles by reaction of acetonitrile with carbonyl compounds catalyzed by KOH is available.[1] The influence of 18-crown-6 on this condensation is still unpredictable.

[1] S. A. DiBase, B. A. Lipisko, A. Haag, R. A. Wolak, and G. W. Gokel, *J. Org.*, **44**, 4640 (1979).

Potassium hydroxide–Dimethyl sulfoxide.

Alkylation of phenols, alcohols, amides, and acids.[1] N-Alkylation of indoles and pyrroles by means of solid KOH in DMSO was reported a few years ago.[2] Actually this method is applicable to a number of substrates. The substrate and alkyl halide are added to powdered KOH and stirred in DMSO, usually at 20°. Methylation of phenols, alcohols, and amides occurs in high yield in about 5–30 minutes. Esterification of acids is slower. Dehydrohalogenation is a competing or predominating reaction when secondary or tertiary halides are used. Another limitation is that amino groups are converted into quaternary salts under these conditions. The general method can be used for permethylation of peptides.

[1] R. A. W. Johnstone and M. E. Rose, *Tetrahedron*, **35**, 2169 (1979).
[2] H. Heaney and S. V. Ley, *J.C.S. Perkin I*, 499 (1973).

Potassium permanganate, 1, 942–952; **2**, 348; **4**, 412–413; **5**, 562–563; **8**, 416–417.

Oxidation of 1,5-dienes. The oxidation of geranyl acetate (**1**) and neryl acetate (**2**) with potassium permanganate proceeds stereospecifically to afford tetrahydrofuranediols **3** and **4** with three chiral centers (equation I).[1] Similarly, the oxidative

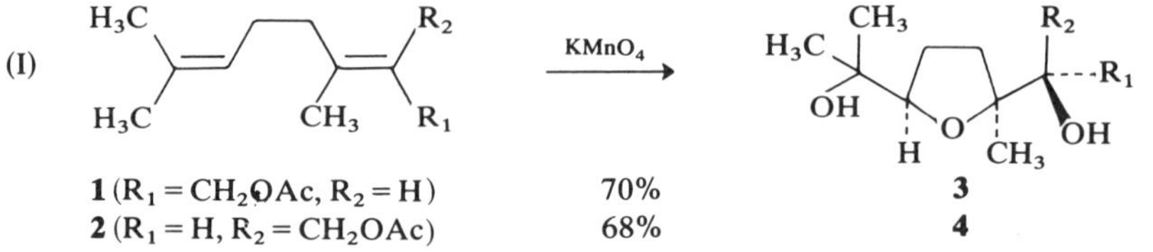

(I)

1 ($R_1 = CH_2OAc$, $R_2 = H$) → ($KMnO_4$, 70%) **3**
2 ($R_1 = H$, $R_2 = CH_2OAc$) → ($KMnO_4$, 68%) **4**

cycloaddition of stereochemically labeled 1,5-hexadienes generates tetrahydrofuranediols with four new chiral centers with complete stereospecificity (equations II and III), but in somewhat low yield. The stereochemistry of these products requires that all new bonds are formed by suprafacial processes. A mechanism has been suggested.[2,3]

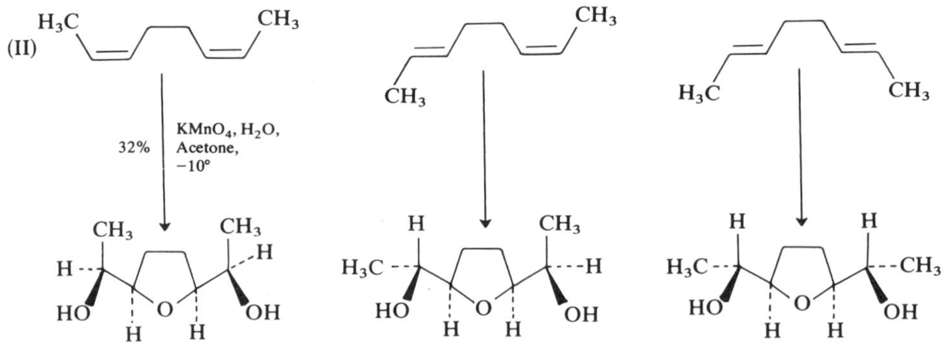

(II)

$KMnO_4$, H_2O, Acetone, −10°; 32%

(III)

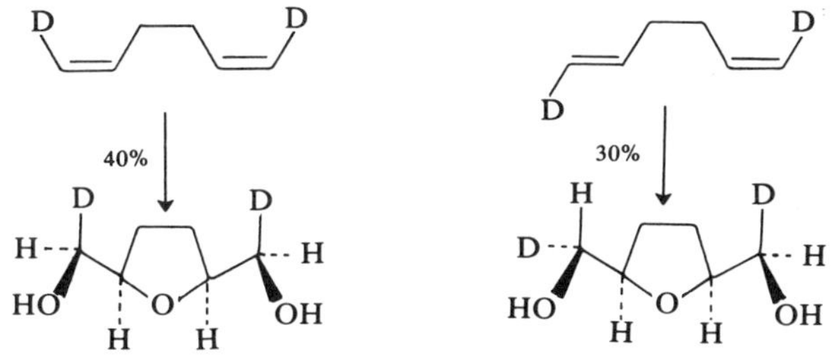

40%

30%

Aromatization.[4] Nonconjugated cyclohexadienes such as γ-terpinene (**1**) can be aromatized in excellent yield at room temperature by exposure to $KMnO_4$ in benzene in the presence of an equimolar amount of dicyclohexyl-18-crown-6.[1] Conjugated cyclohexadienes are recovered unchanged after similar treatment.

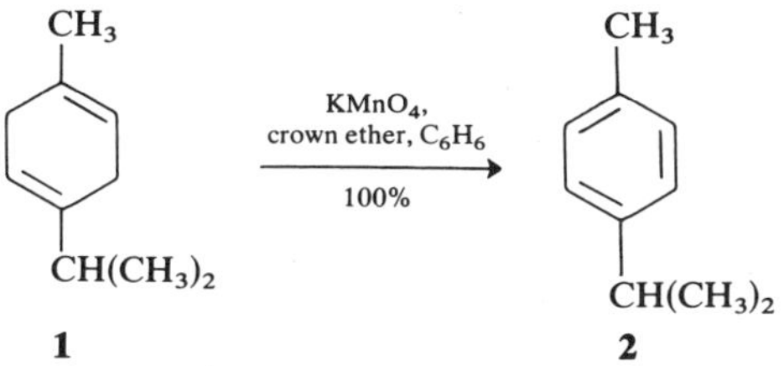

$KMnO_4$, crown ether, C_6H_6 → 100%

1 → **2**

Activation with $CuSO_4 \cdot 5H_2O$.[5] Powdered $KMnO_4$ (1.0 g) and $CuSO_4 \cdot 5H_2O$ (0.50 g) is an effective oxidant for secondary alcohols in benzene. The corresponding ketones are obtained in yields usually over 90% from reactions of 2–10 hours at room temperature. This solid reagent does not react with double bonds and oxidizes primary alcohols at a very slow rate. However, selective oxidation is not feasible experimentally. Apparently one function of the copper salt is to supply water in trace amounts. Powdered $Cu(MnO_4)_2$[6] functions almost as well as the $KMnO_4$–$CuSO_4$ couple.

Oxidation of steroid 5,7-dienes.[7] The main product (about 60% yield) of the oxidation of Δ^7-cholesterol (**1**) or the corresponding acetate with potassium permanganate under neutral or slightly basic conditions is the all-*cis*-epoxydiol **2**. Minor products are the $3\beta,5\alpha,6\alpha$-trihydroxy-Δ^7-steroid and the $3\beta,5\alpha,6\alpha,7\alpha$-tetrahydroxy-$\Delta^8$- and $\Delta^{8(14)}$-steroids, probably formed from **2** by cleavage and dehydration. A precedent for the formation of **2** is the permanganate oxidation of 1,3-cyclohexadiene to an all *cis*-epoxydiol in unspecified yield.[8]

α-Hydroxy ketones. Conditions have been reported for oxidation of internal alkenes by $KMnO_4$ to *vic*-diols (**1**, 948–949), α-diketones (**4**, 412–413; **5**, 563), and α-hydroxy ketones.[9] The last oxidation can also be effected in 75–85% yield by oxidation with $KMnO_4$ in aqueous acetone containing 2–5% acetic acid (to neutralize hydroxide ions).[10]

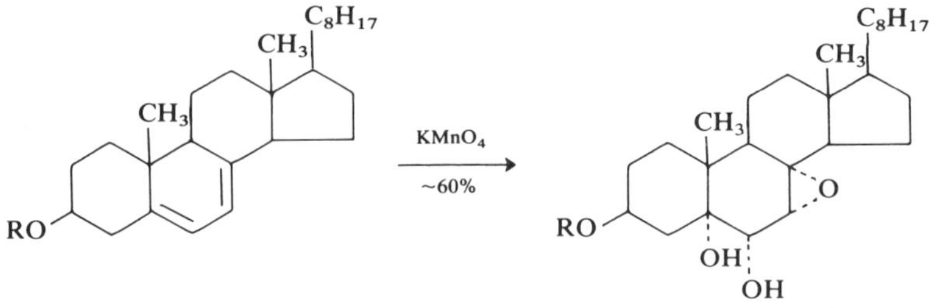

C_8H_{17} CH_3 CH_3 RO $\xrightarrow[\sim 60\%]{KMnO_4}$ C_8H_{17} CH_3 CH_3 RO O OH OH

1 (R = H or Ac) **2**

Oxidation of alkenes. $KMnO_4$ can be solubilized in CH_2Cl_2 by an equimolar amount of benzyltriethylammonium chloride. This solution can be used for homogeneous oxidation of alkenes to intermediates that can be decomposed either to dialdehydes or to *cis*-1,2-diols. In two-phase oxidations with $KMnO_4$ and phase-transfer catalysts, diols or carboxylic acids are obtained.[11]

Example:

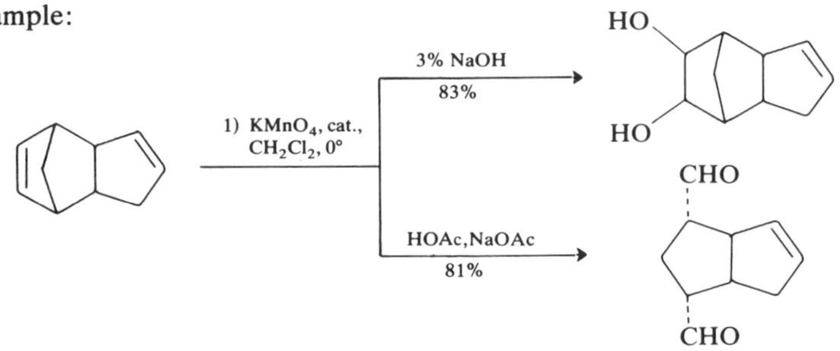

1) $KMnO_4$, cat., CH_2Cl_2, 0° → 3% NaOH, 83% → HO, HO (diol)

→ HOAc, NaOAc, 81% → CHO, CHO (dialdehyde)

Oxidations of alkynes (**1**, 947). A reinvestigation[12] of this reaction indicates that oxidation to diketones is best effected in dry methylene chloride with excess powdered potassium permanganate[13] and a phase-transfer catalyst (Adogen 464). Typical yields are 40–80%. If some water is also present, further oxidative cleavage of the enol of the diketone results in two carboxylic acids. An example is shown in equation (I). Oxidation of terminal alkynes to a carboxylic acid with one less carbon atom may also proceed through a dicarbonyl intermediate.

$$\text{(I)}\quad CH_3(CH_2)_5C{\equiv}C(CH_2)_5CH_3 \xrightarrow{KMnO_4,\ CH_2Cl_2}$$

$$\underset{(54\%)}{CH_3(CH_2)_5\overset{O}{\overset{\|}{C}}-\overset{O}{\overset{\|}{C}}(CH_2)_5CH_3} \rightleftharpoons CH_3(CH_2)_5\overset{O}{\overset{\|}{C}}-\overset{OH}{\overset{|}{C}}{=}CH(CH_2)_4CH_3$$

$$\xrightarrow[-CO_2]{KMnO_4,\ H_2O} \underset{(17\%)}{CH_3(CH_2)_4COOH} + \underset{(25\%)}{CH_3(CH_2)_5COOH}$$

α-Diketones are obtained only in traces from oxidations conducted in water, probably because they are cleaved as rapidly as they are formed.

[1] E. Klein and W. Rojahn, *Tetrahedron*, **21**, 2353 (1965).
[2] D. M. Walba, M. D. Wand, and M. C. Wilkes, *Am. Soc.*, **101**, 4396 (1979).
[3] J. E. Baldwin, M. J. Crossley, and E.-M. M. Lehtonen, *J.C.S. Chem. Comm.*, 918 (1979).
[4] A. Poulose and R. Croteau, *ibid.*, 243 (1979).
[5] F. M. Menger and C. Lee, *J. Org.*, **44**, 3446 (1979).
[6] Mide Chemical Corp.
[7] M. Anastasia, A. Frecchi, and A. Scala, *Tetrahedron Letters*, 3323 (1979).
[8] H. Z. Sable, K. A. Powell, H. Katchian, C. B. Niewoehner, and S. B. Kadlec, *Tetrahedron*, **26**, 1509 (1970).
[9] J. E. Coleman, C. Ricciuti, and D. Swern, *Am. Soc.*, **78**, 5342 (1956).
[10] N. S. Srinivasan and D. G. Lee, *Synthesis*, 520 (1979).
[11] T. Ogino and K. Mochizuki, *Chem. Letters*, 443 (1979).
[12] N. S. Srinivasan and D. G. Lee, *J. Org.*, **44**, 1574 (1979); D. G. Lee and V. S. Chang, *ibid.*, **44**, 2726 (1979).
[13] Cairox M, Carus Chem. Co.

Potassium ruthenate, K_2RuO_4. Mol. wt. 242.27. Supplier: Alfa. This expensive two-electron oxidizing agent can be generated *in situ* from ruthenium trichloride and aqueous alkaline potassium persulfate.

Oxidation of alcohols and aldehydes. Primary alcohols and aldehydes are oxidized to the corresponding acids in good yield by this reagent, which can be used in catalytic amounts in the presence of excess oxidant for regeneration. Secondary alcohols afford ketones; alkenes and tertiary alcohols do not react.[1]

Example:

OH $\xrightarrow[86\%]{}$ COOH

[1] M. Schröder and W. P. Griffith, *J.C.S. Chem. Comm.*, 58 (1979).

Potassium superoxide, 6, 488–490; **7**, 304–307; **8**, 417–419.

Oxidative cleavage of ketones.[1] Potassium superoxide in combination with Aliquat 336 oxidizes ketones in benzene solution at 20°. Ketones lacking an α-hydrogen and 1,3-diketones do not react.

Examples:

=O $\xrightarrow[92\%]{KO_2,\ \text{Aliquat 336},\ C_6H_6}$ COOH, COOH

$$CH_3CH_2\overset{O}{\overset{\|}{C}}CH_2C_6H_5 \xrightarrow[96\%]{} CH_3CH_2COOH + C_6H_5COOH \quad (1:1)$$

$$(CH_3)_2C{=}CHCOCH_3 \xrightarrow[94\%]{} (CH_3)_2CO + CH_3COOH \quad (1:1)$$

Reaction with cyclohexenones.[2] Cyclohexenone (**1**) itself reacts with KO_2–18-crown-6 in benzene to form the dimer (**2**) and a trimer. The presence of two alkyl substituents at the 6-position results in formation of an epoxy ketone (*e.g.*, **4**) in addition. A different reaction is observed if C_4 bears two alkyl substituents. Thus **6** is converted into **7** in fair yield by KO_2. The same products are obtained in lower yields if air/KOH–18-crown-6 is the oxidant.

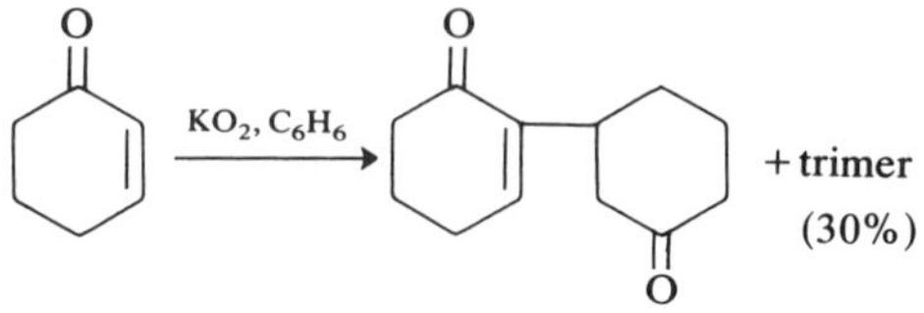

KO_2, C_6H_6 → **1** → **2** (20%) + trimer (30%)

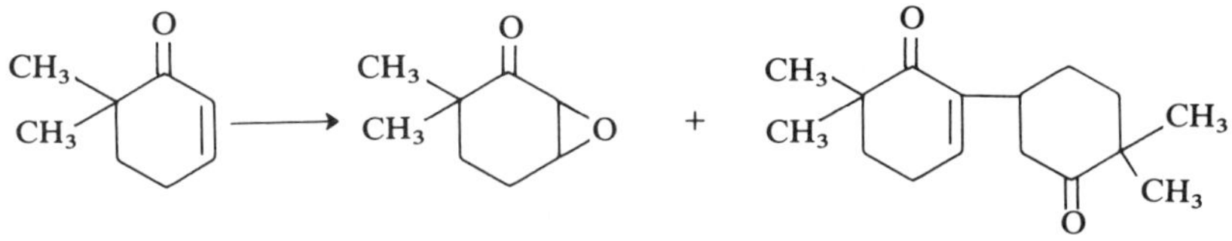

3 → **4** (20%) + **5** (40%)

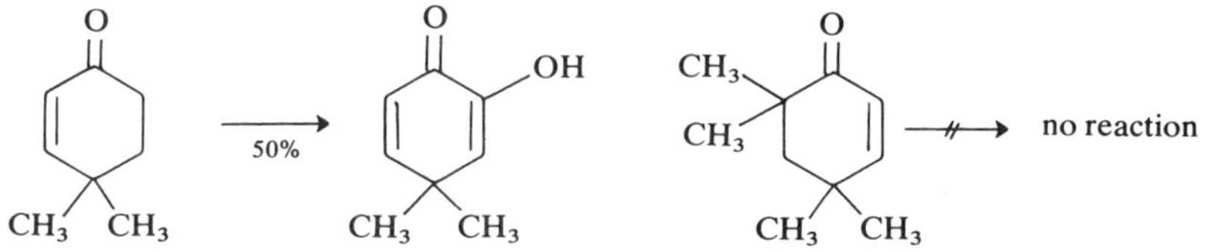

6 → (50%) **7**

8 ↛ no reaction

Cleavage of* t*-butyldimethylsilyl ethers. These ethers can be cleaved by KO_2 and 18-crown-6 in DMSO. In the case of *t*-butyldimethylsilyl 4-phenylbutyl ether the reaction proceeds in quantitative yield in 10 minutes with 3 equiv. of KO_2. This unexpected reaction was first noted in a total synthesis of 11-*epi*-$PGF_{2\alpha}$ accompanying the inversion shown in equation (I).[3]

(I) 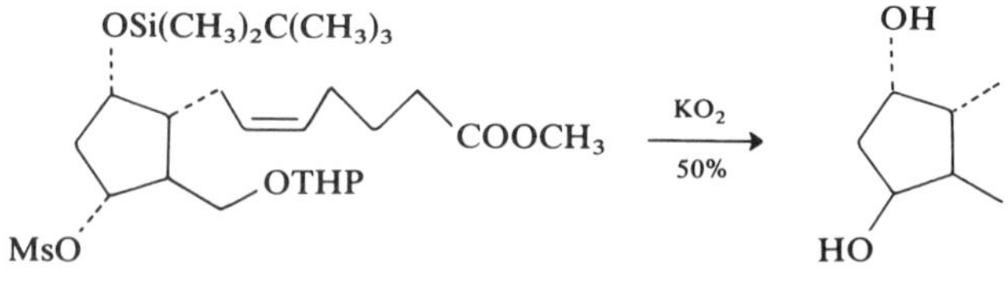 $OSi(CH_3)_2C(CH_3)_3$ … $COOCH_3$, OTHP, MsO $\xrightarrow[50\%]{KO_2}$ OH, HO

Oxidation of aryl amines and phenols. Phenol and aniline are not oxidized by KO_2, but *o*- and *p*-phenylenediamine are oxidized to diaminoazobenzenes and *o*- and *p*-aminophenols are oxidized to dihydroxyazobenzenes. Thiophenol is oxidized to diphenyl disulfide.[4]

Oxidation of *o*-phenylenediamine (**1**) with KO_2 in the presence of 18-crown-6 or with oxygen and KOH complexed with the crown ether results in the products **2–4**.[5] *cis,cis*-Muconitrile, the main product of oxygenation catalyzed by CuCl (**7**, 80), is not observed.

$$\mathbf{1}\ (o\text{-}C_6H_4(NH_2)_2) \xrightarrow[18\text{-crown-6}]{KO_2,} \mathbf{2}\ (o\text{-}NH_2C_6H_4NO,\ 38\%) + \mathbf{3}\ (o\text{-}NH_2C_6H_4NO_2,\ 28\%) + \mathbf{4}\ (H_2N\text{-}C_6H_4\text{-}N{=}N\text{-}C_6H_4\text{-}NH_2,\ 26\%)$$

[1] M. Lissel and E. V. Dehmlow, *Tetrahedron Letters*, 3689 (1978).
[2] A. A. Frimer and P. Gilinsky, *ibid.*, 4331 (1979).
[3] Y. Torisawa, M. Shibasaki, and S. Ikegami, *ibid.*, 1865 (1979).
[4] G. Crank and M. I. H. Makin, *ibid.*, 2169 (1979).
[5] E. Balogh-Hergovich, G. Speier, and E. Winkelmann, *ibid.*, 3541 (1979).

Potassium triisopropoxyborohydride, 5, 565.

Carbonylation of organoboranes (**2**, 60; **3**, 41–42). Potassium triisopropoxyborohydride is superior to other hydrides for this reaction. 9-Alkyl-9-BBN derivatives are excellent substrates, since the 9-alkyl groups are utilized in almost quantitative yield.[1]

$$\text{(9-BBN)}BR + CO + K^+(i\text{-}C_3H_7O)_3BH^- \rightarrow \left[\text{(9-BBN)}B{-}CHR{-}OB(OC_3H_7)_3\right]^- K^+ \xrightarrow[85\text{–}95\%]{H_2O_2,\ OH^-} RCHO$$

$$\xrightarrow[85\text{–}95\%]{NaOH,\ H_2O} RCH_2OH$$

[1] H. C. Brown, J. L. Hubbard, and K. Smith, *Synthesis*, 701 (1979).

Proline, 6, 492–493; **7,** 307; **8,** 421–424.

Asymmetric aldol condensation (**6**, 410–411; **7**, 307). Terashima and coworkers[1] have examined the effect of various (S)-amino acids on asymmetric cyclization of the acyclic triketone **1**. The course of the cyclization is dependent on the solvent. In ether (R,S)-**3** is formed preferentially; in water (R,S)-**2** is the major product. Several amino acids were found to effect asymmetric cyclization, particularly in the cyclization of **1** to **2**. The most interesting result is that addition of (S)-phenylalanine gives (R)-**2** in 56% enantiomeric excess and that addition of (S)-histidine gives (S)-**2** in 54% enantiomeric excess. As expected (S)-proline exerted an effect, but rather slight.

Asymmetric bromolactonization (**8**, 421–423). Detailed papers[2] concerning this reaction have been published. For this reaction L-proline is definitely superior to open-chain L-amino acid derivatives where free rotation of the bond between the asymmetric center and the N atom is possible.

[1] S. Terashima, S. Sato, and K. Koga, *Tetrahedron Letters*, 3469 (1979).
[2] S. Jew, S. Terashima, and K. Koga, *Tetrahedron*, **35**, 2337, 2345 (1979).

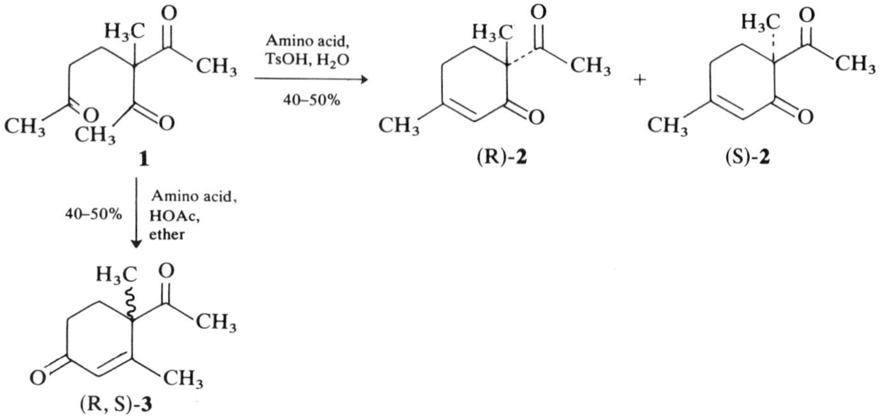

1,3-Propanedithiol, 1, 956; **4,** 413; **6,** 493.

Reduction of azides. Alkyl and aryl azides are reduced to primary amines by the combination of 1,3-propanedithiol and triethylamine (equation I). The method is highly selective, and does not affect double or triple bonds, nitro, nitrile, carboxylic acid, amide, and ester groups.[1]

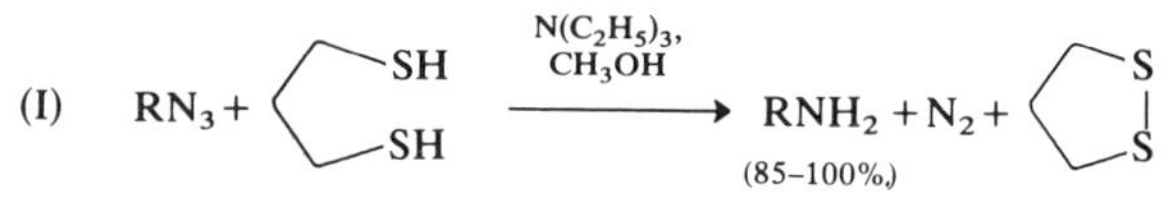

(I) RN_3 + 1,3-propanedithiol $\xrightarrow{N(C_2H_5)_3,\ CH_3OH}$ $RNH_2 + N_2$ + 1,2-dithiolane (85–100%)

This investigation was initiated by the observation that aryl azides are reduced by dithiothreitol (DL-1,4-dimercapto-2,3-butanediol).[2]

[1] H. Bayley, D. N. Standring, and J. R. Knowles, *Tetrahedron Letters*, 3633 (1978).
[2] J. V. Staros, H. Bayley, D. N. Standring, and J. R. Knowles, *Biochem. Biophys. Res. Comm.*, **80**, 568 (1978).

Propargyl alcohol, $HC{\equiv}CCH_2OH$. Mol. wt. 56.06, b.p. 114–115°.

Cyclopentenone annelation.[1] The isomeric adducts **1** of 2-methylcyclohexanone and propargyl alcohol both cyclize regiospecifically to the hydrindanone **2** (70%

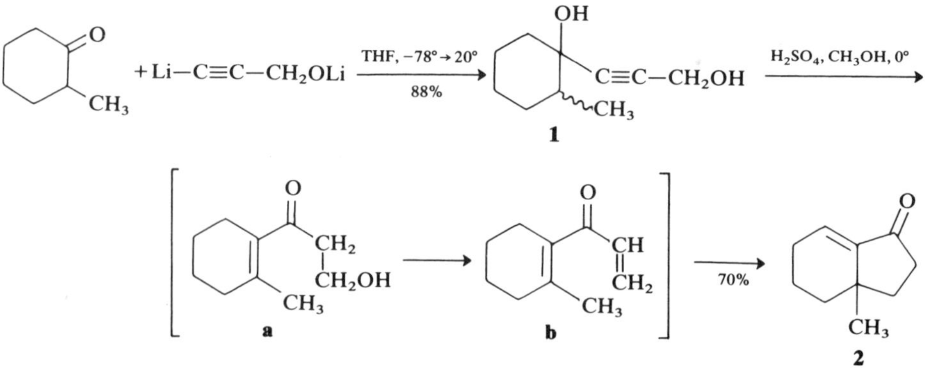

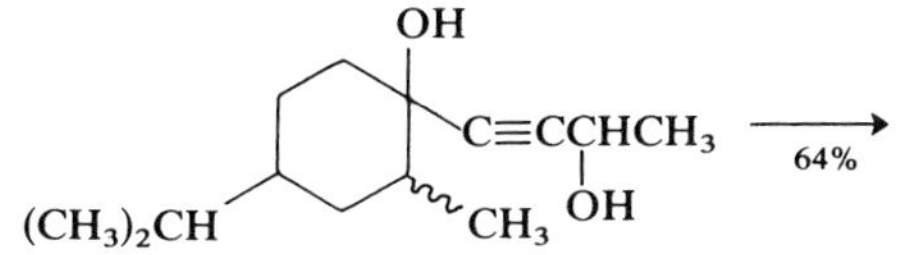

yield). This reaction also exhibits stereoselectivity when applied to alkyl-substituted derivatives of **1**. Thus **3** is cyclized mainly to the hydrindanone **4**. The stereoselectivity probably results from a preference of the alkyl substituents to assume an equatorial configuration in the transition state. The analogous cyclization of **6** to **7** is the key step of a total synthesis of (±) nootkatone (**8**).[2]

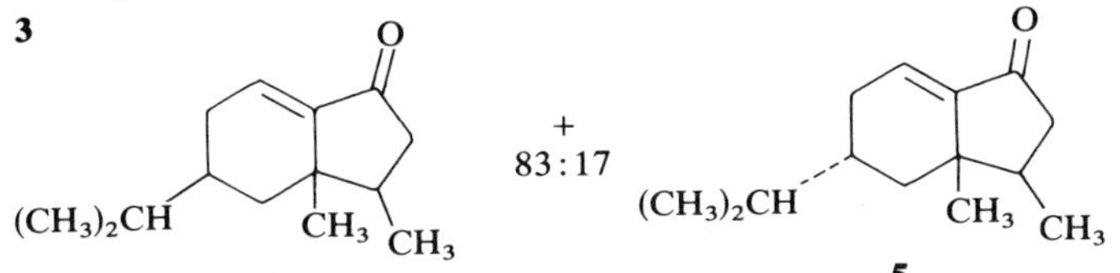

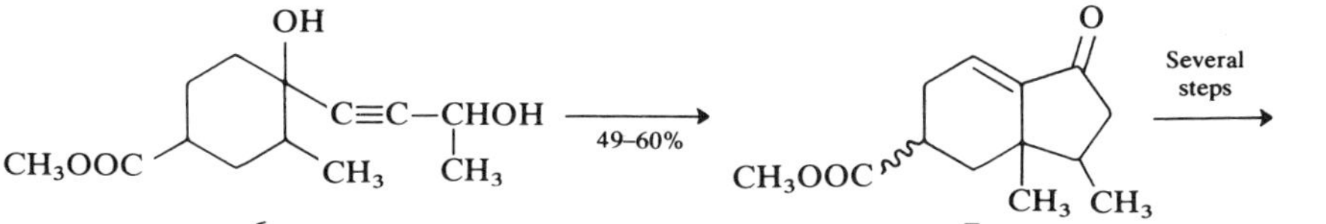

Cyclizations of this type were first reported by Islam and Raphael (**1**, 860–861; equation I).[3]

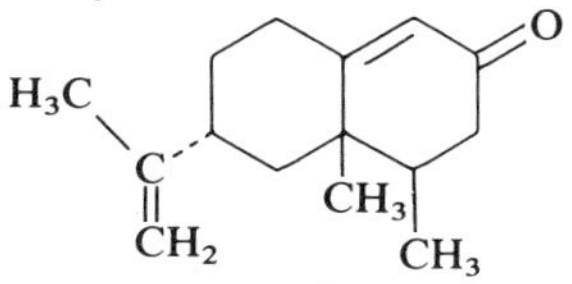

(I) $\xrightarrow[50\%]{H_3PO_4,\ HCOOH}$

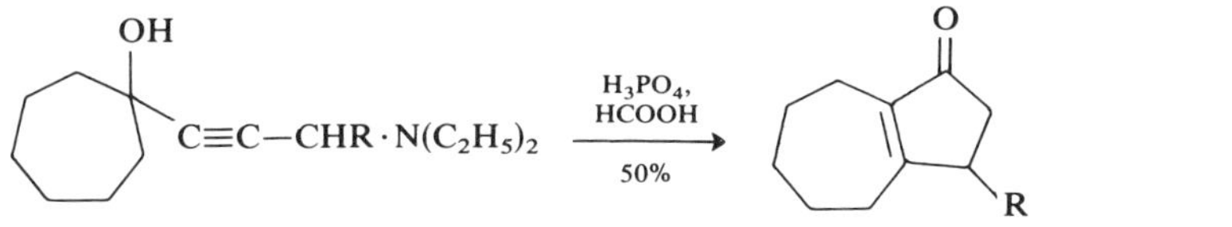

[1] T. Hiyama, M. Shinoda, and H. Nozaki, *Am. Soc.*, **101**, 1599 (1979).
[2] *Idem*, *Tetrahedron Letters*, 3529 (1979).
[3] A. M. Islam and R. A. Raphael, *J. Chem. Soc.*, 2247 (1953).

Pyridazine-1-oxide (1). Mol. wt. 98.09, m.p. 38–40°, very hygroscopic. The oxide is obtained from pyridazine by reaction with 2 equiv. of H_2O_2 (30%) in HOAc (88.6% yield).[1]

Terminal enynes.[2] Reaction of aryl Grignard reagents with pyridazine 1-oxide leads exclusively to terminal (E)-enynes in 50–75% yield. Alkyl Grignard reagents lead to 1,3-dienes in rather low yield.

ArMgX → [H, Ar, N, N, OMgX → Ar, N=N, OMgX] → 50–75% → H, C=C, C≡CH, Ar, H

1

[1] T. Itai and S. Natsume, *Chem. Pharm. Bull. Japan*, **11**, 83 (1963).
[2] L. Crombie, N. A. Kerton, and G. Pattenden, *J.C.S. Perkin I*, 2136 (1979).

Pyridine–Sulfur trioxide complex.

Reduction of α-halo ketones.[1] This complex in combination with sodium iodide reduces α-halo ketones to the parent ketone in 80–90% yield. Sodium iodide and triethylamine–sulfur dioxide is more selective in that only aryl α-haloalkyl ketones are reduced.

[1] G. A. Olah, Y. D. Vankar, and A. P. Fung, *Synthesis*, 59 (1979).

Pyridine N-oxide, 1, 966; **2,** 353; **5,** 567.

α,β-Dehydrogenation of carboxamides. Imide chlorides derived from tertiary amides are dehydrogenated to α,β-unsaturated amides when treated with pyridine N-oxide and triethyl amine in $CHCl_3$. Diphenyl sulfoxide can also be used as oxidant, but separation of the unsaturated amides from sulfur-containing products is troublesome. The method appears to be limited to α-branched amides, since the sole product from N,N-dimethylbutyramide is N,N-dimethyl-α-chlorobutyramide, but it is useful for the synthesis of dehydro amino acid derivatives.[1]

Examples:

$CON(CH_3)_2$, CH_3 → $COCl_2$ → [$C{=}\overset{\oplus}{N}(CH_3)_2$, CH_3, Cl, Cl^-] → (pyridine N-oxide) 75% → $CON(CH_3)_2$, CH_3

$CH_3{-}CH{=}C(CH_3){-}CON(CH_3)_2$ → 40% → $CON(CH_3)_2$, CH_2, CH_2

O, CH_3, $N{-}CH{-}CON$, O → 85% → O, CH_2, N, CON, O

[1] R. Da Costa, M. Gillard, J. B. Falmagne, and L. Ghosez, *Am. Soc.*, **101**, 4381 (1979).

2-Pyridinesulfenyl chloride (PS—Cl), [structure: 2-pyridyl–SCl]. PS—Cl is obtained by reaction of di(2-pyridyl) disulfide in petroleum ether with dry chlorine. It is obtained as a yellow powder after evaporation of the solvent. It is immediately destroyed by water; it is stable in anhydrous acetic acid for hours.

PS—Cl deprotects cysteine residues protected with trityl, diphenylmethyl, acetamidomethyl, *t*-butyl, and *t*-butylsulfenyl groups; a mixed disulfide, CyS—S—PS, is obtained, which can form a disulfide with another SH group. Thus it can also be used to form disulfide bonds between two cysteine-containing peptides.[1]

[1] J. V. Castell and A. Tun-Kyi, *Helv.*, **62**, 2507 (1979).

Pyridinium chlorochromate (PCC) 6, 498–499; **7,** 308–309; **8,** 425–427.

Stoichiometry.[1] Oxidation of alcohols by PCC involves a change of two electrons, rather than the change of three electrons observed in chromic acid oxidations. Under the standard conditions an excess is used, but actually 1 molar equiv. is sufficient for oxidation of 1-octanol (94% yield).

Oxidative rearrangement of cyclopropylcarbinols to β,γ-enones.[2] Tertiary cyclopropylcarbinols (**1**), which can be prepared by addition of cyclopropyl organometallic reagents to ketones, are oxidized by PCC (5 equiv.) mainly to β,γ-enones (**2**). The related reagent $C_5H_5\overset{+}{N}HCrO_3BF_4^-$ is equally effective. Traces

$$\text{(cyclopropyl-}R^1\text{)}-C(R^2)(R^3)OH \ (\mathbf{1}) \xrightarrow[40\text{–}60\%]{PCC,\ CH_2Cl_2,\ 20^\circ} R^1\overset{O}{\overset{\|}{C}}CH_2CH=C(R^2)R^3 \ (\mathbf{2})$$

of water improve the yield. The overall result is a 1,4-transposition of a carbonyl group. The reaction proceeds in rather low yield when applied to cyclopropylcarbinols of type **3**.

$$\mathbf{3} \xrightarrow[20\text{–}30\%]{PCC} \mathbf{4}$$

Oxidation of 1,4-dienes.[3] Oxidation of the 1,4-diene **1** with PCC in CH_2Cl_2 affords dienones **2** and **3** in the ratio of 9 : 1 (70% total yield). Complementary regioselectivity (**2**/**3** = 1 : 3) is achieved with $CrO_3 \cdot 2Py$ or *tert*-butyl chromate as the

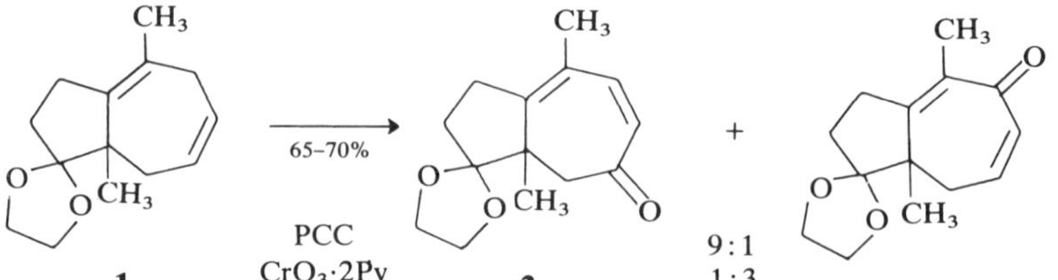

$$\mathbf{1} \xrightarrow[\text{PCC} \quad \text{CrO}_3\cdot 2\text{Py}]{65\text{–}70\%} \mathbf{2} + \mathbf{3} \qquad 9:1 \quad 1:3$$

oxidant (~65% yield). Isolated double bonds and compounds such as diphenylmethane and allylbenzene are inert to PCC.

Oxidation of organoboranes.[4] This oxidant is superior to chromic acid for oxidation of organoboranes derived from cycloalkenes to the corresponding ketones. For example, cyclohexanone is obtained in 81% yield from cyclohexene.

Oxidation of trialkyl borates[5] ***and of trialkoxyboroxines.***[6] Trialkyl borates, $B(OR)_3$, prepared by reaction of primary and secondary alcohols with borane–dimethyl sulfide, are oxidized by PCC to aldehydes and ketones in good yield. This indirect oxidation of alcohols does not involve formation of water, which could be detrimental in some cases. Of even greater interest, carboxylic acids can be converted into aldehydes by reaction with borane–dimethyl sulfide to form a trialkoxyboroxine followed by oxidation with PCC (equation I).

$$\text{(I)}\quad 3\text{RCOOH} \xrightarrow{H_3B:S(CH_3)_2} (\text{RCH}_2\text{OBO})_3 \xrightarrow[70\text{–}80\%]{\text{PCC}} 3\text{RCHO}$$

Oxidation of carbohydrates.[7] The oxidation of isolated secondary hydroxyl groups of carbohydrates with CrO_3 complexed with pyridine is often unsatisfactory. This oxidation can be effected generally in 70–85% yield with PCC; reaction in CH_2Cl_2 is very slow, but proceeds readily in refluxing benzene. The various oxidants based on DMSO are not useful. PCC is also effective for oxidation of primary hydroxyl groups of carbohydrates; Collins reagent is also effective in this case.

γ-Hydroxybutenolides.[8] 5-Bromo-2-furylcarbinols (**1**) are oxidized to γ-hydroxybutenolides (**2**) by PCC in CH_2Cl_2 at room temperature. The reaction is unusual in that only the furane ring is oxidized even though a secondary alcohol group is also present.

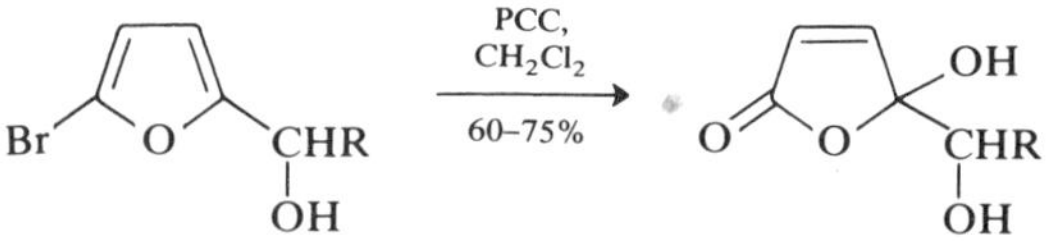

$$\mathbf{1} \xrightarrow[60\text{–}75\%]{\text{PCC},\ CH_2Cl_2} \mathbf{2}$$

[1] H. C. Brown, C. G. Rao, and S. U. Kulkarni, *J. Org.*, **44**, 2809 (1979).
[2] E. Wada, M. Okawara, and T. Nakai, *ibid.*, **44**, 2952 (1979).
[3] P. A. Wender, M. A. Eissenstat, and M. P. Filosa, *Am. Soc.*, **101**, 2196 (1979).
[4] V. V. R. Rao, D. Devaprabhakara, and S. Chandrasekaran, *J. Organometal. Chem.*, **162**, C9 (1978).

[5] H. C. Brown, S. U. Kulkarni, and C. Gundu Rao, *Synthesis*, 702 (1979).
[6] *Idem, ibid.*, 704 (1979).
[7] D. H. Hollenberg, R. S. Klein, and J. J. Fox, *Carbohydrate Res.*, **67**, 491 (1978).
[8] G. Piancatelli, A. Screttri, and M. D'Auria, *Tetrahedron Letters*, 1507 (1979).

Pyridinium dichromate (PDC), $(C_5H_5\overset{+}{N}H)_2Cr_2O_7^{2-}$. Mol. wt. 376.21, m.p. 144–146°, bright orange; soluble in H_2O, DMF, DMSO; sparingly soluble in CH_2Cl_2 or acetone; insoluble in ether, toluene. Supplier: Aldrich.

This solid is obtained by dissolving CrO_3 in a minimum of water, adding pyridine, and collecting the precipitate. PDC is probably present in Sarett and Cornforth oxidants. Unlike pyridinium chlorochromate, it is nearly neutral.

When used in DMF, PDC oxidizes aldehydes and primary alcohols to carboxylic acids. However, allylic primary and secondary alcohols are oxidized only to the α,β-unsaturated carbonyl compounds. All these oxidations proceed in high yield.

The behavior of PDC in CH_2Cl_2 is considerably different. Primary alcohols are oxidized only to aldehydes. Oxidation of secondary alcohols is also satisfactory and can be catalyzed by addition of pyridinium trifluoroacetate. Allylic alcohols are oxidized readily.[1]

CH_3 / COOH / $C(CH_3)_2$ ← PDC, DMF (83%) ← CH_3 / CH_2OH / $C(CH_3)_2$ → PDC, CH_2Cl_2 (92%) → CH_3 / CHO / $C(CH_3)_2$

[1] E. J. Corey and G. Schmidt, *Tetrahedron Letters*, 399 (1979).

Pyridinium hydrobromide perbromide, 1, 967–970; **5**, 568; **6**, 499–500.

Selective monoprotection of dienes.[1] Since the rate of addition of bromine to a double bond increases with substitution,[2] monoprotection is possible. The selectivity is higher with pyridinium hydrobromide perbromide than with bromine, especially at low temperatures (−60°). The dibromide can be deprotected by cathodic reduction at the mercury cathode in DMF.

[1] U. J. U. Husstedt and H. J. Schäfer, *Synthesis*, 964, 966 (1979).
[2] The relative rates for mono-/di-/tri-/tetrasubstituted are 1:60:2000:20,000.

Pyridinium poly(hydrogen fluoride), 5, 538–539; **6**, 473–474; **7**, 294. Supplier: Aldrich.

Aromatic fluorination.[1] A new method of aromatic fluorination involves treatment of aryltriazenes, readily prepared from aryldiazonium ions and dialkylamines,[2] with 70% hydrogen fluoride in pyridine. The yields of product from this reaction are usually higher than those obtained by the reaction of HF–pyridine with a diazonium ion (**6**, 285); *o*-methoxy, iodo-, bromo- and nitro-substituted aryltriazenes generally give unsatisfactory yields. This method may be useful for the synthesis of ^{18}F-labeled compounds.

HF, Py, HOAc, 18°

97%

95%

85%

Fluorination of alkenes. The reagent adds in the Markownikoff fashion to alkenes to yield alkyl fluorides (35–90% yield). It converts secondary and tertiary alcohols into the corresponding alkyl fluorides. *In situ* diazotization followed by dediazotization converts α-amino acids into α-fluorocarboxylic acids (38–98% yield). The reagent converts acid chlorides and anhydrides into acid fluorides.[3]

[1] M. N. Rosenfeld and D. A. Widdowson, *J.C.S. Chem. Comm.*, 914 (1979).
[2] D. Wallach, *Ann.*, **235**, 233 (1886).
[3] G. A. Olah, J. T. Welch, Y. D. Vankar, M. Nojime, I. Kerekes, and J. A. Olah, *J. Org.*, **44**, 3872 (1979).

Pyridinium *p*-toluenesulfonate, 8, 427–428.

Acetalization.[1] Pyridinium tosylate catalyzes the reaction of ketones with ethanediol to form ketals (90–95% yield); it also catalyzes the transacetalization of the ketals with acetone (90–95% yield).

[1] R. Sterzycki, *Synthesis*, 724 (1979).

4-Pyrrolidinopyridine, . Pyrrolidine can be alkylated by 4-bromopyridine using phase-transfer conditions (aqueous NaOH, benzyl triethylammonium bromide). The method is applicable to various secondary cyclic amines. Yields are 65–70%.[1]

[1] K. M. Patel and J. T. Sparrow, *Syn. Comm.*, **9**, 251 (1979).

1-Pyrroline-1-oxide, (**1**). Mol. wt. 85.10, b.p. 74–76°/0.1 mm.

Preparation:[1,2]

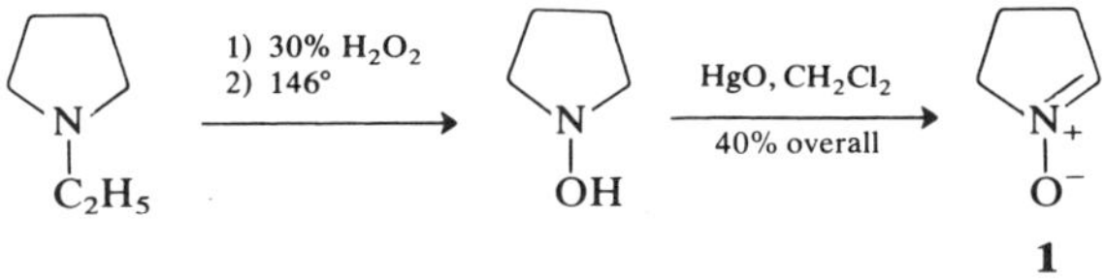
1) 30% H_2O_2
2) 146°
C_2H_5
OH
HgO, CH_2Cl_2
40% overall

1

1,3-Dipolar cycloadditions. Nitrones such as **1** combine with alkenes to form isoxazolidines. Tufariello has employed these 1,3-dipolar cycloadditions as key steps in the synthesis of several alkaloids.[3] Monosubstituted olefins react with nitrones regioselectively to afford 5-substituted isoxazolidines; thus addition of **2** to 1-pyrroline-1-oxide produces the cycloadduct **3**, which serves as an intermediate in a total synthesis of elaeocarpine (**4**).[4]

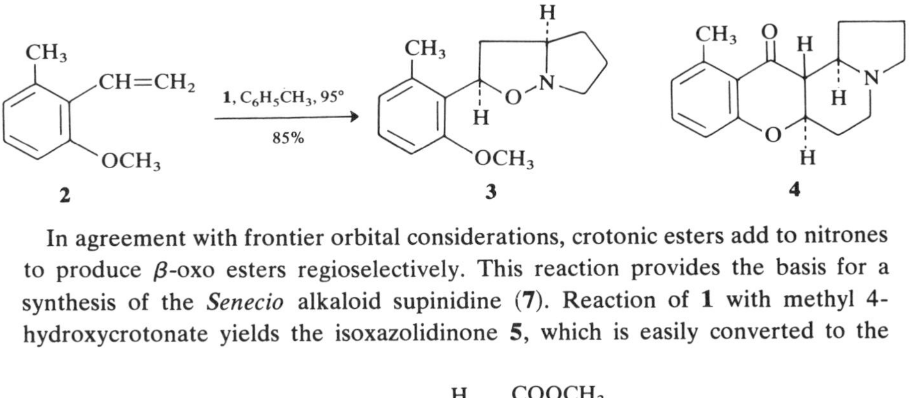
CH_3
CH=CH_2
OCH_3
1, $C_6H_5CH_3$, 95°
85%

2 **3** **4**

In agreement with frontier orbital considerations, crotonic esters add to nitrones to produce β-oxo esters regioselectively. This reaction provides the basis for a synthesis of the *Senecio* alkaloid supinidine (**7**). Reaction of **1** with methyl 4-hydroxycrotonate yields the isoxazolidinone **5**, which is easily converted to the

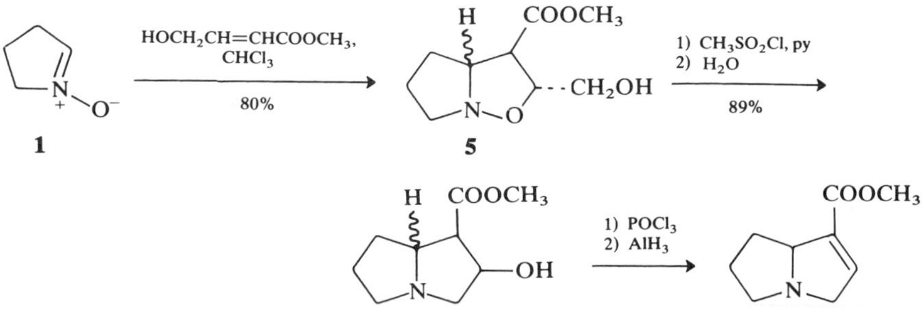
$HOCH_2CH{=}CHCOOCH_3$,
$CHCl_3$
80%
$COOCH_3$
CH_2OH
1) CH_3SO_2Cl, py
2) H_2O
89%
1) $POCl_3$
2) AlH_3
OH

1 **5** **6** **7**

pyrrolizidine alkaloid. Tufariello has also employed the related nitrone (**8**) in alkaloid synthesis.[3]

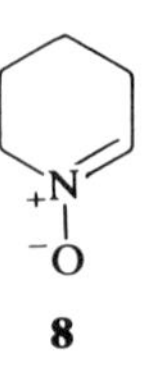

8

[1] J. Thesing and W. Sirrenberg, *Ber.*, **92**, 1748 (1959).
[2] J. J. Tufariello and J. P. Tette, *J. Org.*, **40**, 3866 (1975).
[3] J. J. Tufariello, *Acts. Chem. Res.*, **12**, 396 (1979).
[4] J. J. Tufariello and S. A. Ali, *Am. Soc.*, **101**, 7114 (1979).

Q

Quinine, 6, 501; **7**, 311; **8**, 430–431.

Quinine–acrylonitrile copolymer.[1] Cinchona alkaloids can be copolymerized with another vinyl monomer such as acrylonitrile with AIBN as initiator. The highest yield of polymer in the case of quinine is obtained when the quinine/acrylonitrile ratio is 1:20. This method was used to obtain a polymeric form of the alkaloid in which the crucial part of the molecule for asymmetric reactions—the amino alcohol unit—is free. The polymers are stable, light yellow solids, soluble in polar aprotic solvents (DMF and DMSO), but insoluble in common organic solvents.

These cinchona copolymers are efficient catalysts for some asymmetric Michael reactions. Examples are shown in equations (I) and (II). In the second example, use of quinidine–acrylonitrile polymer as catalyst leads to a product with $\alpha_D + 36.3°$; when quinidine itself is used, the α_D of the product is $+3.9°$. The figures imply that the copolymer is more stereoselective than the monomer.

(I) [indanone with H, $COOCH_3$, =O] $+ CH_2{=}CHCOCH_3 \xrightarrow[98\%]{\text{Quinine copolymer}}$ [indanone with $CH_2CH_2COCH_3$, *, $COOCH_3$, =O] (30% ee)

(II) $C_6H_5CH_2SH + C_6H_5CH{=}CHNO_2 \xrightarrow{\text{Cat.}} C_6H_5CH_2S\overset{*}{C}H(C_6H_5)CH_2NO_2$

Michael reaction of selenophenols. Selenophenols (and other selenides) undergo enantioselective 1,4-addition to cyclohexenone in the presence of catalytic amounts of cinchona alkaloids. Chemical yields are high; optical yields are 10–43%. Usually the optical yield can be enhanced by crystallization. In one case the addition product was converted into an optically active allylic alcohol by hydride reduction followed by selenoxide fragmentation.[2]

$CH_3-C_6H_4-SeH$ + [cyclohexenone] $\xrightarrow[>95\%]{C_6H_5CH_3,\ 20°,\ \text{cinchonidine}}$ [3-($SeC_6H_4CH_3$-*p*)cyclohexanone] (43% ee)

[1] N. Kobayashi and K. Iwai, *Am. Soc.*, **100**, 7071 (1978).
[2] H. Pluim and H. Wynberg, *Tetrahedron Letters*, 1251 (1979).

Quinoline, 1, 975; **2**, 356.

Ethynylchlorosilanes. Quinoline effects dehydrobromination of (α-bromovinyl)chlorosilanes to ethynylchlorosilanes in moderate yields.[1]

Examples:

$$CH_2{=}CBrSi(CH_3)_2Cl \xrightarrow[77\%]{C_9H_7N,\ 180\text{–}200^\circ} CH{\equiv}CSi(CH_3)_2Cl$$

$$CH_2{=}CBrSiCl_2C_2H_5 \xrightarrow[36\%]{} CH{\equiv}CSiCl_2C_2H_5$$

[1] H. Matsumoto, T. Kato, I. Matsubara, Y. Hoshino, and Y. Nagai, *Chem. Letters*, 1287 (1979).

R

Raney nickel, 1, 723–731; **2,** 293–294; **5,** 570–571; **6,** 502; **7,** 312; **8,** 433.

α-Alkylation of ketones.[1] The reaction can be carried out indirectly by α-alkylthiocarbonylation of a ketone to give a β-keto thioester.[2] The products undergo facile α-alkylation even with relatively unreactive alkyl halides. The thioester group is then reductively removed with Raney nickel at room temperature.

Examples:

1) NaH 2) $C_6H_5CH_2Br$, 88%; Raney Ni, C_6H_6, 25°, 92%

$COSC_2H_5$ → $CH_2C_6H_5$, $COSC_2H_5$ → $CH_2C_6H_5$

$COSCH_3$, O, OCH_3 → 88% → CH_3, $COSCH_3$, O, OCH_3 → 72% → CH_3, O, OCH_3

Nitrile → aldehyde (**1,** 725–726; **2,** 293). The nitriles **1** and **2** are reduced with W2 Raney nickel in a slightly acidic buffer system to give the corresponding aldehydes in 72–74% yield.[3] Additional reducing agents (sodium hypophosphite, formic acid, or H_2), which are commonly needed in similar reactions,[4] are not required.

CH_3O, CH_2R, CH_3O, O — **1** (R = CN), **3** (R = CHO)

CH_3O, CH_2R, CH_3O, O, O — **2** (R = CN), **4** (R = CHO)

Inversion of 3β-hydroxy steroid. Reduction of the dihydroxy keto cholanate **1** with Raney nickel (no. 28) results in partial epimerization of the axial 3β-hydroxy group (but not the axial 7α-group). This inversion step must be fairly slow because it is not observed when reduction of a 12-ketocholanic acid group is effected within 4 hours. This reaction is the mildest method known for inversion of an axial C_3-steroidal alcohol.[5]

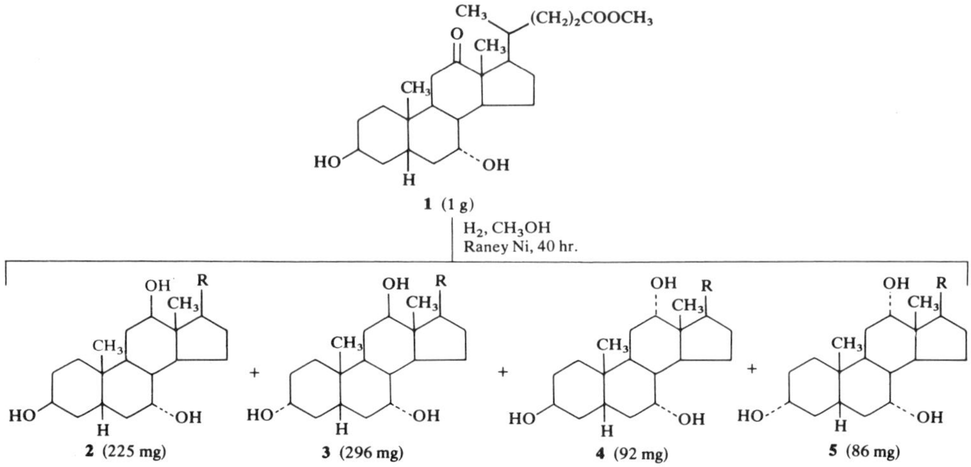

1 (1 g)

H_2, CH_3OH
Raney Ni, 40 hr.

2 (225 mg) + **3** (296 mg) + **4** (92 mg) + **5** (86 mg)

Asymmetric hydrogenation. Raney nickel that has been soaked at 100° in an aqueous solution of tartaric acid and sodium bromide can effect asymmetric hydrogenation of methyl acetoacetate to methyl 3-hydroxybutyrate[6] and of acetylacetone to 2,4-pentanediol.[7] In both cases optical yields near 90% can be achieved.

[1] H.-J. Liu, H. K. Lai, and S. K. Attah-Poku, *Tetrahedron Letters*, 4121 (1979).
[2] *Idem.*, *Syn. Comm.*, **9**, 883 (1979).
[3] B. Glatz, G. Helmchen, H. Muxfeldt, H. Porcher, R. Prewo, J. Senn, J. J. Stezowski, R. J. Stojda, and D. R. White, *Am. Soc.*, **101**, 2171 (1979).
[4] I. T. Harrison and S. Harrison, *Compendium of Organic Synthetic Methods*, Vol. 1, Wiley-Interscience, New York, 1971, p. 161.
[5] F. C. Chang, *J. Org.*, **44**, 4567 (1979).
[6] T. Harada and Y. Izumi, *Chem. Letters*, 1195 (1978).
[7] K. Ito, T. Harada, A. Tai, and Y. Izumi, *ibid.*, 1049 (1979).

Rhodium(II) carboxylates, 5, 571–572; **7,** 313; **8,** 434–435.

Cyclopropanation.[1] Rhodium(II) pivalate is superior to copper catalysts for the cyclopropanation of 1,1-dihalo-4-methylpenta-1,3-dienes (**1**) with ethyl diazoacetate. The yield is higher than that obtained with copper catalysts (48%), and the *cis/trans* ratio is considerably higher. Similar contrathermodynamic product ratios were observed with other substrates.

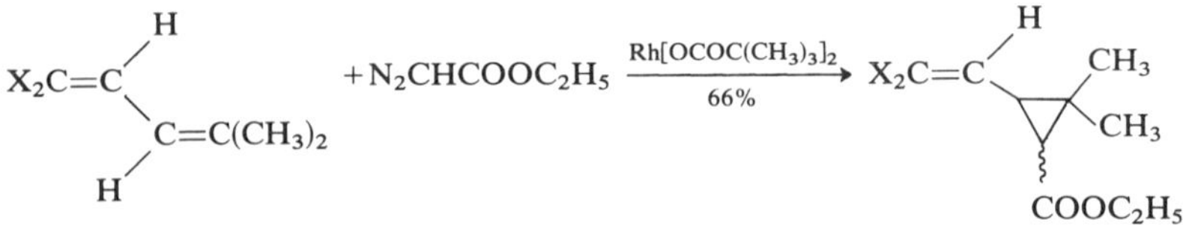

$$X_2C{=}CH{-}CH{=}C(CH_3)_2 + N_2CHCOOC_2H_5 \xrightarrow[66\%]{Rh[OCOC(CH_3)_3]_2} X_2C{=}CH{-}\text{cyclopropane}(CH_3)_2(COOC_2H_5)$$

1 (X = Br, Cl) **2** (*cis/trans* = 65 : 35)

β-Lactams.[2] Diazo compounds **1** and **2** are cyclized to β-lactams **3** and **4**, respectively, by $Rh_2(OAc)_4$ in dichloromethane (75% yield). With Cu in toluene (90°) the yields are lower (25%). Irradiation of **1** affords **3** (55%), whereas similar treatment of **2** affords **4** in 73% yield. Ketone **3** is converted into **5** on $NaBH_4$ reduction (1:1 mixture of epimers). This intermediate is a known precursor of thienamycin (**6**), an unusually potent β-lactam antibiotic.[3]

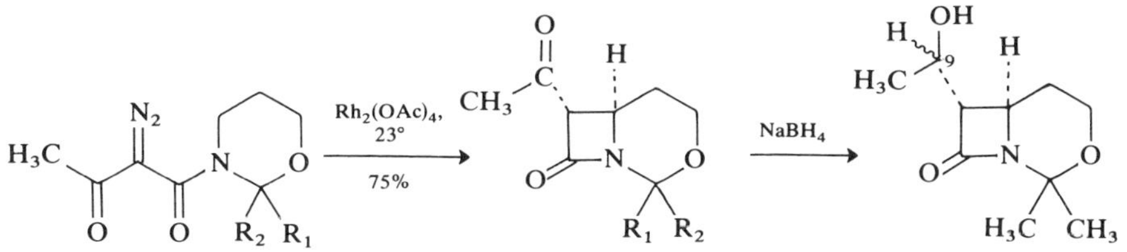

$Rh_2(OAc)_4$, 23°, 75%; $NaBH_4$

1 ($R_1 = R_2 = CH_3$)
2 [($R_1, R_2 = -(CH_2)_5-$)]

3
4

5

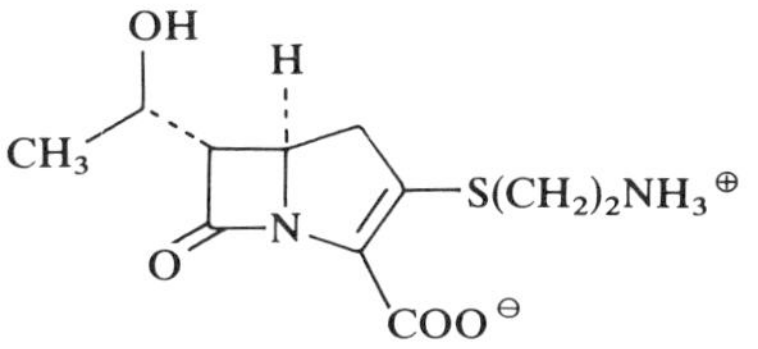

$-S(CH_2)_2NH_3^{\oplus}$, $COO^{\ominus}$

6

β-Keto esters from α-diazo-β-hydroxy esters.[4] α-Diazo-β-hydroxy esters, conveniently prepared by addition of LDA to a mixture of an aldehyde and ethyl diazoacetate in THF at −78°,[5] are rearranged to β-keto esters with catalysis by rhodium(II) acetate (equation I; yields are nearly quantitative). This rearrangement is also conducted at reflux in the presence of HCl or by vacuum thermolysis. The latter two methods are unsatisfactory when sensitive functionality is present in the R

$$\text{(I)}\quad R{-}CHO + N_2CH{-}CO_2C_2H_5 \xrightarrow{\text{LDA}} R{-}CH(OH){-}C(=N_2){-}CO_2C_2H_5 \xrightarrow{Rh_2(OAc)_4(\text{cat.})} R{-}C(=O){-}CH_2{-}CO_2C_2H_5$$

group. It is noteworthy of the Rh(II)-catalyzed reaction that α-diazo-β-hydroxy-γ,δ-unsaturated esters rearrange smoothly to β-keto esters. These highly reactive compounds give complex reaction mixtures when treated with acid, and afford β-keto esters in low yield when pyrolyzed (285°, 0.25 mm.).[6]

Examples:

$$(CH_3)_2CHCH(OH)C(=N_2)CO_2C_2H_5 \xrightarrow[100\%]{\text{Rh(II) cat.}} (CH_3)_2CHC(=O)CH_2CO_2C_2H_5$$

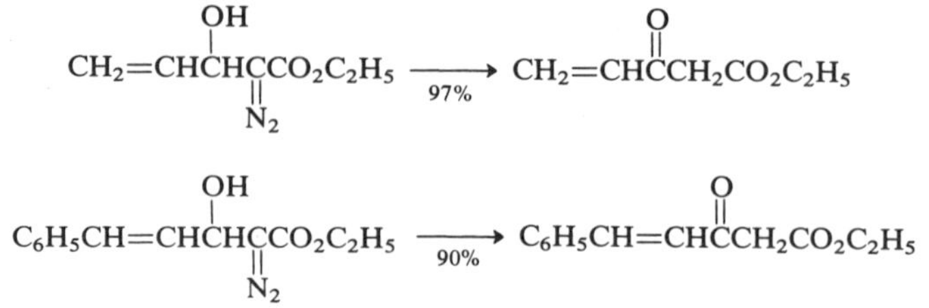

$$CH_2{=}CHCH(OH)C(=N_2)CO_2C_2H_5 \xrightarrow[97\%]{} CH_2{=}CHC(=O)CH_2CO_2C_2H_5$$

$$C_6H_5CH{=}CHCH(OH)C(=N_2)CO_2C_2H_5 \xrightarrow[90\%]{} C_6H_5CH{=}CHC(=O)CH_2CO_2C_2H_5$$

[1] D. Holland and D. J. Milner, *J. Chem. Res.* (*M*), 3734 (1979).
[2] R. J. Ponsford and R. Southgate, *J.C.S. Chem. Comm.*, 846 (1979).
[3] D. B. R. Johnston, S. M. Schmitt, F. A. Bouffard, and B. G. Christensen, *Am. Soc.*, **100**, 313 (1978).
[4] R. Pellicciari, R. Fringuelli, P. Ceccherelli, and E. Sisani, *J.C.S. Chem. Comm.*, 959 (1979).
[5] R. Pellicciari, E. Castagnino, and S. Corsano, *J. Chem. Res.* (*S*), 76 (1979).
[6] E. Wenkert, P. Ceccherelli, and A. A. Fugiel, *J. Org.*, **43**, 3982 (1978).

Rhodium oxide–Platinum oxide (Nishimura catalyst).

Hydrogenation of aromatic rings. This catalyst is prepared by fusion of a mixture of rhodium(III) chloride and chloroplatinic acid (3:1 by weight) with sodium nitrate. It is a particularly effective catalyst for hydrogenation of aromatic rings without hydrogenolysis of C—O bonds. For example, acetophenone is reduced with this catalyst to cyclohexylmethylcarbinol in 80% yield.[1] Corey *et al.*[2] used this reduction to convert **1** to **2** as the major product in a total synthesis of the sesquiterpene 9-isocyanopupukeanane (**3**). The hydrogenation step results in *cis*-ring fusion with a *trans*-relationship between the hydrogens at the ring fusion and the lactone bridge.

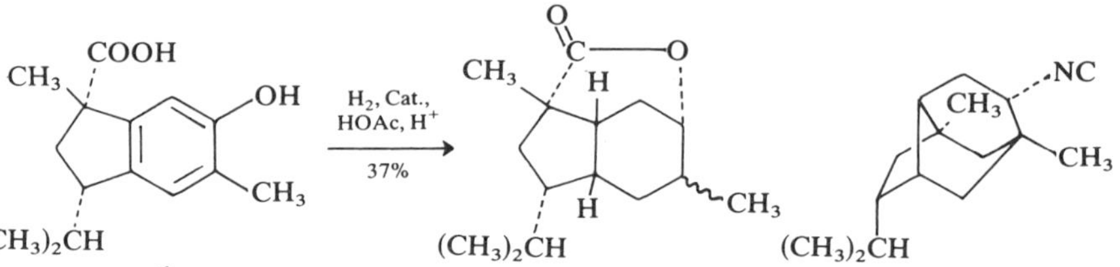

COOH, CH_3, OH, CH_3, $(CH_3)_2CH$ — **1** $\xrightarrow[37\%]{H_2,\ Cat.,\ HOAc,\ H^+}$ **2** (CH_3, H, H, CH_3, $(CH_3)_2CH$, C=O lactone) ; **3** (NC, CH_3, CH_3, $(CH_3)_2CH$)

[1] S. Nishimura, *Bull. Chem. Soc. Japan*, **33**, 566 (1960); **34**, 32 (1961).
[2] E. J. Corey, M. Behforouz, and M. Ishiguro, *Am. Soc.*, **101**, 1608 (1979).

S

Selenium dioxide, 1, 992–1000; **2**, 360–362; **3**, 245–247; **4**, 422–424; **5**, 575–576; **6**, 509–510; **7**, 319; **8**, 439–440.

Nitriles.[1] Aldoximes are dehydrated to nitriles by selenium dioxide in chloroform. The reaction can be used in a one-flask conversion of aldehydes to nitriles. For example, the aldehyde is allowed to react with hydroxylamine (1 equiv.) and pyridine (1 equiv.) in refluxing DMF for 2–3 hours. Selenium dioxide (1 equiv.) is added in portions (exothermic reaction). If the water is removed efficiently, 10 mole % of SeO_2 is sufficient, although yields are somewhat lower (69–80%).

$$\text{RCHO} \xrightarrow[-\text{H}_2\text{O}]{\text{HONH}_2} \text{RCH=NOH} \xrightarrow{\text{SeO}_2} \text{RCH=NOSeOH (Se=O)} \xrightarrow[70\text{–}89\%]{\Delta} \text{RC}\equiv\text{N} + \text{H}_2\text{SeO}_3$$

Oxidation of pyrazolines.[2] Hydroxyaryl pyrazolines (**1**) are oxidized by selenium dioxide in aqueous dioxane to 2′-hydroxychalcones (**2**) in about 50% yield. Use of other common oxidants (LTA, MnO_2, $KMnO_4$, etc.) results in the corresponding pyrazoles.

1 (R-substituted 2-hydroxyphenyl pyrazoline, N–NH, Ar) $\xrightarrow[\sim 50\%]{\text{SeO}_2,\ -\text{N}_2}$ **2** (R-substituted 2-hydroxyphenyl–C(=O)–CH=CHAr)

1,2,3-Selenadiazoles. These substances have been prepared as precursors to cycloalkynes, such as cyclooctyne (equation I).[3]

(I) Cyclooctanone $\xrightarrow{\text{H}_2\text{NNHCNH}_2\ (\text{C=O})}$ cyclooctanone semicarbazone (=N–NH–C(=O)NH₂) $\xrightarrow[81\%]{\text{SeO}_2}$ cycloocta-1,2,3-selenadiazole $\xrightarrow[99\%]{\Delta}$ cyclooctyne

The method has now been extended to preparation of *cis*-cyclooctenyne-3 (equation II)[4] and 5-substituted cyclooctynes (equation III)[5]

(II)

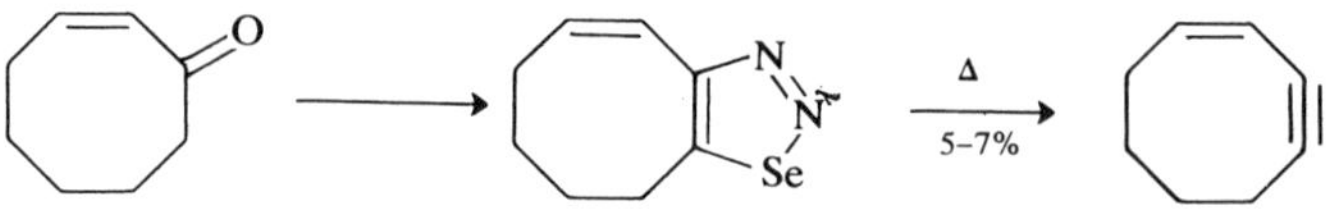

Δ

5–7%

(III)

Δ

22%

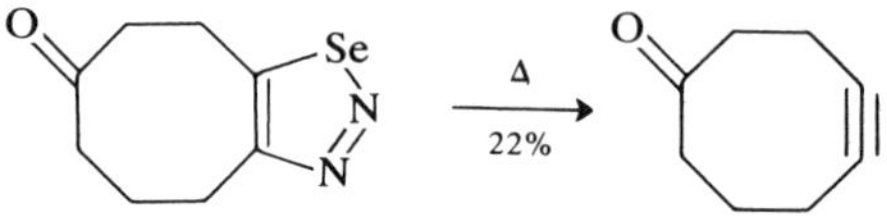

[1] G. Sosnovsky and J. A. Krogh, *Synthesis*, 703 (1978); G. Sosnovsky, J. A. Krogh, and S. G. Umhoefer, *ibid.*, 722 (1979).
[2] D. D. Berge and A. V. Kale, *Chem. Ind.*, 662 (1979).
[3] H. Meier and I. Menzel, *J.C.S. Chem. Comm.*, 1059 (1971); H. Meier and E. Voight, *Tetrahedron*, **28**, 187 (1972).
[4] H. Petersen, H. Kolshorn, and H. Meier, *Angew. Chem., Int. Ed.*, **17**, 461 (1978).
[5] H. Meier and H. Petersen, *Synthesis*, 596 (1978).

Silica gel, SiO_2. Suppliers: Alfa, Merck, Mallinckrodt, and others.

Dehydrochlorination. Exposure of the chlorides **1**[1] and **3**[2] to silica gel results in selective dehydrochlorination to the isopropenyl derivatives **2** and **4**, respectively. Eliminations carried out with other reagents produce mixtures contaminated with the isopropylidene isomers.

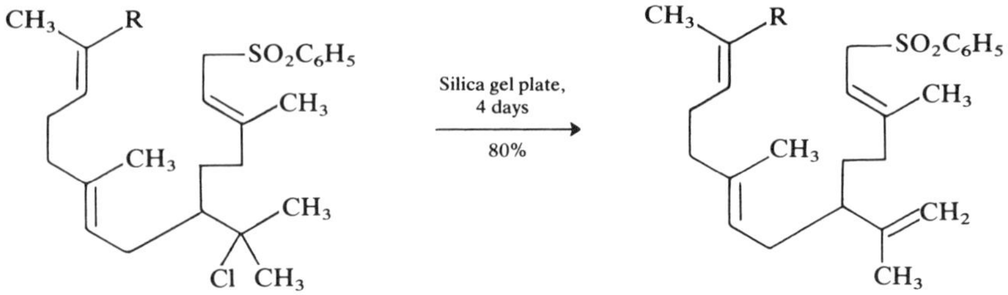

Silica gel plate, 4 days

80%

1 **2**

SiO_2, hexane

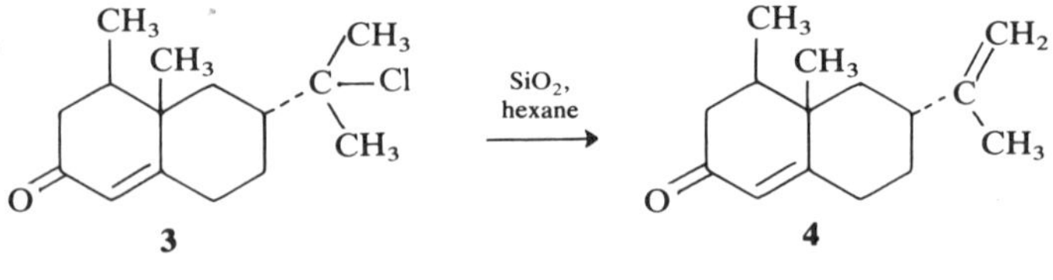

3 **4**

[1] H. Takayanagi, T. Uyehara, and T. Kato, *J.C.S. Chem. Comm.*, 359 (1978).
[2] T. Yanami, M. Miyashita, and A. Yoshikoshi, *ibid.*, 525 (1979).

Silver iododibenzoate, 1, 1007–1008.

Prévost reaction. This reaction constitutes an important step in a synthesis of *trans*-7,8-dihydroxy-*anti*-9,10-epoxy-7,8,9,10-tetrahydrobenzo[*a*]pyrene (**3**), considered to be the ultimate carcinogenic metabolite of benzo[*a*]pyrene. To obtain satisfactory yields, the silver benzoate should be freshly prepared (method of Newman and Beal, **1**, 1004), and the biphase reaction should be stirred efficiently (Vibromixer with a large stirring blade).[1]

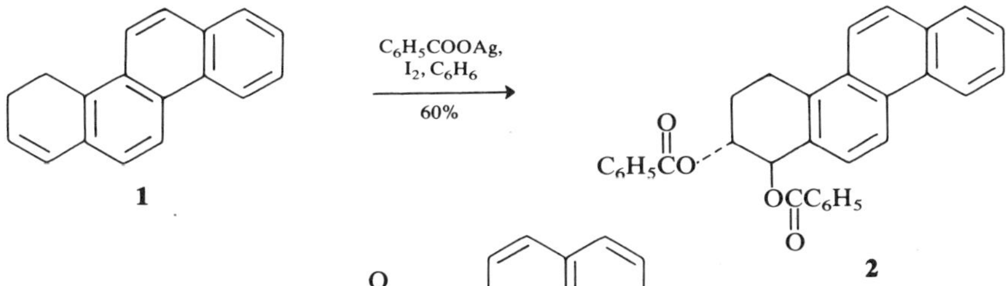

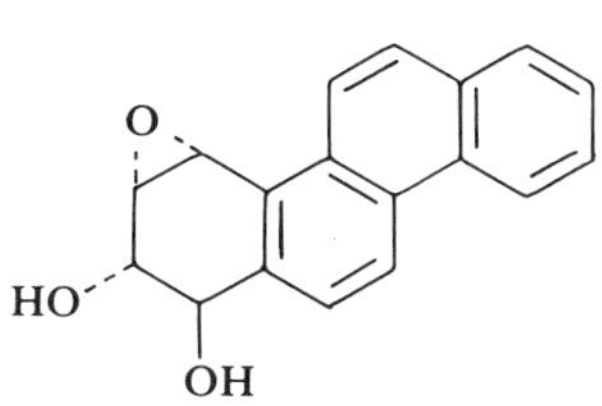

C_6H_5COOAg, I_2, C_6H_6 → 60%

1 → **2** ($C_6H_5\overset{O}{\overset{\|}{C}}O$, $O\underset{O}{\underset{\|}{C}}C_6H_5$)

3 (O, HO, OH)

[1] B. C. Pal and R. P. Iyer, *Org. Syn.*, submitted (1979).

Silver nitrate, 1, 1008–1011; **2,** 366–368; **3,** 252; **4,** 429–430; **5,** 582; **7,** 321.

N,N-Methyltosylhydrazones. Ketones react only slowly with N,N-methyltosylhydrazine (**1**); but hydrazones of this type can be prepared by N-alkylation of tosylhydrazones with phase-transfer catalysis.[1] The more reactive thioketones also react sluggishly with **1**, but in this case, the reaction can be catalyzed by soft Lewis acids.[2] Thus the reaction of **2** with **1** proceeds in high yield at room temperature in the presence of 1 equiv. of silver nitrate. Mercuric acetate also promotes this reaction, but the yield of **3** is only 50% because of formation also of **4** in 44% yield.

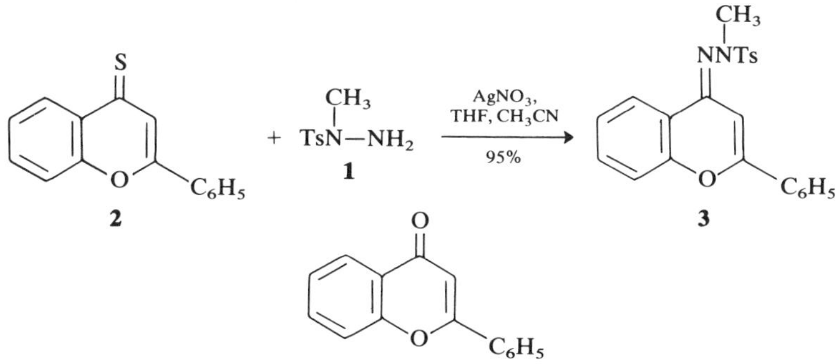

2 (S, O, C_6H_5) + $TsN(CH_3)—NH_2$ **1** → ($AgNO_3$, THF, CH_3CN; 95%) **3** (CH_3, NNTs, O, C_6H_5)

4 (O, O, C_6H_5)

[1] S. Cacchi, F. La Torre, and D. Misiti, *Synthesis*, 301 (1977).
[2] *Idem*, *Chem. Ind.*, 669 (1978).

Silver(II) oxide, 1, 1011–1015; **2**, 368; **3**, 252–254; **4**, 430–431; **5**, 583–585; **6**, 515–518; **7**, 321–322.

Intramolecular oxidative coupling of phenols (**6**, 516).[1] A biomimetic synthesis of natural silybin (**3**), an interesting antihepatotoxic agent, involves oxidation of a mixture of optically active **1** and coniferyl alcohol (**2**) with AgO in a benzene–acetone mixture at 55°. This reaction affords a mixture of silybin (**3**) and isosilybin (**4**) in 78% yield. The products **3** and **4** are diastereomeric mixtures at $C_{2'}$ and $C_{3'}$ relative to C_2 and C_3.[2]

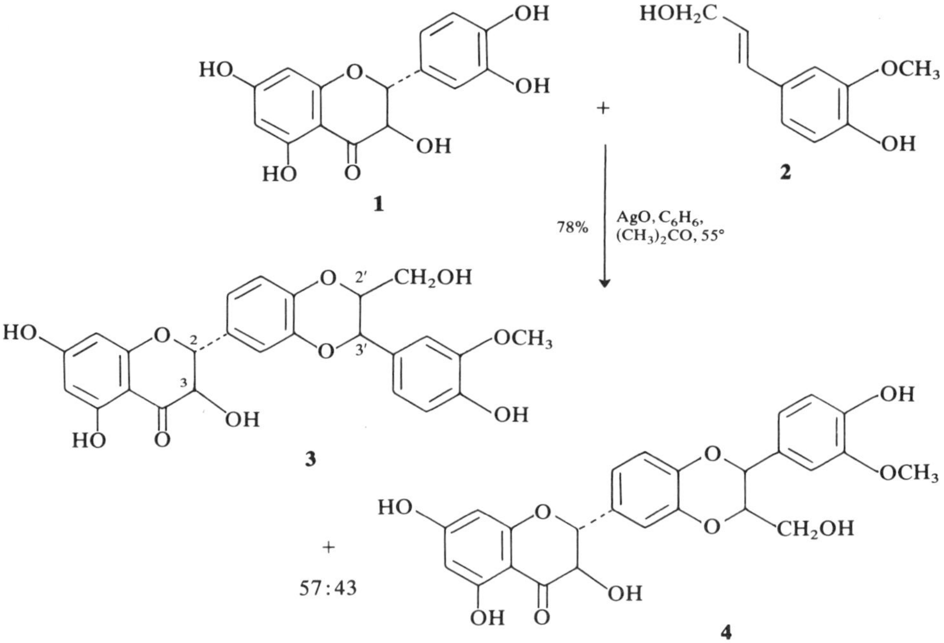

Oxidative demethylation.[3] A new approach to indoloquinones such as **5** involves Michael addition of ethyl acetoacetate to the quinone monoimide **1** to give **2**, which is dehydrated to the indole **3** in 73% overall yield. The latent quinone ring is then modified to give the *p*-methoxyaniline **4**. The final step involves the oxidative demethylation reaction of Rapoport (**4**, 431–432) to give an intermediate quinone imine, which is hydrolyzed to **5**.

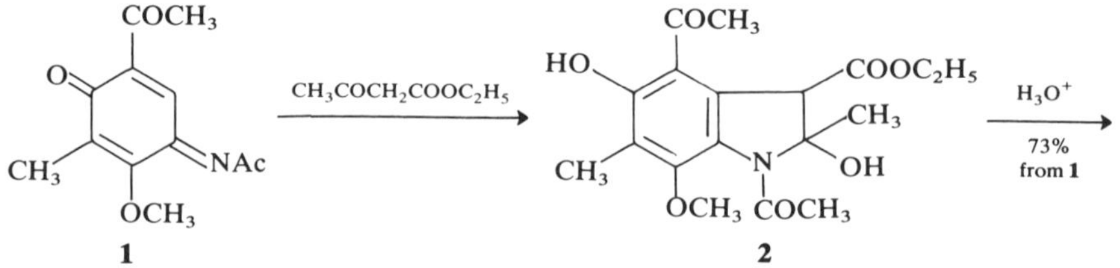

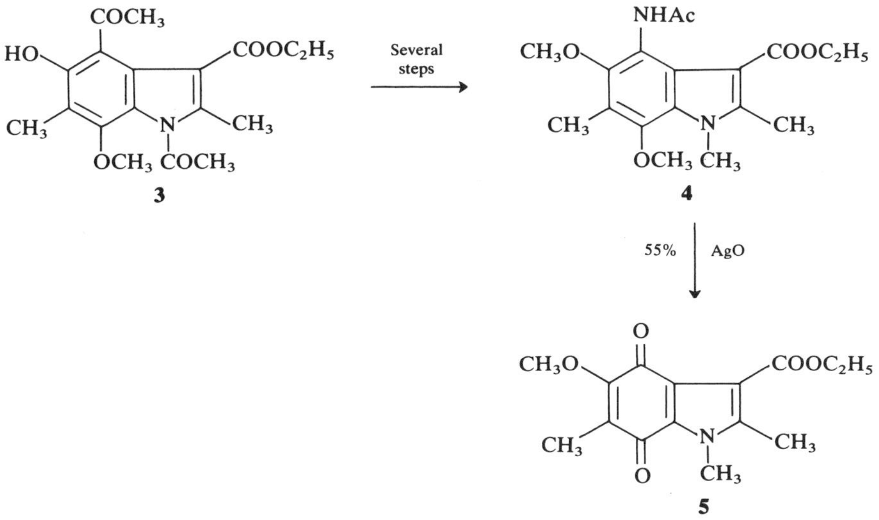

COCH$_3$ HO COOC$_2$H$_5$ CH$_3$ N CH$_3$ OCH$_3$ COCH$_3$ **3** —Several steps→ NHAc CH$_3$O COOC$_2$H$_5$ CH$_3$ N CH$_3$ OCH$_3$ CH$_3$ **4** —55%, AgO→ O CH$_3$O COOC$_2$H$_5$ CH$_3$ N CH$_3$ O CH$_3$ **5**

[1] L. Merlini, A. Zanarotti, A. Pelter, M. P. Rochefort, and R. Hänsel, *J.C.S. Chem. Comm.*, 695 (1979).
[2] A. Arnone, L. Merlini, and A. Zanarotti, *ibid.*, 696 (1979).
[3] K. A. Parker and S.-K. Kang, *J. Org.*, **44**, 1536 (1979).

Silver perchlorate, 2, 369–370; **4**, 432–435; **5**, 585–587; **6**, 518–519; **7**, 322–323.

Lactonization of ω-hydroxy esters to macrolides. Gerlach and co-workers have reported that silver salts can catalyze the cyclization of ω-hydroxy-2-pyridinethiol esters to macrolides (**6**, 246; **7**, 142). Nimitz and Wollenberg[1] have found that silver perchlorate has an even more pronounced and consistent effect on the cyclization of ω-hydroxy thiolesters of 2-amino-4-mercapto-6-methylpyrimidine. Optimal yields are obtained with use of 1.1 equiv. of the silver salt. In addition, high dilution is not necessary. In the example cited, the yield of the lactone in the absence of the catalyst is 41%.

Example:

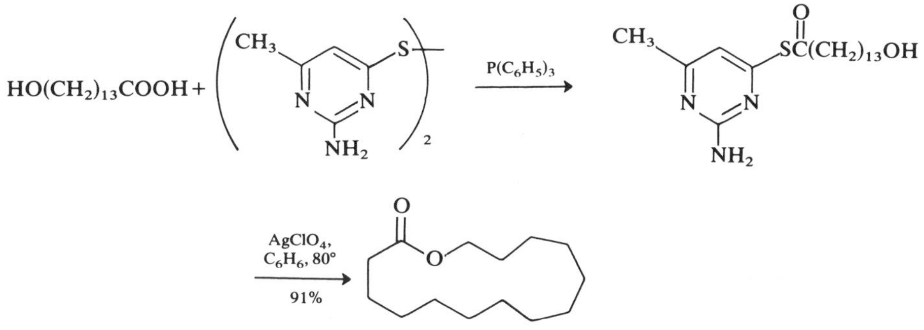

HO(CH$_2$)$_{13}$COOH + (CH$_3$, N, N, NH$_2$, S)$_2$ —P(C$_6$H$_5$)$_3$→ CH$_3$, N, N, NH$_2$, SC(=O)(CH$_2$)$_{13}$OH —AgClO$_4$, C$_6$H$_6$, 80°, 91%→ O, O

Dehydration.[2] Treatment of **1** with this salt gives a mixture of **2** and **3**, which is isomerized to the more stable **3** by acid.[3]

OH, $(CH_2)_6COOCH_3$ (**1**) $\xrightarrow[66\%]{AgClO_4, C_6H_6, 23°}$ $CH(CH_2)_5COOCH_3$ (**2**) + $(CH_2)_6COOCH_3$ (**3**) 1:2

2 $\xrightarrow[80\%]{HCl, n\text{-BuOH}, \Delta}$ **3**

[1] J. S. Nimitz and R. H. Wollenberg, *Tetrahedron Letters*, 3523 (1978).
[2] T. Mukaiyama, S. K. Kobayashi, K. Kamino, and H. Takei, *Chem. Letters*, 237, 287 (1972).
[3] T. Wakamatsu, K. Hashimoto, M. Ogura, and Y. Ban, *Syn. Comm.*, **8**, 319 (1978).

Silver tetrafluoroborate, 1, 1015–1018; **2**, 365–366; **3**, 250–251; **4**, 428–429; **5**, 587–588; **6**, 519–520; **8**, 443–444.

α-Fluoro aldehydes and ketones.[1] These compounds can be prepared by reaction of tertiary α-bromo aldehydes or ketones with silver tetrafluoroborate in ether. Examples:

$$CH_3COC(Br)(CH_3)_2 \xrightarrow[87\%]{AgBF_4} CH_3COC(F)(CH_3)_2 + AgBr + BF_3$$

$$(CH_3)_2C(Br)CHO \xrightarrow[70\%]{} (CH_3)_2C(F)CHO$$

[1] A. J. Fry and Y. Migron, *Tetrahedron Letters*, 3357 (1979).

Silver(I) trifluoromethanesulfonate, 6, 520–521; **7**, 324; **8**, 444–445.

Iminium ions. N-Arylsulfonyliminium ions are generated from N-arylsulfonyl-α-amino acid chlorides on treatment with $AgSO_3CF_3$. These iminium ions are very reactive and readily undergo nucleophilic addition or intramolecular Friedel–Crafts reactions.[1]

$ArSO_2N$ (COCl) $\xrightarrow[-CO, -AgCl]{AgSO_3CF_3}$ $[ArSO_2-N^{\oplus}\ \ {}^{\ominus}OSO_2CF_3]$ $\xrightarrow[40\%]{NH_3}$ $ArSO_2-N$ (NH_2)

SO_2-N (COCl) $\xrightarrow[84\%]{CCl_4, \Delta}$ SO_2, N

[1] E. K. Adesogan and B. I. Alo, *J.C.S. Chem. Comm.*, 673 (1979).

Simmons–Smith reagent, 1, 1019–1022; **2**, 371–372; **3**, 255–258; **4**, 436–437; **5**, 588–589; **6**, 521–523; **8**, 445.

α-Methylene ketones.[1] Treatment of enol acetates or trimethylsilyl enol ethers with the Simmons–Smith reagent (Zn–Cu couple, CH_2I_2) and an α-chloromethyl ether in CH_2Cl_2 affords β-alkoxy carbonyl compounds in 55–90% yield. Cyclopropanes are not formed in the reaction. Although the Simmons–Smith reagent gives the highest yields, iodine, iodoform or iodomethane can be substituted for CH_2I_2, and pure Zn can be used instead of Zn–Cu. Treatment of the resulting α-methoxymethyl ketones with $KHSO_4$ at 160–180° affords the corresponding α-methylene derivatives in 65–89% yield.

Examples:

CH_3, $OSi(CH_3)_3$ (cyclohexene) $\xrightarrow[69\%]{CH_2I_2,\ Zn\text{–}Cu,\ ClCH_2OCH_3}$ CH_3, O, CH_2OCH_3 (cyclohexanone) $\xrightarrow[84\%]{KHSO_4,\ \Delta}$ CH_3, O, $=CH_2$ (cyclohexanone)

$C_6H_5C(OAc)=CHCH_3$ $\xrightarrow[63\%]{CH_2I_2,\ Zn\text{–}Cu,\ C_2H_5OCH(Cl)CH_3}$ $C_6H_5C(=O)CH(CH_3)CH(OC_2H_5)CH_3$ $\xrightarrow{84\%}$ $C_6H_5C(=O)C(=CHCH_3)CH_3$

[1] T. Shono, I. Nishiguchi, T. Komamura, and M. Sasaki, *Am. Soc.*, **101**, 984 (1979).

Sodium–Ammonia, 1, 1041; **2**, 374–376; **3**, 259; **4**, 438; **5**, 589–591; **6**, 523; **7**, 324–325.

Cleavage of benzyl ethers and esters. In a recent liquid-phase synthesis of β-endorphin,[1] a peptide containing 31 amino acids that possesses potent morphine-like analgesic activity,[2] the CBZ group was used for protection of lysine units and the benzyl group was used for protection of hydroxyl groups of serine and threonine and for protection of the carboxyl group of glutamic acid. The final step involved cleavage of the protecting groups. Use of liquid hydrogen fluoride (**2**, 216) resulted in extensive degradation, and the free peptide was obtained in only 13% yield at most. Actually one of the earliest methods of deprotection, sodium in liquid ammonia,[3] was found most satisfactory, but a tenfold excess of sodium was required to prevent ammonolysis of the benzyl ester (OBzl) groups. After cleavage and a single reversed-phase preparative liquid chromatographic step, β-endorphin was obtained in 27.5% yield.

[1] C. Tzougraki, R. C. Makofse, T. F. Grabiel, S.-S. Wang, R. Kutney, J. Meienhofer, and C. H. Li, *Am. Soc.*, **100**, 6248 (1978).
[2] C. H. Li, *Arch. Biochem. Biophys.*, **183**, 592 (1977).
[3] R. H. Sifferd and V. du Vigneaud, *J. Biol. Chem.*, **108**, 753 (1935).

Sodium–Ethanol.

Dechlorination. Sodium in ethanol[1] is reported to be superior to either Li, *t*-butyl alcohol, and THF (**1**, 604–606) or Na, *t*-butyl alcohol, and THF (**2**, 378–379; **6**, 523–524)[2] for dehalogenation of polychlorinated alicyclic compounds.

[1] B. V. Lap and M. N. Paddon-Row, *J. Org.*, **44**, 4979 (1979).
[2] P. G. Gassman and J. L. Marshall, *Org. Syn.*, **48**, 68 (1968).

Sodium–Hexamethylphosphoric triamide.

Deoxygenation of alcohols. Esters are reduced to alkanes by photolysis in HMPT–H_2O in high yield. This reaction has been used to prepare deoxy sugars from O-acetylpyranoses and O-acetylfuranoses.[1] The reaction can also be effected with sodium in HMPT containing *t*-butyl alcohol as a proton source. Deoxygenation of tertiary alcohols to alkanes is quantitative; primary and secondary esters are reduced to mixtures of alkanes and alcohols.[2]

Examples:

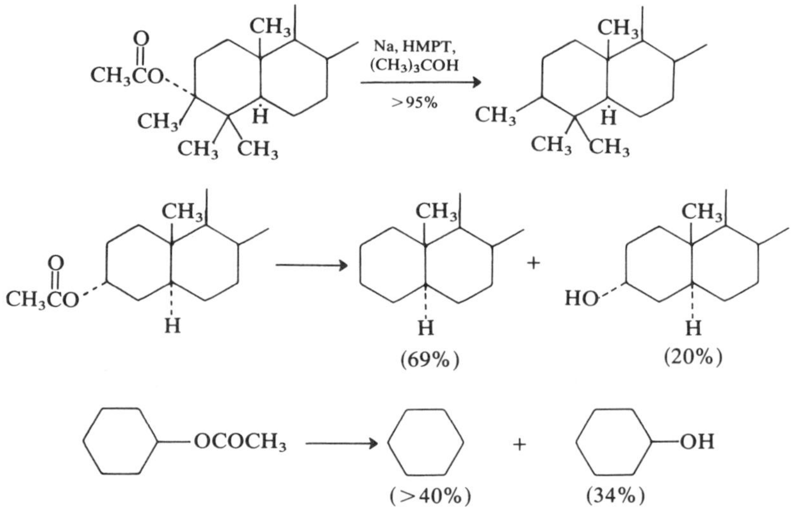

$$CH_3(CH_2)_5CH_2OCOCH_3 \rightarrow CH_3(CH_2)_5CH_3 + CH_3(CH_2)_5CH_2OH$$

(56%) (30%)

[1] J.-P. Pete, C. Portella, C. Monneret, J.-C. Florent, and Q. Khuong-Huu, *Synthesis*, 774 (1977).
[2] H. Deshayes and J.-P. Pete, *J.C.S. Chem. Comm.*, 567 (1978).

Sodium amalgam, 1, 1030–1033; **2**, 373; **3**, 259; **7**, 326; **8**, 171, 296.

Dialkyl alkynes. A recently reported route[1] to these substances involves reductive elimination of enol phosphates of β-oxo sulfones with sodium amalgam in DMSO–THF at 0°.

Examples:

$$\text{1-cyclohexenyl}-C(SO_2C_6H_5)=C(C_6H_5)OP(O)(OC_2H_5)_2 \xrightarrow[72\%]{Na(Hg),\ DMSO\text{-}THF} \text{1-cyclohexenyl}-C\equiv C-C_6H_5$$

$$(CH_3)_2C=CH-C(SO_2C_6H_5)=C(OP(O)(OC_2H_5)_2)CH=C(CH_3)_2 \xrightarrow{69\%} (CH_3)_2C=CH-C\equiv C-CH=C(CH_3)_2$$

[Structure: polyene enol phosphate bearing $(C_2H_5O)_2P(O)O$ and $SO_2C_6H_5$ groups, with CH_3 substituents]

↓ 78.5%

[Structure: corresponding alkyne with CH_3 substituents]

↓ H_2, Lindlar cat.

$$(CH_3)_2C=CH(CH_2)_2C(CH_3)=CH-CH=CH-CH=C(CH_3)(CH_2)_2CH=C(CH_3)_2$$

The last example illustrates use of this elimination as a route to an almost pure conjugated (E,Z,E)-triene.

A similar synthesis of alkynes by reductive elimination of enol phosphates of β-oxosulfones with sodium in liquid ammonia has been reported.[2]

[1] B. Lythgoe and I. Waterhouse, *Tetrahedron Letters*, 2625 (1978); *idem, J.C.S. Perkin I*, 2429 (1979).
[2] P. A. Bartlett, F. R. Green III, and E. H. Rose, *Am. Soc.*, **100**, 4852 (1978).

Sodium amide–Sodium *t*-butoxide, 4, 439–440; **5,** 593; **6,** 526; **7,** 327.

syn-*Elimination* (**5,** 593). The unusual order of leaving group reactivity F > Cl > Br is observed in the *syn*-elimination promoted by this complex base. Lee and

Bortsch[1] suggest that the strength of interaction between the reagent and the halogen is the dominant leaving group effect in these cases, rather than the usual dominance of the carbon–halogen bond strength.

Examples:

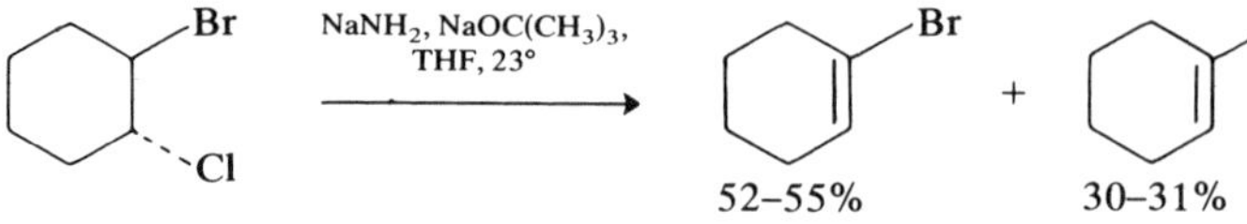

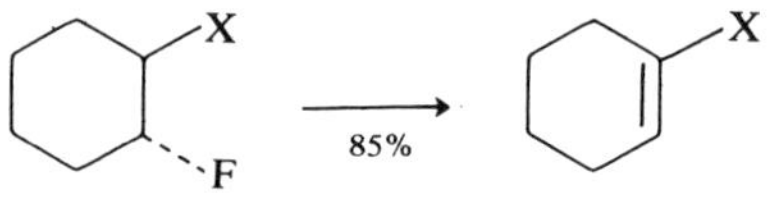

$NaNH_2$, $NaOC(CH_3)_3$, THF, 23°

52–55% 30–31%

85%

X = Cl, Br

[1] J. G. Lee and R. A. Bortsch, *Am. Soc.*, **101**, 229 (1979).

Sodium azide–Triphenylphosphine.

Aziridines.[1] Epoxides can be converted into aziridines with overall retention of configuration by reaction with NaN_3 to form an azido alcohol (inversion), which is converted into an aziridine by reaction with a tertiary phosphine (inversion), equation (I).

(I) NaN_3 88% $P(C_6H_5)_3$ 68%

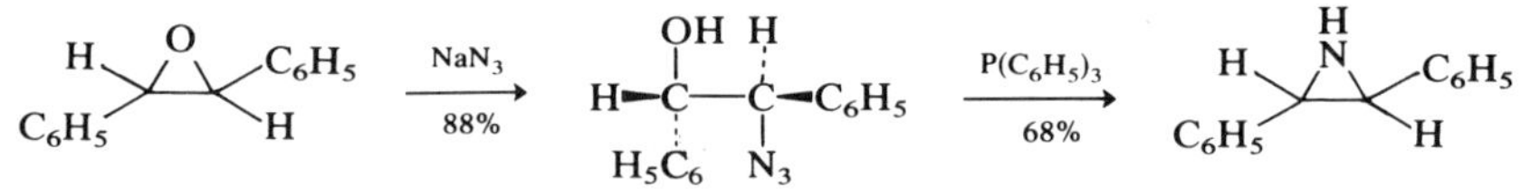

$+ N_2 + (C_6H_5)_3P{=}O$

This method was used to convert phenanthrene-9,10-oxide into the 9,10-imine in 72% yield. This imine is thermally stable at temperatures up to 190°.

[1] Y. I. Hah, Y. Sasson, I. Shahak, S. Tsaroom, and J. Blum, *J. Org.*, **43**, 4271 (1978).

Sodium bis(2-methoxyethoxy)aluminum hydride (SMEAH), 3, 260–261; **4,** 441–442; **5,** 596; **6,** 528–529; **7,** 327–329; **8,** 448–449.

Reduction of $ArCOOCH_3$ to $ArCH_3$. Saucy and co-workers[1] were able to effect the direct reduction of diester **1** to **2** using this reagent (*see also* **3**, 261). The phenol **2** is an intermediate in a total synthesis of vitamin E.

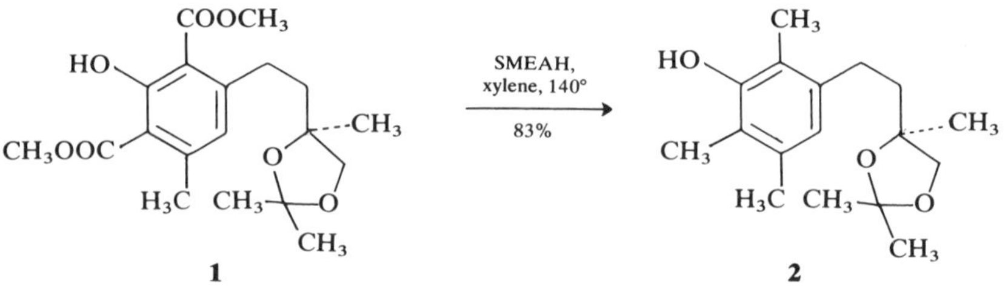

SMEAH, xylene, 140° 83%

1 **2**

Reductive cleavage. Selective scission of the oxymethylene bridge of **1** occurs on exposure to SMEAH in THF.[1] None of the alternate benzoxepin product (**3**) is formed, since only the β-oxygen is disposed perpendicularly to the cyclohexadienone moiety. The product is an intermediate in a synthesis of vitamin E (**4**).

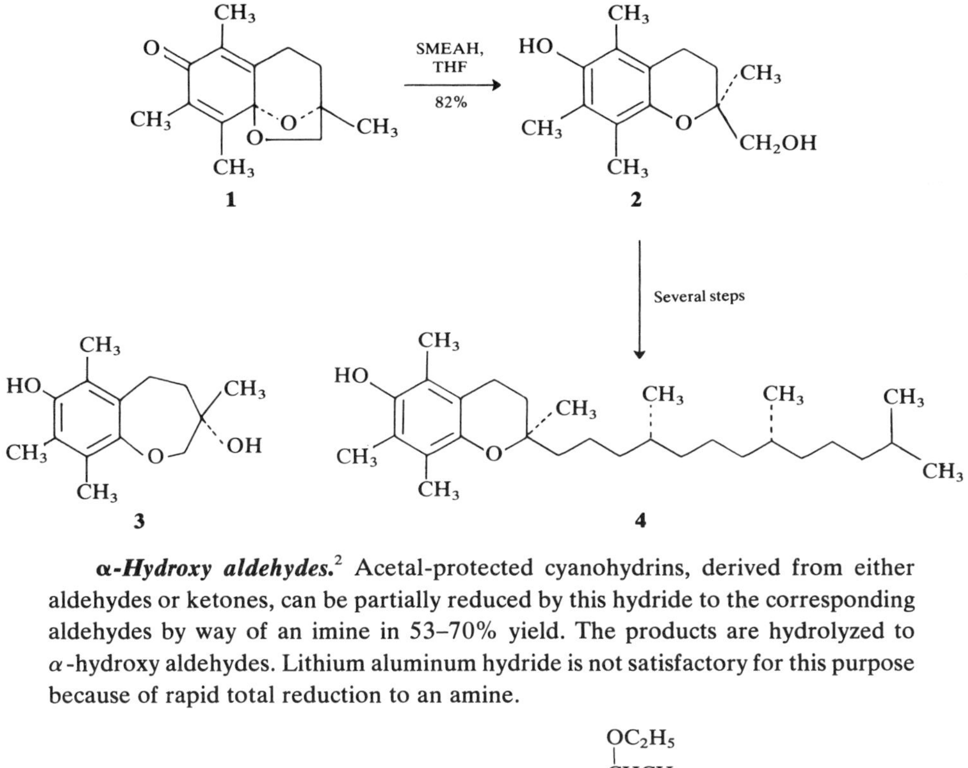

α-Hydroxy aldehydes.[2] Acetal-protected cyanohydrins, derived from either aldehydes or ketones, can be partially reduced by this hydride to the corresponding aldehydes by way of an imine in 53–70% yield. The products are hydrolyzed to α-hydroxy aldehydes. Lithium aluminum hydride is not satisfactory for this purpose because of rapid total reduction to an amine.

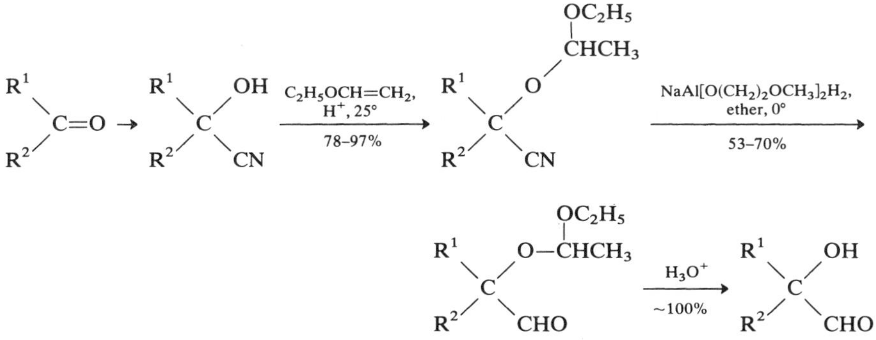

Reduction of α-formyl ketones. This reagent chemoselectively reduces the sodium salt of **1** to the α-hydroxymethyl ketone **2**.[3] Similar reductions have been effected with aluminum hydride (**4**, 233).

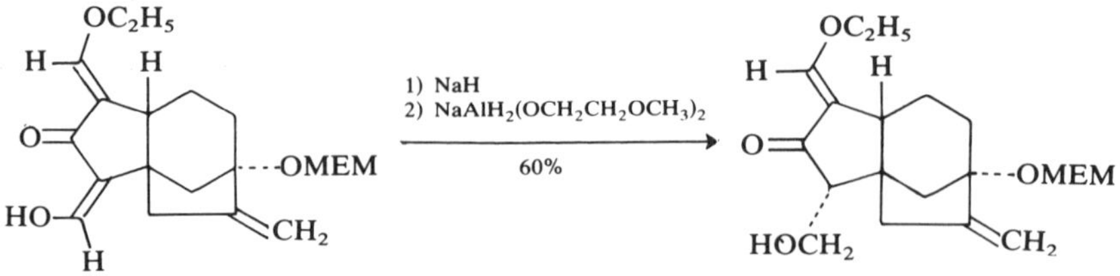

1) NaH
2) $NaAlH_2(OCH_2CH_2OCH_3)_2$
60%

[1] N. Cohen, R. J. Lopresti, and G. Saucy, *Am. Soc.*, **101**, 6710 (1979).
[2] M. Schlosser and Z. Brich, *Helv.*, **61**, 1903 (1978).
[3] E. J. Corey and J. G. Smith, *Am. Soc.*, **101**, 1038 (1979).

Sodium borohydride, 1, 1049–1055; **2**, 377–378; **3**, 262–264; **4**, 443–444; **5**, 597–601; **6**, 530–534; **7**, 329–331; **8**, 449–451.

Reduction of* ortho-*quinones. Conventional reduction of non-K-region *o*-quinones such as **1** with $LiAlH_4$ affords the corresponding dihydrodiols in very low yield. Kundu[1] has found that reduction of the dibromo derivatives of these quinones directly yields mixtures of *cis*- and *trans*-dihydrodiols.

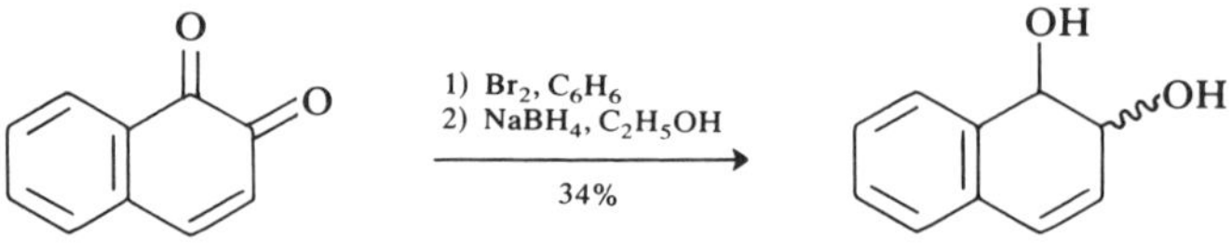

1) Br_2, C_6H_6
2) $NaBH_4, C_2H_5OH$
34%

β-Amino alcohols. Some years ago Yamada *et al.*[2] reported that cyano groups adjacent to a *tert*-amino group are reduced in high yield by sodium borohydride in an alcoholic medium. This reaction coupled with the recent finding that α-cyano derivatives of *tert*-amines can be alkylated at the α-position (*see* N,N-Diethylaminoacetonitrile, this volume) is used as a general route to both *sec*- and *tert*-β-amino alcohols.[3] An example of the method is shown in equation (I). The sequence involves an N to O shift of the benzoyl group via an intermediate oxazolidine. One noteworthy feature is that the product is formed with high stereoselectivity; in this case the *erythro* to *threo* ratio is 85:15.

(I) C_6H_5COCl, 82%; 1) LDA, −78°, 2) C_6H_5CHO; $NaBH_4$, 0°; CH_3OH, 70% overall; 85:15

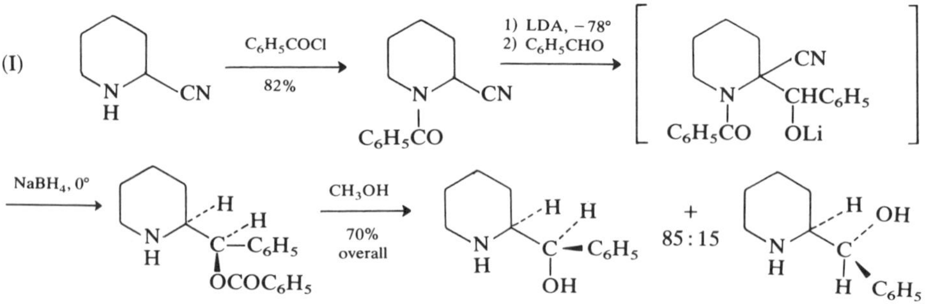

The reaction is also convenient for synthesis of acyclic β-amino alcohols. Thus reaction of **1** with benzaldehyde followed by reduction gives a mixture of (±)-ephedrine (**2**, *erythro*) and (±)-ψ-ephedrine (**3**, *threo*) in the ratio 3.3:1. The opposite stereoselectivity obtains when the reaction is used to prepare a tertiary amino alcohol. Thus the same sequence applied to **4** leads mainly to the *threo*-isomer (**5**, N-methyl-ψ-ephedrine).

$$\underset{\mathbf{1}}{CH_3CH(CN)N(CH_3)COC_6H_5} \xrightarrow[75\%]{\text{1) LDA; 2) } C_6H_5CHO\text{; 3) } NaBH_4, CH_3OH} \underset{\mathbf{2}}{CH_3CH(NHCH_3)CH(OH)C_6H_5\ (erythro)} \overset{3.3:1}{+} \underset{\mathbf{3}}{CH_3CH(NHCH_3)CH(OH)C_6H_5\ (threo)}$$

$$\underset{\mathbf{4}}{CH_3CH(CN)N(CH_3)_2} \xrightarrow[93\%]{} \underset{\mathbf{5}}{CH_3CH(N(CH_3)_2)CH(OH)C_6H_5\ (threo)} \overset{5:1}{+} \underset{\mathbf{6}}{CH_3CH(N(CH_3)_2)CH(OH)C_6H_5\ (erythro)}$$

Reduction of diaryl ketones. Diaryl ketones or diarylmethanols are reduced to diarylmethanes by $NaBH_4$ in trifluoroacetic acid (equation I.)[4]

$$\text{(I)} \qquad ArCOAr' \xrightarrow[75\text{–}95\%]{NaBH_4,\ CF_3COOH} ArCH_2Ar'$$

[1] N. G. Kundu, *J.C.S. Chem. Comm.*, 564 (1979).
[2] S. Yamada and H. Akimoto, *Tetrahedron Letters*, 3105 (1969).
[3] G. Stork, R. M. Jacobson, and R. Levitz, *ibid.*, 771 (1979).
[4] G. W. Gribble, R. M. Reese, and B. E. Evans, *Synthesis*, 172 (1977); G. W. Gribble, W. J. Kelly, and S. E. Emery, *ibid.*, 763 (1978).

Sodium borohydride–Alumina.

Reduction of acid chlorides.[1] Sodium borohydride impregnated on alumina reduces acid chlorides to primary alcohols (equation I).

$$\text{(I)} \qquad RCOCl \xrightarrow[80\text{–}90\%]{NaBH_4, Al_2O_3} RCH_2OH$$

[1] E. Santaniello, C. Farachi, and A. Manzocchi, *Synthesis*, 912 (1979).

Sodium borohydride–Bis(2,4-pentanedionato)copper.

Reduction of nitroarenes.[1] Sodium borohydride can reduce nitroarenes to the corresponding amines in 80–90% isolated yield in the presence of transition metal acetylacetonates, of which $Cu(acac)_2$ is the most effective.

[1] K. Hanaya, T. Muramatsu, H. Kudo, and Y. L. Chow, *J.C.S. Perkin I*, 2409 (1979).

Sodium borohydride–Cerium(III) chloride, 8, 451–452.

Selective reduction of ketones. Luche and Gemal[1] have reported selective reduction of ketones in the presence of aliphatic aldehydes using sodium borohydride in combination with cerium(III) chloride. This reagent also effects selective reduction of conjugated aldehydes in the presence of saturated ones.

$NaBH_4$, $CeCl_3 \cdot 6H_2O$, H_2O, C_2H_5OH; 75%

[1] J.-L. Luche and A. L. Gemal, *Am. Soc.*, **101**, 5848 (1979).

Sodium borohydride–1,2;5,6-Di-O-isopropylidene-α-D-glucofuranose. This reagent is prepared *in situ* from 1 equiv. $NaBH_4$ and 2 equiv. of the glucofuranose in THF.

Asymmetric reduction.[1] Propiophenone is reduced by this reagent to 1-phenylpropanol in 25% enantiomeric excess. When the reduction is performed in the presence of 0.5 equiv. of $ZnCl_2$, the (S)-alcohol is formed in 53% enantiomeric excess. The optical yield increases to 88% if the reagent–$ZnCl_2$ mixture is aged for 48 hours prior to use. Other Lewis acids such as $MgCl_2$, $BaCl_2$, and $AlCl_3$ can be used, but optical yields from propiophenone do not exceed 54%. In these cases (R)-alcohols predominate. Acetophenone and butyrophenone are reduced to the corresponding (S)-alcohols in 41–44% optical yield ($ZnCl_2$, without aging), whereas the optical yield of (S)-alcohol from isobutyrophenone is only 5%.

[1] A. Hirao, S. Nakahama, D. Mochizuki, S. Itsumo, M. Ohowa, and N. Yamazaki, *J.C.S. Chem. Comm.*, 807 (1979).

Sodium chloride–Dimethylformamide.

Decarbomethoxylation.[1] A recent stereoselective synthesis of coccinelline (**4**), an alkaloid used by the ladybug for defense, has been achieved by use of a variant of the classical Robinson–Schopf synthesis of tropinone. The reaction of **1** with acetonedicarboxylic ester at pH 5.5 yields a single product in 75% yield, even though five chiral centers are produced. The next step, conversion of **2** to **3**, requires neutral conditions in order to avoid retro-Mannich or retro-Michael reactions. This decarbomethoxylation was accomplished most satisfactorily by NaCl in refluxing wet

pH 5.5; 75% — NaCl, H_2O, DMF, Δ; 56%

1 **2**

1) $(C_6H_5)_3P{=}CH_2$ (82%)
2) H_2, Pd–C; $ClC_6H_4CO_3H$ (69%)

3 → 4

DMF (4 hours). These conditions were more satisfactory than NaCl in wet DMSO (**4**, 445), NaCN in HMPT, KCl in DMF, or KI in DMSO.

[1] R. V. Stevens and A. W. M. Lee, *Am. Soc.*, **101**, 7032 (1979).

Sodium chlorite, 5, 603.

Oxidation of aldehydes.[1] Sodium chlorite proved far more selective than MnO_2 or AgO–NaCN (**2**, 369) for oxidation of the α,β-unsaturated aldehyde **1** to the acid **2** without attack on the secondary hydroxyl group. The lactonization of **2** also presented a problem because of the presence of the enone group. The pyridyl thiolester method (**6**, 246–247) was not successful. Lactonization with triphenylphosphine and diethyl azodicarboxylate (**7**, 405–406) gave the desired lactone in about 1% yield. However, the yield was improved to 8% when **2** was treated with the preformed betaine **3** from $P(C_6H_5)_3$ and $C_2H_5OOCN{=}NCOOC_2H_5$.

$$\underset{\mathbf{1}}{CH_3CH(OH)(CH_2)_9CH(OC(O)(CH_2)_2COOCH_3)C(O)CH{=}CHCHO} \xrightarrow[69\%]{NaClO_2} \underset{\mathbf{2}}{CH_3CH(OH)(CH_2)_9CH(OC(O)(CH_2)_2COOCH_3)C(O)CH{=}CHCOOH}$$

$$\underset{\mathbf{3}}{C_2H_5OOC{-}N(P^+(C_6H_5)_3){-}\bar{N}COOC_2H_5}$$

[1] T. A. Hase and E.-L. Nylund, *Tetrahedron Letters*, 2633 (1979).

Sodium cyanide–Alumina.

Aryl cyanation.[1] Reaction of iodobenzene with sodium cyanide in the presence of 10 mole % of Pd(0) affords only a 5% yield of benzonitrile. If the sodium cyanide is impregnated on alumina the reaction proceeds in quantitative yield. In the same reaction with aryl bromides higher yields obtain when alumina is present as a co-catalyst rather than as a support.

[1] J. R. Dalton and S. L. Regen, *J. Org.*, **44**, 4443 (1979).

Sodium cyanide–Dimethyl sulfoxide.

Demethylation of aryl methyl ethers.[1] This cleavage can be carried out with NaCN in DMSO at 125–180° for 1–24 hours. DMSO cannot be replaced by DMF. Yields are 45–90%. The method is particularly useful for aromatic methoxy nitriles.

[1] J. R. McCarthy, J. L. Moore, and R. J. Cregge, *Tetrahedron Letters*, 5183 (1978).

Sodium cyanoborohydride, 4, 448–451; **5**, 607–609; **6**, 537–538; **7**, 334–335; **8**, 454–455.

Oxirane cleavage. Isobe and co-workers[1] effected the stereoselective reduction of **1** to **2** by sequential treatment of the epoxide with sodium cyanoborohydride in anhydrous HMPT and diborane in THF. The authors suggest that this reaction involves internal hydride transfer in the hypothetical intermediate **3**. Reduction in DME or THF results in products with the opposite configuration at C_7.

$HC{=}CH_2$, OH, $CH[COOC(CH_3)_3]_2$, H, $COOC_2H_5$ **1** $\xrightarrow[73\%]{1)\ NaBH_3CN,\ HMPT;\ 2)\ BH_3,\ THF}$ $HC{=}CH_2$, OH, 7, $CH[COOC(CH_3)_3]_2$, H, OH, $COOC_2H_5$ **2**

O, $COOC_2H_5$, $COOC(CH_3)_3$, $COOC(CH_3)_3$, H, Na^+, B, H, $CH_2{=}CH$

3

Reduction of tosylhydrazones.[2] Tosylhydrazones of benzoin derivatives (**1**) are reduced by $NaBH_3CN$ in the presence of TsOH stereoselectively to *erythro*-diastereomers (**2**) in excellent yield.

$$Ar^1CH(OH(R))\text{—}C(=NNHTs)Ar^2 \ \xrightarrow[80\text{–}95\%]{NaBH_3CN,\ THF,\ H^+}\ Ar^1CH(OH(R))\text{—}CH(NHNHTs)Ar^2$$

1 → *erythro*-**2**

Tetrahydropyridines.[3] This reagent selectively reduces various pyridinium salts substituted at the 4-position to 1,2,5,6-tetrahydropyridines in 50–90% yield. The reaction probably occurs in two steps. Nitro, amido, cyano, ethoxycarbonyl, and even keto groups are not affected.

$$Z\text{–}C_5H_4\overset{+}{N}\text{—}CH_2Ar\ X^- \xrightarrow{NaBH_3CN,\ H_2O\text{–}C_2H_5OH,\ 25^\circ} \left[Z\text{–(dihydropyridine)}N\text{—}CH_2Ar \xrightarrow{H^+} Z\text{–}\overset{+}{N}\text{—}CH_2Ar \right] \xrightarrow[50\text{–}90\%]{NaBH_3CN} Z\text{–(tetrahydropyridine)}NCH_2Ar$$

Reduction of $-CH_2N(CH_3)_2$ to $-CH_3$.[4] The quaternary ammonium salts of Mannich bases are reduced to the corresponding methyl compounds by $NaBH_3CN$ in HMPT.

Example:

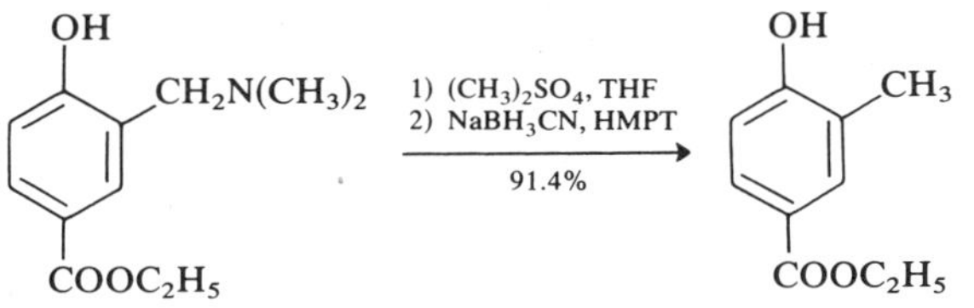

$$\text{4-HO-3-}(CH_2N(CH_3)_2)C_6H_3COOC_2H_5 \xrightarrow[\text{91.4\%}]{\text{1) }(CH_3)_2SO_4\text{, THF; 2) }NaBH_3CN\text{, HMPT}} \text{4-HO-3-}CH_3C_6H_3COOC_2H_5$$

Pyrroloindoles.[5] A simple route to this system, which is present as the corresponding quinone in mitomycins, can be obtained by reaction of N-arylhydroxylamines (**1**) with ethyl 6-oxo-2-hexynoate (**2**) to form a nitrone that is reduced without isolation with $NaBH_3CN$ (1.2 equiv.).

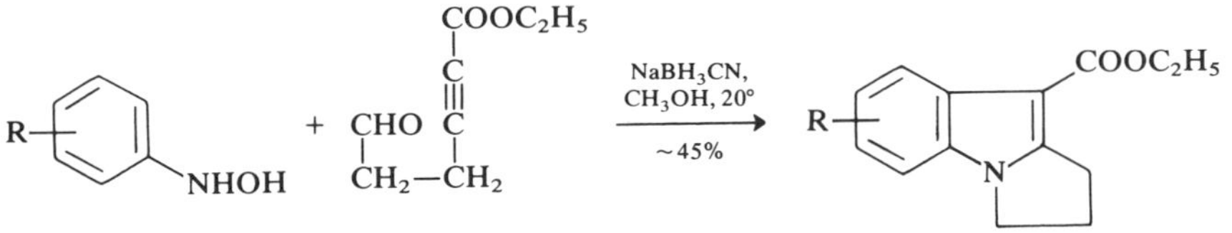

$$R\text{-}C_6H_4NHOH\ (\mathbf{1}) + OHC\text{-}CH_2CH_2\text{-}C{\equiv}C\text{-}COOC_2H_5\ (\mathbf{2}) \xrightarrow[\sim 45\%]{NaBH_3CN,\ CH_3OH,\ 20^\circ} \mathbf{3}$$

Reduction of the oxazolidine ring.[6] Sodium cyanoborohydride (CH_3OH, pH 6–7, 25°) reduces the oxazolidine ring of various C_{20}-diterpene alkaloids almost quantitatively and completely selectively in the presence of a keto or an α,β-unsaturated keto group.

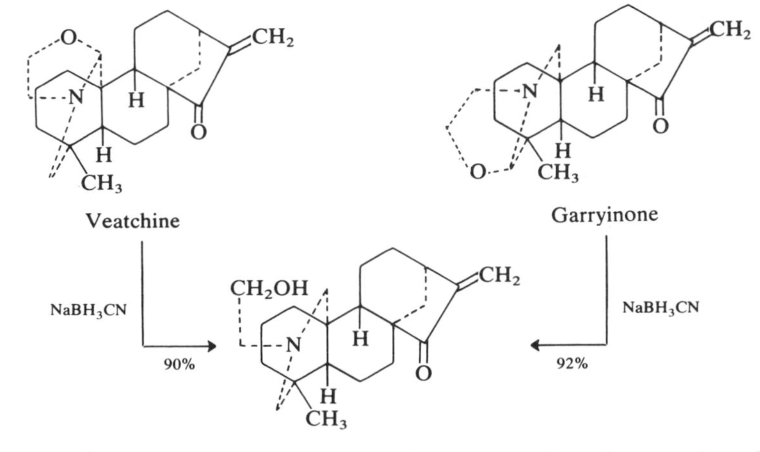

Veatchine $\xrightarrow[90\%]{NaBH_3CN}$ product (CH_2OH) $\xleftarrow[92\%]{NaBH_3CN}$ Garryinone

Review.[7] The various uses of this reducing agent have been reviewed (172 references, including unpublished results).

[1] H. Iio, M. Isobe, T. Kawai, and T. Goto, *Am. Soc.*, **101**, 6076 (1979).
[2] G. Rosini, A. Medici, and M. Soverini, *Synthesis*, 789 (1979).
[3] R. O. Hutchins and N. R. Natale, *ibid.*, 281 (1979).
[4] K. Yamada, N. Itoh, and T. Iwakuma, *J.C.S. Chem. Comm.*, 1089 (1978).
[5] R. M. Coates and C. W. Hutchins, *J. Org.*, **44**, 4742 (1979).
[6] S. W. Pelletier, N. V. Mody, A. P. Venkov, and H. K. Desai, *Tetrahedron Letters*, 4939 (1979).
[7] R. O. Hutchins and N. R. Natale, *Org. Prep. Proc. Int.*, **11**, 201 (1979).

Sodium dicarbonylcyclopentadienyliron, $Na^+[C_5H_5Fe(CO)_2]^-$ (**1**), **5**, 610; **6**, 538–539; **8**, 454–455.

1-Alkenes. This reagent reacts with an alkyl bromide (or tosylate) to form an iron complex of type **2**.[1,2] Treatment with trityl tetrafluoroborate in CH_2Cl_2 at 0° abstracts a hydride ion from a β-carbon atom of **2** to form an alkene iron complex (**3**).[2,3] The alkene **4** is liberated quantitatively and without isomerization by treatment with sodium iodide in acetone.[4] Because of the size, $(C_6H_5)_3C^+$ abstracts a hydride ion preferentially from a methyl rather than a methylene group. In addition, electronic factors may be involved. This elimination reaction is therefore useful for preparation of 1-alkenes.

$$1 + CH_3(CH_2)_2\underset{\displaystyle Br}{CH}CH_3 \xrightarrow[30\%]{} CH_3(CH_2)_2\underset{\displaystyle C_5H_5Fe(CO)_2}{CH}CH_3 \ (\mathbf{2}) \xrightarrow[82\%]{(C_6H_5)_3C^+BF_4^-}$$

$$BF_4^-\left[\begin{matrix} CH_3(CH_2)_2CH{=}CH_2 \\ C_5H_5\overset{+}{Fe}(CO)_2 \end{matrix}\right] (\mathbf{3}) \xrightarrow[quant.]{NaI,\ acetone} CH_3(CH_2)_2CH{=}CH_2 \ (\mathbf{4})$$

Other examples:

$$CH_3(CH_2)_3CH_2Br \xrightarrow[96\%]{1} CH_3(CH_2)_3CH_2Fe(CO)_2C_5H_5 \xrightarrow[50\%]{1)\ (C_6H_5)_3C^+BF_4^-\ \ 2)\ NaI} CH_3(CH_2)_2CH{=}CH_2$$

$$C_6H_{11}\text{-}CH_2OTs \xrightarrow[85\%]{} C_6H_{11}\text{-}CH_2Fe(CO)_2C_5H_5 \xrightarrow[50\%]{} C_6H_{10}{=}CH_2$$

[1] T. S. Piper and G. Wilkinson, *J. Inorg. Nucl. Chem.*, **3**, 104 (1956).
[2] M. L. H. Green and P. L. I. Nagy, *J. Organometal. Chem.*, **1**, 58 (1963).
[3] D. Slack and M. C. Baird, *J.C.S. Chem. Comm.*, 701 (1974).
[4] D. E. Laycock and M. C. Baird, *Tetrahedron Letters*, 3307 (1978).

Sodium dithionite, 1, 1081–1083; **5**, 615–617; **7**, 336; **8**, 456–458.

Reduction of ketones (**8**, 757).[1] Reduction of methylcyclohexanones with sodium dithionite in a two-phase benzene–H_2O mixture with Adogen 464 as a phase-transfer catalyst affords isomeric mixtures of methylcyclohexanols in somewhat higher yield (70–82%) than that obtained when the reduction is conducted in aqueous solution only (15–68%). The stereoselectivity observed in the phase-transfer procedure is comparable to that reported with sodium borohydride.

Examples:

$$\text{2-methylcyclohexanone} \xrightarrow[76\%]{Na_2S_2O_4} \text{cis-2-methylcyclohexanol} + \text{trans} \quad 33:67$$

$$\text{3-methylcyclohexanone} \xrightarrow[81\%]{} \text{cis} + \text{trans} \quad 79:21$$

$$\text{4-methylcyclohexanone} \xrightarrow[82\%]{} \text{cis} + \text{trans} \quad 77:23$$

[1] F. Camps, J. Coll, and M. Riba, *J.C.S. Chem. Comm.*, 1080 (1979).

Sodium hydride, 1, 1075–1081; **2,** 380–383; **4,** 452–455; **5,** 610–614; **6,** 591–592; **8,** 458–459.

Dieckmann condensation.[1] Dithiolesters of dicarboxylic acids undergo Dieckmann condensation under markedly milder conditions than ordinary diesters. Thus di-S-ethyl adipate (**1**) undergoes condensation to S-ethyl 2-cyclopentanone-carboxylthioate (**2**) in 91% yield when treated at 20° with a slight excess of sodium hydride and a catalytic amount of ethanethiol in DME. Six-membered β-keto thiol esters are also formed in high yield, but the method fails in attempted extension to larger rings. The thiol ester group is easily removed by brief treatment with W-2 Raney nickel.

$$\underset{\mathbf{1}}{(CH_2CH_2COSC_2H_5)_2} \xrightarrow[91\%]{NaH,\ DME,\ C_2H_5SH,\ 20^\circ} \underset{\mathbf{2}}{\text{2-(C}_2\text{H}_5\text{SCO)cyclopentanone}} \xrightarrow[\sim 85\%]{Raney\ Ni} \underset{\mathbf{3}}{\text{cyclopentanone}}$$

α,α′-Dihydroxy ketones.[2] A route to these ketones (**5**) from an α-hydroxy carboxylic acid (**1**) involves first conversion to a lactide (**2**) by reaction with an α-bromo carboxylic acid chloride followed by dehydrobromination. On treatment with sodium hydride, **2** undergoes contraction to **a**, which on alkylation affords an α-alkoxytetronic acid (**3**). On basic hydrolysis of **3** a β-keto acid (**b**) is formed, which loses CO_2 to give **4**. When R^4 is CH_3OCH_2 or $C_2H_5OCH_2$, hydrolysis of **4** to **5** is possible with HCl in methanol at 25°. These dihydroxy ketones are unstable.

$$R^1-\underset{R^2}{\overset{OH}{C}}-COOH + R^3\overset{Br}{C}HCOCl \xrightarrow[60\text{–}85\%]{1)\ 80^\circ;\ 2)\ N(C_2H_5)_3} \mathbf{2} \xrightarrow{NaH,\ DMF,\ 25^\circ} [\mathbf{a}]$$

1 **2** **a**

$$\xrightarrow[60\text{–}90\%]{R^4X} \mathbf{3} \xrightarrow{Na_2CO_3,\ H_2O,\ CH_3OH} \left[R^1-\underset{R^2}{\overset{HO}{C}}-\overset{O}{\overset{\|}{C}}-\underset{COOH}{\overset{OR^4}{C}}-R^3\right] \xrightarrow[50\text{–}90\%]{-CO_2} R^1-\underset{R^2}{\overset{HO}{C}}-\overset{O}{\overset{\|}{C}}-\overset{OR^4}{C}HR^3$$

3 **b** **4**, **5** ($R^4 = H$)

β-*Lactams* (**7**, 335). Cyclization of β-halopropionamides to N-alkyl unsubstituted β-lactams is usually not an attractive method because of competing β-elimination. However, dilution favors the desired intramolecular displacements. Under favorable conditions this route can be useful. β-Bromo amides are cyclized in higher yields than β-chloro amides. The structure of the R group on nitrogen also can influence the course of the reaction.[3]

Example:

$$BrCH_2CH_2\overset{O}{\overset{\|}{C}}NHCH_2-C_6H_4-OCH_3 \xrightarrow{NaH,\ DMF,\ CH_2Cl_2} \beta\text{-lactam (N-R)} + CH_2{=}CH\overset{O}{\overset{\|}{C}}NHR$$

(60%) (6%)

A recent synthesis of 3-aminonocardicinic acid (**3**, the nucleus of nocardicins, antibiotics from actinomycetes) involved this cyclization of a β-halopropionamide. Thus treatment of **1** with NaH in dilute solution (DMF–CH_2Cl_2) led to the β-lactam (**2**) in 76% yield.[4]

$$(CH_3OOC)_2C(NHCOCH_2CH_2Cl)-C_6H_4-OCH_2C_6H_5\ (\mathbf{1}) \xrightarrow[76\%]{NaH,\ DMF,\ CH_2Cl_2} \mathbf{2}\ (CH_3OOC,\ COOCH_3,\ OCH_2C_6H_5) \xrightarrow{\text{Several steps}} \mathbf{3}\ (H_2N,\ OH,\ H,\ COOH)$$

[1] H.-J. Liu and H. K. Lai, *Tetrahedron Letters*, 1193 (1979).
[2] U. Schöllkopf, W. Hartwig, U. Sprotte, and W. Jung, *Angew. Chem., Int. Ed.*, **18**, 310 (1979).
[3] H. H. Wasserman, D. J. Hlasta, A. W. Tremper, and J. S. Wu, *Tetrahedron Letters*, 549 (1979).
[4] H. H. Wasserman and D. J. Hlasta, *Am. Soc.*, **100**, 6780 (1978).

Sodium hydride–*n*-Butyllithium.

Sulfenylation of 1,3-dicarbonyl compounds.[1] Dimetalated 1,3-dicarbonyl compounds, prepared by use of 1 equiv. each of NaH and n-C_4H_9Li or with 2 equiv. of LDA, are converted into 4-phenylthio-1,3-dicarbonyl compounds by reaction with C_6H_5SCl, $C_6H_5SSC_6H_5$, or $C_6H_5SO_2SC_6H_5$. Yields are 50–80%. and are not improved by added HMPT.

[1] K. Hiroi, Y. Matsuda, and S. Sato, *Synthesis*, 621 (1979).

Sodium hydrosulfide, NaHS. Mol. wt. 56.07. Very hygroscopic. Preparation.[1]

Thionolactones.[2] A method for conversion of lactones into the corresponding thionolactones involves first conversion of the lactone into the N,N-dimethyliminolactonium tetrafluoroborate (**4**) by a method used by Deslongchamps *et al.*[3] Treatment of the salt **4** with sodium hydrosulfide in acetone at −78° for about 1 hour and then with acetyl chloride and pyridine (to trap the dimethylamine) results in the thionolactone **5**. The low yield of **5** when n = 4 results mainly from cleavage to $AcO(CH_2)_4C(=S)N(CH_3)_2$.

$(CH_2)_n$ lactone (**1**, n = 3, 4, 5) $\xrightarrow{HN(CH_3)_2}$ **2** $\xrightarrow{SOCl_2}$ **3** $\xrightarrow{AgBF_4}$ **4** ($C=\overset{+}{N}(CH_3)_2$ BF_4^-) $\xrightarrow[\text{n = 3, 78\%; n = 4, 43\%; n = 5, 84\%}]{\text{1) NaSH, }(CH_3)_2C=O, -78°\text{; 2) AcCl, Py, }-78°}$ **5** ($C=S$)

[1] R. E. Eibeck, *Inorg. Syn.*, **7**, 128 (1963).
[2] R. B. Nader and M. K. Kaloustian, *Tetrahedron Letters*, 1477 (1979).
[3] P. Deslongchamps, S. Dube, C. Lebreux, D. R. Patterson, and R. J. Taillefer, *Can. J. Chem.*, **53**, 2791 (1975).

Sodium hydroxide–Potassium carbonate, $NaOH–K_2CO_3$.

Solid–liquid phase-transfer catalysts.[1] Diphenylphosphinic hydrazide (**1**) is not alkylated efficiently under usual phase-transfer conditions, but is alkylated by use of solid $NaOH–K_2CO_3$ with benzene as solvent. The reaction is strongly accelerated by tetra-n-butylammonium hydrogen sulfate. The role of K_2CO_3 is not clear. The products are hydrolyzed by 15% HCl to pure monoalkylhydrazines.

$$\underset{\mathbf{1}}{(C_6H_5)_2P(=O)-NH-NH_2} \xrightarrow[30\text{–}95\%]{RX,\ NaOH,\ K_2CO_3,\ C_6H_6,\ cat.} \underset{\mathbf{2}}{(C_6H_5)_2P(=O)-N(R)-NH_2}$$

[1] B. Młotkowska and A. Zwierzak, *Tetrahedron Letters*, 4731 (1978).

Sodium hydroxymethyl sulfoxylate ("Rongalite"), $HOCH_2SO_2Na$. Mol. wt. 118.09, m.p. 63–64°. Suppliers: Fluka, Pfaltz and Bauer, K and K.

Preparation of sodium selenophenolate. Reich *et al.*[1] recommend the use of this reducing agent in place of sodium borohydride for preparation of sodium selenophenolate.

$$PhSeSePh \xrightarrow[H_2O,\ C_2H_5OH]{HOCH_2SO_2Na,\ NaOH,} 2PhSeNa$$

Rongalite has been used as an inexpensive and convenient reducing agent for the conversion of Se(0) to Na_2Se_2[2] and Na_2Se,[2,3] and Te(0) to Na_2Te.[4]

[1] H. J. Reich, F. Chow, and S. K. Shah, *Am. Soc.*, **101**, 6638 (1979).
[2] M. L. Bird and F. Challenger, *J. Chem. Soc.*, 570 (1942).
[3] A. Fredga, *Acta Chem. Scand.*, **17**, 551 (1963).
[4] H. K. Spencer and M. P. Cava, *J. Org.*, **42**, 2937 (1977).

Sodium hypochlorite, 1, 1084–1087; **2**, 67; **3**, 45, 243; **4**, 456; **5**, 617; **6**, 543; **7**, 337–338; **8**, 461–463.

Oxidative cleavage of α-diketones. Corey and Pearce[1] used this reagent to effect oxidative cleavage of the hindered diketone **1** in their synthesis of the sesquiterpene picrotoxinin (**3**).

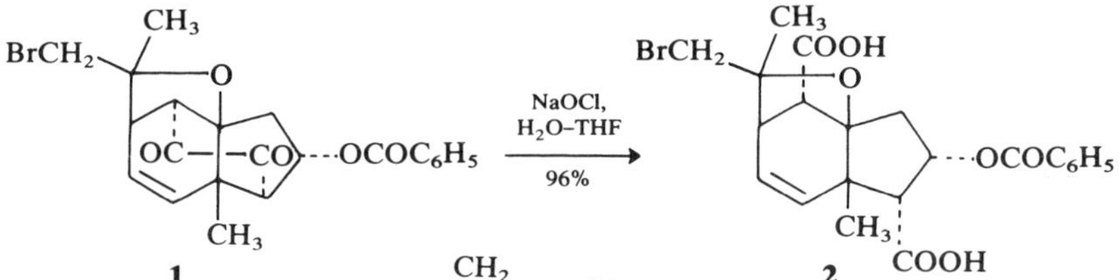

CH_3 $BrCH_2$ O OC— —CO ---$OCOC_6H_5$ CH_3 **1** $\xrightarrow[96\%]{NaOCl,\ H_2O\text{-}THF}$ CH_3 $BrCH_2$ COOH O ---$OCOC_6H_5$ CH_3 COOH **2**

CH_2 CH_3 OH C=O O O CH_3 O—C=O **3**

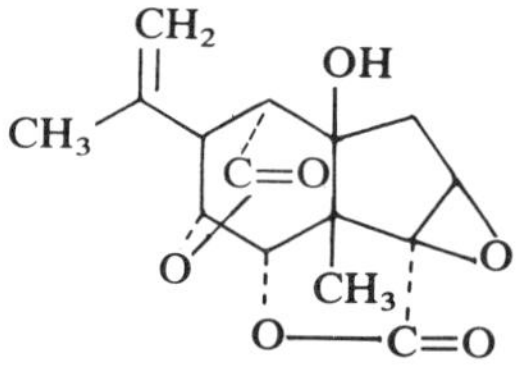

[1] E. J. Corey and H. L. Pearce, *Am. Soc.*, **101**, 5841 (1979).

Sodium methoxide, 1, 1091–1094; **2**, 385–386; **3**, 259–260; **4**, 457–459; **5**, 617–620; **8**, 463.

Cyclopentene-1,3-diones.[1] Only a few natural products contain this system; one is calythrone (**2**). A consideration of the biosynthesis of cyclopentene-1,3-diones suggested that this system could arise from the isomeric 4-ylidenebutenolides, which are also natural products. And indeed treatment of the butenolide **1** with sodium methoxide in methanol followed by acidic work-up does lead to calythrone (**2**) in 80% yield.

This cleavage–recyclization reaction was successfully applied to three other 4-ylidenebutenolides.[2]

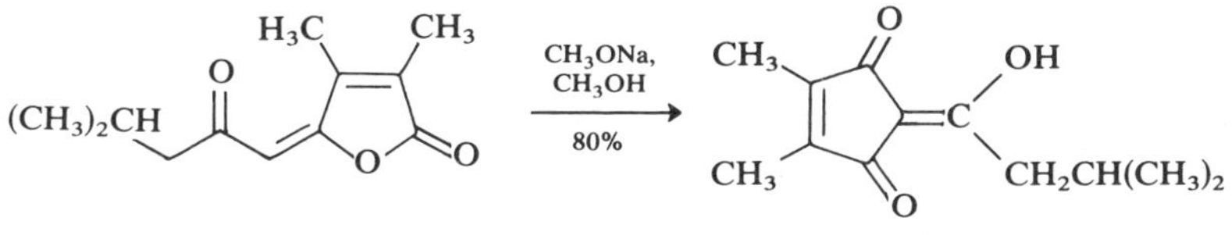

$$(CH_3)_2CHCH_2C(=O)CH{=}\mathbf{1}\ \xrightarrow[80\%]{CH_3ONa,\ CH_3OH}\ \mathbf{2}$$

1 **2**

[1] G. Pattenden, *Prog. Chem. Org. Natural Prod.*, **35**, 133 (1978).
[2] D. R. Gedge and G. Pattenden, *J.C.S. Chem. Comm.*, 880 (1978).

Sodium methylsulfinylmethylide (Dimsylsodium), 1, 310–313; **2**, 166–169; **3**, 123–124; **4**, 195–196; **5**, 621; **6**, 546–547; **7**, 338–339.

α,β-Unsaturated sulfides. The reaction of aromatic aldehydes and DMSO with metallic sodium leads to 2-arylethenyl methyl sulfides with the (E)-configuration. Presumably the actual reagent is dimsylsodium; the final product is then formed by reduction of an intermediate sulfoxide by sodium.[1]

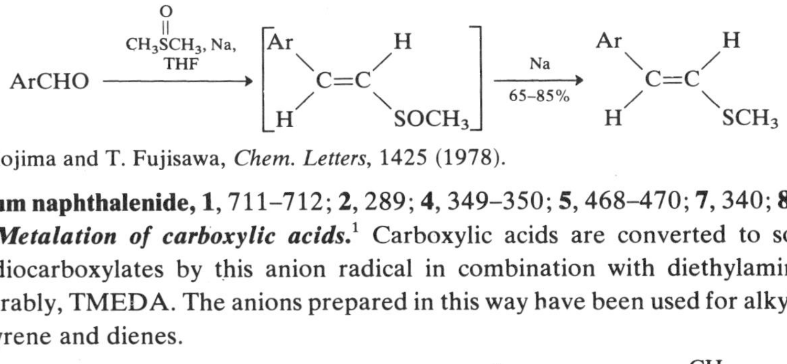

$$ArCHO \xrightarrow[THF]{CH_3S(O)CH_3,\ Na,} \left[\begin{matrix} Ar & & H \\ & C{=}C & \\ H & & SOCH_3 \end{matrix} \right] \xrightarrow[65\text{–}85\%]{Na} \begin{matrix} Ar & & H \\ & C{=}C & \\ H & & SCH_3 \end{matrix}$$

[1] T. Kojima and T. Fujisawa, *Chem. Letters*, 1425 (1978).

Sodium naphthalenide, 1, 711–712; **2**, 289; **4**, 349–350; **5**, 468–470; **7**, 340; **8**, 464.

Metalation of carboxylic acids.[1] Carboxylic acids are converted to sodium α-sodiocarboxylates by this anion radical in combination with diethylamine or, preferably, TMEDA. The anions prepared in this way have been used for alkylation of styrene and dienes.

$$C_6H_5CH{=}CH_2 + (CH_3)_2CHCOOH \xrightarrow[57\%]{C_{10}H_8Na^{\cdot},\ TMEDA,\ THF} C_6H_5CH_2CH_2C(CH_3)_2COOH$$

$$CH_2{=}C(CH_3)CH{=}CH_2 + (CH_3)_2CHCOOH \xrightarrow[66\%]{} (CH_3)_2C{=}CHCH_2C(CH_3)_2COOH$$

Dechlorination.[2] Polychlorinated biphenyl (PCB) can be completely dechlorinated by sodium naphthalenide; about 0.5 equiv. are required per chlorine atom since biphenyl itself forms a radical anion with sodium. Potassium naphthalenide is more effective but less economical. The resulting dechlorinated products (mol. wt. 400–600) can be burned safely for disposal.

[1] T. Fujita, S. Watanabe, K. Suga, and H. Nakayama, *Synthesis*, 310 (1979).
[2] A. Oku, K. Yasufuku, and H. Kataoka, *Chem. Ind.*, 841 (1978).

Sodium nitrite, 1, 1097–1101; **2,** 386–387; **4,** 459–460; **6,** 547.

Oxidative cleavage of hydrazones.[1] Carbonyl compounds can be regenerated from 2,4-dinitrophenylhydrazones, tosylhydrazones, or N-methyl-N-tosylhydrazones in 80–95% yield by oxidation with sodium nitrite in TFA or HOAc (0–5°, 1–2 hours). In the case of tosylhydrazones, the other product of cleavage has been identified as tosyl azide, the product of reaction of tosylhydrazine with nitrous acid.

$$R^1R^2C{=}NNHTs \xrightarrow{NaNO_2,\ TFA} R^1R^2C{=}O + TsN_3$$

(80–95%)

[1] L. Caglioti, F. Gasparrini, D. Misiti, and G. Palmieri, *Synthesis*, 207 (1979).

Sodium ruthenate, 5, 622.

Oxidation. This reagent effects the oxidation of primary alcohols to carboxylic acids under alkaline conditions. Exposure of the hydroxy carboxylate salt derived from **1** to sodium ruthenate in 1 *N* sodium hydroxide solution afforded the diacid **2**, in which the substituent at C_6 had epimerized to the thermodynamically favored β-configuration.[1]

1 $\xrightarrow{1)\ KOH;\ 2)\ Na_2RuO_4,\ NaOH}$ **2**

[1] E. J. Corey, R. L. Danheiser, S. Chandrasekaran, G. E. Keck, B. Gopalan, S. D. Larsen, P. Siret, and J.-L. Gras, *Am. Soc.*, **100**, 8034 (1978).

Sodium selenophenolate, 5, 272–273; **6,** 548–549; **7,** 341.

Phenyl trimethylsilyl selenide; benzeneselenol.[1] The reaction of this anion, generated *in situ* from diphenyl diselenide and sodium in THF, with chlorotrimethylsilane gives phenyl trimethylsilyl selenide in 72% yield. The product is sensitive to oxygen, but is stable under N_2 indefinitely. It reacts rapidly with an alcohol to form benzeneselenol in high yield.

$$C_6H_5SeNa \xrightarrow[72\%]{ClSi(CH_3)_3,\ THF} C_6H_5SeSi(CH_3)_3 \xrightarrow{CH_3OH} C_6H_5SeH + CH_3OSi(CH_3)_3$$

(94%)

Benzeneselenol, generated *in situ* in this way, is useful for selenenylation of α,β-unsaturated ketones (equation I).

$$\text{(I)}\quad CH_3\overset{O}{\overset{\|}{C}}CH{=}CH_2 \xrightarrow[80\%]{C_6H_5SeSi(CH_3)_3,\ CH_3OH,\ CCl_4} CH_3\overset{O}{\overset{\|}{C}}CH_2CH_2SeC_6H_5$$

Epoxide cleavage.[2] Sodium selenophenolate reacts with the furanose **1**, prepared from D-glucose, to form the diselenide **3** in quantitative yield. This reaction involves formation and cleavage of the epoxide **2**, which can be also prepared by treatment of **1** with NaH. The epoxide is cleaved regiospecifically by attack from the less hindered side at C_4. The phenylseleno group can be removed reductively by Raney nickel. The product was used to prepare isoepiallomuscarine (**5**).

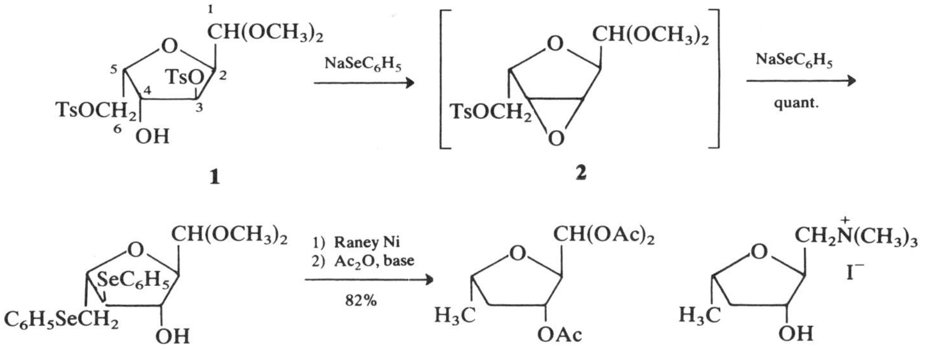

1 $\xrightarrow{NaSeC_6H_5}$ [2] $\xrightarrow[\text{quant.}]{NaSeC_6H_5}$ 3 $\xrightarrow[82\%]{\text{1) Raney Ni; 2) } Ac_2O\text{, base}}$ 4 ; 5

The reduction of **1** with lithium aluminum hydride is not so selective; it results in a 2:3 mixture of diols **6** and **7**, also formed probably through the epoxide **2**.[3] The major product (**7**) was used for synthesis of epiallomuscarine (**8**).

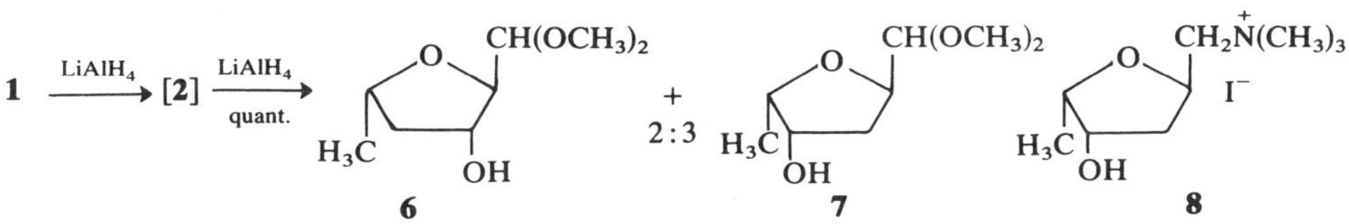

1 $\xrightarrow{LiAlH_4}$ [2] $\xrightarrow[\text{quant.}]{LiAlH_4}$ 6 + 7 (2:3) ; 8

Reaction with alkylidene-γ-lactones. Hoye and Caruso[4] have prepared the reagent by reaction of diphenyl diselenide with sodium sand (Alfa) in THF at 80° in a resealable sealed tube. The reaction is complete after 3 hours. Potassium selenophenolate can be prepared in the same way. The two reagents seem to be comparable

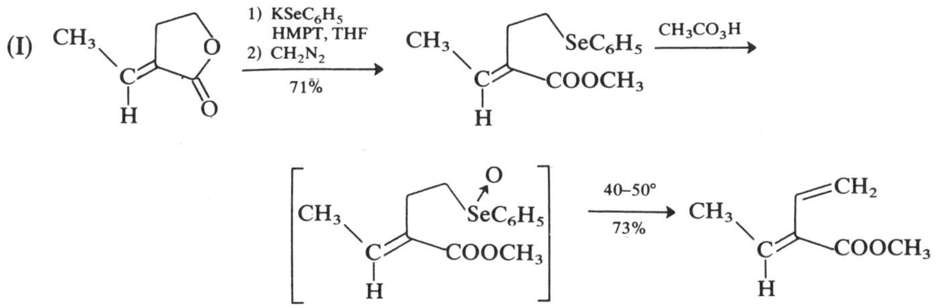

(I) $\xrightarrow[71\%]{\text{1) } KSeC_6H_5\text{, HMPT, THF; 2) } CH_2N_2}$ $CH_3CH{=}C(COOCH_3)CH_2CH_2SeC_6H_5$ $\xrightarrow{CH_3CO_3H}$ [$CH_3CH{=}C(COOCH_3)CH_2CH_2Se(O)C_6H_5$] $\xrightarrow[73\%]{40\text{–}50°}$ $CH_3CH{=}C(COOCH_3)CH{=}CH_2$

in reactivity with alkylidene-γ-lactones. A typical example is shown in equation (I) for the conversion to a 1,3-diene-2-carboxylic acid.

[1] N. Miyoshi, H. Ishii, K. Kondo, S. Murai, and N. Sonoda, *Synthesis*, 300 (1979).
[2] P.-C. Wang, Z. Lysenko, and M. M. Joullié, *Heterocycles*, **9**, 753 (1978).
[3] *Idem, Tetrahedron Letters*, 1657 (1978).
[4] T. R. Hoye and A. J. Caruso, *ibid.*, 4611 (1978).

Sodium sulfide–Sulfur.

—NO₂ → —SH. The combination of reagents is prepared by mixing $Na_2S{\cdot}5H_2O$ and sulfur (1 atom equiv.) in DMSO until the initial blue-green color turns to yellow. A tertiary nitro group is reduced by the reagent to the corresponding thiol and a polysulfide; the latter compound is reduced to the thiol by aluminum amalgam at 0°.[1]

Examples:

$$C_6H_5C(CH_3)_2NO_2 \xrightarrow[92\%]{1)\ Na_2S,\ S;\ 2)\ Al/Hg} C_6H_5C(CH_3)_2SH$$

$$NC{-}C(CH_3)_2{-}C(CH_3)_2{-}NO_2 \xrightarrow[80\%]{} NC{-}C(CH_3)_2{-}C(CH_3)_2{-}SH$$

$$(CH_3)_2C(O_2N){-}C(CH_3)_2{-}COOC(CH_3)_3 \xrightarrow[85\%]{} (CH_3)_2C(SH){-}C(CH_3)_2{-}COOC(CH_3)_3$$

[1] N. Kornblum and J. Widmer, *Am. Soc.*, **100**, 7086 (1978).

Sodium sulfide–Thiophenol.

Reduction of nitroalkenes.[1] Nitroalkenes substituted with at least one aryl group are reduced to alkenes rapidly at room temperature by reaction with $Na_2S \cdot 9H_2O$ and thiophenol in DMF. Thiophenol is essential as a proton source for this reduction; it is converted into diphenyl disulfide during the reaction.

$$C_6H_5(H)C{=}C(C_6H_5)NO_2 \xrightarrow[93\%]{Na_2S\cdot 9H_2O,\ C_6H_5SH,\ DMF} C_6H_5(H)C{=}C(C_6H_5)H$$

[1] N. Ono, S. Kawai, K. Tanaka, and A. Kaji, *Tetrahedron Letters*, 1733 (1979).

Sodium superoxide, NaO_2. Supplier: Alfa.

RC≡N → RCONH₂. Sodium superoxide in DMSO converts nitriles to amides (2–30 hours) in reasonable yields. The paper also reports cleavage of esters by NaO_2 in DMSO at 20° in 1–8 hours.[1]

[1] N. Kornblum and S. Singaram, *J. Org.*, **44**, 4727 (1979).

Sodium tetracarbonylhydridoferrate, 4, 268–269; **5**, 357–358; **6**, 483–486; **8**, 419–420.

Reduction of imidoyl chlorides (1) to imines (2) and amines (3).[1] This reaction can be conducted with sodium tetracarbonylhydridoferrate at 0–25°. Use of excess reagent permits conversion of **2** to **3**. The paper presents a possible mechanism.

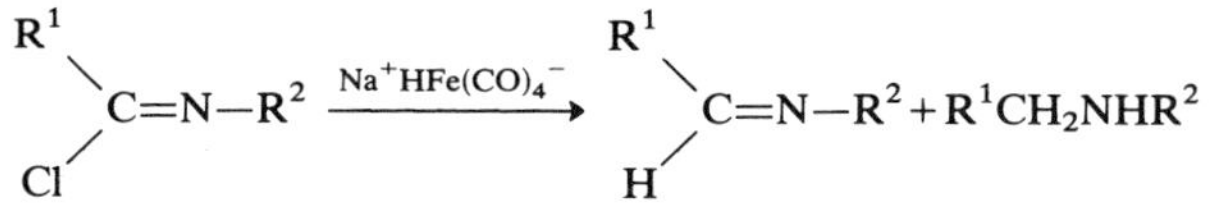

$$\underset{\mathbf{1}}{R^1(Cl)C{=}N{-}R^2} \xrightarrow{Na^+HFe(CO)_4^-} \underset{\mathbf{2}\ (30\text{–}65\%)}{R^1(H)C{=}N{-}R^2} + \underset{\mathbf{3}\ (50\text{–}65\%)}{R^1CH_2NHR^2}$$

[1] H. Alper and M. Tanaka, *Synthesis*, 7891 (1978).

Sodium tetrachloroaluminate (Sodium aluminum chloride), 1, 1027–1029; **2**, 372; **4**, 438; **6**, 524–525.

Cyclodehydration.[1] Several cyclodehydrations have been reported with $NaAlCl_4$ at 170–200° in yields higher than those obtained with the more conventional reagent PPA. Some of the products (with yields) obtained in this way are formulated.

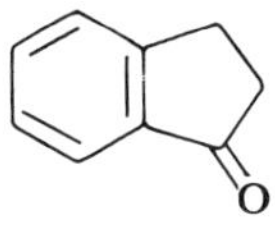

(96%) 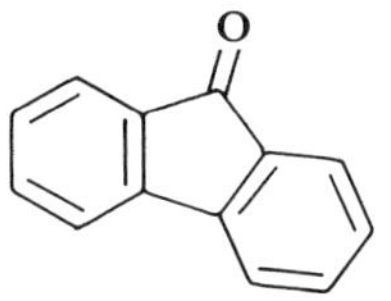 (100%) 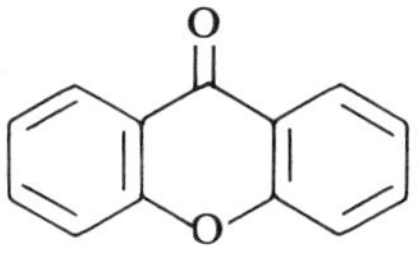 (99%)

[1] L. G. Wade, Jr., K. J. Ackere, R. A. Earl, and R. A. Osteryoung, *J. Org.*, **44**, 3724 (1979).

Sodium trithiocarbonate, Na_2CS_3. Mol. wt. 154.21. This reagent is prepared by the reaction of sodium sulfide and carbon disulfide.[1]

endo-*Disulfides.* Crabbé and co-workers[2] have prepared the thio analog of PGH_2 (**7**) from the diester (**1**) of PGA_2. Conjugate addition of thiolacetic acid to **1** gives **2**, which was reduced to a mixture of **3** and **4**. Treatment of the mesylate of **3** with sodium trithiocarbonate gives the *endo*-trithiocarbonate **5**. This novel derivative has biological properties similar to those of prostaglandin *endo*-peroxides. It is reduced by sodium in alcohol to **6**. The last step, oxidation of **6** to **7**, proved to be unexpectedly difficult, but was accomplished, after numerous attempts, by oxygen in an alkaline medium. The *endo*-disulfide mimics the physiological effects of the

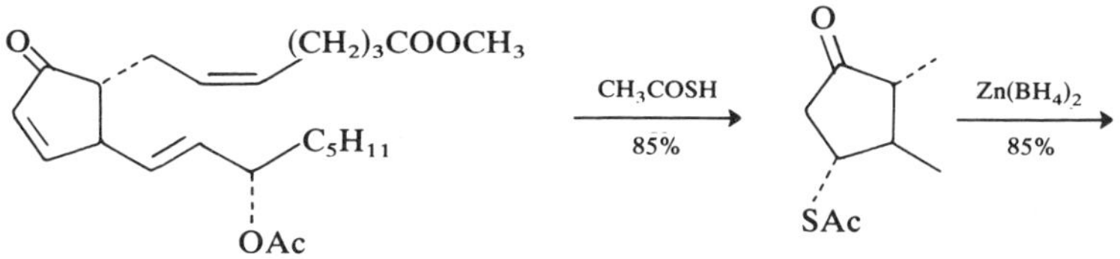

(CH₂)₃COOCH₃ → $(CH_2)_3COOCH_3$; C_5H_{11}; OAc — **1** $\xrightarrow[85\%]{CH_3COSH}$ SAc — **2** $\xrightarrow[85\%]{Zn(BH_4)_2}$

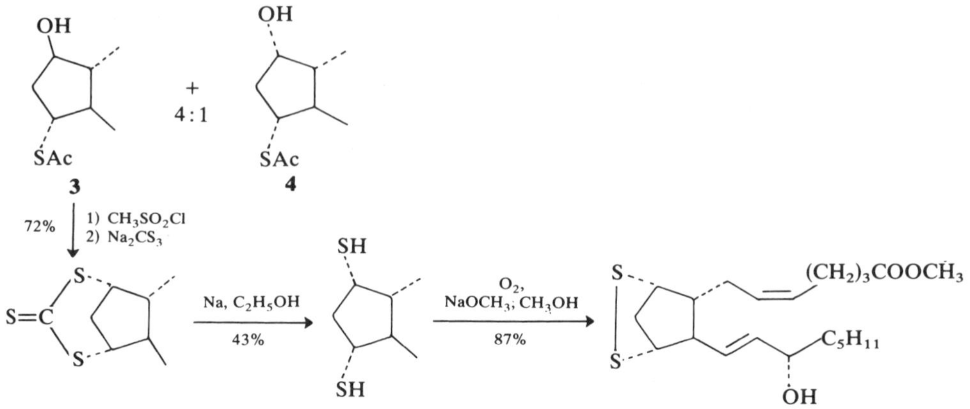
OH
OH
4:1
SAc
SAc
72%
1) CH_3SO_2Cl
2) Na_2CS_3
SH
Na, C_2H_5OH
43%
O_2, $NaOCH_3$, CH_3OH
87%
$(CH_2)_3COOCH_3$
C_5H_{11}
SH
OH

3 **4** **5** **6** **7**

corresponding *endo*-peroxide. A considerably longer synthesis of **7** has been reported by Japanese chemists.[3]

[1] D. J. Martin and C. C. Greco, *J. Org.*, **33**, 1275 (1968).
[2] A. E. Greene, A. Padilla, and P. Crabbé, *ibid.*, **43**, 4377 (1978).
[3] H. Miyake, S. Iguchi, H. Itoh, and M. Hayashi, *Am. Soc.*, **99**, 3536 (1977).

Stannic chloride, 1, 1111–1113; **3,** 269; **5,** 627–631; **6,** 553–554; **7,** 342–345.

Addition of* p*-quinone ketals to olefins. The ketal **1** reacts with 1,2-dimethylcyclopentene in the presence of stannic chloride to afford, after reduction with sodium borohydride, **2** and its diastereomer **3.**[1] The major alcohol (**2**) was used by Büchi and Chu[1] in a total synthesis of the sesquiterpene gymnomitrol (**4**). Several other syntheses of this substance have recently been reported.[2–4]

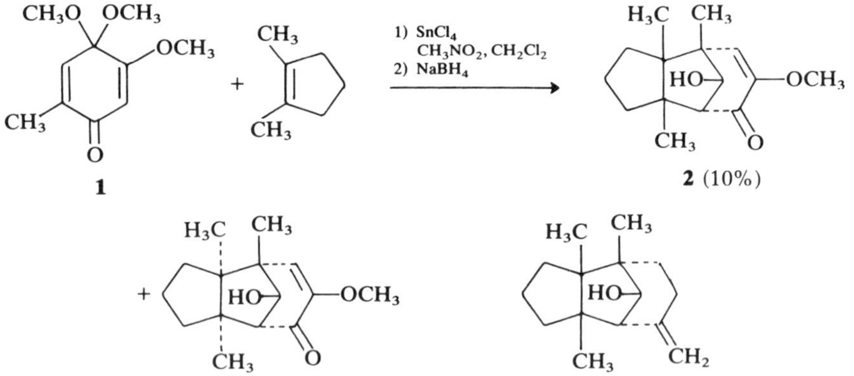
CH_3O OCH_3
OCH_3
CH_3
O
CH_3
CH_3
1) $SnCl_4$ CH_3NO_2, CH_2Cl_2
2) $NaBH_4$
H_3C CH_3
HO
OCH_3
CH_3
O
H_3C CH_3
HO
OCH_3
CH_3
O
H_3C CH_3
HO
CH_3
CH_2

1 **2** (10%) **3** **4**

Cation-π cyclization. Exposure of **1** to stannic chloride in methylene chloride provides **2**, which is an intermediate for synthesis of isodehydroabietenolide (**3**).[5]

This polyene cyclization is the first reported example that employs the β-keto ester unit as initiator. Substances related to **3** exhibit anti-tumor properties.

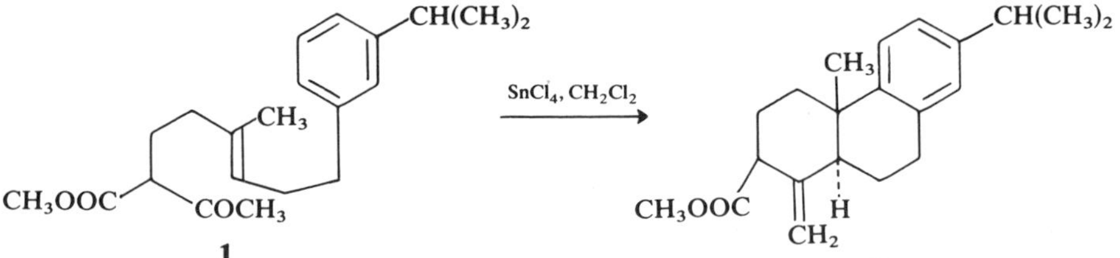

CH(CH$_3$)$_2$ CH$_3$ CH$_3$OOC COCH$_3$ **1** — $SnCl_4$, CH_2Cl_2 → CH(CH$_3$)$_2$ CH$_3$ CH$_3$OOC CH$_2$ H **2**

CH(CH$_3$)$_2$ CH$_3$ O O H **3**

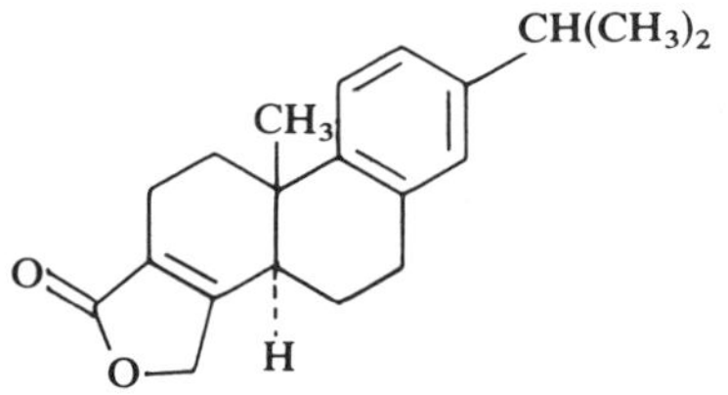

Intramolecular ene reaction. The closure of **1** to **2** with $SnCl_4$ proceeds in high yield and with exceptionally high regiochemical control.[6]

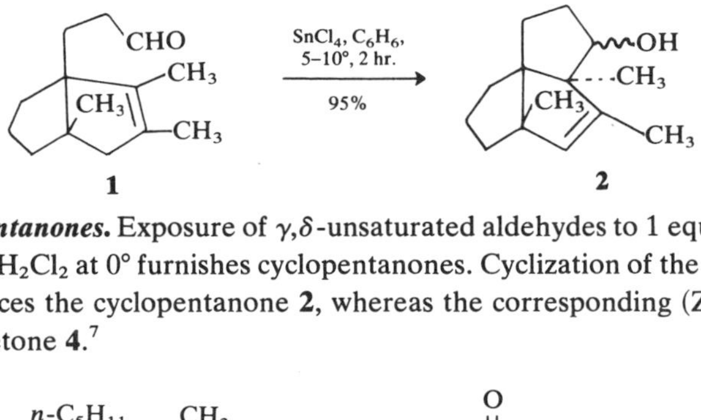

CHO CH$_3$ CH$_3$ CH$_3$ **1** — $SnCl_4$, C_6H_6, 5–10°, 2 hr., 95% → OH CH$_3$ CH$_3$ CH$_3$ **2**

Cyclopentanones. Exposure of γ,δ-unsaturated aldehydes to 1 equiv. of stannic chloride in CH_2Cl_2 at 0° furnishes cyclopentanones. Cyclization of the (E)-aldehyde **1** thus produces the cyclopentanone **2**, whereas the corresponding (Z)-aldehyde **3** cyclizes to ketone **4**.[7]

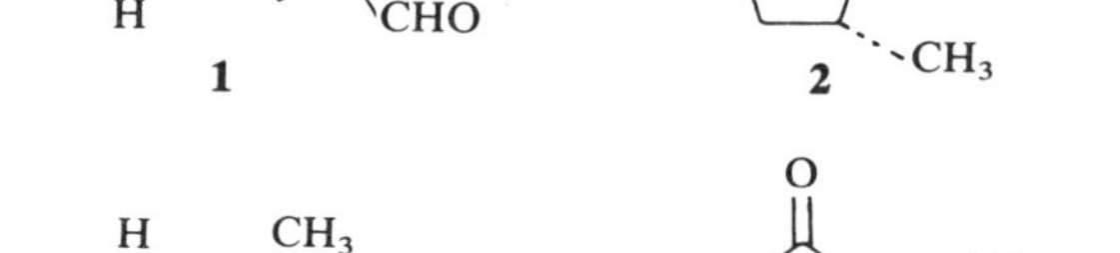

n-C$_5$H$_{11}$ CH$_3$ H CHO **1** — $SnCl_4$, 60% → O C$_5$H$_{11}$-n CH$_3$ **2**

H CH$_3$ n-C$_5$H$_{11}$ CHO **3** — 59% → O C$_5$H$_{11}$-n CH$_3$ **4**

β-*Keto esters.* [8] When ethyl acetoacetate (**1**) is heated with a nitrile in benzene or chlorobenzene in the presence of stannic chloride a tin(IV) complex (**2**) is formed. When **2** is hydrolyzed by acid, the acyl group is cleaved preferentially to give a new β-keto ester (**3**).

Example:

$$C_2H_5CN + CH_3\overset{O}{\overset{\|}{C}}CH_2\overset{O}{\overset{\|}{C}}OC_2H_5 \ (\mathbf{1}) \xrightarrow[86\%]{SnCl_4} CH_3\overset{O}{\overset{\|}{C}}\underset{C(=NH)C_2H_5}{\underset{|}{C}H}\overset{O}{\overset{\|}{C}}OC_2H_5\cdot SnCl_4 \ (\mathbf{2}) \xrightarrow[85\%]{HCl,\ H_2O,\ CHCl_3} C_2H_5\overset{O}{\overset{\|}{C}}CH_2\overset{O}{\overset{\|}{C}}OC_2H_5 \ (\mathbf{3})$$

When **2** is heated with hydroxylamine hydrochloride the isoxazole **4** is obtained in 72% yield.

$$\mathbf{2} \xrightarrow[72\%]{1)\ NH_2OH\cdot HCl\quad 2)\ NaOH} \mathbf{4}$$

(**4**: isoxazole bearing C_2H_5 at C-3, COOH at C-4, and CH_3 at C-5)

[1] G. Büchi and P.-S. Chu, *Am. Soc.*, **101**, 6767 (1979).
[2] Y.-K. Han and L. A. Paquette, *J. Org.*, **44**, 3731 (1979).
[3] R. M. Coates, S. K. Shah, and R. W. Mason, *Am. Soc.*, **101**, 6765 (1979).
[4] S. C. Welch and S. Chayabunjonglerd, *ibid.*, **101**, 6768 (1979).
[5] E. E. van Tamelen, E. G. Taylor, T. M. Leiden, and A. F. Kreft III, *ibid.*, **101**, 7423 (1979).
[6] L. A. Paquette and Y. K. Han, *J. Org.*, **44**, 4014 (1979).
[7] R. C. Cookson and S. A. Smith, *J.C.S. Chem. Comm.*, 145 (1979).
[8] B. Singh and G. Y. Lesher, *Synthesis*, 829 (1978).

Stannic chloride–Sodium borohydride.

Alkene → alcohol. [1] The reagent prepared from $SnCl_4$ (1 equiv.) and $NaBH_4$ (4 equiv.) in THF reacts with olefins to afford anti-Markownikoff alcohols after aqueous work-up. Examples include 2-phenylpropane-1-ol (58%) from α-methylstyrene, isopinocampheol (40%) from α-pinene, *cis*-myrtanol (43%) from β-pinene, 1,2-diphenylethanol (53%) from *trans*-stilbene, and 1-hydroxyacenaphthene (60%, together with acenapthene, 15%) from acenaphthylene. The reaction with dienes such as limonene and 1,5-cyclooctadiene gave complicated results. The mechanism of this hydroxylation is uncertain.

This reagent also reduces nitro, nitrile, and amide derivatives to amines, and carboxylic acids and anhydrides to alcohols. The yields of these reductions are nearly quantitative. Esters and halobenzene derivatives are not reduced.

[1] S. Kano, Y. Yuasa, and S. Shibuya, *J.C.S. Chem. Comm.*, 796 (1979).

Succindialdehyde (1). Mol. wt. 86.09, b.p. 59–60°/12 mm. Preparation[1]:

C_2H_5O–(tetrahydrofuran-2,5-diyl)–OC_2H_5 $\xrightarrow[43\%]{\text{1) HOAc, } H_2O\text{, 98°; 2) Extract with ether}}$ $OHCCH_2CH_2CHO$ (**1**)

Annelation of hydroxyanthraquinones.[2] Leucoquinizarin (**2**), the reduction product of quinizarin with sodium dithionite, reacts with succindialdehyde (excess) to give, after oxygenation, the naphthacenequinone derivative **3** in 62% yield. This annelation is related to the Marschalk reaction[3] by which an alkyl group is introduced into the 2-position of a 1-hydroxyanthraquinone by reaction of an aldehyde with the leuco derivative of the quinone.

Quinizarin $\xrightarrow{Na_2S_2O_4}$ **2** $\xrightarrow[62\%]{\text{1) 1; 2) } O_2}$ **3**

[1] J. Fakstorp, D. Raleigh, and C. E. Schniepp, *Am. Soc.*, **72**, 869 (1950); P. M. Hardy, A. C. Nicholls, and H. N. Rydon, *J.C.S. Perkin II*, 2270 (1972).
[2] M. J. Morris and J. R. Brown, *Tetrahedron Letters*, 2937 (1978).
[3] C. Marschalk, F. Koenig, and N. Ourousoff, *Bull. soc.*, 1545 (1936).

N-Sulfinylbenzenesulfonamide, $C_6H_5SO_2N{=}S{=}O$ (**1**). Mol. wt. 203.24, m.p. 70°. The reagent is prepared by reaction of thionyl chloride with $C_6H_5SO_2NH_2$ (80% yield).

Ene reactions. This reagent undergoes ene reactions at room temperature (equation I).[1]

(I) $R^1R^2C{=}C(R^3)CH_2R^4$ $\xrightarrow[15\text{–}50\%]{1}$ $C_6H_5SO_2NH{-}S(O){-}CR^1R^2{-}C(R^3){=}CHR^4$

This reaction has been used to convert calarene (**2**) into aristolene (**4**).[2]

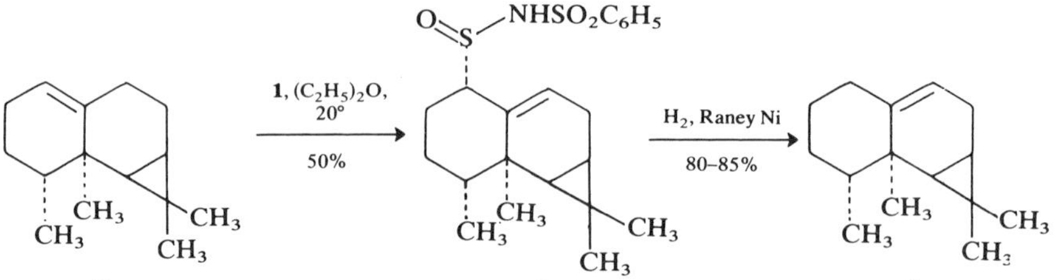

$\xrightarrow[50\%]{\mathbf{1},\ (C_2H_5)_2O,\ 20°}$ $\xrightarrow[80\text{–}85\%]{H_2,\ \text{Raney Ni}}$

$O{=}S{-}NHSO_2C_6H_5$; CH_3; CH_3; CH_3; CH_3

2 **3** **4**

[1] G. Deleris, J. Kowalski, J. Dunogues, and R. Calas, *Tetrahedron Letters*, 4211 (1977).
[2] G. Deleris, J. Dunogues, and R. Calas, *ibid.*, 4835 (1979).

Sulfur dioxide, 1, 1122; **2**, 392; **4**, 469; **5**, 633; **6**, 558; **7**, 346–347; **8**, 464–466.

Addition to vitamin D_3 (1).[1] Vitamin D_3 (**1**) reacts with SO_2 to form the two 1,4-adducts **2** and **3**, which decompose slowly at 20° with loss of SO_2. Thermally induced loss of SO_2 results in either isotachysterol (**4**) and/or isovitamin D_3 (**5**).

Extrusion of SO_2 with KOH–CH_3OH or by contact with alumina affords vitamin D_3 (**1**).

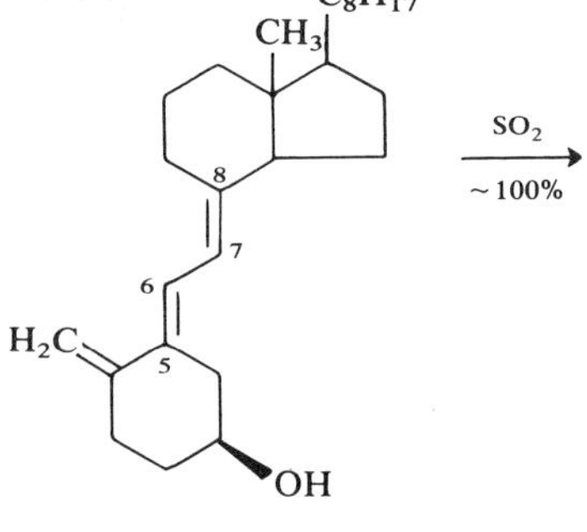

$\xrightarrow[\sim 100\%]{SO_2}$ **2** + **3** (1.6 : 1)

C_8H_{17}; CH_3; H_2C; OH; HO; SO_2; 8; 7; 6; 5

1 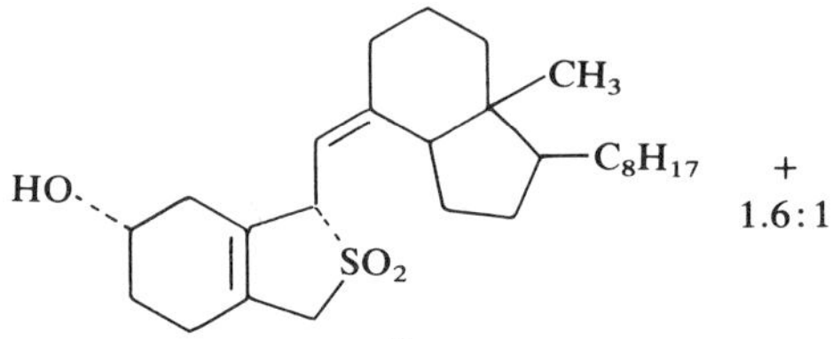 **2** 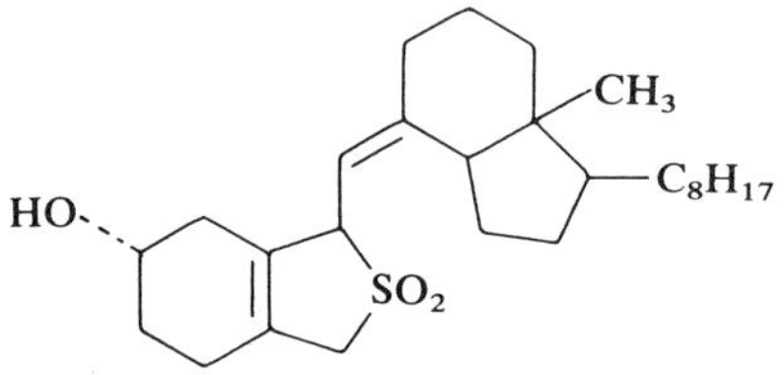 **3** 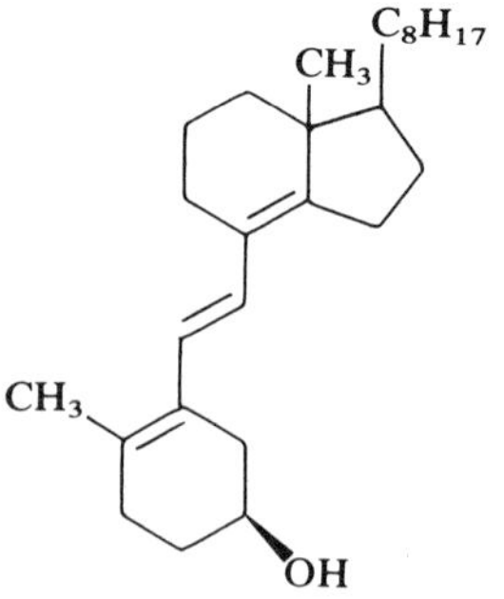 **4** 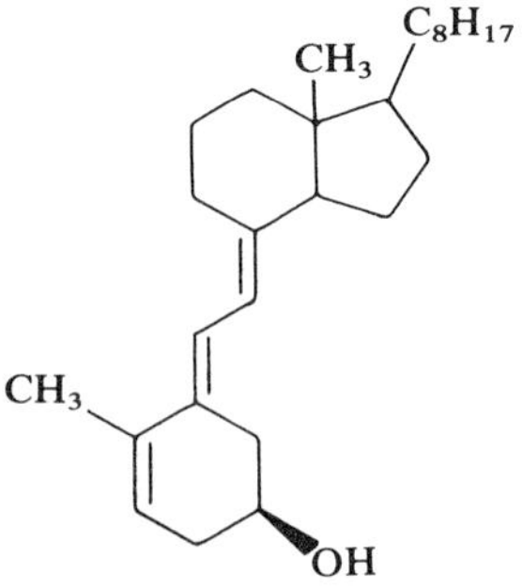 **5**

***Reaction with bicyclo*[5.1.0]*octa-2,5-diene* (1).** This hydrocarbon reacts with SO_2 (dried) in toluene at 150° (5 days) to give **2** in moderate yield. A mechanism for the reaction is suggested.[2]

SO_2, 150°

1 **2**

[1] W. Reischl and E. Zbiral, *Helv.*, **62**, 1763 (1979).
[2] J. Dalling, J. H. Gall, and D. D. MacNicol, *Tetrahedron Letters*, 4789 (1979).

Sulfuric acid, 1, 470–472; **5**, 633–639; **6**, 558–560; **7**, 347–349.

Lactonization of γ,δ-unsaturated acids.[1] Treatment of acid **1a** or **1b**, or of a mixture, with 98% H_2SO_4 for 2 hours at 25° results only in the *cis*-γ-lactone **3** in nearly quantitative yield. The lactone **2** is formed under kinetic control, but rearranges to **3**, the most stable of the four possible γ-lactones. Indeed use of other acid catalysts (HCOOH, $SnCl_4$) results in a mixture of **3** and **2** in the ratio 3:2. Neither of the two possible *trans*-lactones is observed in any case. The stereochemistry of **3** is a feature of many triterpenes.

98% H_2SO_4 ; ~100%

1a ($R^1 = CH_2COOH$, $R^2 = H$)
1b ($R^1 = H$, $R^2 = CH_2COOH$)

2 **3**

Dehydrogenation and dehydration. Some years ago Wells[2] clarified the structure of a known compound **2** obtained by treatment of the dibromolevulinic acid (**1**) with conc. sulfuric acid. The reaction involves dehydration and loss of two hydrogens. Recently a series of halogenated lactones related to **2**, known as fimbrolides,

H_2SO_4, 100° ; 23%

1 **2**

have been isolated from the marine algae *Delisea fimbriata*.[3,4] This unusual reaction of H_2SO_4 was used in the first synthesis of a fimbrolide (**4**) from 3,5-dibromo-2-butyllevulinic acid (**3**).[5] Both **2** and **4** are obtained in improved yield when 100% H_2SO_4 is used.[6]

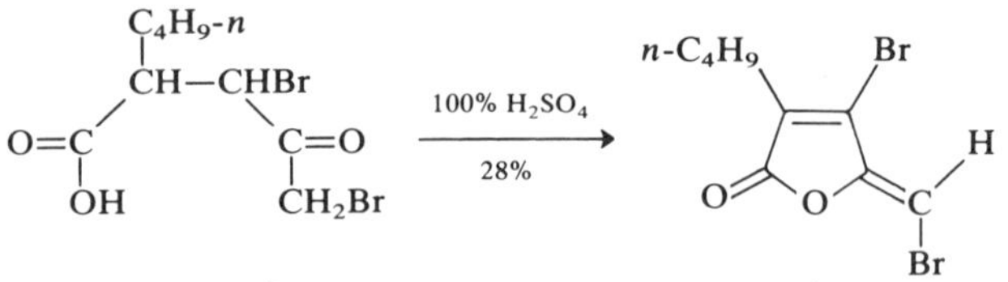

$$\text{n-C}_4\text{H}_9\text{CH(COOH)–CHBr–CO–CH}_2\text{Br} \xrightarrow[28\%]{100\%\ H_2SO_4} \mathbf{4}$$

3 4

[1] F. Rouessac and H. Zamarlik, *Tetrahedron Letters*, 3421 (1979).
[2] P. R. Wells, *Aust. J. Chem.*, **16**, 165 (1963).
[3] J. A. Pettus, Jr., R. M. Wing, and J. J. Sims, *Tetrahedron Letters*, 41 (1977).
[4] R. Kazlauskas, P. T. Murphy, R. J. Quinn, and R. J. Wells, *ibid.*, 37 (1977).
[5] C. M. Beechan and J. J. Sims, *ibid.*, 6149 (1979).
[6] Acid of this strength is prepared by addition of conc. H_2SO_4 to fuming H_2SO_4 until fumes are no longer perceptible.

Sulfur monochloride, 1, 1122–1123; **6**, 560.

Thiones.[1] Hydrazones of ketones react with S_2Cl_2 and $N(C_2H_5)_3$ in benzene at 0–25° to form thiones in moderate to high yield, possibly via a $R_2C{=}S{=}S$ intermediate.

Examples:

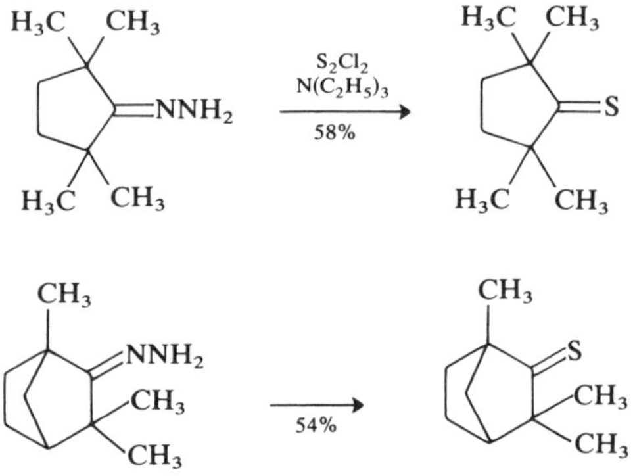

$$\text{2,2,5,5-tetramethylcyclopentanone hydrazone }(=NNH_2) \xrightarrow[58\%]{S_2Cl_2,\ N(C_2H_5)_3} \text{2,2,5,5-tetramethylcyclopentanethione }(=S)$$

$$\text{fenchone hydrazone }(CH_3, =NNH_2, CH_3, CH_3) \xrightarrow[54\%]{} \text{thiofenchone }(CH_3, =S, CH_3, CH_3)$$

[1] R. Okazaki, K. Inoue, and N. Inamoto, *Tetrahedron Letters*, 3673 (1979).

T

Tellurium chloride, $TeCl_4$. Mol. wt. 269.41, m.p. 224°.

(Z)-Iodo- and bromochlorination of $C_6H_5C{\equiv}CR$. This reaction can be conducted in two steps via (Z)-chlorotelluration (equation I).[1]

$$\text{(I)}\quad C_6H_5C{\equiv}CR + TeCl_4 \xrightarrow[75\text{–}90\%]{CCl_4} (C_6H_5)(Cl)C{=}C(R)(TeCl_3)$$

$$(C_6H_5)(Cl)C{=}C(R)(TeCl_3) \xrightarrow[60\text{–}95\%]{I_2,\ CH_3CN} (C_6H_5)(Cl)C{=}C(R)(I)$$

$$(C_6H_5)(Cl)C{=}C(R)(TeCl_3) \xrightarrow[60\text{–}70\%]{NBS,\ AlCl_3,\ CCl_4} (C_6H_5)(Cl)C{=}C(R)(Br)$$

[1] S. Uemura, H. Miyoshi, and M. Okano, *Chem. Letters*, 1357 (1979).

Tetra-*n*-butylammonium acetate, $(C_4H_9)_4\overset{+}{N}\overset{-}{O}COCH_3$. Mol. wt. 301.50, m.p. 114–115°. The reagent is obtained by neutralization of tetrabutylammonium hydroxide in methanol with acetic acid.

This salt is superior to sodium acetate in the reduction of tosylhydrazones with catecholborane (**6**, 98; **7**, 54). The effect results from the increased solubility in organic solvents.[1]

[1] G. W. Kabalka and J. H. Chandler, *Syn. Comm.*, **9**, 275 (1979).

Tetra-*n*-butylammonium chromate, $HCrO_4^{-}\overset{+}{N}(C_4H_9\text{-}n)_4$. Mol. wt. 359.47, yellow-orange solid, soluble in chloroform.

This oxidation reagent is obtained by addition of tetra-*n*-butylammonium chloride to an aqueous solution of chromium trioxide. It oxidizes primary and secondary alcohols to carbonyl compounds in generally good yield. The main advantages are the homogeneous conditions and the requirement for only a small excess of oxidant.[1]

[1] S. Cacchi, F. La Torre, and D. Misiti, *Synthesis*, 356 (1979).

Tetra-*n*-butylammonium di-*t*-butyl phosphate, $[(CH_3)_3CO]_2\overset{O}{\overset{\|}{P}}\overset{-}{O}\overset{+}{N}[C_4H_9\text{-}n]_4$ (**1**). Mol. wt. 451.7, m.p. 108–110°. The salt is prepared by reaction of potassium di-*t*-butyl phosphate with tetra-*n*-butylammonium hydrogen sulfate in aqueous $NaOH\text{–}CH_2Cl_2$ (96% yield).

Phosphorylation of alkyl bromides. Alkyl bromides react with **1** in refluxing DME to form an alkyl di-*t*-butyl phosphate, which is converted by TFA in benzene to a monoalkyl phosphate, usually isolated as the anilinium salt.[1]

$$1 + RBr \xrightarrow[60-85\%]{DME,\ \Delta} [(CH_3)_3CO]_2\overset{O}{\overset{\|}{P}}OR \xrightarrow[50-75\%]{TFA} (HO)_2\overset{O}{\overset{\|}{P}}OR$$

[1] M. Kluba and A. Zwierzak, *Synthesis*, 770 (1978).

Tetra-*n*-butylammonium fluoride, 4, 477–478; **5,** 645; **7,** 353–354; **8,** 467–468. This hygroscopic reagent can be conveniently generated *in situ* by reaction of tetra-*n*-butylammonium chloride and potassium fluoride dihydrate in acetonitrile.[1]

Nitro-aldol reaction; β-amino alcohols.[2] Primary nitro compounds form silyl nitronates (**1**) when treated in sequence with LDA (THF, −78°) and then a silylating reagent. These silyl nitronates undergo aldol condensation with aldehydes in the presence of tetra-*n*-butylammonium fluoride (there is no reaction in the absence of the catalyst). The products **2** are reduced to *β*-amino alcohols (**3**) in good yield by lithium aluminum hydride (equation I).[3] Secondary nitroalkanes undergo the same reaction sequence, but the silyl nitronates are less stable and are obtained in only

$$\text{(I)}\quad R^1CH_2NO_2 \xrightarrow[75\%]{1)\ LDA,\ THF,\ -78^\circ;\ 2)\ ClSi(CH_3)_3} \underset{\mathbf{1}}{R^1CH{=}\overset{+}{N}(O^-)OSi(CH_3)_3} \xrightarrow[60-90\%]{R^2CHO,\ (C_4H_9)_4NF} \underset{\mathbf{2}}{R^2CH(OH){-}CH(NO_2)R^1} \xrightarrow[50-85\%]{LiAlH_4,\ ether} \underset{\mathbf{3}}{R^2CH(OH){-}CH(NH_2)R^1}$$

30–40% yield. Ketones do not react with **1**, but do react with the dianions of nitroalkanes (generated with *n*-butyllithium and THF–HMPT) without catalysis.[4] Silyl ethers of the resulting aldols are also reduced by $LiAlH_4$ to amino alcohols.

$$\left[R^1\overset{-}{C}{=}\overset{+}{N}(O^-)O^-\right]2Li^+ \xrightarrow[40-65\%]{1)\ R^2COR^3;\ 2)\ ClSi(CH_3)_2C(CH_3)_3} R^2R^3C(OSi(CH_3)_2C(CH_3)_3){-}CH(NO_2)R^1 \xrightarrow{LiAlH_4} R^2R^3C(OH){-}CH(NH_2)R^1$$

Terminal alkenes.[5] *β*-Silyl sulfones on treatment with fluoride ion undergo elimination to the corresponding alkene. This reaction, coupled with *α*-alkylation of sulfones, is useful for synthesis of terminal olefins and 1,3-dienes.

Examples:

$$C_6H_5SO_2CH_2CH_2Si(CH_3)_3 \xrightarrow[79\%]{1)\ n\text{-BuLi};\ 2)\ n\text{-}C_8H_{17}Br} n\text{-}C_8H_{17}CH(SO_2C_6H_5)CH_2Si(CH_3)_3 \xrightarrow[80\%]{F^-} n\text{-}C_8H_{17}CH{=}CH_2$$

$$p\text{-}CH_3OC_6H_4C(=CH_2)CH_2SO_2C_6H_5 \xrightarrow[96\%]{1)\ n\text{-BuLi};\ 2)\ (CH_3)_3SiCH_2I} p\text{-}CH_3OC_6H_4C(=CH_2)CH(SO_2C_6H_5)CH_2Si(CH_3)_3 \xrightarrow{79\%} p\text{-}CH_3OC_6H_4C(=CH_2)CH{=}CH_2$$

CH_3, $C_6H_5SO_2$ (decalin) $\xrightarrow{97\%}$ $(CH_3)_3SiCH_2$, $C_6H_5SO_2$, CH_3 (decalin) $\xrightarrow{70\%}$ $CH_2{=}$, CH_3 (decalin)

Alkylation.[6] This salt catalyzes alkylation of purine and pyrimidine bases by alkyl halides, trialkyl phosphates, and dialkyl sulfates. For example, uracil can be converted into 1,3-dimethyluracil in 99% yield with dimethyl sulfate in THF containing 10 equiv. of tetrabutylammonium fluoride. The alkylation of nucleotides under these conditions results in phosphotriesters, a reaction believed to be involved in mutagenic and lethal effects of alkylating reagents on DNA and RNA. An example is the alkylation of thymidine 3′-phosphate (**1**).

Thymidine 3′-phosphate, $O{=}P(OH)_2$ (**1**) $\xrightarrow[90\%]{(CH_3O)_2SO_2,\ (C_4H_9)_4NF}$ N-3-methyl, $O{=}P(OCH_3)_2$ (**2**)

Homoallylic alcohols.[7] In the presence of this salt, allylsilanes (for example, **1**) react with aldehydes or ketones to form, after hydrolysis, homoallylic alcohols (**2**).

$$\underset{\mathbf{1}}{(CH_3)_3SiCH_2CH{=}CH_2} \xrightarrow{n\text{-}Bu_4\overset{+}{N}F^-} [\underset{\mathbf{a}}{(CH_3)_3SiF} + \underset{\mathbf{b}}{n\text{-}Bu_4N^+CH_2\overset{...}{=}\overset{-}{C}H\overset{...}{=}CH_2}] \xrightarrow{R^1COR^2}$$

$$\left[\underset{\mathbf{c}}{R^1R^2C(OSi(CH_3)_3)CH_2CH{=}CH_2}\right] \xrightarrow{H^+} \underset{\mathbf{2}}{R^1R^2C(OH)CH_2CH{=}CH_2}$$

KF is ineffective in the reaction, even when activated by a crown ether. This allylation apparently involves cleavage to **a** and **b**, which adds to the carbonyl group to give **c**, and is possible because of the high Si—F bond energy. Aldehydes are more reactive than ketones.

Examples:

$$1 + C_6H_5CH_2CH_2CHO \xrightarrow[86\%]{} C_6H_5CH_2CH_2\underset{OH}{\underset{|}{C}}HCH_2CH{=}CH_2$$

$$1 + CH_3COCH_2CH_2COOCH_3 \xrightarrow[71\%]{} \text{(γ-lactone: } O{=}C\text{ ring O, C bearing } CH_3 \text{ and } CH_2CH{=}CH_2)$$

$$(CH_3)_3SiCH_2CH{=}CHCH_3 + CH_3(CH_2)_2CHO \longrightarrow$$

$$\underset{(37\%)}{CH_3(CH_2)_2\underset{OH}{\underset{|}{C}}HCH_2CH{=}CHCH_3} + \underset{(23\%)}{CH_3(CH_2)_2\underset{OH}{\underset{|}{C}}H\overset{CH_3}{\overset{|}{C}}HCH{=}CH_2}$$

Isomerization of methyl-substituted allylsilanes.[8] Substrates of this type when heated with fluoride ion in THF rearrange to allyl silanes in which the silicon atom is bonded to the less-substituted carbon atom of the allylic group.

Examples:

$$CH_2{=}CH\overset{CH_3}{\overset{|}{C}}HSi(CH_3)_3 \xrightarrow{F^-,\,\Delta,\,THF} CH_3CH{=}CHCH_2Si(CH_3)_3 \quad (E/Z = 34:66)$$

$$CH_2{=}CH\underset{CH_3}{\underset{|}{\overset{CH_3}{\overset{|}{C}}}}Si(CH_3)_3 \rightarrow (CH_3)_2C{=}CHCH_2Si(CH_3)_3$$

[1] L. A. Carpino and A. C. Saw, *J.C.S. Chem. Comm.*, 514 (1979).
[2] E. W. Colvin and D. Seebach, *ibid.*, 689 (1978).
[3] Unprotected nitro alcohols are cleaved by $LiAlH_4$.
[4] D. Seebach and F. Lehr, *Angew. Chem., Int. Ed.*, **15**, 505 (1976).
[5] P. J. Kocienski, *Tetrahedron Letters*, 2649 (1979).
[6] K. K. Ogilvie, S. L. Beaucage, and M. F. Gillen, *ibid.*, 1663, 3203 (1978).
[7] A. Hosomi, A. Shirahata, and H. Sakurai, *ibid.*, 3043 (1978).
[8] *Idem.*, *Chem. Letters*, 901 (1978).

Tetra-*n*-butylammonium fluoride–Silica.

Anhydrous reagent.[1] Tetraalkylammonium fluorides are useful catalysts in various synthetic reactions, but their use is hampered by their extreme ability to retain water, which reduces the effectiveness. Clark has found that silica gel (60–120 mesh) can give completely anhydrous salts. Thus a 20% aqueous solution of tetra-*n*-butylammonium fluoride is shaken with silica gel; water is then partially removed under reduced pressure. Methanol is added and evaporation is continued to remove solvents. The product is then washed with ether and dried at 100°. This material (**1**) is nonhygroscopic and stable in air for several days, and can be used for fluoride-assisted reactions such as Michael addition, sulfenylation, and C-alkylation. Some examples are formulated below.

$$C_6H_5CH{=}CHCOC_6H_5 + CH_3NO_2 \xrightarrow[75\%]{\mathbf{1},\ THF,\ 20^\circ} C_6H_5CH(CH_2NO_2){-}CH_2COC_6H_5$$

$$CH_3COCH_2COCH_3 + C_6H_5SH \xrightarrow[60\%]{\mathbf{1},\ DMF,\ 20^\circ} CH_3COCH(SC_6H_5)COCH_3$$

$$C_6H_5COCH_2Br + CH_3COOH \xrightarrow[75\%]{\mathbf{1},\ THF,\ 20^\circ} C_6H_5COCH_2OCOCH_3$$

[1] J. H. Clark, *J.C.S. Chem. Comm.*, 789 (1978).

Tetra-*n*-butylammonium iodide, 5, 646–647; **6**, 566–567; **7**, 355.

Deoxyiodo sugars. Preparation of these compounds by displacement reactions usually has limitations because of the vigorous conditions commonly required. Replacement of a triflate group (**4**, 533) by iodine using tetra-*n*-butylammonium iodide, however, proceeds under relatively mild conditions (refluxing benzene, 15 hours). In six reported examples, overall yields were 85–95%.[1]

Examples:

[Scheme: di-O-isopropylidene furanose with OH $\xrightarrow{(CF_3SO_2)_2O,\ Py}$ corresponding OTf $\xrightarrow[87\%]{Bu_4\overset{+}{N}I^-}$ corresponding iodide (I)]

[Scheme: di-O-isopropylidene pyranose with CH_2OH $\xrightarrow{94\%}$ corresponding CH_2I]

[1] R. W. Binkley and D. G. Hehemann, *J. Org.*, **43**, 3244 (1978).

Tetra-*n*-butylammonium octamolybdate, $[(C_4H_9)_4N]_4Mo_8O_{26}$. Mol. wt. 2153.39. The amine salt is prepared in 80% yield by reaction of sodium molybdate, $Na_2MoO_4{\cdot}2H_2O$ (Alfa), with a 40% solution of tetra-*n*-butylammonium hydroxide in a carefully controlled ratio (*cf.* Klemperer and Shum[1]).

cis-*Dichlorination of alkenes.*[2] This reaction when conducted with $MoCl_5$ (**6**, 413–414) suffers from low yields when applied to polysubstituted alkenes. This new reagent in combination with acetyl chloride leads to improved yields (85–95%); less than 2% of *trans*-dichloroalkenes are formed, if any are formed at all. The stoichiometry is shown in equation (I). The reaction is somewhat faster when acetyl chloride is replaced by BCl_3. The actual chlorinating reagent is probably a polychloromolybdenum species.

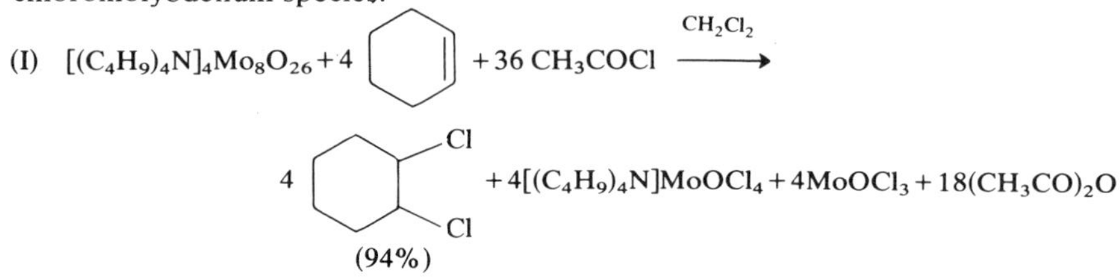

$$\text{(I)}\quad [(C_4H_9)_4N]_4Mo_8O_{26} + 4\ \text{(cyclohexene)} + 36\ CH_3COCl \xrightarrow{CH_2Cl_2}$$

$$4\ \text{(1,2-dichlorocyclohexane, Cl, Cl)} + 4[(C_4H_9)_4N]MoOCl_4 + 4MoOCl_3 + 18(CH_3CO)_2O$$

(94%)

[1] W. A. Klemperer and W. Shum, *Am. Soc.*, **98**, 8291 (1976).
[2] W. A. Nugent, *Tetrahedron Letters*, 3427 (1978).

Tetraethylammonium fluoride, 8, 470–471.

Alkyl aryl ethers.[1] This fluoride can serve as base for the alkylation of phenols with alkyl halides (DMF, 20°). Yields are in the range 65–85%. In the case of 2,6-disubstituted phenols the reaction is conducted at 100°. The method is even satisfactory for alkylation of 2- and 4-nitrophenols. The success of this method probably is a result of the ability of F^- to serve as a hydrogen-bond electron donor.

[1] J. M. Miller, K. H. So, and J. H. Clark, *Can. J. Chem.*, **57**, 1887 (1979).

Tetraethylammonium periodate, $(C_2H_5)_4\overset{+}{N}IO_4^-$. Mol. wt. 321.16, m.p. 176–177°; soluble in H_2O and in many common organic solvents, namely CH_3COOH, acetone, pyridine, $CHCl_3$, DMF; insoluble in ether and benzene. The material was first prepared by Qureski and Sklarz,[1] who observed that the reagent was equivalent to $NaIO_4$ for oxidation of N-alkylhydroxylamines to nitroso compounds. They suggested that the solubility in both water and many organic solvents could be useful for certain oxidations.

Kirby *et al.*[2] have used tetraethylammonium periodate for oxidation of hydroxamic acids, RCONHOH, to highly reactive acyl nitroso compounds, $R\overset{O}{\overset{\|}{C}}-N=O$, which could be trapped by cycloaddition to dienes.

Example:

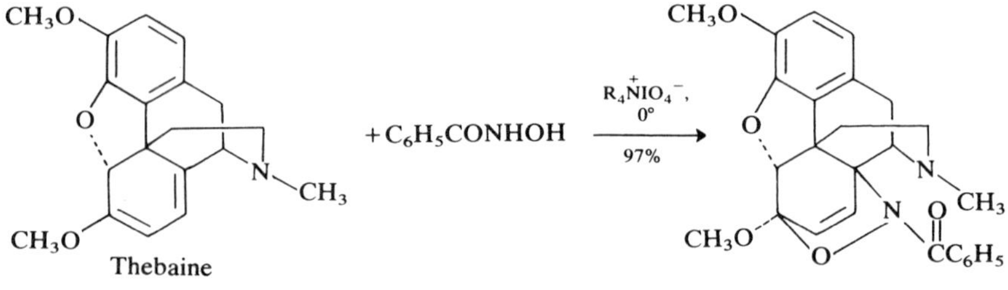

$$\text{Thebaine} + C_6H_5CONHOH \xrightarrow[97\%]{R_4\overset{+}{N}IO_4^-,\ 0^\circ} \text{adduct}$$

CH₃O, O, N, CH₃, CH₃O — Thebaine; product: CH₃O, O, N, CH₃, N, O, CH₃O, $\overset{O}{\overset{\|}{C}}C_6H_5$

Keck and co-workers[3] have shown that this Diels–Alder reaction offers promise for the synthesis of certain alkaloids and other nitrogen-containing products since the adducts are selectively cleaved to hydroxy amides by aluminum or sodium amalgam.

Examples:

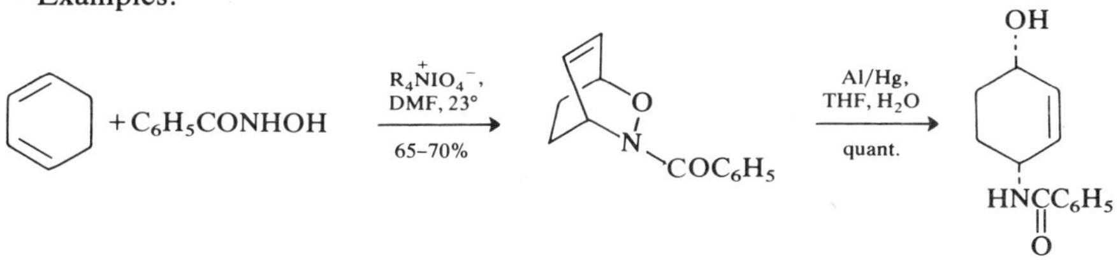

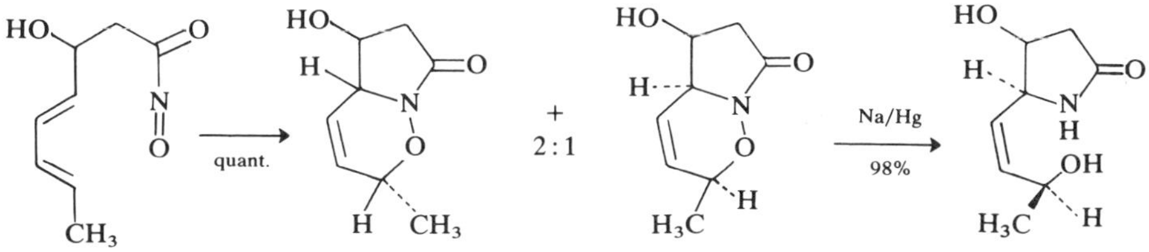

[1] A. K. Qureski and B. Sklarz, *J. Chem. Soc.* (C), 412 (1966).

[2] G. W. Kirby and J. C. Sweeny, *J.C.S. Chem. Comm.*, 704 (1973); J. F. T. Corrie, G. W. Kirby, and R. T. Sharma, *ibid.*, 915 (1975).

[3] G. E. Keck and S. A. Fleming, *Tetrahedron Letters*, 4763 (1978); G. E. Keck, *ibid.*, 4767 (1978); G. E. Keck, S. Fleming, D. Nickell, and P. Weider, *Syn. Comm.*, 281 (1979).

Tetrafluoropyrrolidine, 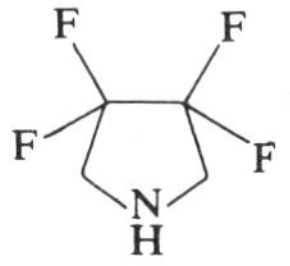 This pyrrolidine is prepared by reduction of 3,3,4,4-tetrafluorosuccinimide with diborane.

This weakly basic amine (pK_a 4.05) is a superior catalyst for reactions involving deprotonation α to a carbonyl group. It is more effective than other weakly basic amines because of ease of iminium ion formation. Thus the decalone **1** is converted into **2** more rapidly by this amine via **a** than by other weakly basic nucleophilic amines.[1]

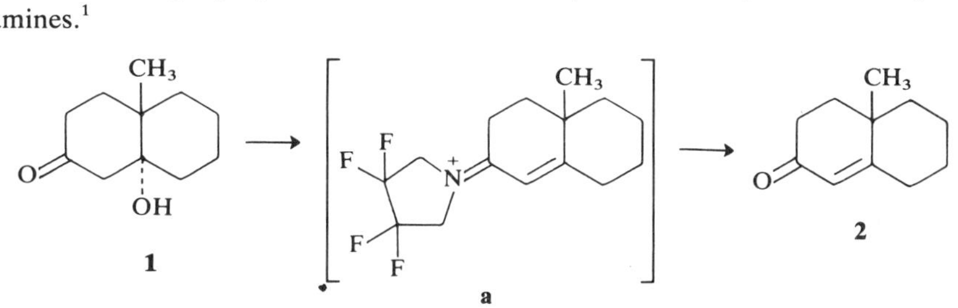

[1] R. D. Roberts and T. A. Spencer, *Tetrahedron Letters*, 2557 (1978).

2-Tetrahydrothienyl diphenylacetate (1). Mol. wt., 298.40, m.p. 82–83°.

Preparation:

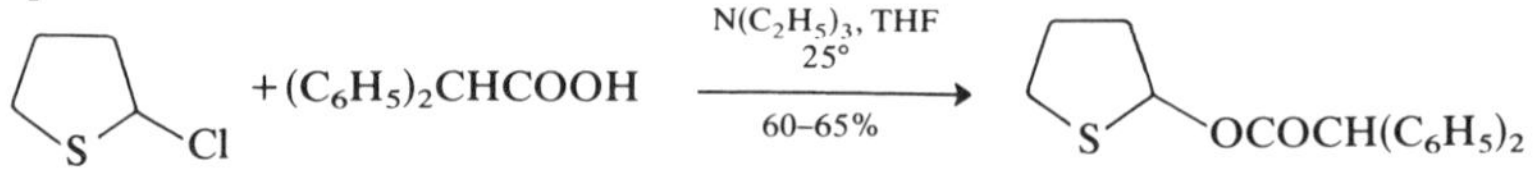

$$\text{2-chlorotetrahydrothiophene} + (C_6H_5)_2CHCOOH \xrightarrow[60-65\%]{N(C_2H_5)_3,\ THF,\ 25°} \text{2-}OCOCH(C_6H_5)_2\text{-tetrahydrothiophene}\ (\mathbf{1})$$

Protection of alcohols.[1] Primary and secondary alcohols are converted into the 2-tetrahydrothienyl (THT) ethers by acid-catalyzed exchange with **1**. The ethers can be cleaved quantitatively by $HgCl_2$ in aqueous acetonitrile; in fact the THT group is cleaved faster than the methylthiomethyl group (**6**, 302).

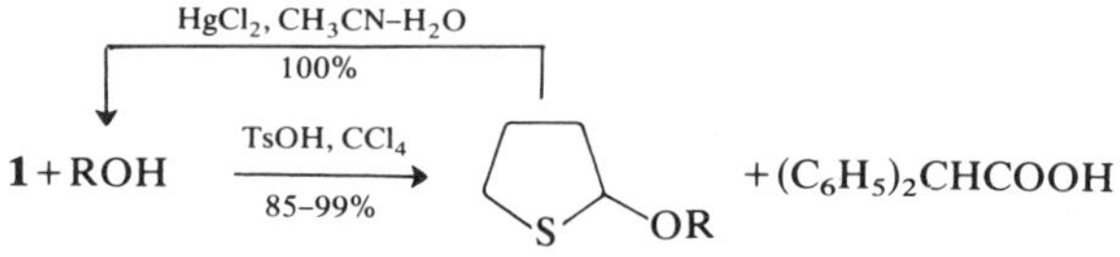

$$\mathbf{1} + ROH \xrightarrow[85-99\%]{TsOH,\ CCl_4} \text{2-}OR\text{-tetrahydrothiophene}\ (\mathbf{2}) + (C_6H_5)_2CHCOOH$$

($\mathbf{2} \rightarrow \mathbf{1} + ROH$: $HgCl_2$, CH_3CN-H_2O, 100%)

[1] C. G. Kruse, E. K. Poels, F. L. Jonkers, and A. van der Gen, *J. Org.*, **43**, 3548 (1978).

Tetrakis(triphenylphosphine)nickel(0), 6, 570; **7**, 357.

Phenanthrene synthesis.[1] A new phenanthrene synthesis employs the nickel(0)-catalyzed condensation of the imine **1** with **2** to give, after hydrolysis, the aldehyde **3**. This product was converted in several steps to the phenanthrene **4**, which is a key intermediate in a synthesis of the lignan steganone (**5**).[2]

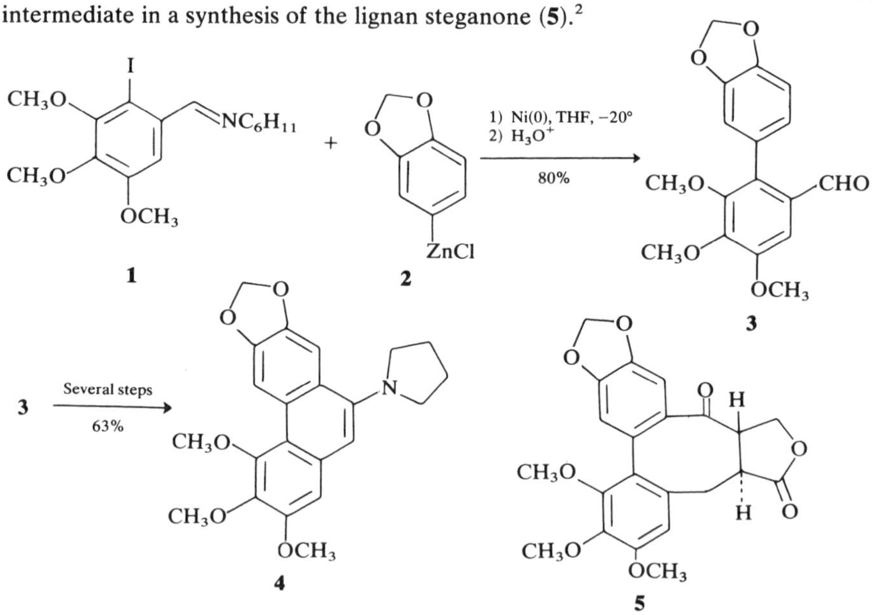

$$\mathbf{1} + \mathbf{2} \xrightarrow[80\%]{1)\ Ni(0),\ THF,\ -20°;\ 2)\ H_3O^+} \mathbf{3}$$

(**1**: 2-iodo-3,4,5-trimethoxybenzaldehyde N-C_6H_{11} imine; **2**: 3,4-methylenedioxyphenylzinc chloride, ZnCl; **3**: biaryl aldehyde with CH_3O, CH_3O, OCH_3, CHO)

$$\mathbf{3} \xrightarrow[63\%]{\text{Several steps}} \mathbf{4}$$

(**4**: methylenedioxy trimethoxy phenanthrene bearing a pyrrolidinyl N; **5**: steganone, CH_3O, CH_3O, OCH_3)

[1] E. R. Larson and R. A. Raphael, *Tetrahedron Letters*, 5041 (1979).
[2] D. Becker, L. R. Hughes, and R. A. Raphael, *J.C.S. Perkin I*, 1674 (1977).

Tetrakis(triphenylphosphine)palladium(0), 6, 571–573; **7,** 357–358; **8,** 472–476.

Cyclization to medium-sized lactones. The nucleophilic substitution of π-allylpalladium compounds[1] can be used to effect intramolecular ring closure.[2] The particular value is that, although two sites are available for cyclization, the kinetically less favored ring size can be obtained predominately. Thus formation of a 9-membered ring can occur exclusively rather than closure to a more stable 7-membered ring (equation I); the preference for formation of an 8-rather than a 6-membered ring (equation II) is even more surprising. The last example (equation III) shows that only a 10-membered ring can be formed when cyclization to an 8-membered ring is also possible.

(I)

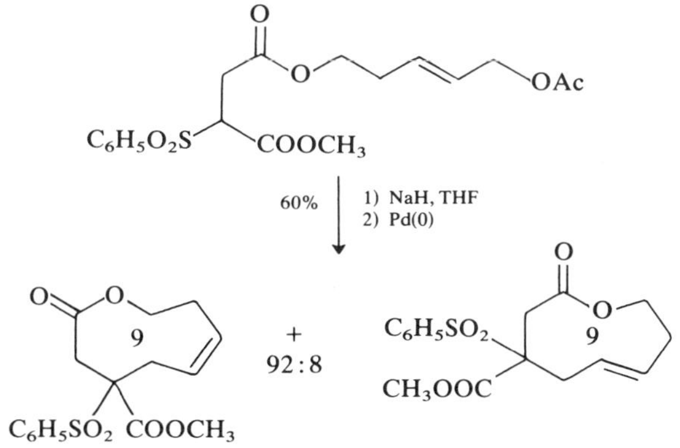

O; O; OAc; $C_6H_5O_2S$; $COOCH_3$

60% 1) NaH, THF 2) Pd(0)

O; O; 9; $C_6H_5SO_2$ $COOCH_3$ + 92:8 $C_6H_5SO_2$; O; O; 9; CH_3OOC

(II)

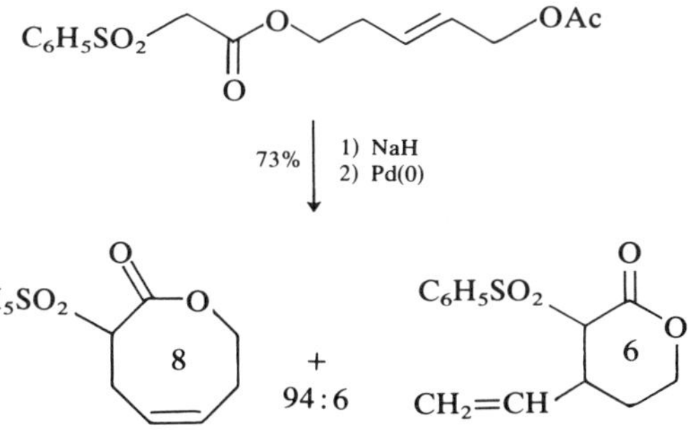

$C_6H_5SO_2$; O; O; OAc

73% 1) NaH 2) Pd(0)

$C_6H_5SO_2$; O; O; 8 + 94:6 $C_6H_5SO_2$; O; O; 6; $CH_2{=}CH$

(III)

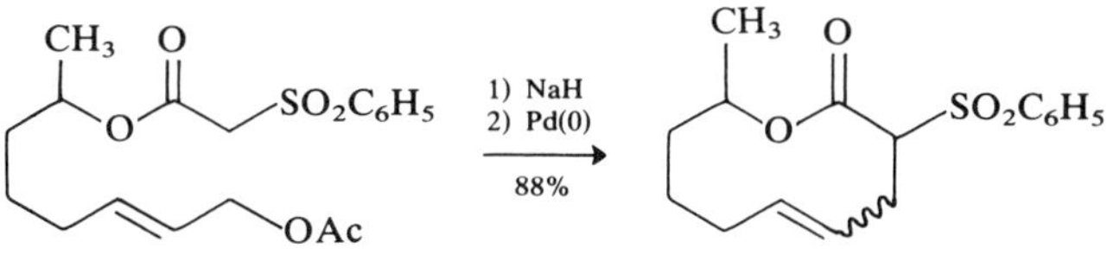

CH_3 O; O; $SO_2C_6H_5$; OAc 1) NaH 2) Pd(0) → 88% CH_3 O; O; $SO_2C_6H_5$

This preference for one site of cyclization over another, however, is not fixed, but depends on the substitution in the ring, the structure of the reacting centers, and the presence of ligands such as bis(1,2-diphenylphosphino)ethane.

***Chiral methyl chiral lactic acid* (5).**[3] This labeled molecule, useful for study of stereospecificity of enzymic reactions, has been prepared in a way that allows for synthesis of all 12 possible isomers. One key step is the stereospecific debromination of **1**, accomplished by conversion to the vinyl-palladium σ-complex **2** followed by cleavage with CF_3COOT to give the tritium-labeled **3**. The next step is the catalytic deuteration of **3**, accomplished with a rhodium(I) catalyst complexed with the ligands norbornadiene and (R)-1,2-bis(diphenylphosphino)propane.[4] This reaction gives **4** with an optical purity of 81%. The product is hydolyzed to **5**, which is obtained optically pure by crystallization.

H(Br)C=C(COOC₂H₅)(OAc) **1** $\xrightarrow{Pd[P(C_6H_5)_3]_4}$ H(Br)([(C₆H₅)₃P]₂Pd)C=C(COOC₂H₅)(OAc) **2** $\xrightarrow{CF_3COOT}$

H(T)C=C(COOC₂H₅)(OAc) **3** $\xrightarrow{D_2,\ RhL_1{}^*L_2ClO_4^-}$ H(D)(T)C–C(D)(COOC₂H₅)(OAc) **4** (81% ee) $\xrightarrow{NaOH}$ H(D)(T)C–C(D)(COOH)(OH) **5** (100% ee)

Isomerization of* 1,3-*diene epoxides. Noyori *et al.*[5] have reported three types of isomerization of allylic oxides in the presence of a catalytic amount of this palladium(0) complex. One class, exemplified by the isomerization formulated in equation (I), involves transfer of a hydrogen atom from the C_2-alkyl group and results in a dienol of the type obtained in the ene reaction of singlet oxygen with 1,3-dienes.

(I) $CH_2{=}CHC(CH_3)\text{–O–}CHR$ (epoxide) $\xrightarrow[80\%]{Pd(0)}$ $CH_2{=}CHC({=}CH_2){-}CH(OH)R$

A second class involves transfer of a C_5-hydrogen and is characteristic of allylic epoxides that do not have an alkyl substituent at C_2 (equation II).

(II) $R^1CH_2CH{=}CHCH\text{–O–}CHR^2$ (epoxide) $\xrightarrow[95\%]{Pd(0)}$ $R^1CH{=}CHCH{=}CHCH(OH)R^2$

The third type is common to monoepoxides of cyclic 1,3-dienes in five- to eight-membered rings and results in rearrangement to β,γ-unsaturated ketones in 55–80% yield. This isomerization can be utilized as one step in a route to 4-hydroxy-2-cycloenones. An example is a synthesis of 4-hydroxy-2-cyclopentenone formulated in equation (III).

(III) 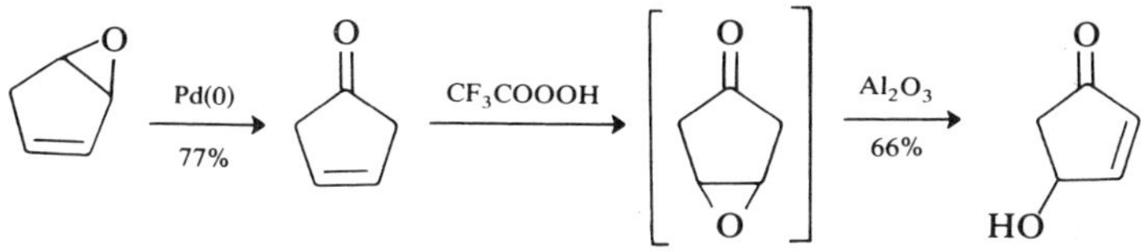 O → Pd(0), 77% → O → CF_3COOOH → [O, O] → Al_2O_3, 66% → O, HO

Alkylation of vinyl lactones. Trost and Klun[6] have reported that alkylation of π-allylpalladium complexes derived from vinyl lactones provides a stereocontrolled approach to acyclic systems such as the side chains of α-tocopherol and vitamin K. The alkylation of lactones **1–4** proceeds with greater than 95% stereoselectivity.

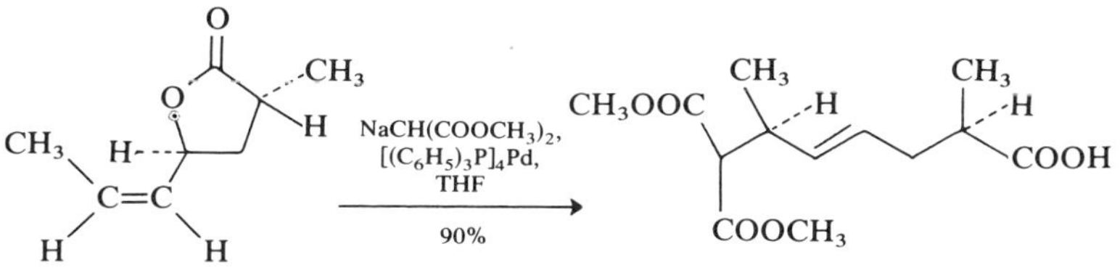

O, O, CH_3, H, CH_3, H, C=C, H, H

1

$NaCH(COOCH_3)_2$, $[(C_6H_5)_3P]_4Pd$, THF, 90%

CH_3OOC, CH_3, H, CH_3, H, COOH, $COOCH_3$

5

O, O, H, CH_3, CH_3, H, C=C, H, H

2

90%

CH_3OOC, CH_3, H, H, CH_3, COOH, $COOCH_3$

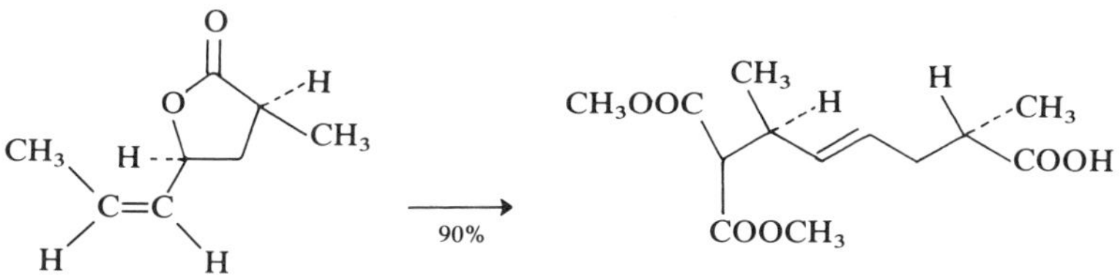

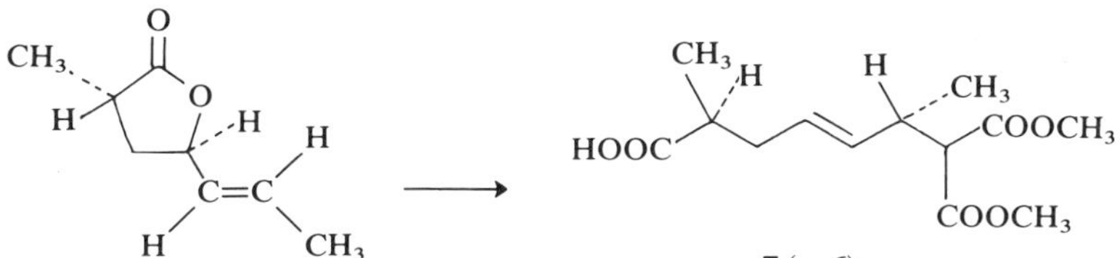

6

O, CH_3, H, O, H, H, C=C, H, CH_3

3

CH_3, H, H, CH_3, HOOC, $COOCH_3$, $COOCH_3$

7 (= **6**)

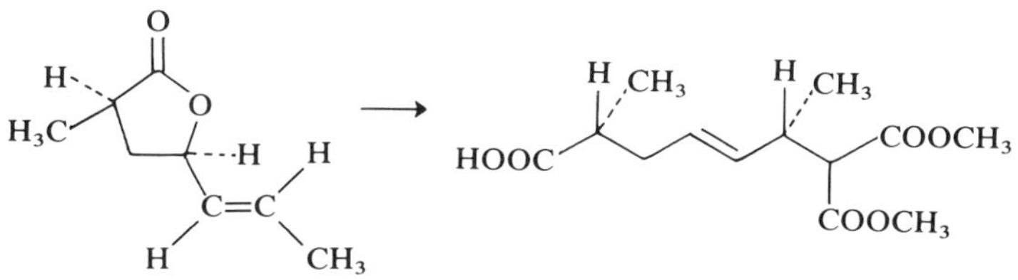

O, H, H_3C, O, H, H, C=C, H, CH_3

4

H, CH_3, H, CH_3, HOOC, $COOCH_3$, $COOCH_3$

8 (= **5**)

Reductive acylation of alkoxy-substituted allylic acetates.[7] Usual procedures do not permit alkylation of allylic acetates substituted with an alkoxyl group. A new method involves treatment of a substrate such as **1** with isopropyl benzenesulfonylacetate and DBU as base in the presence of 10–15 mole% of tetrakis(triphenylphosphine)palladium. A single product, **2**, is obtained. The carbonyl group is unmasked by acid treatment to give **3**. Treatment of **3** with base eliminates $C_6H_5SO_2H$ to give **4**, the product of reductive fumaroylation. The paper reports several applications of this methodology.

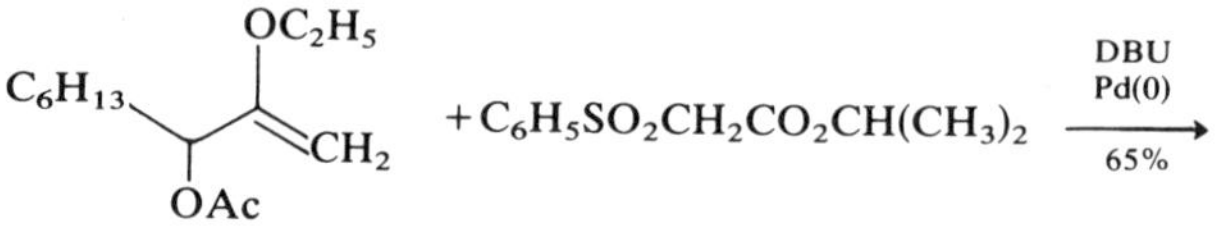

OC_2H_5, C_6H_{13}, CH_2, OAc **1** + $C_6H_5SO_2CH_2CO_2CH(CH_3)_2$ $\xrightarrow[65\%]{\text{DBU, Pd(0)}}$

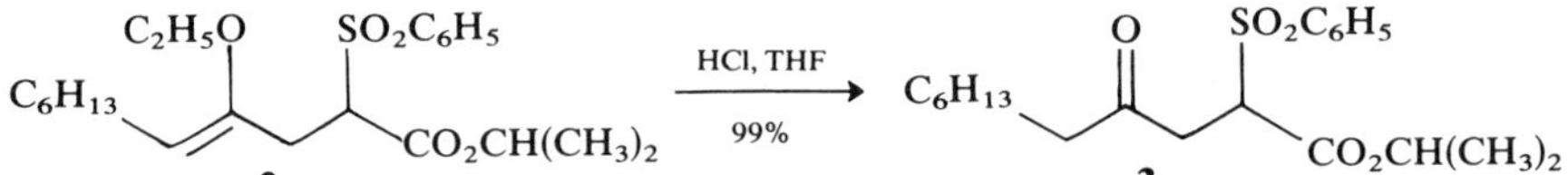

C_2H_5O, $SO_2C_6H_5$, C_6H_{13}, $CO_2CH(CH_3)_2$ **2** $\xrightarrow[99\%]{\text{HCl, THF}}$ O, $SO_2C_6H_5$, C_6H_{13}, $CO_2CH(CH_3)_2$ **3**

$\xrightarrow[81\%]{N(C_2H_5)_3,\ THF,\ 25°}$ O, C_6H_{13}, $CO_2CH(CH_3)_2$ **4**

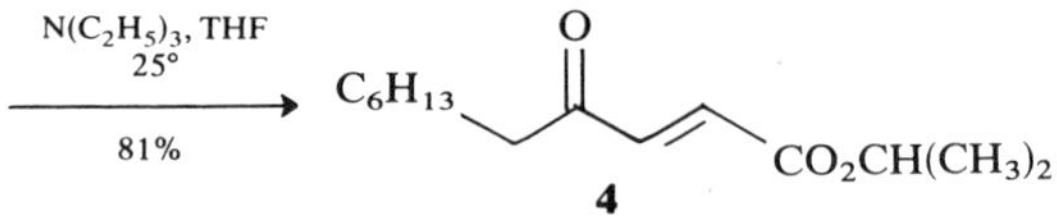

Methylenecyclopentanes. 2-Acetoxymethyl-3-allyltrimethylsilane (**1**) adds to a variety of electron-deficient alkenes in the presence of catalytic amounts of tetrakis(triphenylphosphine)palladium(0) and 1,2-bis(diphenylphosphino)ethane to afford methylenecyclopentanes.[8] The allylsilane **1** is prepared in four steps from methallyl alcohol.

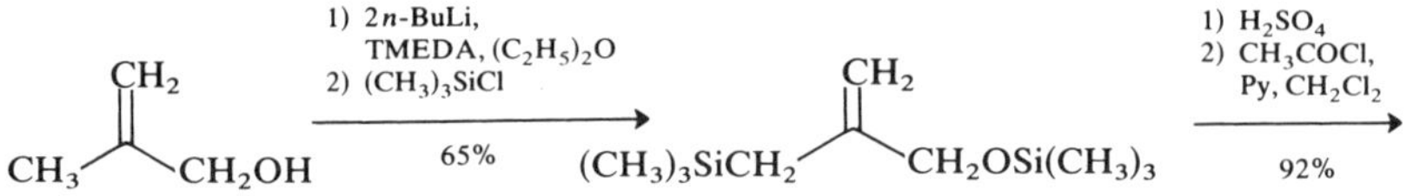

CH_2, CH_3, CH_2OH $\xrightarrow[65\%]{\text{1) 2}n\text{-BuLi, TMEDA, }(C_2H_5)_2O\text{; 2) }(CH_3)_3SiCl}$ CH_2, $(CH_3)_3SiCH_2$, $CH_2OSi(CH_3)_3$ $\xrightarrow[92\%]{\text{1) }H_2SO_4\text{; 2) }CH_3COCl,\ Py,\ CH_2Cl_2}$ CH_2, $(CH_3)_3SiCH_2$, CH_2OCOCH_3 **1**

Examples:

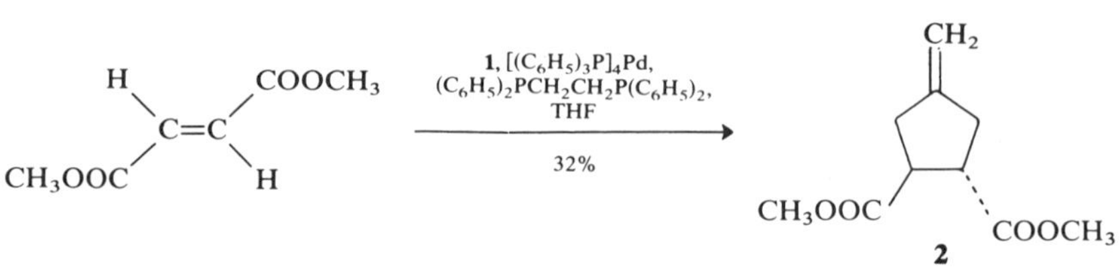

H, $COOCH_3$, C=C, CH_3OOC, H $\xrightarrow[32\%]{\mathbf{1},\ [(C_6H_5)_3P]_4Pd,\ (C_6H_5)_2PCH_2CH_2P(C_6H_5)_2,\ THF}$ CH_2, CH_3OOC, $COOCH_3$ **2**

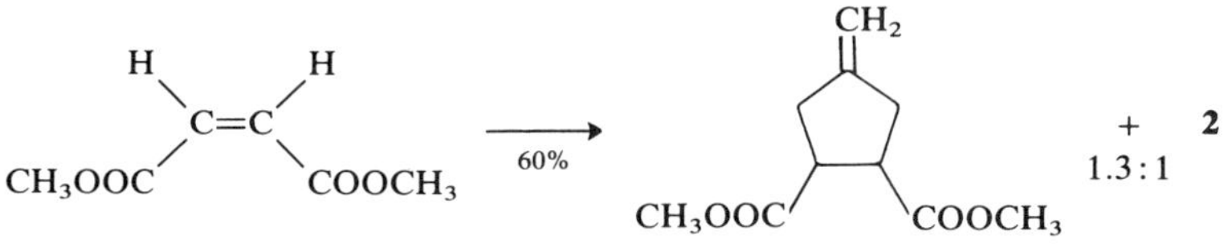

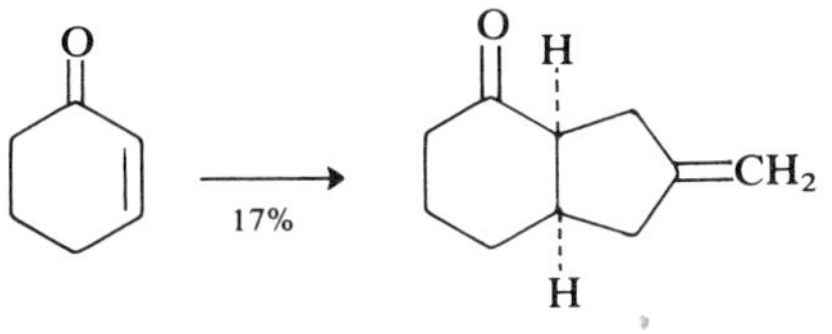

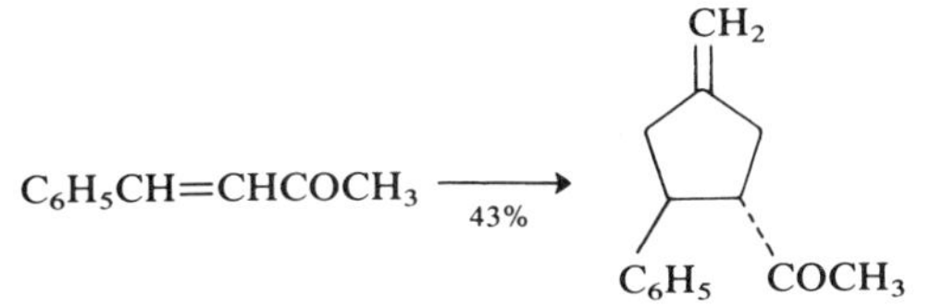

$$\text{cis-}CH_3OOC\text{-}CH{=}CH\text{-}COOCH_3 \xrightarrow[60\%]{} \text{3-methylene-cyclopentane-1,2-}(COOCH_3)_2 + \mathbf{2} \quad (1.3:1)$$

$$\text{2-cyclohexen-1-one} \xrightarrow[17\%]{} \text{methylene-hydrindanone (cis-H)}$$

$$C_6H_5CH{=}CHCOCH_3 \xrightarrow[43\%]{} \text{3-methylene-cyclopentane with } C_6H_5,\ COCH_3$$

Stereospecific synthesis of alkenes. This Pd(0) complex is an efficient catalyst for the cross-coupling of vinylic halides with a variety of Grignard reagents. The resulting alkenes are obtained in 97% isomeric purity and in yields of 75–90%. This reaction is particularly useful for preparation of pure 1,3-dienes (equation I).[9]

$$\text{(I)}\quad n\text{-}C_6H_{13}CH{=}CHI + IMgCH{=}CHCH_3 \xrightarrow[87\%]{Pd(0)} n\text{-}C_6H_{13}CH{=}CH\text{-}CH{=}CHCH_3$$

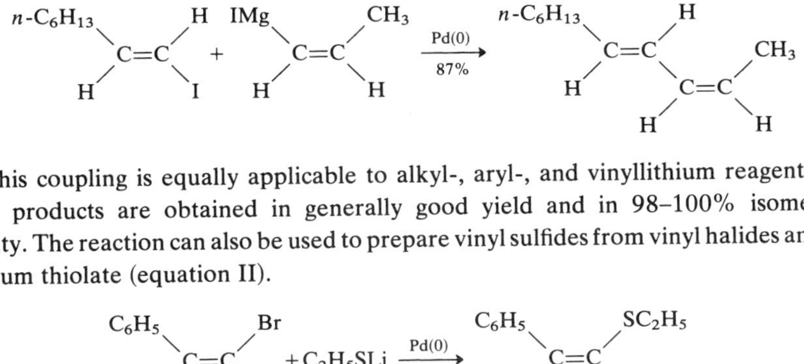

This coupling is equally applicable to alkyl-, aryl-, and vinyllithium reagents.[10] The products are obtained in generally good yield and in 98–100% isomeric purity. The reaction can also be used to prepare vinyl sulfides from vinyl halides and a lithium thiolate (equation II).

$$\text{(II)}\quad C_6H_5CH{=}CHBr + C_2H_5SLi \xrightarrow[93\%]{Pd(0)} C_6H_5CH{=}CHSC_2H_5$$

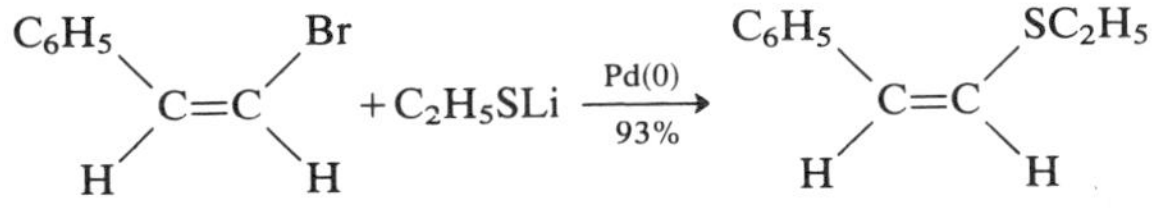

Cross-coupling of aryl halides and 1-alkenylboranes. Aryl iodides and bromides couple efficiently with 1-alkenylboranes (prepared from acetylenes and catecholborane, **4**, 70) in the presence of this Pd(0) complex and a base ($NaOC_2H_5$) (equation I). The reaction proceeds with retention of configuration with respect to the alkenylborane to form only (E)-alkenes. Yields are typically 81–100%. One example is reported where the alkenylborane was prepared with disiamylborane (yield 41%).[11]

(I) $ArX + R^1(H)C{=}C(R^2)(B\text{-catecholborane}) \xrightarrow[50\text{–}100\%]{1)\ Pd(0),\ base,\ \Delta;\ 2)\ H_2O_2,\ OH^-} R^1(H)C{=}C(R^2)Ar$

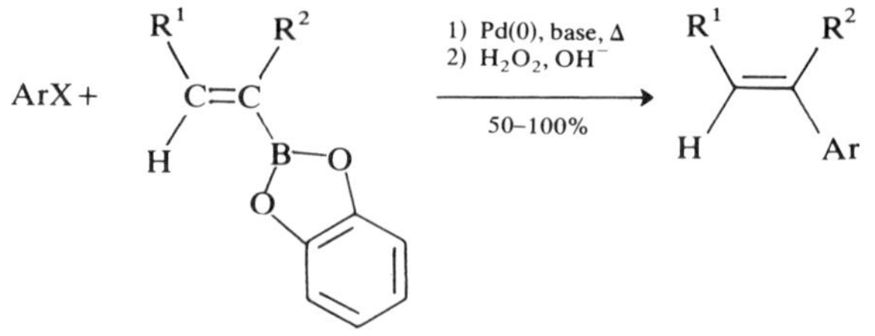

Iboga alkaloids. Trost's strategy for synthesis of the basic skeleton of this alkaloid family by way of organopalladium complexes (**7**, 357–358) has led to a synthesis of cantharanthine (**6**), a prominent member of this group.[11] A key step involved the intramolecular allylic alkylation of the amino group of **1**, which proceeded with fair selectivity for one acyloxy group as desired to give **2**. This product was cyclized to **3** by an internal olefin arylation with a solubilized form of $PdCl_2$ assisted with $AgBF_4$. Remaining steps involved hydrolysis to the alcohol and Moffatt oxidation to the ketone **4**. Reaction of **4** with ethylmagnesium bromide introduced the ethyl side chain to give **5**. This product had been previously converted into **6** by carbomethoxylation and dehydration.[12]

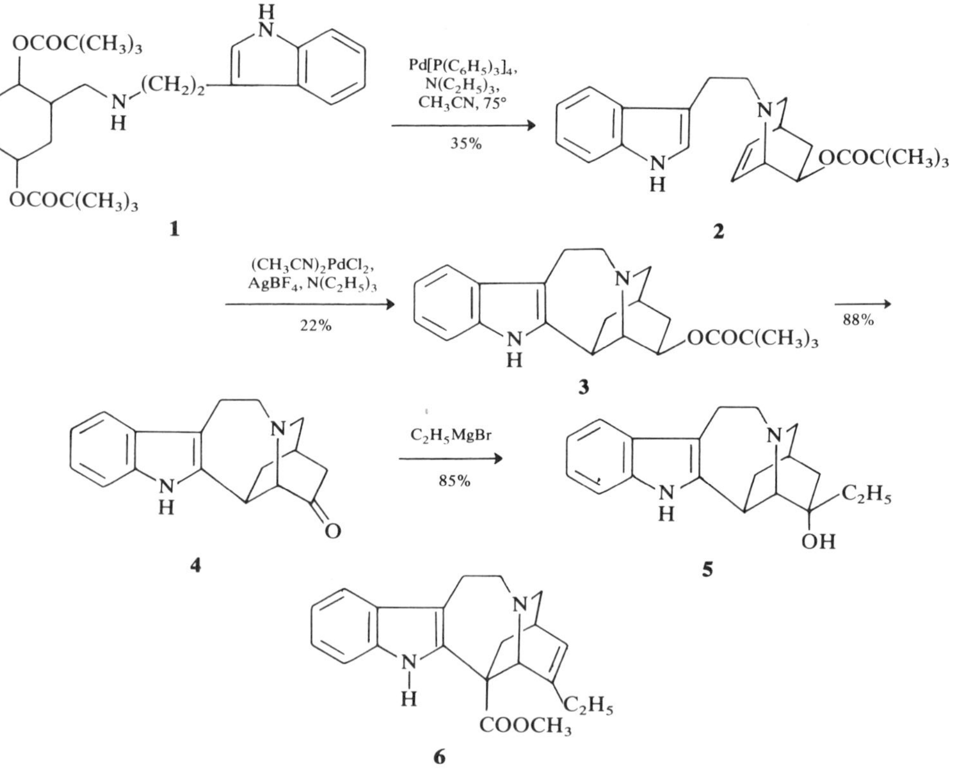

1 (with $OCOC(CH_3)_3$ groups, $N(H)$, $(CH_2)_2$) $\xrightarrow[35\%]{Pd[P(C_6H_5)_3]_4,\ N(C_2H_5)_3,\ CH_3CN,\ 75°}$ **2** ($OCOC(CH_3)_3$)

$\xrightarrow[22\%]{(CH_3CN)_2PdCl_2,\ AgBF_4,\ N(C_2H_5)_3}$ **3** ($OCOC(CH_3)_3$) $\xrightarrow{88\%}$

4 (O) $\xrightarrow[85\%]{C_2H_5MgBr}$ **5** (C_2H_5, OH)

6 (C_2H_5, $COOCH_3$)

Isomerization and elimination of allylic acetates.[13] In the presence of this Pd(0) species, allylic acetates undergo stereochemical isomerization and, more slowly, elimination to dienes (equation I). The conversion to dienes is general and can be effected usually in high yield, particularly when triethylamine is added to react with the acetic acid formed.

(I) [cyclohexene with $COOCH_3$ and OAc] $\xrightarrow{Pd(0)}$ [cyclohexene with $COOCH_3$ and OAc (isomerized)] + [cyclohexadiene with $COOCH_3$]

The isomerization of allylic acetates can be a problem during palladium-catalyzed alkylation of these substrates, but can be circumvented by use of the lactone of the corresponding hydroxy acid (equation II).

(II) [bicyclic lactone] $\xrightarrow[2)\ CH_2N_2]{1)\ Na^+\bar{C}H(SO_2C_6H_5)_2,\ Pd[P(C_6H_5)_3]_4}$ CH_3OOC–[cyclohexene]–$CH(SO_2C_6H_5)_2$ $\xrightarrow[2)\ Na/Hg]{1)\ H_2}$ CH_3OOC–[cyclohexane]–CH_3

Supported reagent.[14] Trost and Keinan have prepared two supported forms of this palladium(0) species: one supported on silica gel, named Ⓢ—Pd, and the other supported on polystyrene and named Ⓟ—Pd. Both are fairly stable to air, in contrast to tetrakis(triphenylphosphine)palladium.

Another advantage of these supported catalysts is an increase in selectivity in allylic alkylation as compared to the solubilized form of Pd. Two examples are shown in equations (I) and (II). Of course, the supported catalysts have the usual advantage of ease of removal and recovery.

(I) CH_3–CH=CH–CH=CH–CH_2OAc $\xrightarrow[Pd[P(C_6H_5)_3]_4,\ Ⓟ\text{-}Pd]{1)\ Na^+CH(SO_2C_6H_5)(COOCH_3)\quad 2)\ 6\%\ Na/Hg}$

CH_3–CH=CH–CH=CH–$(CH_2)_2COOCH_3$ + CH_3–CH=CH–CH(CH_2COOCH_3)–CH=CH_2

69%	31%
80%	20%

(II) [cyclohexenyl acetate with $COOCH_3$, OAc] $\xrightarrow{HN(C_2H_5)_2}$ [product with $COOCH_3$, $N(C_2H_5)_2$] + [product with $COOCH_3$, $N(C_2H_5)_2$ (dashed)]

Catalyst	Yield	Ratio
$Pd[P(C_6H_5)_3]_4$	85%	67:33
Ⓟ-Pd	83%	100:0
Ⓢⓘ-Pd	72%	100:0

[1] Review: B. M. Trost, *Tetrahedron*, **33**, 2615 (1977).
[2] B. M. Trost and T. R. Verhoeven, *Am. Soc.*, **101**, 1595 (1979).
[3] M. D. Fryzuk and B. Bosnich, *ibid.*, **101**, 3043 (1979).
[4] *Idem, ibid.*, **100**, 5491 (1978).
[5] M. Suzuki, Y. Oda, and R. Noyori, *ibid.*, **101**, 1623 (1979).
[6] B. M. Trost and T. P. Klun, *ibid.*, **101**, 6756 (1979).
[7] B. M. Trost and F. N. Gowland, *J. Org.*, **44**, 3448 (1979).
[8] B. M. Trost and D. M. T. Chan, *Am. Soc.*, **101**, 6429 (1979).
[9] H. P. Dang and G. Linstrumelle, *Tetrahedron Letters*, 191 (1978).
[10] S.-I. Murahashi, M. Yamamura, K. Yanagisawa, N. Mita, and K. Kondo, *J. Org.*, **44**, 2408 (1979).
[11] B. M. Trost, S. A. Godleski, and J. L. Belletire, *J. Org.*, **44**, 2052 (1979).
[12] G. Büchi, P. Kulsa, K. Ogasawara, and R. L. Rosati, *Am. Soc.*, **92**, 999 (1970).
[13] B. M. Trost, T. R. Verhoeven, and J. M. Fortunak, *Tetrahedron Letters*, 2301 (1979).
[14] B. M. Trost and E. Keinan, *Am. Soc.*, **100**, 7779 (1978).

1,1,3,3-Tetramethylbutyl isocyanide, 3, 279–280; 4, 480.

Review.[1] The syntheses and uses of isocyanides have been reviewed (34 references).

[1] M. P. Periasamy and H. M. Walborsky, *Org. Prep. Proc. Int.*, **11**, 295 (1979).

Thallium(I) acetate–Iodine, 7, 359–360.

cis,vic-Diols.[1] The reaction of iodine and a thallium(I) carboxylate with an alkene gives a *trans*-iodocarboxylate (**2**) in high yield (**5**, 654–655). Oxidation of **2** by *m*-chloroperbenzoic acid results in displacement of the iodine by a hydroxyl group with inversion to give the monocarboxylate of a *cis*-diol (**3**) in high yield. The *cis*-diol is obtained by reduction with $LiAlH_4$. This reaction is an alternative to the Prévost reaction.

Simple primary and secondary alkyl iodides are also converted into alcohols by peracid oxidation in high yield.

1 $\xrightarrow[90\%]{I_2,\ TlOCOCH_3}$ **2** (I, $OCOCH_3$) $\xrightarrow[89\%]{ClC_6H_4CO_3H}$ **3** (OH, $OCOCH_3$) $\xrightarrow[>95\%]{LiAlH_4}$ **4** (OH, OH)

[1] R. C. Cambie, B. G. Lindsay, P. S. Rutledge, and P. D. Woodgate, *J.C.S. Chem. Comm.*, 919 (1978).

Thallium(III) acetate, 1, 1150–1151; **2,** 406; **3,** 286; **4,** 492; **5,** 655; **7,** 360–361.

Enamine oxidation (**6,** 406); ***synthesis of vinblastine*** (**3**). Treatment of enamine (**2**) with thallium(III) acetate followed by $NaBH_4$ reduction affords vinblastine (**3**) in 30% overall yield from the N-oxide **1**. In contrast, treatment of **2** with OsO_4 followed by $NaBH_4$ affords leurosidine (**4**) in 25% overall yield. Potier *et al.*[1] argue that the transformation **2** → **3** is governed by stereoelectronic factors, whereas steric approach control governs **2** → **4**. These transformations may have biogenetic implications.

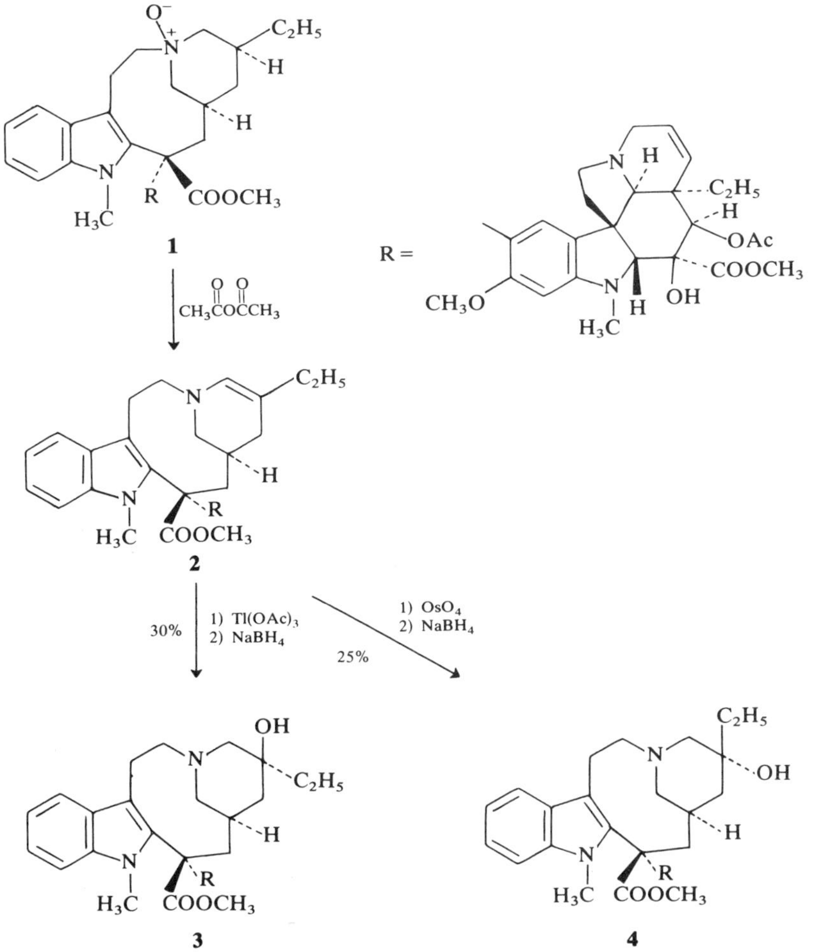

O^- N^+ C_2H_5 H H R $COOCH_3$ H_3C N

1

CH_3COCCH_3 (O O)

R = CH_3O N H C_2H_5 H OAc $COOCH_3$ N H OH H_3C

N C_2H_5 H N R H_3C $COOCH_3$

2

30% 1) $Tl(OAc)_3$ 2) $NaBH_4$

25% 1) OsO_4 2) $NaBH_4$

OH N C_2H_5 H N R H_3C $COOCH_3$

3

C_2H_5 N OH H N R H_3C $COOCH_3$

4

[1] P. Mangeney, R. Z. Andriamialisoa, N. Langlois, Y. Langlois, and P. Potier, *Am. Soc.*, **101,** 2243 (1979).

Thallium(III) nitrate (TTN), 4, 492–497; **5,** 656–657; **6,** 578–579; **7,** 362–365; **8,** 476–478.

α-Carbonyl dimethyl acetals. *β*-Carbonyl sulfides, which can be obtained in several ways, react with TTN in methanol at room temperature to form *α*-carbonyl acetals and ketals.[1,2]

Examples:

$$CH_3OCOCH_2Br \xrightarrow[73\%]{TlSC_2H_5} CH_3OCOCH_2SC_2H_5 \xrightarrow[87\%]{1)\ LDA,\ THF\text{–}HMPT;\ 2)\ C_6H_5CH_2Br} CH_3OCOCH(SC_2H_5)CH_2C_6H_5 \xrightarrow[85\%]{TTN,\ CH_3OH} CH_3OCOC(OCH_3)_2CH_2C_6H_5$$

$$\text{5-}CH_3(CH_2)_3\text{-}\gamma\text{-lactone} \xrightarrow[53\%]{C_2H_5SSC_2H_5} \alpha\text{-}SC_2H_5\text{ lactone} \xrightarrow[68\%]{TTN,\ CH_3OH} \alpha,\alpha\text{-}(OCH_3)_2\text{ lactone}$$

$$CH_3OCOCH_2SC_2H_5 \xrightarrow[53\%]{TTN,\ CH_3OH} CH_3OCOCH(OCH_3)_2$$

p-Quinone monoketals.[3] Oxidation of 5-methoxy-1-naphthol (**1**) with TTN and trimethyl orthoformate (**7**, 363) in ethylene glycol at −40° gives the 4-monoketal of juglone methyl ether (**2**) in 27% yield. The corresponding *p*-quinone is also formed, but can be separated from **2** by formation of the bisulfite addition complex.

$$\mathbf{1} \xrightarrow[27\%]{TTN,\ HC(OCH_3)_3,\ HOCH_2CH_2OH} \mathbf{2}$$

Oxidation of flavylium salts to flavones.[4] The highest yield previously reported for this direct oxidation is about 10% (O_2, Na_2CO_3 or NaOH, 20°).[5] The oxidation of **1** to **2** can be effected in about 70% yield with TTN in methanol.

$$\mathbf{1}\ (R = H,\ OCH_3,\ CH_3),\ ClO_4^- \xrightarrow{Tl(NO_3)_3,\ CH_3OH} [\mathbf{a}] \xrightarrow{\sim 70\%} \mathbf{2}$$

Certain 2-hydroxychalcones (**3**) are also converted into flavones under the same conditions and in comparable yield.

$$\mathbf{3}\ (R^1 \text{ or } R^2 = H, OCH_3, CH_3) \xrightarrow[\sim 70\%]{Tl(NO_3)_3,\ CH_3OH} \mathbf{4}\ (2\text{-}C_6H_3R^1R^2\text{ flavone})$$

Oxythallative cyclization (**6**, 578–579; **7**, 365). The reaction of elemol acetate (**1**) with TTN in HOAc gives as the main product a diacetate, which is reduced by lithium aluminum hydride to the guaiene diol **2**. On the other hand, elemol is cyclized by mercury(II) acetate under similar conditions without rearrangement to crytomeridiol (**3**). These results may have biogenetic significance, since they suggest that enzymatic reactions can vary with different metal-containing enzymes.[6]

$$\mathbf{1} \xrightarrow[63.5\%]{(R = Ac)\ 1)\ TTN, HOAc;\ 2)\ LiAlH_4} \mathbf{2}$$

$$\mathbf{1} \xrightarrow[69\%]{(R = H)\ 1)\ Hg(OAc)_2;\ 2)\ LiAlH_4} \mathbf{3}$$

α-Nitrato ketones. Thallium(III) nitrate in acetonitrile oxidizes enolizable ketones to α-nitrato ketones (60–80°, 12 hours).[7]

Examples:

$$RC_6H_4COCH_3 \xrightarrow[85\text{–}95\%]{Tl(NO_3)_3,\ CH_3CN} RC_6H_4COCH_2ONO_2$$

$(R = H, CH_3, OCH_3, Br, NO_2)$

$$CH_3CH_2COCH_2CH_3 \xrightarrow{85\%} CH_3CH(ONO_2)COCH_2CH_3$$

$$C_6H_5COCH_2CH_3 \xrightarrow{90\%} C_6H_5COCH(ONO_2)CH_3$$

Hydrolysis of thioacetals (**7**, 364). TTN has been used for selective dethioacetalization of the bis thioketal **1** to the mono thioketal **2**. The paper includes examples of hydrolysis of simpler thioacetals.[8]

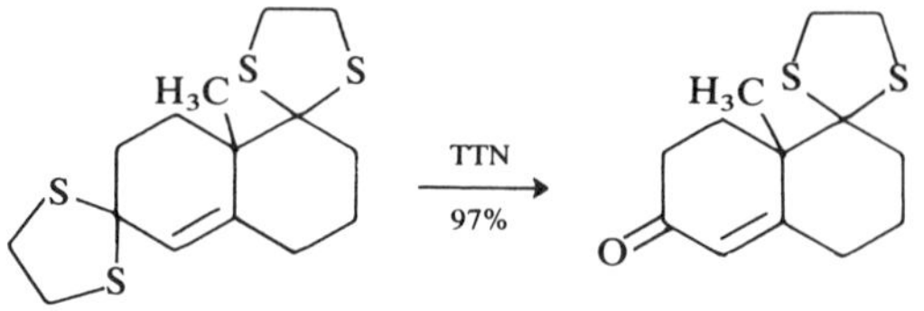

$$\mathbf{1} \xrightarrow[97\%]{\text{TTN}} \mathbf{2}$$

[1] Y. Nagao, M. Ochiai, K. Kaneko, A. Maeda, K. Watanabe, and E. Fujita, *Tetrahedron Letters*, 1345 (1977).
[2] Y. Nagao, K. Kaneko, and E. Fujita, *ibid.*, 4115 (1978).
[3] D. J. Crouse and D. M. S. Wheeler, *ibid.*, 4797 (1979).
[4] M. Meyer-Dayan, B. Bodo, C. Deschamps-Vallet, and D. Molho, *ibid.*, 3359 (1978).
[5] D. W. Hill and R. R. Melhuish, *J. Chem. Soc.*, 1161 (1935).
[6] W. Renold, G. Ohloff, and T. Norin, *Helv.*, **62**, 985 (1979).
[7] A. McKillop, D. W. Young, M. Edwards, R. P. Hug, and E. C. Taylor, *J. Org.*, **43**, 3773 (1978).
[8] R. A. J. Smith and D. J. Hannah, *Syn. Comm.*, **9**, 301 (1979).

Thallium(III) trifluoroacetate (TTFA), 3, 286–289; **4,** 496–501; **5,** 658–659; **7,** 365; **8,** 478–481.

Oxidative dimerization of cinnamic acids.[1] Cinnamic acid itself is stable to TTFA in the presence of BF_3 etherate; but 4-alkoxycinnamic acids are converted by these two reagents in TFA–CH_2Cl_2 to fused bislactones by a nonphenolic oxidative dimerization (equation I). A few natural products of this type are known.

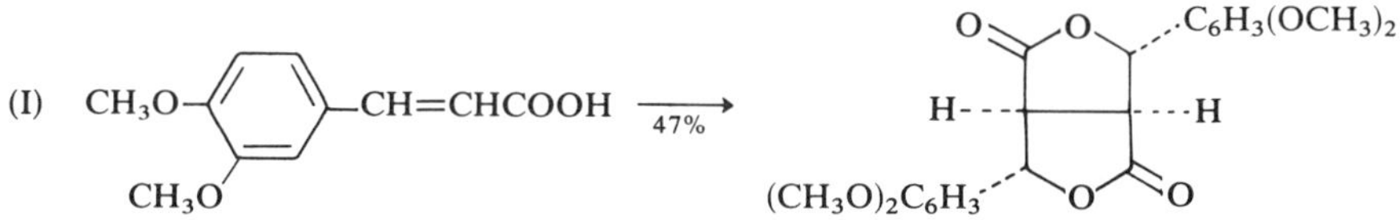

(I) $(CH_3O)_2C_6H_3CH{=}CHCOOH \xrightarrow{47\%}$ bislactone with two $C_6H_3(OCH_3)_2$ groups

Diaryl sulfones.[2] Arylthallium bistrifluoroacetates can be converted into diaryl sulfones by reaction with an aqueous solution of sodium benzenesulfinate and copper(II) sulfate (4 : 1) under mild conditions. The copper salt is essential.

$$ArTl(OCOCF_3)_2 + C_6H_5SO_2Na \xrightarrow[60\%]{CuSO_4} ArSO_2C_6H_5$$

Oxidative* para-ortho *phenol coupling (**6**, 580–581). Oxidation of the N-acylsecoreticulines **1a** and **1b** with TTFA (1 equiv., CH_2Cl_2, 25°) results in N-acylsecosalutaridines **2a** and **2b** accompanied by some minor oxidation products. Hydrolysis of the $-COCF_3$ group of **2a** was accompanied by spontaneous cyclization to the hasubanan derivative **3** in 75% yield. Reduction of **2b** ($NaBH_4$) followed by sequential treatment with $SOCl_2$–pyridine and then hot aqueous NaOH affords the secothebaine **4** in 57% overall yield. This product was converted into the secocodeine **5** by reaction with singlet oxygen and then with sodium borohydride.

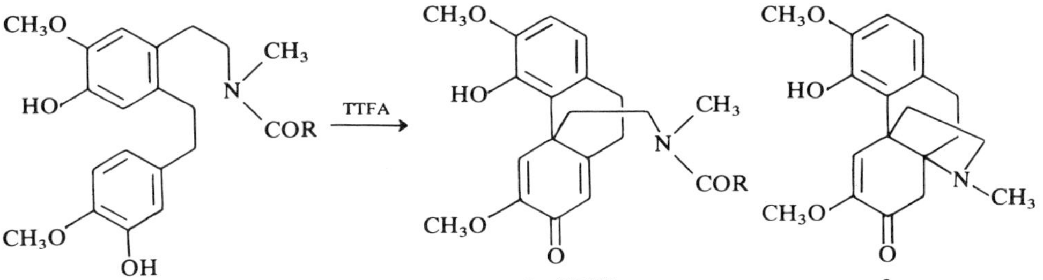

1a ($R = CF_3$)
1b ($R = OCH_3$)

2a (15%)
2b (24%)

3

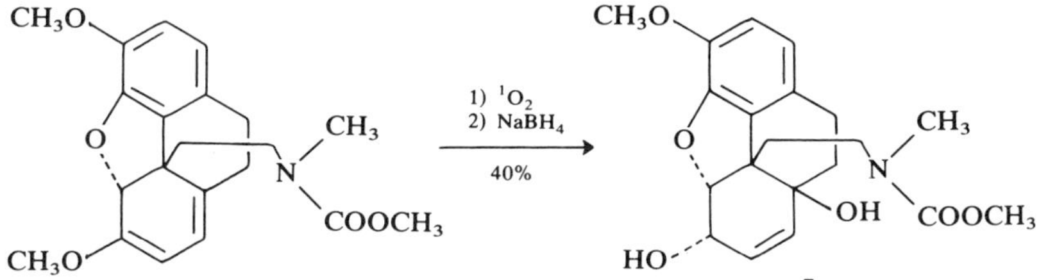

1) $^1O_2$
2) $NaBH_4$
40%

4

5

The oxidation of **1a** with $VOCl_3$ (**3**, 331–332; **5**, 744) does not give **2a**, but instead **6** and **7** in about equal amounts. These are minor products (3 and 5%, respectively) in the oxidation of **1a** with TTFA.[3]

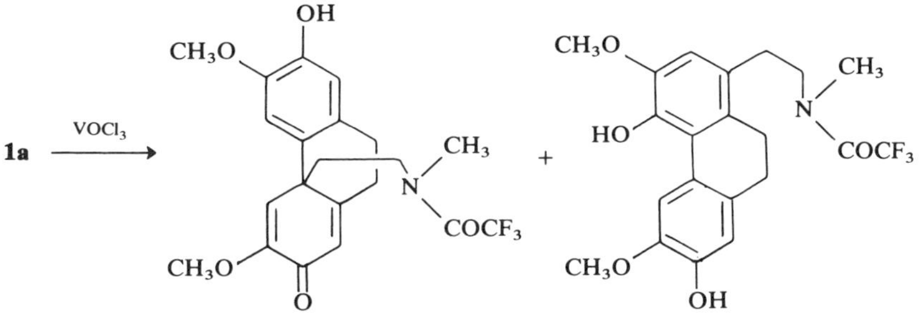

1a $\xrightarrow{VOCl_3}$

6 (45%)

7 (46%)

Oxidation of 3-(alkoxyaryl)propionic acids. This reaction of TTFA can be used for synthesis of dihydrocoumarins and spirohexadienone lactones and of *p*-benzoquinones. An example of both sequences is formulated in equation (I).[4]

[1] E. C. Taylor, J. G. Andrade, G. J. H. Rall, and A. McKillop, *Tetrahedron Letters*, 3623 (1978).
[2] R. A. Hancock and S. T. Orszulik, *ibid.*, 3789 (1979).
[3] M. A. Schwartz and R. A. Wallace, *ibid.*, 3257 (1979).
[4] E. C. Taylor, J. G. Andrade, G. J. H. Rall, and A. McKillop, *J. Org.*, **43**, 3632 (1978).

(I) [Reaction scheme: 3-(3,4-dimethoxyphenyl)propanoic acid with TTFA, TFA, $BF_3 \cdot (C_2H_5)_2O$ (57%) gives the 6,7-dimethoxy lactone + spirodienone lactone (3:1); with 1) TTFA, 2) CH_3OH gives the methyl ester phenol, which with TTFA (73%) gives the methoxy-p-quinone ester $COOCH_3$]

Thexylborane–N,N-Diethylaniline (TBDA), $(CH_3)_2CH-C(CH_3)_2-BH_2 \cdot C_6H_5N(C_2H_5)_2$. Stable at 0°, soluble in hydrocarbon or ethereal solvents.

Hydroboration.[1] Thexylborane stabilized as the triethylamine complex is not useful for hydroboration, because 2,3-dimethyl-2-butene is displaced with formation of $RBH_2-N(C_2H_5)_3$. However, TBDA is a useful reagent for hydroboration and for various reductions. Thus it reacts with 1-octene to form di-*n*-octylthexylborane in quantitative yield. It is comparable to thexylborane–THF for reduction of aldehydes and ketones. Carboxylic acids are reduced to the corresponding alcohol. 10-Undecenoic acid is reduced selectively to undecanoic acid (90% yield). Tertiary amides are reduced very rapidly to *t*-amines. Acid chlorides and nitriles are reduced very slowly.

[1] A. Pelter, D. J. Ryder, and J. H. Sheppard, *Tetrahedron Letters*, 4715 (1978).

Thioanisole–Trifluoromethanesulfonic acid.

Deprotection of O-methyltyrosine.[1] O-Methylation of tyrosine (**1**) has represented an irreversible protection of the phenol group, since cleavage is accomplished only under drastic conditions. A new, mild method for deprotection involves treatment of **1** with thioanisole and trifluoromethanesulfonic acid in trifluoroacetic acid. This method was used in the final step of a synthesis of a pentapeptide (43% yield).

$$\underset{\mathbf{1}}{p\text{-}CH_3OC_6H_4CH_2CH(\overset{\oplus}{N}H_3)COO^{\ominus}} + C_6H_5SCH_3 \xrightarrow[\sim 100\%]{CF_3SO_3H,\ CF_3COOH} \underset{\mathbf{2}}{p\text{-}HOC_6H_4CH_2CH(\overset{\oplus}{N}H_3)CO_2H} + C_6H_5\overset{\oplus}{S}(CH_3)_2 + 2CF_3SO_3^-$$

[1] Y. Kiso, S. Nakamura, K. Ito, K. Ukawa, K. Kitagawa, T. Akita, and H. Moritoki, *J.C.S. Chem. Comm.*, 971 (1979).

Thionyl chloride, 1, 1158–1163; **2,** 412; **3,** 290; **4,** 503–505; **5,** 663–667; **6,** 585; **7,** 366–367; **8,** 481.

Pyrazoles.[1] Aryl-substituted pyrazoles (**3**) are available in 68–80% overall yield from enimines (**1**) by condensation with thionyl chloride in pyridine followed by thermal extrusion of SO to form the N—N bond.

1, ($R = H, CH_3$) $+ SOCl_2$ —(Py, 25°; 77–87%)→ **2**

2 —(Toluene, 90°; 84–93%)→ **3**

1 $+ SOCl_2$ —(Py, 90°; 68–80%)→ **3**

[1] J. Barluenga, J. F. López-Ortiz, and V. Groton, *J.C.S. Chem. Comm.*, 891 (1979).

Thionyl chloride–Dimethylformamide.

Benzal chlorides.[1] Benzaldehyde or the bisulfite addition product reacts with thionyl chloride and DMF at $-10 \rightarrow 20°$ to form benzal chloride, $C_6H_5CHCl_2$, in $\sim 88\%$ yield. Newman favors structure **1** for the Vilsmeier complex involved in this reaction. This reaction is general for aromatic aldehydes and α,β-unsaturated aldehydes. The reaction is not observed with diaryl ketones.

$$[(CH_3)_2N \cdots CHOSOCl]^+Cl^-$$

1

[1] M. S. Newman and P. K. Sujeeth, *J. Org.*, **43**, 4367 (1978).

Thiophenol, 4, 505; **5,** 585–586; **6,** 458–459; **7,** 367–368.

Reduction of aryl diazonium salts.[1] Aryl diazonium tetrafluoroborates are reduced to arenes by thiophenol in 85–100% yield. Replacement by deuterium can be effected with C_6H_5SD.

$$ArN_2^+BF_4^- + C_6H_5SH \xrightarrow[85\text{–}100\%]{H_2O,\ C_5H_{12},\ 3\ hr,\ 20°} ArH$$

[1] T. Shono, Y. Matsumura, and K. Tsubata, *Chem. Letters*, 1051 (1979).

Thiophosgene, 5, 667–668.

Review.[1] Recent developments in the use of thiophosgene for synthesis of various heterocycles have been reviewed (226 references).

[1] S. Sharma, *Synthesis*, 803 (1978).

Thiophosphoryl bromide, $PSBr_3$. Mol. wt. 302.76, m.p. 38°. Supplier: Alfa. The reagent is prepared from red phosphorus, bromine, and phosphorus pentasulfide (67% yield).[1]

Reduction of sulfoxides.[2] Sulfoxides are reduced to sulfides in generally high yield by reaction with 1 equiv. of this reagent in CH_2Cl_2 at room temperature. The reaction times vary from 5–10 minutes to 24 hours. Carbonyl groups do not react. One selenoxide has been reduced in this way.

This reagent is superior to phosphorus pentasulfide[3] in respect to yields, although reaction times are more variable. Thiophosphoryl chloride, $PSCl_3$, is ineffective for this reduction.

[1] H. S. Booth and C. A. Seabright, *Inorg. Syn.*, **2**, 153 (1946).
[2] I. W. J. Still, J. N. Reed, and K. Turnbull, *Tetrahedron Letters*, 1481 (1979).
[3] I. W. J. Still, S. K. Hasan, and K. Turnbull, *Can. J. Chem.*, **56**, 1423 (1978).

2-Thiopyridyl chloroformate, (2-pyridyl)—SCOCl (**1**). Mol. wt. 174.62, very unstable to water and silica gel; can be stored for 1 month at −25°. The reagent is prepared by reaction of 2-pyridinethiol and phosgene (96% yield).

2-Pyridinethiol esters. In combination with triethylamine **1** converts acids into 2-pyridinethiol esters in over 95% yield. The method is superior to the usual earlier route involving treatment of an acid with 2,2′-dipyridyl disulfide and triphenylphosphine (**5**, 286; **6**, 246).[1]

[1] E. J. Corey and D. A. Clark, *Tetrahedron Letters*, 2875 (1979).

Titanium(0), 7, 368–369; **8**, 481–482.

McMurry reductive carbonyl coupling (**7**, 368–369). The complete paper on reductive carbonyl coupling to alkenes is available. Ti(0) can be prepared by reduction of $TiCl_3$ with either potassium or lithium, the former reagent being somewhat more effective. The optimum ratio of $TiCl_3$/K/ketone is 4:14:1. Although yields are highest with two identical ketones, unsymmetrical coupling is efficient when one ketone is a diaryl ketone. Intramolecular dicarbonyl coupling to cyclic alkenes is possible; in this case the combination $TiCl_3$ with Zn–Cu is the most efficient reagent. The reductions of some 1,2-diols to alkenes with $TiCl_3$–K is also possible.[1]

Reductive coupling of carbonyl compounds to diols. Activated Ti(0) (**2**, 368–369) effects the reductive coupling of carbonyl compounds to olefins and also converts 1,2-diols to alkenes (**7**, 368–369). Corey *et al.*[2] employed this reagent for the intramolecular coupling of **1** in their synthesis of gibberellic acid. Reductive deoxygenation of this diol, which would produce a strained bridgehead olefin, was not observed under their conditions.

H H

Ti(0) 55%

O CHO THPOCH$_2$ **1** ---OH OH THPOCH$_2$ **2**

Deoxygenation of phenols.[3] The reduction of enol phosphates to alkenes by titanium metal (**8**, 482) has been extended to reduction of aryl diethyl phosphates to arenes. Yields are in the range 75–95%; reduction with lithium in liquid ammonia (**1**, 248) usually proceeds in low yield.

Example:

OH 1) NaH 2) $ClPO(OC_2H_5)_2$ 91% $OPO(OC_2H_5)_2$ Ti, THF, Δ 74%

[1] J. E. McMurry, M. P. Fleming, K. L. Kees, and L. R. Krepski, *J. Org.*, **43**, 3255 (1978).
[2] E. J. Corey, R. L. Danheiser, S. Chandrasekaran, P. Siret, G. E. Keck, and J.-L. Gras, *Am. Soc.*, **100**, 8031 (1978).
[3] S. C. Welch and M. E. Walters, *J. Org.*, **43**, 4797 (1978).

Titanium(III) chloride, 2, 415; **4**, 506–508; **5**, 669–671; **6**, 587–588; **7**, 369; **8**, 482–483.

4-Methylaminomethylindole.[1] The reaction of 2-methyl-5-nitroisoquinolinium iodide (**1**) with $TiCl_3$ (7.7 equiv.) gives 4-methylaminomethylindole (**2**) directly in 20% yield. The conversion proceeds via reduction of the nitro group to an amino group, ring fission, and recyclization.

NO_2 I^- $\overset{+}{N}$ CH_3 **1** $TiCl_3$, H_2O, 19° 20% CH_2NHCH_3 N H **2**

[1] M. Somei, F. Yamada, and C. Kaneko, *Chem. Letters*, 943 (1979).

Titanium(III)chloride–Diisobutylaluminum hydride.

Oxime → imine.[1] The low valent titanium reagent prepared from $TiCl_3$ and diisobutylaluminum hydride (1:3) in THF reduces the oxime **1** to the very labile imine **2**. Reduction of **2** to the corresponding amine with diisobutylaluminum hydride followed by formylation and dehydration affords (±)-9-isocyanopupukeanane (**3**) in ~53% overall yield.

H_3C, N—OH, CH_3, $CH(CH_3)_2$ **1** —DIBAH–$TiCl_3$ (3:1), THF→ H_3C, NH, CH_3, $CH(CH_3)_2$ **2** —1) DIBAH; 2) $CH_3\overset{O}{\overset{\|}{C}}OCHO$; 3) MsCl, pyridine; ~53% overall→ H_3C, NC, CH_3, $CH(CH_3)_2$ **3**

Corey *et al.*[2] have also synthesized **3** from **1**. Their route involves reduction of **1** with a rhodium oxide–platinum oxide catalyst (this volume) to the corresponding amine, formylation, and dehydration to give **3**.

[1] H. Yamamoto and H. L. Sham, *Am. Soc.*, **101**, 1611 (1979).
[2] E. J. Corey, M. Bekforouz, and M. Ishiguro, *ibid.*, **101**, 1608 (1979).

Titanium(III) chloride–Lithium aluminum hydride, 6, 588–589; **7**, 369–370.

3,3-Disubstituted cyclopropenes.[1] The McMurry coupling reaction is convenient for preparation of 3,3-disubstituted cyclopropenes of type **3** and **4**, even though these products are considerably strained (equation I).

(I) $C_6H_5COC(R)_2COC_6H_5$ —$TiCl_3$, $LiAlH_4$, THF, Δ→ 3,3-R,R-1,2-(C_6H_5)$_2$-cyclopropene

1 ($R = CH_3$) → **3** (46%)
2 ($R = C_2H_5$) → **4** (40%)

[1] A. L. Baumstark, C. J. McCloskey, and K. E. Witt, *J. Org.*, **43**, 3609 (1978).

Titanium(IV) chloride, 1, 1169–1171; **2**, 414–415; **3**, 291; **4**, 507–508; **5**, 671–672; **6**, 590–596; **7**, 370–372; **8**, 483–486.

Pseudoguaianes.[1] Aldol condensation has not been particularly useful in synthesis of hydroazulenes. A new approach to pseudoguaianes employs a related intramolecular condensation of a trimethylsilyl enol ether and a ketal for this purpose. Thus conjugate addition of lithium dimethylcuprate to the enone group of **1** followed by reaction with chlorotrimethylsilane gives **2** in over 90% yield. Treatment of **2** with $TiCl_4$ at 0° gives the hydroazulenone **3** in 55% overall yield. Base-catalyzed condensation of the unprotected diketone corresponding to **2** does not produce a hydroazulene.

Reaction of **4** with lithium dimethylcuprate (2 equiv.) in THF effects β-addition and intramolecular alkylation to **5**, but in only 15% yield. Use of methylmagnesium chloride (2 equiv.) and the dimethyl sulfide complex of copper(I) bromide (1 equiv.) increases the yield to 25%. The product is easily dehydrated by TsOH to two isomeric pseudoguaianes (70% yield).

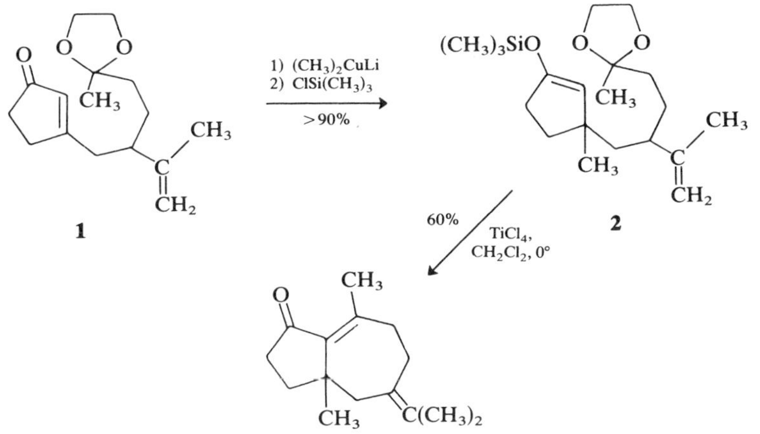

Acylation of tetramic acids.[2] This Lewis acid is generally superior to BF_3 etherate or $SnCl_4$ for acylation of tetramic acids (pyrrolidine-2,4-diones), particularly when the acyl group is unsaturated. An example is shown in equation (I).

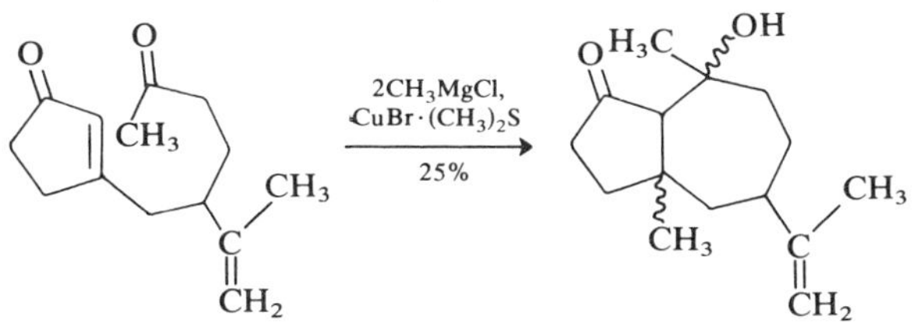

(I)

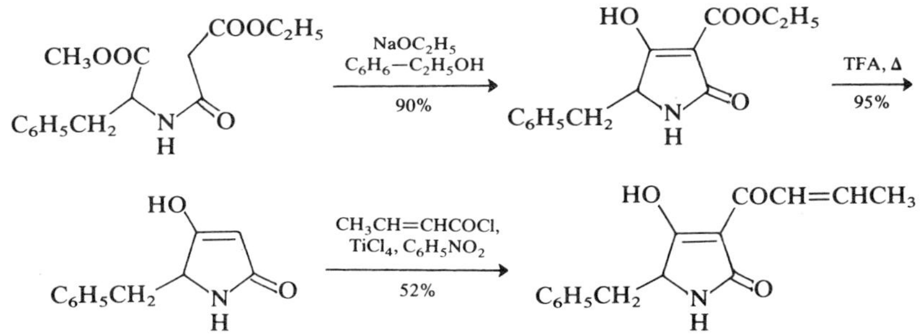

Aza–Claisen rearrangements. $TiCl_4$ has been reported to catalyze enamine formation.[3] It has now been found to catalyze also sigmatropic rearrangement (aza–Claisen rearrangement) of enamines. This rearrangement without a catalyst requires temperatures near 250°, but proceeds at 25–75° in the presence of $TiCl_4$. Simple ketones do not react under comparable conditions.[4]

Examples:

$$(CH_3)(C_6H_5)CHCHO + C_6H_5NHCH(CH_3)CH{=}CH_2 \xrightarrow{TiCl_4} [\,\mathbf{a} \rightarrow \mathbf{b}\,] \xrightarrow[47\%]{H_3O^+} CH_3{-}C(CHO)(C_6H_5){-}CH_2CH{=}CHCH_3 \quad (E/Z=9:1)$$

$$\text{3-cyclohexene-CHO} + C_6H_5NHCH(CH_3)CH{=}CH_2 \xrightarrow[31\%]{TiCl_4} \text{1-CHO-1-}(CH_2CH{=}CHCH_3)\text{-3-cyclohexene}$$

Furanes. γ-Sulfenyl-β,γ-unsaturated ketones react with $TiCl_4$ in acetonitrile under anhydrous conditions to afford furanes (equation I).[5]

$$\text{(I)}\quad \text{cyclohexenyl-CH(R)COCH}_3\text{ (2-SCH}_3) \xrightarrow[60\text{–}77\%]{TiCl_4,\ CH_3CN} \text{tetrahydrobenzofuran (3-R, 2-CH}_3)$$

[1] A. Alexakis, M. J. Chapdelaine, G. H. Posner, and A. W. Runquist, *Tetrahedron Letters*, 4205 (1978).
[2] R. C. F. Jones and S. Sumaria, *ibid.*, 3173 (1978).
[3] W. A. White and H. Weingarten, *J. Org.*, **32**, 213 (1967).
[4] R. K. Hill and H. N. Khatri, *Tetrahedron Letters*, 4337 (1978).
[5] S. Kano, Y. Tanaka, S. Hibino, and S. Shibuya, *J.C.S. Chem. Comm.*, 238 (1979).

Titanium(IV) chloride–N-Methylaniline, $TiCl_4–C_6H_5NHCH_3$ (**1**). The two reagents combine at 0° in CH_2Cl_2 to form a stable complex.

Cyclization of (Z)-allylic alcohols.[1] Cyclization of several (Z)-allylic alcohols to cyclohexenes and cycloheptenes has been observed with this complex in CH_2Cl_2. Complexes of $TiCl_4$ and 2,2,6,6-tetramethylpiperidine, $VCl_4–C_6H_5NHCH_3$, and $AlCl_3–C_6H_5NHCH_3$ are less active than **1**.

Examples:

$$(CH_3)_2C{=}CHCH_2CH_2C(CH_3){=}CHCH_2OH \xrightarrow[67\%]{\mathbf{1},\ CH_2Cl_2} \text{1-methyl-4-}(ClC(CH_3)_2)\text{cyclohexene}$$

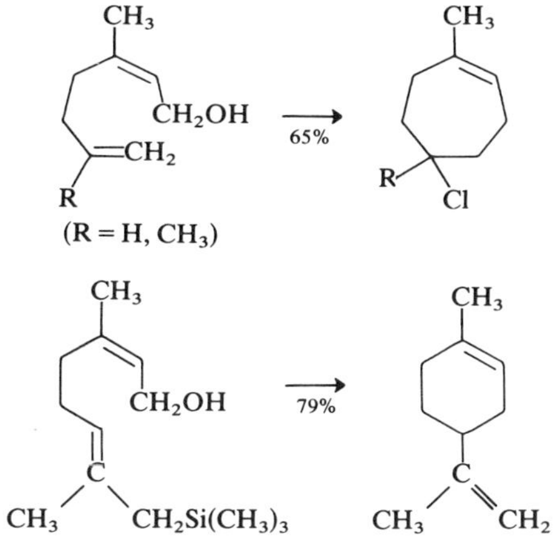

$$\text{(R = H, CH}_3\text{)} \xrightarrow{65\%}$$

$$\xrightarrow{79\%}$$

[1] T. Saito, A. Itoh, K. Oshima, and H. Nozaki, *Tetrahedron Letters*, 3519 (1979).

Titanocene dichloride–Dialkylaluminum chloride, $TiCp_2Cl_2$–R_2AlCl.

Carbotitanation of alkynylsilanes.[1] The reaction of an alkynylsilane with 1 equiv. each of titanocene dichloride and dimethylaluminum chloride followed by hydrolysis with aqueous sodium bicarbonate or sodium hydroxide results mainly in *cis*-carbometalation (equation I). Addition of triethylamine prior to quenching results in loss of stereochemistry.[2] Use of trimethylaluminum results in similar *cis*-addition, but this reaction is slower than that with $(CH_3)_2AlCl$. Surprisingly, alkynylsilanes do not react with zirconocene dichloride and dimethylaluminum chloride.

$$\text{(I)}\quad n\text{-}C_6H_{13}C{\equiv}CSi(CH_3)_3 \xrightarrow[75\%]{TiCp_2Cl_2,\ (CH_3)_2AlCl} (C_6H_{13})(CH_3)C{=}C(Si(CH_3)_3)(H) + (C_6H_{13})(CH_3)C{=}C(H)(Si(CH_3)_3) \quad 95:5$$

[1] B. B. Snider and M. Karras, *J. Organometal. Chem.*, **179**, C37 (1979).
[2] J. J. Eisch, R. J. Manfre, and D. A. Komar, *ibid.*, **159**, C13 (1978).

***p*-Toluenesulfonic acid, 1,** 1172–1178; **4,** 508–510; **5,** 673–675; **6,** 596; **7,** 374–375; **8,** 488–489.

Wagner–Meerwein rearrangement. Exposure of **2** to 0.3 equiv. of TsOH in refluxing benzene effects rearrangement to the tricyclic sesquiterpene isocomene (**3**).[1]

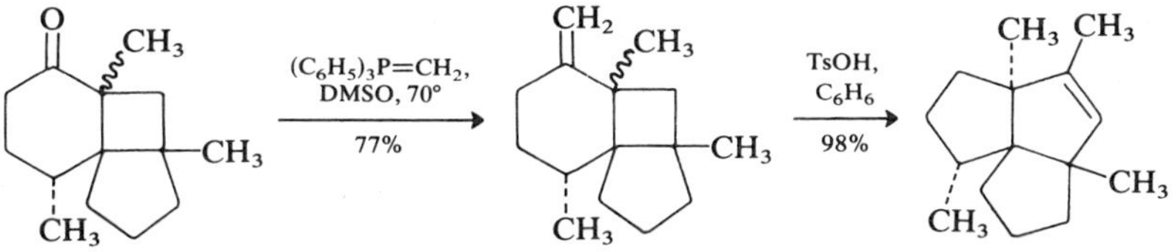

$$\mathbf{1} \xrightarrow[77\%]{(C_6H_5)_3P{=}CH_2,\ DMSO,\ 70^\circ} \mathbf{2} \xrightarrow[98\%]{TsOH,\ C_6H_6} \mathbf{3}$$

[1] M. C. Pirrung, *Am. Soc.*, **101**, 7130 (1979).

p-Toluenesulfonyl azide, 1, 1178–1179; **2**, 415–417; **3**, 291–292; **4**, 510; **5**, 675; **6**, 597.

Reduction of hydrazines.[1] Monosubstituted hydrazines are converted into hydrocarbons by reaction with the reagent under phase-transfer conditions (equation I). Yields are poor to moderate (6–40%) for arylhydrazines, but often fairly high for heterocyclic substrates (7–95%).

$$\text{(I)} \qquad ArNHNH_2 + TsN_3 \xrightarrow[NaOH,\ C_6H_6]{(C_2H_5)_4NBr,} ArH + 2N_2 + TsNH_2$$

[1] B. Stanovnik, M. Tišler, M. Kunaver, D. Gabrijelčič, and M. Kočevar, *Tetrahedron Letters*, 3059 (1978).

p-Toluenesulfonyl chloride, 1, 1179–1185; **3**, 292; **4**, 510–511; **6**, 598; **8**, 489.

Beckmann rearrangement. The standard Beckmann rearrangement of Δ^4-3-keto steroids is complicated by the fact that the corresponding oximes consist of *syn*- and *anti*-isomers (**1b** and **1a**), but the former isomers undergo rearrangement in the presence of acid much more readily than the *anti*-isomers. However, the tosylates of the isomeric oximes equilibrate in methanol very readily in the presence of acid. Thus treatment of the tosylated oxime mixture with conc. HCl at 50° results in 3-azalactams (**2**) in 85–95% yield.[1]

TsO–N= ... CH_3 ⇌ (HCl, CH_3OH) ... CH_3 ... N–OTs → (50°, 85–95%) HN ... O ... CH_3

1a (*anti*) **1b** (*syn*) **2**

[1] K. Oka and S. Hara, *J. Org.*, **43**, 3790 (1978).

p-Toluenesulfonylhydrazine, 1, 1185–1187; **2**, 417–423; **3**, 293; **4**, 511–512; **5**, 678–681; **6**, 598–600, **7**, 375–376; **8**, 489–493.

Olefin synthesis.[1] Condensation of aldehyde tosylhydrazones with stabilized carbanions affords alkenes by an addition–fragmentation process. The overall sequence is formulated in equation (I). Functional groups such as sulfides, sulfones, thioacetals, hemithioacetals, and nitriles can serve as Y, the anion stabilizing group in **2** and the leaving group in **3**. When $Y = SO_2R$, the anions **1** and **2** are conveniently prepared by treatment of a mixture of the tosylhydrazone and sulfone with LDA in

$$\text{(I)} \quad \underset{\substack{|\\ H}}{RC}{=}NNHTs \xrightarrow[THF,\ -20^\circ]{n\text{-BuLi or LDA,}} \underset{\substack{|\\ H}}{RC}{=}NN(Li)Ts \;(\mathbf{1}) \xrightarrow[THF,\ -20^\circ]{R'CH(Li)Y\ (\mathbf{2}),}$$

$$R{-}\underset{H}{\overset{N=N-Li}{C}}{-}\underset{Y}{CHR'} \;(\mathbf{3}) \xrightarrow{-N_2,\ LiY} RCH{=}CHR' \;(\mathbf{4})$$

THF at $-20°$. All other anions should be prepared separately, and **2** should be added to **1** at $-20°$. By these procedures mono- and disubstituted olefins, vinyl sulfides, vinyl ethers, and allylsilanes are available in 35–80% yield, usually as *cis–trans* mixtures. Trisubstituted olefins are best prepared by similar routes from α-branched nitrilė anions.

Example:

$$C_6H_5CH_2CH_2CH{=}NNHTs \xrightarrow[\text{2) } CH_3OCH(Li)SC_6H_5,\ -20°]{\text{1) LDA, THF, } -78°} C_6H_5CH_2CH_2CH{=}CHCH_2OCH_3$$

(*cis/trans* = 1 : 3)

1,2-Transposition of a carbonyl group. Two groups[2,3] have used the Shapiro reaction to transpose a carbonyl group. Thus treatment of the tosylhydrazone of an α-sulfenylated ketone (**1**) with *n*-butyllithium and DABCO at $-78 \rightarrow 20°$ gives **2**, which is hydrolyzed to the ketone **3** by mercuric chloride.[2]

$NNHSO_2C_7H_7$ / SC_6H_5 (cyclopentane) **1** $\xrightarrow[90\%]{n\text{-BuLi, DABCO}}$ cyclopentene-SC_6H_5 **2** $\xrightarrow[88\%]{HgCl_2}$ cyclopentanone **3**

Another example:[3]

CH_3-cyclohexanone NNHTs $\xrightarrow{\text{1) 2LDA; 2) } CH_3SSCH_3}$ [CH_3, $\overset{Li}{N}NTs$, SCH_3 cyclohexane $\xrightarrow{n\text{-BuLi}}$ CH_3, $\overset{Li}{N}NTs$, Li, SCH_3 cyclohexane] $\xrightarrow[92\%]{\Delta}$ CH_3, SCH_3 cyclohexene $\xrightarrow[78\%]{HgCl_2}$ CH_3-cyclohexanone

Alkylative 1,2-transposition of a carbonyl group is also possible (equation I).[2]

(I) 2-(SC_6H_5)cyclopentanone $\xrightarrow[87\%]{CH_2{=}P(C_6H_5)_3}$ CH_2=, SC_6H_5 cyclopentane $\xrightarrow{n\text{-BuLi}}$ CH_3, SC_6H_5 cyclopentene $\xrightarrow[85\%]{HgCl_2}$ CH_3-cyclopentanone

[1] E. Vedejs, J. M. Dolphin, and W. T. Stolle, *Am. Soc.*, **101**, 249 (1979).
[2] S. Kano, T. Yokomatsu, T. Ono, S. Hibino, and S. Shibuya, *J.C.S. Chem. Comm.*, 414 (1978).
[3] T. Nakai and T. Mimura, *Tetrahedron Letters*, 531 (1979).

p-Tolylthionochlorocarbonate (O-4-Methylphenyl chlorothioformate), 4, 342–343; **5**, 457.

Selective esterification.[1] This reagent (**1**) was used to convert both of the epimeric diols **2** and **3** into 14α-hydroxyhibaene-15 (**6**). Elimination of the hydroxyl group of **6** afforded (−)-hibaene-15 (**7**) in fair yield.[1]

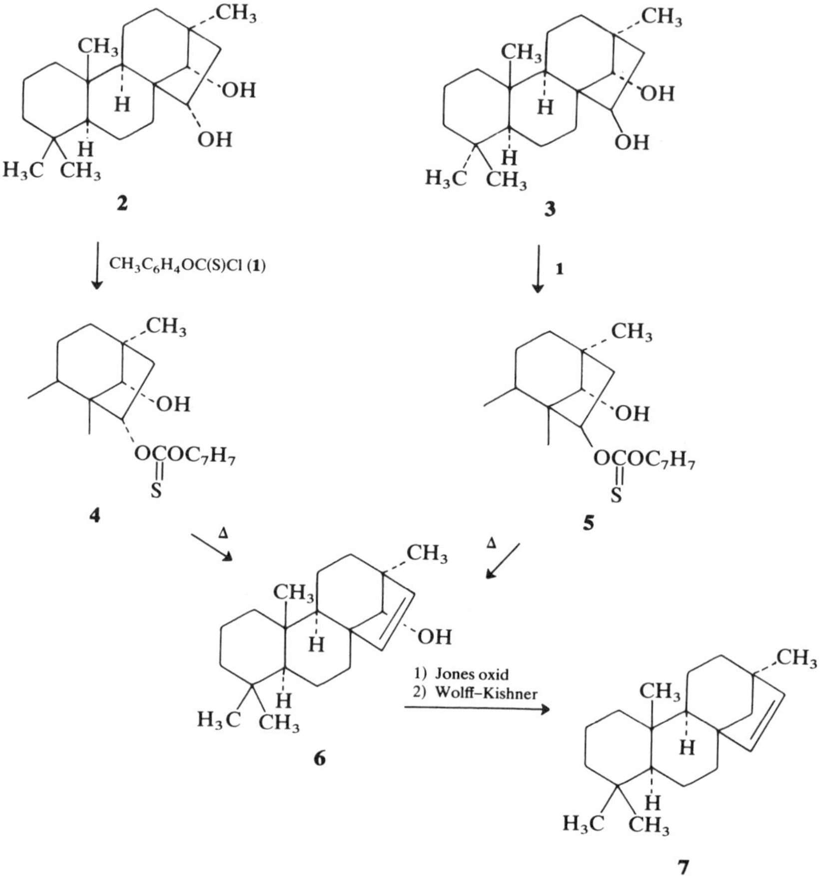

$CH_3C_6H_4OC(S)Cl$ (**1**)

1) Jones oxid
2) Wolff–Kishner

[1] D. K. M. Duc, M. Fetizon, and S. Lazare, *Tetrahedron*, **34**, 1207 (1978).

(S)-p-Tolyl p-tolylthiomethyl sulfoxide (1). Mol. wt. 276.41, m.p. 78–79°.

Preparation:[1]

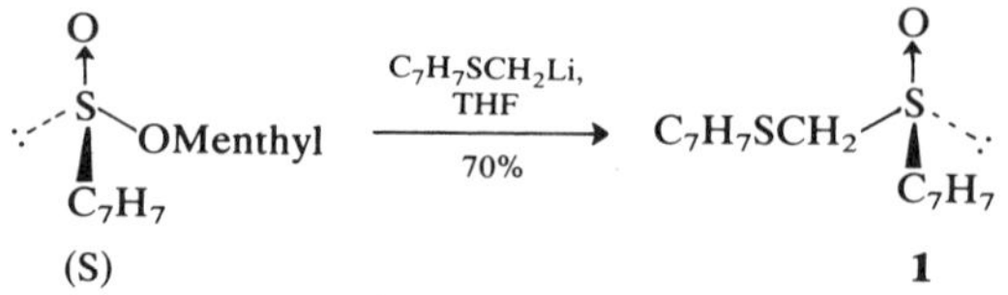

$C_7H_7SCH_2Li$, THF, 70%

Asymmetric synthesis of α-methoxyaldehydes. The anion from this chiral sulfoxide combines with benzaldehyde to produce **2**, which can be transformed into (R)-α-methoxytolualdehyde (**3**) in ≥70% optical yield.[2]

$$\mathbf{1} \xrightarrow[100\%]{1)\ n\text{-BuLi, THF};\ 2)\ C_6H_5CHO} C_6H_5CH(OH)\text{—}CH(SC_7H_7)\text{—}S(O)C_7H_7\ (\mathbf{2}) \xrightarrow[67\%]{1)\ n\text{-}Bu_4NOH, (CH_3)_2SO_4;\ 2)\ NaI, I_2, (C_6H_5)_3P;\ 3)\ I_2, NaHCO_3, H_2O} H\text{—}C(OCH_3)(C_6H_5)\text{—}CHO\ (\mathbf{3})$$

[1] L. Colombo, C. Gennari, and E. Narisano, *Tetrahedron Letters*, 3861 (1978).
[2] L. Colombo, C. Gennari, C. Scolastico, G. Guanti, and E. Narisano, *J.C.S. Chem. Comm.*, 591 (1979).

Tri-*n*-butyliodomethyltin. $(n\text{-}C_4H_9)_3SnCH_2I$ (**1**). Mol. wt. 430.95, b.p. 100–110°/0.01 mm. The reagent is prepared by reaction of ICH_2ZnI with $(n\text{-}C_4H_9)_3SnCl$ in THF for 18 hours (96% yield).[1]

(Z)-Trisubstituted alkenes. Still and Mitra[2] have described an efficient synthesis of alkenes of this type from allylic alcohols by a [2.3] sigmatropic Wittig rearrangement.[3] The alcohol **2** is converted into the allyl stannylmethyl ether (**3**), which can be isolated if desired. Treatment with *n*-butyllithium results in tin–lithium exchange and rearrangement to the homoallylic alcohol **4** in 95% overall yield. When **3** is transmetalated and immediately quenched with cyclohexanone, **5** is obtained in 73% yield.

$$n\text{-}C_4H_9CH(OH)\text{—}C(CH_3){=}CH_2\ (\mathbf{2}) \xrightarrow{1)\ KH, THF;\ 2)\ \mathbf{1}} n\text{-}C_4H_9CH(OCH_2Sn(C_4H_9\text{-}n)_3)\text{—}C(CH_3){=}CH_2\ (\mathbf{3}) \xrightarrow[95\%]{n\text{-BuLi},\ -78^\circ} (n\text{-}C_4H_9)(H)C{=}C(CH_2CH_2OH)(CH_3)\ (\mathbf{4})$$

$$\mathbf{3} \xrightarrow[73\%]{1)\ n\text{-BuLi};\ \text{cyclohexanone}} n\text{-}C_4H_9CH(OCH_2\text{—}C_6H_{10}OH)\text{—}C(CH_3){=}CH_2\ (\mathbf{5})$$

This rearrangement can be used in cyclic systems, for example, for the conversion of **6** into **7**.

$$\mathbf{6}\ (\text{2-cyclohexen-1-ol}) \xrightarrow{93\%} \mathbf{7}\ (HOCH_2\text{-cyclohexene})$$

This rearrangement provided the key step in a synthesis of the sex attractant (**8**) of the California red scale.

$$(CH_3)_2C{=}CHCOOC_2H_5 \xrightarrow[80\%]{1)\ LDA;\ 2)\ Br(CH_2)_2CH{=}CH_2} CH_2{=}CH(CH_2)_2CH(C(=CH_2)CH_3)COOC_2H_5 \xrightarrow[93\%]{1)\ LiAlH_4;\ 2)\ NBS,\ P(C_6H_5)_3}$$

$$CH_2{=}CH(CH_2)_2CH(C(=CH_2)CH_3)CH_2Br \xrightarrow[52\%]{1)\ Mg;\ 2)\ OHC{-}C(CH_3){=}CH_2} CH_2{=}CH(CH_2)_2CH(C(=CH_2)CH_3)CH_2CH(OH)C(CH_3){=}CH_2 \xrightarrow[83\%]{1)\ KH,\ \mathbf{1};\ 2)\ n\text{-}BuLi;\ 3)\ Ac_2O,\ Py}$$

$$CH_2{=}CH(CH_2)_2CH(C(=CH_2)CH_3)CH_2(H)C{=}C(CH_3)CH_2CH_2OAc$$

8

The same general scheme provided a synthesis of the C_{15} *Cecropia* juvenile hormone.[4]

[1] W. C. Still, *Am. Soc.*, **100**, 1481 (1978).
[2] W. C. Still and A. Mitra, *ibid.*, **100**, 1927 (1978).
[3] G. Wittig, *Experientia*, 14, 389 (1958).
[4] W. C. Still, J. H. McDonald III, D. B. Collum, and A. Mitra, *Tetrahedron Letters*, 593 (1979).

Tri-*n*-butyltin hydride, 1, 1192–1193; **2**, 424; **3**, 294; **4**, 518–520; **5**, 685–686; **6**, 604; **7**, 379–380; **8**, 497–498.

Desulfonylation. Allylic sulfones with a terminal double bond (**1**) react with tri-*n*-butyltin hydride (azobisisobutyronitrile initiation or irradiation) to form allyltin derivatives (**2**), which on protonolysis give terminal alkenes (**3**).[1] The reaction can be conducted without isolation of **2** in 80–87% overall yield.

$$p\text{-}CH_3C_6H_4SO_2{-}CH(R){-}CH{=}CH_2\ (\mathbf{1}) \xrightarrow[65\text{–}75\%]{Bu_3SnH,\ AIBN\ or\ h\nu} (R)(H)C{=}C(H)CH_2SnBu_3\ (\mathbf{2}) \xrightarrow[80\text{–}87\%\ from\ \mathbf{1}]{H^+} RCH_2CH{=}CH_2\ (\mathbf{3})$$

Another reagent for effecting desulfonylation is potassium–graphite, but in this case (Z)- and (E)-2-alkenes are formed, with the latter usually predominating.[2]

$$\mathbf{1} \xrightarrow[55\text{–}80\%]{C_8K} RCH{=}CHCH_3$$

$$(Z/E = 30\text{–}65:70\text{–}35)$$

Stereocontrolled reduction of 6-halopenicillanates.[3] This reaction is a useful route to 6β-substituted penicillanates.

Examples:

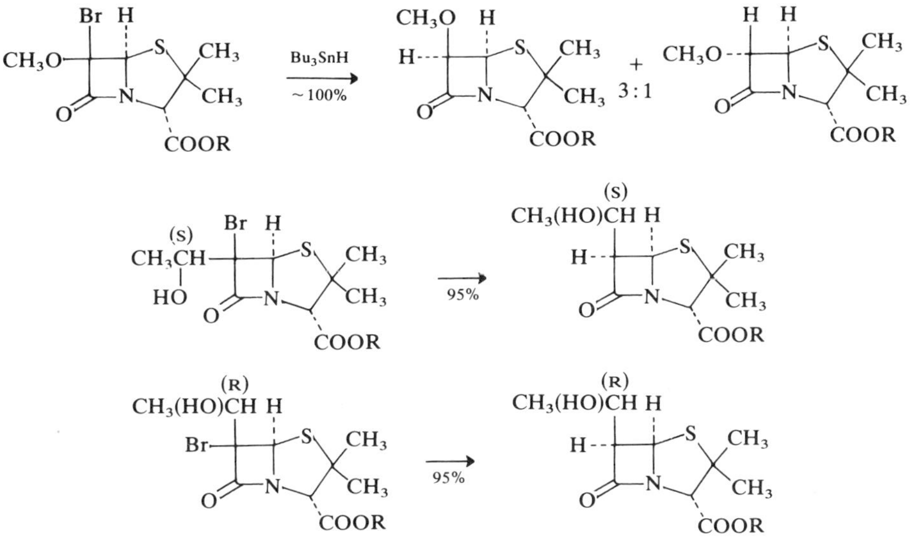

Bu_3SnH, ~100%; 3 : 1

95%

95%

Regardless of the stereochemistry of the bromine, hydrogen is transferred from the less hindered side of the substrate.

[1] Y. Ueno, S. Aoki, and M. Okawara, *Am. Soc.*, **101**, 5414 (1979).
[2] D. Savoia, C. Trombini, and A. Umani-Ronchi, *J.C.S. Perkin I*, 123 (1977).
[3] J. A. Aimetti, E. S. Hamanaka, D. A. Johnson, and M. S. Kellogg, *Tetrahedron Letters*, 4631 (1979).

Tributyltin hydride–Silica gel.

Selective reduction of aldehydes.[1] In the absence of radical initiators, tributyltin hydride does not ordinarily reduce carbonyl groups. However, when slurried in cyclohexane with dried silica gel (activated by heating at 220° under reduced pressure), this hydride reduces aldehydes and ketones to alcohols in high yield. The rate of reduction is aldehydes > dialkyl ketones > aryl alkyl ketones > diaryl ketones. Thus it is possible to reduce aldehydes selectively. The function of SiO_2 apparently is that of a mild acid catalyst.

[1] N. Y. M. Fung, P. de Mayo, J. H. Schauble, and A. C. Weedon, *J. Org.*, **43**, 3977 (1978).

Tri-μ-carbonylhexacarbonyldiiron, 1, 259–260; **2**, 139–140; **3**, 101; **4**, 157–158; **5**, 221–224; **6**, 195–198; **7**, 110–111; **8**, 498–499.

Intramolecular [3+2]cycloadditions.[1] The intramolecular [3 + 2]cycloaddition of olefinic dibromo ketones proceeds smoothly in benzene at 80–110° for 1.5–3 hours in the presence of 1.2 equiv. of this iron carbonyl. The reaction is sensitive to the substitution pattern of the olefinic double bond as well as to the length of the chain separating the olefin and the bromo ketone moiety.

Examples:

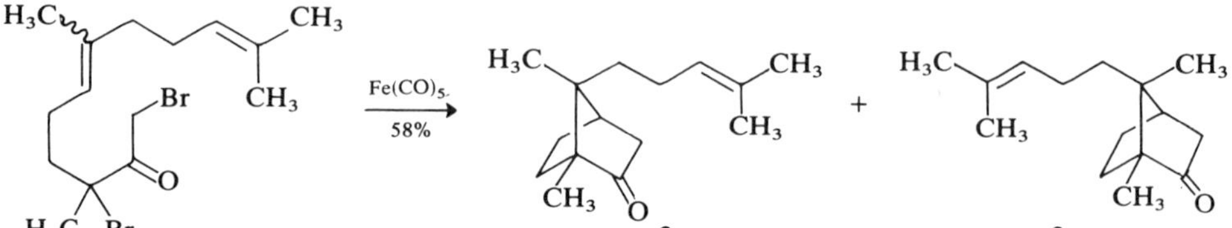

$\xrightarrow[58\%]{Fe(CO)_5}$ 2 + 3

1 (E) 2:1
(Z) 1:2

R = H (41%)
R = CH_3 (38%)

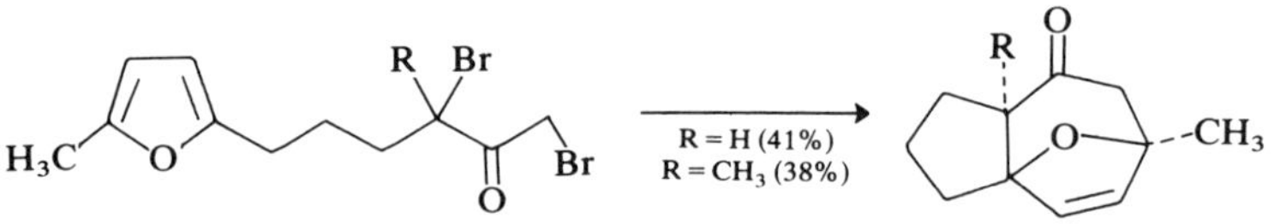

[1] R. Noyori, M. Nishizawa, F. Shimizu, Y. Hayakawa, K. Maruoka, S. Hashimoto, H. Yamamoto, and H. Nozaki, *Am. Soc.*, **101**, 220 (1979).

2,4,6-Trichlorobenzoyl chloride, (**1**)

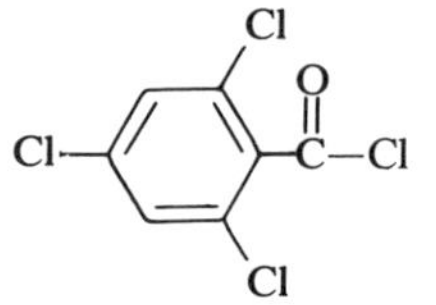

Lactonization. Esters and lactones can be prepared using a trichlorobenzoic mixed anhydride, prepared by reaction of an acid with trichlorobenzoyl chloride.[1] Thus the anhydride reacts with an alcohol in the presence of 4-dimethylamino-pyridine to form an ester in yields usually >90%. The same method can be used for lactonization.[2] Nine- to thirteen-membered rings can be prepared in 35–65% yield. This method was used for synthesis of methynolide (**6**), a macrolide antibiotic, from a sequence shown in equation (I).

(I) **2** $\xrightarrow{1,\ DMAP}$ $\xrightarrow[87\%]{MnO_2}$

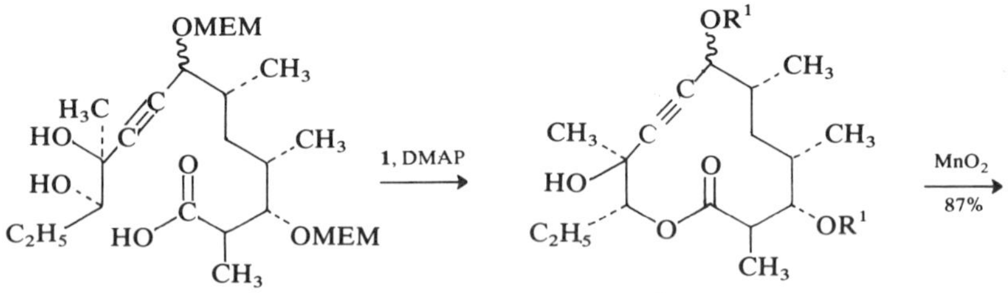

2

3, R^1 = MEM, 42%
4, R^1 = H, 67%

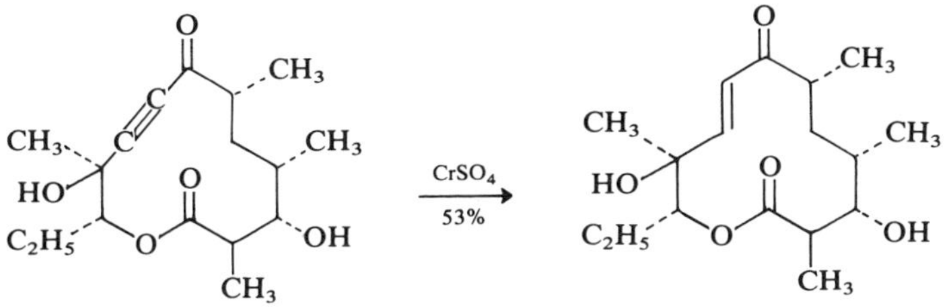

5 → (CrSO₄, 53%) → 6

[1] J. Inanaga, K. Hirata, H. Sacki, T. Katsuki, and M. Yamaguchi, *Bull. Chem. Soc. Japan*, **52**, 1989 (1979).
[2] J. Inanaga, T. Katsuki, S. Takimoto, S. Ouchida, K. Inone, A. Nakano, N. Okukuda, and M. Yamaguchi, *Chem. Letters*, 1021 (1979).

2,2,2-Trichloroethyl *trans*-1,3-butadiene-1-carbamate, $CH_2{=}CH{-}CH{=}CH{-}NHCO_2CH_2CCl_3$ (**1**). Mol. wt. 244.51, m.p. 70–71°. The carbamate is prepared from *trans*-2,4-pentadienoic acid and 2,2,2-trichloroethanol (63% yield).

***Diels–Alder reaction with ethyl atropate* (2).** The diene reacts with **2** to form a single cycloadduct **3**. In contrast, when the $—CH_2CCl_3$ group is replaced by benzyl, the addition yields two crystalline adducts **4** and **5**. In general 1-(acylamino)-1,3-dienes show favored *endo*-stereoselectivity, whereas 1-(dialkylamino)-1,3-dienes show favored *exo*-stereoselectivity.[1]

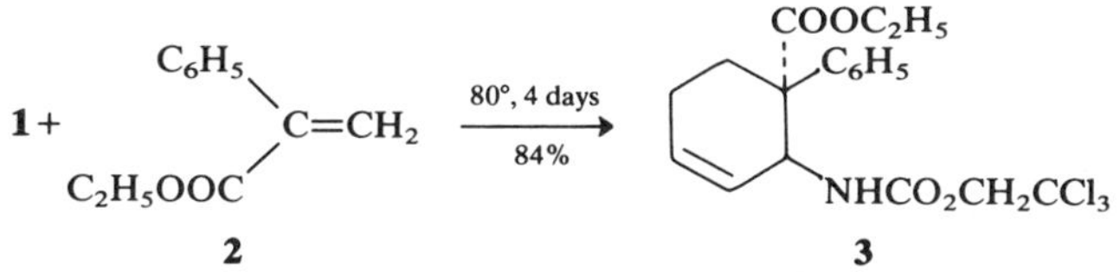

$$\mathbf{1} + \underset{\mathbf{2}}{(C_6H_5)(C_2H_5OOC)C{=}CH_2} \xrightarrow[84\%]{80°,\ 4\ \text{days}} \mathbf{3}$$

3: $COOC_2H_5$, C_6H_5, $NHCO_2CH_2CCl_3$

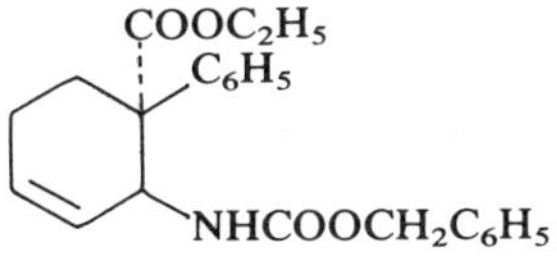

4 (70%): $COOC_2H_5$, C_6H_5, $NHCOOCH_2C_6H_5$

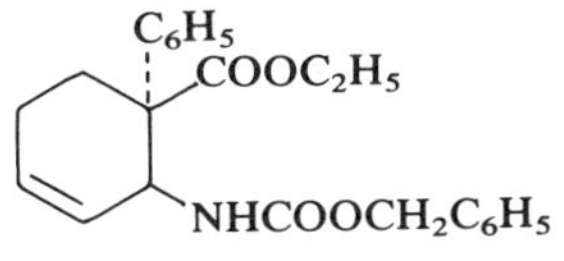

5 (20%): C_6H_5, $COOC_2H_5$, $NHCOOCH_2C_6H_5$

[1] L. E. Overman, C. B. Petty, and R. J. Doedens, *J. Org.*, **44**, 4183 (1979).

Trichloroethylene, $CHCl{=}CCl_2$ (**1**). Mol. wt. 131.39, b.p. 87°. Supplier: Aldrich.

Dichlorovinylation.[1] Enolates of ketones react with **1** to form α-dichlorovinyl ketones, which can be converted into α-acetylenic ketones (equations I and II).

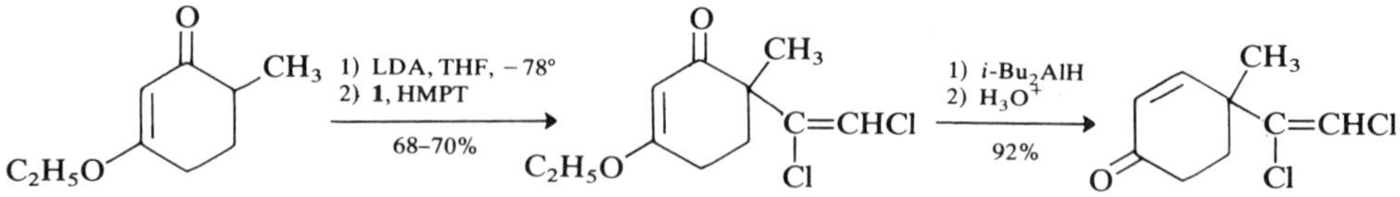

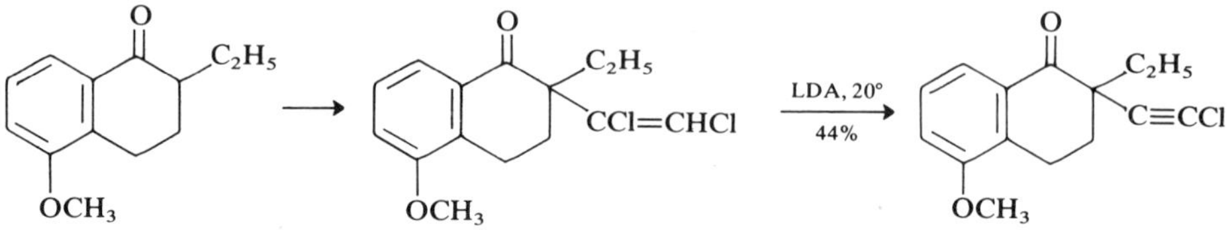

[1] A. S. Kende, M. Benechie, D. P. Curran, P. Fludzinski, W. Swenson, and J. Clardy, *Tetrahedron Letters*, 4513 (1979).

Trichloro-1,2,4-triazine (1). Mol. wt. 184.42, m.p. 60–62°.

Preparation[1,2]:

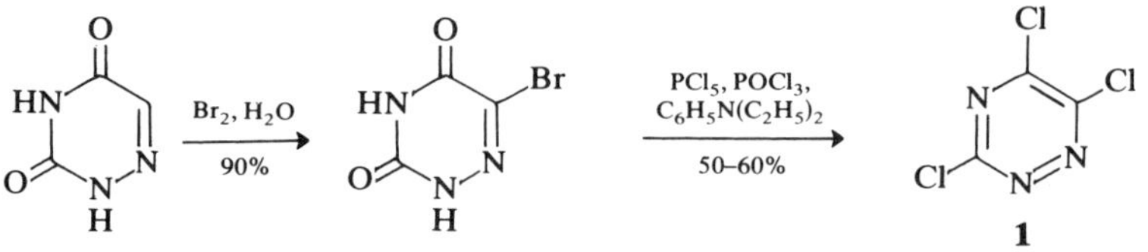

Diels–Alder reactions. This triazine reacts with unactivated disubstituted olefins to produce dichloropyridines (**4**).[3] This reaction appears to involve a [1,5] hydrogen migration in the initial adduct (**2**) followed by loss of hydrogen chloride.

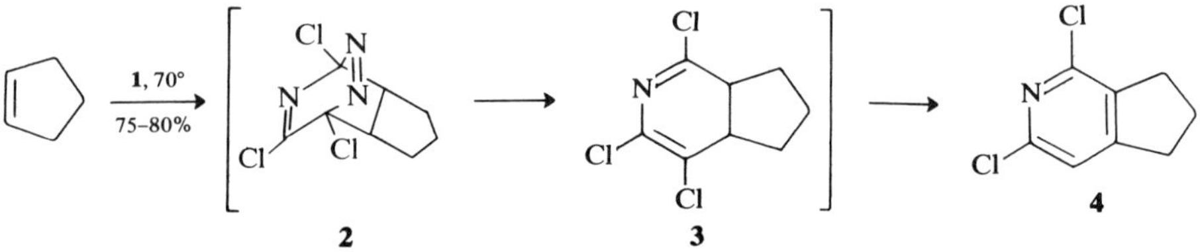

[1] B. A. Loving, C. E. Snyder, G. L. Whittier, and K. R. Fountain, *J. Heterocyclic Chem.*, **8**, 1095 (1971).

[2] P. K. Chary and T. L. V. Ulbricht, *Am. Soc.*, **80**, 976 (1958).

[3] M. G. Barlow, R. N. Haszeldine, and D. J. Simpkin, *J.C.S. Chem. Comm.*, 658 (1979).

Triethoxydiiodophosphorane, $(C_2H_5O)_3PI_2$ (**1**). Mol. wt. 389.0. The reaction of triethyl phosphite with iodine in CH_3CN, CH_2Cl_2, or ether at 0° results in a complex formulated as **1**. Substances of this type are intermediates in the Arbuzov reaction. The phosphorane is stable in solution for several days. In combination with 1 equiv. of triethylamine it can dehydrate amides and aldoximes to nitriles. It effects condensation of acids and amines to amides. The paper cites one example of use for synthesis of a dipeptide but with complete racemization, which was partially inhibited by 1-hydroxybenzotriazole.[1]

[1] D. Cooper and S. Trippett, *Tetrahedron Letters*, 1725 (1979).

Triethylamine, 1, 1198–1203; **2**, 427–429; **4**, 526; **5**, 689–690; **7**, 385.

Catechol to phenol rearrangement. Treatment of 2-acetoxy-4-(methylthio)-3,5-xylenol (**1**) with triethylamine (benzene, reflux) leads to **2**, also obtained from **3** under the same conditions. This migration of an acetoxyl group is probably related to the Pummerer rearrangement.[1]

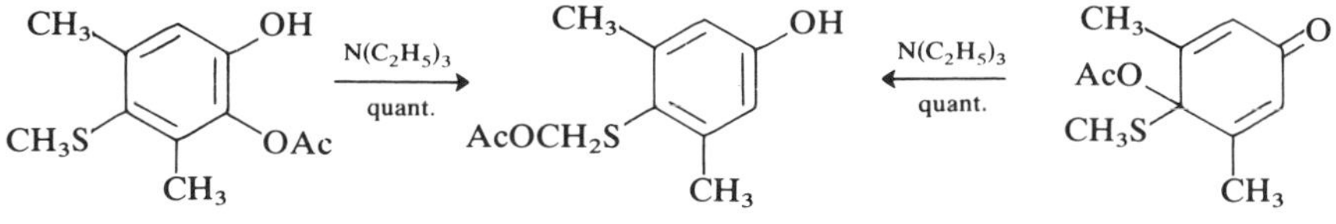

Cine rearrangement.[2] Triethylamine catalyzes the "cine rearrangement" of α-chlorocyclobutanones;[3] the α'-chloro isomers (**2**) predominate only when R is an electron-withdrawing substituent such as CH_2CCl_3 or Cl.

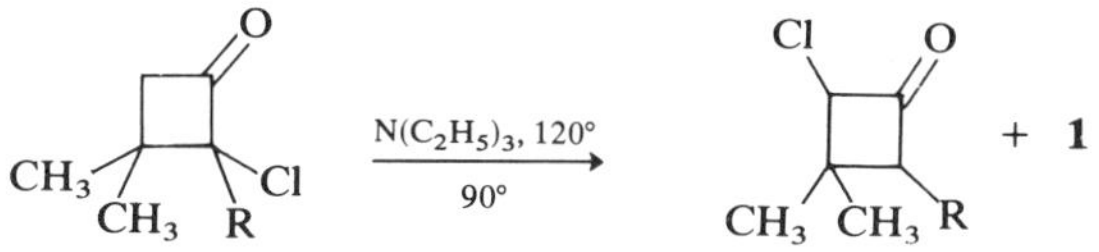

R	2 : 1
$R = CH_2CCl_3$	100:0
$R = Cl$	100:0
$R = CH_3$	43:57

[1] R. R. King, *J. Org.*, **43**, 3784 (1978); **44**, 4194 (1979).
[2] P. Martin, H. Greuter, and D. Belluš, *Am. Soc.*, **101**, 5853 (1979).
[3] For preparation *see* Chlorotrichloroethylketene, this volume

Triethylammonium formate–Palladium.

Reduction of enones.[1] Palladium-catalyzed reductions with trialkylammonium formates have been used to reduce α,β-unsaturated aldehydes, ketones, and esters to saturated carbonyl compounds.

Examples:

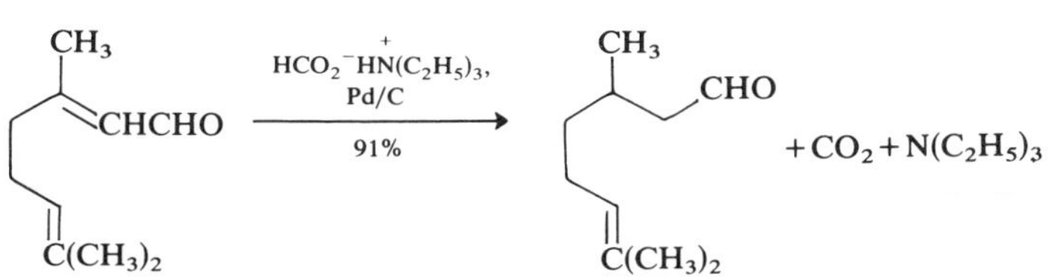

$$\xrightarrow[91\%]{HCO_2^-\overset{+}{H}N(C_2H_5)_3,\ Pd/C} \quad + CO_2 + N(C_2H_5)_3$$

$$CH_3OOC\text{-}CH(CH_3)\text{-}CH{=}CH\text{-}CH{=}C(CH_3)\text{-}COOCH_3 \xrightarrow{65\%} CH_3OOC\text{-}CH(CH_3)\text{-}CH_2\text{-}CH{=}CH\text{-}C(CH_3)\text{-}COOCH_3$$

Simple 1,3-dienes are reduced mainly to monoenes. Some alkynes are reduced satisfactorily to *cis*-alkenes.

[1] N. A. Cortese and R. F. Heck, *J. Org.*, **43**, 3985 (1978).

Triethylborane, $B(C_2H_5)_3$. Mol. wt. 98, gas. Supplied as a 1 *M* solution in hexane or THF (Aldrich).

Selective α-alkylation of ketones.[1] Potassium enolates of ketones and an unhindered trialkylborane such as triethylborane form a potassium enoxytriethylborate, which undergoes selective α-monoalkylation with alkyl halides in high yield. Lithium enolates do not form the corresponding borates. In the absence of triethylborane dialkylated products are formed and some of the original ketone is recovered.

$$\text{Cyclohexanone} \xrightarrow[2)\ B(C_2H_5)_3]{1)\ KH,\ THF} \left[\text{1-}\bar{O}B(C_2H_5)_3K^+\text{-cyclohexene}\right] \xrightarrow[\substack{2)\ NaOH,\ H_2O_2 \\ 90\%}]{1)\ BrCH_2CH{=}CH_2} \text{2-}CH_2CH{=}CH_2\text{-cyclohexanone}$$

$$C_6H_5COCH_3 \rightarrow \xrightarrow[98\%]{CH_3I} C_6H_5COCH_2CH_3$$

$$C_6H_5COCH_3 \rightarrow \xrightarrow[63\%]{BrCH_2C{\equiv}CH} C_6H_5COCH_2CH_2C{\equiv}CH$$

[1] E. Negishi, M. J. Idacavage, F. DiPasquale, and A. Silveira, Jr., *Tetrahedron Letters*, 845 (1979).

Triethyloxonium tetrafluoroborate, 1, 1210–1212; **2,** 430–431; **3,** 303; **4,** 527–529; **6,** 691–693; **7,** 386–387; **8,** 500–501.

Esterification (**4,** 528).[1] Complete details are available for preparation of methyl and ethyl esters of carboxylic acids with trialkyloxonium salts. The reaction is rapid even with hindered acids and yields generally are in the range 80–95%.

Hydrolysis of 2-alkyl-1,3-dithanes.[2] A recent method[3] for this reaction involves oxidation with periodate to the monosulfoxide followed by alkylation with triethyloxonium tetrafluoroborate to form a 1-ethoxy-1,3-dithianium tetrafluoroborate. Salts of this type are hydrolyzed by water at 5°.

$$\text{2-H-2-R-1,3-dithiane} \xrightarrow[85\text{–}95\%]{NaIO_4} \text{1-oxide} \xrightarrow[50\text{–}80\%]{(C_2H_5)_3O^+BF_4^-} \text{1-}OC_2H_5\text{-dithianium}\ BF_4^- \xrightarrow[>60\%]{H_2O}$$

$$RCHO + C_3H_6S_2 + C_2H_5OH + HBF_4$$

[1] D. J. Raber, P. Gariano, Jr., A. O. Brod, A. Gariano, W. C. Guida, A. R. Guida, and M. D. Herbst, *J. Org.*, **44**, 1149 (1979); D. J. Raber, P. Gariano, A. O. Brod, and W. C. Guida, *Org. Syn.*, **56**, 59 (1977).
[2] Review: B.-T. Gröbel and D. Seebach, *Synthesis*, 357 (1977).
[3] I. Stahl, J. Opel, R. Manske, and J. Gosselek, *Angew. Chem., Int. Ed.*, **18**, 165 (1979).

Triethylsilane–Boron trifluoride hydrate ($BF_3 \cdot H_2O$). $BF_3 \cdot H_2O$, m.p. 6°, is unstable above 15°. This reagent is obtained by bubbling BF_3 into water. It is a strong acid, $H^+[BF_3 \cdot OH]^-$, comparable to HF.[1]

Some reductions that are not possible with triethylsilane and trifluoroacetic acid (**5**, 695; **6**, 616) can be effected with triethylsilane and boron trifluoride hydrate. Although benzene, naphthalene, and phenanthrene are not reduced, anthracene and naphthacene are reduced to tetrahydro derivatives in high yield. 1- or 2-Hydroxynaphthalene is reduced to tetralin in moderate yield.[2]

$HSi(C_2H_5)_3$, $BF_3 \cdot H_2O$, 88%

OH → 40–50%

[1] N. N. Greenwood and R. L. Martin, *J. Chem. Soc.*, 1915 (1951).
[2] J. W. Larsen and L. W. Chang, *J. Org.*, **44**, 1168 (1979).

Trifluoroacetic acid (TFA), 1, 1219–1221; **2**, 433–434; **3**, 305–308; **4**, 530–532; **5**, 695–700; **6**, 613–615; **7**, 388–389; **8**, 503.

trans-Fused α-methylene-γ-lactones. A useful new strategy for synthesis of this system (**2**) involves cleavage of cyclopropyl acrylates (**1**), prepared as shown in equation (I).[1]

(I) $ClC(=O)-COOC(CH_3)_3 \rightarrow N_2CHC(=O)COOC(CH_3)_3$ → (cyclohexadiene, $Rh(OAc)_2$) → $COOC(CH_3)_3$ (ketone)

$(C_6H_5)_3P{=}CH_2$ → **1** (CH_2, $COOC(CH_3)_3$) → TFA → [**a** (CH_2, $COOC(CH_3)_3$)] → 89% → **2**

[1] L. G. Mueller and R. G. Lawton, *J. Org.*, **44**, 4741 (1979).

Trifluoroacetic anhydride, 1, 1221–1226; **3,** 308; **5,** 701; **6,** 616–617; **7,** 389–390.

Polonovski intramolecular cyclization (**7**, 389). Treatment of the N-oxide of desmethylhirsutine (**1**) with trifluoroacetic anhydride at 0° gives dihydromancunine (**2**) in 33% yield.[1] Intramolecular cyclizations of this type are believed to be involved in biosynthesis of indole alkaloids.

$(CF_3CO)_2O$, CH_2Cl_2, 0°; 33%

1 → **2**

A similar Polonovski reaction with the N-oxide of desmethylhirsuleine (**3**) leads to two heteroyohimbine alkaloids, 3-isorauniticine (**4**) and akuammigine (**5**), and desmethylcorynantheine (**6**), formed by epimerization of **3**.[2]

1) $(CF_3CO)_2O$ 2) $NaCNBH_3$

3 → **4** (C_{19} αH, 4%), **5** (C_{19} βH, 11%) +[**6**, C_3 epimer of **3**, 18%)]

Acyl trifluoroacetates, $RCOOCCF_3$ (RC(=O)OC(=O)CF₃).[3] Mixed anhydrides of this type can be obtained by reaction of a carboxylic acid in CH_3CN with trifluoroacetic anhydride (3 equiv.) in the presence of 85% phosphoric acid (1 equiv.). Isolation of the mixed anhydride is not necessary for use in Friedel–Crafts acylation of arenes. The method is applicable even to long-chain carboxylic acids and yields are generally satisfactory.

[1] S. Sakai and N. Shinma, *Chem. Pharm. Bull. Japan*, **25**, 842 (1977).
[2] *Idem, ibid.*, **26**, 2596 (1978).
[3] C. Galli, *Synthesis*, 303 (1979).

Trifluoroacetic anhydride–Pyridine, 8, 504.

Nitriles from aldoximes.[1] The system trifluoroacetic anhydride and pyridine, which converts primary amides to nitriles, also converts aldoximes to nitriles. The geometrical configuration of the substrate determines the ease of dehydration: (E)-Aldoximes are less reactive than (Z)-aldoximes. However, high yields can also be obtained from the former substrates if the amount of pyridine is increased or if the reaction is conducted at higher temperatures (60–65°).

[1] A. Carotti, F. Campagna, and R. Ballini, *Synthesis*, 56 (1979).

Trifluoroacetyl triflate, $CF_3\overset{O}{\overset{\|}{C}}OSO_2CF_3$ (**1**). Mol. wt. 258.11, b.p. 62.5°. This stable anhydride is prepared by treatment of an equimolar mixture of trifluoroacetic acid and triflic acid with P_2O_5 at reflux.

This mixed anhydride, in combination with a base such as 2,6-di-*t*-butyl-4-methylpyridine, is a potent reagent for trifluoroacetylation. The reaction of **1** with anthracene results in formation of the 9-trifluoroacetyl derivative (81% yield). Cyclohexanone reacts with **1** to form the enol trifluoroacetate (72% yield). The reaction of trityl chloride and **1** leads to trityl triflate.[1]

[1] T. R. Forbus, Jr., and J. C. Martin, *J. Org.*, **44**, 313 (1979).

Trifluoromethanesulfonic acid, 4, 533; **5,** 701–702; **6,** 617–618; **8,** 504.

Koch–Haaf carboxylation.[1] This acid is much superior to the previously used 95% H_2SO_4 for carboxylation of olefins, alcohols, and esters with CO at atmospheric pressure. The beneficial effect appears to be the higher solubility of CO in this acid.

Another general advantage of CF_3SO_2OH is that quantitative regeneration is possible by precipitation as the barium salt followed by reaction with 100% H_2SO_4.[2]

[1] W. Bird, *Chem. Rev.*, **62**, 283 (1962).
[2] B. L. Booth and T. A. El-Fekky, *J.C.S. Perkin I*, 2441 (1979).

Trifluoromethanesulfonyl chloride, CF_3SO_2Cl. Mol. wt. 168.52, b.p. 29–32°. Supplier: Aldrich.

Chlorination.[1] Fairly acidic compounds such as $CH_3COCH_2COCH_3$, $CH_2(COOCH_3)_2$, and $Cl_2CHCOOCH_3$ are chlorinated in high yield by CF_3SO_2Cl in combination with $N(C_2H_5)_3$ or DBU. Cyclohexanone, methyl phenylacetate, and acetophenone do not react.

Tetrahydrofuranes; tetrahydropyranes.[2] Use of $CF_3SO_2Cl–N(C_2H_5)_3$ for synthesis of some oxygen heterocycles is formulated in equations (I)–(III).

(I) $(C_2H_5OOC)_2CH(CH_2)_3CH_2OH$ $\xrightarrow[N(C_2H_5)_3]{CF_3SO_2Cl,}$ $(C_2H_5OOC)_2CCl(CH_2)_3CH_2OH$ + $(C_2H_5OOC)_2$-tetrahydropyran (50:50)

(II) $(C_2H_5OOC)_2CH(CH_2)_2CH_2OH$ $\xrightarrow[DBU]{CF_3SO_2Cl,}$ 2,2-bis($COOC_2H_5$)tetrahydrofuran

(III) $(C_2H_5OOC)_2CHCH_2CH_2OH$ $\xrightarrow[N(C_2H_5)_3]{CF_3SO_2Cl}$ 2-Cl-2-(C_2H_5OOC)-γ-butyrolactone

[1] G. H. Hakimelahi and G. Just, *Tetrahedron Letters*, 3643 (1979).
[2] *Idem, ibid.*, 3645 (1979).

Trifluoromethylcopper, CF_3Cu. Mol. wt. 132.55. The reagent is prepared[1] from CF_3I and copper powder[2] in HMPT.

Trifluoromethylation.[1] This reagent converts aliphatic and aromatic halides into the corresponding trifluoromethyl compounds in moderate yield (~10–65%).

[1] Y. Kobayashi, K. Yamamoto, and I. Kumadaki, *Tetrahedron Letters*, 4071 (1979).
[2] Prepared according to R. Q. Brewster, and T. Groening, *Org. Syn. Coll. Vol.*, **2**, 445 (1943).

Triisobutylaluminum, 1, 260; **4,** 535; **7,** 391.

Reduction of α-keto epoxides.[1] Epoxidation of cyclic allylic alcohols results mainly in the epoxide *syn* to the hydroxyl group (**4**, 76). Schlessinger *et al.*[1] have reported a method for isomerization of the alcohol group by oxidation to the ketone and reduction to the *anti*-alcohol. In a model case, **2** → **4**, pyridinium chlorochromate buffered with sodium acetate was found to be the most satisfactory oxidant. Stereoselective reduction to **4** was found to be a more difficult problem, but eventually triisobutylaluminum was found to effect this reduction in high yield.

CH3, CH3, OH — **1** —($ClC_6H_4CO_3H$, 88%)→ CH3, CH3, OH, O — **2** —($C_5H_5\overset{+}{N}HCrO_3Cl^-$, 80%)→ CH3, CH3, O, O — **3** —($Al(i\text{-}Bu)_3$, 99%)→ CH3, CH3, OH, O — **4**

[1] M. R. Roberts, W. H. Parsons, and R. H. Schlessinger, *J. Org.*, **43**, 3970 (1978).

2,4,6-Triisopropylbenzenesulfonyl hydrazine, 4, 535; **7,** 392. Supplier: Mallinckrodt.

Vinyllithium reagents. Bond *et al.*[1] have modified the Shapiro reaction to permit generation of vinyllithium reagents for use in reactions with electrophiles. For this purpose the trisylhydrazones, which lack α-hydrogens in the aryl group, are preferable to tosylhydrazones. Thus treatment of the hydrazone (**1**) with 2 equiv. of *n*-butyllithium–TMEDA in hexane generates the anion **2** by removal of a primary α-hydrogen. The anion is a useful precursor to allylic alcohols, vinylsilanes, and vinyl bromides, acrylic acids, and acrylic aldehydes. In general the ease of removal of the α-hydrogen is primary > secondary > tertiary; generation of the anion can be facilitated by use of excess base or of *sec*-butyllithium.

$$\underset{\mathbf{1}\,[(E)\text{ and }(Z)]}{CH_3C(=NNHSO_2C_6H_2(i\text{-}Pr)_3)C_6H_{13}\text{-}n} \xrightarrow[C_6H_{14}]{n\text{-BuLi,}} \left[\underset{\mathbf{2}}{CH_2{=}C(Li)C_6H_{13}}\right] \xrightarrow[62\%]{C_6H_5CHO} CH_2{=}C(CHOHC_6H_5)C_6H_{13}\text{-}n$$

$$\mathbf{2} \xrightarrow[71\%]{ClSi(CH_3)_3} CH_2{=}C(Si(CH_3)_3)C_6H_{13}\text{-}n$$

$$\mathbf{2} \xrightarrow[\sim 60\%]{RCH_2Br} CH_2{=}C(CH_2R)C_6H_{13}\text{-}n$$

On the other hand, the pure (Z)-isomer of **1** on treatment with *sec*-butyllithium in THF generates the more susbstituted vinyllithium reagent **3**. Thus by use of either sequence stereoselective routes are possible to alkenes, allylic alcohols, and aldehydes.[2]

$$\underset{(Z)\text{-}\mathbf{1}}{CH_3C(=NNHSO_2C_6H_2(i\text{-}Pr)_3)C_6H_{13}\text{-}n} \xrightarrow[THF]{sec\text{-BuLi,}} \underset{\mathbf{3}}{(Li)(CH_3)C{=}C(H)(C_5H_{11}\text{-}n)} \xrightarrow[47\%]{CH_3CH_2I} (C_2H_5)(CH_3)C{=}C(H)(C_5H_{11}\text{-}n)$$

$$\mathbf{3} \xrightarrow[52\%]{HCHO} (HOCH_2)(CH_3)C{=}C(H)(C_5H_{11}\text{-}n)$$

$$\mathbf{3} \xrightarrow[56\%]{(CH_3)_2NCHO} (OHC)(CH_3)C{=}C(H)(C_5H_{11}\text{-}n)$$

α-Methylene-γ-lactones.[3] The trisylhydrazone of acetone (**1**) has been used in a synthesis of α-methylene-γ-lactones (**2**). The dianion is generated with n-BuLi and treated with a ketone or an aldehyde in DME and then with n-BuLi and CO_2. Although several intermediates are involved, the whole sequence (equation I) can be conducted in one pot with overall yields of 40–75%.

(I)
$$\underset{\mathbf{1}}{(CH_3)_2C{=}NNHSO_2Ar} \xrightarrow[R^1COR^2,\ -70^\circ]{2n\text{-BuLi,}} \left[CH_3C(=N\bar{N}SO_2Ar)CH_2C(O^-)R^1R^2 \xrightarrow{n\text{-BuLi}} CH_2{=}C(Li)CH_2C(O^-)R^1R^2\right] \xrightarrow[40\text{–}75\%]{1)\ CO_2\ \ 2)\ H^+} \underset{\mathbf{2}}{CH_2{=}\text{(γ-lactone)}R^1R^2}$$

[1] A. R. Chamberlin, J. E. Stemke, and F. T. Bond, *J. Org.*, **43**, 147 (1978); A. R. Chamberlin, E. L. Liotta, and F. T. Bond, *Org. Syn.*, submitted (1977).
[2] A. R. Chamberlin and F. T. Bond, *Synthesis*, 44 (1979).
[3] R. M. Adlington and A. G. M. Barrett, *J.C.S. Chem. Comm.*, 1071 (1978).

Trimethylaluminum–Titanocene dichloride (1).

Trisubstituted alkenes. Negishi *et al.*[1] have extended the reaction of alkynes with trimethylalane and zirconocene dichloride (this volume) to a corresponding addition reaction with trimethylalane–titanocene dichloride. This latter reagent is somewhat more reactive. It reacts with diphenylacetylene to form (Z)-α-methylstilbene (**2**) in high yield. The yield of **2** is $<30\%$ when **1** is replaced by the corresponding zirconocene reagent.

(I) $$C_6H_5C{\equiv}CC_6H_5 \xrightarrow{Al(CH_3)_3\text{–}TiCp_2Cl_2,\ CH_2Cl_2,\ 20^\circ} \xrightarrow[84\%]{H_2O} (C_6H_5)(CH_3)C{=}C(C_6H_5)(H)$$

2

Unexpectedly the reaction of **1** with 5-decyne under the same conditions leads to the allene **3**, formed by dehydrometalation of the initial addition product (equation II).

(II) $$CH_3(CH_2)_3C{\equiv}C(CH_2)_3CH_3 \xrightarrow{1} \xrightarrow[92\%]{H_2O} (CH_3(CH_2)_3)(CH_3)C{=}C{=}C((CH_2)_2CH_3)(H)$$

3

Yields are low in reactions of **1** with 1-alkynes unless $ZnCl_2$ is added to form an intermediate alkynylzinc chloride. Under these conditions 1-alkenes can be obtained in 75–85% yield.

[1] D. E. Van Horn, L. F. Valente, M. J. Idacavage, and E. Negishi, *J. Organometal. Chem.*, **156**, C20 (1978).

Trimethylamine–Sulfur dioxide, $(CH_3)_3\overset{+}{N}{-}SO_2^-$. Trimethylamine and sulfur dioxide form a stable complex melting at 77°.[1]

Nitriles. Aldoximes[2] and primary nitro compounds[3] can be converted into nitriles by dehydration with this complex in CH_2Cl_2. Both reactions proceed under mild conditions and in satisfactory yields.

[1] D. van der Helm, J. D. Childs, and S. D. Christian, *J.C.S. Chem. Comm.*, 887 (1969).
[2] G. A. Olah and Y. D. Vankar, *Synthesis*, 702 (1978).
[3] G. A. Olah, Y. D. Vankar, and B. G. B. Gupta, *ibid.*, 36 (1979).

Trimethylamine N-oxide, 1, 1230–1231; **2,** 434; **3,** 309–310; **6,** 624–625; **7,** 392; **8,** 507.

Removal of —Fe(CO)₃ (**6,** 624). The cation of cycloheptadiene has found limited use in synthesis because of limited stability. However, the iron tricarbonyl complex of the cation reacts readily with a variety of nucleophiles [H_2O, $NaOCH_3$, $HN(CH_3)_2$, $(CH_3)_2CuLi$] to form 5-substituted derivatives of the complex. Of the known reagents for removal of the $—Fe(CO)_3$ group, namely, CAN (**4**, 72), $FeCl_3$, and $(CH_3)_3N \rightarrow O$, the last reagent was found to be the most versatile.[1]

Oxidation of organoaluminum compounds. Organoaluminum compounds can be oxidized conveniently to the corresponding alcohols in nearly quantitative yield by anhydrous trimethylamine oxide (4 hours, 140°).[2]

$$R_3Al \xrightarrow{3(CH_3)_3NO} (RO)_3Al \xrightarrow{H_3O^+} 3ROH$$

[1] B. Y. Shu, E. R. Biehl, and P. C. Reeves, *Syn. Comm.*, **8**, 523 (1978).
[2] G. W. Kabalka and R. J. Newton, Jr., *J. Organometal. Chem.*, **156**, 65 (1978).

2,4,6-Trimethylphenoxymagnesium bromide, 2,4,6-$(CH_3)_3C_6H_2OMgBr$ (**1**). Mol. wt. 239.41. The base is prepared by reaction of 2,4,6-trimethylphenol and ethylmagnesium bromide in ether.

Condensation of carbonyl compounds.[1] This magnesium phenolate catalyzes the self-condensation of linear aliphatic aldehydes. When the reaction is conducted in HMPT, 1,3-diol monoesters (**3** and **4**) are formed (equation I). When the reaction is conducted in benzene, 2,3-dialkylacrylaldehydes (**5**) are formed, also in high yield (equation II). Magnesium ions ($MgBr^+$ and Mg^{2+}) are much more effective in these reactions than Li^+ or Na^+.

$$\text{(I)} \quad \underset{\mathbf{2}}{3RCH_2CHO} \xrightarrow[90\text{–}95\%]{1,\ HMPT} \underset{\mathbf{3}}{RCH_2\underset{OH}{CH}\overset{R}{C}HCH_2OCOCH_2R} + \underset{\mathbf{4}}{RCH_2\underset{OCOCH_2R}{CH}\overset{R}{C}HCH_2OH}$$

$$\text{(II)} \quad 2RCH_2CHO \xrightarrow{1,\ C_6H_6} \underset{\mathbf{5}}{RCH_2CH{=}\overset{R}{C}CHO}$$

The reagent also catalyzes cross-condensation of α,β-unsaturated aldehydes and methyl ketones (equation III). The products **8** predominate when $R^1{=}CH_3$, and are the only products when $R^1 \geq C_2H_5$ and $R^2 = CH_3$, C_6H_5.[2]

$$\text{(III)} \quad \underset{\mathbf{6}}{CH_3COCH_2R^1} + \underset{\mathbf{7}}{R^2CH{=}CHCHO} \xrightarrow{1,\ C_6H_6} \underset{\mathbf{8}}{R^2CH{=}CHCH{=}CHCOCH_2R^1}$$

$$\text{and/or} \quad \underset{\mathbf{9}}{R^2CH{=}CHCH{=}CHR^1COCH_3}$$

[1] G. Casnati, A. Pochini, G. Salerno, and R. Ungaro, *J.C.S. Perkin I*, 1527 (1975).
[2] E. Dradi, A. Pochini, G. Salerno, and R. Ungaro, *Gazz. Chim. Ital.*, 109 (1979).

Trimethyl 3-(phenylseleno)orthopropionate, 1. Mol. wt. 289.22, b.p. 140°/ 0.02 mm.; stable below 175°.

Preparation:

$$C_6H_5SeNa + BrCH_2CH_2CN \xrightarrow[97\%]{} C_6H_5SeCH_2CH_2CN \xrightarrow[65\%]{HCl,\ CH_3OH}$$

$$C_6H_5SeCH_2CH_2CCl{=}NH{\cdot}HCl \xrightarrow[92\%]{1)\ CH_3OH;\ 2)\ N(C_2H_5)_3} C_6H_5SeCH_2CH_2C(OCH_3)_3$$

1

2-Substituted acrylic acids; α-methylene-γ-butyrolactones. The reaction of **1** with the allylic alcohol **2** at 170° gives the product (**3**) of Claisen ortho ester rearrangement. Oxidative elimination of C_6H_5SeOH gives the 2-substituted acrylic ester **4**.

$$\mathbf{1} + HOCH_2CH{=}CHC_6H_5 \ (\mathbf{2}) \xrightarrow[84\%]{(CH_3)_3COH,\ 170^\circ} \mathbf{3} \xrightarrow[80\%]{H_2O_2,\ 25^\circ} \mathbf{4}$$

The reagent has also been used for synthesis of α-methylene-γ-butyrolactones (equation II).[1]

(II) $\xrightarrow[\sim 35\%]{1)\ \mathbf{1},\ 160^\circ;\ 2)\ LiI}$ $\xrightarrow[60\%]{C_6H_5SeCl}$ $\xrightarrow[60\%]{H_2O_2}$

[1] S. Raucher, K.-J. Hwang, and J. E. Macdonald, *Tetrahedron Letters*, 3057 (1979).

Trimethyl phosphite, 1, 1233–1235; **2**, 439–441; **3**, 315–316; **4**, 541–542; **5**, 717; **7**, 393–394.

Cleavage of penicillins.[1] Reaction of a methyl penicillinate 1-oxide (**1**) with a carboxylic acid and trimethyl phosphite in benzene or benzene–THF gives an azetidinone (**2**) in 65–86% yield. The reaction involves replacement of sulfur with an acyloxy group with inversion at C_4. The 6-epimer (**3**) of **1** under the same conditions gives two epimeric azetidinones, **4** and **5**.

$$\mathbf{1} \xrightarrow[65\text{–}86\%]{R^1COOH,\ P(OCH_3)_3} \mathbf{2}$$

3 → **4** + **5**

The reaction provides a route to optically active azetidinones.

[1] A. Suarato, P. Lombardi, C. Galliani, and G. Franceschi, *Tetrahedron Letters*, 4059 (1978).

Trimethylselenonium hydroxide (1). Mol. wt. 141.07. The crystalline solid is fairly stable below 10°.

Preparation:

$$(CH_3)_2Se \xrightarrow[\text{quant.}]{CH_3I} (CH_3)_3SeI \xrightarrow[82\%]{Ag_2O,\ CH_3OH,\ H_2O} (CH_3)_3Se^+OH^-$$

1

Methylation. The reagent methylates carboxyl, sulfhydryl, and phenolic hydroxyl groups rapidly in essentially quantitative yield. The by-products are dimethyl selenide (which can be recycled) and water. It does not methylate aliphatic hydroxyl groups or amino groups, but does methylate aromatic heterocyclic NH groups in high yield.[1]

A number of related reagents have also been prepared, but of these **1** is the most reactive, as judged by reaction of uridine to give 3-methyluridine. One closely related reagent with similar reactivity is trimethylsulfonium hydroxide, $(CH_3)_3S^+OH^-$, prepared by reaction of trimethylsulfonium iodide with silver oxide.[2]

[1] K. Yamauchi, K. Nakamura, and M. Kinoshita, *Tetrahedron Letters*, 1787 (1979).
[2] K. Yamauchi, T. Tanabe, and M. Kinoshita, *J. Org.*, **44**, 638 (1979).

1-Trimethylsilylbutadiene, 8, 395–397. The reagent (**1**) has also been obtained by the route shown in equation (I).[1]

(I) $\xrightarrow[67\%]{C_5H_5\overset{+}{N}HCrO_3Cl^-}$ $\xrightarrow[52\%]{(C_6H_5)_3P{=}CH_2}$ **1**

[1] M. E. Jung and B. Gaede, *Tetrahedron*, **35**, 623 (1979).

Trimethylsilyl dichloroacetate, $CHCl_2COOSi(CH_3)_3$. Mol. wt. 201.13.

α-Chloro-γ-butyrolactones.[1] These useful precursors to $\Delta^{\alpha,\beta}$-butenolides and α-methylene-γ-butyrolactones can be prepared by reaction of an alkene with a trimethylsilyl α-chlorocarboxylate in toluene catalyzed by dichlorotris(triphenylphosphine)ruthenium(II). Isolated yields are 70–80%.

Examples:

$$C_6H_5CH{=}CH_2 + CHCl_2COOSi(CH_3)_3 \xrightarrow[76\%]{Ru(II)} \text{(5-phenyl-3-chloro-γ-butyrolactone)} + (CH_3)_3SiCl$$

(*cis/trans* = 4:6)

$$CH_2{=}C(CH_3)COOCH_3 + CCl_3COOSi(CH_3)_3 \xrightarrow[73\%]{} \text{(5-methyl-5-carbomethoxy-3,3-dichloro-γ-butyrolactone)}$$

$$C_6H_5CH{=}CH_2 + CH_3CCl_2COOSi(CH_3)_3 \xrightarrow[81\%]{} \text{(5-phenyl-3-methyl-3-chloro-γ-butyrolactone)}$$

(two isomers, 1:1)

[1] H. Matsumoto, K. Ohkawa, S. Ikemori, T. Nakano, and Y. Nagai, *Chem. Letters*, 1011 (1979).

β-Trimethylsilylethylidenephosphorane (1), Mol. wt. 362.53.

Preparation[1]:

$$(C_6H_5)_3P{=}CH_2 + (CH_3)_3SiCH_2I \xrightarrow[85\%]{THF} (C_6H_5)_3\overset{+}{P}CH_2CH_2Si(CH_3)_3I^- \xrightarrow{n\text{-BuLi}} (C_6H_5)_3P{=}CHCH_2Si(CH_3)_3 \; \mathbf{1}$$

gem-*Alkylation of ketones.*[2] As expected **1** reacts readily with carbonyl compounds to form allylsilanes. The reaction can be conducted as a one pot procedure with yields of 60–85%. In the presence of Lewis acids, allylsilanes react with various electrophiles (alkyl halides, acid chlorides, ethylene oxide). A typical sequence is shown in equation (I).

$$\text{(I)} \quad \text{cyclohexanone} \xrightarrow[86\%]{\mathbf{1}} C_6H_{10}{=}CHCH_2Si(CH_3)_3 \xrightarrow[85\text{–}95\%]{RX,\ TiCl_4,\ CH_2Cl_2,\ -78°} C_6H_{10}(R)CH{=}CH_2$$

$$C_6H_{10}{=}CHCH_2Si(CH_3)_3 \xrightarrow[99\%]{BF_3\cdot HOAc,\ CHCl_3,\ 20°} C_6H_{11}CH{=}CH_2$$

Allylsilanes undergo protodesilylation in high yield with the $BF_3\cdot$acetic acid complex at 20° in 5 minutes.

[1] D. Seyferth, K. R. Wursthorn, and R. E. Mammarella, *J. Org.*, **42**, 3104 (1977).
[2] I. Fleming and I. Paterson, *Synthesis*, 446 (1979).

Trimethylsilyl 2-lithio-2-carboethoxyacetate, 5, 722–723.

β-Keto esters. Taylor and Turchi[1] have extended the original use to a general synthesis of β-keto esters and bis-β-keto esters.

Examples:

$$\underset{\mathbf{1}}{LiCH(COOSi(CH_3)_3)(COOC_2H_5)} + (CH_3)_3CCH_2COCl \xrightarrow[93\%]{1)\ DME,\ -78 \rightarrow 20°;\ 2)\ H_2O} (CH_3)_3CCH_2COCH_2COOC_2H_5$$

$$2\,\mathbf{1} + \text{cyclobutane-1,2-}(COCl)_2 \xrightarrow{79\%} \text{cyclobutane-1,2-}(COCH_2COOC_2H_5)_2$$

[1] E. C. Taylor and I. J. Turchi, *Org. Prep. Proc. Int.*, **10**, 221 (1978).

Trimethylsilyllithium, $(CH_3)_3SiLi$, **7**, 400.

Modified Peterson reaction.[1] This reagent was used in an unusual application of the Peterson reaction. The α-trimethylsilyloxy aldehyde (**1**) was converted in 80% yield to the *bis*-trimethylsilyl compound (**2**) by treatment at −35° with 0.95 equiv. of trimethylsilyllithium in HMPT. Nucleophilic attack at the carbonyl carbon was followed by silicon migration to the secondary alkoxide. The reaction of **2** with 3 equiv. of lithium diisopropylamide in THF containing 5% HMPT at 23° gave the saturated aldehyde **3** in 80% yield. The reaction was used to convert a hindered >C=O into >CHCH₂OH.

$$\mathbf{1}\ (\text{C–OTMS, CHO}) \xrightarrow[80\%]{(CH_3)_3SiLi,\ HMPT,\ -35°} \mathbf{2}\ (\text{C(OH)–CH(OTMS)TMS}) \xrightarrow[\sim 80\%]{LDA} \mathbf{3}\ (\text{CH–CHO})$$

[1] E. J. Corey, M. A. Tius, and J. Das, *Am. Soc.*, **102**, 1742 (1980).

2-Trimethylsilylmethyl-1,3-butadiene (1). Mol. wt. 140.30, b.p. 70°/80 mm.

Preparation:

$$(CH_3)_3SiCH_2MgCl + CH_2{=}C(Cl)CH{=}CH_2 \xrightarrow[90\%]{Cl_2NiL_2} \underset{\mathbf{1}}{CH_2{=}C(CH_2Si(CH_3)_3)-CH{=}CH_2}$$

$L = (C_6H_5)_2P(CH_2)_3P(C_6H_5)_2$

Isoprenylation.[1] In the presence of $TiCl_4$ or $AlCl_3$, this isoprenylsilane (**1**) reacts with acid chlorides, acetals, and carbonyl compounds to form isoprenylated compounds. Reactions with the first two electrophiles proceed in higher yield than those with carbonyl compounds. Isoprenylation provides simple syntheses of ipsenol (**2**) and ipsdienol (**3**), components of the aggregation pheromone of a bark beetle.

$(CH_3)_2CHCH_2COCl + $ **1** $\xrightarrow[77\%]{TiCl_4,\ CH_2Cl_2}$ ketone $\xrightarrow[62\%\ overall]{DIBAH}$ **2**

CH2 CH2 CH2 CH2 O CH(CH3)2 HO CH(CH3)2 2

$(CH_3)_2C{=}CHCOCl \xrightarrow[71\%]{}$ ketone $\xrightarrow[75\%]{}$ **3**

CH2 CH2 CH2 CH2 O C CH3 CH3 HO C CH3 CH3 3

[1] A. Hosomi, M. Saito, and H. Sakurai, *Tetrahedron Letters*, 429 (1979).

2-Trimethylsilylmethylenecyclobutane (1). Mol. wt. 140.3, b.p. 130°/760 mm.
Preparation:[1]

Methylenecyclobutane $\xrightarrow[77\%]{n\text{-}BuLi/TMEDA,\ (CH_3)_3SiCl,\ hexane,\ -78°}$ **1** + 1-(trimethylsilylmethyl)cyclobutene (60:40)

CH2 CH2 Si(CH3)3 Si(CH3)3 1

$\downarrow$ Δ, $CH_2{=}CHCOOCH_3$

Si(CH3)3 CH3OOC
(b.p. 150°/20 mm.)

Isoprenoid synthesis.[1] The reagent reacts with aldehydes in the presence of $TiCl_4$ to afford substituted cyclobutenes. The reagent functions as an isoprene synthon since cyclobutenes are converted into 1,3-dienes on thermolysis. This method is preferable to use of 1-cyclobutenylmethyllithium (**6**, 151), since reactions of **1** with electrophiles are regiospecific. A typical application of this synthon is illustrated by synthesis of tagetol (**3**).

[1] S. R. Wilson, L. R. Phillips, and K. J. Natalie, Jr., *Am. Soc.*, **101**, 3340 (1979).

$$\text{(CH}_3)_2\text{CHCH}_2\text{CHO} \xrightarrow[86\%]{\mathbf{1},\ \text{TiCl}_4,\ \text{CH}_2\text{Cl}_2} \mathbf{2} \xrightarrow[\sim 100\%]{150°} \mathbf{3}$$

Trimethylsilylmethylene dimethylsulfurane, $(CH_3)_3SiCH{=}S(CH_3)_2$ (**1**). Mol. wt. 172.36.

Preparation:

$$(CH_3)_3SiCH_2SCH_3 \xrightarrow{CH_3I} (CH_3)_3SiCH_2\overset{+}{S}(CH_3)_2I^- \xrightarrow{sec\text{-BuLi}} \mathbf{1}$$

Silylcyclopropanation.[1] This ylide reacts with α,β-unsaturated cyclohexenones and cyclopentenones to give silylcyclopropyl ketones in fair yields (35–65%). The method by which **1** is generated is extremely important for this reaction. Neither *t*- nor *n*-butyllithium can be used as base. Epoxides, which would have resulted from desilylation to dimethylsulfonium methylide, are not formed.

Examples:

$$CH_3CH{=}CHCOCH_3 \xrightarrow[65\%]{\mathbf{1}} \text{1-COCH}_3\text{-2-CH}_3\text{-3-Si(CH}_3)_3\text{cyclopropane}$$

$$\text{2-cyclohexenone} \xrightarrow[55\%]{} \text{7-Si(CH}_3)_3\text{bicyclo[4.1.0]heptan-2-one}$$

[1] F. Cooke, P. Magnus, and G. L. Bundy, *J.C.S. Chem. Comm.*, 714 (1978).

Trimethylsilylmethyllithium, 6, 635–636.

α-Trimethylsilyl ketones.[1] Trimethylsilylmethyllithium (**1**) reacts with carboxylic acids to form α-trimethylsilyl ketones in rather low yield. However, the reaction of **1** with methyl esters, particularly of secondary and tertiary carboxylic acids, is a useful route to α-trimethylsilyl ketones and to methyl ketones. The intermediate enolate can also be alkylated regioselectively.

Example:

$$(CH_3)_2CHCOOCH_3 \xrightarrow{2LiCH_2Si(CH_3)_3} (CH_3)_2CH\overset{O^-\ Li^+}{C}{=}CHSi(CH_3)_3$$

$$\xrightarrow[82\%]{H_2O} (CH_3)_2CHCOCH_2Si(CH_3)_3$$

$$\xrightarrow[83\%]{CH_3OH} (CH_3)_2CHCOCH_3$$

$$\xrightarrow[70\%]{CH_2{=}CHCH_2Cl} (CH_3)_2CHCOCH(Si(CH_3)_3)CH_2CH{=}CH_2$$

[1] M. Demuth, *Helv.*, **61**, 3136 (1978).

Trimethylsilyl nonaflate, $(CH_3)_3SiOSO_2C_4F_9$-*n* (**1**). Mol. wt. 372.29. The reagent can be generated *in situ* from chlorotrimethylsilane and potassium nonaflate (*n*-$C_4F_9SO_3K$).

Enol silyl ethers.[1] Ketones can be converted into enol silyl ethers by reaction with **1** and triethylamine at reflux temperature. Typical yields are 70–80%.

[1] H. Vorbrüggen and K. Krolikiewicz, *Synthesis*, 34 (1979).

Trimethylsilyl phenyl selenide, $C_6H_5SeSi(CH_3)_3$ (**1**). Mol. wt. 229.25, b.p. 63–65°/0.9 mm. The silane is prepared (68% yield) by reaction of C_6H_5SeNa with $ClSi(CH_3)_3$.

Deoxygenation.[1] The reagent (2 equiv.) reduces sulfoxides, selenoxides, and telluroxides to the corresponding sulfides, selenides, and tellurides in 85–95% yield.

Phenylselenenylation of acetates and lactones.[2] In the presence of zinc iodide, **1** reacts with alkyl acetates to form alkyl phenyl selenides. The reaction is particularly facile with acetates of tertiary, allylic, and benzylic alcohols. Acetates of primary and secondary alcohols react only at 110°.

Under the same conditions, **1** converts lactones into ω-phenylselenocarboxylic acids (equation I).

(I) [δ-valerolactone] $\xrightarrow{\mathbf{1},\ ZnI_2,\ 110°}$ $C_6H_5Se(CH_2)_4C(=O)OSi(CH_3)_3$ $\xrightarrow[\text{70\% overall}]{CH_3OH}$ $C_6H_5Se(CH_2)_4C(=O)OH$

Reaction with epoxides.[3] The reagent (**1**) in the presence of ZnI_2 as catalyst converts epoxides into β-siloxyalkyl phenyl selenides in 70–90% yield (*cf.* cleavage of epoxides with $NaSeC_6H_5$, **5**, 272–273). The facility of C—O bond cleavage is tertiary > primary > secondary. The selectivity follows the reverse order if *n*-butyllithium is used as the catalyst.

Examples:

[1-methylcyclohexene oxide] $\xrightarrow[79\%]{\mathbf{1},\ CH_2Cl_2,\ ZnI_2}$ [1-methyl-1-SeC_6H_5-2-$OSi(CH_3)_3$-cyclohexane] + [1-methyl-1-$OSi(CH_3)_3$-2-SeC_6H_5-cyclohexane] (59:41)

[styrene oxide, C_6H_5] $\xrightarrow[70\%]{}$ $C_6H_5CH(SeC_6H_5)CH_2OSi(CH_3)_3$ + $C_6H_5CH(OSi(CH_3)_3)CH_2SeC_6H_5$ (90:10)

[1] M. R. Delty, *J. Org.*, **44**, 4528 (1979).
[2] N. Miyoshi, H. Ishii, S. Murai, and N. Sonoda, *Chem. Letters*, 873 (1979).
[3] N. Miyoshi, K. Kondo, S. Murai, and N. Sonoda, *ibid.*, 909 (1979).

2-Trimethylsilyl-2-propenol, $CH_2{=}C(Si(CH_3)_3)CH_2OH$ (**1**). Mol. wt. 130.26, b.p. 74–75°/30 mm. The alcohol is prepared by reaction of α-lithiovinyltrimethylsilane with formaldehyde (65% yield).[1]

Protection of phosphoric acids.[2] 2-Trimethylsilyl-2-propenyl esters of phosphoric acids are relatively stable to acid and to base. They are cleaved selectively by either tetraethylammonium fluoride or by catalytic hydrogenation. The former method is particularly useful because one of two protective groups can be cleaved selectively by use of 1 equiv. of the fluoride ion (equation I).

$$\text{(I)}\quad C_2H_5OP(=O)\left[OCH_2C(Si(CH_3)_3){=}CH_2\right]_2 \xrightarrow[\sim 90\%]{(C_2H_5)_4\overset{+}{N}F^-,\ CH_3CN} C_2H_5OP(=O)(^-ON^+(C_2H_5)_4){-}OCH_2C(Si(CH_3)_3){=}CH_2 + FSi(CH_3)_3 + CH_2{=}C{=}CH_2$$

[1] T. H. Chan, W. Mychajlowskij, B. S. Ong, and D. N. Harpp, *J. Org.*, **43**, 1526 (1978); T. H. Chan and B. S. Ong, *ibid.*, **43**, 2994 (1978).
[2] T. H. Chan and M. Di Stefano, *J.C.S. Chem. Comm.*, 761 (1978).

Trimethylsilyl trifluoromethanesulfonate (1), 8, 514–516.

Epoxide → allylic alcohol.[1] Treatment of an oxirane with equimolar amounts of **1** and DBU in an aromatic solvent affords allylic trimethylsilyl ethers in moderate yield. 2,2-Di-, tri-, and tetrasubstituted oxiranes, as well as oxides of cycloalkenes, react at 23° or below. 2,3-Di- and monosubstituted oxiranes do not react at this temperature; these species react with **1** and DBU at 70–80° to give trimethylsilyl enol ethers. The reaction of epoxycyclooctane gives a product of transannular cyclization. In the case of epoxycyclohexane, the intermediate **2** has been isolated.

$$\text{cyclohexene oxide} \xrightarrow{(CH_3)_3SiOTf\ (\mathbf{1})} \underset{\mathbf{2}}{\text{2-OTf-cyclohexyl-}OSi(CH_3)_3} \xrightarrow[87\%]{DBU} \underset{\mathbf{3}}{\text{cyclohex-2-enyl-}OSi(CH_3)_3}$$

[1] S. Murata, M. Suzuki, and R. Noyori, *Am. Soc.*, **101**, 2738 (1979).

Ethers from acetals.[2] Reduction of acetals with trimethylsilane (or triethylsilane) catalyzed by trimethylsilyl trifluoromethanesulfonate gives ethers in 75–95% isolated yield (equation I).

$$\text{(I)}\quad R^1R^2C(OCH_3)_2 + HSi(CH_3)_3 \xrightarrow[75\text{–}95\%]{(CH_3)_3SiOTf,\ CH_2Cl_2,\ 0\rightarrow 28^\circ} R^1R^2CHOCH_3 + (CH_3)_3SiOCH_3$$

[2] T. Tsunoda, M. Suzuki, and R. Noyori, *Tetrahedron Letters*, 4679 (1979).

α-Trimethylsilylvinyllithium, $CH_2{=}C(Si(CH_3)_3)Li$ (**1**).

Tetrasubstituted alkenes. A new route to alkenes of this type involves as the initial step reaction of ketones with **1** to form the tertiary alcohols **2**. These alcohols are not acetylated under conventional methods, but are acetylated in satisfactory yields by acetyl chloride in the presence of silver cyanide. Reaction of **3** with organocopper reagents gives tetrasubstituted alkenes as a mixture of (E)- and (Z)-isomers, **4** and **5**. The stereoselectivity of this reaction depends on the relative size of R^1 and R^2. When R^2 is more bulky than R^1, **5** is formed predominately. Both products can be halodesilylated to vinyl iodides **6** or **7** or to vinyl bromides by $BrCN\text{–}AlCl_3$.

$$R^1R^2C{=}O + \underset{\mathbf{1}}{(CH_3)_3Si(Li)C{=}CH_2} \rightarrow \underset{\mathbf{2}}{R^1R^2C(OH)\text{–}C(Si(CH_3)_3){=}CH_2} \xrightarrow[80\%]{CH_3COCl,\ AgCN} \underset{\mathbf{3}}{R^1R^2C(OAc)\text{–}C(Si(CH_3)_3){=}CH_2}$$

$$\xrightarrow{(R^3)_2CuLi} \underset{\mathbf{4}}{R^1R^2C{=}C(CH_2R^3)Si(CH_3)_3} + \underset{\mathbf{5}}{R^2R^1C{=}C(CH_2R^3)Si(CH_3)_3}$$

$$\downarrow I_2$$

$$\underset{\mathbf{6}}{R^1R^2C{=}C(CH_2R^3)I} \qquad \underset{\mathbf{7}}{R^2R^1C{=}C(CH_2R^3)I}$$

[1] R. Amouroux and T. H. Chan, *Tetrahedron Letters*, 4453 (1978).

Trimethylvinylsilane, $CH_2{=}CHSi(CH_3)_3$. Mol. wt. 100.24. Supplier: Petrarch Systems.

Cyclopentenones. Trimethylvinylsilane can function as an ethylene equivalent in an aliphatic Friedel–Crafts type of acylation with α,β-unsaturated acid chlorides.

For example, reaction of cyclohexenecarboxylic acid chloride (**1**) in CH_2Cl_2 with trimethylvinylsilane in the presence of stannic chloride at $-30°$ and finally at $20°$ leads to the bicyclic enone **2** in 46% yield. The reaction involves an intermediate dienone (**a**) that cyclizes under acidic conditions (Nazarov reaction). Although yields are not high, this one step synthesis of bicyclo[n.3.0]enones is very convenient.[1]

COCl **1** $+ CH_2{=}CHSi(CH_3)_3$ $\xrightarrow{SnCl_4, CH_2Cl_2}$ [**a** (CH_2)] $\xrightarrow{46\%}$ **2**

Acyclic α,β-unsaturated acid chlorides do not undergo this acylation. However, cyclic vinylsilanes also undergo Friedel–Crafts acylation with α,β-unsaturated acid chlorides to give dienones of type **a**, which are cyclized by $SnCl_4$ to cyclopentenones.[2] A typical example is formulated for the reaction of 5-methyl-1-trimethylsilylcyclopentene (**3**) with β,β-dimethylacryloyl chloride. The first step is conducted with $AlCl_3$ and gives **4**, which after isolation is cyclized with $SnCl_4$ to a mixture of **5** and **6**. The cyclopentenone **6** is obtained in 55% overall yield after isomerization with $RhCl_3$.

3 (CH_3, $Si(CH_3)_3$) $+ (CH_3)_2C{=}CHCOCl$ $\xrightarrow{AlCl_3, -78°, CH_2Cl_2}$ **4** (CH_3, O, $C(CH_3)_2$)

$\xrightarrow{SnCl_4, CH_2Cl_2, \Delta}$ **5** (CH_3, O, CH_3 CH_3) + **6** (CH_3, O, CH_3CH_3) ~3:2

5 $\xrightarrow{RhCl_3, C_2H_5OH}$ **6** (55% overall)

[1] F. Cooke, J. Schwindeman, and P. Magnus, *Tetrahedron Letters*, 1995 (1979).
[2] W. E. Fristad, D. S. Dime, T. R. Bailey, and L. A. Paquette, *ibid.*, 1999 (1979).

Trimethylstannyllithium, $(CH_3)_3SnLi$.

Polyprenylation of quinones. Trimethylstannyllithium converts polyprenyl halides (e.g., geranyl chloride, phytyl chloride) into polyprenyltrimethylstannyl compounds in high yield with retention of stereochemistry of the allylic double bond. In the presence of a Lewis acid (BF_3 etherate) these tin compounds couple regio- and stereoselectively with 2-methyl-1,4-naphthoquinone. The reaction to form vitamin K (**2**) is typical (equation I).[1]

The same sequence is applicable to synthesis of coenzyme Q (ubiquinones, involved in respiratory and photosynthetic electron transport). A representative reaction is the preparation of *trans*-coenzyme Q_2 (**5**, equation II) from geranyltrimethyltin (**4**).[2]

(I) Phytyl chloride $\xrightarrow[\text{quant.}]{(CH_3)_3SnLi,\ THF,\ -60\rightarrow 20^\circ}$ **1**

2-methyl-1,4-naphthoquinone $\xrightarrow{1)\ \mathbf{1},\ BF_3\cdot(C_2H_5)_2O;\ 2)\ Ag_2O}$ **2** (70%) + **3** (17%)

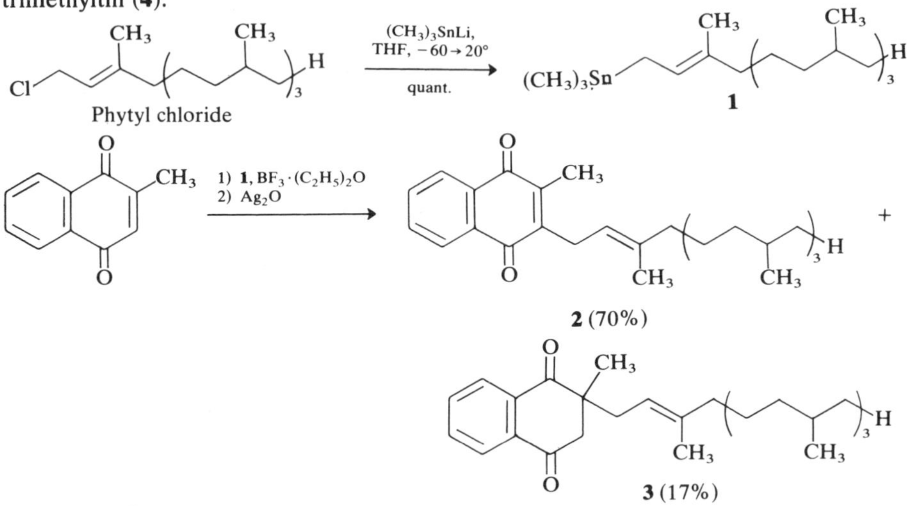

(II) 2,3-dimethoxy-5-methyl-1,4-benzoquinone + **4** $\xrightarrow[\text{65\% (isolated)}]{1)\ BF_3\cdot(C_2H_5)_2O,\ \sim 60^\circ;\ 2)\ H_3O^+;\ 3)\ Ag_2O}$ **5**

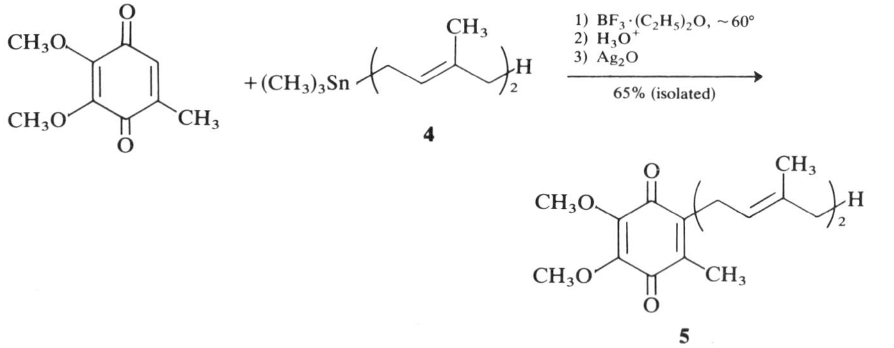

Cleavage of 2-bromoethyl esters.[3] These esters can be cleaved by treatment with Still's reagent in THF and then with tetra-*n*-butylammonium fluoride (cf. cleavage of 2-trimethylsilylethyl esters, **8**, 510).

$$RCOOCH_2CH_2Br \xrightarrow{LiSn(CH_3)_3} [RCOOCH_2CH_2Sn(CH_3)_3] \xrightarrow[80\text{–}85\%]{F^-} RCOOH$$

[1] Y. Naruta and K. Maruyama, *Chem. Letters*, 881 (1979).
[2] *Idem, ibid.*, 885 (1979).
[3] T.-L. Ho, *Syn. Comm.*, **8**, 359 (1978).

Trimethylstannylsodium, $(CH_3)_3SnNa$ (**1**). Mol. wt. 186.78. The reagent is prepared in tetraglyme or THF by reaction of sodium metal with hexamethylditin (90–95% yield).

Dehalogenation of vic-dihalides.[1] The reaction of these substrates with **1** proceeds by *anti*-elimination to give alkenes in high yield. Thus double bonds can be protected by stereospecific *anti*-addition of halogen and stereospecific *anti*-dehalogenation with **1**. Inversion of (E)- and (Z)-alkenes can be achieved by stereospecific *anti*-addition of methyl hypobromite followed by *syn*-demethoxybromination by **1** in THF.

Aromatic substitution.[2] The reagent (**1**) reacts with aryl halides by halogen–metal exchange as the initial step to give $(CH_3)_3SnAr$. ArH is formed when proton donors are present.

[1] H. G. Kuivila and Y. M. Choi, *J. Org.*, **44**, 4774 (1979).
[2] K. R. Wursthorn, H. G. Kuivila, and G. F. Smith, *Am. Soc.*, **100**, 2779 (1978).

Triphenylbismuth carbonate (**1**). Mol. wt. 500.31, m.p. 164–165° dec.

Preparation[1]:

$$(C_6H_5)_3BiCl_2 \xrightarrow{K_2CO_3,\ H_2O,\ \text{acetone}} \underset{\mathbf{1}}{(C_6H_5)_3BiCO_3}$$

Oxidation.[2] Triphenylbismuth carbonate suspended in CH_2Cl_2 is a heterogeneous oxidant for a variety of functional groups. Allylic alcohols are efficiently oxidized to the corresponding unsaturated aldehydes or ketones, even in the presence of a thiol, which is itself oxidized by this reagent to a disulfide. *cis*- and *trans*-1,2-Glycols are cleaved to dialdehydes; hydrazones are oxidized to diazocompounds; oximes are cleaved to ketones; and 1,2-dialkylhydrazines are oxidized to azo compounds. Phenylhydrazones, semicarbazones, tosylhydrazones, aromatic and aliphatic amines, enamines, and enol ethers are inert to **1**.

Examples:

CH₂OH (geraniol-type allylic alcohol, CH₃, H₃C, CH₃) $\xrightarrow[95\%]{\mathbf{1},\ CH_2Cl_2,\ 40°}$ CHO (CH₃, H₃C, CH₃)

Cyclohexane-1,2-diol (OH, OH) $\xrightarrow[100\%]{}$ CHO, CHO

$$(C_6H_5)_2C{=}NNH_2 \xrightarrow[97\%]{} (C_6H_5)_2CN_2 \qquad C_6H_5NH{-}NHC_6H_5 \xrightarrow[90\%]{} C_6H_5N{=}NC_6H_5$$

Oxime (HO–N=, H₃C, H) $\xrightarrow[60\%]{}$ ketone (O=, H₃C, H)

$$2C_6H_5SH \xrightarrow[70\%]{} C_6H_5SSC_6H_5$$

[1] R. G. Goel and H. S. Prosad, *Can. J. Chem.*, **49**, 2529 (1971).
[2] D. H. R. Barton, D. J. Lester, W. B. Motherwell, and M. T. B. Papoula, *J.C.S. Chem. Comm.*, 705 (1979).

Triphenylmethylphosphonium permanganate, $(C_6H_5)_3\overset{+}{P}CH_3MnO_4^-$. Mol. wt. 396.3, violet crystals, decomposes explosively about 70°. The salt is prepared by reaction of $(C_6H_5)_3\overset{+}{P}CH_3Br^-$ and $KMnO_4$.

cis-Hydroxylation.[1] Olefins can be oxidized to *cis*-diols by this reagent in anhydrous CH_2Cl_2 at −70°. *This reaction should not be carried out on a large scale.* Yields of diols are generally higher than those obtained with $KMnO_4$ in the presence of a phase-transfer catalyst or a crown ether.

[1] W. Reischl and E. Zbiral, *Tetrahedron*, **35**, 1109 (1979).

Triphenylmethylpotassium (Tritylpotassium), $(C_6H_5)_3CK$. Mol. wt. 282.43. The base is prepared by reaction of triphenylmethane with potassium hydride.

Dieckmann cyclization.[1] Esters of a dicarboxylic acid substituted at the α,α'-positions with alkyl groups (**1**) do not undergo cyclization under the usual conditions, but can be cyclized by tritylpotassium (4 equiv.) in good yield. The product, **2**, can also be alkylated by means of the same base.

$$CH_3CH(COOCH_3)-(CH_2)_3-CH(COOCH_3)CH_3 \ (\mathbf{1}) \xrightarrow[80\%]{(C_6H_5)_3CK,\ DME} \mathbf{2}\ \text{(two isomers)} \xrightarrow[65\%]{1)\ (C_6H_5)_3CK\quad 2)\ CH_3I} \mathbf{3}$$

(**2**: 2-methoxycarbonyl-2,6-dimethylcyclohexanone, CH_3, CH_3, $COOCH_3$, C=O; **3**: 2-methoxycarbonyl-2,6,6-trimethylcyclohexanone, CH_3, CH_3, CH_3, $COOCH_3$, C=O)

[1] G. Nee and B. Tchoubar, *Tetrahedron Letters*, 3717 (1979).

Triphenyl(1-phenylthiovinyl)phosphonium iodide (1). Mol. wt. 512.79
Preparation:

$$(C_6H_5)_3\overset{+}{P}CH_2SC_6H_5I^- + CH_2{=}\overset{+}{N}(CH_3)_2Cl^- \xrightarrow[90\%]{} CH_2{=}C(SC_6H_5)\overset{+}{P}(C_6H_5)_3I^-\ (\mathbf{1})$$

Cyclopentanones.[1] The reaction of the sodium enolate of the keto diester **2** with **1** in THF at 20° leads to the cyclopentene **3** in 90% yield. The product is readily converted to the ketone **4** by treatment with TFA.

$$CH_3COCH_2CH(COOC_2H_5)_2\ (\mathbf{2}) \xrightarrow[90\%]{1)\ NaH\quad 2)\ \mathbf{1}} \mathbf{3} \xrightarrow[56\%]{TFA} \mathbf{4}$$

(**3**: 4,4-bis(ethoxycarbonyl)-1-phenylthio-2-methylcyclopentene, C_6H_5S, CH_3, C_2H_5OOC, $COOC_2H_5$; **4**: 4,4-bis(ethoxycarbonyl)-2-methylcyclopentanone, O, CH_3, C_2H_5OOC, $COOC_2H_5$)

[1] A. T. Hewson, *Tetrahedron Letters*, 3267 (1978).

Triphenylphosphine–Carbon tetrachloride, 1, 1247; **2,** 445; **3,** 320; **4,** 551–552; **5,** 727; **6,** 644–645; **7,** 404; **8,** 516.

Amides. Barstow and Hruby[1] have published two methods for synthesis of amides. In one, the two reagents are heated in THF to form the adduct; the carboxylic acid and, after 10 minutes, the amine (2 equiv.) are added. The amide is obtained after a reflux period of about 1 hour. Yields are generally 80–90%. The second method involves triphenylphosphine, bromotrichloromethane, the acid, and the amine. The reported yields are somewhat higher than those obtained with the first method. One example of peptide synthesis by this method was reported (85% yield).

This reaction provided an important step in a recent stereospecific synthesis of **4**, which contains the basic skeleton of lycorine (**5**).[2]

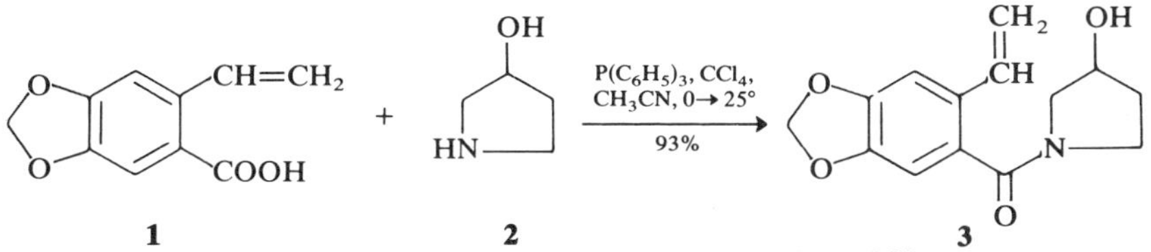

$CH{=}CH_2$, COOH (**1**) + OH, HN (**2**) $\xrightarrow[93\%]{P(C_6H_5)_3,\ CCl_4,\ CH_3CN,\ 0\rightarrow 25°}$ CH_2, CH, OH, N, O (**3**)

H, H, O, N, O (**4**) 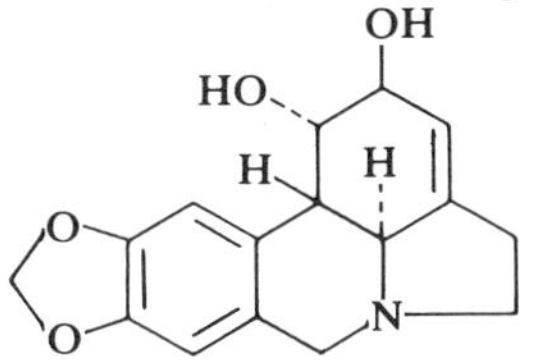   OH, HO, H, H, O, N (**5**)

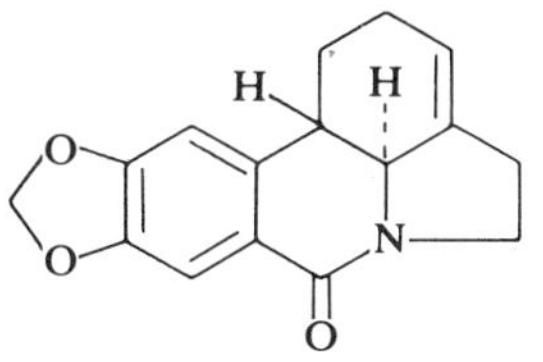

Alkyl chlorides; acid chlorides (**1,** 1247).[3] The conversion of alcohols into alkyl chlorides and of acids into acid chlorides with $(C_6H_5)_3P$ and CCl_4 proceeds more rapidly when the phosphine is polymer supported. One factor contributing to the rate increase may be the close proximity of the phosphorus-containing residues.

Azetidines.[4] The cyclization of β-amino alcohols to aziridines by $P(C_6H_5)_3$ and CCl_4 (**5,** 727) can be used for a similar synthesis of azetidines from 3-methylaminopropanols.

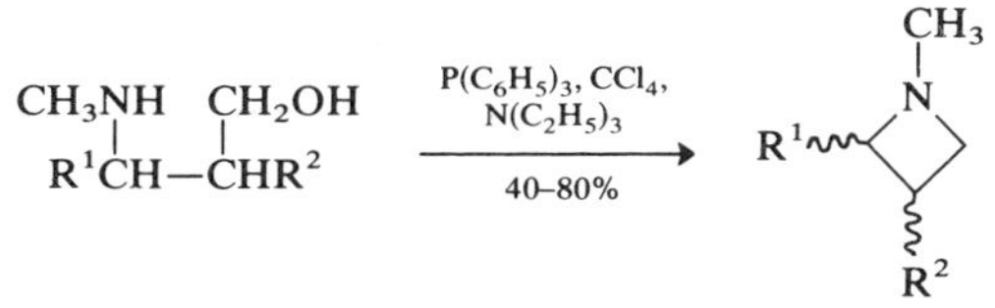

CH_3NH CH_2OH / $R^1CH{-}CHR^2$ $\xrightarrow[40\text{–}80\%]{P(C_6H_5)_3,\ CCl_4,\ N(C_2H_5)_3}$ CH_3, N, R^1, R^2

[1] L. E. Barstow and V. J. Hruby, *J. Org.*, **36**, 1305 (1971).
[2] G. Stork and D. J. Morgans, Jr., *Am. Soc.*, **101**, 7110 (1979).
[3] C. R. Harrison and P. Hodge, *J.C.S. Chem. Comm.*, 813 (1978).
[4] V. Stoilova, L. S. Trifonov, and A. S. Orahovats, *Synthesis*, 105 (1979).

Triphenylphosphine–Diethyl azodicarboxylate, 1, 245–247; **4**, 553–555; **5**, 727–728; **6**, 645; **7**, 404–406; **8**, 517.

Decarboxylative dehydration of 3-hydroxycarboxylic acids. *threo*-3-Hydroxycarboxylic acids (**1**) undergo *anti*-elimination when treated with this reagent to give (*Z*)-olefins (**2**) (97–99% stereoselectivity). The corresponding (E)-olefins can be prepared by thermal *syn*-elimination of the β-lactones derived from **1**.[1]

COOH; H, R^1; H, R^2; OH (**1**) —[$P(C_6H_5)_3$, $C_2H_5OOCN{=}NCOOC_2H_5$; 68–77%]→ H, H; R^2, R^1 (**2**)

N-Hydroxy-2-azetidinones.[2] A biomimetic synthesis of the β-lactam **4** from BOC-L-serine (**1**) has been reported. The protected serine derivative is converted into the hydroxamic acid **2** by condensation with O-benzylhydroxylamine mediated by the water-soluble 1-ethyl-3(3′-dimethylaminopropyl)carbodiimide (**1**, 371). The product is cyclized directly in high yield to the β-lactam **3** by treatment with diethyl azodicarboxylate and triphenylphosphine. No racemization is observed. Deprotection by catalytic hydrogenation gives the N-hydroxy-β-lactam **4**. Previous biomimetic syntheses of 2-azetidinones have involved cyclization of β-chloroamides with sodium hydride (e.g., **7**, 335).

BOCHN–CH(CH_2OH)–C(=O)–OH (**1**) + $H_2NOCH_2C_6H_5$ —[80–90%]→ BOCHN–CH(CH_2OH)–C(=O)–$NHOCH_2C_6H_5$ (**2**)

—[$C_2H_5OOCN{=}NCOOC_2H_5$, $P(C_6H_5)_3$, THF; 80–90%]→ BOCHN, H, O, N, $OCH_2C_6H_5$ (**3**) —[H_2, Pd/C; ~100%]→ BOCHN, H, O, N, OH (**4**)

cis-*Rheadane to* trans-*rheadane.* The *cis*-rheagenine **1** has been transformed into the less stable *trans*-isomer **3** by hydrolysis to the hydroxy acid **2** followed by dehydration with triphenylphosphine and diethyl azodicarboxylate.[3]

O, O, H, N–CH_3, H, CH_3O, O, O, CH_3O, OCH_3 (**1**) —[NaOH, CH_3OH]→

$$\textbf{2} \xrightarrow[\text{26\%}]{(C_6H_5)_3P,\ C_2H_5OOCN{=}NCOOC_2H_5} \textbf{3}$$

Reaction with 1,2-glycols.[4] This redox system converts *trans*-1,2-cyclohexanediol into cyclohexane epoxide with no indication of a phosphorus-containing intermediate. The reaction with *cis*-1,2-cyclohexanediol results in a phosphorane. This differing behavior was first observed in reactions of the nucleosides **1** and **3**.

$$\textbf{1} \xrightarrow[\text{78\%}]{P(C_6H_5)_3,\ C_2H_5OOCN{=}NCOOC_2H_5} \textbf{2}$$

$$\textbf{3} \xrightarrow[\text{79\%}]{} \textbf{4}$$

Cyclic ethers.[5] This system converts 1,3-propanediol into oxetane in near quantitative yield. The system is equally effective when applied to 1,2-, 1,3-, 1,4-, 1,5-, and 1,6-diols. 2,2-Dimethylaziridine was prepared in high yield in this way from 2-amino-2-methyl-1-propanol.

$$HOCH_2CH_2CH_2OH + (C_6H_5)_3P + C_2H_5OOCN{=}NCOOC_2H_5 \xrightarrow{C_6H_5CH_3,\ 0^\circ} \text{oxetane}\ (98\%) + (C_6H_5)_3P{=}O + C_2H_5OOCNH{-}NHCOOC_2H_5$$

$$(CH_3)_2C(NH_2){-}CH_2OH \xrightarrow[\text{quant.}]{\text{Tetralin, } 20^\circ} \text{2,2-dimethylaziridine}$$

Dehydroamino acids.[6] Dehydroamino acids can be prepared from protected forms of serine and threonine with this reagent. Methyl esters are used; the amino protecting group can be phthaloyl, CBZ, or BOC. Yields range from 63 to 69%.

$$\underset{\displaystyle\text{NHBOC}}{\text{HOCH}_2\text{CHCOOCH}_3} \xrightarrow[69\%]{\text{P(C}_6\text{H}_5)_3,\ \text{C}_2\text{H}_5\text{OOCN}{=}\text{NCOOC}_2\text{H}_5} \underset{\displaystyle\text{NHBOC}}{\text{CH}_2{=}\text{CCOOCH}_3}$$

[1] J. Mulzer, A. Pointner, A. Chucholowski, and G. Brüntrup, *J.C.S. Chem. Comm.*, 52 (1979).
[2] P. G. Mattingly, J. F. Kerwin, Jr., and M. J. Miller, *Am. Soc.*, **101**, 3983 (1979).
[3] R. Hohlbrugger and W. Klötzer, *Ber.*, **112**, 3486 (1979).
[4] R. Mengel and M. Bartke, *Angew. Chem., Int. Ed.*, **17**, 679 (1978).
[5] J. T. Carlock and M. P. Mack, *Tetrahedron Letters*, 5153 (1978).
[6] H. Wojciechowska, R. Pawlowicz, R. Andruszkiewicz, and J. Grzybowska, *ibid.*, 4063 (1978).

Triphenylphosphine–Diethyl azodicarboxylate–Methyl iodide.

Deoxyiodo sugars.[1] Iodo sugars have been prepared by reaction of isolated hydroxyl groups with this combination of reagents to form alkoxyphosphonium iodides in one operation. When the salts are heated in toluene at 110°, iodo sugars are formed with inversion. A typical example is formulated in equation (I) for the reaction of 1,2;5,6-diisopropylideneglucose.

(I)

$$\text{(1,2;5,6-diisopropylideneglucose, OH)} \xrightarrow[70\%]{\text{P(C}_6\text{H}_5)_3,\ \text{DEAD},\ \text{CH}_3\text{I}} \text{(alkoxyphosphonium salt, O}\overset{+}{\text{P}}\text{(C}_6\text{H}_5)_3\ \text{I}^-) \xrightarrow[90\%]{110^\circ} \text{(3-deoxy-3-iodo sugar, I)}$$

[1] H. Kunz and P. Schmidt, *Tetrahedron Letters*, 2123 (1979).

Triphenylphosphine–Selenocyanogen.

Alkyl selenocyanates; diselenides.[1] Primary alcohols are converted by the reagent into alkyl selenocyanates in about 80% yield. If the reaction mixture is treated with 5% KOH in CH_3OH for 2 hours the alkyl selenocyanates are transformed to dialkyl diselenides (about 70% yield). The reagent converts secondary alcohols into selenocyanates and to a lesser extent into isoselenocyanates. It does not react with tertiary alcohols.

[1] Y. Tamura, M. Adachi, T. Kawasaki, and Y. Kita, *Tetrahedron Letters*, 2251 (1979).

Triphenylphosphine–Thiocyanogen, $(C_6H_5)_3P(SCN)_2$ (**1**), **8**, 518–519.

Thiocyanation of epoxy ketones.[1] The reagent (**1**) converts α,β-epoxy ketones into 2-thiocyanate enones (**2**) in 60–90% yield. A typical example is formulated, together with two transformations of the product.

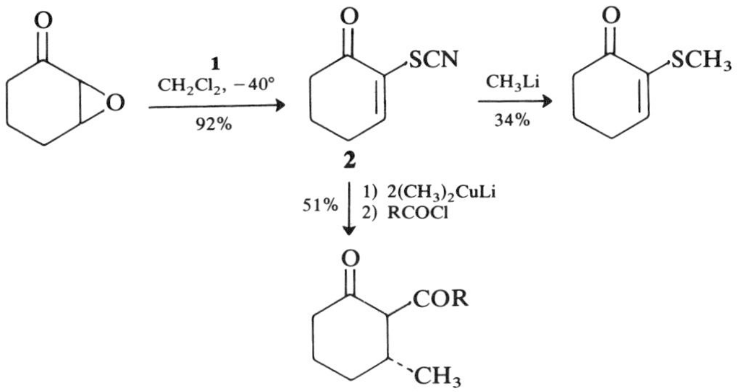

Thioamides.[2] The reaction of Grignard reagents with **1** results, after acid hydrolysis, in N-unsubstituted thioamides (**3**) in fair yields (equation I). In one case, $R = C_6H_5CH_2$, **2** was isolated in 31% yield.

$$\text{(I)}\quad \underset{\mathbf{1}}{(C_6H_5)_3P(SCN)_2} + RMgX \xrightarrow[\text{40–90\% overall}]{C_6H_6,\ \text{ether},\ THF} \underset{\mathbf{2}}{(C_6H_5)_3P{=}N\overset{\overset{S}{\|}}{C}R} \xrightarrow{HCl} \underset{\mathbf{3}}{H_2N\overset{\overset{S}{\|}}{C}R}$$

[1] Y. Tamura, T. Kawasaki, N. Gohda, and Y. Kita, *Tetrahedron Letters*, 1129 (1979).
[2] Y. Tamura, T. Kawasaki, M. Adachi, and Y. Kita, *Synthesis*, 887 (1979).

Triphenylphosphine–Triiodoimidazole (1). 2,4,5-Triiodoimidazole (m.p. 190–192°) is prepared by reaction of imidazole in water with iodine (3 equiv.) in hexane and sodium hydroxide (76% yield).[1]

Alkenes from* vic-*diols.[2] Hexopyranoside diols are converted into alkenes by reaction with **1** and triphenylphosphine in refluxing toluene. Addition of imidazole improves the yield. The most satisfactory molar proportion of diol, triphenylphosphine, **1**, and imidazole is 1:4:1.6:2. Yields are high with *vic,trans*-diols, either diequatorial or diaxial.

Examples:

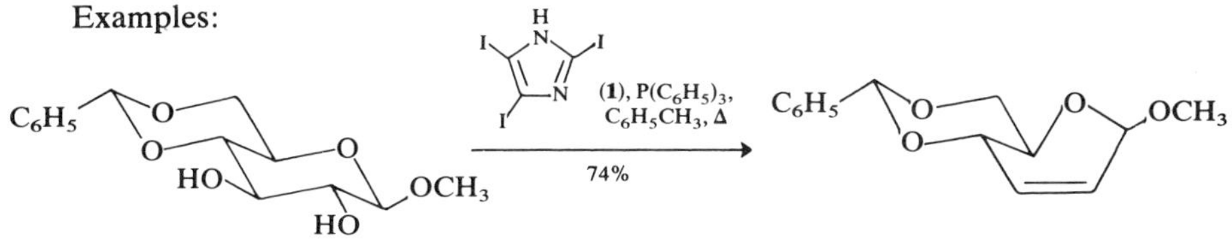

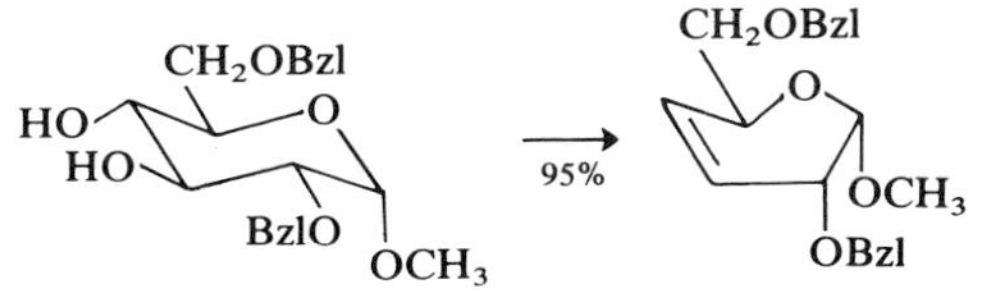

This reaction is an improved modification of an earlier method that uses triphenylphosphine, iodine, and imidazole in refluxing toluene.[3]

ROH → RI.[4] The combination of triphenylphosphine and 2,4,5-triiodoimidazole in toluene at elevated temperature (120°) effects replacement of primary and secondary hydroxyl groups of carbohydrates by iodine with inversion of configuration. The reaction is heterogeneous and the carbohydrate need not be soluble in toluene. Triphenylphosphine, iodine, and imidazole can also be used, but the yields are generally lower.

Examples:

HO, OCH_3, O, O, O, H_3C, CH_3 → (97%) I, OCH_3, O, O, O, H_3C, CH_3

H_3C, H_3C, O, O, O, O, O, CH_3, CH_3, HO → (78%) H_3C, H_3C, O, O, O, O, O, CH_3, CH_3, I

[1] H. Pauly and K. Gundermann, *Ber.*, **41**, 3999 (1908).
[2] P. J. Garegg and B. Samuelsson, *Synthesis*, 813 (1979).
[3] *Idem, ibid.*, 469 (1979).
[4] *Idem, J.C.S. Chem. Comm.*, 978 (1979).

2,4,6-Triphenylpyrylium fluoride (1). Mol. wt. 328.39, m.p. 112°.

Preparation[1]:

C_6H_5, OH, H_5C_6, O, C_6H_5 → (40% HF, >90%) C_6H_5, H_5C_6, O^+, C_6H_5, F^-

1

Alkyl and benzyl fluorides.[1] Pyridinium fluorides (**2**) result from reaction of primary amines and **1** with removal of water by azeotropic distillation with benzene–ethanol. These pyridinium fluorides decompose smoothly at 80–120° to afford the corresponding alkyl and benzyl fluorides (**3**).

[1] A. R. Katritzky, A. Chermprapai, and R. C. Patel, *J.C.S. Chem. Comm.*, 238 (1979).

$$RCH_2NH_2 \xrightarrow[82\%]{1,\ -H_2O} \text{2} \xrightarrow[40\text{–}65\%]{80\text{–}120°} RCH_2F$$

(2: N-(CH_2R)-2,4,6-triphenylpyridinium fluoride, H_5C_6, C_6H_5, C_6H_5, N^+, F^-, CH_2R)

2 3

Triphenylsilyl hydroperoxide (1). Mol. wt. 292.40, m.p. 110–112°, stable at 20°.

Preparation[1]:

$$ClSi(C_6H_5)_3 \xrightarrow[56.2\%]{98\%\ H_2O_2,\ NH_3,\ ether} (C_6H_5)_3SiOOH$$

1

Epoxidation.[2] This hydroperoxide (2 equiv.) epoxidizes double bonds at room temperature in satisfactory yield. It can also be used for Baeyer–Villiger oxidations.

Examples:

$$(CH_3)_2C{=}C(CH_3)_2 \xrightarrow[70\%]{1,\ CH_2Cl_2,\ 25°} (CH_3)_2C\overset{O}{—}C(CH_3)_2$$

Cyclohex-2-enol –OH $\xrightarrow[77\%]{}$ epoxy alcohol ~OH (Syn/anti = 19 : 1)

Cyclopentanone $\xrightarrow[56\%]{}$ δ-valerolactone

[1] R. Dannley and G. Talies, *J. Org.*, **30**, 2417 (1965).
[2] J. Rebek, Jr. and R. McCready, *Tetrahedron Letters*, 4337 (1979).

Triphenylstannylmethyllithium, $(C_6H_5)_3SnCH_2Li$ (**1**). Mol. wt. 370.98. The reagent is obtained in almost quantitative yield from (triphenylstannyl)methyl iodide by halogen–metal exchange with *n*-butyllithium.[1]

Carbonyl methylenation.[2] The reagent reacts with carbonyl compounds to form, after hydrolysis, 2-triphenylstannylethanols (**2**), which form olefins (**3**) on pyrolysis at 110–175°. Overall yields are improved if the alcohols (**2**) are not isolated.

Examples:

$$R^1R^2C{=}O \xrightarrow{1} \xrightarrow{H_2O} (C_6H_5)_3SnCH_2C(OH)R^1R^2 \xrightarrow[-(C_6H_5)_3SnOH]{\Delta} CH_2{=}CR^1R^2$$

2 3

$$C_6H_5CHO \xrightarrow[87\%]{1)\ 1;\ 2)\ 110°} C_6H_5CH{=}CH_2$$

Cyclohexanone (=O) $\xrightarrow[86\%]{}$ methylenecyclohexane (=CH_2)

[1] D. Seyferth and S. B. Andrews, *J. Organometal. Chem.*, **30**, 151 (1971).
[2] T. Kauffmann, R. Kriegesmann and A. Woltermann, *Angew. Chem., Int. Ed.*, **16**, 862 (1977).

Tris(p-bromophenyl)ammoniumyl hexachloroantimonate, $(p\text{-}BrC_6H_4)_3\dot{N}^+SbCl_6^-$ (**1**). Mol. wt. 816.48, m.p. 141–142° dec., stable in the dark, fairly soluble in organic solvents.[1]

Cleavage of* p*-methoxybenzyl ethers.[2] The p-methoxybenzyl ether group is a common protecting group for alcohols. It is generally removed by hydrogenation or, less frequently, by sodium in liquid ammonia. A newer method is oxidative cleavage with the cation radical $(BrC_6H_4)_3\dot{N}^+$ in moist acetonitrile, used either as the salt (**1**) or generated electrochemically from the amine. Both methods give good to excellent yields of the alcohol.

Benzyl ethers are not cleaved by **1**. Removal of the benzyl ether function can, however, be achieved with the cation radicals of amines **2** or **3**.[3] Thus selective deprotection of primary and secondary hydroxyl groups is possible. These deprotections can be performed in the presence of primary or secondary alkyl bromides, which are unaffected by **1** or the radical cations of **2** or **3**.

2 (tris(2,4-dibromophenyl)amine) **3** (bis(2,4-dibromophenyl)(4-bromophenyl)amine)

$$C_3H_7-CH(OH)-CH(C_2H_5)-CH_2-OH \;(\mathbf{4}) \xrightarrow[83\%]{1)\ Na;\ 2)\ CH_3O-C_6H_4-CH_2Cl} C_3H_7-CH(OH)-CH(C_2H_5)-CH_2OCH_2C_6H_4OCH_3 \;(\mathbf{5})$$

$$\mathbf{5} \xrightarrow[75\%]{NaH,\ C_6H_5CH_2Br} C_3H_7-CH(OCH_2C_6H_5)CH(C_2H_5)-CH_2OCH_2C_6H_4OCH_3 \;(\mathbf{6}) \xrightarrow[95\%]{\mathbf{1}} C_3H_7CH(OCH_2C_6H_5)-CH(C_2H_5)-CH_2OH \;(\mathbf{7}) \xrightarrow[74\%]{\mathbf{3}} \mathbf{4}$$

[1] F. A. Bell, A. Ledwith, and D. C. Sherrington, *J. Chem. Soc.* (*C*), 2719 (1969).
[2] W. Schmidt and E. Steckhan, *Angew. Chem., Int. Ed.*, **17**, 673 (1978).
[3] *Idem, ibid.*, **18**, 801, 802 (1979).

Tris(phenylseleno)borane, $B(SeC_6H_5)_3$. Mol. wt. 479.02, m.p. 152–153°, decomposes slowly in air. The borane is prepared in 84% yield by the reaction of benzeneselenol with boron tribromide at 25°.[1]

Phenylseleno acetals. This borane converts both aldehydes and ketones into diphenyl diselenoacetals. Addition of TFA is usually advantageous.[2] An alternative route, which sometimes is more efficient, is acid-catalyzed exchange from oxygen acetals.[3] The selenoacetals are precursors to useful selenium-stabilized carbanions.

$$R^1R^2C{=}O \xrightarrow[70\text{–}90\%]{B(SeC_6H_5)_3} R^1R^2C(SeC_6H_5)_2 \xleftarrow[70\text{–}90\%]{B(SeC_6H_5)_3,\ TFA} R^1R^2C(OR^3)_2$$

These acetals are reduced by triphenyltin hydride to the corresponding hydrocarbons in high yield. Raney nickel is less efficient for this purpose. This reaction constitutes a useful alternative to Wolff–Kishner reduction of carbonyl compounds. Phenyl selenides are also reduced by tin hydrides.[4]

Deoxygenation of sulfoxides.[5] Tris(phenylseleno)borane is an effective reagent for the reduction of sulfoxides under mild conditions. Although this reagent converts ketones to selenoketals, selective deoxygenation of keto sulfoxides is possible. Deoxygenation of vinyl sulfoxides also proceeds smoothly.

Examples:

$$(C_6H_5CH_2)_2SO \xrightarrow[91\%]{(C_6H_5Se)_3B,\ CHCl_3,\ -30°} (C_6H_5CH_2)_2S$$

$$CH_3(CH_2)_{14}C({=}O)CH_2S({=}O)CH_3 \xrightarrow[90\%]{} CH_3(CH_2)_{14}C({=}O)CH_2SCH_3$$

$$n\text{-}C_4H_9CH{=}CHS({=}O)CH_3 \xrightarrow[86\%]{} n\text{-}C_4H_9CH{=}CHSCH_3$$

[1] M. Schmidt and H. D. Block, *J. Organometal. Chem.*, **25**, 17 (1970).
[2] D. L. J. Clive and S. M. Menchen, *J.C.S. Chem. Comm.*, 356 (1978).
[3] *Idem, J. Org.*, **44**, 1883 (1979).
[4] D. L. J. Clive, G. Chittattu, and C. K. Wong, *J.C.S. Chem. Comm.*, 41 (1979).
[5] D. L. J. Clive and S. M. Menchen, *ibid.*, 168 (1979).

Tris(phenylthio)methyllithium, 6, 650–651.

γ-Keto aldehydes. Cohen and Nolan[1] have used this reagent for synthesis of γ-keto aldehydes because of its ability to undergo conjugate addition to enones. Remaining steps involve conversion of the adducts (**2**) to **4**. Raney nickel was not useful for monodesulfurization, but this reaction was realized in high yield with chromous chloride (6 equiv.) in DMF–H_2O at 100°. The last step, hydrolysis of the thioacetal group, was accomplished with trichloroisocyanuric acid (**6**, 605–606) and silver nitrate in $CH_3CN–H_2O$ containing $CdCO_3$.

$$R^1CH{=}CHCOR^2 + (C_6H_5S)_3CLi \rightarrow (C_6H_5S)_3C\overset{R^1}{\overset{|}{C}}HCH_2COR^2 \xrightarrow[95\%]{CrCl_2}$$

1 2

$$(C_6H_5S)_2CH\overset{R^1}{\overset{|}{C}}HCH_2COR^2 \xrightarrow[70\text{–}90\%]{Cl^+,\ Ag^+,\ H_2O} OCH\overset{R^1}{\overset{|}{C}}HCH_2COR^2$$

3 4

[1] T. Cohen and S. M. Nolan, *Tetrahedron Letters*, 3533 (1978).

Tris(trimethylsilyloxy)ethene, 8, 523.

α,β-Dihydroxy carboxylic acids.[1] The reagent reacts with an aldehyde in the presence of tin(IV) chloride as catalyst to form an intermediate (**a**), which is converted on acid hydrolysis into an α,β-dihydroxy carboxylic acid.

Example:

$$n\text{-}C_5H_{11}CHO + \overset{OSi(CH_3)_3}{\overset{|}{C}}H{=}C[OSi(CH_3)_3]_2 \xrightarrow{SnCl_4}$$

$$n\text{-}C_5H_{11}\overset{OSi(CH_3)_3}{\overset{|}{C}}H\underset{OSi(CH_3)_3}{\underset{|}{C}}HCOOSi(CH_3)_3 \xrightarrow[80\%]{H_3O^+} n\text{-}C_5H_{11}\overset{OH}{\overset{|}{C}}H\underset{OH}{\underset{|}{C}}HCOOH$$

a

[1] A. Wissner, *Synthesis*, 27 (1979).

Trityl tetrafluoroborate, 1, 1256–1258; **2**, 454; **4**, 565–567; **6**, 657; **7**, 414–415; **8**, 524–525.

Lactonization.[1] Reaction of **1** with trityl tetrafluoroborate not only effects oxidative debenzylation as expected (**4**, 567), but also cyclization to **2**, *dl*-aplysistatin, a brominated sesquiterpene. Attempted hydrogenolysis of the benzyl ether resulted only in hydrogenation of the double bond.

CH₃ H CH₂OCH₂C₆H₅ O COOCH₃ Br H H₃C CH₃ **1** $\xrightarrow[51\%]{(C_6H_5)_3C^+BF_4^-}$ H₃C H O O O Br H H₃C CH₃ **2**

[1] T. R. Hoye and M. J. Kurth, *Am. Soc.*, **101**, 5065 (1979).

V

Vanadium(II) perchlorate, $V(ClO_4)_2$. Mol. wt. 249.85. The salt is prepared by reaction of V_2O_5 in aqueous $HClO_4$ with H_2O_2 followed by electrolysis.

Pinacol-type reduction.[1] α-Keto acids are reduced to substituted tartaric acids (either the *dl*- or *meso*-form) by the reagent. Cr(II) and Eu(II) ions give mixtures of products of reduction.

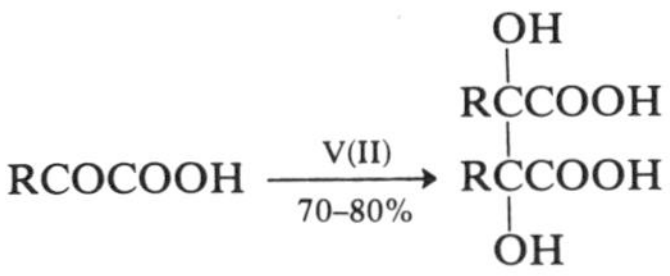

$$\text{RCOCOOH} \xrightarrow[70\text{–}80\%]{\text{V(II)}} \begin{array}{c} \text{OH} \\ | \\ \text{RCCOOH} \\ | \\ \text{RCCOOH} \\ | \\ \text{OH} \end{array}$$

[1] N. Katsaros, E. Vrachnou-Astra, J. Konstantatos, and C. I. Stassinopoulou, *Tetrahedron Letters*, 4319 (1979).

Vanadyl trifluoride, 5, 745–746; **6**, 660; **7**, 418–421; **8**, 528–529.

Monophenolic oxidative coupling. Kupchan *et al.*[1] have reported use of the VOF_3–TFA/TFAA system for synthesis of aporphines by monophenolic intramolecular oxidative coupling.

Examples:

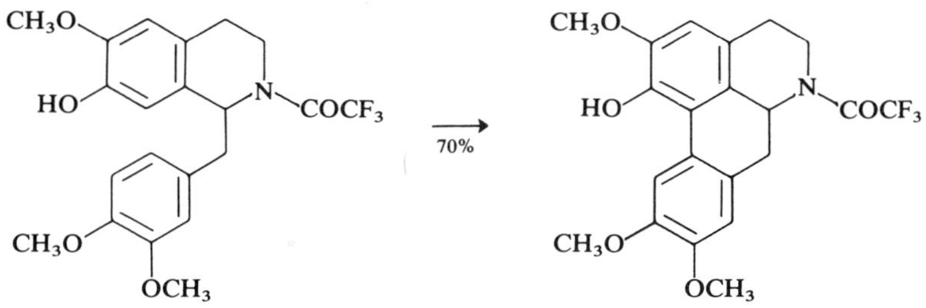

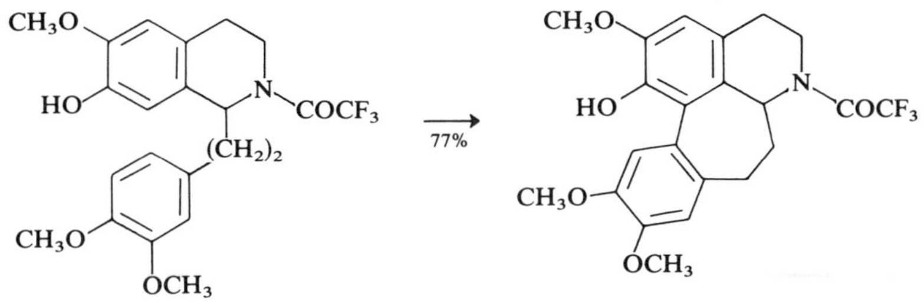

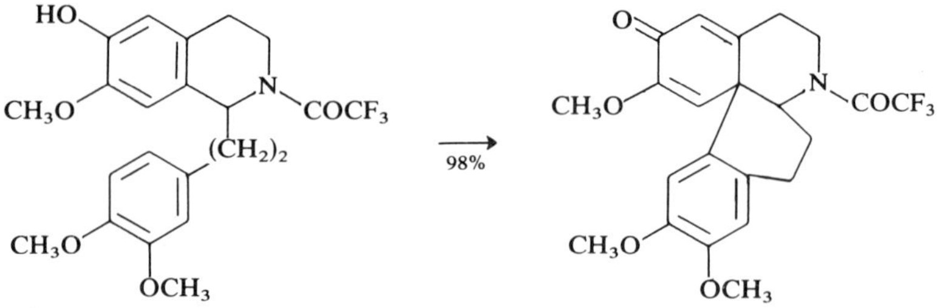

HO, CH_3O, N, $COCF_3$, $(CH_2)_2$, CH_3O, OCH_3 $\xrightarrow{98\%}$ O, CH_3O, N, $COCF_3$, CH_3O, OCH_3

[1] S. M. Kupchan, O. P. Dhingra, and C.-K. Kim, *J. Org.*, **43**, 4076 (1978).

Vilsmeier reagent, 1, 284–289; **2**, 154; **3**, 116; **4**, 186; **5**, 251; **6**, 220; **7**, 422–424; **8**, 529–530.

Esterification.[1] Dimethylchloroformiminium chloride, generated *in situ* from DMF and oxalyl chloride (**5**, 251), converts carboxylic acids into an activated derivative, which reacts with alcohols or phenols in the presence of pyridine to form esters in 70–90% yield. The method is also useful for preparation of active esters from N-protected amino acids, since no racemization is observed.

$$R^1COOH \xrightarrow[CH_3CN,\ -20°]{ClCH{=}\overset{+}{N}(CH_3)_2Cl^-,} \left[R^1\overset{O}{\overset{\|}{C}}O{-}CH{=}\overset{+}{N}(CH_3)_2Cl^- \right] \xrightarrow[70\text{–}90\%]{R^2OH,\ Py,\ 0\text{–}20°} R^1COOR^2$$

a

Formylation of limonene. (+)-(4R)-Limonene (**1**) is formylated by the Vilsmeier reagent (**2**) stereoselectively to give mainly the (E)-α,β-unsaturated aldehyde **3**.[2] The product was used for the synthesis of (4R),(E)-α-atlantone (**5**), a sesquiterpene from *Cedrus* oils.

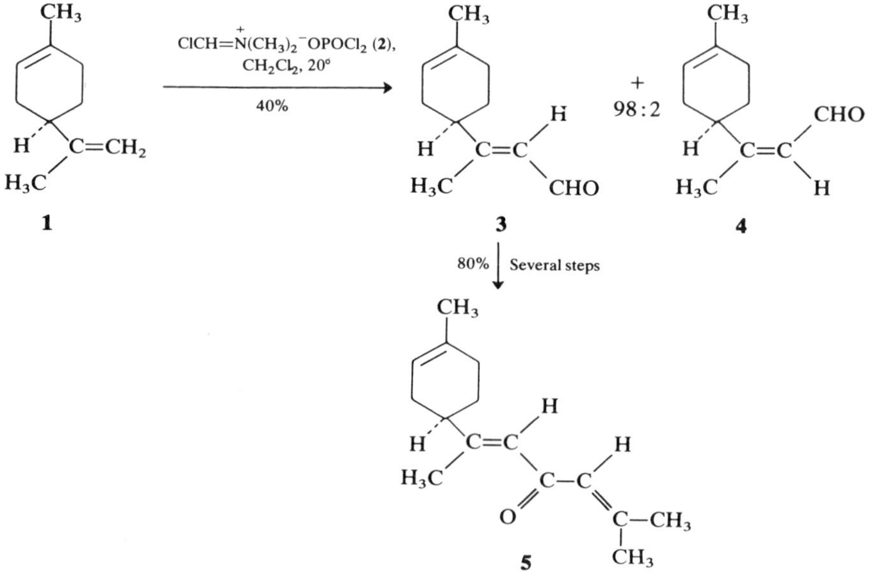

1 (CH_3, H, $C{=}CH_2$, H_3C) $\xrightarrow[40\%]{ClCH{=}\overset{+}{N}(CH_3)_2\ {}^-OPOCl_2\ (\mathbf{2}),\ CH_2Cl_2,\ 20°}$ **3** (CH_3, H, $C{=}C$, H, CHO, H_3C) + **4** (CH_3, H, $C{=}C$, CHO, H, H_3C), 98 : 2

3 $\xrightarrow[80\%]{\text{Several steps}}$ **5** (CH_3, H, $C{=}C$, H, H_3C, $C{-}C$, H, O, $C{-}CH_3$, CH_3)

Quinolines. A new synthesis of quinolines is formulated in equation (I). Yields are satisfactory except in the reaction of **1** when R = Cl. A new route to quinoxazolines utilizing N-nitrosodimethylamine is shown in equation (II).[3]

(I) **1** (R[1]-substituted acetanilide, N–H, C=O, CH₂R) $\xrightarrow[75\text{–}95\%]{DMF,\ POCl_3}$ **2** (2-chloro-3-R-quinoline, R[1]) $\xrightarrow[70\text{–}98\%]{Zn,\ HOAc}$ **3** (3-R-quinoline, R[1])

(II) **4** (3,5-dimethoxyphenyl propionamide: OCH_3, CH_3O, N–H, C=O, CH_3) $\xrightarrow[11\%]{(CH_3)_2NNO,\ POCl_3}$ **5** (5,7-dimethoxy-2-chloro-3-methylquinoxaline: OCH_3, CH_3O, N, N, CH_3, Cl)

[1] P. A. Stadler, *Helv.*, **61**, 1675 (1978).
[2] G. Dauphin, *Synthesis*, 799 (1979).
[3] O. Meth-Cohn, S. Rhouati, and B. Tarnowski, *Tetrahedron Letters*, 4885 (1979).

β-Vinylbutenolide, ($CH_2{=}CH$–butenolide) (**1**). Mol. wt. 110.11. This compound is prepared most conveniently by sulfenylation of β-vinylbutyrolactone followed by sulfoxide elimination. It is somewhat unstable.

Lactone annelation. Enolates of 1,3-dicarbonyl compounds undergo 1,6-Michael addition to **1**; subsequent cyclization results in lactone annelation (equation I).[1]

(I) 2-methylcyclohexane-1,3-dione (CH_3, O, O) $\xrightarrow[47\%]{1)\ Na,\ THF;\ 2)\ \mathbf{1},\ \Delta}$ tricyclic lactone (O, CH_3, HO, O, O) + tricyclic lactone epimer (O, CH_3, HO, O, O), 11 : 1

This reaction has been used in a total synthesis of the eudesmanolide frullanolide (**4**), equation (II)[2,3].

(II) **2** (methyl 3-methyl-2-oxocyclohexanecarboxylate: $COOCH_3$, O, CH_3) + **1** ($H_2C{=}$CH–butenolide, O, O) $\xrightarrow[51\%]{Na,\ THF}$

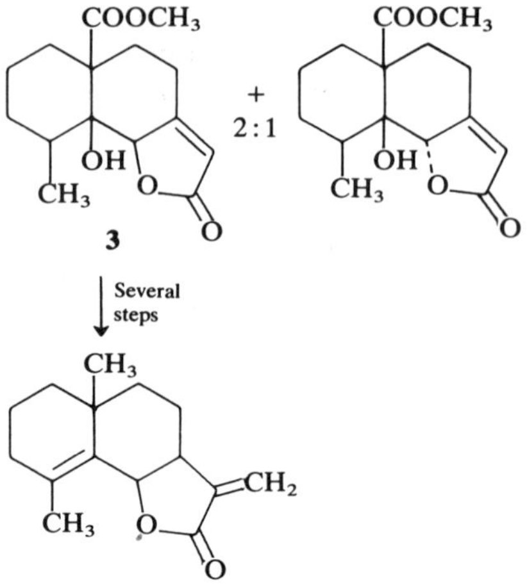

$COOCH_3$ $COOCH_3$

OH CH_3 + 2:1 OH CH_3

3

Several steps

CH_3

CH_2

CH_3

4

[1] F. Kido, T. Fujishita, K. Tsutsumi, and A. Yoshikoshi, *J.C.S. Chem. Comm.*, 337 (1975).
[2] F. Kido, R. Maruta, K. Tsutsumi, and A. Yoshikoshi, *Chem. Letters*, 311 (1979).
[3] F. Kido, K. Tsutsumi, R. Maruta, and A. Yoshikoshi, *Am. Soc.*, **101**, 6420 (1979).

Y

Ytterbium, Yb. At. wt. 173.04. Very pure (99.9%) metal is available from Research Chemicals, P.O. Box 14588, Phoenix, Ariz. 85063.

Ytterbium–ammonia reductions.[1] Ytterbium possesses reducing properties similar to those of lithium and sodium. It dissolves in liquid ammonia to form a blue solution that is stable for several hours at −33°. This solution reduces arenes to 1,4-dihydroarenes, enones to ketones, and alkynes to *trans*-alkenes.

Examples:

$$-OCH_3 \xrightarrow[80\%]{Yb\text{–}NH_3} -OCH_3$$

$$\xrightarrow[88\%]{}$$

$$(CH_3)_2C{=}CHCOCH_3 \xrightarrow[40\%]{} (CH_3)_2CHCH_2COCH_3$$

$$-COCH_3 \xrightarrow[48\%]{} -COCH_3$$

$$C_6H_5C{\equiv}CC_6H_5 \xrightarrow[75\%]{} C_6H_5\overset{H}{C}{=}\underset{H}{C}C_6H_5$$

[1] J. D. White and G. L. Larson, *J. Org.*, **43**, 4555 (1978).

Z

Zinc–Hydrochloric acid-*t*-Butyl alcohol.

1-Oxacephems. These substances (**6**) generally have higher antibacterial activity than the related sulfur analogs, the cephalosporins. A synthesis from penicillins has been reported by Japanese chemists (scheme I).[1] The 6-aminopenicillanate **1** is converted in several steps into the oxazolidine **2**, which is reductively cleaved by activated zinc, hydrochloric acid, and *t*-butyl alcohol in C_6H_6 or CH_2Cl_2 mainly to **3**. Aluminum amalgam–TFA–$HOC(CH_3)_3$ effects the same cleavage, but in lower yield. The methyl ketone **3** is brominated only at the terminal methyl group by $CuBr_2$ (**1**, 161–162) and $HC(OC_2H_5)_3$ in ethanol at 60°. Subsequent hydrolysis gives the desired bromo ketone **5**. This product is converted into 1-oxacephems such as **6** by reactions previously reported by the same laboratory.[2]

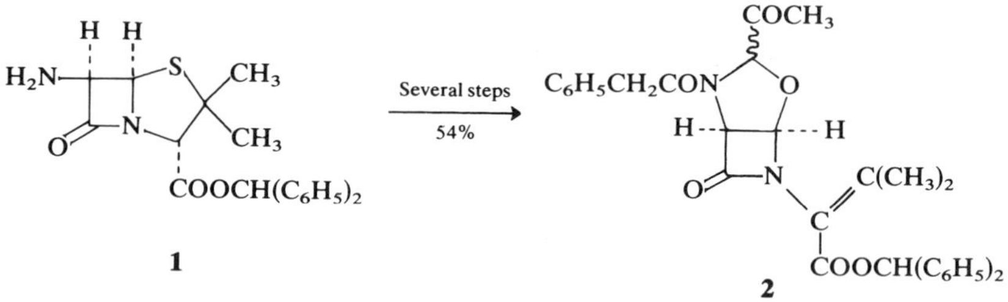

H H
H_2N S CH_3
N CH_3
O
$COOCH(C_6H_5)_2$

1

Several steps
54%

$COCH_3$
$C_6H_5CH_2CON$ O
H H
O N $C(CH_3)_2$
C
$COOCH(C_6H_5)_2$

2

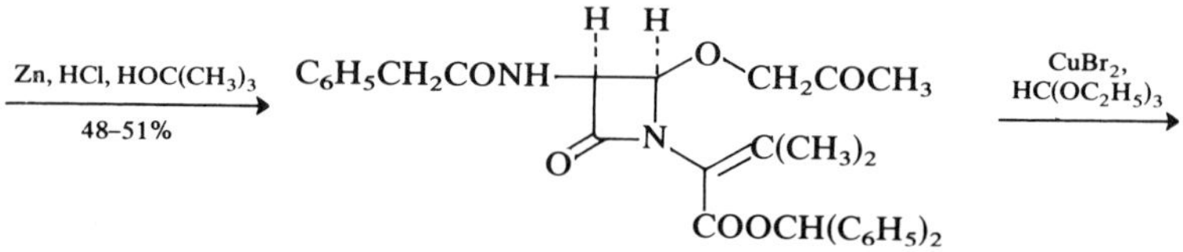

Zn, HCl, $HOC(CH_3)_3$
48–51%

H H
$C_6H_5CH_2CONH$ O CH_2COCH_3
O N $C(CH_3)_2$
$COOCH(C_6H_5)_2$

3

$CuBr_2$, $HC(OC_2H_5)_3$

H H $(OC_2H_5)_2$
$C_6H_5CH_2CONH$ O CH_2CCH_2Br
O N $C(CH_3)_2$
$COOCH(C_6H_5)_2$

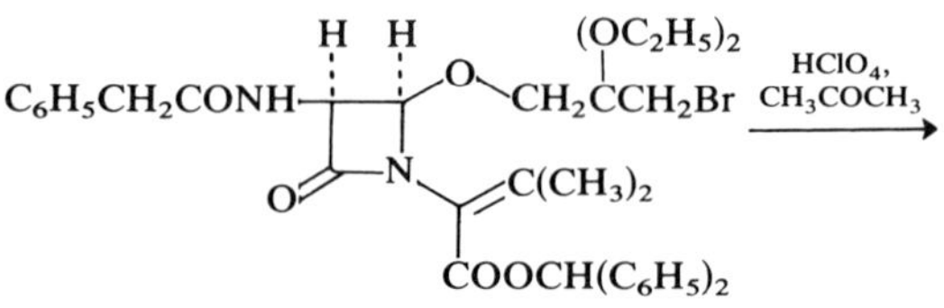

4

$HClO_4$, CH_3COCH_3

$C_6H_5CH_2CONH$ ... OCH_2COCH_2Br ... $C(CH_3)_2$ ... $COOCH(C_6H_5)_2$

5

Several steps →

$C_6H_5CH_2CONH$ ... CH_2Br ... $COOCH(C_6H_5)_2$

6

Scheme (I)

[1] M. Yoshioka, I. Kikkawa, T. Tsuji, Y. Nishitani, S. Mori, K. Okada, M. Murakami, F. Matsubara, M. Yamaguchi, and W. Nagata, *Tetrahedron Letters*, 4287 (1979).
[2] S. Uyeo, I. Kikkawa, Y. Hamashima, H. Ona, Y. Nishitani, K. Okada, T. Kubota, K. Ishikura, Y. Ide, K. Nakano, and W. Nagata, *Am. Soc.*, **101**, 4403 (1979).

Zinc–Silver couple, 4, 436; **5**, 760.

[3+4]Cyclocoupling of polybromo ketones and furane (**5**, 222–223; **6**, 195–196). This reaction has usually been promoted by tri-μ-carbonylhexacarbonyldiiron. Zinc–silver couple is also effective. Although yields are not so high, this variation is more economical for large scale preparations. An example is the preparation of 8-oxabicyclo[3.2.1]octene-6-one-3 (**1**) from $\alpha,\alpha,\alpha',\alpha'$-tetrabromoacetone and furane.[1]

Zn–Ag, THF; Zn–Cu, 55% overall

1

Chiral 4-hydroxy-2-cyclopentenones. Both (R)- and (S)-4-hydroxy-2-cyclopentenones can be obtained from phenol. The first step is alkaline hypochlorite oxidation[2] to the acid **1**, which is resolved with brucine. Oxidative decarboxylation of **1** gives **2**, which is partially dechlorinated and protected as the silyl ether (**3**).[3] The last step to **4**[4] is reduction with zinc–silver (**5**, 760) or zinc–copper.[5]

1) NaOCl (60%) 2) Resolution (75%) → HOOC, OH, Cl, Cl, HO, Cl **1** → $Pb(OAc)_4$, 98% → O, Cl, Cl, HO, Cl **2** → 1) $CrCl_2$ 2) $ClSi(CH_3)_2C(CH_3)_3$, 61% →

$$\underset{\mathbf{3}}{(CH_3)_3CSi(CH_3)_2O\text{-}(\text{3-chlorocyclopent-2-enone})} \xrightarrow[94\%]{Zn(Ag),\ CH_3OH} \underset{\mathbf{4}}{(CH_3)_3CSi(CH_3)_2O\text{-}(\text{cyclopent-2-enone})}$$

[1] T. Sato and R. Noyori, *Bull. Soc. Japan*, **51**, 2745 (1978).
[2] A. W. Burgstahler, T. B. Lewis, and M. O. Abdel-Rahman, *J. Org.*, **31**, 3516 (1966); C. J. Moye and S. Sternhell, *Aust. J. Chem.*, **19**, 2107 (1966).
[3] M. Gill and R. W. Richards, *J.C.S. Chem. Comm.*, 121 (1979).
[4] *Idem, Tetrahedron Letters*, 1539 (1979).
[5] R. M. Blankenship, K. A. Burdett, and J. S. Swenton, *J. Org.*, **39**, 2300 (1974).

Zinc bromide, 2, 463–464; **8**, 535.

Alkylation of silyl enol ethers. Regioselective monoalkylation of silyl enol ethers with *t*-alkyl chlorides in the presence of $TiCl_4$ has been reported recently (**8**, 485). The method is suitable for alkylation of kinetic enolates of ketones at the less substituted α-position.[1]

This reaction has been extended to a similar alkylation with more reactive primary and secondary alkyl halides, such as benzylic and allylic halides. For this purpose the milder Lewis acid zinc bromide is generally preferable to titanium(IV) chloride as catalyst.[2]

Examples:

$$C_6H_5COCH(CH_3)CH_2CH{=}C(CH_3)_2 \xleftarrow[]{(CH_3)_2C{=}CHCH_2Br,\ ZnBr_2} C_6H_5C(OSi(CH_3)_3){=}CHCH_3 \xrightarrow[83\%]{C_6H_5CH(CH_3)Br,\ ZnBr_2} C_6H_5COCH(CH_3)CH(CH_3)C_6H_5$$

$$\text{1-OSi(CH}_3)_3\text{-2-CH}_3\text{-cyclohexene} \xrightarrow[71\%]{C_6H_5CH(CH_3)Br,\ ZnBr_2} \text{2-CH}_3\text{-2-(CH(CH}_3)C_6H_5)\text{cyclohexanone}$$

$$C_6H_5CH{=}C(OSi(CH_3)_3)(OC_2H_5) \xrightarrow[72\%]{CH_3OCH_2Cl,\ ZnBr_2} C_6H_5CH(CH_2OCH_3)COC_2H_5$$

$$(CH_3)_2C{=}CHCOCH_3 \xrightarrow[96\%]{} (CH_3)_2C{=}CHC(OSi(CH_3)_3){=}CH_2 \xrightarrow[83\%]{p\text{-}CH_3C_6H_4CH(CH_3)Cl,\ ZnBr_2} p\text{-}CH_3C_6H_4CH(CH_3)CH_2COCH{=}C(CH_3)_2$$

The last example formulates a two step synthesis of ar-turmerone from mesityl oxide.

γ-Alkylation of O-silylated dienolates. The trimethylsilyl ethers of unsubstituted α,β- or β,γ-unsaturated esters and ketones undergo alkylation in the presence of zinc bromide about equally at the α- and γ-positions. A β-substituent increases the regioselectivity significantly to favor γ-alkylation. In the case of esters, a bulky alkoxy group suppresses α-alkylation. This regioselectivity applies to a number of electrophiles such as $ClCH_2SC_6H_5$, AcCl, CH_3OCH_2Cl, all catalyzed by zinc bromide.[3]

Examples:

$$[(CH_3)_2CH]_2CHOC(=O)CH{=}C(CH_3)CH_2CH(SC_6H_5)C_3H_7\text{-}n \xleftarrow[90\%]{ClCH(SC_6H_5)C_3H_7\text{-}n,\ ZnBr_2} [(CH_3)_2CH]_2CHOC(OSi(CH_3)_3){=}CHC(CH_3){=}CH_2$$

$$[(CH_3)_2CH]_2CHOC(OSi(CH_3)_3){=}CHC(CH_3){=}CH_2 \xrightarrow[80\%]{AcCl,\ ZnBr_2} [(CH_3)_2CH]_2CHOC(=O)CH{\cdots}C(CH_3){\cdots}CHC(=O)CH_3$$

$$\xrightarrow[71\%]{CH(OCH_3)_3,\ ZnBr_2} [(CH_3)_2CH]_2CHOC(=O)CH{=}C(CH_3)CH_2CH(OCH_3)_2 \quad (E/Z = 63:37)$$

Phenylthioalkylation of silyl enol ethers.[4] Silyl enol ethers of ketones, aldehydes, esters, and lactones can be alkylated regiospecifically by α-chloroalkyl phenyl sulfides in the presence of a Lewis acid. Zinc bromide and titanium(IV) chloride are the most effective catalysts. The former is more satisfactory for enol ethers derived from esters and lactones. $ZnBr_2$ and $TiCl_4$ are about equally satisfactory for enol ethers of ketones. The combination of $TiCl_4$ and $Ti(O\text{-}i\text{-}Pr)_4$ is more satisfactory for enol ethers of aldehydes. Since the products can be desulfurized by Raney nickel, this reaction also provides a method for alkylation of carbonyl compounds. Of more interest, sulfoxide elimination provides a useful route to α,β-unsaturated carbonyl compounds.

The α-chloroalkyl phenyl sulfides are prepared by reaction of the corresponding alkyl bromide with sodium thiophenoxide to give an alkyl phenyl sulfide, which is then chlorinated by NCS.

Examples:

$$\text{(bicyclic silyl ketene acetal)}\text{-}OSi(CH_3)_3 \xrightarrow[96\%]{C_6H_5SCH_2Cl,\ ZnBr_2} \text{lactone with } CH_2SC_6H_5 \xrightarrow[96\%]{NaIO_4,\ \Delta} \text{α-methylene lactone } (=CH_2)$$

$$\text{octalin-}OSi(CH_3)_3 \xrightarrow[80\%]{C_6H_5SCH_2Cl,\ TiCl_4} \text{decalone with } C_6H_5SCH_2 \xrightarrow[92\%]{\text{Raney Ni}} \text{decalone with } CH_3$$

$$(CH_3)_3SiO\text{-cyclohexenyl-}CH_3 \xrightarrow[71\%]{C_6H_5SCH(Cl)CH_2CH_2CH_3\ (\mathbf{1}),\ ZnBr_2} \text{2-methyl-2-[CH(SC}_6H_5)CH_2CH_2CH_3]\text{cyclohexanone} \xrightarrow{\text{Raney Ni}} \text{2-methyl-2-}(CH_2)_3CH_3\text{-cyclohexanone}$$

$$CH_3(CH_2)_5C(OSi(CH_3)_3){=}CH_2 \xrightarrow{\mathbf{1},\ TiCl_4} CH_3(CH_2)_5COCH_2CH(SC_6H_5)CH_2CH_2CH_3 \xrightarrow[87\%]{NaIO_4,\ \Delta} CH_3(CH_2)_5COCH{=}CHCH_2CH_2CH_3$$

$$CH_3(CH_2)_5CH{=}CHOSi(CH_3)_3 \xrightarrow[86\%]{\mathbf{1},\ TiCl_4\text{-}Ti(OR)_4} CH_3(CH_2)_5CH(CHO)CH(SC_6H_5)CH_2CH_2CH_3 \xrightarrow{97\%} CH_3(CH_2)_5C(CHO){=}CHCH_2CH_2CH_3$$

$$CH_3CH_2CH{=}C(OC_2H_5)(OSi(CH_3)_3) \xrightarrow[98\%]{\mathbf{1},\ ZnBr_2} CH_3CH_2CH(COOC_2H_5)CH(SC_6H_5)CH_2CH_2CH_3 \xrightarrow{97\%} CH_3CH_2C(COOC_2H_5){=}CHCH_2CH_2CH_3$$

[1] M. T. Reetz and W. F. Maier, *Angew. Chem., Int. Ed.*, **17**, 48 (1978); M. T. Reetz, I. Chatziiosifidis, W. Löwe, and W. F. Maier, *Tetrahedron Letters*, 1427 (1979).
[2] I. Paterson, *ibid.*, 1519 (1979).
[3] I. Fleming, J. Goldhill, and I. Paterson, *ibid.*, 3209 (1979).
[4] I. Paterson and I. Fleming, *ibid.*, 993 (1979); *idem, ibid.*, 995 (1979); *idem, ibid.*, 2179 (1979).

Zinc chloride, 1, 1289–1292; **2**, 464; **3**, 338; **5**, 763–764; **6**, 676; **7**, 430; **8**, 536–537.

Cyclic orthoesters. The cycloaddition of ketene acetals with epoxides can be effected by catalysis with $ZnCl_2$, which does not promote polymerization. Some useful transformations of the resulting cyclic orthoesters are formulated.[1]

Examples:

$$C_6H_5CH\overset{O}{-}CH_2 + HClC{=}C(OC_2H_5)_2 \xrightarrow[50\%]{ZnCl_2,\ 60^\circ} \text{2,2-(OC}_2H_5)_2\text{-3-Cl-4-}C_6H_5\text{-tetrahydrofuran} \xrightarrow[55\%]{1)\ KOC(CH_3)_3\ \ 2)\ H_3O^+} \text{4-}C_6H_5\text{-2(5H)-furanone}$$

$$ClCH_2\overset{O}{\widehat{CH-CH_2}} + CH_3CH{=}C(OCH_3)_2 \xrightarrow{60\%} \text{2-}ClCH_2\text{-5,5-}(OCH_3)_2\text{-4-}CH_3\text{-tetrahydrofuran} \xrightarrow[65\%]{KOC(CH_3)_3} \text{2-}CH_2{=}\text{-5,5-}(OCH_3)_2\text{-4-}CH_3\text{-tetrahydrofuran}$$

$$C_6H_5\overset{O}{\widehat{CH-CH_2}} + CH_3CH{=}C(OCH_3)_2 \xrightarrow{50\%} \text{2-}C_6H_5\text{-5,5-}(OCH_3)_2\text{-4-}CH_3\text{-tetrahydrofuran} \xrightarrow{Al[OC(CH_3)_3]_3} \text{2-}C_6H_5\text{-5-}OCH_3\text{-4-}CH_3\text{-2,3-dihydrofuran}$$

Chloro- and bromoacylation of carbonyl compounds.[2] This reaction (equation I) has been known in the literature since 1900, but has attracted little attention because of poor yields. The unsatisfactory yields are a consequence of the fact that the reaction is an equilibrium and that the reactants can revert to starting materials if heated in the presence of the Lewis acid catalyst. Modern research has established that the products are favored in the case of aliphatic, α,β-unsaturated, and aromatic aldehydes and aliphatic ketones. Highest yields are obtained at low temperatures ($\sim -10°$). By use of optimum conditions yields >95% are obtainable in some cases.

$$\text{(I)} \qquad R^1R^2C{=}O + CH_3COCl(Br) \underset{}{\overset{ZnCl_2}{\rightleftharpoons}} R^1R^2C(OCOCH_3)Cl(Br)$$

[1] H. W. Scheeren, F. J. M. Dahman, and C. G. Bakker, *Tetrahedron Letters*, 2925 (1979); *see also* H. W. Scheeren, R. W. N. Aben, P. H. J. Ooms, and R. J. F. Nivard, *J. Org.*, **42**, 3128 (1977).

[2] M. Neuenschwander, P. Bigler, K. Christen, R. Iseli, R. Kyburz, and H. Mühle, *Helv.*, **61**, 2047 (1978); P. Bigler, S. Schonholzer, and M. Neuenschwander, *ibid.*, **61**, 2059 (1978); P. Bigler and M. Neuenschwander, *ibid.*, **61**, 2165 (1978).

INDEX OF REAGENTS ACCORDING TO TYPES

AUTHOR INDEX

SUBJECT INDEX

Page numbers referring to reagents are indicated in **boldface.**